Biotoxins

Majeti Narasimha Vara Prasad
Sarada Devi Tetali • Catherine Bennetau-Pelissero

Editors

Biotoxins

Biotechnological and Therapeutic Applications

Springer

Editors
Majeti Narasimha Vara Prasad
Retired, Department of Plant Sciences
School of Life Sciences
University of Hyderabad
Hyderabad, Telangana, India

Sarada Devi Tetali
Department of Plant Sciences
School of Life Sciences
University of Hyderabad
Hyderabad, Telangana, India

Catherine Bennetau-Pelissero
Department of Feed & Food
Bordeaux Sciences Agro
Gradignan, France

ISBN 978-3-031-75308-4 ISBN 978-3-031-75309-1 (eBook)
https://doi.org/10.1007/978-3-031-75309-1

This Springer imprint is published by the registered company Springer Nature Switzerland AG
The registered company address is: Gewerbestrasse 11, 6330 Cham, Switzerland

The manufacturer's authorised representative in the EU is Springer Nature Customer Service Centre GmbH, Europaplatz 3, 69115 Heidelberg, Germany. If you have any concerns regarding our products, please contact ProductSafety@springernature.com

Printed and bound by CPI Group (UK) Ltd, Croydon, CR0 4YY
30/09/2026
02225563-0016

Canavalia brasiliensis and *Cratylia argentea* in a mice model of Salmonella infection. Int Immunopharmacol 31:233–238
Sodhi A, Tarang S, Kesherwani V (2007) Concanavalin A induced expression of toll-like receptors in murine peritoneal macrophages in vitro. Int Immunopharmacol 7:454–463
Souza MA, Carvalho FC, Ruas LP, Ricci-Azevedo R, Roque-Barreira MC (2013) The immunomodulatory effect of plant lectins: a review with emphasis on ArtinM properties. Glycoconj J 30:641–657
Spadoro-Tank JP, Etzler ME (1988) Heath shock enhances the synthesis of a lectin-related protein in Dolichos biflorus cell suspension cultures. Plant Physiol 88(4):1131–1135
Sumner JB, Howell SF (1936) Identification of hemagglutinin of jack bean with concanavalin A. J Bacteriol 32(2):227–237
Takahashi F, Kuromori T, Urano K, Yamaguchi-Shinozaki K, Shinozaki K (2020) Drought stress responses and resistance in plants: from cellular responses to long-distance intercellular communication. Front Plant Sci 11:556972
Timofeeva OA, Nevmerzhitskaya YY, Moscowkina MA (2010) Lectin activity and composition in the wheat cell walls under the effects of low temperature and inhibitors of calcium signaling pathway. Russ J Plant Physiol 57:198–204
Tsaneva M, Van Damme EJ (2020) 130 years of plant lectin research. Glycoconj J 37:533–551
Tyagi N, Tyagi M, Pachauri M, Ghosh PC (2015) Potential therapeutic applications of plant toxin-ricin in cancer: challenges and advances. Tumour Biol 36:8239–8246
Van Damme EJM (2014) History of plant lectin research. In: Hirabayashi J (ed) Lectins: methods and protocols, methods in molecular biology, vol 1200. Humana Press, New York, NY
Van Damme EJ (2020) Purification of GNA-related lectins from natural sources. In: Hirabayashi J (ed) Lectin purification and analysis: methods and protocols, pp 413–419
Van Damme EJM, Allen AK, Peumans WJ (1987) Isolation and characterization of a lectin with exclusive specificity towards mannose from snowdrop (*Galanthus nivalis*) bulbs. FEBS Lett 215(1):140–144
Van Hove J, De Jaeger G, De Winne N, Guisez Y, Van Damme EJ (2015) The Arabidopsis lectin EULS3 is involved in stomatal closure. Plant Sci 238:312–322
Vandenborre G, Smagghe G, Van Damme EJ (2011) Plant lectins as defense proteins against phytophagous insects. Phytochemistry 72(13):1538–1550
Varki NM, Varki A (2007) Diversity in cell surface sialic acid presentations: implications for biology and disease. Lab Investig 87:851–857
Wang H, Ng TB, Ooi VE, Liu WK (2000) Effects of lectins with different carbonhydrate-binding specifities on hepatoma, chriocarcinoma, melanoma and osteosarcomacell lines. Int J Biochem Cell Biol 32(3):365–372
Wang SY, Yu QJ, Bao JK, Liu B (2011) Polygonatum cyrtonema lectin, a potential antineoplastic drug targeting programmed cell death pathways. Biochem Biophys Res Commun 406(4):497–500
Watkins WM, Morgan WTJ (1952) Neutralization of the anti-H agglutinin in eel serum by simple sugars. Nature 169(4307):825–826
Wu L, Bao JK (2013) Anti-tumor and anti-viral activities of Galanthus nivalis agglutinin (GNA)-related lectins. Glycoconj J 30:269–279
Xiang Y, Song M, Wei Z, Tong J, Zhang L, Xiao L et al (2011) A jacalin-related lectin-like gene in wheat is a component of the plant defence system. J Exp Bot 62(15):5471–5483
Ye T, Wang T, Yang X, Fan X, Wen M, Shen Y, Xi X, Men R, Yang L (2018) Comparison of Concanavalin a-induced murine autoimmune hepatitis models. Cell Physiol Biochem 46(3):1241–1251

Preface

Here is an up-to-date review of Biotoxins, their occurrence, their effects, and their new applications. Toxic compounds that could be either natural (microbial, plant, or animal) or anthropic origin fall under international regulations that can differ from place to place. Local authorities carefully claim on these compounds giving recommendations on exposure limits, usages, and dosage for therapeutic purposes. These recommendations require regular updating based on general knowledge progression on their incidence in exposed human populations. A follow-up of exposure is also of major importance in explaining health adverse events giving rise to new recommendations.

The biotoxins represent a scientific field in constant evolution, thanks to methodological, computational, and regulatory improvements. Biotoxins are essentially defined by their deleterious effects on human beings, but their properties on living organisms including humans largely go beyond these toxic effects.

Biotoxins of natural secondary metabolites, produced by living entities including bacteria, fungi, plants, or animals present in the air, water, and soil at all depths and altitudes, can have adverse impacts on the environment. They are generally of small size and are present/secreted in small amounts in/by the producing organisms, respectively. Due to these characteristics, their analysis and discovery require specific and highly sophisticated tools. Indeed, the improvement of analytical techniques including all the omics gives rise to the discovery of new substances among complex mixtures. These methods also provide new data on biotoxins' mode of action. Artificial Intelligence will be of great help in the systematic analysis of meta-data generated by the omics techniques.

Humans can be exposed to biotoxins due to their natural occurrence in foods of plant and animal origins or can be isolated and used as chemical defenses against predators, insects, or microorganisms. Various types of biotoxins such as cyanogenic glycosides, lectins (modified proteins), mycotoxins, pyrrolizidine alkaloids, etc. can be harmful when consumed in sufficient quantities. Some of the biotoxins produced by cyanobacteria and microalgae are known to have numerous adverse effects on the environment and can reach humans in various ways, also *via* fishes, which get exposed. Unfortunately, biotoxins are also exploited as biological

weapons in the form of bacteria, viruses, or toxins either to kill people, agricultural crops, or herds of farm animals.

The role of traditional herbal medicines in primary healthcare systems is known to be globally important. Several countries have been involved in the identification of medicinal plants and also developed criteria for quality assurance and quality control of phytomedicines. Further, the chemicals recognized as phytotoxins of various origins categorized such as terpenes, alkaloids, glycosides, phenolics, cyanotoxins, coumarins, tannins, quinones, and many more can be exploited for their pharmacological activities.

The main role of biotoxins is to ensure interactions between living organisms with their surrounding environment. They are often produced as defense compounds against abiotic or biotic stressors, the latter being viruses, pests, competitors, or predators. However, the same compound can also be positive leading, for instance, to symbiotic associations between different species such as microorganisms or pollinators.

Biotoxins are so potent that small doses can be lethal in humans and as such can also be used as biological weapons. Scientific works are still ongoing in state laboratories on these issues although biological weapons are officially forbidden by international agreements. However, deciphering their mode of action at the cell level led to new medical discoveries and therapeutical applications in many medicinal fields. Thanks to new biotechnological tools, it is now possible to modify them for greater efficacy or to find antidotes. These tools allow us also to produce them in great amounts either for direct use as drugs or for phytosanitary applications.

Because of their potential deleterious effects, biotoxins are regulated and the exposure of the living populations should be carefully monitored leading to officially recognized analytical techniques. The regulations also evolve considering the discovery of new substances and their use for human applications.

The main readership groups of this book are intended to be medical practitioners, food scientists, chefs, environmental scientists, toxicologists, government bodies governing the safety of medicinal and food plant or animal-based products, academicians of teaching and research field as well as higher education students and research scholars who are interested and/or whose work is comprised of toxicity and uses of biotoxins. For the aforementioned professional interests and requirements, the Editors thank all the authors of the book chapters. In fact, the authors have been invited to fulfill the requirement of information included in various chapters of the book.

The following contents are covered in this work with the new applications around biotoxins:

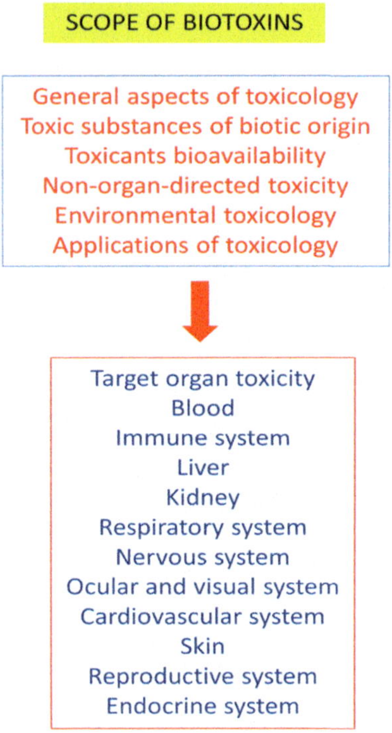

Hyderabad, Telangana, India — Majeti Narasimha Vara Prasad
Hyderabad, Telangana, India — Sarada Devi Tetali
Gradignan, France — Catherine Bennetau-Pelissero

References

Clark GC, Casewell NR, Elliott CT, Harvey AL, Jamieson AG, Strong PN, Turner AD (2019) Friends or foes? Emerging impacts of biological toxins. Trends Biochem Sci 44(4):365–379.

Gurib-Fakim A (2006) Medicinal plants: traditions of yesterday and drugs of tomorrow. Mol Asp Med 27(1):1–93.

Contents

About the Editors

Majeti Narasimha Vara Prasad earned his B.Sc. and M.Sc. degrees from Andhra University, Waltair; and PhD degree from the Lucknow University, Lucknow (1975–79 (Research conducted at Birbal Sahni Institute of Palaeosciences, An autonomous Inst. Under the Dept. of Sci & Tech, Govt. of India).

Prasad served as Lecturer, June 1980–85, Dept. of Botany, North Eastern Hill University, Shillong, Meghalaya State, India; Lecturer (Senior Scale), University of Hyderabad, Sch. of Life Sci., Lecturer (1985–86), Reader 1986–90; Feb. 1990–98, and Full Professor of Environmental Biotechnology from 1st Aug 1998 to Feb 2018 in the Dept of Plant Sciences, University of Hyderabad, Sch. of Life Sciences.

Prasad supervised 19 PhD, several post-doctoral and M.Phil. students and hosted numerous visiting foreign scientists and students from Kazakhstan, Nigeria, Poland, Switzerland, Thailand, Turkey, Ukraine, the USA, etc.

Published research articles in peer-reviewed journals 224, book chapters 205, edited books 49 published by, viz., Elsevier (23), John Wiley & Sons (10), Springer (8), One each by Fizmatlit Russia, Kluwer Academic, M/o of Environment & Forests, GOI, New Delhi. Marcel Dekker, Narosa, Russian Academy of Sciences, CRC and Taylor & Francis.

Awards and honors

- XIX International Botanical Congress **Excellent Scholar** Award Shenzhen, China, July 23–29, 2017
- Pitambar Pant **National Environment Fellow 2007** awarded by the **Ministry of Environment, Forests & Climate Change, Government of India**
- Recipient of Prof KS Bilgrami Memorial Award 2015 by the Soc. for Plant Research, India
- Served as COST action 859 (Phytotechnologies) working group member, European Science Foundation

- Elected Fellow—Linnean Society of London, UK
- Elected Fellow—National Institute of Ecology, New Delhi

Visiting assignments in various universities

- Univ. Québec INRS-Eau, **Canada, 1994** NSERC foreign res awardee
- Jagiellonian Univ. Krakow, **Poland 1996** Dept. Plant Physiol. & Biochem.
- Univ. of Coimbra, **Portugal 1999**
- Stockholm University, Inst. of Botany, **Sweden, 2000**
- Univ. Oulu, Oulu by Finnish Academy, **Finland 2002**
- Univ. of South Australia. Adelaide, **Australia 2005**
- Al-Farabi Kazakh National Univ. Botany Dept Almaty, **Kazakhstan 2006**
- Ural Federal University, Ekaterinburg, **Russia 2007–2015**
- Ghent University Faculty of Bioscience Engineering, Gent, **Belgium 2011**
- Mahasarakham University, **Thailand 2013–2014**
- Univ of Santa Cruz, Ilhéus-Bahia, **2015 Brazil**
- MHRD, Govt. of India deputation as visiting Professor, to Asian Institute of Technology, **Thailand** 2017 Jan–May semester

Sarada D. Tetali did M.Sc. Life Sciences (Plant Sciences) from University of Hyderabad, 1987–89; M.Phil. and Ph.D. (Plant Sciences), University of Hyderabad, Hyderabad 1989–95. Worked as postdoctoral research scientist at Iowa State University (Ames, Iowa), University of California (Berkeley and Davis) during 1996 to 2007.

Teaching experience:

- Reader, May 2007–10, Associate Professor (2007–16), Professor since 6th Dec 2016, Department of Plant Sciences, School of Life Sciences, University of Hyderabad, Telangana, India.

- Made significant contributions to the fields of "Photosynthesis, Algal Biotechnology and Molecular basis of phytomedicine for chronic inflammatory diseases.
- Research articles in peer-reviewed journals 41, book chapters 2
- https://scholar.google.com/citations?user=qg9kaYsAAAAJ&hl=en

Academic honors/awards

1. 2006. The California Clinical Nutrition Research Unit and the Richard C. Woodard Award
2. 2010–11. DBT-CREST Award (Research Area—Nanobiotechnology) Host Institute: Prof. J. Jankowski, Charité, Campus Benjamin Franklin (CBF), Berlin, Germany
3. May 2014. Visiting Professor at University of California, Davis, USA
4. Feb–Mar 2020. Visiting Professor at University of Tokyo, Tokyo, Japan

Research publications (selected)

- 2021 Reddi KK, Li H, Li W, **Tetali SD**. **Molecules.** 26(16):4733. doi:10.3390/molecules26164733.
- 2021 Kulyal P, Acharya S, Ankari AB, Kokkiripati PK, **Tetali SD**, **Frontiers in Pharmacolology.** 12:659546. doi:10.3389/fphar.2021.659546 .
- 2021 **Tetali SD**, Acharya S, Ankari AB, Nanakram V, Raghavendra AS. **Journal of Ethnopharmacology.** 16: 113469
- 2021. Sapuri-Butti, A.R., Wang, L., **Tetali, S.D.,** Rutledge, J.C., Parikh, A.N. **Biochimica et Biophysica Acta—Biomembranes.** 1863(3) 183535.
- 2019 Reddi KK, **Tetali** SD. **Phytomedicine**. 2019 Aug;61:152831. doi: 10.1016/j.phymed.2019.152831.
- 2018 Kumar RK, Heuchel M, Kratz K, Lendlein A, Jankowski J, **Tetali SD**. **Clin Hemorheol Microcirc.** 2018;69(1–2):175–185. doi: 10.3233/CH-189112.
- 2016 **Tetali SD**, Jankowski V, Luetzow K, Kratz K, Lendlein A, Jankowski J. **Clin Hemorheol Microcirc.** 2016;61(4):657–65. doi: 10.3233/CH-152026. PMID: 26639769.

- 2016 Kumar RK, Basu S, Lemke HD, Jankowski J, Kratz K, Lendlein A, **Tetali SD**. **Clin Hemorheol Microcirc**. 2016;61(4):667–80. doi: 10.3233/CH-152027. PMID: 26639770.

- 2016 Kumar RK, Basu S, Lemke HD, Jankowski J, Kratz K, Lendlein A, **Tetali SD**. (HAECs). **Clin Hemorheol Microcirc**. 2016 64(4):931–940

- 2014 Choudhury SS, Bashyam L, Manthapuram N, Bitla P, Kollipara P, **Tetali SD**. **Journal of Ethnopharmacology.** 2014 May 28;154(1):148–55.

- 2012 den Hartigh LJ, Altman R, Hutchinson R, Petrlova J, Budamagunta MS, **Tetali SD**, Lagerstedt JO, Voss JC, Rutledge JC. **PLoS One. 2012**;7(11):e50513. doi: 10.1371/journal.pone.0050513.

- 2010 **Tetali SD**, Budamagunta MS, Simion C, Hatters DM, Higgins LJ, Weisgraber KH, Voss JC and Rutledge, JC, MD. **Journal of Lipid Research**. 51: 1273–1283

- 2007 Mullick AE, Powers AF, Kota RS, **Tetali SD**, Arteriosclerosis Thrombosis Vascular Biology (ATVB). 2007 27(2):339–45

- 2006 **Tetali, S. D.**, M. S. Budamagunta, J. C. Voss and J. C. Rutledge. **Journal of Lipid Research**. 47(7): 1358–65.

Catherine Bennetau-Pelissero earned her first PhD in Marine Biology in Bordeaux University in 1988. Her research focused on the study of the reproductive endocrinology of a Siberian fish imported in France. She discovered that the fish were contaminated by soy phytoestrogens that were present in the fish farm food. She earned a second PhD still in Bordeaux University in Physiology of Interactions and Adaptations in 1990. She had a post-doctoral fellowship of 2 years at Brunel University in the Biology and Biochemistry Laboratory where she studied the impact of soy phytoestrogens in trout and set up an *in vitro* test on trout liver cells for xenoestrogens. This test was retained by OECD to classify endocrine disruptors. She came back to France where she got a first position in Mont de Marsan in 1992. Then she moved to Bordeaux Sciences Agro in 1993 where she is still working. She also worked as a professor in the University of Bordeaux where she developed a master on Plant Biomolecules of Health Interest.

She developed her research first in the Marine Biology Lab of the University of Bordeaux 1993–2000. Then she held a research team named Micronutrients Reproduction and Health Unit in Bordeaux Sciences Agro from 2000 to 2010. Afterwards she worked in neurosciences from 2010 to 2017 in the Magendie Neurocenter of Bordeaux in the Pathophysiology of Declarative Memory team. Finally she joined the ARNA unit in Bordeaux where she worked as a senior scientist from 2018 to 2023. During this period she passed from fish endocrinology to

human health still working on plant estrogenic molecules. Her research focused on the detection of these substances in food and in biological fluids. With her team she also looked at the bioavailability of these substances in human beings as well as on their health effects. The approach was based on studies held *in vitro*, in animal models and also in clinical approaches. She thus explored the exposure of consumers to plant estrogenic molecules first showing a dramatic increase of this exposure in the French population and second showing that soy phytoestrogens could easily be reduced in food by following traditional water treatments.

Production

Professor Catherine Bennetau-Pelissero wrote 102 scientific articles in peer-reviewed journals and 9 book chapters in English. She managed 82 students at different study levels from bachelor degree to post-doctoral fellowships.

Expertise

- Catherine Bennetau-Pelissero worked as an expert in food safety since 1995.
- In 1996 she proposed a full report on plant phytoestrogens to the French Superior Council of Hygiene.
- From 2002 to 2005 she took part in the working group of Afssa (French Food Safety Agency) and Afssaps (French Drug Safety Agency) on benefits and risks of phytoestrogens in the French population, Recommendations.
- From 2010 to 2022, she worked as a Nutrition expert for Anses (French Food Safety Agency) and took part in several working groups on toxic exposures.
- From 2008 until 2015, she was a member on the National Program on Endocrine Disruptors, scientific commitee
- From 2010 to 2018 she was a member on the French Society of Nutrition as a member of the scientific committee and the administration committee.
- In 2015 she was involved as an invited scientist in a working group of the European Food Safety Authority dealing with the risk and benefits of isoflavones in peri- and post-menopausal women.

Visiting assignments in various universities

- Tuft University, Boston, USA, 11/2015
- Moncton University, Canada, 11/2015

- University of Tsukuba, Japan, 09/2017
- University of Hyderabad, India 02/2018
- University of Taipei, Taiwan, 03/2018
- University of Trans-Disciplinary Health Sciences and Technology, Bangalore, India, 03/2022
- University of Mumbai 03/2022

Part I
General Insights on the Chemistry and Analysis of Biotoxins

Chapter 1
Chemistry of Biotoxins and Mode of Action/Antidotes

Gyanajeet Parida, Madhabi Madhusmita Bhanjadeo, and Nihar Ranjan Singh

Abstract Organisms are known to produce compounds that play vital roles in metabolism, sustenance, growth, and development, along with toxic chemicals that have deleterious effects on living organisms. Biotoxins are toxic compounds originating from microbes, fungi, plants, and animals and pose health risks to other organisms. Biotoxins can be classified based on their chemical structures, source of origin, and mode of action. In recent years, a good amount of interest has been shown in toxinology for its multilevel role in clinical, agricultural, and even bioterrorism. Understanding the underlying mechanism of biotoxin action and its toxicity can reveal a plethora of pragmatic knowledge on the predator–prey relationship, defence mechanisms, drug discovery and therapies, food safety, and human health. The chemistry behind biotoxin reactions and remediation is sparingly known. The present knowledge of most biotoxin's action is that it is known to exert toxicity by blocking cellular ion channels, disrupting cellular membranes, altering signalling pathways, and inhibiting proteins and enzymes. An enthralling list of biotoxins and their chemistry advances our understanding of these natural toxins and paves the way for developing novel preventive methods in the form of antidotes for better utilisation towards human health and global security. In this chapter, broad categories of biotoxins and their chemistry are evaluated, and their modes of action with implications are presented for a better understanding of biotoxins for the future beneficial role of humankind.

Keywords Biotoxins · Toxin structure · Toxin chemistry · Bioterrorism · Health · Toxin antidotes · Toxinology

---

G. Parida · N. R. Singh (✉)
Department of Botany, Ravenshaw University, Cuttack, Odisha, India
e-mail: nrsingh@ravenshawuniversity.ac.in

M. M. Bhanjadeo
KIIT School of Biotechnology, KIIT University, Bhubaneswar, Odisha, India

M. N. V. Prasad et al. (eds.), *Biotoxins*,
https://doi.org/10.1007/978-3-031-75309-1_1

1.1 Introduction

The statement of famous Renaissance physician Paracelsus (1493–1541), often referred to as the "Father of Toxicology" translated from his German writing that roughly says "*What is there that is not poison? All things are poison and nothing is without poison. Solely the dose determines that a thing is not a poison*" paved way for the definition of toxicology as the science of poison or poisoning (Grandjean 2016). The poison is nothing but the toxin that can cause harmful reaction or effect on biological system and can be broadly configured as biotoxins. A study of toxins that is produced naturally by a species can be called as toxinology and it differs from its superset toxicology as the study of adverse effect of toxins or any chemicals, including its source, effect, and treatment on human beings (Duffull and Isbister 2022). Even extremely small amounts of biotoxins can induce disease symptoms or malafide effects on human beings when they enter the body by physical contact, food ingestion, or air inhalation. It is therefore not inaccurate to say that the powerful poisons are harvested from living organisms. Based on the source of biotoxin, it can be categorised broadly into microtoxins, mycotoxins, phytotoxins, and zootoxins and their mode of action vary with their origin, chemical composition, and properties (Nwaji et al. 2022). The biotoxins are also chemically diverse and can be structurally classified into steroids, active peptide, and secondary metabolites. Also, these biotoxins affect human organ system from circulatory system to nervous systems and can affect specifically body parts like liver, intestine, and blood (Foster et al. 2002). The modes of action of these biotoxins also vary and they act by inactivating or inhibiting enzymes or by blocking ion channels. Due to such interactions, the biotoxins can adversely affect human health and cause paralysis, painful infection, hepatotoxicity, and cardiovascular disorders (Parak et al. 2024). It is imperative to evaluate an individual group of biotoxins and understand the modalities of its mode of action. The main class of biotoxins is described in Figure 1.1, along with a few significant species that fall within each category.

1.2 Microtoxins

Microtoxins are the group of toxins synthesised from microbes or bacterial pathogens and cause a range of diseases in human beings (Schmitt et al. 1999). They are classified into exotoxins and endotoxins. Exotoxins are secreted by microbes and affect the host cells in its vicinity, contributing to pathogenicity. In contrast, endotoxins are toxic substances bound to microbial membranes that are released upon the rupture of the microbes (Cavaillon 2018). Exotoxins are synthesised within the microbial cells and secreted actively with the expense of energy or after the lysis of cell. They can recognise specific receptors on the host cell surface and modulate the host cell leading to its disruptions (Sakari et al. 2022). Exotoxins are highly antigenic proteins and can elicit potent antibody responses on host cells (Ghazaei 2022).

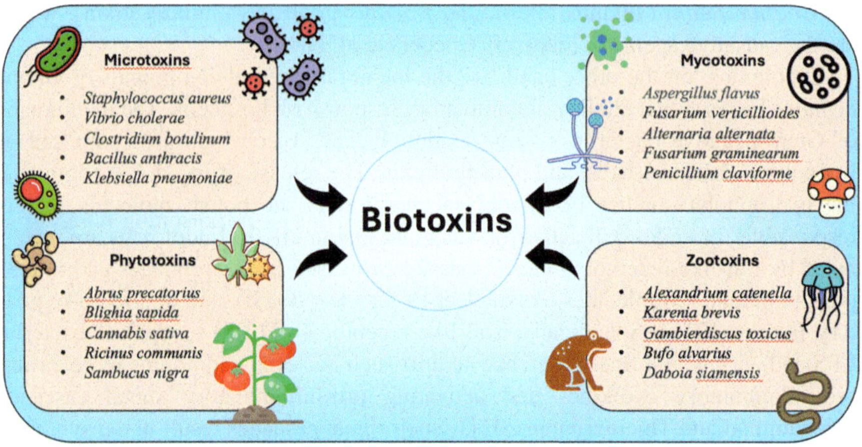

Fig. 1.1 Major groups of biotoxins and some commonly known species belonging to them

They are further categorised into type-1exotoxins, also called as superantigens and their mode of action is through binding major histocompatibility complex class II (MHC II) of macrophages and the receptors on T helper cells. The non-specific binding enhances the release of cytokines into the bloodstream, including Interleukin-2 (IL-2), Interferon-γ (INF-γ), and Tumour Necrosis Factor α (TNF-α) that can lead to diseases like diarrhoea, fever, shock, and potentially death. Superantigens include bacterial toxins such as staphylococcal enterotoxin A (SEA) and toxic shock syndrome toxin (TSST) coming from toxic microbes like *Staphylococcus aureus* and *Staphylococcus pyogenes* (Fraser 2011). The type-2 exotoxins disrupt the membranes of host cells and attenuate their immune system. Such membrane piercing toxins can either form protein channels to enter into the host cells or disrupt the integrity of host cells by altering membrane bound lipopolysaccharides (Ulhuq and Mariano 2022). The targets of these toxins are red blood cells, white blood cell, or cell membrane and therefore can be regarded as haemolysin, leukocidins, or cytolysins (Prévost et al. 2006). Numerous diseases, including sepsis, pneumonia, and infections of the skin and soft tissues, are brought on by the common class of these pore-forming toxins, which are also known as alpha, beta, and gamma toxins. Some microbial species belonging to these group of toxins are *Streptococcus pneumoniae*, *Mycobacterium tuberculosis*, and *Staphylococcus aureus* causes a wide-range of diseases such as pneumonia, sepsis, and skin and soft tissue infections (Seilie and Bubeck Wardenburg 2017). The third type of exotoxin also called as multi-subunit AB toxins are under intense investigation, which has been named based on two peptide chains. While the B-chain is in charge of attaching to host cell receptors, the A-chain is an active enzyme that has a cytotoxic effect because of its Adenosine diphosphate ribosylation activity (Odumosu et al. 2010). The well-known AB-type protein toxins secreted by many bacterial species like *Clostridium difficile* (toxins A and B), *Vibrio cholerae* (cholera toxin), enterohaemorrhagic *Escherichia coli* (Shiga toxin), *Clostridium botulinum* (Botulin toxin),

Clostridium tetani (Tetanus toxin), and *Bacillus anthracis* (anthrax toxin), which are the causative agents of diseases (Goeders et al. 2016).

Endotoxins, on the other hand, are the membrane bound compounds of Gram-negative bacteria that elicit an inflammatory response on host cells. The membranes of Gram-negative bacteria possess peptidoglycans layer which are composed of lipopolysaccharides (LPS) and phospholipids. The release of Lipid A (endotoxin), the hydrophobic anchor of lipopolysaccharides, is the potent molecule that is responsible for endotoxin activity (Wang and Quinn 2010). Endotoxins are recognised by toll-like receptors (TLRs) that are expressed in macrophages and induce pro-inflammatory molecules like nuclear factor-κB (NF-κB), and interferon-regulatory factors (IRFs). Specifically, Toll-like receptor 4 (TLR4) recognises bacterial LPS and acts as first line of defence against such bacterial endotoxins by releasing pro-inflammatory cytokines and activating pro-inflammatory signal cascades (Mohammad and Thiemermann 2021). Such signal cascades result in the synthesis of interleukin-1 (IL-1), interleukin-6 (IL-6), and tumour necrosis factor (TNF) (Galanos and Freudenberg 1993). Among the common microorganisms in this group are *Klebsiella pneumoniae, Pseudomonas species, Heliobacter* species, and *Enterobacter* species (Bonhomme et al. 2024).

1.3 Mycotoxins

Mycotoxins are group of biotoxins produced by fungus as secondary metabolites. Such mycotoxins are capable to cause pharmacological effects and diseases on human beings ranging from inflammatory responses to even death (Bennett and Klich 2003). Mycotoxins have also been linked to numerous financial losses in the fields of public health, animal husbandry, and agriculture. Additionally, mycotoxins frequently infect food and drink items, resulting in significant losses for the food industry (Janik et al. 2020). Alternatively, mycotoxins can also be promoted as antibiotics, growth promoters, pharmacological drugs or even as chemical warfare agents (Lootens et al. 2022). Mycotoxins are structurally diverse groups and usually with low molecular weight. Over 400 harmful mycotoxins from the genera *Aspergillus, Penicillim, Alternaria*, and *Fusarium* have been discovered. Such toxins are normally hepatotoxic, neurotoxic, nephrotoxic, and immunotoxic and possibly lead to cancer and collectively its toxic effects are known as mycotoxicosis (Ülger et al. 2020). The major groups of mycotoxins are trichothecenes (T-2 toxins), ochratoxins, aflatoxins, zearalenone, fumonisins, patulin, ergots, and citrinin, and the basic chemistry of each of these mycotoxins is given in Table 1.1.

The primary mode of action of mycotoxins is by modifying DNA template to impair the transcription or translation process. Each of the mycotoxins group has its own mechanism of action (Ueno 1991). T-2 toxin at cellular level inhibits protein translation process by inactivating the peptidyl transferase and targeting 60S ribosomal unit and preventing polypeptide chain initiation (Janik et al. 2021b). T-2 toxin exposure also leads to the production of ROS, dysregulation of antioxidant defences,

Table 1.1 Major groups of mycotoxins and their chemical structure with a range of pathogenicity

Mycotoxin	Mould species	Structure	Empirical formula	Affected food commodity	Pathological effects in humans	References
T-2 toxin	*Fusarium* spp.	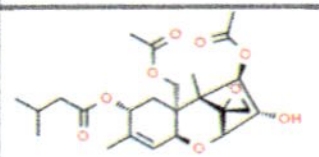	$C_{24}H_{34}O_9$	• Barley • Oats • Wheat	• Dermal toxicity • Immunotoxicity • Hepatotoxicity • Neurotoxicity	• (Meneely et al. 2023) • (Janik et al. 2021a)
Ochratoxin A	*Aspergillus* spp. *Penicillium* spp.	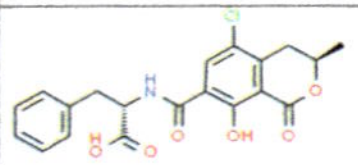	$C_{20}H_{18}ClNO_6$	• Coffee beans • Maize • Dried fruits • Spices • Eggs • Meat	• Nephrotoxicity • Hepatotoxicity • Genotoxicity • Teratogenicity • Immunotoxicity • Neurotoxicity	• (El Khoury and Atoui 2010) • (Klingelhöfer et al. 2020) • (Więckowska et al. 2023)
Aflatoxins	*Aspergillus* spp.	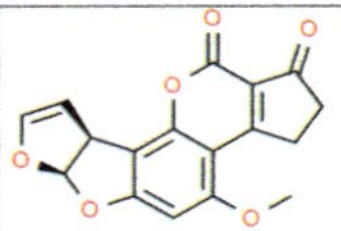	$C_{17}H_{12}O_6$	• Wheat • Walnut • Maize • Peanuts • Milk • Eggs	• Hepatotoxicity • Teratogenicity • Immunotoxicity • Carcinogenicity	• (Pickova et al. 2021) • (Kumar et al. 2016) • (Benkerroum 2020)
Zearalenone	*Fusarium* spp.	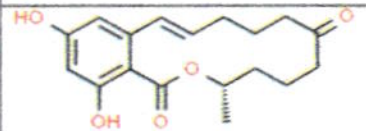	$C_{18}H_{22}O_5$	• Maize • Barley • Oats • Sorghum • Wheat	• Reproductive disorders • Hepatotoxicity • Immunotoxicity • Genotoxicity	• (Rogowska et al. 2019)
Fumonisins	*Fusarium* spp. *Aspergillus* spp.	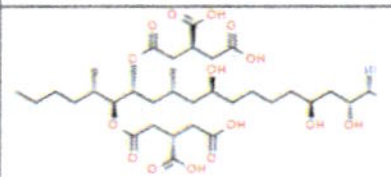	$C_{34}H_{59}NO_{15}$	• Maize • Barley • Rye • Millets • Wheat	• Neurotoxicity • Hepatotoxicity • Nephrotoxicity • Cytotoxicity	• (Kamle et al. 2019)

(continued)

Table 1.1 (continued)

Mycotoxin	Mould species	Structure	Empirical formula	Affected food commodity	Pathological effects in humans	References
Patulin	*Penicillium* spp. *Aspergillus* spp. *Byssochlamys* spp. *Paecilomyces* spp.	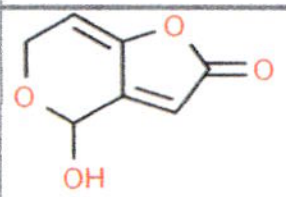	$C_7H_6O_4$	• Apples • Grapes • Plums • Peaches • Pears • Tomatoes	• Hepatotoxicity • Nephrotoxicity • Immunotoxicity • Genotoxicity • Neurotoxicity • Hepatotoxicity	• (Saleh and Goktepe 2019) • (Ramalingam et al. 2019) • (Wei et al. 2020)
Ergots	*Claviceps* spp.	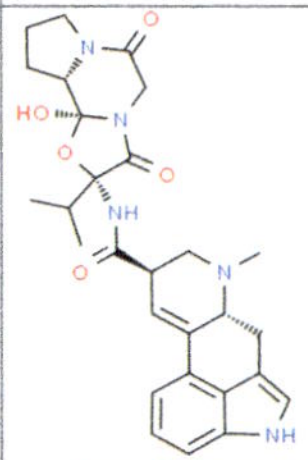	$C_{28}H_{33}N_5O_5$	• Rye • Wheat • Barley • Oats	• Nausea • Vomiting • Itching • Gangrene • Convulsions	• (Reddy et al. 2020)
Citrinin	*Penicillium* spp. *Aspergillus* spp. *Monascus* spp.	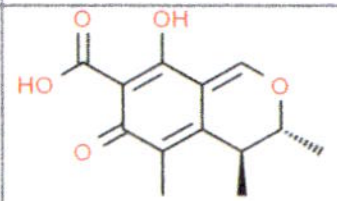	$C_{13}H_{14}O_5$	• Rice • Wheat • Oats • Rye • Corn • Barley	• Diarrhoea • Teratogenic • Nephrotoxicity • Hepatotoxicity	• (He and Cox 2016) • (Zhang et al. 2021) • (Zargar and Wani 2023)

mitochondrial dysfunction, the activation of PPAR-γ/TGF-β1/Smad2/3 pathway, endoplasmic reticulum (ER) stress, and autophagy and inflammatory response in cardiac cells (Dai et al. 2022). In another study, T-2 toxin is found to enhance the expression of inflammatory cytokines and the activity of inducible nitric oxide synthase (iNOS), which increases the production of nitric oxide (NO) and reduces the production of growth hormone (GH). The toxin also upregulates the expression of Fas protein on cell surfaces, upregulates p53 protein, and activates caspase-3, thereby leading to cell apoptosis. In addition, T-2 toxin suppresses the Nrf2/HO-1 pathway, promotes mitochondrial dysfunction, and increases the production of ROS (Li et al. 2022).

Mycotoxin Ochratoxin A when ingested is known to absorb rapidly into the blood circulation and is known to bind the subdomain IIA of human serum albumin. It purportedly targets and accumulates primarily in the kidney, liver, adipose tissue, skeletal muscle, gastrointestinal tissues, and brain (Obafemi et al. 2023). This toxin has shown detrimental activity by inhibiting phenylalanine t-RNA synthase and thus inhibits protein translation, alters cellular energy production by decreasing the activity of phosphoenolpyruvate-carboxykinase (PEPCK), binding covalently to DNA causing mutation, producing pro-oxidants and leading to cell arrest and apoptosis (Kőszegi and Poór 2016).

Aflatoxins are processed in the liver cells and converted into aflatoxin epoxides that bind to guanine residue in DNA leading to depurination. This toxin intercalates into the DNA helix and is believed to cause genetic mutation, activating Ras oncogenes and inactivating p53 tumour suppressing gene leading to increasing risk of cancer (Okechukwu et al. 2024). Aflatoxin epoxides also react with cellular macromolecules, including proteins like cytochrome p450, albumin, and phospholipids, to induce various genetic, metabolic, signalling, and cell structure disruptions (Benkerroum 2020).

Zearalenone is known to bind oestrogenic receptors and hence alters the transcription of oestrogen-responsive elements. They are therefore regarded as endocrine disruptor (Fink-Gremmels and Malekinejad 2007). Zearalenone are known to activate the pro-proliferative Wnt/β catenin signalling pathway and inhibits immunosuppressive TGF-β signalling. To induce apoptosis, they increase the expression of cleaved caspase-3 and caspase-9, the ratio of Bax/Bcl-2, and mitogen activated protein kinases (MAPKs), such as extracellular regulated protein kinases (ERK) and Jun N-terminal kinase (JNK) (Zheng et al. 2018). According to Bulgaru et al. (2021), exposure to zearalenone raises nuclear factor erythroid 2-related factor 2 (Nrf2) and decreases NF-κB, which increases oxidative stress and modifies TLRs, which in turn promotes pro-inflammatory responses.

Fumonisins are also called as sphingosine-like mycotoxins. They have structural similarity with cellular sphingolipids, and this similarity disturbs the metabolism of sphingolipids by inhibiting the enzyme ceramide synthase leading to accumulation of sphinganine in cells and tissues (Stockmann-Juvala and Savolainen 2008). Fumonisins also mediate cytotoxicity through the induction of oxidative stress (Qu et al. 2022). In another study, it has been seen that this biotoxin induces a

dose-dependent increase in the expression of tumour necrosis factor (TNFα) and IL-1β suggesting cytokine production (Chen et al. 2021).

Patulin is another group of mycotoxins that interacts covalently with sulfur containing biomolecules like glutathione (GSH), cysteine, and thioglycolate and induces oxidative stress. Studies have revealed that patulin can induce apoptosis by decreasing superoxide dismutase (SOD), catalase (CAT), GSH, promoting CHOP transcription and ROS-mediated mitochondrial dysfunction (Wei et al. 2020). Patulin induces cell cycle arrest and significantly affects proteins of cell (tight) junctions, membrane potential, protein synthesis, cell signalling, and gene expression (Ramalingam et al. 2019).

Being one of the oldest known disease, Ergot alkaloids cause hallucinations, convulsions, agalactia, burning sensation, vasoconstriction, and gangrenous loss of limbs in humans after consuming foods contaminated with ergot alkaloids (Agriopoulou 2021). Ergot alkaloids have the ability to modify neurotransmitter receptors of dopamine, serotonin, and noradrenaline due to the structural homology of their ergoline ring with biogenic amine receptors (Sharma et al. 2016).

Another group of mycotoxin is citrinin that imparts toxic effects through oxidative stress and modifying antioxidative enzymatic defences (e.g. GSH and transhydrogenase). In yeast cells, it upregulates the expression of oxidative stress response genes, *AADs*, *FLR1*, *OYE3*, *GRE2*, and *MET17 and also* arrests cell cycle at G0/G1 phase, leading to apoptotic cell death (De Oliveira Filho et al. 2017). It is evident from the above that mycotoxins act on host cells in varied way. The toxicity extends from acute to chronic level mycotoxicosis. To curtail the adverse effect of mycotoxin, a better understanding of its mode of action and the chemistry of such toxins is required that can help finding better antidotes against their toxic effects.

1.4 Phytotoxins

Phytotoxins are poisonous compounds produced by plants and plant pathogens. There are numerous variations in their composition, occurrence, fatal effects, and mechanism of action. They are produced as defensive agents against predators (pests, insects, stress, etc.) and also cause harm to organism if taken into the cells and cause diseases. Chemically, they are secondary metabolites and have detrimental effects on both humans and animals, but they can also be advantageous to plants in situations like providing adaptation and protecting from predators or pathogens (Chen et al. 2022). Broadly, plant toxins can be classified into lectins, ribosome-inactivating protein, protease inhibitors, α-amylase inhibitors, ureases, arcelins, antimicrobial peptides, and pore-forming toxins and they have variability on their toxicity and mode of action. Most of action of these toxins is mediated through carbohydrate binding, glycosidase activity, protease inhibitor, and pore-forming activity (Dang and Van Damme 2015). Table 1.2 highlights the mode of action by phytotoxins.

Table 1.2 Mode of action of phytotoxins and major plant groups belonging to such groups (Kocyigit et al. 2023)

Mode of action	Phytotoxins	Plants	References
Carbohydrate binding and glycosidase activity	• Lectins • Ribosome-inactivating proteins	• *Ricinus communis* • *Sambucus nigra* • Cucurbitaceae • Euphorbiaceae • Rosaceae	• (Iglesias et al. 2022) • (Hartley and Lord 2004)
Protease inhibitor	• Protease inhibitors	• Solanaceae • Leguminosae (Fabaceae) • Gramineae (Poaceae)	• (Cotabarren et al. 2020)
Pore-forming activity	• Acrelins • Thionins • Cyclotides	• *Phaseolus* spp. • Many monocots and dicots	• (Dang and Van Damme 2015)

Carbohydrate-binding phytotoxins like lectins are the well-studied plant toxins and are first discovered in plants (ricin from *Ricinus communis* and abrin from *Abrus precatorius*). Lectins are known to bind the carbohydrate moieties on the surface of erythrocytes and agglutinate the erythrocytes, without altering the properties of the carbohydrates (Sharon and Lis 2004). Lectins are hugely diverse and can be classified into *Agaricus bisporus* lectin, Amaranthin, Chitinase-like lectins, Cyanovirin, *Euonymus europaeus* lectin (EUL), *Galanthus nivalis* lectin (GNA), Hevein, Jacalin, legume lectin, LysM, *Nicotiana tabacuum* lectin (Nictaba), and Ricin-B (Tsaneva and Van Damme 2020). Based on their carbohydrate-binding specificity lectins have been classified as Mannose/glucose binding, Galactose/*N*-acetyl-D-galactosamine binding, *N*-acetyl-D-glucosamine binding, L-Fucose binding lectin, and Sialic acid binding (Katoch and Tripathi 2021). On the basis of binding sites, the lectins can be merolectin (monovalent), hololectin (bivalent), chimerolectin (consisting of one or more substrate binding sites together with another independent catalytic domain), and superlectin (multivalent carbohydrate-binding sites binding with two structurally unrelated carbohydrates) (Van Damme 2022). Lectins can also be classified on the basis of their function as they can act as toxins, adhesins, and haemagglutinins (Chettri et al. 2021). Considering the mode of action of lectins towards human health, it is found that they are extremely resistant to digestion when consumed and can conjugate with intestinal cells receptors, glycolipoids, and glycoproteins leading to severe damages to the cellular linings (Adamcová et al. 2021).

Ribosome-inactivating proteins (RIPs) are the members of *N*-glycosidases produced by organisms including bacteria, fungi, and plants, and known for inactivating ribosomes. RIPs can inactivate prokaryotic and eukaryotic ribosomes by removing a specific adenine residue from the highly conserved and surface-exposed sarcin/ricin loop (SRL: GAGA) of 26S or 28S rRNAs (Wang et al. 2016). As the depurinated 50S/60S subunits are incapable of binding to corresponding elongation factors (EF-2/GTP complex), the protein syntheses are halted leading to cell death

(Abbes et al. 2021). The best studies lectin and RIPs is the toxin Ricin from Ricinus communis plant, that has a known history of toxic use and was awarded the dubious title of "poisonous plant of the year 2018". The Ricin toxin has two polypeptide chains A and B that are connected through disulfide bond. The ricin toxin A-chain is responsible for the cause of ricin toxicity and can inactivate ribosomes by hydrolysing the N-glycosidic bond of an adenosine residue in the 28 S ribosomal RNA, whereas toxin B-chain binds to the surface of cells and facilitates entry of toxin into the cell (Bolt and Hengstler 2023).

Another class of phytotoxins are protease inhibitors and such plant protease inhibitors are classified into different families such as Bowman-Birk, Kunitz, Potato I, Potato II, Serpin, Cereal, Rapeseed, Mustard, Squash, etc. Although protease inhibitors are known to inhibit catalytic action of enzymes and are considered non-nutritional compounds but also protease inhibitors like Bowman-Birk inhibitors (BBI) and Kunitz-type inhibitors (KTI) have shown promising therapy against human illness like autoimmune disease, cardiovascular disease, and cancer (Cid-Gallegos et al. 2022).

The pore-forming toxins from plants can be broadly classified as Arcelins, thionins, and cyclotides. The arcelin sequences belong to the arcelin/phytohaemagglutinin/α-amylase inhibitor (APA) family and have sequence similarity with α-amylase inhibitors. The primary effect of thionin toxicity is by increasing membrane permeability. Such change in permeability provokes membrane depolarisation, increase in Ca^{2+} and K^{+} ion permeability and also activation of some enzymes. The biological activities of cyclotides are most probably related to their ability to form pores in the host membrane. All these secondary events might strengthen the initial toxicity and lead to final cell destruction (Dang and Van Damme 2015).

Apart from these phytotoxins, a vast diverse group of plant secondary metabolites generally called as phytochemicals exist in almost all plants which have may adverse effects on animals when ingested (Madariaga-Mazón et al. 2019). Such phytochemicals include alkaloids, terpenes, amines, glucosinolates, cyanogenic glucosides, sterols, quinones, phenolics, peptides, and polyacetylenes and these compounds do not have direct role in the basic life processes of plants, but they play a vital role in adaptation and defence against herbivores (Divekar et al. 2022). Phytochemicals vary in their chemistry but can be divided into hydrophilic and hydrophobic compounds, of which a wide variety of polyphenolic and terpenoid compounds, as well as alkaloids, carbohydrates, and non-protein amino acids, invoke special interest.

1.4.1 Zootoxins

The term "zootoxin" itself denotes an animal poison or venomous chemicals that organisms produce for defence and predation (Gwaltney-Brant 2017). Zootoxins with varying compositions and activity are produced because of several

circumstances, such as food, age, sex, populations, and geographic dispersion. Depending on where they originated, various animal species contain toxins unique to that area. The venomous and poisonous animal species responsible for producing zootoxins mostly belong to arthropods (insects like scorpions), amphibians (like toads), reptiles (like vipers), or marine organisms (like jelly fish) (Yu et al. 2020). Being most of animal toxins are peptide, they act on the nervous system, affecting ion channels, perturbing glutamatergic transmissions, stimulating transmitter release, etc. Others disrupt host cell membrane, behaving as antibacterial, cytolytic, or haemolytic peptides. In addition, a wide variety of enzymes, such as metalloproteinase, lectins, phospholipase A2, proteases, oxidases, sphingomyelinases, etc., are found in animal venom in different taxonomic groups (Jungo et al. 2010). When this toxin penetrates the bodies of other non-poisonous creatures, it produces a variety of biological and physiological complications by exhibiting effects such as haematotoxicity, cytotoxicity, cardiotoxicity, neurotoxicity, and enzyme activity (Gwaltney-Brant et al. 2018). Figure 1.2 provides a graphical representation of toxic effects of biotoxin on human body.

The effects of zootoxins on animals and people that arise from direct contact (oral intake and cutaneous exposure) include neurotoxic, coagulotoxic, necrotoxic, pro-inflammatory, and vasoactive effects. Its systemic and localised effects harm tissues and cells, which disrupts the body's haemostatic and signalling systems. Zootoxins are used in extracted, recombinant, and synthetic forms for the treatment of humans as well. While there is no specific treatment for these toxins, non-specific supportive care (such as breathing and cardiovascular support), antibiotics, analgesics, and, if feasible, species-specific antivenoms can all be used to treat them (Bolon et al. 2023).

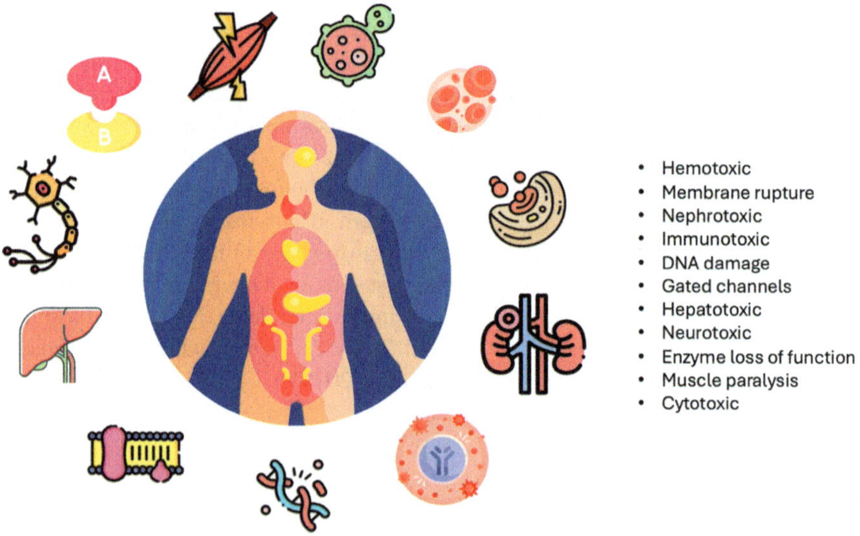

Fig. 1.2 Illustrates the major impact that biotoxins have on human health

As mentioned earlier, animal toxins can be broadly classified into marine toxins, insect toxins, zootoxins from *Arachnida* (spider/scorpions), zootoxins from toads and snake venoms. Table 1.3 provides a list of zootoxins and its mode of action.

1.4.2 Biotoxin as Bioweapon

It is well said that the "*powerful poisons are harvested from living organisms*". There is a genuine chance that biological weapons will be employed in acts of terrorism against people, animals, and crops in the form of microorganisms, phytotoxins, and zootoxins (Atlas 2002). Biotoxins can thus be a source for bioterrorism. Ricin, botulinum, *Clostridium perfringens*, *Epsilon* toxin, conotoxins, Shigatoxins, saxitoxins, tetrodotoxins, mycotoxins, anthrax, and nicotine are a few biotoxins that are pertinent to bioterrorism. The Center for Disease Control and prevention (CDC) in the United States classified biotoxins into three groups based on their bioterrorism activity. The factors in Category A are highly pathogenic and mortal, as well as having a high ease of dissemination. Among them are *Bacillus anthracis*, *Francisella tularensis*, *Yersinia pestis*, botulinum neurotoxins, haemorrhagic fever viruses (*Ebola virus*), and Variola major. Category B pathogens are moderately easy to spread, with low morbidity and mortality rates, but require specific monitoring and improvement of diagnostic capabilities. This category includes: ricin, abrin, *Brucella* species, *Clostridium perfringens* E toxin, and *Burkholderia mallei*. Category C includes emerging pathogens against which the population has little or no immunity and which could be engineered as bioweapons. These include the Hanta and Nipah viruses (Janik et al. 2019). Anthrax toxin when inhaled can disrupt the endocytosis pathway, and Biotoxin like botulinum causes acetylcholine to be inhibited in the peripheral nervous system, which paralyses muscles. Ricin blocks the synthesis of proteins because it is harmful to multiple organs. Epsilon toxin damages cell membranes. Conotoxins, which is a venom from marine snails can inhibit the potassium and sodium channels found in neurons. Mycotoxins can be carcinogenic and can promote cancer development in normal cells. Trichothecenes prevent the synthesis of proteins and nucleic acids. Numerous nicotinic actions are produced by nicotine in the neurological system. Many toxins affect the nervous systems of mammals by interfering with the transmission of nerve impulses, which gives them their high potential in bioterrorist attacks. Others are responsible for blocking the main cellular metabolism, causing cellular death. Moreover, most toxins act quickly and are lethal in low doses ($LD50 < 25$ mg/kg), which are frequently lower than chemical warfare agents (Janik et al. 2019).

Table 1.3 Major groups of zootoxins

Zootoxin	Animal species	Structure/composition	Empirical formula	Mode of action	References
Saxitoxin (paralytic shellfish toxins)	Dinoflagellates of the *Alexandrium* genus	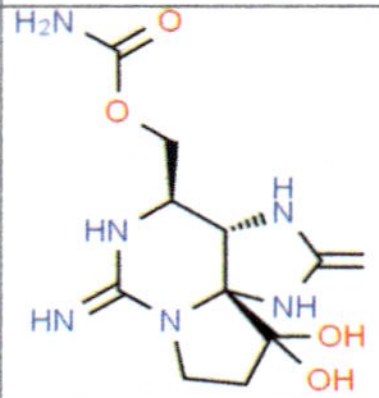	$C_{10}H_{17}N_7O_4$	• Inhibition of the voltage-gated sodium channel • Reduced action potential	• (Christensen and Khan 2020) • (Barzegari Naeeni et al. 2022)
Domoic acid	• Marine alga. • Diatoms.	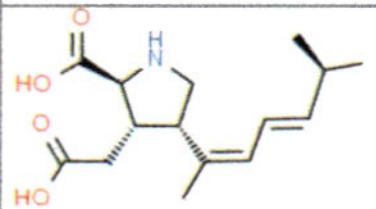	$C_{15}H_{21}NO_6$	• Neurotoxin by binding glutamate receptors. • Opening of the membrane channels (permeable to sodium). • Increased sodium influx and membrane depolarisation.	• (Pulido 2014) • (Pérez-Gómez and Tasker 2022)
Ciguatoxins	Epibenthic dinoflagellates of the genera *Gambierdiscus* and *Fukuyoa*	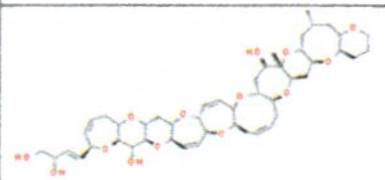	$C_{60}H_{86}O_{19}$	• Activate voltage-dependent Na^+ channels • Opening of the ion pore • Plasma membrane depolarisation • Triggers spontaneous action potentials.	• (Vilariño et al. 2018)

(continued)

Table 1.3 (continued)

Zootoxin	Animal species	Structure/composition	Empirical formula	Mode of action	References
Brevetoxins	Dinoflagellate *Karenia brevis*.	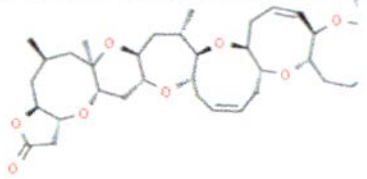	$C_{49}H_{70}O_{13}$	• Binding voltage gated sodium channels • Nerve and muscle depolarisation	• (Baden and Adams 2000) • (Furey et al. 2007)
Tetrodotoxins	• Pufferfish. • Marine animals including goby fish, gastropods, crabs.	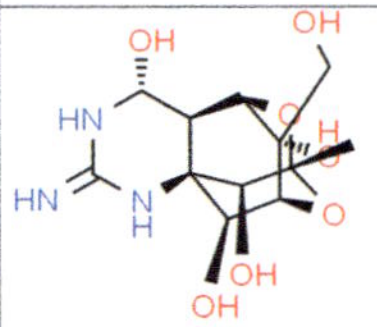	$C_{11}H_{17}N_3O_8$	• Binding voltage-gated sodium channels • Blockade of the nerve conduction • Muscle paralysis	• (Bane et al. 2014) • (Katikou et al. 2022)
Okadaic acid group	• Marine dinoflagellates. • *Dinophysis* spp. • *Prorocentrum* spp.	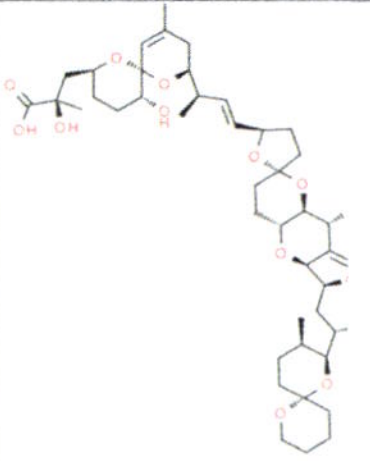	$C_{44}H_{68}O_{13}$	• Protein phosphatase (PP) inhibitors. • Decrease in sodium secretion • Loss of solute permeability in the cell	• (Fu et al. 2019)
Azaspiracid	• Dinoflagellates • Genera *Azadinium* and *Amphidoma*.	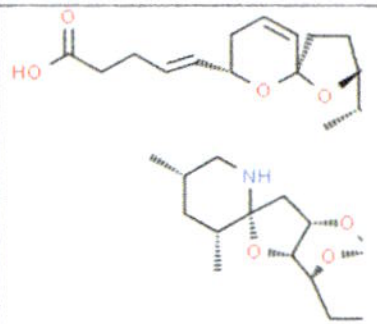	$C_{47}H_{71}NO_{12}$	• Gastrointestinal symptoms • Nausea • Vomiting • Stomach cramps	• (Fu et al. 2019)

Palytoxin	• Soft corals • Dinoflagellates of the genus *Ostreopsis*	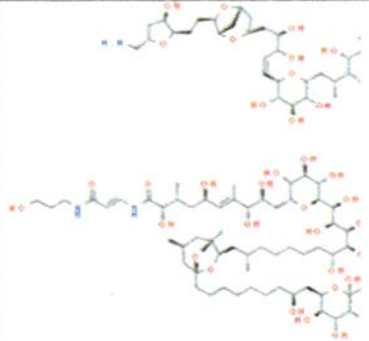	$C_{129}H_{223}N_3O_{54}$	• Alters Na^+/K^+-ATPase pump • Disrupts membrane polarity	• (Carlin et al. 2022)
Bufotoxin	• *Bufo* spp. • Toads	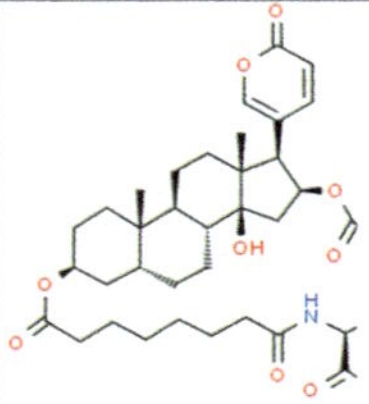	$C_{40}H_{60}N_4O_{10}$	• Inhibition of Na^+, K^+-adenosine triphosphatase	• (Asrorov et al. 2023)
Cantharidin	Order *Coleoptera*: *Meloidae*, *Oedemeridae*, and *Staphylinidae*	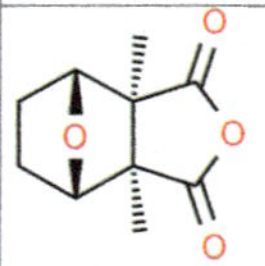	$C_{10}H_{12}O_4$	• Cantharidin activation of the serine/threonine protein phosphatases • Detachment of tonofilaments from desmosomes	• (Langenegger et al. 2019)
Thaumetopoein	• *Thaumetopoea pityocampa* (family Notodontidae)	• 2 protein subunits with 13 & 15 kDa	• Peptide	• IgE-mediated hypersensitivity response • Anaphylactic reactions	• (Seldeslachts et al. 2020)
Scorpion venom	• Scorpion spp.	• Cysteine-stabilised (CS) α/β motifs • Calcins • Non-disulfide bridged peptides (NDBPs)	• Highly complex and heterogeneous	• Inactivation of voltage-gated sodium channels • Causing a persistent Na^+ influx.	• (Ahmadi et al. 2020)

(continued)

Table 1.3 (continued)

Zootoxin	Animal species	Structure/composition	Empirical formula	Mode of action	References
Snake venom (α-Bungarotoxin and γ-Bungarotoxin), Cardiotoxins, Dendrotoxins	• *Bungarus multicinctus.* • Cobra snake poisoning. • Mamba snake poisoning.	• 5′Nucleotidases (5′NUCs) • 3 finger toxins (3FTXs) • Cysteine-rich secretory proteins (CRISPs) • C-type lectins • Disintegrins • Hyaluronidases • Metalloproteinases • Serine proteases • Waprin/Kunitz type inhibitors	• Highly complex and heterogeneous	• α-Neurotoxins—Non-enzymatic nAChRs binding proteins • β-Neurotoxins—Presynaptic phospholipase A2 (PLA2)	• (Oliveira et al. 2022)
Spider toxin (alpha-latrotoxin)	• Spiders spp.	• Concoction of enzymes • Polyamines • Nucleic acids, inorganic salts, and disulfide-rich peptides (DRPs)	• Highly complex and heterogeneous	• Forming cation permeable pores • Leads to massive influx of Ca^{2+} • Stimulating the release of all known types of neurotransmitters from presynaptic cells • Clinical symptoms such as muscular paralysis	• (Langenegger et al. 2019)

1.4.3 Antidotes of Biotoxins

The quote by Suzy Kassem that "*He who creates a poison, also has the cure. He who creates a virus, also has the antidote*" manifests that the nature carries the biotoxin along with its antidote. The International Programme on Chemical Safety broadly defines an antidote as a therapeutic agent that counteracts the toxic actions of a drug/toxin. Antidotes can act directly on toxin, or altering the toxin binding site, or decreasing the toxic metabolites, or by counteracting the effects of toxins (Chacko and Peter 2019). Natural antidotes like turmeric, milk of magnesia, sodium salicylate, sodium bicarbonate, etc., mainly form an insoluble complexes with the toxins and make the complex inactive (Magowska 2021). Certain microbial toxins, such as cholera, botulin, shiga, and tetanus, have antidotes that can reduce their severity. These antidotes are in the form of copper complexes, 3,4-diaminopyridine, pterins, human monoclonal antibodies (mAbs) produced from B cells, and meta-nitrophenyl α-galactoside (MNPG). Since there is no recognised counterdote for phytotoxins such as abrin, ricin, and cannabinol, they are typically more harmful. Fortunately, there are some established counteragents for zootoxins and mycotoxin. Table 1.4 listed some selected biotoxins with their antidotes.

Table 1.4 Selected biotoxins and their antidotes

Toxin name	Source organism	Antidote	Reference
Botulin toxin	*Clostridium botulinum*	Copper complexes and 3,4-diaminopyridine	(Hefter and Samadzadeh 2022)
Shiga toxin	*E. Coli*	Pterins	(Wahome et al. 2012)
Tetanus toxin	*Clostridium tetani*	Human monoclonal antibodies (mAbs) derived from B cells	(Aliprandini et al. 2019)
Cholera toxin	*Vibrio cholerae*	Meta-nitrophenyl α-galactoside (MNPG)	(Haksar et al. 2019)
Clostridium perfringens toxins	*Clostridium perfringens*	• Siastatin B (SB) • N-acetyl-2,3-dehydro-2-deoxyneuraminic acid (NADNA)	(Wang 2020)
Aflatoxins	*Aspergillus flavus*	Charcoal, *Camellia sinensis*	(Adam and Oh 2022)
Ochratoxin A	*Aspergillus ochraceus*	Bentonite	(Bhatti et al. 2017)
T-2 toxin	*Fusarium sporotrichioides*	Hydrated sodium calcium aluminosilicate (HSCAS)	(Wei et al. 2019)
Zearalenone	*Fusarium graminearum*	Hydrated sodium calcium aluminosilicate (HSCAS)	(Xu et al. 2021)
Cicutoxin	Water hemlock	Gastric suction, administration of activated charcoal	(Poison 2014)
Urushiol	Poison ivy	*Bhallataka* (*Semecarpus anacardium* Linn.)	(Nandakumar 2022)
Thaumetopoein	Pine processionary caterpillar	• Affected area is cleaned with soap and water. • Cooling icepacks are applied. • Topical or systemic antihistamines.	(Galip et al. 2022)

1.5 Conclusion

Biotoxins pose a significant threat to humanity due to their toxicogenicity. Each year many health risks and pathophysiological effects including paralysis and mortality are reported due to biotoxins. Modern advancements in toxinology pave way for finding new toxins from many living organisms from microbes to animal's kingdom. Understanding the chemistry and pathophysiological effects of these toxins will be boon for the healthcare industries as new structures and new mechanism of action will be evaluated. The great diversity of biotoxins is a significant barrier in their investigation. Microbial toxins, fungal toxins, plant based toxins, and animal based toxins (including marine and insect toxins) contribute huge list of toxins with diverse pathophysiological effects. It will be a huge opportunity and possibility to identify such toxins with application in food safety, environment regulation, predator–prey relationship, and public health with its novel drugs or beneficial chemicals. On contrary, the biotoxins can be misused and may be a source for bioweapon and can be used in bioterrorism. The toxin inoculum of anthrax has already been used as bioweapon. With so diverse availability of natural biotoxins, alteration in its chemical structure or peptide conformation can build more potent and effective bioweapon that can terrorise humankind. Hence, it is the need of the hour to find appropriate antidote against each of these biotoxins to neutralise them and to counter them against the use of human civilisation. Finding appropriate antidote against biotoxin is the biggest blockade as many toxins do not have any antidote molecule. The future course of action on biotoxin must involve better understanding of its chemistry and mechanism of action, so that antidotes can be developed to counter the adverse effects of biotoxins.

References

Abbes M, Montana M, Curti C, Vanelle P (2021) Ricin poisoning: a review on contamination source, diagnosis, treatment, prevention and reporting of ricin poisoning. Toxicon 195:86–92

Adam GO, Oh H-G (2022) Mycotoxins and antidotes of herbal, binding and enzymatic nature. Novel Res Microbiol J 6:1583–1600

Adamcová A, Laursen KH, Ballin NZ (2021) Lectin activity in commonly consumed plant-based foods: calling for method harmonization and risk assessment. Food Secur 10:2796

Agriopoulou S (2021) Ergot alkaloids mycotoxins in cereals and cereal-derived food products: characteristics, toxicity, prevalence, and control strategies. Agronomy 11:931

Ahmadi S, Knerr JM, Argemi L, Bordon KCF, Pucca MB, Cerni FA, Arantes EC, Çalişkan F, Laustsen AH (2020) Scorpion venom: detriments and benefits. Biomedicines 8:118

Aliprandini E, Takata DY, Lepique A, Kalil J, Boscardin SB, Moro AM (2019) An oligoclonal combination of human monoclonal antibodies able to neutralize tetanus toxin in vivo. Toxicon X(2):100006

Asrorov AM, Kayumov M, Mukhamedov N, Yashinov A, Mirakhmetova Z, Huang Y, Yili A, Aisa HA, Tashmukhamedov M, Salikhov S (2023) Toad venom bufadienolides and bufotoxins: an updated review. Drug Dev Res 84(5):815–838

Atlas RM (2002) Bioterrorism: from threat to reality. Ann Rev Microbiol 56:167–185

Baden DG, Adams DJ (2000) Brevetoxins: chemistry, mechanism of action, and methods of detection. In: Food science and technology. Marcel Dekker, New York, pp 505–532
Bane V, Lehane M, Dikshit M, O'riordan A, Furey A (2014) Tetrodotoxin: chemistry, toxicity, source, distribution and detection. Toxins 6:693–755
Barzegari Naeeni R, Amani J, Nobakht M, Mirhosseini SA, Mahmoodzadeh Hosseini H (2022) Human risk of saxitoxins poisoning: incidence, source, toxicity and diagnostic methods. J Mar Med 4:16–25
Benkerroum N (2020) Chronic and acute toxicities of aflatoxins: mechanisms of action. Int J Environ Res Public Health 17:423
Bennett JW, Klich M (2003) Mycotoxins. Clin Microbiol Rev 16:497–516
Bhatti SA, Khan MZ, Saleemi MK, Saqib M, Khan A, Ul-Hassan Z (2017) Protective role of bentonite against aflatoxin B1-and ochratoxin A-induced immunotoxicity in broilers. J Immunotoxicol 14:66–76
Bolon B, Heinz-Taheny K, Yeung KA, Oguni J, Erickson TB, Chai PR, Goldfine CE (2023) Animal toxins. In: Haschek and Rousseaux's handbook of toxicologic pathology, vol 3. Elsevier
Bolt HM, Hengstler JG (2023) Ricin: an ancient toxicant, but still an evergreen. Arch Toxicol 97:909–911
Bonhomme D, Cavaillon J-M, Werts C (2024) The dangerous liaisons in innate immunity involving recombinant proteins and endotoxins: examples from the literature and the Leptospira field. J Biol Chem 300
Bulgaru CV, Marin DE, Pistol GC, Taranu I (2021) Zearalenone and the immune response. Toxins 13:248
Carlin M, Pelin M, Ponti C, Sosa S, Tubaro A (2022) Functional and structural biological methods for palytoxin detection. J Mar Sci Eng 10:916
Cavaillon J-M (2018) Exotoxins and endotoxins: inducers of inflammatory cytokines. Toxicon 149:45–53
Chacko B, Peter JV (2019) Antidotes in poisoning. Indian J Crit Care Med 23:S241–s249
Chen J, Wen J, Tang Y, Shi J, Mu G, Yan R, Cai J, Long M (2021) Research progress on fumonisin B1 contamination and toxicity: a review. Molecules 26:5238
Chen H, Singh H, Bhardwaj N, Bhardwaj SK, Khatri M, Kim K-H, Peng W (2022) An exploration on the toxicity mechanisms of phytotoxins and their potential utilities. Crit Rev Environ Sci Technol 52:395–435
Chettri D, Boro M, Sarkar L, Verma AK (2021) Lectins: biological significance to biotechnological application. Carbohydr Res 506:108367
Christensen VG, Khan E (2020) Freshwater neurotoxins and concerns for human, animal, and ecosystem health: a review of anatoxin-a and saxitoxin. Sci Total Environ 736:139515
Cid-Gallegos MS, Corzo-Ríos LJ, Jiménez-Martínez C, Sánchez-Chino XM (2022) Protease inhibitors from plants as therapeutic agents - a review. Plant Foods Hum Nutr 77:20–29
Cotabarren J, Lufrano D, Parisi MG, Obregón WD (2020) Biotechnological, biomedical, and agronomical applications of plant protease inhibitors with high stability: a systematic review. Plant Sci 292:110398
Dai C, Das Gupta S, Wang Z, Jiang H, Velkov T, Shen J (2022) T-2 toxin and its cardiotoxicity: new insights on the molecular mechanisms and therapeutic implications. Food Chem Toxicol 167:113262
Dang L, Van Damme EJM (2015) Toxic proteins in plants. Phytochemistry 117:51–64
De Oliveira Filho JWG, Islam MT, Ali ES, Uddin SJ, Santos JVDO, De Alencar MVOB, Júnior ALG, Paz MFCJ, De Brito MDRM, Sousa JMDC, Shaw S, De Medeiros MDGF, Dantas SMMDM, Rolim HML, Ferreira PMP, Kamal MA, Pieczynska MD, Das N, Gupta VK, Mocan A, Dos Santos Andrade TDJA, Singh BN, Mishra SK, Atanasov AG, Melo-Cavalcante AADC (2017) A comprehensive review on biological properties of citrinin. Food Chem Toxicol 110:130–141

Divekar PA, Narayana S, Divekar BA, Kumar R, Gadratagi BG, Ray A, Singh AK, Rani V, Singh V, Singh AK, Kumar A, Singh RP, Meena RS, Behera TK (2022) Plant secondary metabolites as defense tools against herbivores for sustainable crop protection. Int J Mol Sci 23:2690

Duffull S, Isbister G (2022) Challenges faced when modeling clinical toxicology and toxinology events. CPT Pharmacometrics Syst Pharmacol 11:532–534

El Khoury A, Atoui A (2010) Ochratoxin a: general overview and actual molecular status. Toxins 2:461–493

Fink-Gremmels J, Malekinejad H (2007) Clinical effects and biochemical mechanisms associated with exposure to the mycoestrogen zearalenone. Anim Feed Sci Technol 137:326–341

Foster J, Kane P, Speight N (2002) The Detoxx system: detoxification of biotoxins in chronic neurotoxic syndromes. Townsend Lett Doctors Patients 232:77–83

Fraser JD (2011) Clarifying the mechanism of superantigen toxicity. PLoS Biol 9:e1001145

Fu L-L, Zhao X-Y, Ji L-D, Xu J (2019) Okadaic acid (OA): toxicity, detection and detoxification. Toxicon 160:1–7

Furey A, García J, O'callaghan K, Lehane M, Amandi MF, James KJ (2007) Brevetoxins: structure, toxicology, and origin. In: Phycotoxins: chemistry and biochemistry, pp 19–46

Galanos C, Freudenberg MA (1993) Bacterial endotoxins: biological properties and mechanisms of action. Mediat Inflamm 2:S11–S16

Galip N, Şanlidağ B, Babayiğit A, Bahçeciler NN (2022) Cutaneous allergic reactions to pine processionary caterpillar (Thaumetopoea Pityocampa): A complicated cutaneous reaction in an infant and review of the literature. Turk J Pediatr 64:389–393

Ghazaei C (2022) Advances in the study of bacterial toxins, their roles and mechanisms in pathogenesis. Malays J Med Sci 29:4–17

Goeders N, Chai R, Chen B, Day A, Salmond GPC (2016) Structure, evolution, and functions of bacterial type III toxin-antitoxin systems. Toxins 8:282

Grandjean P (2016) Paracelsus revisited: the dose concept in a complex world. Basic Clin Pharmacol Toxicol 119:126–132

Gwaltney-Brant SM (2017) Chapter 49 - Zootoxins. In: Gupta RC (ed) Reproductive and developmental toxicology, 2nd edn. Academic Press

Gwaltney-Brant SM, Dunayer E, Youssef H (2018) Terrestrial zootoxins. Elsevier, Veterinary toxicology

Haksar D, De Poel E, Van Ufford LQ, Bhatia S, Haag R, Beekman J, Pieters RJ (2019) Strong inhibition of cholera toxin B subunit by affordable, polymer-based multivalent inhibitors. Bioconjug Chem 30:785–792

Hartley MR, Lord JM (2004) Cytotoxic ribosome-inactivating lectins from plants. Biochim Biophys Acta 1701:1–14

He Y, Cox RJ (2016) The molecular steps of citrinin biosynthesis in fungi. Chem Sci 7:2119–2127

Hefter H, Samadzadeh S (2022) Clinical relevance of neutralizing antibodies in botulinum neurotoxin type A. Intech

Iglesias R, Russo R, Landi N, Valletta M, Chambery A, Di Maro A, Bolognesi A, Ferreras JM, Citores L (2022) Structure and biological properties of ribosome-inactivating proteins and lectins from elder (Sambucus nigra L.) leaves. Toxins 14:611

Janik E, Ceremuga M, Saluk-Bijak J, Bijak M (2019) Biological toxins as the potential tools for bioterrorism. Int J Mol Sci 20:1181

Janik E, Niemcewicz M, Ceremuga M, Stela M, Saluk-Bijak J, Siadkowski A, Bijak M (2020) Molecular aspects of mycotoxins-a serious problem for human health. Int J Mol Sci 21:8187

Janik E, Niemcewicz M, Podogrocki M, Ceremuga M, Stela M, Bijak M (2021a) T-2 toxin-the most toxic trichothecene mycotoxin: metabolism, toxicity, and decontamination strategies. Molecules 26:6868

Janik E, Niemcewicz M, Podogrocki M, Ceremuga M, Stela M, Bijak M (2021b) T-2 toxin—the Most toxic trichothecene mycotoxin: metabolism, toxicity, and decontamination strategies. Molecules 26:6868

Jungo F, Estreicher A, Bairoch A, Bougueleret L, Xenarios I (2010) Animal toxins: how is complexity represented in databases? Toxins 2:262–282
Kamle M, Mahato DK, Devi S, Lee KE, Kang SG, Kumar P (2019) Fumonisins: impact on agriculture, food, and human health and their management strategies. Toxins 11:328
Katikou P, Gokbulut C, Kosker AR, Campàs M, Ozogul F (2022) An updated review of tetrodotoxin and its peculiarities. Mar Drugs 20:47
Katoch R, Tripathi A (2021) Research advances and prospects of legume lectins. J Biosci 46:104
Klingelhöfer D, Braun M, Schöffel N, Oremek GM, Brüggmann D, Groneberg DA (2020) Ochratoxin—characteristics, influences and challenges of global research. Food Control 114:107230
Kocyigit E, Kocaadam-Bozkurt B, Bozkurt O, Ağagündüz D, Capasso R (2023) Plant toxic proteins: their biological activities, mechanism of action and removal strategies. Toxins 15:356
Kőszegi T, Poór M (2016) Ochratoxin A: molecular interactions, mechanisms of toxicity and prevention at the molecular level. Toxins 8:111
Kumar P, Mahato DK, Kamle M, Mohanta TK, Kang SG (2016) Aflatoxins: a global concern for food safety, human health and their management. Front Microbiol 7:2170
Langenegger N, Nentwig W, Kuhn-Nentwig L (2019) Spider venom: components, modes of action, and novel strategies in transcriptomic and proteomic analyses. Toxins 11:611
Li J, Wang Y, Deng Y, Wang X, Wu W, Nepovimova E, Wu Q, Kuca K (2022) Toxic mechanisms of the trichothecenes T-2 toxin and deoxynivalenol on protein synthesis. Food Chem Toxicol 164:113044
Lootens O, Vermeulen A, Croubels S, De Saeger S, Van Bocxlaer J, De Boevre M (2022) Possible mechanisms of the interplay between drugs and mycotoxins—is there a possible impact? Toxins 14:873
Madariaga-Mazón A, Hernández-Alvarado RB, Noriega-Colima KO, Osnaya-Hernández A, Martinez-Mayorga K (2019) Toxicity of secondary metabolites. Phys Sci Rev 4:20180116
Magowska A (2021) The natural history of the concept of antidote. Toxicol Rep 8:1305–1309
Meneely J, Greer B, Kolawole O, Elliott C (2023) T-2 and Ht-2 toxins: toxicity, occurrence and analysis: a review. Toxins 15:481
Mohammad S, Thiemermann C (2021) Role of metabolic endotoxemia in systemic inflammation and potential interventions. Front Immunol 11:594150
Nandakumar A (2022) Ayurvedic management of Urushiol-induced contact dermatitis with Agadayogas: a case report. J Ayurveda Case Re 5:80–83
Nwaji AR, Arieri O, Anyang AS, Nguedia K, Abiade EB, Forcados GE, Oladipo OO, Makama S, Elisha IL, Ozele N, Gotep JG (2022) Natural toxins and one health: a review. Sci One Health 1:100013
Obafemi BA, Adedara IA, Rocha JBT (2023) Neurotoxicity of ochratoxin A: molecular mechanisms and neurotherapeutic strategies. Toxicology 497-498:153630
Odumosu O, Nicholas D, Yano H, Langridge W (2010) Ab toxins: A paradigm switch from deadly to desirable. Toxins 2:1612–1645
Okechukwu VO, Adelusi OA, Kappo AP, Njobeh PB, Mamo MA (2024) Aflatoxins: occurrence, biosynthesis, mechanism of action and effects, conventional/emerging detection techniques. Food Chem 436:137775
Oliveira AL, Viegas MF, Da Silva SL, Soares AM, Ramos MJ, Fernandes PA (2022) The chemistry of snake venom and its medicinal potential. Nat Rev Chem 6:451–469
Parak M, Asgari A, Nourian YH, Ghanei M (2024) A review of poisoning with various types of biotoxins and its common clinical symptoms. Toxicon:107629
Pérez-Gómez A, Tasker RA (2022) Domoic acid as a neurotoxin. In: Kostrzewa RM (ed) Handbook of neurotoxicity. Springer, Cham
Pickova D, Ostry V, Toman J, Malir F (2021) Aflatoxins: history, significant milestones, recent data on their toxicity and ways to mitigation. Toxins 13:399
Poison PY (2014) The chemist. Chemist:19

Prévost G, Mourey L, Colin D, Monteil H, Dalla Serra M, Menestrina G (2006) Alpha-helix and beta-barrel pore-forming toxins (leucocidins, alpha-, gamma-, and delta-cytolysins) of Staphylococcus aureus. Elsevier
Pulido OM (2014) Domoic acid: biological effects and health implications. Toxins Biol Active Comp Microalgae 2:219–252
Qu L, Wang L, Ji H, Fang Y, Lei P, Zhang X, Jin L, Sun D, Dong H (2022) Toxic mechanism and biological detoxification of fumonisins. Toxins 14:182
Ramalingam S, Bahuguna A, Kim M (2019) The effects of mycotoxin patulin on cells and cellular components. Trends Food Sci Technol 83:99–113
Reddy P, Hemsworth J, Guthridge KM, Vinh A, Vassiliadis S, Ezernieks V, Spangenberg GC, Rochfort SJ (2020) Ergot alkaloid mycotoxins: physiological effects, metabolism and distribution of the residual toxin in mice. Sci Rep 10:9714
Rogowska A, Pomastowski P, Sagandykova G, Buszewski B (2019) Zearalenone and its metabolites: effect on human health, metabolism and neutralisation methods. Toxicon 162:46–56
Sakari M, Laisi A, Pulliainen AT (2022) Exotoxin-targeted drug modalities as antibiotic alternatives. ACS Infect Dis 8:433–456
Saleh I, Goktepe I (2019) The characteristics, occurrence, and toxicological effects of patulin. Food Chem Toxicol 129:301–311
Schmitt CK, Meysick KC, O'brien, A. D. (1999) Bacterial toxins: friends or foes? Emerg Infect Dis 5:224
Seilie ES, Bubeck Wardenburg J (2017) Staphylococcus aureus pore-forming toxins: the interface of pathogen and host complexity. Semin Cell Dev Biol 72:101–116
Seldeslachts A, Peigneur S, Tytgat J (2020) Caterpillar venom: a health hazard of the 21st century. Biomedicines 8:143
Sharma N, Sharma VK, Manikyam HK, Krishna AB (2016) Ergot alkaloids: a review on therapeutic applications. Eur J Med Plants 14:1–17
Sharon N, Lis H (2004) History of lectins: from hemagglutinins to biological recognition molecules. Glycobiology 14:53R–62R
Stockmann-Juvala H, Savolainen K (2008) A review of the toxic effects and mechanisms of action of fumonisin B1. Hum Exp Toxicol 27:799–809
Tsaneva M, Van Damme EJM (2020) 130 years of plant Lectin research. Glycoconj J 37:533–551
Ueno Y (1991) Biochemical mode of action of mycotoxins. In: Mycotoxins and animal foods, pp 437–453
Ülger TG, Uçar A, Çakiroğlu FP, Yilmaz S (2020) Genotoxic effects of mycotoxins. Toxicon 185:104–113
Ulhuq FR, Mariano G (2022) Bacterial pore-forming toxins. Microbiology 168:001154
Van Damme EJM (2022) 35 years in plant lectin research: a journey from basic science to applications in agriculture and medicine. Glycoconj J 39:83–97
Vilariño N, Louzao MC, Abal P, Cagide E, Carrera C, Vieytes MR, Botana LM (2018) Human poisoning from marine toxins: unknowns for optimal consumer protection. Toxins 10:324
Wahome PG, Robertus JD, Mantis NJ (2012) Small-molecule inhibitors of ricin and shiga toxins. In: Mantis N (ed) Ricin and shiga toxins: pathogenesis, immunity, vaccines and therapeutics. Springer, Berlin
Wang Y-H (2020) Sialidases from Clostridium perfringens and their inhibitors. Front Cell Infect Microbiol 9:462
Wang X, Quinn PJ (2010) Endotoxins: lipopolysaccharides of gram-negative bacteria. In: Endotoxins: structure, function and recognition, pp 3–25
Wang S, Li Z, Li S, Di R, Ho C-T, Yang G (2016) Ribosome-inactivating proteins (Rips) and their important health promoting property. RSC Adv 6:46794–46805
Wei J-T, Wu K-T, Sun H, Khalil MM, Dai J-F, Liu Y, Liu Q, Zhang N-Y, Qi D-S, Sun L-H (2019) A novel modified hydrated sodium calcium aluminosilicate (Hscas) adsorbent can effectively reduce T-2 toxin-induced toxicity in growth performance, nutrient digestibility, serum biochemistry, and small intestinal morphology in chicks. Toxins 11:199

Wei C, Yu L, Qiao N, Zhao J, Zhang H, Zhai Q, Tian F, Chen W (2020) Progress in the distribution, toxicity, control, and detoxification of patulin: a review. Toxicon 184:83–93

Więckowska M, Szelenberger R, Niemcewicz M, Harmata P, Poplawski T, Bijak M (2023) Ochratoxin A—the current knowledge concerning hepatotoxicity, mode of action and possible prevention. Molecules 28:6617

Xu Y, Sun MH, Li XH, Ju JQ, Chen LY, Sun YR, Sun SC (2021) Modified hydrated sodium calcium aluminosilicate-supplemented diet protects porcine oocyte quality from zearalenone toxicity. Environ Mol Mutagen 62:124–132

Yu C, Yu H, Li P (2020) Highlights of animal venom research on the geographical variations of toxin components, toxicities and envenomation therapy. Int J Biol Macromol 165:2994–3006

Zargar S, Wani TA (2023) Food toxicity of mycotoxin citrinin and molecular mechanisms of its potential toxicity effects through the implicated targets predicted by computer-aided multidimensional data analysis. Life 13:880

Zhang H, Ahima J, Yang Q, Zhao L, Zhang X, Zheng X (2021) A review on citrinin: its occurrence, risk implications, analytical techniques, biosynthesis, physiochemical properties and control. Food Res Int 141:110075

Zheng W, Wang B, Li X, Wang T, Zou H, Gu J, Yuan Y, Liu X, Bai J, Bian J, Liu Z (2018) Zearalenone promotes cell proliferation or causes cell death? Toxins 10:184

Chapter 2
Biosensors for the Detection of Biotoxins in Finfish and Shellfish

Tiasha Dutta, Sushil Kumar Mandal, and Jayanta Kumar Biswas

Abstract Fish is a vital entity of aquatic biodiversity contributing immensely to ecosystem functions and ecological services including nutritional and economical support to mankind. Currently, fishes have been exposed to biotoxins such as ciguatoxins, microcystins, okadaic acid, etc., produced by harmful algal blooms (HABs) occurring in various degraded aquatic systems. These toxins when ingested by finfishes and shellfishes get accumulated at hazardous levels in their tissues which eventually penetrate into the food chain, affecting a wide spectrum of species. Biotoxins are quite persistent in the environment because of their high stability and low degradability. It is predicted that with climate change, the occurrence of HABs and production of biotoxins will increase which demands monitoring of such outbreaks and assessment of associated ecotoxicological risk. Hence, fast detection, identification, and quantification of biotoxins affecting finfish and shellfish are of utmost importance. Biosensors have proved to be rapid, cost-effective, specific, sensitive, and real-time devices for detecting biotoxins as compared to the classical Mouse Based Assay (MBA), chromatographic, and spectroscopic methods. This chapter aims to elucidate the application of biosensors in detecting biotoxins affecting finfish and shellfish, their limitations and future prospects as a durable sensing device to tackle and mitigate the environmental, health, and economic implications, globally.

Keywords Biosensors · HABs · Biotoxins · Fish · Physiology · Environment · Health · Food safety · Economy

---

T. Dutta · S. K. Mandal · J. K. Biswas (✉)
Department of Ecological Studies & International Centre for Ecological Engineering, University of Kalyani, Kalyani, West Bengal, India
e-mail: jkbiswas@klyuniv.ac.in

M. N. V. Prasad et al. (eds.), *Biotoxins*,
https://doi.org/10.1007/978-3-031-75309-1_2

2.1 Introduction

Consumption of fish and shellfish contributes significantly to meeting the socio-economic and nutritional needs of the global population. They hold a key position in the ecological framework. But these finfishes and shellfishes face a serious risk of contamination with harmful algal blooms (HABs) releasing biotoxins. HABs occur in various aquatic habitats such as freshwater, marine, and brackish systems. These produce a plethora of harmful biotoxins such as ciguatoxin, cyanotoxins, etc., which when ingested by fishes such as pufferfish or shellfishes (e.g., oysters, clams) get accumulated in their tissues to hazardous levels. These biotoxins are quite stable and cannot be easily broken down subject to heating or cooking. As a result, they can be easily transferred via the food chain to higher trophic levels ranging from small herbivorous fishes to large predatory fishes eventually entering human population, potentially endangering a wide spectrum of aquatic life and humans. It is anticipated that with climate change, the occurrence of HABs and production of biotoxins are going to increase, and hence monitoring of such outbreaks and assessment of associated ecotoxicological risk are urgently warranted as a local to global concern (Wu et al. 2019; Bertani and Lu 2021; McPartlin et al. 2016).

Consequently, fast detection, identification, and quantification of biotoxins affecting finfish and shellfish are of utmost importance to combat socioeconomic and environmental threats (Cunha et al. 2018). Over the past few decades, a multitude of chemical and biological instrumental analyses have been established to detect the presence of biotoxins (Wu et al. 2019). The process is quite challenging because of the immensity of their chemical structures, properties, and toxicity mechanism. Due to insufficient evidence on their prevalence and technical difficulties involved in detection, most countries do not have any regulations to monitor their toxic effects (Vasconcelos 2015). Detection of biotoxins with Mouse Based Assay (MBA) was practiced conventionally. Due to ethical concerns, the assay was discontinued and banned in many countries. Parallelly, several other instrumental analyses such as Ultra-Performance Liquid Chromatography-Fluorescence Detection (UPLC-FLD), Liquid Chromatography with tandem Mass Spectrometry (LC-MS-MS), Liquid Chromatography-High Resolution Mass Spectrometry (LC-HRMS), Enzyme-Linked Immunosorbent Assay (ELISA), High-Performance Liquid Chromatography (HPLC), etc., are implemented for toxin detection. In spite of their efficiency, the processes were expensive, labor intensive, and time consuming. Extensive efforts are being made by scientists and institutions for developing alternative detection technology like biosensors which would be highly specific, sensitive, and cost-friendly devices for rapid screening and analysis of biotoxins affecting aquatic organisms (Ye et al. 2019; Reverté et al. 2023).

This chapter elucidates the application of biosensors in detecting aquatic biotoxins seriously affecting finfishes and shellfishes, their limitations and future prospects in the field.

2.2 Biotoxins in Aquatic Habitat

Biotoxins are natural toxins produced by living organisms as a defense mechanism. Aquatic habitats suffer from serious biotoxin contamination owing to harmful algal blooms (HABs). Several cyanobacteria, marine algae, dinoflagellates, etc., produce toxins contaminating freshwater, seawater, and brackish water. Freshwater systems (e.g., lakes, ponds, rivers) are generally affected by cyanobacteria producing cyanotoxins including microcystins, cylindrospermopsins, etc., while marine systems are affected by phycotoxins such as ciguatoxins, domoic acid, okadoic acid, etc., produced by diatoms and dinoflagellates. These toxins accumulate in the tissues of fishes and bivalve mollusks, which endangers their life, inflicts damage to the biodiversity, and poses severe risk to human health (Pinto et al. 2023). Often harmful algal blooms comprise a combination of toxins generated by multiple genera of cyanobacterial and algal species (Wu et al. 2019). Factors such as temperature, light, turbidity, salinity, eutrophication, hydrologic and meteorological conditions affect the production of biotoxins (Wu et al. 2019). These biotoxins cause neurotoxic, hepatotoxic, and dermatotoxic effects in organisms (Bertani and Lu 2021; Cunha et al. 2018). Fishes and bivalves are susceptible to biotoxin contamination majorly due to their feeding habits or through contact on their epithelium surfaces. Planktivorous filter feeders are potent carriers of marine biotoxins during harmful algal blooms (HABs). Intense feeding on mollusks and shellfishes contaminated with various marine biotoxins affect fishes dangerously (Corriere et al. 2021). Both exposure routes exhibit histopathological effects in gills, intestine, liver, and kidney of fishes. Other physiological dysfunctions comprising cardiac, hematological, immunological, biochemical alterations, ionic imbalance, etc., have also been reported. Besides, performances of fishes in terms of foraging, swimming, escaping from predators, etc., are also impacted through biotoxin exposure.

2.2.1 Freshwater Toxins

Freshwater habitats are generally contaminated by cyanotoxins. Cyanobacteria are known to produce secondary metabolites which are categorized as cyclic peptides, lipopolysaccharides, and alkaloids (Pinto et al. 2023). The most common cyanotoxins affecting the fish and shellfish are discussed in the following section and summarized in Table 2.1.

Table 2.1 Classification of aquatic toxins affecting finfish and shellfish

Sl No.	Type of toxins	Organisms	Nature of toxicity	Chemical structure	Reference
1.	*Freshwater toxins*				
	Microcystin (MC)	*Microcystis* sp., *Nostoc* sp., *Anabaena* sp., *Planktothrix* sp., *Phormidium* sp.	Hepatotoxic	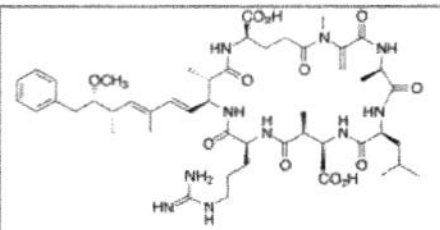	Cunha et al. (2018), Wu et al. (2019), Pinto et al. (2023), image source: Wikipedia.
	Anatoxin-A (ATX)	*Arthrospira* sp., *Oscillatoria* sp., *Raphidiopsis* sp., *Anabaena* sp.	Neurotoxic	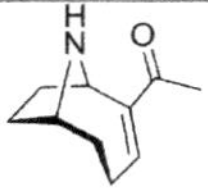	Cunha et al. (2018), image source: Wikipedia.
	Cylindrospermopsin (CYN)	*Cylindrospermopsis* sp., *Raphidiopsis* sp., *Lyngbya* sp., *Aphanizomenon* sp., *Umezakia* sp., and *Anabaena* sp.	Cytotoxic, dermatoxic, genotoxic, tumor promoter	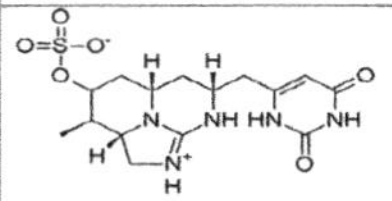	Cunha et al. (2018), image source: Wikipedia
	Saxitoxin (STX)	*Aphanizomenon* sp., *Scytonema* sp., *Lyngbya* sp., *Anabaena* sp., *Cylindrospermopsis* sp. and *Planktothrix* sp.	Neurotoxins causing paralytic shellfish poisoning (PSP)	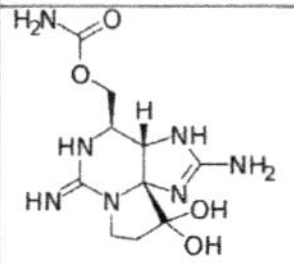	Pinto et al. (2023), image source: Wikipedia
2.	*Marine toxins*				

	Brevetoxin (BTX)	*Karenia* sp., *Chattonella* sp., *Fibrocapsa japonica*, *Heterosigma akashiwo*	Neurotoxins causing neurotoxic shellfish poisoning (NSP)	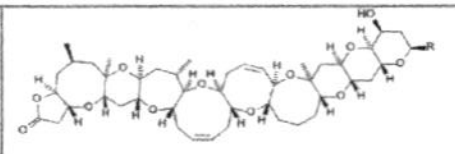	Cunha et al. (2018), Murk et al. (2019), Wu et al. (2019), Pinto et al. (2023), image source: Wikipedia
	Saxitoxin (STX)	*Coolia* sp., *Protogonyaulax* sp., *Alexandrium* sp., *Prorocentrum* sp. *Gymnodinium* sp., and *Pyrodinium* sp.	Neurotoxins causing paralytic shellfish poisoning (PSP)	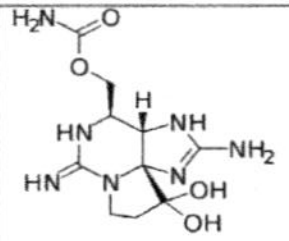	Cunha et al. (2018), Murk et al. (2019), Wu et al. (2019), image source: Wikipedia
	Okadaic acid (OA) and its analog dinophysis toxins (DTXs)	*Dinophysis* sp., *Prorocentrum* sp., and *Phalacroma* sp.	Gastrointest-inal problems causing diarrhetic shellfish poisoning (DSP)	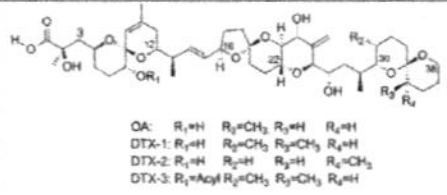	Corriere et al. (2021), Cunha et al. (2018), Murk et al. (2019), Wu et al. (2019), image source: Wikipedia
	Domoic acid (DA)	*Chondria armata* Pseudo-nitzschia	Neurotoxins causing amnesic shellfish poisoning (ASP)	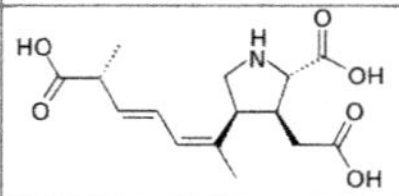	Murk et al. (2019), Wu et al. (2019), Pinto et al. (2023), image source: Wikipedia
	Tetrodotoxin (TTX)	*Pseudoalteromonas tetraodonis*, *Pseudomonas* spp., and *Vibrio* spp.	Neurotoxic	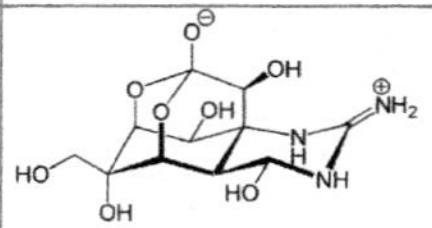	Cunha et al. (2018), Murk et al. (2019), image source: Wikipedia
	Ciguatoxin (CTX)	*Gambierdiscus toxicus* and *Fukuyoa* sp.	Neurotoxins causing ciguatera poisoning (CP)	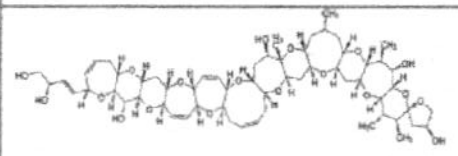	Murk et al. (2019), Wu et al. (2019), Pinto et al. (2023), image source: Wikipedia
	Palytoxin (PTX)	*Ostreopsis* sp., *Trichodesmium* sp.	Neurotoxic	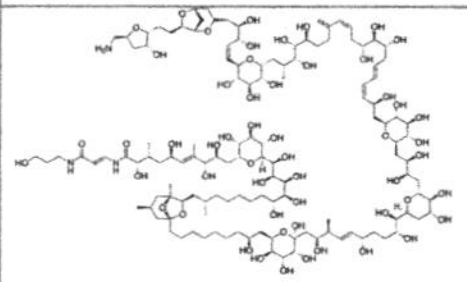	Cunha et al. (2018); Pinto et al. (2023), image source: Wikipedia
	β-N-Methylamino-L-alanine (BMAA)	*G. catenatum*	Neurotoxic	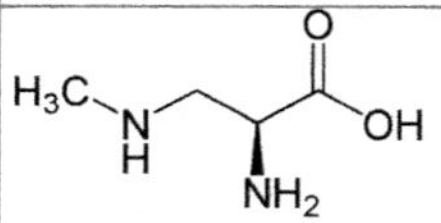	Pinto et al. (2023), image source: Wikipedia.

2.2.1.1 Microcystins (MCs)

Microcystins are secondary metabolites released by multiple species of cyanobacteria exemplified by *Microcystis*, *Nostoc*, *Dolichospermum* (previously *Anabaena*), *Planktothrix*, *Phormidium* spp., etc., found in freshwater habitat when the cells undergo lysis due to stress or aging. MCs are monocyclic heptapeptides consisting of a peculiar aromatic amino acid named ADDA (3-amino-9-methoxy-2, 6, 8-trimethyl-10-phenyldeca-4, 6-dienoic acid) along with two other variable amino acids, leucine (L) and arginine (R). MC-LR is the most prevalent MC congeners exhibiting structural variations. They also differ based on the extent of hydroxylation, epimerization, methylation, etc., which serve as key determinants of their toxicity. Microcystins are water-soluble, heat-resistant, and stable compounds. As a result of which their natural degradation is a slow process and easily persist in warm and cold conditions in aquatic habitat (Wu et al. 2019). Another toxin nodularins, having analogous chemical structure and toxicity mechanism exist in the environment (Bittner et al. 2021). MC-LR is hepatotoxic in nature. These toxins are responsible for hyperphosphorylation of proteins which can cause liver damage, disruption of DNA damage repair, cell division processes, mitochondrial and endoplasmic reticulum activities. They also possess carcinogenic properties acting as tumor promoters and also induce oxidative stress in organisms (Cunha et al. 2018; Pinto et al. 2023; Kholafazad Kordasht et al. 2020; Bittner et al. 2021; Chen et al. 2016; Vichi et al. 2016; Wu et al. 2019).

2.2.1.2 Anatoxin-a (ATX-a)

Anatoxin has several analogs in freshwater among which Anatoxin-a (ATX-a) is the most prevalent one. Cyanobacteria such as *Arthrospira*, *Oscillatoria*, *Raphidiopsis*, *Anabaena* spp. produce it as a secondary amine (Cunha et al. 2018). The toxin is relatively stable in acidic conditions with pH less than 3 and 1–2 h of an average half-life under normal light conditions with pH ranging from 8 to 9. Anatoxins exhibit higher persistence and longer half-life under lower light conditions and temperature (under 20 °C) (Christensen and Khan 2020; Pinto et al. 2023; Chorus and Welker 2021). Anatoxins are neurotoxins which disrupt the synaptic hydrolysis of acetylcholine causing excitation of nerves and muscles leading to muscle exhaustion, fatigue, convulsions and in extreme cases leading to respiratory arrest and death (Cunha et al. 2018; Fiore et al. 2020; Pinto et al. 2023). Since the nerves and muscles are severely affected, fishes undergo several motor and metabolic issues, thereby increasing their vulnerability to predators. Decrease in mobility and behavioral changes pose a detrimental effect on fish populations affecting their reproductive potential and predator–prey interactions (Pinto et al. 2023).

2.2.1.3 Cylindrospermopsin (CYN)

This freshwater toxin is produced by some cyanobacterial species, for example *Aphanizomenon*, *Cylindrospermopsis*, *Raphidiopsis*, *Anabaena*, *Lyngbya*, and *Umezakia* spp. It is a polycyclic uracil derivative alkaloid and highly stable in water because of its zwitterionic property. The presence of uracil group determines its toxicity (Ye et al. 2019; Cunha et al. 2018).

Cylindrospermopsin (CYN) is a potent cytotoxin, genotoxin, dermatoxin, and promotes tumors. This toxic alkaloid obstructs synthesis of proteins and causes modification of DNA and/or RNA covalently resulting in genomic instability. These toxins affect the liver and hamper the functions of other organ systems in the body (Cunha et al. 2018; Bittner et al. 2021; Pinto et al. 2023).

2.2.2 *Marine Toxins*

Marine environment is dominated by toxins majorly released by diatoms and dinoflagellates such as *Karenia, Pseudo-nitzschia, Alexandrium*, *Dinophysis*, and *Gymnodinium* spp. (Farabegoli et al. 2018; Pinto et al. 2023; Visciano et al. 2016). These toxins accumulate and persist in the tissues of bivalve mollusks, crustaceans, and fishes for several months eventually entering the food chain (Visciano et al. 2016; Nicolas et al. 2017). They are highly persistent in the shellfish because of their poor natural detoxification mechanism (Botelho et al. 2019). Sometimes, the marine organisms may also be exposed to a mixture of toxins which can produce synergistic effects. Marine biotoxins are generally classified into 10 groups based on their chemical structure. The most common marine biotoxins have been elaborated in the following section (as depicted in Table 2.1) (Pinto et al. 2023; Cunha et al. 2018).

2.2.2.1 Brevetoxins (BTX)

They are lipophilic cyclic polyether compounds released by different species of selected dinoflagellates (e.g., *Karenia* and *Chattonella* spp.) (Visciano et al. 2016; Farabegoli et al. 2018; Murk et al. 2019; Pinto et al. 2023; Ye et al. 2019). *Fibrocapsa japonica* and *Heterosigma akashiwo* are some other algal species producing compounds similar to BTX (Farabegoli et al. 2018; Pinto et al. 2023). They are stable toward heat and acidic conditions. BTX are potent neurotoxins binding and activating Na^+ channels in neuronal and muscle cell membranes hence disrupting the normal voltage processes. Cells are depolarized affecting the transmission of neuromuscular impulses. This toxin is fatal to aquatic fishes, mammals, and many other aquatic invertebrates as well. Ingestion of fish or shellfish by humans contaminated with BTX leads to Neurotoxic Shellfish Poisoning (NSP). It is released by *Karenia brevis* which is responsible for several physiological anomalies leading to

respiratory and myocardial dysfunction. Furthermore, BTX cause gastrointestinal problems, embryotoxicity, developmental toxicity, immunotoxicity, genotoxicity, and carcinogenic effects (Pinto et al. 2023; Ye et al. 2019; Cunha et al. 2018; Murk et al. 2019).

2.2.2.2 Saxitoxin (STX)

It is a light carbamate alkaloid—a tetrahydropurine derivative. In marine environment, dinoflagellates such as *Protogonyaulax* sp., *Coolia* sp., *Gymnodinium* sp., *Pyrodinium* sp., *Prorocentrum* sp., and *Alexandrium* sp. produce STX, whereas filamentous cyanobacteria species such as *Cylindrospermopsis* sp., *Anabaena* sp., *Planktothrix* sp., *Aphanizomenon* sp., *Scytonema* sp., and *Lyngbya* sp. are responsible for STX production in freshwater. Lobster, carnivorous snails and crabs are potent vectors. Bivalve mollusks serve as the prime vectors to humans (Cunha et al. 2018; Pinto et al. 2023; Ye et al. 2019). Saxitoxins generally accumulate in the viscera and/or digestive glands of the contaminated marine organisms including shellfish, octopus, squid, etc. (Pinto et al. 2023; Chorus and Welker 2021). These toxins are accumulated in the livers, digestive tract, and muscular tissues of finfishes and shellfishes (Murk et al. 2019).

As reported by de Morais Calado et al. (2020), saxitoxins cause DNA damage, oxidative stress and alter osmoregulatory functions in a freshwater fish like *Geophagus brasiliensis* (cichlid fish). They modulate and promote genes such as VEGFC and DLL4 responsible for vascular development. They also alter the functions of genes related to apoptosis (P53, BAX, CASPASE 3, BCL-2) in zebrafish (*Danio rerio*) embryos (Chen et al. 2020). STXs are potent neurotoxins causing paralytic shellfish poisoning (PSP) in humans. Voltage-gated Na^+ and Ca^- channels in neurons are blocked like other neurotoxins eventually affecting neuromuscular junctions leading to paralysis (Christensen and Khan 2020; Cunha et al. 2018; Pinto et al. 2023; Ye et al. 2019). Saxitoxin activity is dependent on heat, light and pH. High temperature, extracellular concentration of Na^+/Cl^-, and light intensity upregulate the activity of stx gene cluster producing saxitoxin whereas it undergoes downregulation under dark conditions and in the presence of excess nitrogen concentration (Boopathi and Ki 2014). It is a great threat to humans because they cannot be destroyed by the digestive system as it is easily absorbed in the gastrointestinal tract. Gonyautoxins (GTX 1/4) are saxitoxin analogs released by several marine dinoflagellate species and detected in echinoderms, bivalves, and some gastropods (Cunha et al. 2018; Pinto et al. 2023; Ye et al. 2019).

2.2.2.3 Okadaic Acid (OA)

Okadaic acid (OA) along with its analogs dinophysistoxins (DTXs) is a complex group of polyketide polyether lipophilic compounds which are secondary metabolites released by marine dinoflagellates species belonging to the genus *Dinophysis*,

Prorocentrum, and *Phalacroma*, and was also found in black sponges (*Halichondria okadai* and *Halichondria melanodocia*) and mussels (*Mytilus* sp.) (Cunha et al. 2018; Corriere et al. 2021).

Understanding the toxicity mechanism of OA is quite complex. Both OA and DTX have permeability through cell membranes due to which they can easily alter intracellular processes and inhibit protein (serine/threonine) and alkaline phosphatases such as PP1 and PP2A leading to intestinal mucosal damages. They cause diarrhetic shellfish poisoning (DSP) in humans leading to vomiting, nausea, and abdominal pain. The causative agents for DSP are both okadaic acid and dinophysis toxin-1 and -2 majorly carried by bivalves. Apart from OA and DTX, pectenotoxins and yessotoxins are also associated with DSP (Wu et al. 2019). They also alter gene expression, metabolic processes, exert neurotoxic, carcinogenic, cytotoxic, immunotoxic, and genotoxic effects (Cunha et al. 2018; Corriere et al. 2021; Ye et al. 2019). These toxins also play a role as tumor promoters. Since they are lipophilic, they can be easily accumulated in the fatty tissues of fishes and shellfishes, hence entering the food chain.

2.2.2.4 Domoic Acid (DA)

It is a cyclic tricarboxylic amino acid which was primarily detected in *Chondria armata*, a red algal species (Murk et al. 2019; Pinto et al. 2023). Later, it was discovered that *Pseudo-nitzschia* (a marine diatom) releases domoic acid and gets accumulated in the digestive tracts of organisms like bivalves, crabs, fish, etc., during filtration or grazing of benthic algae (Murk et al. 2019; Ye et al. 2019; Pinto et al. 2023).

Domoic acid (DA) is neurotoxic and induces excitatory toxicity with a tendency to bind competitively with the receptor of glutamic neurotransmitter. ROS is produced due to the over-excitation of these glutamate receptors causing cell death, via activation of N-methyl-D-aspartate (NMDA) receptors. It also leads to the imbalance of sodium osmotic pressure (Ye et al. 2019; Pinto et al. 2023). DA induces behavioral and developmental anomalies as observed in zebrafish embryos. DNA fragmentation was observed in the digestive glands of other organisms like *Mytilus edulis* (Pinto et al. 2023). DA produced from pennate diatom *Pseudo-nitzschia pungens* is known to cause amnesic shellfish poisoning (ASP). Major symptoms include neurological and gastrointestinal disorders which might ultimately lead to memory loss, coma or even death (Wu et al. 2019).

2.2.2.5 Tetrodotoxin (TTX)

Tetrodotoxin is an exclusively lethal marine neurotoxin found in nature composed of an amino perhydroquinazoline compound mostly isolated from Pufferfish and amphibians such as Salamander, Xenopus, etc. It has also been identified in gastropods, newts, star fishes, crabs, sea slugs ribbon worms, blue-ringed octopuses, and

bacteria. They are released by endosymbiotic bacteria such as *Pseudoalteromonas tetraodonis*, *Vibrio* spp., and *Pseudomonas* spp. which naturally inhabit in the host organism body. Its mechanism is similar to other neurotoxins leading to paralysis of respiratory muscles, hence respiratory failure (Cunha et al. 2018; Murk et al. 2019; Ye et al. 2019; Reverté et al. 2023).

2.2.2.6 Ciguatoxin (CTX)

Ciguatoxins (CTXs) are lipophilic polyether compounds released by benthic dinoflagellates like *Fukuyoa* sp. and *Gambierdiscus toxicus*. They generally occur in predatory fishes like moray eels, Spanish mackerels, snappers, and barracuda (Visciano et al. 2016; Murk et al. 2019). They too possess thermal stability and oxygenated characteristics.

They are neurotoxic causing ciguatera poisoning (CP). The nerve cells get depolarized due to blockage in the voltage-gated sodium channels (Reverté et al. 2023; Pinto et al. 2023; Wu et al. 2019; Murk et al. 2019). Overall, they induce severe neurological, cardiological, and gastrointestinal issues in the organisms.

2.2.2.7 Palytoxin (PTX)

Palytoxins are another class of biotoxins released by cyanobacteria (*Trichodesmium* sp.). and marine dinoflagellates (*Ostreopsis* sp.). Initially, they were also reported in diverse corals such as *Palythoa* sp. Major vectors of this toxin are crabs, goldspot herring, parrotfish, etc. They are made of complex fatty alcohols. They induce toxic effects on dermal, ocular, and inhalation exposure. Binding of these toxins to extracellular K^+ and Na^{2+} channels leads to overstimulation of the axons inhibiting active transportation of the ions across the membranes. Overall, it induces a neurotoxic effect on the organisms (Cunha et al. 2018; Pinto et al. 2023).

2.2.2.8 β-N-Methylamino-L-Alanine (BMAA)

BMAA is another group of biotoxin produced by freshwater cyanobacteria as well as dinoflagellates such as *G. catenatum* and marine diatoms (Pierozan et al. 2020; Christensen and Khan 2020; Rodgers et al. 2018; Pinto et al. 2023). It generally accumulates in bivalves, fishes, and crustaceans (Chorus and Welker 2021; Lance et al. 2018). It induces glutamate receptor-mediated excitotoxicity hence, affecting the motor neurons (Pulido 2016; Rodgers et al. 2018).

2.2.3 Evolutionary Significance

Biotoxins are natural toxins produced by organisms and have unique roles in the environment. The prime role is to provide chemical defense against other organisms in a competitive environment, for example, cyanotoxins are produced to combat other algal species, cellular nitrogen source, chemical signaling, improving physiological processes like photosynthesis, growth rates, etc. They are also released as a result of cell death and senescence (Pinto et al. 2023; Walls et al. 2018; Rodgers et al. 2018; Christensen and Khan 2020). In some cases, anatoxin is released during the growth phase of the organism particularly under stressed conditions like low light conditions, nitrogen starvation, presence of competition (Christensen and Khan 2020; Boopathi and Ki 2014; Pinto et al. 2023). The mystery behind the production of these biotoxins lies in the specific genetic makeup of the organisms which is controlled by environmental parameters like water temperatures, nutrient (nitrate and phosphate availability), light, salinity, etc. (Christensen and Khan 2020; Pinto et al. 2023; Macario et al. 2021; Rodgers et al. 2018). Production of saxitoxin is an ancient trait of cyanobacteria dating back to millions of years ago. The inheritance of the stx gene cluster is the backbone of the survival of these species (O'Neill et al. 2016). In case of marine saxitoxins produced by dinoflagellates, it is believed that saxitoxin biosynthesis is dependent on a different cluster of stx genes. Sometimes it is also produced after chemical and enzymatic reactions occurring in the tissues of vector species (Pinto et al. 2023; Cembella and Durán-Riveroll 2021; Akbar et al. 2020). The evolutionary background and functional significance of these toxins are not clearly known yet and are under active research.

2.2.4 Biotechnological and Therapeutic Applications

Currently, researchers are interested to explore the brighter side of these fatal biotoxins in the pharmaceutical and biotechnological fields. Modern day lifestyle and rapid urbanization have given rise to several dangerous diseases such as cancer, obesity, physiological issues, etc., which demand vigorous treatments leading to harmful side effects. In this aspect, it is anticipated that these toxins could serve as a potent natural source for drug and clinical development as reviewed by Pradhan and Ki (2022). Biosensors could act as an essential tool for detection, quantification, and analysis of these toxins for further therapeutic development. Cyanotoxins like microcystin (MC-LR), nodularins, etc., and domoic acid (DA) produced by certain diatom species have been known to possess anti-tumor or anti-cancer properties effective through apoptosis. It has been studied that toxins produced by dinoflagellates have wider applications due to their interesting structure and function. Saxitoxins (STX) are known to be potent neurotoxins but they have been identified as promising anesthetic agent and exhibit effective antibacterial, antifungal, antialgal, and antiprotozoal activity in vitro. Tetrodotoxin (TTX) has a clinical

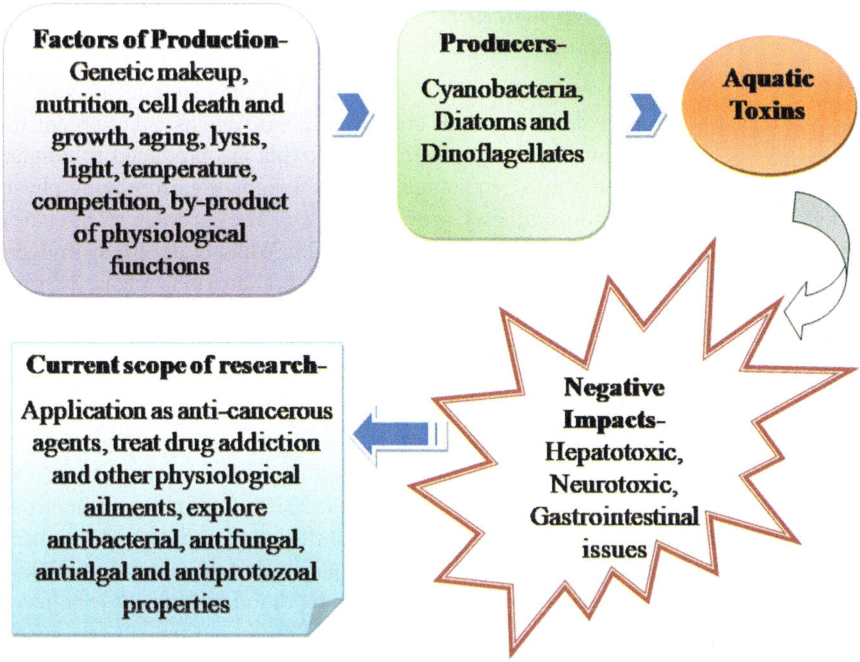

Fig. 2.1 Origin and role of aquatic toxins in the environment

application similar to STX in addition to treatment of drug addiction. OA and its analog DTX have been used to explore cellular processes related to phosphorylation, anti-tumor properties, immunological responses, fungicidal effects, and neurodegenerative illnesses. Ciguatoxin (CTX) has been found to provide a therapeutic agent for cardiovascular issues, including epilepsy. Brevetoxins (BTX) are known to reduce certain respiratory issues, cystic fibrosis, and brain related ailments. Palytoxins (PLTX) have also been found to possess anti-tumor properties. All these toxins direct toward a wide scope in therapeutic and commercial development but the major limitation lies in their availability for active testing and research. They are still in their pre-clinical and clinical stage of testing and yet to be explored further for pharmacological applications, hence combating harmful and fatal diseases in humans. The role of aquatic toxins in the environment has been depicted in Fig. 2.1.

2.3 Biosensors: Detection of Aquatic Toxins

Analytical devices composed of bio-recognition components including enzyme, cell, microorganism, or antibody combined with a physical or chemical transducer element (optical, piezoelectric, etc.) are known as biosensors. Till date, aquatic toxins such as DA, OA, PTXs, BTXs, STXs, microcystins, etc., were primarily detected

using conventional instrumental analysis methods (e.g., LC–MS) or biological assays such as mouse bioassay, lateral flow assays using antibodies, and ELISA kits (Wu et al. 2019; Pinto et al. 2023; Merel et al. 2013; Nicolas et al. 2017; Morabito et al. 2018; Vogiazi et al. 2019; Cruz et al. 2018; Cunha et al. 2018; Murk et al. 2019; Raposo et al. 2020; Liang et al. 2022; Tang et al. 2023; Tian et al. 2021). Due to their limitations regarding sensitivity, cost-effectiveness, specificity, labor, and time factor, they are slowly being replaced by biosensors (Murk et al. 2019; Morabito et al. 2018; Nicolas et al. 2017; Merel et al. 2013). Biosensors are currently serving as a potential tool for detecting, quantifying, and monitoring aquatic toxins. Categorization of biosensors and their efficiency is determined by the following four factors (Tang et al. 2023).

- Biorecognition elements like cells, receptors, antibodies, and aptamers which are used as sensing elements.
- Types of physicochemical transducers like electrochemical, optical, photothermal, etc., which sense the recognition reactions between analytes and biomolecule and provide the biochemical information.
- Materials used for signal amplification like nanoparticles for high sensitivity.
- Type of device used for signal processing, acquisition, and display of results.

Several kinds of novel biosensors have been designed which have been discussed further in the chapter (Fig. 2.2) (Table 2.2).

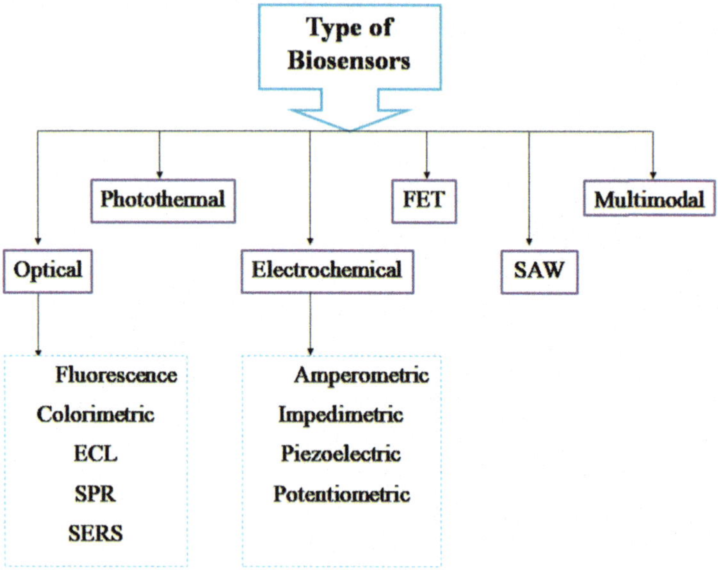

Fig. 2.2 Types of biosensors for detecting aquatic biotoxins

Table 2.2 Summary of biosensors to detect aquatic toxins

Sl. No.	Types of biosensor	Mode of action	Toxins detected	Remarks		References
				Positive	Negative	
1.	*Optical*					
	Fluorescence	Fluorescent signals produced using materials like traditional fluorescence proteins, dyes, carbon dots, advanced fluorescent nanomaterials, etc.	Okadaic acid (OA), MC-LR, saxitoxin (STX), and domoic acid (DA)	Multiple analysis, nondamaging, highly sensitive	Manufacturing cost, sample autofluorescence	Tang et al. (2023), Wu et al. (2019), Tian et al. (2021)
	Colorimetric	Absorption of matter with respect to light following the Lambert–beer law for detecting analyte concentration	Domoic acid (DA), saxitoxin (STX), DSP, okadaic acid (OA), MC-LR	Rapid analysis, naked-eye readout, easy operation	Less stable	Tang et al. (2023), Wu et al. (2019), Tian et al. (2021)
	ECL	Electrode driven luminescence process	Okadaic acid (OA), palytoxin (PLTX)	Easy observation, high sensitivity and selectivity, wide LOD, stable, high luminosity, low-background interference, easy operation, cost-effective, and non-damaging characteristics	Less portable, sample matrix interference	Peng et al. (2020), Tang et al. (2023), Wu et al. (2019), Tian et al. (2021)
	SPR	Measurement of the excitation of surface plasmons using refractometric sensing technique	Saxitoxin (STX), MC-LR, PSTs, okadaic acid (OA), domoic acid (DA)	Stable and portable	Less sensitive, cannot detect very low concentrations	Tang et al. (2023), Wu et al. (2019), Tian et al. (2021)
	SERS	Raman scattering	Domoic acid (DA), MC-LR, okadaic acid (OA)	High speed, sensitivity, simple operation	Can be damaging and suitable functionalization required	Tang et al. (2023), Wu et al. (2019), Tian et al. (2021)

2.	*Electrochemical*					
	Amperometric	Measurement of current produced at the electrode surface using an enzymatic or bioaffinity reaction at a stable working potential coordinated with the reference electrode	Domoic acid (DA), MC-LR, okadaic acid (OA)	Easy to fabricate, disposable electrode, selective, accurate, and stable	Expensive metal electrodes need to be altered, better portability, sensitivity and time consumption to be improved	Wu et al. (2019), Tian et al. (2021)
	Impedimetric	Measuring the resistance and capacitance properties of materials and recording the impedance spectrum with respect to the shift in electric impedance as a function of the probed frequencies	Saxitoxin (STX), cylindrospermopsin (CYN), okadaic acid (OA)			Wu et al. (2019), Tian et al. (2021)
	Piezoelectric	Application of an alternating electric potential to the piezoelectric crystal hence the generation of acoustic waves in solids via piezoelectric effect	MC-LR, okadaic acid (OA)			Tang et al. (2023), Wu et al. (2019)
	Potentiometric	Measurement of differences in the electrical and potential intensity due to aggregated electrical charges over the dielectric layer	Saxitoxin (STX)			Tian et al. (2021)
3.	*Photothermal*	Absorption and conversion of light energy to heat energy in the presence of a specific laser	–	High resolution	Expensive	Tang et al. (2023)

(continued)

Table 2.2 (continued)

4.	*Field-effect transistor (FET)-based*	Biological sensing element and FET transducer	Okadaic acid (OA)	High sensitivity, selectivity, stability, low cost, short time span	For field detection, compaction of the equipment and improvement in detection flux are required	Tian et al. (2021)
5.	*Surface acoustic wave (SAW)*	Recording the propagation of a series of elastic waves on shallow solid surface using devices such as oscillator, interface membrane material, and oscillation circuit	Okadaic acid (OA)	No-contact sensing, speed, easy, and unpowered operation, non-interference, simple coding, excellent secrecy, and cheap	Sensitivity and measurement time need to be improved	Tian et al. (2021)
6.	*Multimodal*	Combination of colorimetric, fluorescence, electrochemistry, SERS, and photothermal signal elements	–	High LOD, accurate, stable	–	Tang et al. (2023)

2.3.1 *Optical Biosensor*

Optical biosensors are a class of sensors which recognizes analytes and furnishes results based on optical changes. These sensors are versatile and popular in their application because they provide quick and real-time determination of the toxins. The measurements are generally recorded based on changes in the frequency signal, phase or polarized light and amplitude, directly proportionate to the toxin concentration (Wu et al. 2019). Some of the common optical biosensors used for detection of biotoxins in aquatic habitats have been discussed in further sections (Tang et al. 2023; Wu et al. 2019; Kholafazad Kordasht et al. 2020; Bertani and Lu 2021; Tian et al. 2021).

2.3.1.1 Fluorescence Biosensor

Fluorescence optical biosensors are integrated with elements that mimic biological samples capable of recognizing target analyte and conversion into a detectable fluorescent optical signal, directly proportional to the analyte concentration. These biosensors are label-based devices ranging from traditional fluorescence proteins, dyes, carbon dots, advanced fluorescent nanomaterials, etc. (Tang et al. 2023). These biosensors are known for their nondamaging mode of detection, high sensitivity, cost-effectiveness, and less polluting functional nature. Fluorescence biosensors developed for shellfish toxin detection is mainly based on aptamers and immunoassays which incorporate different techniques including microfluids, flow cytometry, chromatography, etc., to enhance the detection performance (Sun et al. 2018). For example, Lian et al. (2021) developed a dual-emission ratiometric fluorescence biosensor labeled with carbon dots for detecting OA of about 25 ng/l (LOD) in marine water (Tang et al. 2023). Another researcher developed a sensitive waveguide based fluorescent immunosensor for detecting MC-LR in lakes (Wu et al. 2019). Several aptasensor-based fluorescent probes have also been designed by scientists for OA detection. Multiplex microarray sensors are also used for detecting several harmful cyanobacterial and algal toxins such as STX and its analogs, MCs and its analogs, OA, CYN, and DA, found in freshwater, brackish water, and marine environments (Kholafazad Kordasht et al. 2020).

2.3.1.2 Colorimetric Biosensor

Colorimetric biosensors are popularly used for detecting aquatic toxins. Based on the principle of Lambert–Beer law, sensors like aptasensors, immunosensors, cell-based sensors, etc., selectively absorb matter with respect to light for the detection of concentration of analyte (Tian et al. 2021). Recently, aptamers and nanomaterials (e.g., AgNPs, AuNPs, etc.) have gained a lot of popularity in designing colorimetric biosensors because they are more stable, effective, economical, and less time

consuming. For example, colorimetric aptasenors designed as per the peroxidase-like catalytic action of AuNPs have been used to detect saxitoxins (STX) (Zhao et al. 2021a). Similarly ELISA based colorimetric biosensors have been designed for detecting other important algal toxins like MC-LR, OA, DA, STX, etc. (Tang et al. 2023; Tian et al. 2021).

2.3.1.3 Electrochemiluminescence Biosensor

Electrochemiluminescence (ECL) biosensors are based on electrode-driven luminescence process in which modified electrode surface produces luminescence of excited states by controlling the electrode potential (Tang et al. 2023). It is a combination of the principle of chemiluminescence and electrochemistry. Its cost-efficiency, high selectivity, luminosity, low-background noise, easy operation, and non-damaging characteristics have aroused research interest as compared to other optical biosensors. Several ECL biosensors have been implemented for detection of biotoxins such as OA, PLTX, etc. (Wu et al. 2019). Peng et al. (2020) is known to have designed an ECL biosensor with carboxylated multiwalled carbon nanotubes (MWCNTs) and poly (diallyldimethylammonium) chloride-Au nanocluster conjugated with a monoclonal antibody against okadaic acid and also an ECL immunosensor having graphene quantum dots co-doped with uniform-sized phosphorus and sulfur (P, S-GQDs) dispersible in water in combination with monoclonal antibody (McAb) for detection and quantification of OA.

2.3.1.4 Surface Plasmon Resonance (SPR) Biosensor

This biosensor is made up of a chip whose surface is coated with thin films of gold conducting material providing an interface for binding and immobilization of specific ligands. The signals are detected with an optical detector and analyzed through an integrated flow-through system (Wu et al. 2019). The contact between the immobilized molecule and analyte molecule excites the surface plasmon polaritons and hence produces a signal which is detected by the device (Tang et al. 2023). The analysis of toxin concentration is dependent on the shift in the refractive index of the label-free sensor caused by the SPR phenomenon which is proportionate to the toxin content. These biosensors do not require the use of harmful solvents or dangerous labels; hence they are fit to be designed as portable biosensors for quick monitoring. Further, the sensitivity of these biosensors can be increased using nanomaterials. SPR biosensors are known to be the first among the optical biosensors to detect saxitoxins and its analogs effectively. Apart from saxitoxin, SPR biosensors have also been reported to detect various aquatic toxins in fishes and shellfishes like MC-LR, OA, DTX, YTX, PSTs, etc. For detecting okadaic acid in mussels, Garibo et al. (2014) fabricated an immunosensor based on SPR where the antibody carriers are designed with magnetic particles. Similarly, several such SPR based biosensors can be designed on the basis of surface immobilization of the toxin onto the surface

of the sensor chip along with varieties of recognition materials or binders like sodium channel receptors or antibody elements for SPR inhibition assay (Wu et al. 2019; Tang et al. 2023; Tian et al. 2021).

2.3.1.5 Surface-Enhanced Raman Scattering (SERS) Biosensor

The process known as Raman scattering occurs when a light wave's frequency changes as a result of scattering. An analytical tool, i.e. SERS, was developed to magnify the signals produced by molecules present at incredibly low concentrations to an extent of single-molecule level. It is known to detect the increase in analyte vibration spectrum intensity as it approaches nanometallic surfaces (Wu et al. 2019). It was based on Raman scattering. Recent years have seen the introduction of SERS biosensors by researchers to detect more pertinent data on the biotoxins' functional groups. It is a reliable tool in food industry because of its ease of use, promptness, and responsiveness. Label-free and labeled SERS detection are the two main categories of current detection technologies. The label-free SERS detection which is a simple and convenient method is intended to get the spectra immediately. An indirect detection technique called labeled SERS detection is used to find and identify tiny compounds like biotoxins. It depends on a powerful electromagnetic field produced by metal-surface Localized Surface Plasmon Resonance (LSPR). Instead of the target analyte, organic molecules with high Raman activity, such as Raman dye, typically come into contact with the metal surface (Tang et al. 2023). Several techniques have been used to detect toxins such as DA, MC-LR, OA, and STX with SERS probes conjugated with amino-functionalized Ag nanoparticles and 4-aminothiophenol as SERS tag (Wu et al. 2019) and aptamer-based SERS paper swabs (Zhao et al. 2021b).

2.3.2 *Electrochemical Biosensor*

Electrochemical biosensors offer an appealing method for detecting algal toxins because the toxin concentration is directly converted to an electric signal. A particular recognition element based on chemical/biological substances and an electrode which converts the recognition activity into a detectable electronic signal are typically included in a basic electrochemical sensor design (Wu et al. 2019). The analytical signal produced by the sensor is converted by the electrode conversion element into an electronic circuit, while the chemical or biological recognition element primarily indicates the toxin occurrence and determines the sensitivity of biosensors. Different kinds of nanomaterials can be added to their sensing surface to greatly increase their sensitivity. Currently, graphene has gained popularity due to its catalytic ability, biocompatibility, surface area, and electronic property (Tang et al. 2023). Most of the electrochemical biosensors developed are immunotypes, i.e. using antibodies as biorecognition elements. Other methods like protein

phosphatase 2A inhibition and use of aptamers, enzymes, etc., are popular in fabricating various electrochemical biosensors. Thanks to its high sensitivity, easy miniaturization, cost-effectiveness, and dependability, electrochemical biosensors have many uses in the food industry, healthcare, and agriculture as well as environmental monitoring. Based on the electrochemical approach, there are three types of electrochemical biosensors for detecting shellfish toxins: amperometric, potentiometric, and impedimetric (Tian et al. 2021). For algal toxins, impedimetric as well as amperometric biosensors are mostly used, especially immunosensors that rely on the precise activity between antigens and antibodies (Wu et al. 2019).

2.3.2.1 Amperometric Sensor

The measurement of current produced at the electrode surface by bioaffinity or enzymatic reaction at a stagnant working potential in relation to the electrode used as reference is the foundation of the amperometric sensor's operation. These sensors are one of the most reliable sensing technologies for general toxin analysis because the redox potential determines the characteristic of the analytes. Its simplicity in manufacturing processes, disposability of electrodes, and enhanced selectivity make it a reasonable device for use. A small number of amperometric sensors take advantage of the suppression of enzyme processes by particular toxins such as OA. The tactic often relies on the suppression of toxins on fixed protein phosphatase and uses the proper enzyme substrates to detect the current produced by variations in enzyme activity. For identifying algal toxins, such MC-LR and DA amperometric immunosensors are mostly popular. The primary benefit of amperometric biosensors is that they make a variety of helpful detection techniques possible such as cyclic voltammetry (CV), square wave voltammetry (SWV), differential pulse voltammetry (DPV), and chronoamperometry (Wu et al. 2019; Tian et al. 2021).

2.3.2.2 Impedimetric Sensor

When a sinusoidal AC excitation signal (2–10 mV) with small amplitude at different frequencies disrupts the electrochemical cell system, electrochemical impedance spectroscopy (EIS) evaluates the resistance and capacitance of the materials. By calculating the variations in the electric impedance with respect to the tested frequencies, the impedance spectrum is recorded (Wu et al. 2019). Ever since the first findings on impedimetric immunosensors were put forth, EIS-based sensors are becoming more valuable for examining various biomolecular interactions, such as antibody–antigen interactions. The impedance signal is often produced by the analytes' and recognition elements' specific binding to the electrode surface, which obstructs the electron transfer's path. When CYN binds to aptamers, the lower EIS signal produced by the aptamer's conformational shift is measured to identify CYN in lake water using EIS. The aptasensor exhibits excellent selectivity and sensitivity. Because biological cells have dielectric characteristics, impedimetric biosensors are

typically combined with cell technology for detecting toxins in shellfish (Li HongBo et al. 2018). By adding the antagonistic effect, Zou et al. (2015) suggested an impedance biosensor for STX detection based on a neuroblastoma cell element. Neuroblastoma cells when treated with veratridine and ouabain, they swelled up and eventually perished because ouabain hindered the function of Na^{+}/K^{+}-ATPases and veratridine induced a prolonged input of Na^{+}. However, because STX blocks sodium channels, it reduced the cytotoxic effects of veratridine and ouabain. Later, another impedance biosensor based on the excitation-contraction property of cardiomyocyte cells was designed by Wang et al. (2015b) for STX detection. This biosensor analyzed the cytotoxic potential of shellfish toxins with wide dimensions because of the multiple characteristics of cardiomyocyte beating (Wu et al. 2019; Tian et al. 2021).

2.3.2.3 Piezoelectric Sensor

Known for their simplicity, portability, and sensitivity as gravimetric sensors, piezoelectric immunosensors made up of piezoelectric crystal that are mass-sensitive and are used as microbalances have become the most extensively used transducers in biosensors (Wu et al. 2019). The bulk acoustic wave (BAW) based PZ device is incorporated in this kind of piezoelectric (PZ) sensor. Based on the variations in acoustic wave propagation caused by changes in the physical properties of interfacial thin films or by particular antibody–antigen interaction coated on the crystal, concentration of analytes captured at the sensor's surface can be determined. While the majority of biosensors require certain chemicals and labeling in order to record the affinity of contacts, piezoelectric biosensors may identify targets using readily available materials. PZ quartz crystals' ease of automation, simplicity, portability, affordability, and ease of miniaturization have drawn a lot of interest. The mass loaded on the crystal surface changes in a proportionate amount that changes the quartz crystal's resonance frequency. The PZ biosensors were utilized to identify different biotoxins by combining quartz crystals with aptamers, antibodies, and receptor proteins, among other substances. Tian et al. (2021) created label-free AuNP amplification-based PZ aptasensor for detection of OA.

2.3.2.4 Potentiometric Biosensors

The electro-readout method known as potentiometry is based on the variations in electrical potential when electric charges are accumulated over a dielectric layer. For a number of years, PSP detection was primarily conducted using enzymes, cells, and immunization in combination with potentiometric biosensors. A microelectrode array (MEA)-based portable cardiomyocyte-based potential biosensor was created (Wang et al. 2015a) for the quick detection of STX. Plasticized polyvinyl chloride (PVC) membrane has been used for the measurement of GTX5 based on the carbamoylase hydrolysis (Raposo et al. 2020). A miniature potentiometric

biosensor composed of graphene nanosheets and polymer stabilized lipid films was created for quick detection of STX (5–20 min) (Bratakou et al. 2017). The lipid membrane provided a biologically suitable setting at the interface of biosensing surface which facilitated the precise adsorption of cationic and anionic species, hence reducing background constituent interference (Tian et al. 2021).

2.3.3 Photothermal Biosensors

In addition to the traditional techniques discussed earlier, various novel sensing approaches have been created lately. When exposed to a certain type of laser radiation, photothermal nanomaterials absorbed light energy and transformed it into thermal energy. The photothermal conversion efficiency is found to be high. There is, however, little research on the application of nanomaterials possessing photothermal properties to be used in biosensors for detecting biotoxins. The majority of the research so far have focused on creating photothermal probes using metal complexes, noble metals, and carbon-based photothermal nanomaterials (Tang et al. 2023).

2.3.4 FET Biosensors

FET biosensors have a FET acting as a transducer and a sensitive biological membrane for molecule determination at the sensor element. The exceptional selectivity and sensitivity of its sensor parts give FET biosensors significant advantages in real-time and online analysis and characterization. Antunes et al. (2018) proposed a FET-containing graphene immunosensor (label-free) for OA detection. Because of its atomic thickness, graphene's electrical characteristics were receptive to interactions between the adsorbed molecules and graphene surface enhancing Gr-FET sensitivity. FET biosensors are quite sensitive and stable providing a quick detection and are cost friendly. However, there is still scope for improvement with respect to equipment integration and miniaturization and detection flux for field detection (Tian et al. 2021).

2.3.5 Surface Acoustic Wave (SAW) Biosensors

There are various modes of SAWs—Love, Lamb waves, Rayleigh waves, etc. SAWs are elastic in nature which are propagated on the shallow solid surface and measured with the help of an interface membrane material, oscillator, and oscillation circuit. They are an integration of material science and electronic technology (Tian et al. 2021). Due to its numerous significant features, including contactless sensing,

unpowered operation, speed, non-interference, simple coding, cryptic nature, and cost efficiency, they are being extensively incorporated in several domains, including the recognition of shellfish toxins (Deng et al. 2019). An immunosensor for the detection of OA utilizing a Love wave sensor was described by Zou et al. (2017) using an immunogold staining approach. In that work, staphylococcal protein A (SPA) on the sensor surface bound to bovine serum albumin-monoclonal antibody (BSA-MAb) was used to collect okadoic acid-bovine serum albumin (OA-BSA), and the phase signal was enhanced using AuNPs. To achieve sensitive and real-time OA detection, Zhang et al. (2016) created a biosensor based on Love Wave incorporating HepG2 cells which allowed real-time tracking of the cell attachment process. OA can cause aberrant morphological changes in cell or even their death. Hence, OA-induced morphological alterations in the cells affected the cells' adherence to the sensor.

2.3.6 Multimodal Biosensors

The linear range of detection (LOD) can be expanded by multimodal biosensors' ability to produce a variety of signals. Furthermore, the accuracy of the detection results can be increased by these signals validating one another. The signal components used are mostly electrochemistry, SERS, colorimetric, fluorescence, photothermal, and so on. An integrated fluorescence-colorimetric method-based biosensor possesses an extremely sensitive fluorescence signal in addition to a visual colorimetric signal. To fulfill the increasing requirement for precision and stability, multimodal biosensors have been used more frequently in recent years. In the future, this could be a useful method for on-site biotoxin identification (Tang et al. 2023).

2.4 Gaps/Limitations

Biosensing technology has gained a lot of significance in the past decade because of its portability and sensitivity. But there are certain challenges which need to be addressed for better functioning of these devices with respect to detection of aquatic toxins. In case of nanomaterial-based biosensors, their storage and stability act as a hindrance under field conditions. The major limitations of optical, electrochemical, surface exciton polariton resonance (SEPR), SPR, ECL, SAW, and FET biosensors lie in their stability, processing cost, sensitivity, flux, and portability which need to be further improved. The pretreatment techniques used for fish and shellfish nowadays are basically manual, multi-link, or single-flux, which requires long time and laborious preparation procedures. Pretreatment of liquids, like seawater, usually involves a filtering technique. Also, biotoxin accumulations in fishes and shellfishes occur at such low concentrations that it is quite difficult to determine their toxicity for sample matrices under real-time environmental scenario. Detection of saxitoxin

and its analogs is quite challenging because their molecular-weight is very low. Issues such as environmental and sample autofluorescence also need to be considered. The biggest challenges biosensors will face are developing regulations and norms for their utilization so non-technical workers may utilize them, although new technology will gradually seep into the commercial and agency sectors. Furthermore, a lot of studies present great results that normally compete with ELISA and other commercial procedures; yet, comparisons with more established technologies are frequently omitted. Before biosensors are widely used, this needs to be resolved.

2.5 Conclusions and Future Prospects

Population increase is undoubtedly pushing the world toward urbanization, hence producing more wastewater and contaminating aquatic environments (Mondal and Bordoloi 2023). Biotoxins produced by different freshwater and marine water algae species impacting fish and shellfish are a matter of concern. Up until now, research and detection efforts have focused on toxins such as cylindrospermopsins, saxitoxins, microcystins, nodularins, domoic acid, brevetoxins, ciguatoxins, and BMAA; however, details regarding certain other neurotoxins remain unresolved (Bittner et al. 2021; Christensen and Khan 2020). Novel nanostructural elements, aptamers, molecularly imprinted polymers (MIPs), receptors, and recombinant or modified proteins are examples of novel target recognition techniques recommended for the specific detection of known/unknown toxins and their respective analogs. Although these devices cater to the general issues of poor stability, excess background noise, interference, and photobleaching, certain areas still need to be investigated, such as the creation of fluorescent nanomaterials in the near-infrared (NIR) range to prevent autofluorescence, more straightforward extraction and cleanup techniques for quick analysis of biotoxins, improving the efficiency of biomaterial and target recognition, use of spectroscopy and other techniques for active on-spot biotoxin detection in samples excluding sample pretreatment; and finally, creation of portable and compact instruments for field based detections. Smart biosensors may continue to be essential in the detection of biotoxins for satisfying market demands and regulatory requirements with the advancements in detection technologies, better portability and data analysis (Tang et al. 2023; Wu et al. 2019).

References

Akbar MA, Mohd Yusof NY, Tahir NI, Ahmad A, Usup G, Sahrani FK, Bunawan H (2020) Biosynthesis of saxitoxin in marine dinoflagellates: an omics perspective. Mar Drugs 18(2):103

Antunes J, Justino C, da Costa JP, Cardoso S, Duarte AC, Rocha-Santos T (2018) Graphene immunosensors for okadaic acid detection in seawater. Microchem J 138:465–471

Bertani P, Lu W (2021) Cyanobacterial toxin biosensors for environmental monitoring and protection. Med Nov Technol Devices 10:100059

Bittner M, Štern A, Smutná M, Hilscherová K, Žegura B (2021) Cytotoxic and genotoxic effects of cyanobacterial and algal extracts — microcystin and retinoic acid content. Toxins 13(2):107

Boopathi T, Ki JS (2014) Impact of environmental factors on the regulation of cyanotoxin production. Toxins 6(7):1951–1978

Botelho MJ, Vale C, Ferreira JG (2019) Seasonal and multi-annual trends of bivalve toxicity by PSTs in Portuguese marine waters. Sci Total Environ 664:1095–1106

Bratakou S, Nikoleli GP, Siontorou CG, Nikolelis DP, Karapetis S, Tzamtzis N (2017) Development of an electrochemical biosensor for the rapid detection of saxitoxin based on air stable lipid films with incorporated anti-STX using graphene electrodes. Electroanalysis 29(4):990–997

Cembella AD, Durán-Riveroll LM (2021) Marine guanidinium neurotoxins: biogenic origins and interactions, biosynthesis and pharmacology. In: Advances in neurotoxicology, vol 6. Academic Press, pp 1–47

Chen L, Chen J, Zhang X, Xie P (2016) A review of reproductive toxicity of microcystins. J Hazard Mater 301:381–399

Chen G, Jia Z, Wang L, Hu T (2020) Effect of acute exposure of saxitoxin on development of zebrafish embryos (*Danio rerio*). Environ Res 185:109432

Chorus I, Welker M (2021) Toxic Cyanobacteria in Water: A Guide to Their Public Health Consequences, Monitoring and Management. Taylor & Francis, p 858

Christensen VG, Khan E (2020) Freshwater neurotoxins and concerns for human, animal, and ecosystem health: a review of anatoxin-a and saxitoxin. Sci Total Environ 736:139515

Corriere M, Soliño L, Costa PR (2021) Effects of the marine biotoxins okadaic acid and dinophysistoxins on fish. J Mar Sci Eng 9(3):293

Cruz MG, Ferreira NS, Gomes MTS, Botelho MJ, Costa ST, Vale C, Rudnitskaya A (2018) Determination of paralytic shellfish toxins using potentiometric electronic tongue. Sensors Actuators B Chem 263:550–556

Cunha I, Biltes R, Sales MGF, Vasconcelos V (2018) Aptamer-based biosensors to detect aquatic phycotoxins and cyanotoxins. Sensors 18(7):2367

de Morais Calado SL, Santos GS, Vicentini M, Bozza DC, Prodocimo V, de Magalhaes VF, Cestari MM, de Assis HCS (2020) Multiple biomarkers response in a Neotropical fish exposed to paralytic shellfish toxins (PSTs). Chemosphere 238:124616

Deng Y, Zheng H, Yi X, Shao C, Xiang B, Wang S, Zhao Z, Zhang X, Hui G (2019) Paralytic shellfish poisoning toxin detection based on cell-based sensor and non-linear signal processing model. Int J Food Prop 22(1):890–897

Farabegoli F, Blanco L, Rodríguez LP, Vieites JM, Cabado AG (2018) Phycotoxins in marine shellfish: origin, occurrence and effects on humans. Mar Drugs 16(6):188

Fiore MF, de Lima ST, Carmichael WW, McKinnie SM, Chekan JR, Moore BS (2020) Guanitoxin, re-naming a cyanobacterial organophosphate toxin. Harmful Algae 92:101737

Garibo D, Campbell K, Casanova A, De La Iglesia P, Fernández-Tejedor M, Diogène J, Elliott CT, Campàs M (2014) SPR immunosensor for the detection of okadaic acid in mussels using magnetic particles as antibody carriers. Sensors Actuators B Chem 190:822–828

Kholafazad Kordasht H, Hassanpour S, Baradaran B, Nosrati R, Hashemzaei M, Mokhtarzadeh A, de la Guardia M (2020) Biosensing of microcystins in water samples; recent advances. Biosens Bioelectron 165:112403

Lance E, Arnich N, Maignien T, Biré R (2018) Occurrence of β-N-methylamino-l-alanine (BMAA) and isomers in aquatic environments and aquatic food sources for humans. Toxins 10(2):83

Li H, Wei X, Gu C, Su K, WH, Hu N, Wang P (2018) A dual functional cardiomyocyte-based hybrid-biosensor for the detection of diarrhetic shellfish poisoning and paralytic shellfish poisoning toxins. Anal Sci 34(8):893–900

Lian Z, Zhao M, Wang J, Yu RC (2021) Dual-emission ratiometric fluorescent sensor based molecularly imprinted nanoparticles for visual detection of okadaic acid in seawater and sediment. Sensors Actuators B Chem 346:130465

Liang Y, Li A, Chen J, Tan Z, Tong M, Liu Z, Qiu J, Yu R (2022) Progress on the investigation and monitoring of marine phycotoxins in China. Harmful Algae 111:102152

Macario IP, Ventura SP, Goncalves FJ, Torres-Acosta MA, Pereira JL (2021) The "bright side" of cyanobacteria: revising the nuisance potential and prospecting innovative biotechnology-based solutions to integrate water management programs. ACS Sustain Chem Eng 9(21):7182–7197

McPartlin DA, Lochhead MJ, Connell LB, Doucette GJ, O'Kennedy RJ (2016) Use of biosensors for the detection of marine toxins. Essays Biochem 60(1):49–58

Merel S, Walker D, Chicana R, Snyder S, Baurès E, Thomas O (2013) State of knowledge and concerns on cyanobacterial blooms and cyanotoxins. Environ Int 59:303–327

Mondal B, Bordoloi N (2023) Metal in water: an assessment of toxicity with its biogeochemistry. In: Metals in Water. Elsevier, pp 71–91

Morabito S, Silvestro S, Faggio C (2018) How the marine biotoxins affect human health. Nat Prod Res 32(6):621–631

Murk AJ, Nicolas J, Smulders FJ, Bürk C, Gerssen A (2019) Marine biotoxins: types of poisoning, underlying mechanisms of action and risk management programmes. In: Chemical Hazards in Foods of Animal Origin. Wageningen Academic Publishers, pp 398–402

Nicolas J, Hoogenboom RL, Hendriksen PJ, Bodero M, Bovee TF, Rietjens IM, Gerssen A (2017) Marine biotoxins and associated outbreaks following seafood consumption: prevention and surveillance in the 21st century. Glob Food Sec 15:11–21

O'Neill K, Musgrave IF, Humpage A (2016) Low dose extended exposure to saxitoxin and its potential neurodevelopmental effects: a review. Environ Toxicol Pharmacol 48:7–16

Peng J, Zhao Z, Zheng M, Su B, Chen X, Chen X (2020) Electrochemical synthesis of phosphorus and sulfur co-doped graphene quantum dots as efficient electrochemiluminescent immunomarkers for monitoring okadaic acid. Sensors Actuators B Chem 304:127383

Pierozan P, Cattani D, Karlsson O (2020) Hippocampal neural stem cells are more susceptible to the neurotoxin BMAA than primary neurons: effects on apoptosis, cellular differentiation, neurite outgrowth, and DNA methylation. Cell Death Dis 11(10):910

Pinto A, Botelho MJ, Churro C, Asselman J, Pereira P, Pereira JL (2023) A review on aquatic toxins-do we really know it all regarding the environmental risk posed by phytoplankton neurotoxins? J Environ Manag 345:118769

Pradhan B, Ki JS (2022) Phytoplankton toxins and their potential therapeutic applications: a journey toward the quest for potent pharmaceuticals. Mar Drugs 20(4):271

Pulido OM (2016) Phycotoxins by harmful algal blooms (HABS) and human poisoning: an overview. Int Clin Pathol J 2(6):145–152

Raposo M, Botelho MJ, Costa ST, Gomes MTS, Rudnitskaya A (2020) A carbamoylase-based bioassay for the detection of paralytic shellfish poisoning toxins. Sensors 20(2):507

Reverté J, Alkassar M, Diogène J, Campàs M (2023) Detection of ciguatoxins and tetrodotoxins in seafood with biosensors and other smart bioanalytical systems. Food Secur 12(10):2043

Rodgers KJ, Main BJ, Samardzic K (2018) Cyanobacterial neurotoxins: their occurrence and mechanisms of toxicity. Neurotox Res 33:168–177

Sun A, Chai J, Xiao T, Shi X, Li X, Zhao Q, Li D, Chen J (2018) Development of a selective fluorescence nanosensor based on molecularly imprinted-quantum dot optosensing materials for saxitoxin detection in shellfish samples. Sensors Actuators B Chem 258:408–414

Tang X, Zuo J, Yang C, Jiang J, Zhang Q, Ping J, Li P (2023) Current trends in biosensors for biotoxins (mycotoxins, marine toxins, and bacterial food toxins): principles, application, and perspective. TrAC Trends Anal Chem 165:117144

Tian Y, Du L, Zhu P, Chen Y, Chen W, Wu C, Wang P (2021) Recent progress in micro/nano biosensors for shellfish toxin detection. Biosens Bioelectron 176:112899

Vasconcelos V (2015) Global changes and the new challenges in the research on cyanotoxin risk evaluation. Limnetica 34(1):149–158

Vichi S, Buratti FM, Testai E (2016) Microcystins: toxicological profile. Mar Freshw Toxins:219–238

Visciano P, Schirone M, Berti M, Milandri A, Tofalo R, Suzzi G (2016) Marine biotoxins: occurrence, toxicity, regulatory limits and reference methods. Front Microbiol 7:1051

Vogiazi V, de la Cruz A, Mishra S, Shanov V, Heineman WR, Dionysiou DD (2019) A comprehensive review: development of electrochemical biosensors for detection of cyanotoxins in freshwater. ACS Sens 4(5):1151–1173

Walls JT, Wyatt KH, Doll JC, Rubenstein EM, Rober AR (2018) Hot and toxic: temperature regulates microcystin release from cyanobacteria. Sci Total Environ 610:786–795

Wang Q, Fang J, Cao D, Li H, Su K, Hu N, Wang P (2015a) An improved functional assay for rapid detection of marine toxins, saxitoxin and brevetoxin using a portable cardiomyocyte-based potential biosensor. Biosens Bioelectron 72:10–17

Wang Q, Su K, Hu L, Zou L, Wang T, Zhuang L, Hu N, Wang P (2015b) A novel and functional assay for pharmacological effects of marine toxins, saxitoxin and tetrodotoxin by cardiomyocyte-based impedance biosensor. Sensors Actuators B Chem 209:828–837

Wu X, Hou L, Lin X, Xie Z (2019) Application of novel nanomaterials for chemo-and biosensing of algal toxins in shellfish and water. In: Novel Nanomaterials for Biomedical, Environmental and Energy Applications. Elsevier, pp 353–414

Ye W, Liu T, Zhang W, Zhu M, Liu Z, Kong Y, Liu S (2019) Marine toxins detection by biosensors based on aptamers. Toxins 12(1):1

Zhang X, Fang J, Zou L, Zou Y, Lang L, Gao F, Hu N, Wang P (2016) A novel sensitive cell-based love wave biosensor for marine toxin detection. Biosens Bioelectron 77:573–579

Zhao Y, Li L, Ma R, Wang L, Yan X, Qi X, Wang S, Mao X (2021a) A competitive colorimetric aptasensor transduced by hybridization chain reaction-facilitated catalysis of AuNPs nanozyme for highly sensitive detection of saxitoxin. Anal Chim Acta 1173:338710

Zhao P, Liu H, Zhu P, Ge S, Zhang L, Zhang Y, Yu J (2021b) Multiple cooperative amplification paper SERS aptasensor based on AuNPs/3D succulent-like silver for okadaic acid quantization. Sensors Actuators B Chem 344:130174

Zou L, Wu C, Wang Q, Zhou J, Su K, Li H, Hu N, Wang P (2015) An improved sensitive assay for the detection of PSP toxins with neuroblastoma cell-based impedance biosensor. Biosens Bioelectron 67:458–464

Zou L, Tian Y, Zhang X, Fang J, Hu N, Wang P (2017) A competitive love wave immunosensor for detection of okadaic acid based on immunogold staining method. Sensors Actuators B Chem 238:1173–1180

Chapter 3
Nanomaterials, Magnetic Beads, and Microfluidic Systems for Detecting Emerging Biotoxins

H. K. I. S. Vishwajith, Nadun H. Madanayake, and Nadeesh M. Adassooriya

Abstract Living organisms produce certain metabolites that can be toxic to other organisms. Biotoxins produced by microalgae and cyanobacteria have numerous adverse effects on the environment. Therefore, it is crucial to recognize and rule out biotoxins that could be present in different environments. Numerous techniques have been developed to detect biotoxins. Nanotechnology has improved precision and accuracy in the detection of biotoxins, and nanomaterials (NMs) have been employed in sensor configurations to boost their effectiveness. Furthermore, there has been a significant surge in investigation into the application of magnetic beads in biotoxin detection. Magnetic beads are implied to target pre-concentration, signal amplifiers, and the immobilization of toxins, which leads to sensitive, reliable, and robust analytical systems. Recently, microfluidic systems have been harnessed to detect these noxious agents due to their minimal sample and reagent consumption, automation and integration possibilities, and optimal range of portability. Hence, nanotechnology, along with microfluid systems can be ideal candidates to detect biogenic toxins in the environment. Therefore, this book chapter elaborates on the potential applications of nanomaterials, magnetic beads, and microfluidic systems for emerging environmental biotoxin detection.

Keywords Biotoxins · Microalgae · Cyanobacteria · Nanomaterials · Magnetic beads · Microfluidic systems · Analytical systems

---

H. K. I. S. Vishwajith · N. M. Adassooriya (✉)
Department of Chemical and Process Engineering, Faculty of Engineering, University of Peradeniya, Peradeniya, Sri Lanka
e-mail: nadeeshm@eng.pdn.ac.lk

N. H. Madanayake
Department of Botany, Faculty of Science, University of Peradeniya, Peradeniya, Sri Lanka

M. N. V. Prasad et al. (eds.), *Biotoxins*,
https://doi.org/10.1007/978-3-031-75309-1_3

3.1 Introduction

A wide range of hazardous compounds produced by microorganisms, plants, and animals are generally referred as biotoxins. These comprise of substances made toxic by metabolic pathways of living organisms and residues from decomposing organisms. Different organisms including bacteria, algae, fungi, and plants can produce biotoxins, and these can have numerous effects on metabolic processes. These toxins pose significant hazards to both humans and the other organisms and have a wide range of adverse reactions which leads to short- and long-term detrimental effects. These effects might range from gastrointestinal upset to brain impairment and ultimately up to organ failure and death (Parak et al. 2024). Most of the time, biotoxins found in marine environments include amnesic shellfish poisoning (ASP) and ciguatera fish poisoning (CFP). Dinoflagellates transfer these poisons, which additionally result in the corrosive properties of domoic acid and ciguatoxins (Bahramian 2020; Parak et al. 2024). Also, dinoflagellates produces saxitoxins that have been connected to paralytic shellfish poisoning (PSP), bacterial poisons are the root cause of botulism (Bahramian 2020). Among the microbes which generate botulinum poison, a strong neurotoxin that causes botulism is *Clostridium botulinum* (Gupta and Gupta 2020). The chemical structures, mode of action, and origins of biotoxins display a significant variation. Therefore, examining and understanding the sources and related hazards associated with biotoxins is essential to comprehend their nature and outcomes.

Biotoxins can be detected using various strategies such as biochemical assays, molecular techniques, biosensors, etc. In the last few years, the detection of biotoxins has been more effective due to the integration of nanotechnology, magnetic bead technology, and microfluidic technologies. The integration of these contemporary technologies offers previously unseen opportunities for sensitive, rapid, and precise detection methods, as it is critical to safeguard the public's health and the quality of the environment. Employing nanomaterials, magnetic beads, and microfluidic devices enables to monitor pollutants in the environment and track down potential biological threats. Therefore, this book chapter will elaborate on the potential of the application of nanomaterials, magnetic beads, and microfluidic systems in the detection of emerging biotoxins.

3.2 Biotoxins and Their Sources

Biotoxins can come in various forms, ranging from microorganisms to venomous animals. Understanding such bio-harms and their connected risks is important to alleviate anticipated harm and develop effective countermeasures. Bacteria, being among the most ancient and versatile organisms, have created an enormous assortment of toxins for both offensive and defensive uses. Among the worst neurotoxins known to man is botulinum toxin, which is produced by the well-known bacterium

Clostridium botulinum. This toxin can cause botulism, a dangerous and occasionally deadly disease that paralyzes muscles, even when minute amounts are consumed. Apart from their aesthetic appeal and nutritional value, plants are also known for their diverse spectrum of toxins that act as barriers against illnesses. For example, ricin is a vigorous poison present in the seeds of the castor oil plant (*Ricinus communis*) (Alshannaq and Yu 2017). If swallowed or injected, ricin inhibits cells from generating new proteins, thereby resulting in cell death and systemic toxicity. Different types of poisons are produced by living organisms including reptiles, insects, scorpions, and marine life, especially to knock prey unconscious or safeguard themselves from predators. For instance, snake venoms are composed of a mixture of proteins, peptides, and enzymes that are meant to interfere with the victims' metabolic functions. Neurotoxic venoms, such as those produced by cobras and mambas, target the nervous system and trigger respiratory failure and paralysis. Besides the biotoxins found in nature, humans have fabricated a wide range of chemical compounds possessing hazardous characteristics for diverse applications including industrial operations, control of pests, and military. Also, studies have shown that natural hydrocarbon raw materials and processed products can harbor biotoxins, highlighting the need for comprehensive analysis and monitoring to prevent potential health and environmental risks (Vinh et al. 2022). Table 3.1 summarizes different sources of biotoxins and their potential risks to living organisms.

3.3 Detection of Biotoxins

As mentioned elsewhere in the text, biotoxins show serious effects on different living organisms. Furthermore, emerging biotoxins pose a major health danger and upset the environment by poisoning food supply, water sources, and ecosystems, which can emerge from different sources, including microbial, algal, fungal, and plant-based organisms (Still and Mohapatra 2020). Hence, effective strategies to detect and identify biotoxins are utterly important. This will enable the authorities to mitigate health hazards and guarantee the safety of food and water and other important consumables. Recently, nanotechnological-based approaches have been developed in biotoxin detection. Therefore, incorporation of NMs enables to develop detection systems with higher sensitivity, specificity, and speed. For example, biosensors based on gold nanoparticles (NPs) have become effective instruments for fast pathogen identity, providing great sensitivity and selectivity in biotoxin sensing applications to detect botulinum toxin, aflatoxins, etc. (Hegde et al. 2022). Magnetic nanoparticles integrated into microfluidic platforms have revolutionized biotoxin detection, enabling efficient transport and detection processes with enhanced sensitivity and reduced assay time (Khizar et al. 2020). Furthermore, understanding the environmental impacts and biotoxin activities associated with emerging biotoxins is essential for assessing risks and formulating effective management strategies.

Table 3.1 Sources of biotoxins and their risks

Organism	Source	Biotoxin	Effect	Reference
Algal blooms	Marine environments	Saxitoxin	Paralysis (PSP)	Murk et al. (2019)
Dinoflagellates	Marine environments	Ciguatoxin	Gastrointestinal symptoms, neurological effects	Murk et al. (2019)
Shellfish	Contaminated waters	Domoic acid	Amnesic shellfish poisoning (ASP)	Landsberg et al. (2007)
Algal blooms	Marine environments	Ciguatoxin	Gastrointestinal symptoms, neurological effects	Murk et al. (2019)
Algae	Marine environments	Algal toxins	Neurotoxicity, paralysis	Landsberg et al. (2007)
Shellfish	Contaminated waters	Tetrodotoxin	Paralysis, respiratory failure	Parak et al. (2024)
Fish	Marine environments	Ciguatoxins	Neurological symptoms	Morabito et al. (2018)
Bacteria	Food	Botulinum toxin	Muscle paralysis, respiratory failure	Gupta and Gupta (2020)
Tetrodotoxin	Shellfish	Tetrodotoxin	Neurotoxicity, paralysis	Shkembi et al. 2021
Botulinum	Food	Botulinum toxin	Muscle paralysis	Aghamohammadi et al. 2023
Aflatoxin B1	Food	Aflatoxin B1	Hepatotoxicity, carcinogenicity	Yadav et al. 2021
Ochratoxin A	Food	Ochratoxin A	Nephrotoxicity, carcinogenicity	Jiang et al. 2018
Diclofenac	Water contamination	Diclofenac	Cardiovascular and renal toxicity	Ecke et al. 2022
SARS-CoV-2	Biological fluids (e.g., blood)	SARS-CoV-2 antibodies	Detection of COVID-19 infection	Huergo et al. 2021
Mycotoxins	Food	Various mycotoxins	Hepatotoxicity, nephrotoxicity	Fattahi and Hasanzadeh 2022
Cholera toxin	Fecal samples	Cholera toxin	Diarrhea, dehydration	Bunyakul et al. 2015
Algal toxins	Shellfish and water	Various algal toxins	Neurotoxicity, gastrointestinal issues	Wu et al. 2019

3.4 Nanomaterials in Biotoxin Detection

Unique physical and chemical properties of NMs above those of their bulk counterparts make them promising candidates in biotoxin detection systems. NMs are particles with at least one measurement within the nanometer scale typically between 1 and 100 nm (Kaliva and Vamvakaki 2020). Higher surface area-to-volume ratio and reactivity of NMs permit effective interactions with target analytes in biotoxin

detection techniques (Barhoum et al. 2022). Moreover, quantum confinement effects results in changes to electrical, optical, and magnetic characteristics of NPs enabling them to utilize in precise detection methods (Barhoum et al. 2022).

Carbon-based materials are getting attention because of their attractive electrochemical properties and biocompatibility (Hwang et al. 2020). Carbon nanotubes (CNTs) and graphene-based materials have been comprehensively analyzed for their possible applications in biosensing because of their exceptional electron transport limits and sweeping surface reach (Sanati et al. 2019). The astoundingly fragile and accurate area of biotoxins is made conceivable by the functionalization of carbon nanomaterials with specific receptors. In addition, Su et al. (2019) report that two-layered (2D) nanomaterials seem to ensure sensible decisions for biosensing applications. Graphene, phosphorus, and change metal dichalcogenides are events of materials with fascinating electrical and optical attributes that have phenomenal affectability and particularity for recognizing various kinds of biotoxins. The enormous surface-to-volume extent of 2D nanomaterials increases the extent in biosensing stages and adds the sufficiency of biomolecule immobilization. Nanomaterials are therefore vital for the unique verification of biotoxin production since they provide enhanced affectability, selectivity, and reconciliation capabilities. The execution of biosensing tests is empowered to gain ground by the mix of carbon-based and 2D nanomaterials with magnetic beads and microfluidic devices, which has massive potential for biosensing applications. Through advanced examination and improvement around here, moves in the biotoxin area should be extended and potential health and environmental risks can be minimized.

The development of aflatoxin B1, a strong mycotoxin that causes disease, by specific molds is a serious threat to the security of food. A broad assessment of nanomaterial-based biosensors for aflatoxin B1 (AFB1) detection was done by Yadav et al. (2021). The review stressed the utilization of a few nanomaterials to work on the responsiveness and selectivity of biosensors, including metal nanoparticles, graphene, and carbon nanotubes. Analysts have effectively and immediately recognized AFB1 in food tests by using the exceptional characteristics of nanomaterials, which lessens the risks implied in devouring it.

To identify harmful bacteria and microbial toxins in food matrices, Gupta et al. (2021) investigated the developments in nanomaterial-based electrochemical biosensors. They gave an example of how well-performing biosensors may be improved by using nanomaterials such as magnetic nanoparticles, quantum dots, and metal oxides. Researchers have developed sensitive and accurate techniques for detecting bacteria and microbial toxins through the integration of nanomaterials into sensor systems. This enables prompt actions to guarantee the safety and quality of food. Both human health and aquatic ecosystems are severely impacted by algae poisons. Wu et al. (2019) examined the use of new nanomaterials in the chemo- and biosensing of algal toxins in water and shellfish. The paper outlines in detail how to detect algae toxins selectively using nanomaterials such as metal nanoparticles, carbon nanomaterials, and polymers possessing molecular imprints. Using nanomaterials' distinct characteristics, scientists have developed portable and sensitive sensing

platforms that may detect algae toxins instantly, enabling the invention of early warning systems and efficient management approaches. The studies mentioned above illustrate the significant function of nanomaterials in augmenting the identification of biotoxins, presenting unparalleled prospects for increasing food safety, environmental surveillance, and public health.

3.5 Magnetic Beads for Biotoxin Detection

Because of their great sensitivity, speed, and adaptability, magnetic bead-based assays have grown into highly effective instruments in the field of biotoxin detection. These assays effectively collect, separate, and detect target analytes by utilizing the unique qualities of magnetic beads. The basic ideas behind magnetic bead-based assays will be covered in this part, along with an exploration of how they might be utilized for recognizing newly developing biotoxins. Small, solid particles known as magnetic beads could bind to biomolecules including proteins, DNA, RNA, and antibodies with specificity. Magnetic beads are composed of magnetic minerals like magnetite or iron oxide, and these beads respond to external magnetic fields. This makes it simple for researchers to organize and separate samples while analyzing them. The concept behind assays with magnetic beads is the selective binding interactions that occur between target molecules and capture molecules that are deposited on the surface of the beads. As a result of their strong affinity and selectivity for the target analyte, these capture molecules—which are frequently antibodies or aptamers—ensure accurate detection even at low concentrations. Several important steps are usually involved in the assay workflow using magnetic beads.

1. Target Capture: The magnetic beads functionalized with capture molecules are introduced into the sample containing the target biotoxin. The target molecules precisely attach to the capture molecules on the bead surface when the right circumstances are met.
2. Magnetic Separation: After capturing the target, an external magnetic field is provided, which causes the unbound components to be washed away and the magnetic beads to move in the direction of the magnet. This minimizes background interference by effectively separating the target-bound magnetic beads from the sample matrix.
3. Recognition: After the attractive beads have been isolated, various techniques, including colorimetry, electrochemical strategies, and fluorescence, can be utilized to distinguish the objective analyte. The signal quickly correspond with the convergence of the objective biotoxin in the example.
4. Quantification: Accurate assessment of biotoxin levels is made possible by quantitative analysis of the target analyte concentration, which is accomplished by comparing the signal intensity with established standards or calibration curves (Manea et al. (2024); Campàs et al. (2020); Moro et al. (2017)).

Figure 3.1 illustrates how magnetic beads are used in biotoxin separation.

A significant progression has shown lately in distinguishing the proof of biotoxins with the guide of attractive beads, nanomaterials, and microfluidic frameworks to improve the novel detection methodologies. Among these, attractive bead-based tests have turned into a fantastic choice on account of the various benefits they offer. The increased sensitivity and selectivity of magnetic bead-based tests is one of their main benefits. The capacity of magnetic beads to concentrate target analytes, hence magnifying the signal and enhancing detection limits, is emphasized by Gaiani et al. (2019). This feature is especially important for the detection of low doses of biotoxins when conventional techniques might not be able to provide accurate findings. Due to its massive surface area-to-volume ratio, magnetic beads allow target analytes to attach quickly and effectively. This property makes it possible to quickly extract biotoxins from intricate sample matrices and immobilize them. Kim et al. (2019) used surface-enhanced Raman scattering (SERS) immunoassays to show how well magnetic bead-based assays detect botulinum toxins A and B. The effective immobilization of antibodies was made possible by the magnetic beads, which also increased sensitivity and repeatability.

Notable benefits of magnetic bead-based assays are their automation platform compatibility and simplicity of handling. Numerous research studies have examined the creation of biosensors to identify bacterial toxins in intricate food matrices, highlighting the significance of automation in achieving high-throughput analysis. Magnetic bead-based assays have a straightforward process and can be integrated

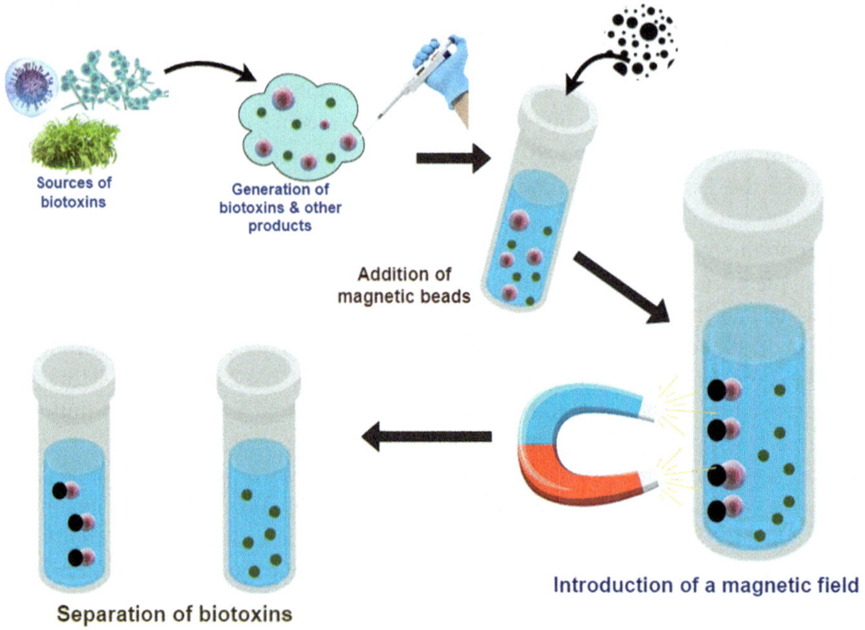

Fig. 3.1 Separation of biotoxins using magnetic beads

with robotic liquid handling systems, and they are ideally suited for automation. This feature increases overall efficiency, decreases the need for manual intervention, and streamlines the detection process. Assays based on magnetic beads frequently need little sample preparation, which makes them appropriate for applications requiring quick screening. Complex samples may be processed immediately, negating the need for labor- and time-intensive sample preparation procedures. Furthermore, certain ligands or antibodies can be used to functionalize magnetic beads, allowing for the selective capture of target analytes with the least amount of interference from non-target elements. These characteristics make magnetic bead-based assays ideal for a wide range of biotoxin detection applications, from food safety studies to environmental monitoring.

Numerous biotoxins, such as chemical toxins, microbial infections, and environmental pollutants, may be detected using magnetic bead-based assays. To detect tetrodotoxins (TTXs) in shellfish, Campàs et al. (2020) created a quick magnetic bead-based colorimetric immunoassay. Specificity of antibodies coupled to magnetic beads can effectively collect TTXs found in intricate matrices. The utilization of magnetic reagents in the toxic sites upgrades the speed and adequacy of screening shellfish tests for TTX tainting. Kim et al. (2019) proposed a surface-upgraded Raman scattering based immunoassay for the location of botulinum poisons utilizing attractive globules. This method achieved high sensitivity and specificity for BoNT detection by immobilizing the antibodies on magnetic beads. Botulinum toxin contamination in food and environmental samples may be monitored with greater ease thanks to the reliable platform provided by the combination of SERS and magnetic beads for the quick and sensitive detection of BoNTs. A functional fluorescence test employing magnetic beads to identify botulinum neurotoxin A (BoNT/A) in complicated matrices was presented by Klisara et al. (2019). This test makes use of surface immobilized antibodies of BoNT/A's on magnetic beads to capture and then quantifies the results using fluorescence detection. By effectively capturing and concentrating BoNT/A from complicated samples, magnetic beads improve the assay's specificity and enable dependable detection in a range of matrices.

Applying magnetic bead-based immunoassays, Ecke et al. (2022) uncovered the technique's true capacity for drug toxin location by exhibiting the touchy distinguishing proof of diclofenac, a non-steroidal mitigating medication. Huergo et al. (2021) developed an appealing bead-based immunoassay that could potentially be used to quickly and accurately identify human SARS-CoV-2 antibodies. Moreover, as surveyed by Manea et al. (2024), attractive bead-based Systematic Evolution of Ligands by Exponential Enrichment (SELEX) advances give an adaptable stage to choose aptamers with high tendency and specificity towards an assortment of target particles, working with the improvement of creative bio poison location examines. Consequently, it has been shown that magnetic bead-based assays offer significant possibilities for the quick, accurate, and focused identification of a range of biotoxins. The examples illustrate the versatility of magnetic beads in biotoxin identification and emphasize how important they are in the advancement of analytical methods for food security, monitoring the environment, and healthcare diagnosis.

3.6 Microfluidic Systems in Biotoxin Detection

Microfluidics has developed as a capable device in a few areas, including the recognizable proof of biotoxins. To handle small amounts of liquids (10^{-9} to 10^{-18} liters), microfluidics uses simple channels that range in size from tens to hundreds of microliters. Microfluidic systems can mix, transport, or separate fluids for analysis. Every one of these approaches has developed over the years into a unique set of strategies for maintaining stable fluid flow. These platforms utilize microchannels, chambers, and valves to precisely control the movement of fluids and particles at the microscale.

Figure 3.2exhibits how microfluidic systems can be used to detect and quantify biotoxins in the environment.

The detection of biotoxins can be performed quickly and sensitively because of the control that microfluidic devices provide over fluid flow and response conditions. Functionalized surfaces and capture probes work together to improve detection selectivity, allowing for the differentiation of closely related toxins and lowering false positive rates (Bunyakul et al. 2015). Multiple biotoxins can be detected simultaneously in a single sample because of the superior multiplexed analytical capabilities of microfluidic devices. In addition to improving efficiency and throughput, multiplexed detection offers detailed information on the kind and amount of toxins in a sample, which improves decision-making and diagnostic accuracy. For the detection of biotoxins, microfluidic devices provide many benefits (Alahmad et al. 2023). One of their primary benefits is that they are very sensitive, which makes it feasible to detect low quantities of biotoxins with tiny sample volumes. This is essential to the early diagnosis and treatment of disorders associated with toxins. Microfluidic devices might be made more versatile and have a wider detection range by combining several analytical methods such as mass spectrometry, electrochemical sensing, and fluorescence detection. Microfluidic devices offer an inexpensive solution for biotoxin detection because of their compact size and ideal reagent and consumable utilization. Because of the smaller sample sizes and reagent use, microfluidic technology may now be used by a broader range of clients, which also means less waste and operating expenses. They enable rapid and sensitive detection with minimal sample volumes, enhancing efficiency and reducing reagent consumption (Aghamohammadi et al. 2023). Also, the integration of microfluidics with bead-based assays allows for multiplexed toxin detection, enabling simultaneous analysis

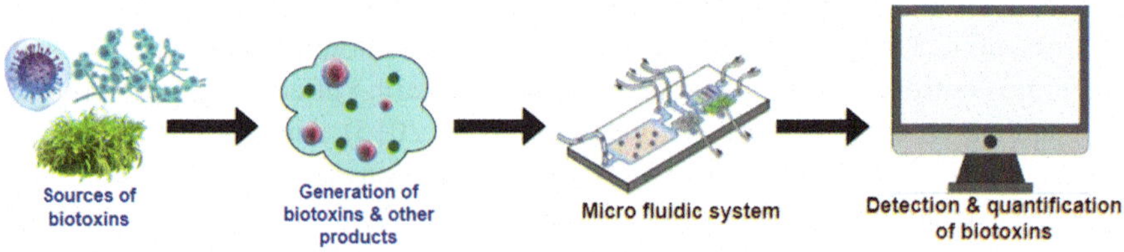

Fig. 3.2 Detection and quantification of biotoxins using microfluidic systems

of multiple targets within a single sample. By using microvalves, microarrays, and parallel channels, these devices can support a variety of detection tests that are specific to certain toxins or biomarkers. In addition to improving efficiency and throughput, multiplexed detection offers detailed compositional and concentration information.

Furthermore, advancements in microfluidic technology facilitate precise control over reaction conditions, enhancing the specificity and sensitivity of detection methods (Bendre et al. 2022). Microfluidic platforms' small size makes it possible to continuously monitor the amounts of biotoxins in saltwater or extracts from shellfish, providing timely warnings to prevent any poisoning situations. Cyanobacterial blooms are a severe environmental issue because they generate cyanotoxins such as microcystins that can pollute water sources and pose a health risk to consumers. Bendre et al. (2022) presented the utilization of magnetic beads functionalized with specific receptors for toxin detection and capture, in addition to the on-chip screening of cyanotoxins employing microfluidic technology. The integration of magnetic manipulation within microfluidic channels enabled efficient concentration and separation of cyanotoxins from complex environmental samples, followed by sensitive detection using fluorescence or electrochemical methods. This integrated approach offers a rapid and reliable solution for monitoring cyanotoxin levels in water bodies, facilitating timely intervention strategies to mitigate health risks.

The creation of extremely selective and sensitive platforms for the detection of shellfish toxins has resulted from recent advancements in micro/nano biosensors (Tian et al. 2021). The crucial necessity to monitor seafood safety has been addressed by researchers who have successfully and quickly detected shellfish poisons by combining nanomaterials with microfluidic devices. In a similar way employing nanomaterial-integrated immunosensors, outstanding sensitivity and specificity of food toxin detection have been established (Bhardwaj et al. 2021). These immunosensors offer effective on-chip detection of several foodborne toxins when combined with microfluidic devices, guaranteeing food items' safety and quality. Certain forms of biotoxins, including botulinum toxins and cholera, were successfully identified using microfluidic biosensors. Selective and sensitive detection is made practical by the combination of microfluidics with certain biorecognition components, helping with early diagnosis and intervention during cholera epidemics. Koh et al. (2015) showcased the application of centrifugal microfluidic technology in the very sensitive identification of botulinum toxin.

For consistent sample dispersion and the best possible interaction between target analytes and recognition components, efficient fluid manipulation and mixing are essential. Using centrifugal forces, centrifugal microfluidic devices provide accurate control over fluid flow and mixing, facilitating effective sample processing and increased detection sensitivity (Koh et al. 2015). Another significant benefit in the design of microfluidic systems is the capacity to detect numerous biotoxins concurrently in a single experiment. In addition to increasing assay throughput, multiplexed detection makes it possible to thoroughly analyze complicated materials. High sensitivity and specificity mycotoxin detection in parallel

has been established using techniques such as microfluidic fluorescence immunoassays and multiplexed capillary microfluidic immunoassays (Machado et al. 2018; Soares et al. 2018).

Building biotoxin detection microfluidic devices using magnetic beads and nanomaterials is crucial. Biosensors may be significantly improved in terms of sensitivity and selectivity by utilizing nanomaterials, which have special qualities including large surface area, biocompatibility, and catalytic activity. However, according to Pinto et al. (2017), magnetic beads offer effective target analyte capture and separation from complicated samples, resulting in improved detection limits and shorter test times. The ability to integrate microfluidic devices with point-of-need platforms and their portability thanks to their miniaturization, makes it easier to detect biotoxins on-site in a variety of environments. Machado et al. (2018) have demonstrated the successful integration of microfluidic immunoassays with smartphone-based data gathering systems to enable remote monitoring and real-time analysis. This integration has the potential to facilitate decentralized testing and early warning systems.

When developing microfluidic systems for biotoxin detection, several factors need to be taken into consideration, such as the integration of magnetic beads and nanomaterials, fluid manipulation techniques, multiplexed detection abilities, and miniaturization for point-of-need applications. Food safety, monitoring of the environment, and biodefense might all benefit from this technology. Microfluidic paper-based analytical devices (μPADs) are among some of the noteworthy developments in the field of biotoxin detection utilizing microfluidic systems; they are cheap, portable, and easy to use. Alahmad et al. (2023) showed how useful μPADs are for identifying aflatoxins when they are present in food samples. Using antibody-based assays in conjunction with the natural capillary action of paper channels, the μPADs allowed for the quick and accurate detection of mycotoxins in a matter of minutes. This method has great potential for on-site food product screening, particularly in situations with limited resources where conventional laboratory methods would not be practical. Marine biotoxins are a serious risk to human health and seafood safety, thus it is important to create quick and accurate detection techniques. The potential of microfluidic biosensors for real-time monitoring of marine biotoxins was demonstrated by Farré and Barceló (2020). Saxitoxin and domoic acid were among the target toxins that the biosensors showed excellent sensitivity and selectivity toward by combining microfluidic channels with receptors, including aptamers or antibodies. By using modern facilities and microfluidic systems, researchers are striving toward more accessible, rapid, and sensitive detection approaches, which will eventually benefit public health and safety.

3.7 Integration of Nanomaterials, Magnetic Beads, and Microfluidic Systems

Biotoxins have several different features and may pose health hazards; therefore detecting them can be practically difficult. Conventional detection techniques frequently lack the speed, specificity, and sensitivity needed for prompt and reliable diagnosis. On the other hand, recent advances in magnetic beads, nanomaterials, and microfluidic devices provide viable paths for improving biotoxin detection. Nanomaterials serve as the sensing elements for toxin detection in this integrated platform, while magnetic beads aid in the efficient extraction and concentration of specific toxins from the sample matrix (Bhardwaj et al. 2021).

Minimal sample sizes and processing durations are possible for on-chip analysis because of the infrastructure for sample manipulation provided by microfluidic devices.

Researchers design portable, user-friendly biosensing devices with these components to identify biotoxins quickly and accurately in a range of settings, including food safety, environmental monitoring, and biomedical diagnostics. Using magnetic beads, nanomaterials, and microfluidic devices together is a potent way to improve biotoxin detection. This novel strategy transforms the way we identify and lessen the risks presented by dangerous poisons in a variety of settings. It also has numerous impressive advantages. Combining magnetic beads with nanomaterials and microfluidic devices provides enhanced biotoxin detection sensitiveness and selectivity. One of the biggest benefits of integrated systems is their capacity to do fast research on biotoxins. Assay periods are shortened, and the detection process is expedited thanks to microfluidic devices, which provide accurate micrometer-scale control over fluid flow and reaction conditions. Integrated systems can provide real-time results in a matter of minutes by utilizing nanomaterial-based detection methods with high sensitivity and magnetic beads for quick target analyte concentration and capture. This allows for the miniaturization of biotoxin detection platforms, which facilitates portability and field applications (Bendre et al. 2022). Because of their small size and great portability, these systems can be used for decentralized testing in places with limited resources or distant locations without access to conventional laboratory equipment. The capacity of integrated systems to perform multiplexed detection of biotoxins, which enables the simultaneous analysis of numerous analytes inside a single sample, is another benefit (Kim et al. 2019). Combining magnetic beads, nanomaterials, and microfluidic devices, it is possible to automate every step of the biotoxin detection process, from sample preparation to analysis. Also, this lowers the possibility of human errors and eliminates the need for manual intervention. These benefits open the door to the creation of cutting-edge biosensing technology with important outcomes. Through the utilization of these combined advantages, scientists have made impressive advancements in a range of fields. The integration of nanomaterials, including magnetic and quantum dots, with microfluidic devices has resulted in notable enhancements in terms of selectivity and sensitivity. As a result of their special optical qualities, quantum dots may be

functionalized for attachment to certain toxins with preference, increasing the sensitivity of the detection process and signal amplification (Moro et al. 2017). Additionally, magnetic particles offer a versatile platform for capturing and isolating target analytes from complex matrices, thus improving selectivity and reducing interference from background noise (Giouroudi and Kokkinis 2017). The integration of magnetic beads with microfluidic systems enables multiplexed detection of biotoxins, allowing simultaneous analysis of multiple targets within a single assay. Aghamohammadi et al. (2023) demonstrated a competitive, bead-based assay combined with microfluidics for multiplexed toxin detection, offering rapid and comprehensive analysis of diverse toxin profiles. This capability is particularly valuable in scenarios where the presence of multiple toxins needs to be assessed simultaneously. Microfluidic systems, with their ability to precisely control fluid flow at the microscale, enable rapid and portable assays for biotoxin detection. When coupled with magnetic beads, these systems facilitate efficient capture and concentration of target analytes, reducing assay time and sample volume requirements (Nakhlband et al. 2022). Miniaturized platforms hold great promise for on-site detection applications, where timely decision-making is crucial. Functionalization of magnetic beads with specific ligands allows for the selective capture of target toxins, while the modular design of microfluidic devices enables easy integration of various detection modalities (Rezvani Jalal et al. 2021). This adaptability facilitates the development of customized assays for different biotoxin targets and sample matrices. Target analytes may be quickly and effectively isolated from complex biological matrices utilizing magnetic beads functionalized with nanomaterials as good capture probes (Pirzada and Altintas 2019). Biosensing tests may be made more selective and sensitive through the elimination of interfering species using magnetic separation methods. Additionally, on-chip sample preparation, detection, and analysis with low sample volume and processing time are made possible by microfluidic systems combined with nanomaterial-based biosensors (Carneiro et al. 2019). Biosensing assays function more effectively when fluid flow and reaction conditions are precisely controlled in microfluidic devices. A microfluidic system functionalized with magnetic beads and graphene oxide nanosheets was created and described by Tian et al. (2021) for the quick identification of paralytic shellfish toxins (PSTs). The magnetic beads were coated with specific antibodies to help PSTs bind to shellfish samples more selectively. emphasizes the significance of cross-disciplinary collaboration between the fields of biosensing and nanotechnology. It also highlights the necessity of constant innovation to meet changing food safety concerns. The synergistic integration of microfluidic systems with nanotechnology support for chemical sensing, biosensing, and bioanalysis is highlighted by Fattahi and Hasanzadeh (2022). If AFB1, a strong carcinogen, is present in consumables more than what is allowed, serious health hazards could result. Yadav et al. (2021) provide a thorough analysis of biosensors based on nanomaterials and demonstrate their effectiveness in detecting AFB1. It is expected that additional developments in fluid dynamics, materials science, and assay optimization will increase the usefulness of these integrated platforms in combating new biotoxin threats.

3.7.1 Challenges and Solutions in Integrating these Technologies

The identification of biotoxins is essential for maintaining environmental sustainability and public health. The integration of nanomaterials, magnetic beads, and microfluidic systems has been identified as a viable strategy for improving the biotoxin detection platforms to enhance sensitivity, selectivity, and efficiency. There are a few issues and hurdles that need to be resolved in order to effectively exploit these technologies. Establishing compatibility and smooth integration of various parts, such as magnetic beads, microfluidic systems, and nanomaterials, constitutes one of the main problems. The distinct qualities and functions of each component might not always be consistent with or easily incorporated into a single detection platform. For early and precise identification, a high sensitivity in biotoxin detection is essential. The integration of magnetic beads, nanomaterials, and microfluidic systems with optimum sensibility is a challenging task. Integrating magnetic beads for sample pre-concentration and purifying into microfluidic systems necessitates precise control over fluid flow, magnetic manipulation, and sample interaction. This is because effective sample preparation and handling are critical to the reliable detection of biotoxins. Another challenge is creating scalable and manufacturable integrated detection platforms for large production. It can be challenging technically to ensure affordability, consistency, and reproducibility when increasing output (Bendre et al. 2022). It is still difficult to identify biotoxins with high selectivity and specificity while reducing interference from background signals and non-specific binding. Enhancing specificity and selectivity requires integrating various recognition elements and adjusting test conditions (Gaiani et al. 2019).

3.7.2 Future of Biotoxin Detection Using Nanomaterials, Magnetic Beads, and Microfluidic Systems

Biotoxin identification has recently advanced recognition of the combination of nanomaterials, magnetic beads, and microfluidic devices, which provide previously unheard-of levels of sensibility, mobility, and effectiveness. Improving the biotoxin detection systems' sensitivity and selectivity is one of the main areas of attention for future study. Through the utilization of nanomaterials' distinctive characteristics, like their elevated surface-to-volume ratio and adjustable surface chemistry, scientists hope to enhance the specificity of detection tests. Toxins can be selectively captured by functionalizing magnetic beads with ligands or antibodies, as Nakhlband et al. (2022) noted in their review of magnetic materials in microfluidic-based sensing systems.

Another exciting path in biotoxin detection is the creation of multiplexed detection systems. Through the integration of diverse nanomaterials and several detection channels in microfluidic devices, researchers can accurately and efficiently detect

multiple poisons at the same time. The potential of nanomaterial-based biosensing platforms for multiplexed detection of marine toxins was demonstrated by Shati et al. (2023), underscoring the significance of modern technology in accomplishing this objective. Microfluidic systems are great candidates for point-of-care (POC) applications because of their miniaturization and portability. To enable quick and on-site detection of biotoxins in a variety of contexts, the future research efforts will probably concentrate on further downsizing and integrating these devices. The use of magnetic nanoparticles in microfluidic sensors for virus detection is discussed by Rezvani Jalal et al. (2021), who highlight possible uses of POC in the monitoring of the environment and healthcare.

Additionally, integrating cutting-edge signal transduction approaches including electrochemical and optical technologies shows potential for enhancing the biotoxin sensing platforms' detection limits and dependability. Capability for real-time data processing and monitoring is crucial for the prompt identification and remediation of biotoxin contamination incidents. Future studies will probably concentrate on creating integrative sensing systems that can automatically analyze data and monitor continuously, allowing for quick decision-making and action. The use of magnetic beads, nanomaterials, and microfluidic technology for biotoxin detection has great promise for addressing novel problems in food safety, public health, and environmental monitoring. By combining the advantages with these state-of-the-art technologies, researchers may be capable of developing extremely portable, sensitive, and selective detection systems that will successfully meet the growing demand for quick and accurate biotoxin identification.

The combination of magnetic beads, nanomaterials, and microfluidic technology has resulted in a significant turning point in the detection of recently identified biotoxins during the past several years. When it concerns strengthening the sensitivity, selectivity, and efficiency of biosensing systems for the detection of marine toxins, nanomaterials are crucial. The addition of nanomaterials, such as carbon nanotubes and graphene oxide, enhances the surface area available for immobilizing proteins and encourages signal amplification, which lowers detection limits and strengthens the dependability of biosensors. Applying nanomaterials including gold nanoparticles and quantum dots enables the growth of ultrasensitive biosensors capable of detecting even trace levels of shellfish toxins. Additionally, functionalizing nanomaterials using receptors improves the selectivity of biosensors, allowing for highly precise differentiation among various toxin kinds.

When creating microfluidic-based sensing systems, magnetic beads have special benefits, especially for sample preparation and analyte enrichment. By integrating magnetic beads, microfluidic devices can handle samples quickly and automatically, cutting down on assay time and increasing overall sensibility. Miniaturized platforms for detailed biochemical investigations with little reagent and sample usage are made accessible by microfluidic devices. The coupling of microfluidic systems with biosensors for rapid and sensitive detection has been addressed in research assessments of current developments in the use of microfluidic chip biosensors to detect foodborne pathogenic bacteria. Biosensing platform downsizing addresses the growing need for convenient and easy-to-use tests by enabling

point-of-care testing and on-site surveillance. The subject of biotoxin detection can be greatly advanced by combining magnetic beads, nanomaterials, and microfluidic technologies. Future studies should concentrate on tackling important issues such as strengthening integrated detection systems' scalability and resilience, developing multiplexed capability for the simultaneous identification of several toxins, and maximizing device performance for practical uses.

Moreover, the creation of new nanomaterials with customized qualities, like enhanced stability and biocompatibility, would increase the range of applications for integrated detection systems that involve the detection of biotoxins. To drive development and turn lab prototypes into commercial goods, cooperative interdisciplinary research combining experts in nanotechnology, materials science, biochemistry, and microfluidics will be crucial. Integrating nanomaterials, magnetic beads, and microfluidic devices is a feasible way to improve the sensitivity, selectivity, and portability of biotoxin detection systems. Technological developments in this area will lead to the creation of future biosensing systems having a variety of applications in nutrition safety, biomedical diagnosis, and environmental tracking.

Solving compatible concerns and creating integrated detecting platforms require cooperation among specialists in nanomaterials, magnetic beads, microfluidics, and biosensing. Diverse component design, optimization, and integration can be made easier with the use of multidisciplinary techniques. Furthermore, precise manufacture and assembly of integrated detection platforms can be made possible by utilizing cutting-edge fabrication techniques including soft lithography, 3D printing, and microfluidic patterning. These methods provide control over device architecture, scalability, and flexibility (Bendre et al. 2022). The specificity and sensitivity of detection can be improved by creating novel assay designs that take advantage of the special characteristics of magnetic beads and nanomaterials. Assay performance can be increased by techniques such as surface functionalization, multiplexing, and signal amplification (Rezvani Jalal et al. 2021; Aghamohammadi et al. 2023). Using robotics and automation for sample handling, preparation, and evaluation can increase reproducibility, decrease errors made by humans, and streamline operations. High precision and throughput complex manipulations can be carried out by integrated robotic systems (Fattahi and Hasanzadeh 2022). For integrating detecting platforms to be dependable and repeatable, defined methods, performance measures, and validating processes must be established.

3.8 Conclusion

The development of cutting-edge technology has transformed biosensors in the past few years, providing creative approaches to the identification and reduction of biotoxins. According to Bazin et al. (2017), one important development is the use of new biorecognition molecules for biosensors. These compounds are essential for improving the assays used for toxin detection in terms of both sensitivity and specificity. Through the utilization of unique characteristics found in different

biomolecules, including enzymes, antibodies, aptamers, and DNA/RNA fragments, biosensors can accurately and precisely identify a broad spectrum of toxins. This discovery is significant for food safety, environmental monitoring, and healthcare applications because it not only increases the accuracy of toxin identification but also makes analysis faster and less expensive. Mycotoxins are a danger to food safety, and awareness of their prevalence and toxicity is increaing globally (Alshannaq and Yu 2017). To detect and quantify mycotoxins in different food matrices, sophisticated analytical methods have been developed, such as mass spectrometry, chromatography, and immunoassays. By enabling thorough screening for many mycotoxin pollutants, these approaches ensure regulatory compliance and protect public health. Furthermore, ongoing improvements in detection sensitivity and specificity improve our capacity to locate traces of mycotoxins, reducing the chance of exposure and associated health threats. A novel method for the detection of biomolecules is the development of wearable materials integrated with synthetic biology sensors (Nguyen et al. 2021). With their real-time monitoring features, these wearables help identify toxins and pathogens early in a variety of settings. Users can receive timely alerts and take proactive steps to avoid dangers by incorporating synthetic biology parts, like modified microbial cells or nucleic acid-based sensors, onto wearable systems. This technology could revolutionize toxin detection and surveillance and has applications in biodefense, environmental monitoring, and customized health care. While advanced technologies are employed to combat biotoxins, there are unprecedented opportunities to improve real-time monitoring capabilities, sensitivity, and detection precision. We may effectively decrease the risks associated with biotoxins, protecting public health and environmental sustainability, by using novel biorecognition molecules, wearable sensor platforms, and analytical techniques.

Acknowledgement The authors would like to thank the University Research Council of University of Peradeniya, Sri Lanka (Multidisciplinary Research Grant-MDRG 390) for the financial support.

References

Aghamohammadi H, Thomas KE, Srikant S, Deglint J, Wong A, Poudineh M (2023) A competitive, bead-based assay combined with microfluidics for multiplexed toxin detection. Lab Chip 23(14):3245–3257

Alahmad W, Varanusupakul P, Varanusupakul P (2023) Recent developments and applications of microfluidic paper-based analytical devices for the detection of biological and chemical hazards in foods: a critical review. Crit Rev Anal Chem 53(2):233–252

Alshannaq A, Yu JH (2017) Occurrence, toxicity, and analysis of major mycotoxins in food. Int J Environ Res Public Health 14(6):632

Bahramian A (2020) Hazards identification and the unit's assessment of a the water treatment plant against pathogenic and biotoxin threats affecting by physicochemical parameters. Iranian Journal of Chemical Engineering 17(4):33–48

Barhoum A, García-Betancourt ML, Jeevanandam J, Hussien EA, Mekkawy SA, Mostafa M, Omran MM, Abdalla M, Bechelany M (2022) Review on natural, incidental, bioinspired, and engineered nanomaterials: history, definitions, classifications, synthesis, properties, market, toxicities, risks, and regulations. Nano 12(2):177

Bazin I, Tria SA, Hayat A, Marty JL (2017) New biorecognition molecules in biosensors for the detection of toxins. Biosens Bioelectron 87:285–298

Bendre A, Bhat MP, Lee KH, Altalhi T, Alruqi MA, Kurkuri M (2022) Recent developments in microfluidic technology for synthesis and toxicity-efficiency studies of biomedical nanomaterials. Mater Today Adv 13:100205

Bhardwaj H, Rajesh, Sumana G (2021) Recent advances in nanomaterials integrated immunosensors for food toxin detection. J Food Sci Technol 59:1–22

Bunyakul N, Promptmas C, Baeumner AJ (2015) Microfluidic biosensor for cholera toxin detection in fecal samples. Anal Bioanal Chem 407:727–736

Campàs M, Reverté J, Rambla-Alegre M, Campbell K, Gerssen A, Diogène J (2020) A fast magnetic bead-based colorimetric immunoassay for the detection of tetrodotoxins in shellfish. Food Chem Toxicol 140:111315

Carneiro P, Morais S, Pereira MC (2019) Nanomaterials towards biosensing of Alzheimer's disease biomarkers. Nano 9(12):1663

Ecke A, Westphalen T, Hornung J, Voetz M, Schneider RJ (2022) A rapid magnetic bead-based immunoassay for sensitive determination of diclofenac. Anal Bioanal Chem 414:1–11

Farré M, Barceló D (2020) Microfluidic devices: biosensors. In: Chemical analysis of food. Academic Press, pp 287–351

Fattahi Z, Hasanzadeh M (2022) Nanotechnology-assisted microfluidic systems for chemical sensing, biosensing, and bioanalysis. TrAC Trends Anal Chem 152:116637

Gaiani G, O'Sullivan CK, Campàs M (2019) Magnetic beads in marine toxin detection: a review. Magnetochemistry 5(4):62

Giouroudi I, Kokkinis G (2017) Recent advances in magnetic microfluidic biosensors. Nanomaterials 7(7):171

Gupta PK, Gupta PK (2020) Biotoxins and venomous organisms. In: Problem solving questions in toxicology: a study guide for the board and other examinations, pp 219–228

Gupta R, Raza N, Bhardwaj SK, Vikrant K, Kim KH, Bhardwaj N (2021) Advances in nanomaterial-based electrochemical biosensors for the detection of microbial toxins, pathogenic bacteria in food matrices. J Hazard Mater 401:123379

Hegde M, Pai P, Shetty MG, Shakuntala K (2022) Gold nanoparticle based biosensors for rapid pathogen detection: a review. Environ Nanotechnol Monit Manag 18:100756. https://doi.org/10.1016/j.enmm.2022.100756

Huergo LF, Selim KA, Conzentino MS, Gerhardt EC, Santos AR, Wagner B et al (2021) Magnetic bead-based immunoassay allows rapid, inexpensive, and quantitative detection of human SARS-CoV-2 antibodies. ACS Sens 6(3):703–708

Hwang HS, Jeong JW, Kim YA, Chang M (2020) Carbon nanomaterials as versatile platforms for biosensing applications. Micromachines 11(9):814

Jiang C, Lan L, Yao Y, Zhao F, Ping J (2018) Recent progress in application of nanomaterial-enabled biosensors for ochratoxin a detection. TrAC Trends Anal Chem 102:236–249

Kaliva M, Vamvakaki M (2020) Nanomaterials characterization. In: Polymer science and nanotechnology. Elsevier, pp 401–433

Khizar S, Ben Halima H, Ahmad NM, Zine N, Errachid A, Elaissari A (2020) Magnetic nanoparticles in microfluidic and sensing: from transport to detection. Electrophoresis 41(13–14):1206–1224

Kim K, Choi N, Jeon JH, Rhie GE, Choo J (2019) SERS-based immunoassays for the detection of botulinum toxins A and B using magnetic beads. Sensors 19(19):4081

Kim JH, Park JY, Lee C, Park JS (2019) Marine biotoxins and shellfish poisoning. Compr Rev Food Sci Food Saf 18(4):1018–1057

Klisara N, Peters J, Haasnoot W, Nielen MW, Palaniappan A, Liedberg B (2019) Functional fluorescence assay of botulinum neurotoxin A in complex matrices using magnetic beads. Sensors Actuators B Chem 281:912–919

Koh CY, Schaff UY, Piccini ME, Stanker LH, Cheng LW, Ravichandran E, Singh BR, Sommer GJ, Singh AK (2015) Centrifugal microfluidic platform for ultrasensitive detection of botulinum toxin. Anal Chem 87(2):922–928

Landsberg JH, Van Dolah FM, Doucette GJ (2007) Algal biotoxins. In: Botana L (ed) Algal Toxins: Nature, Occurrence, Effect, and Detection. Springer, pp 185–216

Machado JM, Soares RR, Chu V, Conde JP (2018) Multiplexed capillary microfluidic immunoassay with smartphone data acquisition for parallel mycotoxin detection. Biosens Bioelectron 99:40–46

Manea I, Casian M, Hosu-Stancioiu O, de los Santos-Álvarez N, Lobo-Castañón MJ, Cristea C (2024) A review on magnetic beads-based SELEX technologies: applications from small to large target molecules. Analy Chim Acta, p 342325

Morabito G, Silvestro S, Faggio C, Guglielmino SPP (2018) Water quality of drinking and recreational water resources: emerging contaminants and cyanotoxins. Int J Environ Res Public Health 15(8):1699. https://doi.org/10.3390/ijerph15081699

Moro L, Turemis M, Marini B, Ippodrino R, Giardi MT (2017) Better together: strategies based on magnetic particles and quantum dots for improved biosensing. Biotechnol Adv 35(1):51–63

Murk AJ, Nicolas J, Smulders FJ, Bürk C, Gerssen A (2019) Marine biotoxins: types of poisoning, underlying mechanisms of action and risk management programmes. In: Chemical hazards in foods of animal origin. Wageningen Academic Publishers, pp 398–402

Nakhlband A, Kholafazad-Kordasht H, Rahimi M, Mokhtarzadeh A, Soleymani J (2022) Applications of magnetic materials in the fabrication of microfluidic-based sensing systems: recent advances. Microchem J 173:107042

Nguyen PQ, Soenksen LR, Donghia NM, Angenent-Mari NM, de Puig H, Huang A, Lee R, Slomovic S, Galbersanini T, Lansberry G, Sallum HM (2021) Wearable materials with embedded synthetic biology sensors for biomolecule detection. Nat Biotechnol 39(11):1366–1374

Parak M, Asgari A, Nourian YH, Ghanei M (2024) A review of poisoning with various types of biotoxins and its common clinical symptoms. Toxicon 107629

Pinto E, Van Nieuwerburgh L, Sampayo MAM (2017) Biotoxins from microalgae. In: Hallegraeff GM, Anderson DM, Cembella AD (eds) Toxins and Biologically Active Compounds from Microalgae, Volume 2: Biological Effects and Risk Management. CRC Press, pp 137–162

Pirzada M, Altintas Z (2019) Nanomaterials for healthcare biosensing applications. Sensors 19(23):5311

Rezvani Jalal N, Mehrbod P, Shojaei S, Labouta HI, Mokarram P, Afkhami A, Madrakian T, Los MJ, Schaafsma D, Giersig M, Ahmadi M (2021) Magnetic nanomaterials in microfluidic sensors for virus detection: a review. ACS Appl Nano Mater 4(5):4307–4328

Sanati A, Jalali M, Raeissi K, Karimzadeh F, Kharaziha M, Mahshid SS, Mahshid S (2019) A review on recent advancements in electrochemical biosensing using carbonaceous nanomaterials. Microchim Acta 186:1–22

Shati AA, Al-dolaimy F, Alfaifi MY, Sayyed RZ, Mansouri S, Aminov Z, Alubady R, Gandla K, Alawady AHR, Alsaalamy AH (2023) Recent advances in using nanomaterials for portable biosensing platforms towards marine toxins application: up-to-date technology and future prospects. Microchem J 195:109500

Shkembi X, Skouridou V, Svobodova M, Leonardo S, Bashammakh AS, Alyoubi AO, Campàs M, O′ Sullivan, C.K. (2021) Hybrid antibody–aptamer assay for detection of tetrodotoxin in pufferfish. Anal Chem 93(44):14810–14819

Soares RR, Santos DR, Pinto IF, Azevedo AM, Aires-Barros MR, Chu V, Conde JP (2018) Multiplexed microfluidic fluorescence immunoassay with photodiode array signal acquisition for sub-minute and point-of-need detection of mycotoxins. Lab Chip 18(11):1569–1580

Still KR, Mohapatra A (2020) Biotoxins. In: Information resources in toxicology. Academic Press, pp 117–132

Su S, Sun Q, Gu X, Xu Y, Shen J, Zhu D, Chao J, Fan C, Wang L (2019) Two-dimensional nanomaterials for biosensing applications. TrAC Trends Anal Chem 119:115610
Tian Y, Du L, Zhu P, Chen Y, Chen W, Wu C, Wang P (2021) Recent progress in micro/nano biosensors for shellfish toxin detection. Biosens Bioelectron 176:112899
Vinh TQ, Huy DTN, Yakutseny S (2022) Environment impacts and composition-biotoxic activity in natural hydrocarbon raw materials & processed products. J Lifesci 3(3):128–132
Wu X, Hou L, Lin X, Xie Z (2019) Application of novel nanomaterials for chemo-and biosensing of algal toxins in shellfish and water. In: Novel nanomaterials for biomedical, environmental and energy applications. Elsevier, pp 353–414
Yadav N, Yadav SS, Chhillar AK, Rana JS (2021) An overview of nanomaterial-based biosensors for detection of Aflatoxin B1 toxicity in foods. Food Chem Toxicol 152:112201

Part II
Data on Biotoxins from Different Sources

Chapter 4
Biotechnological and Therapeutic Applications of Bacterial Toxins

Rasika Jinadasa

Abstract Bacterial protein toxins are highly specific, extremely potent, and highly efficient to translocate into the cytosol of the target cells. These inherent properties make them very good candidates for biotechnological and therapeutic applications. Targeted intracellular drug delivery is a highly promising therapeutic approach, particularly in cancer therapy. This novel therapeutic approach requires highly specific targeting mechanisms capable of passage through the cell membrane. Bacterial toxins have evolved naturally to specifically target cell subtypes and subcellular organelles via receptor binding. Additionally, these toxins penetrate the cell membrane efficiently through the membrane translocation process, therefore making them ideal vehicles to deliver therapeutic agents into the host cytosol.

Keywords Anthrax toxin · Cholera toxin · Diphtheria toxin · Pertussis toxin · Pore-forming toxins · Rho-activating toxins · Synaptic proteins

4.1 Introduction

The "classical" bacterial toxins are proteins, which are released or secreted from the bacteria and act on or in target cells (Hu et al. 2021; Johannes 2022). Some of these toxins act on the cell surfaces by forming pores while others act exclusively in the cytoplasm and have elaborate mechanisms to enter cells (Hu et al. 2021). Many of the classical bacterial toxins covalently bind and modify their target proteins in eukaryotic cells, and alter the functions of their targets, which often results in a permanent inhibition or stimulation of the activities or functions of target proteins. Pore formation and uptake mechanisms of these toxins are frequently utilized as protein delivery systems to translocate proteins into the cytosol (Johannes 2022).

R. Jinadasa (✉)
Department of Veterinary Pathobiology, Faculty of Veterinary Medicine & Animal Science, University of Peradeniya, Peradeniya, Sri Lanka

M. N. V. Prasad et al. (eds.), *Biotoxins*,
https://doi.org/10.1007/978-3-031-75309-1_4

Bacterial protein toxins are highly specific, extremely potent, and highly efficient. Additionally, despite the fairly large size, these toxins can easily translocate into the cytosol of the target cells without causing much damage to the cell membrane (Kumar et al. 2019; Balmforth et al. 2021). These inherent properties make them very good candidates for biotechnological and therapeutic applications. Most of these toxins are used as pharmacological and cell biology tools (Kumar et al. 2019; Balmforth et al. 2021; Kenworthy et al. 2021; White et al. 2022).

The high specificity is the most important property of these toxins making them suitable candidates as pharmacological and cell biology tools, and for biotechnological and therapeutic applications (Kumar et al. 2019; Johannes 2022; Khoshnood et al. 2022). Many bacterial toxins have remarkable specificity to differentiate between highly related and structurally similar members of eukaryotic protein families. For example, actin-ADP-ribosylating clostridial toxins can differentiate between α-and β/γ-actin, which are more than 95% identical in their amino acid sequences (Heber et al. 2023; Nagahama et al. 2023). Toxins/toxin families commonly used as pharmacological and cell biology tools are listed in Table 4.1.

High potency of bacterial protein toxins means that they can act at extremely low concentrations. For example, *Clostridium botulinum* neurotoxins (BoNTs) are the most potent biological agents known. The minimal lethal dose (LD50) of BoNTs for mice is as low as 10 pg/Kg (Rossetto and Montecucco 2019). The high potency of bacterial protein toxins is primarily due to two properties. Many toxins are enzymes (proteases, ADP-ribosyltransferases, glucosyltransferases, N-glycosidases, or deamidases) and act as catalysts (Balmforth et al. 2021; Kenworthy et al. 2021; White et al. 2022; Locht and Antoine 2023). Additionally, these toxins preferentially target cells that are particularly sensitive to the toxin using high-affinity interactions with specific cell surface receptors (Pirazzini et al. 2017, 2022). The high

Table 4.1 Bacterial toxins commonly used as pharmacological and cell biology tools. Modified from Aktories (2006)

Pore-forming toxins	Streptolysin O *Staphylococcus aureus* α-toxin Tetanolysin *Escherichia coli* hemolysin
Toxins targeting heterotrimeric G proteins	Cholera toxin Pertussis toxin *Pasteurella multocida* toxin
Toxins targeting low molecular weight GTPases	C3 exoenzymes *Clostridium difficile* toxin A and B *Escherichia coli* CNF
Toxin targeting actin	*Clostridium botulinum* C2 toxin *Clostridium perfringens* iota toxin
Toxin-based cell delivery systems (intracellular protein delivery)	Diphtheria toxin *Pseudomonas* exotoxin A Anthrax toxin

efficiency of bacterial toxins is due to their specificity and potency, and additionally, due to the choice to attack functionally important eukaryotic cell substrates (Balmforth et al. 2021; Kenworthy et al. 2021; White et al. 2022). The substrates attacked by the bacterial protein toxins are often important switch proteins of cellular signaling. Therefore, the activation or inactivation of these target proteins has profound and rapid effects on cellular functions. Examples include the activation of the Gs proteins by cholera toxin, and inhibition of Gi proteins by pertussis toxin. These toxins can be used for biotechnological and therapeutic applications in their native form, or with modifications (Balmforth et al. 2021; Kenworthy et al. 2021; Johannes 2022; White et al. 2022).

4.2 Pore-Forming Toxins

Formation of pores by bacterial toxins affects the eukaryotic cells by causing cellular electrolyte imbalances, depletion of nucleotides, or other essential low molecular weight factors (Hu et al. 2021). Pore-forming toxins are widely used as pharmacological tools to permeabilize eukaryotic cells and to introduce small molecules, which are membrane impermeable (e.g., nucleotides and ions), peptides or proteins into target cells (Hu et al. 2021).

Streptolysin O from *Streptococcus pyogenes* is probably the most common pore-forming toxin used as a pharmacological tool. It forms very large pores, enabling delivery of large molecules. The toxin does not bind to membranes which do not contain cholesterol. The number of binding sites and the binding affinity are also dependent on the amount of cholesterol present in the membrane. Usually, the permeabilization by streptolysin O is lethal for the cell. However, reversible membrane permeabilization has been attempted. The effect of the toxin is not limited to the cell membrane, it may affect the intracellular organelles also. Certain other pore-forming toxins such as *Staphylococcus aureus* α-toxin affect only the cell membrane. Pore-forming toxins have been widely used to study exocytosis from secretory cells, calcium regulation, signal-contraction coupling in smooth muscle cells, and intracellular membrane trafficking. Other pore-forming toxins used as cell biological tools include tetanolysin from *Clostridium tetani*, *Escherichia coli* hemolysin, and perfringolysin O from *C. perfringens*, and *Staphylococcus aureus* α-toxin.

Pore-forming toxins from *Bacillus thuringiensis,* primarily the three-domain crystal (Cry) toxins are commercially available as insecticides to control insects such as disease-vector mosquitoes and many crop pests (Miranda et al. 2024; Palma et al. 2024; Takahashi et al. 2024; Wang et al. 2024). *Bacillus thuringiensis* is a naturally occurring soil bacteria. These toxins kill insect larvae by pore forming in the midgut cell epithelium and causing an osmotic shock leading to cell lysis (Bravo et al. 2013).

Different strains of *B. thuringiensis* have specific toxicity to certain insects or insect families depending on the type and number of Cry toxins present in the given bacterial strain(Bravo et al. 2013; Palma et al. 2024; Takahashi et al. 2024). Due to

this selectivity and mode of action, *B. thuringiensis* is non-infective and non-toxic to vertebrates, including humans, and the toxins have very little or no effect on vertebrates, non-target aquatic animals and terrestrial invertebrates, making it an environmentally friendly alternative to chemical pesticides (Bravo et al. 2013; Palma et al. 2024; Takahashi et al. 2024). Cry toxins are commonly applied as aerial sprays (prepared toxin or *B. thuringiensis* spores). Genetically modified crop varieties such as soybean, corn, and cotton incorporating *B. thuringiensis* toxin genes have also been developed and widely cultivated (Bravo et al. 2013; Elkins et al. 2024; García Degano et al. 2024; Sakuno et al. 2024; Wang et al. 2024). However, resistance to single Cry toxins is common, particularly in field strains of crop pest insect populations. Efficacy in commercial operations can be achieved by mixing synergistic toxins from different *B. thuringiensis* subspecies that produce several Cry toxins (Palma et al. 2024; Wang et al. 2024). This also prevents the evolution of resistance to ensure its sustainability as a biopesticide (Palma et al. 2024; Wang et al. 2024).

4.3 Toxins Targeting Heterotrimeric G Proteins

4.3.1 Cholera Toxin

Cholera toxin is produced by *Vibrio cholerae.* The ADP-ribosyltransferase activity of the toxin is responsible for the watery diarrhea and possible eventual death due to dehydration in cholera patients (Balmforth et al. 2021; Kenworthy et al. 2021; White et al. 2022). The ~85-kDa toxin is composed of an enzyme subunit of ~28 kDa (A-subunit) and a binding component consisting of five identical B subunits of about 11.6 kDa. The enzyme part consists of a ~ 21-kDa A1 and a ~ 6-kDa A2 subunit linked by a disulfide bridge (Kenworthy et al. 2021; White et al. 2022). A1 is the ADP-ribosyltransferase. The A2 subunit links the enzyme to the B component. The B subunit binds to the ganglioside receptor GMI. Up on binding the enzyme subunit is internalized and translocated to the cytosol by retrograde transport via endoplasmic reticulum (Kenworthy et al. 2021; White et al. 2022).

Cholera toxin is widely used as a cell biology tool to study the structure and function of cell membranes. For example, the toxin is often used to study the membrane (lipid) rafts, which are regions of the cell membranes enriched in sterols, sphingolipids, and saturated phospholipids (Kenworthy et al. 2021). The toxin is also used as a sensor and inducer of cell membrane curvature, a reporter of clathrin-independent endocytosis, and a probe for retrograde trafficking mechanisms (Kenworthy et al. 2021). The toxin has been investigated for potential therapeutic uses in multiple areas as well. For example, its being studied as a means of highly efficient method for intracellular delivery of proteins. It has been shown that cholera toxin B subunit alone could deliver other recombinant proteins into cells through either covalent conjugation or noncovalent interaction (Jia et al. 2021). Interestingly, the advancing knowledge on the biology of cholera toxin has resulted in the

utilization of its cellular targets for therapeutic purposes. For example, cholera toxin-binding proteins have been investigated as vehicles for targeted delivery of proteins to cells (Balmforth et al. 2021).

4.3.2 Pertussis Toxin

Pertussis toxin is produced by *Bordetella pertussis* which causes whooping cough in humans. The toxin is a hexameric toxin of 105 kDa, consisting of the enzyme subunit (~26 kDa, A-subunit) and the binding pentamer (B-subunit) formed of subunits with molecular weights ranging from ~11 to 22 kDa (Locht and Antoine 2023). Pertussis toxin is the most complex bacterial protein toxin currently known, with respect to both the structure and diversity of biological activities. Importantly it is the main protective antigen in pertussis vaccines currently in use worldwide (Ausar et al. 2020; Locht and Antoine 2023). In addition to its role in vaccines, pertussis toxin played an important role in our understanding of many cellular signaling pathways. Historically it was used for the identification of G_i as an inhibitory G protein, compared to the stimulatory G_s protein. Importantly, several G proteins in addition to the G_s are not modified by pertussis toxin. Accordingly, the toxin is frequently used as a cell biology tool to test whether a signaling pathway involves "pertussis toxin-sensitive" G proteins or not (Locht and Antoine 2023).

Since the G protein-coupled receptor family consists of more than 800 members, many functions of the pertussis toxin remain to be discovered. Therefore, it has a huge potential for novel therapeutic applications, such as for the treatment of cancer by inhibiting cancer cell proliferation and metastasis (Lyu et al. 2021; Locht and Antoine 2023).

4.3.3 **Pasteurella multocida** *Toxin*

Pasteurella multocida toxin (PMT) is the major toxin produced by *Pasteurella multocida*, which colonizes the gastrointestinal tract and nasopharynx of several animal species including cats, dogs, and pigs. This toxin causes progressive atrophic rhinitis in pigs, which results in the loss of nasal turbinate bones (Ebner et al. 2019). The toxin deamidates an essential glutamine residue of G_α subunits of heterotrimeric G proteins resulting in glutamic acid. This modification results in the loss of the intrinsic GTPase activity of the G protein rendering the protein constitutively active (Ebner et al. 2019). The constitutive activation of heterotrimeric G proteins alters bone metabolism, leading to bone loss. PMT has been studied as a therapeutic agent to treat rare human disorders causing excessive and uncontrolled bone formation (Ebner et al. 2019).

4.4 Toxins Targeting Low Molecular Weight GTPases

4.4.1 Rho Inactivating Toxins

4.4.1.1 C3 Exoenzymes

C3 exoenzymes are ADP-ribosylating toxins produced by several bacteria including *Clostridium botulinum, Bacillus cereus*, and *Staphylococcus aureus* (Just et al. 2011; Rohrbeck and Just 2017). C3 exoenzymes are valuable cell biology tools for studying their target proteins. The toxins generally induce important changes in function upon modification of the targets. The C3 exoenzymes have been used to understand the physiological role of the small GTP-binding protein Rho, due to its ability to inhibit Rho signaling activity (Cells 2015; Rohrbeck and Just 2017). In recent years, C3 has become a promising drug candidate to inactivate RhoA in neurons (Just et al. 2011).

4.4.1.2 *Clostridium difficile* Toxin A and B

Clostridial glucosylating toxins target Rho GTPases. These toxins modify the GTPases by glucosylation. They are large toxins with molecular masses between 250 and 308 kDa (Chen et al. 2015; Aktories et al. 2017). *C. difficile* toxins A and B are the major virulence factors causing antibiotic-associated diarrhea and are the most studied members of this toxin family (Aktories et al. 2017). These toxins can also be used as tools to study the signaling activity mediated by Rho GTPases (Chen et al. 2015).

4.4.2 Rho-Activating Toxins

4.4.2.1 *Escherichia coli* CNF

Some bacterial protein toxins such as Cytotoxic necrotizing factor 1 (CNF1) & 2 from *E. coli* can form persistently active Rho by deamidation of glutamine 63 of RhoA to glutamic acid. Due to Rho GTPase activation, CNFs induce prominent cellular changes including stress fiber, microspikes, and lamellipodia formation. Multinucleation is also seen as a typical feature of CNF-treated cells (Chaoprasid and Dersch 2021). Due to the extreme potency, high efficacy, and long-lasting and specific activity on Rho GTPases, CNF toxins are being investigated as strong candidates for therapeutic use. They are being tested for the treatment of central nervous system diseases and cancer (Chaoprasid and Dersch 2021). The CNFs are routinely secreted into the environment, are up taken by host cells with a very high efficiency. Therefore, the non-toxic transport fragments of these toxins can be

engineered as carriers to deliver therapeutic agents as cargo into target cells (Chaoprasid and Dersch 2021).

4.4.3 Toxin Targeting Actin

4.4.3.1 *Clostridium botulinum* C2 Toxin

4.4.3.1.1 *Clostridium perfringens* Iota Toxin

Several bacterial toxins including *Clostridium botulinum* C2 toxin, *C. perfringens* iota toxin, *C. spiroforme* toxin, and the ADP-ribosylating toxin from *C. difficile* act primarily by ADP-ribosylating of actin (Heber et al. 2023; Nagahama et al. 2023). All these toxins contain separate enzyme (active) and cell-binding/translocation subunit (Nagahama et al. 2021). C2 toxin ADP-ribosylates nonmuscle β/γ actin isoforms but not the α-actin isoforms. In contrast, *C. perfringens* iota toxin catalyzes the ADP-ribosylation of all mammalian actin isoforms. The ADP-ribosylation is specific to actin (Heber et al. 2023; Nagahama et al. 2023). Accordingly, *Clostridium botulinum* C2 toxin and *Clostridium perfringens* iota toxin are frequently used as pharmacological or cell biological tools. For example, *botulinum* C2 toxin has been shown to function as a vehicle for the delivery of N-terminally His-tagged proteins or untagged proteins with a positively charged N-terminal to cells (Nagahama et al. 2021, 2023; Heber et al. 2023).

4.4.4 Neurotoxins Targeting Synaptic Proteins

4.4.4.1 *Clostridium botulinum* Neurotoxins

The clostridial neurotoxins consisting of *Clostridium botulinum* neurotoxins and tetanus toxin (from *Clostridium tetani*) contain a light chain (~50 kDa) and a heavy chain (~100 kDa) linked by a disulfide bond (Pirazzini et al. 2017, 2022; Rossetto and Montecucco 2019; Megighian et al. 2021). The light chain has the enzyme activity while the heavy chain is responsible for binding and membrane translocation. One isoform of tetanus toxin (TeNT) and at least seven serotypes of botulinum neurotoxins (BoNT/A, B, C1, D, E, F, and G) are currently known. The toxins have metalloendoprotease activity and selectively cleave synaptic proteins (synaptobrevin/VAMP, syntaxin, SNAP-25, and their isoforms). The synaptic proteins targeted by clostridial neurotoxins are involved in the exocytosis and neurotransmitter release from presynaptic nerve endings. Botulinum neurotoxins are the most potent biological toxic substances known (Pirazzini et al. 2017, 2022; Megighian et al. 2021).

The toxins have very high substrate specificity and preference for sites of action. Specifically, synaptobrevin is cleaved by tetanus toxin and BoNT/B, D, F, and G, whereas SNAP25 is cleaved by BoNT/A, C, and E; and syntaxin is cleaved by BoNT/C (Pirazzini et al. 2017, 2022). The botulinum neurotoxins act on peripheral nerves and cause flaccid paralysis, while the tetanus toxin acts on the central nervous system and induces spastic paralysis (Pirazzini et al. 2017). As cell biology tools, clostridial neurotoxins were highly useful for the understanding of the processes involved in neuronal exocytosis and vesicle-cell membrane fusion processes. The toxins were also used to study signal secretion coupling in nonneuronal cells (Pirazzini et al. 2017, 2022).

Tetanus toxin is also applied to induce experimental epilepsies and neuronal degeneration. Tetanus toxin is also the major protective antigen in the tetanus toxoid. Recombinant toxin has shown strong immune response and vaccine potency under experimental conditions (Przedpelski et al. 2020). Botulinum neurotoxins are now formulated to be used widely as therapeutics (Pirazzini et al. 2017). Their therapeutic uses include treating blepharospasm, strabismus, and many different types of dystonias. They are also used to treat hyperhidrosis, pain, autoimmune disorders, and migraine headaches. Recently, the toxin is widely used for cosmetic purposes (Pirazzini et al. 2017). In fact, nearly 50% of the medical grade BoNT produced at present is used for cosmetic purposes. It is primarily used as an injection to smoothen facial lines, mainly to resolve the glabellar lines. Recent studies show that ~20 U total dose of BoNT is remarkably effective and safe for the treatment of glabellar lines, and repeated use has no significant ill effects (Pirazzini et al. 2017). In addition to the therapeutic uses in human medicine, BoNT is being used for multiple applications in veterinary medicine as well (Turin et al. 2023).

4.5 Toxin-Based Cell Delivery Systems (Intracellular Protein Delivery)

Targeted intracellular drug delivery is a highly promising therapeutic approach, particularly in cancer therapy. This novel therapeutic approach requires highly specific targeting mechanisms capable of passage through the cell membrane (Lingwood 2021; Ruschig and Marschall 2023; Yin and Thaker 2023). Bacterial toxins have evolved naturally to specifically target cell subtypes and subcellular organelles via receptor binding. Additionally, these toxins penetrate the cell membrane efficiently through the membrane translocation process, therefore making them ideal vehicles to deliver therapeutic agents into the host cytosol (Lingwood 2021). Some of the candidates currently been investigated include anthrax toxin protective antigen, diphtheria toxin, and *Pseudomonas* exotoxin A (Lingwood 2021; Ruschig and Marschall 2023; Yin and Thaker 2023).

4.5.1 Anthrax Toxin Protective Antigen

Anthrax toxin is the major virulence factor for *Bacillus anthracis* that causes severe disease called anthrax in all warm-blooded animals. This binary toxin consists of the receptor-binding and translocation machinery protective antigen (PA) and enzymatic factors, which are referred to as the lethal factor (LF) and edema factor (EF). As the membrane translocation module, PA facilitates the delivery of LF and EF through the membrane into the cytosol (Ruschig and Marschall 2023; Yin and Thaker 2023). Anthrax toxin PA has been shown experimentally to facilitate the cytosolic delivery of various non-native cargoes (Yin and Thaker 2023).

4.5.2 Diphtheria Toxin

Diphtheria toxin (DT) is a ~ 58 kDa protein secreted by lysogenic strains of *Corynebacterium diphtheriae* that causes diphtheria in humans. The toxin consists of three domains, the receptor (R-), translocation (T-), and catalytic (C-) domains like many other A-B toxins. The toxin inhibits protein synthesis following the entry to the cell via receptor-mediated endocytosis (Rodnin et al. 2020; Ruschig and Marschall 2023; Yin and Thaker 2023). Diphtheria toxin is a readily expressed and extremely potent toxin that has minimal adverse effects on humans. Therefore, it is widely used to selectively target various cancers. Replacing the B domain with various cancer antigen-targeting antibodies, growth factors or other therapeutic agents can successfully achieve tumor cell-specific targeting (Yin and Thaker 2023).

4.5.3 **Pseudomonas** *Exotoxin A*

Pseudomonas exotoxin A is a highly potent toxin secreted by *Pseudomonas aeruginosa*. It is a single chain multidomain AB toxin. The cell-binding B fragment specifically binds to the host cell receptor and the toxin is internalized by clathrin-coated vesicles-mediated endocytosis (Ruschig and Marschall 2023; Yin and Thaker 2023). *Pseudomonas* exotoxin A based immunotoxins for cancer treatment have been successfully achieved. Modified versions of the toxin have been developed to minimize the protein size and reduce immune clearance. The truncated versions were linked to various targeting molecules such as antibodies, antibody fragments, or ligands (Yin and Thaker 2023).

References

Aktories K (2006) Toxins as tools. In: Alouf JE, Popoff MR (eds) The comprehensive sourcebook of bacterial protein toxins, 3rd edn. Elsevier, San Diego, pp 976–990

Aktories K, Schwan C, Jank T (2017) Clostridium difficile toxin biology. Ann Rev Microbiol 8(71):281–307. https://doi.org/10.1146/annurev-micro-090816-093458

Ausar SF et al (2020) Genetically detoxified pertussis toxin displays near identical structure to its wild-type and exhibits robust immunogenicity. Commun Biol 3:427. https://doi.org/10.1038/s42003-020-01153-3

Balmforth MR et al (2021) Piggybacking on the cholera toxin: identification of a CTB-binding protein as an approach for targeted delivery of proteins to motor neurons. Bioconjug Chem 32(10):2205. https://doi.org/10.1021/acs.bioconjchem.1c00373

Bravo A et al (2013) Evolution of bacillus thuringiensis cry toxins insecticidal activity. Microb Biotechnol 6(1):17–26. https://doi.org/10.1111/j.1751-7915.2012.00342.x

Cells., U. of clostridium botulinum C. exoenzyme into intact H. and J (2015) Uptake of clostridium botulinum C3 exoenzyme into intact HT22 and J774A.1 cells. Toxins 2(7(2)):380–395. https://doi.org/10.3390/toxins7020380

Chaoprasid P, Dersch P (2021) The cytotoxic necrotizing factors (CNFs)-a family of rho GTPase-activating bacterial exotoxins. Toxins 15(13(12)):901. https://doi.org/10.3390/toxins13120901

Chen S et al (2015) The role of rho gtpases in toxicity of Clostridium difficile toxins. Toxins 7:5254–5267. https://doi.org/10.3390/toxins7124874

Ebner JK et al (2019) Activation of Gq signaling by Pasteurella multocida toxin inhibits the osteoblastogenic-like actions of Activin A in C2C12 myoblasts, a cell model of fibrodysplasia ossificans progressiva. Bone 127:592–601. https://doi.org/10.1016/j.bone.2019.07.031

Elkins BH et al (2024) Sublethal effects of a commercial Bt product and Bt cotton flowers on the bollworm (Helicoverpa zea) with impacts to predation from a lady beetle (Hippodamia convergens). PLoS One 19(5):e0302941. https://doi.org/10.1371/journal.pone.0302941

García Degano F et al (2024) Susceptibility of Diatraea saccharalis populations (Lepidoptera: Crambidae) to Bt toxins expressed in genetically modified corns for development of transgenic sugarcane in Argentina. Sugar Tech 26:776–785. https://doi.org/10.1007/s12355-024-01392-2

Heber S, Borho J, Stadler N, Wondany F, König I, Michaelis J, Papatheodorou P, Barth H, Fellermann M (2023) The Clostridium botulinum C2 toxin subunit C2IIa delivers enzymes with positively charged N-termini into the cytosol of target cells. Toxins 15:390. https://doi.org/10.3390/toxins15060390

Hu H, Liu M, Sun S (2021) Pore-forming toxins during bacterial infection: molecular mechanisms and potential therapeutic targets. In: Drug design, development and therapy, pp 3773–3781. https://doi.org/10.2147/DDDT.S322393

Jia X et al (2021) Highly efficient method for intracellular delivery of proteins mediated by cholera toxin-induced protein internalization. Mol Pharm 18(1):4067–4078. https://doi.org/10.1021/acs.molpharmaceut.1c00479

Johannes L (2022) Application of protein toxins as cell biological and pharmacological tools. Toxins 14:242. https://doi.org/10.3390/toxins14040242

Just I et al (2011) Therapeutic effects of Clostridium botulinum C3 exoenzyme. Naunyn Schmiedeberg's Arch Pharmacol 383(3):247–252. https://doi.org/10.1007/s00210-010-0589-3

Kenworthy AK et al (2021) Cholera toxin as a probe for membrane biology. Toxins 13:543. https://doi.org/10.3390/toxins13080543

Khoshnood S et al (2022) Bacteria-derived chimeric toxins as potential anticancer agents. Front Oncol 12:953678. https://doi.org/10.3389/fonc.2022.953678

Kumar R et al (2019) Evolutionary features in the structure and function of bacterial toxins. Toxins 11:15. https://doi.org/10.3390/toxins11010015

Lingwood C (2021) Therapeutic uses of bacterial subunit toxins. Toxins 13:378. https://doi.org/10.3390/toxins13060378

Locht C, Antoine R (2023) The history of pertussis toxin. Toxins 13:623. https://doi.org/10.3390/toxins13090623

Lyu C et al (2021) Targeting Gi/o protein-coupled receptor signaling blocks HER2-induced breast cancer development and enhances HER2-targeted therapy. JCI Insight 6(8):e150532. https://doi.org/10.1172/jci.insight.150532

Megighian A et al (2021) Tetanus and tetanus neurotoxin: from peripheral uptake to central nervous tissue targets. J Neurochem 158:1244–1253. https://doi.org/10.1111/jnc.15330

Miranda LS et al (2024) The perpetual vector mosquito threat and its eco-friendly nemeses. Biology 13:182. https://doi.org/10.3390/biology13030182

Nagahama M et al (2021) Internalization of clostridium botulinum C2 toxin is regulated by Cathepsin B released from lysosomes. Toxins 13:272. https://doi.org/10.3390/toxins13040272

Nagahama M et al (2023) Cellular uptake and cytotoxicity of Clostridium perfringens iota-toxin. Toxins 15:695. https://doi.org/10.3390/toxins15120695

Palma L, Sauka DH, Ibarra JE (2024) Bacillus thuringiensis: a broader view of its biocidal activity. Toxins 16:162. https://doi.org/10.3390/toxins16030162

Pirazzini M et al (2017) Botulinum neurotoxins: biology, pharmacology, and toxicology. Pharmacol Rev 69(2):200–235. https://doi.org/10.1124/pr.116.012658

Pirazzini M, Montecucco C, Rossetto O (2022) Toxicology and pharmacology of botulinum and tetanus neurotoxins: an update. Arch Toxicol 96:1521. https://doi.org/10.1007/s00204-022-03271-9

Przedpelski A et al (2020) A novel high-potency tetanus vaccine. mBio 10(11):1128. https://doi.org/10.1128/mbio.01668-20

Rodnin MV et al (2020) Structure of the diphtheria toxin at acidic pH: implications for the conformational switching of the translocation domain. Toxins2 12:704. https://doi.org/10.3390/toxins12110704

Rohrbeck A, Just I (2017) Cell entry of C3 exoenzyme from Clostridium botulinum. Curr Top Microbiol Immunol 406:97–118. https://doi.org/10.1007/82_2016_44

Rossetto O, Montecucco C (2019) Tables of toxicity of botulinum and tetanus neurotoxins. Toxins 11(22):686. https://doi.org/10.3390/toxins11120686

Ruschig M, Marschall ALJ (2023) Targeting the inside of cells with biologicals: toxin routes in a therapeutic context. BioDrugs 37:181–203. https://doi.org/10.1007/s40259-023-00580-y

Sakuno CIR et al (2024) Transgenic sugarcane expressing Bacillus thuringiensis Cry1Ac protein offers new possibilities for controlling the giant borer, Telchin lcus (Drury), a pest difficult to control. Crop Prot 175:106469. https://doi.org/10.1016/j.cropro.2023.106469

Takahashi H et al (2024) Mutational analysis of the transmembrane α4-helix of Bacillus thuringiensis mosquito-larvicidal Cry4Aa toxin. Curr Microbiol 81:80. https://doi.org/10.1007/s00284-023-03602-8

Turin L et al (2023) Therapeutic applications of botulinum neurotoxins in veterinary medicine. Vet Sci 10:460. https://doi.org/10.3390/vetsci10070460

Wang Y et al (2024) A chromosome-level genome assembly of the soybean pod borer: insights into larval transcriptional response to transgenic soybean expressing the pesticidal Cry1Ac protein. BMC Genomics 25:355. https://doi.org/10.1186/s12864-024-10216-2

White C, Bader C, Teter K (2022) The manipulation of cell signaling and host cell biology by cholera toxin. Cell Signal 100:110489. https://doi.org/10.1016/j.cellsig.2022.110489

Yin L, Thaker H (2023) Cancer drug delivery systems using bacterial toxin translocation mechanisms. Bioengineering 10:813. https://doi.org/10.3390/bioengineering10070813

Chapter 5
Cyanotoxins: Industrial Potential

Dhammika N. Magana-Arachchi and H. M. S. A. T. Gunathilaka

Abstract Cyanotoxins are secondary metabolites produced by specific cyanobacterial species, able to grow and bloom in all aquatic and terrestrial biotas. The rationale for cyanotoxin production is still a mystery. There are many records of the adverse impacts of cyanotoxins, harming the organisms, animals, plants, and humans. However, with the continuous global research on these complex biological toxins, scientists realize their potency in industry, which benefits humans. The pharmaceutical industry is always looking for novel drugs from natural sources as most current medications cause adverse side effects, and most drugs/antibiotics have become ineffective because of the resistant nature of the infective microbes. Hence, cyanotoxins are an excellent source to be targeted. This chapter begins with an introduction to cyanobacteria, followed by a description of the diverse cyanobacterial toxins and their chemical structures. The next section will be on their bioactivity and genomics. The following section discusses the potential biomedical applications of various cyanotoxins, such as microcystins, oscillatoxins, anatoxins, and kalkitoxins, including those currently being tested and as future targets for different diseases. The following section discusses the utility of cyanotoxins in other industrial applications, including the production of algaecides, herbicides, and insecticides. The final section will include the conclusions and prospects as these cyanotoxins still need to be fully explored.

Keywords Biomedical applications of cyanotoxins · Cyanotoxin chemical structures · Cyanotoxins and cytotoxic compounds · Cylindrospermopsin · Microcystins · Nodularin · Saxitoxin

D. N. Magana-Arachchi (✉) · H. M. S. A. T. Gunathilaka
Molecular Microbiology and Human Diseases Research Programme, National Institute of Fundamental Studies, Kandy, Sri Lanka
e-mail: dhammika.ma@nifs.ac.lk

M. N. V. Prasad et al. (eds.), *Biotoxins*,
https://doi.org/10.1007/978-3-031-75309-1_5

5.1 Introduction

Natural products from microorganisms are an essential source in the pharmaceutical industry. Cyanobacteria are oxygenic phototrophic prokaryotes ubiquitous in nature and can produce diverse biologically active chemical compounds. These have been found to possess a wide range of potential antimicrobial, anticancer, antiviral, and anti-inflammatory activities. Cyanotoxins are secondary metabolites produced by several cyanobacterial species. These toxins are well-known for their harmful effects on humans and animals, and the rationale for their production is still a mystery. Cyanotoxins comprise a rich source of natural cytotoxic compounds. However, their potential in the pharmaceutical industry and the biotechnology industry as a whole has yet to be exploited. This chapter begins with an introduction to cyanobacteria, followed by a description of the diverse cyanobacterial toxins and their chemical structures. The following section will be on their bioactivity and genomics. The following section discusses the potential biomedical applications of various cyanotoxins, such as microcystins, oscillatoxins, anatoxins, and kalkitoxins, including those currently being tested and as future targets for different diseases. The next section discusses the utility of cyanotoxins in other industrial applications, including the production of algaecides, herbicides, and insecticides. The final section will include the conclusions and prospects as these cyanotoxins still need to be fully explored.

5.1.1 Cyanobacteria, Cyanotoxins, and Cytotoxic Compounds

5.1.1.1 Cyanobacteria

Cyanobacteria is one of the oldest life forms on Earth; It is evidenced by fossilized records dating back 3.5 million years. It has profoundly impacted the planet's evolutionary history. These photoautotrophic bacteria exhibit remarkable adaptability, thriving in diverse environments such as freshwater, brackish water, marine habitats, and wastewater. Cyanobacteria, recognized as pioneers of oxygenic photosynthesis, primarily obtain nutrients through photosynthesis. However, some adapt to hydrogen sulfide-rich environments by switching to anoxygenic photosynthesis, similar to bacteria. Cyanobacteria's photosynthetic machinery includes various pigments which help them adapt to changes in light intensity. Through complementary chromatic adaptation, they adjust to different light wavelengths by synthesizing phycobiliproteins, also shielding cells from harmful radiation (Sanfilippo et al. 2019). Microscopic in nature, cyanobacteria display a diverse morphology, existing in the environment as single cells, in groups, colonized, or as filamentous forms (Magana-Arachchi and Wanigatunge 2023). Many cyanobacteria live independently; some form symbiotic relationships with eukaryotes, playing crucial roles in ecosystems. They serve as natural nitrogen-fixing agents, transforming atmospheric

nitrogen into organic compounds such as ammonia, nitrite, or nitrate (Álvarez et al. 2023).

5.1.1.2 Cyanotoxins

Biotoxins originate from various living organisms, including bacteria, fungi, algae, plants, and animals. They are substances that can induce harm upon other organisms, including humans, through multiple routes of exposure, such as ingestion, inhalation, or direct contact. These toxins exhibit diverse chemical structures and biological activities, significantly contributing to ecological dynamics and human health (Still and Mohapatra 2020). Cyanobacterial toxins or cyanotoxins are a group of secondary metabolites originating from cyanobacteria. These biotoxins negatively impact aquatic ecosystems, harming aquatic life and human populations (Du et al. 2019). The favourable environmental conditions and nutrient availability boost cyanobacterial proliferation, causing blooms in waterbodies (Weralupitiya et al. 2022). Most of these blooms contain cyanobacterial species that are capable of producing cyanotoxins. Typically, these toxins are found within (intracellular) the live cells of cyanobacteria. However, both live and dead cells can release cyanotoxins into the environment (Magana-Arachchi and Wanigatunge 2023). Cyanobacterial species, such as *Cylindrospermopsis*, naturally excrete the toxin cylindrospermopsin (CYN) into water. A study reported the historical presence of the cyanotoxin CYN in sediments of hypereutrophic Lake Griffin, Florida, USA, dating back to 4700 years up to 2015 (Waters 2016).

Not all cyanobacteria produce cyanotoxins. Certain cyanobacterial species belonging to different genera are capable of producing cyanotoxins, and it is interesting to note that diverse genera can produce one particular toxin. Furthermore, a range of cyanobacterial species can produce more than one cyanotoxin, such as *Anabaena* species, which produce microcystin, cylindrospermopsin, and anatoxins (Magana-Arachchi and Wanigatunge 2023) (Fig. 5.1).

5.1.1.3 Cytotoxic Compounds from Cyanobacteria

Cytotoxic compounds of cyanobacteria refer to a broader category of bioactive substances. They exhibit toxicity towards cells, as cyanotoxins are a subset of cytotoxic compounds. This is because cyanotoxin includes compounds with cytotoxic effects, such as anticancer properties or those that disrupt cellular processes. These cytotoxic compounds may not necessarily be classified as cyanotoxins but can still harm biological systems.

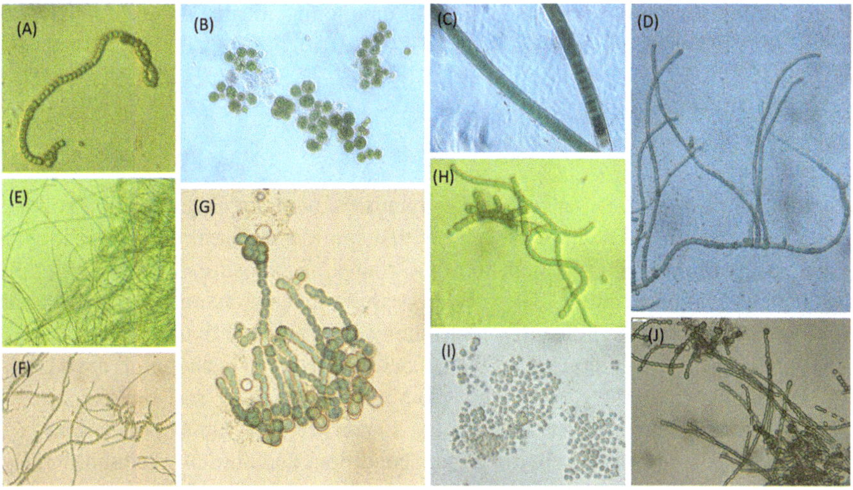

Fig. 5.1 Morphological diversity of cyanobacteria (**a**) *Nostoc* sp. (**b**) *Chroococcidiopsis* sp. (**c**) *Oscillatoria* sp. (**d**) *Anabaena* sp. (**e**) *Leptolyngbya* sp. (**f**) *Mastigocladus* sp. (**g**) *Tolypothrix* sp. (**h**) *Fischerella* sp. (**i**) *Microcystis* sp. (**j**) *Westiellopsis* sp.

5.2 Cyanobacterial Toxins and Their Chemical Structures

The mechanisms of cyanobacterial toxicity show significant diversity, including hepatotoxic, neurotoxic, and dermatotoxic effects, as well as inhibition of protein synthesis (Nugumanova et al. 2023). It is crucial to understand their chemical and physical properties to assess the risk associated with cyanotoxins, the prevalence of cyanotoxins in water sources used by humans, the regulations governing their production, and their environmental fate (Du et al. 2019).

Depending on the chemical structure, the cyanotoxins can be categorized into three groups: cyclic peptides, alkaloids, and lipopolysaccharides (LPS). Major cyanotoxins, including microcystins (MCs) and nodularins (NODs), are peptides, while cylindrospermopsin (CYN), anatoxins, lyngbyatoxins, and saxitoxins are alkaloids. Table 5.1 describes an overview of identified toxic substances originating from various genera of cyanobacteria. Additionally, the table outlines the main target organs in humans affected by these toxins.

5.2.1 Microcystin

Microcystins (MCs) are the most broadly studied hepatotoxins, which are cyclic peptides produced by different cyanobacterial genera such as *Microcystis*, *Planktothrix*, *Anabaena Nostoc*, etc. (Table 5.1). MCs' molecular weights are within the range of 800–1100 Da. The general structure of MC consists of

Table 5.1 The cyanotoxins detected and their corresponding taxa from which they have been isolated, along with their primary targets in mammals. *The dose needed to kill 50% of exposed animals

Toxin	Producer	LD_{50} Intraperitoneal (i.p.) Pure toxin (oral)*mouse F(Churro et al. 2012)	Primary target organ	Mechanism of action (Churro et al. 2012)
Microcystins (~280 variants) Cyclic heptapeptides	*Microcystis* sp. *Planktothrix* sp. *Dolichospermum* sp. *Anabaenopsis* sp. *Aphanocapsa* sp. *Spirulina* sp. *Snowella* sp. *Aphanizomenon* sp. *Oscillatoria* sp. *Nostoc* sp. *Phormidium* sp. *Anabaena* sp. *Rivularia* sp. *Hapalosiphon* sp. *Plectonema* sp.	~50–1000 µg/kg Bw (body weight)	Liver	Hepatotoxic Tumour promoting, inhibition of eukaryotic protein phosphatase PP1, PP2A, and phosphoprotein phosphatases PPP4, PPP5
Nodularin (9 variants) Cyclic pentapeptides	*Nodularia* sp. *Nostoc* sp. (symbiotic) *Spumigena* sp.	~30–50 µg/kg Bw	Liver	Same as MCs. Inhibit the serine/thionine, protein phosphatase type 1/2a. Induce oxidative stress
Cylidrospermopsin (4 variants) Tricyclic guanidine alkaloids	*Lyngbya* sp. *Cylindrospermopsis* sp. *Raphidiopsis* sp. *Oscillatoria* sp. *Umezakia* sp. *Anabaena* sp. *Aphanizomenon* sp.	~200–2100 µg/kg Bw (24 h) 200 µg/kg Bw 5–6 days	Liver Kidneys Lung Heart	Inhibition of glutathione (GSH) and protein synthesis Inhibition of cytochrome P450 Multiple organ toxicity, neurotoxic, genotoxic, protein synthesis inhibitor
Anatoxin-a (5 variants) Atropine-like alkaloid	*Anabaena* sp. *Oscillatoria* sp.	~250 µg/kg Bw/d	Post-synaptic neuromuscular junction	Agonism of muscular and neuronal nicotinic acetylcholine receptor

(continued)

Table 5.1 (continued)

Toxin	Producer	LD_{50} Intraperitoneal (i.p.) Pure toxin (oral)*mouse F(Churro et al. 2012)	Primary target organ	Mechanism of action (Churro et al. 2012)
Homoanatoxin-A (Alkaloid)	*Anabaena* sp. *Oscillatoria* sp. *Planktothrix* sp.	~250 µg/kg Bw/d	Post-synaptic neuromuscular junction	Agonism of muscular and neuronal nicotinic acetylcholine receptor
Anatoxin-A(s) (Guanidine methyl phosphate ester)	*Anabaena* sp. *Aphanizomenon* sp. *Phormidium* sp.	~20–250 µg/kg Bw/d	Post-synaptic neuromuscular junction	Irreversible inhibition of acetylcholinesterase Neurotoxic Binds competitively at acetylcholine receptors
Saxitoxin (~60 variants) Carbamate alkaloids	*Anabaena* sp. *Lyngbya* sp. *Cylindrospermopsis* sp. *Planktothrix* sp. *Aphanizomenon* sp.	~10–30 µg/kg Bw/d	Axons	Blockage of calcium or sodium channels of the nerve axon membrane Neurotoxic Blocks voltage-gated sodium channels
Lyngbyatoxin-A (Alkaloid)	*Lyngbya* sp. *Oscillatoria* sp. *Schizothrix* sp.	~250 µg/kg Bw/d	Skin	Inflammatory agent Tumour promoting, binds to eukaryotic protein kinase C
Aplysiatoxin (Phenolic bislactone)	*Lyngbya* sp. *Planktothrix* sp.	~107–117 µg/kg Bw/d	Skin	Inflammatory agent Protein kinase c activator
Debro-moaplysiatoxin (Phenolic bislactone)	*Lyngbya* sp.	~107–117 µg/kg Bw/d	Skin	Inflammatory agent Protein kinase c activator
Kalkitoxin (Lipopolypeptide) (Shrestha et al. 2023)	*Moorea producens*	LC_{50} 3.86 nM (Morgan et al. 2015)	Neurons	Neuronal necrosis Blocks the electron transport chain complex 1 Block voltage sensitive-sodium channels

cyclo-(D-Ala[1]-X[2]-D-Masp[3]-Z[4]-Adda[5]-D-γ-Glu[6]-Mdha[7]) (Fig. 5.2a) in which X and Z are variable L-amino acids, D-Masp[3] is D-*erythro*-β-methyl-isoaspartic acid, and Mdha is *N*-methyldehydroalanine. Currently, a universal system of nomenclature is being used based on the original term microcystin-XZ, where X and Z denote the variable amino acid residues in positions 2 and 4. MCs induce haemorrhagic shock in the liver, leading to death within 45 min to a few hours and are considered tumour promoters in chronic exposures. Around 300 distinct variants of microcystins have been thoroughly characterized by 2021 (Jones et al. 2021).

5.2.2 *Nodularins*

Nodularins (NOD) are another type of hepatotoxic cyclic peptide. Structural variations are observed across all ten amino acids, with common alterations including the substitution of D-amino acids at positions 2 and 4, the replacement of Mdha with Dhb (dehydrobutyrine) or serine at position 7, (Fig. 5.2b) and the absence of methylation in amino acids at positions 3 and/or 7 (Brezeştean et al. 2022).

5.2.3 *Cylindrospermopsin*

Cylindrospermopsin (CYN) is a cyclic guanidinic alkaloid (Fig. 5.2f). It is also a hepatotoxin but functions as a protein synthesis inhibitor. CYNs are characterized by a tricyclic guanidino moiety, which is connected through a hydroxylated bridging carbon in the C7 position to uracil (WHO 2020). Four distinct structural variants of the CYNs have been identified: 7-epi-cylindrospermopsin (7-epi-CYN), 7-deoxy-cylindrospermopsin (7-deoxy-CYN), 7-deoxy-desulpho-cylindrospermopsin, and 7-deoxy-desulpho-12-acetyl cylindrospermopsin (Wimmer et al. 2014).

The hydrophilic nature of CYNs facilitates the uptake of them into the transport systems and mediates the intestinal absorption of various cell types, including hepatocytes. However, the small size of these molecules suggests limited passive diffusion through biological membranes. At low concentrations, the primary effect of CYN toxicity appears to be the inhibition of protein synthesis mediated by the parent compound (Pichardo et al. 2017).

5.2.4 *Anatoxins*

Neurotoxic anatoxins (ATX) are isolated from the cyanobacterial species *Anabaena*. ATX-a is a neurotoxic bicyclic alkaloid, containing 2-acetyl-9-aza bicycle (4.2.1) non-2-ene (Fig. 5.2c). ATX-a had been initially isolated from *Anabaena flos-aquae*

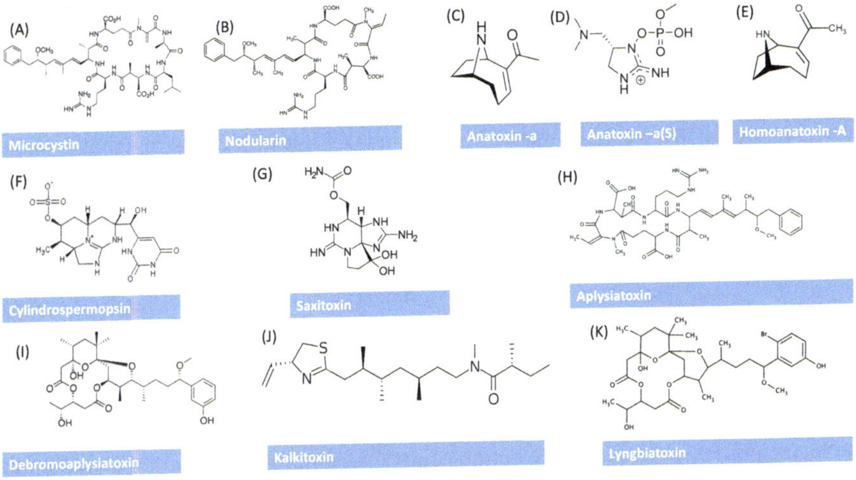

Fig. 5.2 Chemical structures of major cyanotoxins in environment

in the 1970s. Four congeners of this compound have been identified, namely; ATX-a, homoATX-a, dihydroATX-a, and dihydrohomoATX-a. According to Du et al. 2019, no analogues of ATX-a(s) (Fig. 5.2d) have been found.

Homoanatoxin-a (HTX) was isolated from a sample of *Oscillatoria* sp. HTX is a structural variant (Fig. 5.2e) differing from ATX by an ethyl group at the carbonyl-C (Colas et al. 2021). Due to its structural resemblance to ATX, HTX is produced by the same biosynthetic pathway, with the additional carbon derived from L-methionine via S-adenosyl-methionine (Plata-Calzado et al. 2022).

5.2.5 *Saxitoxins*

Saxitoxins (STXs), known as Paralytic Shellfish Poisons, are alkaloid neurotoxins (Oyaneder-Terrazas et al. 2022). These are primarily found in molluscs contaminated with toxic dinoflagellates. They act rapidly, blocking sodium channels in neurons, causing muscle paralysis, and leading to death within 2–30 min in mice. Saxitoxins form 57 analogues (Fig. 5.2g). They comprise a tetrahydropurine group and two guanidine subunits, representing a tricyclic perhydropurine backbone (Akbar et al. 2020).

Based on substitutions in variable positions (Functional Groups), the Saxitoxins family can be categorized into four groups: Carbamate, Decarbamoyl, N-sulfocarbamoyl, Deoxydecarbamoyl. Most known STXs are hydrophilic, except those produced by *L. wollei* in a freshwater environment (Mihali et al., 2008).

Among the roughly 200 natural products associated with tropical "*Lyngbya majuscula*", some have been identified as inducers of irritant contact dermatitis, known as dermatoxins. These include Aplysiatoxin (Nagai et al. 2019),

Table 5.2 Bioactive secondary metabolites produced by various cyanobacteria against human pathogens

Medical property of metabolite	Metabolite	Chemical class	Biological target	References
Antiviral	Lectins	Protein	HIV-1 Simian immunodeficiency virus (SIV) Human herpes virus 6 (HHV-6)	(Shahid et al. 2020) (Huskens et al. 2010)
	Calcium-Spirulan (Ca-SP)	Polysaccharide	HIV-1, herpes simplex virus type 1 (HSV-1) Human cytomegalovirus (HCMV) Measles morbillivirus (MeV) Mumps virus, Influenza virus	(Carpine and Sieber 2021)
Antibacterial	Ambiguines	Alkaloid	*M. Tuberculosis*	(Nowruzi and Porzani 2021)
	Hapalindoles	Alkaloid	*S. aureus* *Pseudomonas aeruginosa* *M. tuberculosis*	(Nowruzi and Porzani 2021)
	Scytoscalarol	Terpene	*Bacillus anthracis* *S. aureus* *E. coli* *Candida albicans*	(Nowruzi and Porzani 2021)
Anticancer/ Antitumoral	Apratoxins	Cyclic depsipeptides	A549 (adenocarcinoma human alveolar basal epithelial cells) cancer cells	(Huang et al. 2016) (Robles-Bañuelos et al. 2022)
	Microcystins	Cyclic heptapeptides	Myeloma cancer cell lines Pancreatic cancer	(Shishido et al. 2019) (Kailash et al. 2022)
	Saxitoxin	Alkaloid	Interacting with site 1 of the voltage-gated Na + channel Suppresses the expression of inflammatory cytokines	(Flores-Holguín et al. 2024)
	Cryptophycins	Alkaloids	Apoptosis in cancer cells through microtubule inhibition	(Eren et al. 2021; Fanale et al. 2015)

(continued)

Table 5.2 (continued)

Medical property of metabolite	Metabolite	Chemical class	Biological target	References
Analgesic/ Anaesthetics	Neosaxitoxin	Alkaloid	Local anaesthesia	(Jeong et al. 2020)
	Gonyautoxins	Alkaloid	Pain relief in tension-type headache	(Jeong et al. 2020)
Antifungal	Laxaphycins	Cyclic heptapeptide	*Aspergillus oryzae* *Candida albicans* *Penicillium notatum* *Saccharomyces cerevisiae* *Trichophyton mentagrophytes*	(Srivastava et al. 2022a) (Bornancin et al. 2019)
	Lyngbyabellin	Lipopeptides	*Moorea* sp. *Okeania* sp. *Perforafilum* sp.	(Dahiya et al. 2020)
	Cryptophycin A	Depsipeptide	*Candida albicans* *Candida glabrata*	(Weiss et al. 2017)
	Tolytoxin	Diterpene lactone	*Saccharomyces pastorianus* *Neurospora crassa* *Candida albicans* *Pythium ultimum* *Rhizoctonia solani*	(Shishido et al. 2015)
Anti-protozoal	Aerucyclamide	Alkaloid	*P. falciparum*	(Niedermeyer and Orst 2015)
Anti-inflammatory	Aeruginosins	Fatty acids	Reduces levels of interleukin-8 (IL-8) and intercellular adhesion molecule-1 (ICAM-1)	(Tabarzad et al. 2020)

5.4.1 Anticancer Activity

5.4.1.1 Microcystin

Microcystin-LR has been identified as inducing toxic effects in the myeloma Sp2/01 cell line. The microcystin-LR delineates their potential therapeutic efficacy and modes of toxicity in myeloma cancer cell lines (Kailash et al. 2022). Consequently, it is proposed that microcystin compounds represent promising candidates for targeted drug development in cancer research and warrant thorough pharmacological evaluation, underscoring their significance in this context.

MC-LR is a model molecule for developing targeted cancer treatments that are effective against pancreatic cancer and other tumours with specific organic anion-transporting polypeptides (OATPs) expression. OATPs, particularly OATP1B1 and 1B3, enable cellular uptake. Increased levels of OATPs in pancreatic cancer show potential for new treatments (Kounnis et al. 2015). However, careful adjustment of

chemicals is necessary to prevent organ-specific side effects. Structural modifications of cyanotoxins can alter their selectivity for specific substrates, rendering them valuable for targeted cancer therapy.

5.4.1.2 Cyanotoxin Analogues

Analogues of cyanotoxins show potential as antitumor agents in biomedicine. Specific cyanotoxins exhibit strong cytotoxic effects against cancer and viral cells, making them promising candidates for antitumour therapy.

Approximately 80% of adult acute leukaemia cases in the United States are classified as acute myeloid leukaemia (AML) (Jani et al. 2023). The current treatment for the disease relies on chemotherapy and anti-leukaemic drugs. Several studies have shown that cyanotoxins contain anti-leukaemic compounds and hold the potential for natural medicines to AML. Some cyanotoxins have shown significant anti-leukaemic activity, with organic extracts displaying ≥70% effectiveness, while some toxins display the properties in hydrophilic extracts.

Cell extracts from *Nostoc* sp., *Cyanobium* sp., and *Oxynema* sp. induced apoptosis in AML cell lines. Aqueous extracts from these species, except *Nostoc* sp., also demonstrated anti-leukaemic activity, particularly at a concentration of 13.3 mg DW mL^{-1}, which exhibited a high potency in inducing cell death in human leukaemia MOLM-13 cells. Some strains of *Fischerella* spp. and *Nostoc* spp. have produced anti-leukaemic compounds with greater potency against MOLM-13 cells than normal rat kidney epithelial cells (Shishido et al. 2019).

5.4.1.2.1 Saxitoxin

Saxitoxin is a group of compounds with a basic structure of trialkyl tetrahydrofuran. This structure constitutes the two permanent guanidinium moieties characterized by NH_2 groups at positions 2 and 8 of the purine ring. The functional groups located at four positions around the purine ring divide this compound into four main categories. Within these toxins, the most potent ones fall under the carbamate division, ranked as saxitoxin (STX), neosaxitoxin (NeoSTX), and gonyautoxin (GTX1) (Flores-Holguín et al. 2024). Certain analogues of STX demonstrate significant pharmaceutical promise due to their capacity to induce anaesthesia by interacting with site 1 of the voltage-gated Na + channel (Wiese et al. 2010). Neosaxitoxin (NeoSTX) suppresses the expression of inflammatory cytokines produced by macrophages by blocking intracytoplasmic voltage-gated sodium channels. The NeoSTX potentially helps treat various conditions characterized by inflammation, such as multiple sclerosis, osteoarthritis, and cancer, as well as disorders affecting the muscles and immune system (Flores-Holguín et al. 2024).

Though cyanotoxin analogues exhibit anaesthetic properties owing to their potent neurotoxicity, showing promise for anaesthetic development, ensuring reversible nerve function suspension is essential to prevent permanent damage.

5.4.1.2.2 Apratoxins

Apratoxins are renowned for their powerful antiproliferative effects against various human cancer cell lines. These compounds are classified as cyclodepsipeptides and are extracted from marine cyanobacteria (Robles-Bañuelos et al. 2022). Apratoxin analogues enhance therapeutic efficacy by enhanced selectivity and robust antitumor activity on cancer cells. Anticancer drugs usually affect organs like the bone marrow, liver, gastrointestinal tract, and kidneys. Interestingly, the pancreas is the primary site affected by apratoxin A, which is a less frequently targeted organ. However, apratoxin A exhibited minimal toxicity in the organs commonly targeted by current therapeutic regimens. This suggests it could be a safe combination partner with standard chemotherapeutic agents. Also, Apratoxin A shows selective activity against A549 (Adenocarcinoma human alveolar basal epithelial cells) cancer cells compared to normal liver cells (Huang et al. 2016).

5.4.1.2.3 Cryptophycins

Cryptophycin-1, a cyanotoxin derived from filamentous cyanobacterium *Nostoc* sp., has been explored as a promising anticancer agent. The antitumor cryptophycins are synthetic derivatives of the cryptophycins. Cryptophycin exhibits selective antitumor properties by inducing apoptosis in cancer cells through microtubule inhibition, similar to Vinca alkaloids (Fanale et al. 2015). Cryptophycins is a depsipeptide with the attachments of four fragments; two ester and two amide linkages make cryptophycin-52 structure, which undergoes phase II clinical trials (Eren et al. 2021).

5.4.2 Antifungal Activity

5.4.2.1 Tolytoxin

Tolytoxin is a macrocyclic lactone isolated from *T. conglutinate* and is an essential antifungal metabolite. The compound demonstrates activity against various fungi, including *Saccharomyces pastorianus*, *Neurospora crassa*, *Candida albicans*, *Pythium ultimum*, *Rhizoctonia solani*, and *Sclerotinia homoeocarpa* (Karpiński 2019).

5.4.2.2 Scytophycins

Scytophycins are macrolides initially isolated from *Scytonema pseudohofmanni*. They are potent cytotoxins and fungicides. Later, Scytophycin was detected from many *Anabaena* strains, including *Anabaena* sp. HAN21/1, *Anabaena* cf. *cylindrica* PH133, *Nostoc* sp. HAN11/1, and *Scytonema* sp. HAN3/2(Karpiński 2019) (Fig. 5.3).

5.5 Cyanobacterial Cytotoxic Metabolites Other Than Cyanotoxins

5.5.1 Anticancer Activity

5.5.1.1 Dolastatin 10

Dolastatin 10 is a significant marine cyanobacterial metabolite initially isolated from sea hare *Dolabella auricularia* and later isolated from *Symploca* sp. It binds to tubulin at the rhizoxin-binding site, disrupting microtubule assembly and inducing cell death in the G2/M phase (Nandi et al. 2024). In the Phase II clinical trial for patients with hormone-refractory prostate cancer, Dolastatin 10 showed excellent tolerance, particularly among the elderly pre-treatment population (Montuori et al. 2023). Dolastatin 10 exhibited notable antiproliferative activity across four distinct human lymphoma cell lines of DB, HT, RL, and SR. Additionally, it effectively triggers apoptosis in non-Hodgkin lymphoma cells (Cragg and Pezzuto 2016).

5.5.1.2 Calothrixin A and B

These two are bioactive molecules whose structural arrangement is assembly to quinoline, quinone, and indole pharmacophores. Calothrixins have been isolated from the cyanobacterial species *Calothrix*. Calothrixin B is commonly recognized as the neutral analogue, whereas calothrixin A is identified as its N-oxide counterpart. It has considerable antiproliferative efficacy against various cell lines like human cervical cancer cells, human colon cancer cells, and human Jurkat cancer cells (Moorthy et al. 2018). Calothrixin A was found to induce cell death through apoptosis in a manner dependent on both time and concentration.

5.5.1.3 Suomilide

Sequencing the complete genome of *Nodularia sphaerocarpa* UHCC 0038 and using bioinformatic analysis of Ahmed et al. 2021 demonstrated that suomilide belongs to the aeruginosin family. Aeruginosins have a complex chemical structure

characterized by the presence of an unusual 2-carboxy-6-hydroxyoctahydroindole (Choi) moiety and the C-terminal arginine derivatives argininal, argininol, agmatine, 1-amidino-2-ethoxy-3-aminopiperidine, or more rarely the 1-amino-2-(*N*-amidino-Δ3-pyrrolinyl)-ethyl (Aeap) moiety (May et al. 2020). Studies have shown that suomilide is a potential lead for developing anti-invasive drugs; in the bigger picture, advocates that aeruginosins are an excellent source of drug leads for developing protease inhibitors (Ahmed et al. 2021).

5.5.2 Antiviral Activity

Antiviral compounds isolated from cyanobacteria typically exhibit bioactivity by blocking viral absorption or penetration and inhibiting replication stages of progeny viruses after penetration into cells. The most common antiviral compounds of cyanobacteria are lectins and carbohydrates.

5.5.2.1 Lectins

Lectins are proteins that can exist as monomers or oligomers. They can bind to carbohydrates specifically and reversibly, such as by including glycoproteins in viral envelopes. The most common cyanobacterial lectins are Cyanovirin-N, Microvirin, Scytovirin, *Microcystis viridis* lectin, and *Oscillatoria agardhii* agglutinins. These lectins are typically isolated from freshwater cyanobacterial cultures; *Nostoc ellipsosporum, Microcystis aeruginosa, Cytonema varium, Microcystis viridis,* and *Oscillatoria agardhii*, respectively (Mazur-Marzec et al. 2021).

Cyanovirin-N (CV-N)

Cyanovirin-N (CV-N) functions by blocking the interaction between the human immunodeficiency virus (HIV) gp120 protein and the CD4 T-cell receptor. In addition to that, it is effective against a range of enveloped viruses such as simian immunodeficiency virus (SIV), chimeric SIV/HIV-1 virus (SHIV89.6P), feline immunodeficiency virus (FIV) (Mazur-Marzec et al. 2021), human herpes virus 6 (HHV-6), measles virus (MeV), Ebola virus, hepatitis virus, and influenza virus. These antiviral effects are related to CV-N's interaction with N-linked high mannose oligosaccharides present on the glycoprotein components of these viruses.

Microvirin (MVN)

Microvirin (MVN) proved highly effective against different strains of HIV-1 and cell types, with IC50 values of 2–12 nM. It also prevented merging infected and uninfected cells in the T-cell line HUT-78 (Shahid et al. 2020).

Scytovirin (SVN)

Scytovirin (SVN) possesses activity against different HIV isolates as well as Zaire Ebola virus (ZEBOV).

***Microcystis viridis* Lectin**

Microcystis viridis lectin actively inhibited the cell conjugation of HIV-I cells and HCV cells. The IC value is around 30 nM (Singh et al. 2023).

***Oscillatoria Agardhii* Agglutinins (OAA)**

Oscillatoria agardhii agglutinins (OAA) block the viral insertion of target cells in HIV-I, HIV-I group O isolates, and HIV-2 strains. They also inhibit HIV replication in MT-4 cells.

Calcium Spirulina (Ca-SP) and Nostoflan

The most common carbohydrate antiviral compound of cyanobacteria is Calcium Spirulina (Ca-SP), isolated from *Spirulina platensis*, and nostoflan, isolated from *Nostoc flagelliforme*. Both carbohydrates are active against enveloped viruses such as HIV-1, herpes simplex virus type 1 (HSV-1), human cytomegalovirus (HCMV), measles morbillivirus (MeV), mumps virus, and influenza virus (Carpine and Sieber 2021).

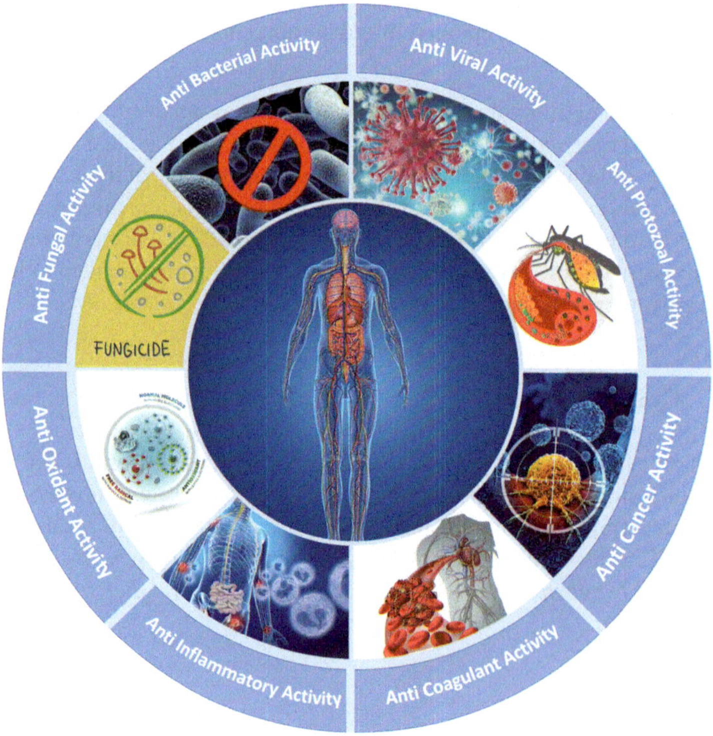

Fig. 5.3 Schematic representation of the diverse biomedical properties associated with cyanobacterial secondary metabolites, highlighting their potential applications in biomedical research

5.5.3 Antibacterial Activity

Hapalindoles

These groups are categorized as tetracyclic and tricyclic compounds. Various hapalindoles, such as Hapalindoles A, B, and T, along with epimers of hapalindoles, are derived from the *Hapalosiphon fontinalis* species. Hapalindole A is described as a chlorinated isonitrile-containing tetracyclic indole alkaloid. Hapalindole T exhibited antibacterial properties, with a MIC value of 2.5 µg/ml against *S. aureus* and 2.0 µg/ml against *Pseudomonas aeruginosa*. Additionally, at a 2.5 µg/ml concentration, Hapalindole T demonstrated activity against *M. tuberculosis* comparable to the anti-TB drug rifampicin (Nowruzi and Porzani 2021).

Ambiguines

Ambiguines were extracted from crude samples of *Fischerella ambigua, H. hibernica*, and *Westiellopsis prolifica*. These compounds bear a structural resemblance to hapalindoles. It features an isoprene unit attached to the C(2) position of the indole moiety. The different toxic compounds of the ambiguine group, such as Ambiguines A–F, K–O, K–N, H–J, P, and Q, are harmful. Ambiguine K isonitrile exhibited moderate activity against *M. tuberculosis*, with a minimum inhibitory concentration (MIC) value of 6.6 µM. Ambiguines H and I isonitriles showed antibacterial activity similar to antibiotics like streptomycin and puromycin (Nowruzi and Porzani 2021).

Scytoscalarol

Scytoscalarol, an antimicrobial cyano-terpene that was obtained from cultured cyanobacterium *Scytonema* sp. (UTEX 1163) through bioassay-guided fractionation. It contains a guanidinium or guanidino group. It displayed antimicrobial activity against *Bacillus anthracis, S. aureus, E. coli*, and *Candida albicans,* with MIC values ranging from 2 to 110 µM (Nowruzi and Porzani 2021). Furthermore, scytoscalarol demonstrated low toxicity (IC_{50} 135 µM) against green monkey kidney cells. Additionally, they indicate weak activity against *M. tuberculosis.*

5.5.3.1 Cyano Peptides

Only a few cyanobacterial peptides exhibit antibacterial properties. For instance, kawaguchi-peptins A and B, isolated from *Microcystis aeruginosa* NIES-88 strain, demonstrate similar minimum inhibitory concentration (MIC) values against *S. aureus* (Parajuli et al. 2016). Additionally, pitipeptolides A, B, C, E, and F, reported from *Lyngbya majuscula*, show activity against *M. tuberculosis* (Cock and Cheesman 2023).

5.5.3.2 Cyano Terpenes

A few terpenes from cyanobacteria have been identified for their antibacterial properties. One example is the diterpenoid noscomin, isolated from *Nostoc commune*, which shows activity against *Bacillus cereus*, *Staphylococcus epidermidis*, and *Escherichia coli* (Srivastava et al. 2022). Another antibacterial terpenoid is cybastacines A and B from *Nostoc* sp., which exhibit more significant activity against the gram-positive bacteria *Bacillus anthracis* and *S. aureus* compared to *E. coli* (Nandagopal et al. 2021a).

5.5.3.3 Cyanobacterial Lipids

Most of the cyanobacterial lipids possess quorum sensing inhibiting activity. Lyngbyoic acid, pitinoic acid A, and doscadecanamide A have been discovered to disrupt quorum sensing (QS) in *Pseudomonas aeruginosa*. The potent antibacterial γ-linolenic acid (GLA) from *Fischerella* sp. was active against *S. aureus*. Polyhalogenated compounds (PHCs) such as Ambigols A, B, C, D, and E are also produced by cyanobacteria. These compounds are effective against methicillin-resistant *S. aureus* (MRSA) (Kar et al. 2022).

5.5.3.4 Malyngolides

The first polyketide antibiotic found in cyanobacteria is malyngolide, which is isolated from *Lyngbya majuscula*. It effectively inhibits the growth of several gram-positive human pathogenic strains like *Mycobacterium smegmatis*, *Streptococcus pyogenes*, *Staphylococcus aureus*, and *Bacillus subtilis*, but is inactive against the gram-negative *Salmonella enteritidis* and *Escherichia coli*. The malyngolide disrupts the quorum sensing (QS) of two gram-negative bacteria, *Chromobacterium violaceum* CV017 and *Pseudomonas aeruginosa*. This disruption can help mitigate the development of antibiotic-resistant strains (Nandagopal et al. 2021b).

5.5.4 *Antifungal Activity*

According to the literature, among the secondary metabolites of cyanobacteria, nearly 106 secondary metabolites contain antifungal properties. These compounds belong to different chemical groups, including peptides, phycobiliproteins, enzymes, carbohydrates, fatty acids, alkaloids, polyketides, macrolides, phenolic compounds, terpenoids, and polymers. The hapalindole-type alkaloids were predominant among the identified antifungal compounds. The genera *Fischerella* and *Scytonema* were pivotal producers of these antifungal compounds (do Amaral et al. 2023). Initially

identified, most of the antifungal agents were later revealed to be highly potent cytotoxins.

5.5.4.1 Cyanotoxin Analogues

Laxaphycins
Laxaphycins (Laxas) are β-amino fatty acid-derived antifungal metabolites from cyanobacterium *Anabaena laxa*. They have been isolated and found to exhibit antifungal activity against various fungal species, including *Aspergillus oryzae*, *Candida albicans*, *Penicillium notatum*, *Saccharomyces cerevisiae*, and *Trichophyton mentagrophytes* (Bornancin et al. 2019).

Lyngbyabellin
Lyngbyabellin B is an antifungal compound containing two thiazole units and one chlorinated β-hydroxy acid. This compound is produced by the *Lyngbya* genus, as well as genera such as *Moorea*, *Okeani*a, and *Perforafilum*. It has been found to suppress *Candida albicans* (Dahiya et al. 2020).

Cryptophycin A
Cryptophycin A is a cyclic depsipeptide that has been documented to exhibit activity against *C. neoformans* and various other species of fungi. This compound is produced by a symbiotic cyanobacterium, *Nostoc* sp. (Weiss et al. 2017).The fungi *Candida albicans, Candida glabrata*, and a fluconazole-resistant strain of *Candida albicans* have their growth suppressed by Cryptophycin A.

Majusculamides
Four Majusculamides (Majusculamides A-D) have been identified. Majusculamides A and B are cytotoxic lipopeptides extracted from the marine cyanobacteria *Lyngbya majuscula*, exhibiting antilarval settlement activities (Du et al. 2019). Majusculamide C was previously discovered in cyanobacteria originating from the Marshall Islands and has been documented to possess antifungal properties (Barzkar et al. 2019).

Schizotrin A
Schizotrin A is an antifungal metabolite derived from the strain *Schizothrix* sp. This compound incorporates β-amino acids into its peptide chain. This compound has been found effective against various fungi, including *Fusarium oxysporum, Colletotrichum gloeosporioides, Saccharomyces cerevisiae, Candida albicans, Candida tropicalis, Rhodotorula mucilaginosa, Sclerotium rolfsii*, and *Rhizoctonia solani*.

Nostofungicidine is a cyclic peptide that is an antifungal compound extracted from the terrestrial cyanobacterium *Nostoc commune*. It exhibits bioactivity against the fungus *Aspergillus candidus* (Nandagopal et al. 2021).

Hassallidin A
Hassallidin A was tested against a panel of 16 Candida strains, including *C. albicans*, *C. krusei*, *C. neoformans*, and *C. parapsilosis*. This compound was extracted from the epilithic cyanobacterium *Hassallia* sp. (Nandagopal et al. 2021).

5.5.5 *Anti-Protozoal Activity*

Against the activity of protozoans such as *Plasmodium*, *Trypanosoma*, *Leishmania*, and *Schistosoma*, cyanobacteria produce a variety of anti-protozoal compounds, including Viridamide A, Symplocamide A, Venturamides, Dragomabin, and Ambigol C.

An alkaloid extracted from *Nostoc* sp., known as nostocarboline, exhibits activity against *Trypanosoma brucei*, *Trypanosoma cruzi*, *Leishmania donovani*, and *Plasmodium falciparum*, with IC_{50} values ranging from 0.5 to 0.194 μM (Nandagopal et al. 2021). Additionally, *Microcystis aeruginosa* releases a compound known as Aerucyclamide C, which is documented for its suppressive effects on *T. brucei* and *P. falciparum*.

5.5.5.1 Antimalarial Properties

5.5.5.1.1 Cyanotoxin Analogues

Aerucyclamide
The ribosomal cyclic peptide aerucyclamide B derived from *M. aeruginosa* is identified as the most potent antiplasmodial compound discovered in cyanobacteria. It demonstrates an IC_{50} value of 0.7 μM against *P. falciparum* with significantly lower cytotoxicity. Another promising lead compound, Balgacyclamides, is obtained from a different strain of *M. aeruginosa* and displays similar activity levels.

5.5.5.1.2 Cyanobacterial Metabolites

Lagunamides A and B
The marine cyanobacterium *L. majuscula* possesses antimalarial properties. It contains cyclodepsipeptides known as Lagunamides A and B. Lagunamide A demonstrated an IC_{50} value of 190 nM, Lagunamide B showed an IC_{50} value of 910 nM against the *P. falciparum* NF54 strain.

Lyngbyabellin
Marine cyanobacteria such as *Moorea producens*, *Moorea bouillonii*, and *Okeania* sp. are also abundant sources of depsipeptides called lyngbyabellins. These compounds exhibit significant antimalarial activities. Lyngbyabellin A, isolated from

Fijian *Moorea producens* that emerged as the most potent cyclic peptide, with an EC_{50} of 1.5 nM against the asexual stages of *P. falciparum*. Furthermore, compounds derived from antimalarial extracts of *Moorea producens* displayed moderate activity against *P. berghei* liver-stage parasites (Sweeney-Jones et al. 2020).

Cell extracts of Calothrix isolates inhibited the growth in vitro of a chloroquine-resistant strain of the malarial parasite *Plasmodium falciparum*.

5.5.6 Anticoagulant and Anti-Inflammatory Activity

Various chemicals with anti-inflammatory activity are excreted from cyanobacteria.

5.5.6.1 Cyanobacterial Metabolites

Aeruginosins

Aeruginosins are comprised of an N-terminal short fatty acid chain, L-Tyr, L-Choi, and L-argininal, and in some cases, a pentose sugar. It is notably found in *Nostoc* sp. but also structural variants in *Nodularia spumigena*. It reduces levels of interleukin-8 (IL-8) and intercellular adhesion molecule-1 (ICAM-1), both proinflammatory mediators (Tabarzad et al. 2020).

Sacran

Sacran from the *Aphanothece sacrum* exhibits anti-inflammatory activity in different animals, such as mice. Additionally, the heteroglycan compound Nc-5-s, isolated from *Nostoc commune*, significantly reduces IL-6 secretion and increases IL-10 secretion in monocytes (Tabarzad et al. 2020).

Coibacins

Coibacin B is a lactone derivative, while Coibacin A is a natural product isolated from *Oscillatoria* sp. Both are significantly affecting the gene transcription of IL-1β and iNOS cytokines (Tabarzad et al. 2020).

Furthermore, C-phycocyanins and exopolysaccharides of some cyanobacteria have significant anticoagulant activity in the human body (Barzkar et al. 2019).

5.6 Utility of Cyanotoxins in Other Industrial Applications

5.6.1 Allelopathic Agents and Biocide

Numerous secondary metabolites synthesized by cyanobacteria demonstrate inhibitory properties against the growth of diverse organisms, such as encompassing large land plants, algae, and microorganisms. Thereby presenting themselves as potential allelochemicals. Compounds such as MC-LR and MC-RR have hindered photoautotrophic organisms' growth by disrupting photosynthetic pigment synthesis. Moreover, compounds such as MC-LR and ATX-a have been found to inhibit the growth of insect larvae, including the stem borer *Chilo agamemnon* and the leaf miner *Hydrellia prosternalis.* This indicates their potential application as insecticides (Li et al. 2023). However, when developing agricultural products, it is crucial to carefully evaluate the possible adverse effects of modified natural cyanotoxins on non-target organisms. Although there is potential for creating valuable products, it is essential to ensure that cyanotoxins or secondary metabolites do not inadvertently harm the environment or non-target organisms.

5.6.1.1 Allelopathic Reactions

Microcystins have been discovered as growth-hindering molecules of numerous macrophyte species and reduce photosynthetic oxygen production. This leads to changes in pigment patterns and content in various macrophytes such as *Chara demersum*, *Myriophyllum spicatum*, and *Vallisneria natans*. Furthermore, exposure to MCs can diminish growth rates, chlorophyll-a concentration, and photosynthetic rates of specific charophyte species. They also act as allelopathic agents, with a pronounced impact on charophyte species compared to angiosperms.

Aquatic plants have an antioxidant system comprising antioxidant enzymes and glutathione (GSH) to defend against oxidative stress. Exposure to elevated levels of MCs has increased the activities of antioxidant enzymes like peroxidase (POD), superoxide dismutase (SOD), glutathione S-transferase (GST), glutathione peroxidase (GPX), and ascorbate peroxidase (APX) in diverse macrophytes. Moreover, MCs can impact GSH content, a crucial element in detoxification, through conjugation with absorbed toxins.

While most of the studies have concentrated on the allelopathic effects of cyanotoxins on aquatic plants at their high concentrations, some investigations have reported minor or detrimental effects even at environmentally relevant concentrations (Pindihama and Gitari 2017). Additionally, cyanotoxins such as cylindrospermopsin can impact aquatic plants by hindering protein and glutathione synthesis, leading to growth inhibition, modulating protease activity, and inducing changes in root histology and microtubule organization (Máthé et al. 2021).

Exposure to anatoxin-a has been associated with reduced photosynthetic oxygen production and heightened peroxidase and glutathione S-transferase activities in

floating aquatic macrophytes like *Lemna minor*. Furthermore, anatoxin-a can disturb homeostasis in submerged aquatic macrophytes such as *Ceratophyllum demersum* by triggering oxidative stress.

5.6.2 Insecticide and Larvicide Activity

The presence of hepatotoxic microcystins and the neurotoxic compound anatoxin-a were found to possess significant larvicidal properties against *Aedes aegypti* (Ayala et al. 2019). Additionally, when cyanobacteria were utilized as a biofertilizer, several strains demonstrated the ability to inhibit the development of mosquito larvae. Methanolic extracts from an isolate of *Westiellopsis* sp. exhibited larvicidal effects on various mosquito species (Rao et al. 1999), containing *Aedes aegypti* (a vector for Dengue Fever), *Anopheles stephensi* (a vector for malaria), and *Culex tritaeniorhynchus* and *C. quinquefasciatus* (vectors of encephalitis).

Cyanobacterial anatoxin-a specifically targets nicotine acetylcholine receptors in arthropods. Simultaneous exposure to natural toxins could impair the challenge for *Daphnia* to withstand cyanobacterial blooms, potentially reducing reproduction rates (Schwarzenberger 2022). The survival of *Daphnia magna* declined progressively over time, with microcystin-LR and other secondary metabolites playing a significant role in the negative impact of *D. magna* survival and stress response (Bojadzija Savic et al. 2021) (Fig. 5.4).

5.7 Conclusion

In conclusion, the chapter has provided an insightful exploration of cyanobacterial toxins and their multifaceted applications in therapeutic and industrial contexts. We have explored the elaborate chemistry between cyanobacteria and their toxin-producing gene clusters to uncover their medicinal uses and industrial potential

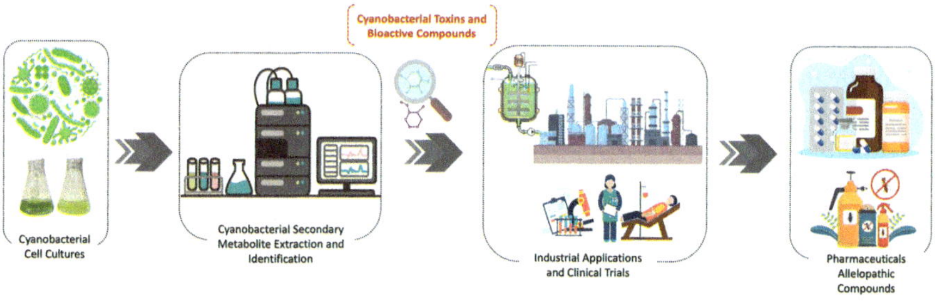

Fig. 5.4 Progression from laboratory-scale research to scaled-up industrial production of cyanobacterial metabolites, for commercially valuable pharmaceutical and other industrial applications

using a diverse array of cyanotoxins and secondary metabolites. As we unravel the mysteries of cyanobacterial toxins, there is great potential for exploiting their therapeutic benefits and industrial applications. The future holds exciting possibilities for unlocking the full potential of cyanobacterial toxins in various fields. Ongoing research and innovation are key to supporting these novel advancements and discoveries.

Acknowledgement The authors are expressing their sincere appreciation to Mr. Yasasa Mendis for the technical support in the preparation of the figures.

References

Ahmed MN, Wahlsten M, Jokela J, Nees M, Stenman UH, Alvarenga DO, Strandin T, Sivonen K, Poso A, Permi P, Metsä-Ketelä M, Koistinen H, Fewer DP (2021) Potent inhibitor of human trypsins from the aeruginosin family of natural products. ACS Chem Biol 16:2537–2546. https://doi.org/10.1021/acschembio.1c00611

Akbar MA, Yusof NYM, Tahir NI, Ahmad A, Usup G, Sahrani FK, Bunawan H (2020) Biosynthesis of saxitoxin in marine dinoflagellates: an omics perspective. Mar Drugs 18(2):103. https://doi.org/10.3390/md18020103

Álvarez C, Jimenez-Ríos L, Iniesta-Pallares M, Jurado-Flores A, Molina-Heredia FP, Ng CKY, Mariscal V (2023) Symbiosis between cyanobacteria and plants: from molecular studies to agronomic applications. J Exp Bot 74(19):6145–6157. https://doi.org/10.1093/jxb/erad261

Ayala FI, Becerra LM, Quintana J, Bayona LM, Ramos FA, Puyana M, Duque F, Castellanos L (2019) Environmental and cultured cyanobacteria as sources of Aedes aegypti larvicides. Univ Sci 24:465–496. https://doi.org/10.11144/Javeriana.SC24-3.eacc

Ballot A, Scherer PI, Wood SA (2018) Variability in the anatoxin gene clusters of Cuspidothrix issatschenkoi from Germany, New Zealand, China and Japan. PLoS One 13(7):e0200774. https://doi.org/10.1371/journal.pone.0200774

Barzkar N, Jahromi ST, Poorsaheli HB, Vianello F (2019) Metabolites from marine microorganisms, micro, and macroalgae: immense scope for pharmacology. Mar Drugs 17(8):464. https://doi.org/10.3390/md17080464

Bojadzija Savic G, Colinet H, Bormans M, Edwards C, Lawton LA, Briand E, Wiegand C (2021) Cell free Microcystis aeruginosa spent medium affects Daphnia magna survival and stress response. Toxicon 195:37–47. https://doi.org/10.1016/j.toxicon.2021.03.009

Bornancin L, Alonso E, Alvariño R, Inguimbert N, Bonnard I, Botana LM, Banaigs B (2019) Structure and biological evaluation of new cyclic and acyclic laxaphycin-A type peptides. Bioorg Med Chem 27(10):1966–1980

Brezeştean IA, Gherman AMR, Colniţă A, Dina NE, Müller Molnár C, Marconi D, Chiş V, David IL, Cîntă-Pînzaru S (2022) Detection and characterization of nodularin by using label-free surface-enhanced spectroscopic techniques. Int J Mol Sci 23(24):15741. https://doi.org/10.3390/ijms232415741

Brown A (2022) Identification and dereplication of lipopeptides from cyanobacteria. University of Illinois at Chicago

Carpine R, Sieber S (2021) Antibacterial and antiviral metabolites from cyanobacteria: their application and their impact on human health. Curr Res Biotechnol 3:65–81. https://doi.org/10.1016/j.crbiot.2021.03.001

Chen Z, Chen N, Fu P, Wang W, Bian S, Zhang H, Shen S, Han B (2023) Structure elucidation of two intriguing neo-debromoaplysiatoxin derivatives from marine cyanobacterium Lyngbya sp.

showing strong inhibition of Kv1.5 potassium channel and differential cytotoxicity. Molecules 28(6):2786. https://doi.org/10.3390/molecules28062786

Churro C, Dias E, Valério E (2012) Risk assessment of cyanobacteria and cyanotoxins, the particularities and challenges of Planktothrix spp. monitoring. In: Novel approaches and their applications in risk assessment. INTECH

Cock IE, Cheesman MJ (2023) A review of the antimicrobial properties of cyanobacterial natural products. Molecules 28(20):7127. https://doi.org/10.3390/molecules28207127

Colas S, Marie B, Lance E, Quiblier C, Tricoire-Leignel H, Mattei C (2021) Anatoxin-a: overview on a harmful cyanobacterial neurotoxin from the environmental scale to the molecular target. Environ Res 193:110590. https://doi.org/10.1016/j.envres.2020.110590

Cragg GM, Pezzuto JM (2016) Natural products as a vital source for the discovery of cancer chemotherapeutic and chemopreventive agents. Med Princ Pract 25:41–59. https://doi.org/10.1159/000443404

Dahiya R, Dahiya S, Fuloria NK, Kumar S, Mourya R, Chennupati SV, Jankie S, Gautam H, Singh S, Karan SK, Maharaj S, Fuloria S, Shrivastava J, Agarwal A, Singh S, Kishor A, Jadon G, Sharma A (2020) Natural bioactive thiazole-based peptides from marine resources: structural and pharmacological aspects. Mar Drugs 18(6):329. https://doi.org/10.3390/md18060329

do Amaral SC, Xavier LP, Vasconcelos V, Santos AV (2023) Cyanobacteria: a promising source of antifungal metabolites. Mar Drugs 21(6):359. https://doi.org/10.3390/md21060359

Du X, Liu H, Yuan L, Wang Y, Ma Y, Wang R, Chen X, Losiewicz MD, Guo H, Zhang H (2019) The diversity of cyanobacterial toxins on structural characterization, distribution and identification: a systematic review. Toxins 11(9):530. https://doi.org/10.3390/toxins11090530

Eren E, Watts NR, Sackett DL, Wingfield PT (2021) Conformational changes in tubulin upon binding cryptophycin-52 reveal its mechanism of action. J Biol Chem 297(4):101138. https://doi.org/10.1016/j.jbc.2021.101138

Fanale D, Bronte G, Passiglia F, Calò V, Castiglia M, Di Piazza F, Barraco N, Cangemi A, Catarella MT, Insalaco L, Listì A, Maragliano R, Massihnia D, Perez A, Toia F, Cicero G, Bazan V (2015) Stabilizing versus destabilizing the microtubules: a double-edge sword for an effective cancer treatment option? Anal Cell Pathol 2015:690916. https://doi.org/10.1155/2015/690916

Filatova D, Jones MR, Haley JA, Núñez O, Farré M, Janssen EML (2021) Cyanobacteria and their secondary metabolites in three freshwater reservoirs in the United Kingdom. Environ Sci Eur 33:41. https://doi.org/10.1186/s12302-021-00472-4

Flores-Holguín N, Salas-Leiva JS, Núñez-Vázquez EJ, Tovar-Ramírez D, Glossman-Mitnik D (2024) Marine toxins as pharmaceutical treasure troves: a focus on saxitoxin derivatives from a computational point of view. Molecules 29(1):275. https://doi.org/10.3390/molecules29010275

Geffroy S, Lechat MM, Le Gac M, Rovillon GA, Marie D, Bigeard E, Malo F, Amzil Z, Guillou L, Caruana AMN (2021) From the sxtA4 gene to saxitoxin production: what controls the variability among Alexandrium minutum and Alexandrium pacificum strains? Front Microbiol 12:613199. https://doi.org/10.3389/fmicb.2021.613199

Gemma S, Molteni M, Rossetti C (2016) Lipopolysaccharides in cyanobacteria: a brief overview. Adv Microbiol 06:391–397. https://doi.org/10.4236/aim.2016.65038

Huang KC, Chen Z, Jiang Y, Akare S, Kolber-Simonds D, Condon K, Agoulnik S, Tendyke K, Shen Y, Wu KM, Mathieu S, Choi HW, Zhu X, Shimizu H, Kotake Y, Gerwick WH, Uenaka T, Woodall-Jappe M, Nomoto K (2016) Apratoxin a shows novel pancreas-targeting activity through the binding of sec 61. Mol Cancer Ther 15:1208–1216. https://doi.org/10.1158/1535-7163.MCT-15-0648

Huskens D, Férir G, Vermeire K, Kehr JC, Balzarini J, Dittmann E, Schols D (2010) Microvirin, a novel α(1,2)-mannose-specific lectin isolated from Microcystis aeruginosa, has anti-HIV-1 activity comparable with that of cyanovirin-N but a much higher safety profile. J Biol Chem 285:24845–24854. https://doi.org/10.1074/jbc.M110.128546

Jacinavicius FR, De Carvalho LR, Carneiro RL, Sant'Anna CL (2018) The influence of temperature on Radiocystis fernandoi strain (cyanobacteria) growth and microcystin production. Rev Bras Bot 41:675–680. https://doi.org/10.1007/s40415-018-0490-8

Jani CT, Ahmed A, Singh H, Mouchati C, Al Omari O, Bhatt PS, Sharma R, Farooq M, Liu W, Shalhoub J, Marshall D, Salciccioli JD, Warner JL, Lam P (2023) Burden of AML, 1990-2019: estimates from the global burden of disease study. JCO Glob Oncol 9:e2300229. https://doi.org/10.1200/go.23.00229

Jeong Y, Cho SH, Lee H, Choi HK, Kim DM, Lee CG, Cho S, Cho BK (2020) Current status and future strategies to increase secondary metabolite production from cyanobacteria. Microorganisms 8(12):1849. https://doi.org/10.3390/microorganisms8121849

Jin X, Miller ES, Lindsey JS (2021) Natural product gene clusters in the filamentous Nostocales cyanobacterium ht-58-2. Life 11(4):356. https://doi.org/10.3390/life11040356

Jones MR, Pinto E, Torres MA, Dörr F, Mazur-Marzec H, Szubert K, Tartaglione L, Dell'Aversano C, Miles CO, Beach DG, McCarron P, Sivonen K, Fewer DP, Jokela J, Janssen EML (2021) CyanoMetDB, a comprehensive public database of secondary metabolites from cyanobacteria. Water Res 196

Kailash J, Ragini G, As Y (2022) Microcystin-LR exhibit cytotoxicity in myeloma Sp2/01 cancer cell line and emerging as a potential anticancer therapeutics. Int J Biotech Trends Technol 12:18–30. https://doi.org/10.14445/22490183/ijbtt-v12i3p603

Kar J, Ramrao DP, Zomuansangi R, Lalbiaktluangi C, Singh SM, Joshi NC, Kumar A, Kaushalendra MS, Yadav MK, Singh PK (2022) Revisiting the role of cyanobacteria-derived metabolites as antimicrobial agent: a 21st century perspective. Front Microbiol 3:1034471. https://doi.org/10.3389/fmicb.2022.1034471

Karpiński TM (2019) Marine macrolides with antibacterial and/or antifungal activity. Mar Drugs 17(4):241. https://doi.org/10.3390/md17040241

Komaki H, Tamura T (2020) Polyketide synthase and nonribosomal peptide synthetase gene clusters in type strains of the genus phytohabitans. Life 10:1–16. https://doi.org/10.3390/life10110257

Kounnis V, Chondrogiannis G, Mantzaris MD, Tzakos AG, Fokas D, Papanikolaou NA, Galani V, Sainis I, Briasoulis E (2015) Microcystin LR Shows Cytotoxic Activity Against Pancreatic Cancer Cells Expressing the Membrane OATP1B1 and OATP1B3 Transporters. Anticancer Res:5857–5865. PMID: 26504008

Li Z, Zhu X, Wu Z, Sun T, Tong Y (2023) Recent advances in cyanotoxin synthesis and applications: a comprehensive review. Microorganisms 11(11):2636. https://doi.org/10.3390/microorganisms11112636

Lopicic S, Svirčev Z, Palanački Malešević T, Kopitović A, Ivanovska A, Meriluoto J (2022) Environmental neurotoxin β-N-methylamino-L-alanine (BMAA) as a widely occurring putative pathogenic factor in neurodegenerative diseases. Microorganisms 10(12):2418. https://doi.org/10.3390/microorganisms10122418

Magana-Arachchi DN, Wanigatunge RP (2023) Cyanotoxin in hydrosphere and human interface. In: One health. Wiley. https://doi.org/10.1002/9781119867333.ch6

Máthé C, Hamvas M, Vasas G, Garda T, Freytag C (2021) Subcellular alterations induced by cyanotoxins in vascular plants—a review. Plan Theory 10(5):984. https://doi.org/10.3390/plants10050984

May DS, Crnkovic CM, Krunic A, Wilson TA, Fuchs JR, Orjala JE (2020) 15N stable isotope labeling and comparative metabolomics facilitates genome mining in cultured cyanobacteria. ACS Chem Biol 15:758–765. https://doi.org/10.1021/acschembio.9b00993

Mazur-Marzec H, Cegłowska M, Konkel R, Pyrć K (2021) Antiviral cyanometabolites—a review. Biomol Ther 11(3):474. https://doi.org/10.3390/biom11030474

Mihali TK, Kellmann R, Muenchhoff J, Barrow KD, Neilan BA (2008) Characterization of the gene cluster responsible for cylindrospermopsin biosynthesis. Appl Environ Microbiol 74:716–722. https://doi.org/10.1128/AEM.01988-07

Montuori E, Hyde CAC, Crea F, Golding J, Lauritano C (2023) Marine natural products with activities against prostate cancer: recent discoveries. Int J Mol Sci 24(2):1435. https://doi.org/10.3390/ijms24021435

Moorthy ND, Ramalingam BM, Iqbal S, Mohanakrishnan AK, Gunasekaran K, Vellaichamy E (2018) Novel isothiacalothrixin B analogues exhibit cytotoxic activity on human colon can-

cer cells in vitro by inducing irreversible DNA damage. PLoS One 13:e0202903. https://doi.org/10.1371/journal.pone.0202903
Moraes MAB, Rodrigues RAM, Schlüter L, Podduturi R, Jørgensen NOG, Calijuri MC (2021) Influence of environmental factors on occurrence of cyanobacteria and abundance of saxitoxin-producing cyanobacteria in a subtropical drinking water reservoir in Brazil. Water 13:1716. https://doi.org/10.3390/w13121716
Morgan JB, Liu Y, Coothankandaswamy V, Mahdi F, Jekabsons MB, Gerwick WH, Valeriote FA, Zhou YD, Nagle DG (2015) Kalkitoxin inhibits angiogenesis, disrupts cellular hypoxic signaling, and blocks mitochondrial electron transport in tumor cells. Mar Drugs 13:1552–1568. https://doi.org/10.3390/md13031552
Nagai H, Sato S, Iida K, Hayashi K, Kawaguchi M, Uchida H, Satake M (2019) Oscillatoxin i: a new aplysiatoxin derivative, from a marine cyanobacterium. Toxins 11:366. https://doi.org/10.3390/toxins11060366
Nandagopal P, Steven AN, Chan LW, Rahmat Z, Jamaluddin H, Noh NIM (2021) Bioactive metabolites produced by cyanobacteria for growth adaptation and their pharmacological properties. Biology 10:1061. https://doi.org/10.3390/biology10101061
Nandi S, Khatua S, Nag A, Sen S, Chakraborty N, Naskar A, Acharya K, Mekky RH, Contreras M d M, Calina D, Dini I, Sharifi-Rad J (2024) Dolastatins and their analogues present a compelling landscape of potential natural and synthetic anticancer drug candidates. Curr Res Biotechnol 7:100167. https://doi.org/10.1016/j.crbiot.2023.100167
Niedermeyer TH, Orst J (2015) Anti-infective Natural Products from Cyanobacteria. Planta Med. https://doi.org/10.1055/s-0035-1546055
Nowruzi B, Porzani SJ (2021) Toxic compounds produced by cyanobacteria belonging to several species of the order Nostocales: a review. J Appl Toxicol 41(4):510–548. https://doi.org/10.1002/jat.4088
Nugumanova G, Ponomarev ED, Askarova S, Fasler-Kan E, Barteneva NS (2023) Freshwater cyanobacterial toxins, cyanopeptides and neurodegenerative diseases. Toxins 15(3):233. https://doi.org/10.3390/toxins15030233
Oyaneder-Terrazas J, Figueroa D, Araneda OF, García C (2022) Saxitoxin group toxins accumulation induces antioxidant responses in tissues of Mytilus chilensis, Ameghinomya antiqua, and Concholepas concholepas during a bloom of Alexandrium pacificum. Antioxidants 11:392. https://doi.org/10.3390/antiox11020392
Parajuli A, Kwak DH, Dalponte L, Leikoski N, Galica T, Umeobika U, Trembleau L, Bent A, Sivonen K, Wahlsten M, Wang H, Rizzi E, De Bellis G, Naismith J, Jaspars M, Liu X, Houssen W, Fewer DP (2016) A unique tryptophan C-prenyltransferase from the kawaguchipeptin biosynthetic pathway. Angew Chem 128:3660–3663. https://doi.org/10.1002/ange.201509920
Pichardo S, Cameán AM, Jos A (2017) In vitro toxicological assessment of cylindrospermopsin: a review. Toxins 9(12):402. https://doi.org/10.3390/toxins9120402
Pindihama GK, Gitari MW (2017) Uptake and growth effects of cyanotoxins on aquatic plants Ludwigia adscendens and Amaranthus hybridus in raw surface waters. Int J Environ Sci Dev 8:93–101. https://doi.org/10.18178/ijesd.2017.8.2.928
Plata-Calzado C, Prieto AI, Cameán AM, Jos A (2022) Toxic effects produced by anatoxin-a under laboratory conditions: a review. Toxins 14(12):861. https://doi.org/10.3390/toxins14120861
Pradhan B, Kim H, Abassi S, Ki JS (2022) Toxic effects and tumor promotion activity of marine phytoplankton toxins: a review. Toxins 14(6):397. https://doi.org/10.3390/toxins14060397
Rao DR, Thangavel C, Kabilan L, Suguna S, Mani TR, Shanmugasundaram S (1999) Larvicidal properties of the cyanobacterium Westiellopsis sp. (blue-green algae) against mosquito vectors. Trans R Soc Trop Med Hyg 93:232. https://doi.org/10.1016/S0035-9203(99)90002-0
Robles-Bañuelos B, Durán-Riveroll LM, Rangel-López E, Pérez-López HI, González-Maya L (2022) Marine cyanobacteria as sources of lead anticancer compounds: a review of families of metabolites with cytotoxic, antiproliferative, and antineoplastic effects. Molecules 27(15):4814. https://doi.org/10.3390/molecules27154814

Sanfilippo JE, Garczarek L, Partensky F, Kehoe DM (2019) Chromatic acclimation in cyanobacteria: a diverse and widespread process for optimizing photosynthesis. Ann Rev Microbiol 73:407. https://doi.org/10.1146/annurev-micro-020518-115738

Schwarzenberger A (2022) Negative effects of cyanotoxins and adaptative responses of Daphnia. Toxins 14(11):770. https://doi.org/10.3390/toxins14110770

Shahid M, Qadir A, Yang J, Ahmad I, Zahid H, Mirza S, Windisch MP, Shahzad-ul-Hussan S (2020) An engineered microvirin variant with identical structural domains potently inhibits human immunodeficiency virus and hepatitis C virus cellular entry. Viruses 12:199. https://doi.org/10.3390/v12020199

Shishido TK, Humisto A, Jokela J, Liu L, Wahlsten M, Tamrakar A, Fewer DP, Permi P, Andreote AP, Fiore MF, Sivonen K (2015) Antifungal compounds from cyanobacteria. Mar Drugs 13(4):2124–2140. https://doi.org/10.3390/md13042124. PMID: 25871291; PMCID: PMC4413203

Shishido TK, Popin RV, Jokela J, Wahlsten M, Fiore MF, Fewer DP, Herfindal L, Sivonen K (2019) Dereplication of natural products with antimicrobial and anticancer activity from Brazilian cyanobacteria. Toxins 12(1):12. https://doi.org/10.3390/toxins12010012

Shrestha SK, Min KH, Kim SW, Kim H, Gerwick WH, Soh Y (2023) Kalkitoxin: a potent suppressor of distant breast cancer metastasis. Int J Mol Sci 24:1207. https://doi.org/10.3390/ijms24021207

Singh U, Gandhi HA, Nikita, Bhattacharya J, Tandon R, Tiwari GL, Tandon R (2023) Cyanometabolites: molecules with immense antiviral potential. Arch Microbiol 205:164. https://doi.org/10.1007/s00203-023-03514-y

Srivastava R, Prajapati R, Kanda T, Yadav S, Singh N, Yadav S, Mishra R, Atri N (2022) Phycochemistry and bioactivity of cyanobacterial secondary metabolites. Mol Biol Rep 49(11):11149–11167. https://doi.org/10.1007/s11033-022-07911-2

Still KR, Mohapatra A (2020) Biotoxins. Inform Resourc Toxicol 1:117–132. https://doi.org/10.1016/B978-0-12-813724-6.00011-6

Sweeney-Jones AM, Gagaring K, Antonova-Koch J, Zhou H, Mojib N, Soapi K, Skolnick J, McNamara CW, Kubanek J (2020) Antimalarial peptide and polyketide natural products from the Fijian marine cyanobacterium Moorea producens. Mar Drugs 18:167. https://doi.org/10.3390/md18030167

Tabarzad M, Atabaki V, Hosseinabadi T (2020) Anti-inflammatory activity of bioactive compounds from microalgae and cyanobacteria by focusing on the mechanisms of action. Mol Biol Rep 47(4):6793. https://doi.org/10.1007/s11033-020-05562-9

Weralupitiya C, Wanigatunge RP, Gunawardana D, Vithanage M, Magana-Arachchi D (2022) Cyanotoxins uptake and accumulation in crops: phytotoxicity and implications on human health. Toxicon:21–35. https://doi.org/10.1016/j.toxicon.2022.03.003. Epub 2022 Mar 11. PMID: 35288171

Vico P, Bonilla S, Cremella B, Aubriot L, Iriarte A, Piccini C (2020) Biogeography of the cyanobacterium Raphidiopsis (Cylindrospermopsis) raciborskii: integrating genomics, phylogenetic and toxicity data. Mol Phylogenet Evol 148:106824. https://doi.org/10.1016/j.ympev.2020.106824

Videau P, Wells KN, Singh AJ, Gerwick WH, Philmus B (2016) Assessment of Anabaena sp. strain PCC 7120 as a heterologous expression host for cyanobacterial natural products: production of Lyngbyatoxin A. ACS Synth Biol 5:978–988. https://doi.org/10.1021/acssynbio.6b00038

Vingiani GM, Stalberga D, De Luca P, Ianora A, De Luca D, Lauritano C (2020) De novo transcriptome of the non-saxitoxin producing Alexandrium tamutum reveals new insights on harmful dinoflagellates. Mar Drugs 18:386. https://doi.org/10.3390/MD18080386

Wang S, Ding P, Lu S, Wu P, Wei X, Huang R, Kai T (2021) Cell density-dependent regulation of microcystin synthetase genes (mcy) expression and microcystin-LR production in Microcystis aeruginosa that mimics quorum sensing. Ecotoxicol Environ Saf 220:112330. https://doi.org/10.1016/j.ecoenv.2021.112330

Waters MN (2016) A 4700-year history of cyanobacteria toxin production in a Shallow Subtropical Lake. Ecosystems 19:426–436. https://doi.org/10.1007/s10021-015-9943-0

Weiss C, Figueras E, Borbely AN, Sewald N (2017) Cryptophycins: cytotoxic cyclodepsipeptides with potential for tumor targeting. J Pept Sci 23:514–531. https://doi.org/10.1002/psc.3015

WHO (2020) Cyanobacterial toxins: cylindrospermopsins Background document for development of WHO

Wiese M, D'Agostino PM, Mihali TK, Moffitt MC, Neilan BA (2010) Neurotoxic alkaloids: Saxitoxin and its analogs. Mar Drugs 8(7):2185–2211. https://doi.org/10.3390/md8072185

Wimmer KM, Strangman WK, Wright JLC (2014) '7-Deoxy-desulfo-cylindrospermopsin and 7-deoxy-desulfo-12-acetylcylindrospermopsin: two new cylindrospermopsin analogs isolated from a Thai strain of Cylindrospermopsis raciborskii. '*Harmful*. Algae 37:203–206. https://doi.org/10.1016/j.hal.2014.06.006

Zhang L, Hoshino S, Awakawa T, Wakimoto T, Abe I (2016) Structural diversification of lyngbyatoxin a by host-dependent heterologous expression of the tleABC biosynthetic gene cluster. Chembiochem 17:1407–1411. https://doi.org/10.1002/cbic.201600229

Chapter 6
Phycotoxins: Compounds of Immense Importance

Bhupinder Dhir

Abstract Phycotoxins are metabolites synthesized naturally by marine algal and cyanobacteria species. They accumulate in the marine organisms and eventually cause poisoning in humans. Phycotoxins include chemical compounds such as amino acids, alkaloids, and polyketides. Most of these chemical compounds are toxic in nature but some of them can be harnessed for useful purposes such as preparation of medicines for certain ailments. Phycotoxins have also found their use in production of beneficial materials such as Botox and other therapeutic agents. Research studies need to be conducted to exploit and harness the potential of these algal products that can be used for benefit of mankind.

Keywords Algae · Phycotoxins · Harmful algal blooms · Cyanobacteria · Cyanotoxins · Human health

6.1 Introduction

Phycotoxins are compounds produced naturally by marine algae mainly dinoflagellates and cyanobacteria species (Rossini and Hess 2010; Karlson et al. 2021). They mainly include secondary metabolites such as alkaloids and polyketides which are toxic in nature (Gallardo-Rodríguez et al. 2012; Pulido 2016; Cousseau et al. 2020; Liang et al. 2022). These compounds have molecular mass in the range of 300–3000 Da. Some of the commonly known phycotoxins include saxitoxin, okadaic acid, yessotoxins, azaspiracids, brevetoxins, pinnatoxins, ciguatoxins, and maitotoxins (Fig. 6.1).

Studies have indicated that synthesis of phycotoxins occurs as a result of unfavourable environmental conditions and/or a defensive mechanism (Selander et al. 2006; Ajuzie 2007; Ianora et al. 2011; Anderson et al. 2012; Kohli et al. 2016; Pires et al. 2023). The synthesis of these compounds has also been linked to processes

B. Dhir (✉)
Department of Chemistry, St. Stephen's College, University of Delhi, Delhi, India

M. N. V. Prasad et al. (eds.), *Biotoxins*,
https://doi.org/10.1007/978-3-031-75309-1_6

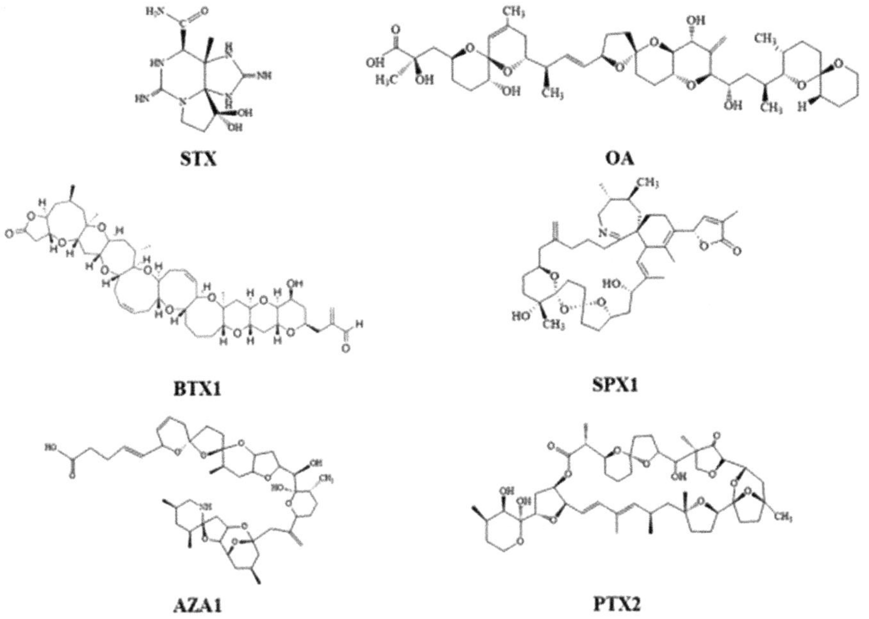

Fig. 6.1 Molecular structure of some useful phycotoxins. Saxitoxin (STX), okadaic acid (OA), brevetoxin type-A (BTX1), spirolides (SPX) type 1, azaspiracid-1 (AZA1), pectenotoxins (PTXs) type 2 (modified from Louzao et al. 2022)

such as photosynthesis, nitrogen and carbon metabolism (Cassell et al. 2015; Glibert et al. 2016). Alexandrium, Gymnodinium, Dinophysis, and Pseudo-nitzschia are some of the major dinoflagellates or diatom species producing phycotoxins.

High concentrations of phycotoxins get accumulated in tissues of aquatic animals (Zohdi and Abbaspour 2019). Shellfish such as mussels, oysters, clams and bivalve molluscs such as echinoderms, tunicates, marine gastropods or crustaceans accumulate these biotoxins in high quantities (Rossini and Hess 2010; Rasmussen et al. 2016). Phycotoxins that get accumulated within the tissues of shellfish are called as "shellfish toxins" (Farabegoli et al. 2018). These compounds get metabolized to more toxic metabolites in marine organisms. Poisoning in animals and humans occurs due to consumption of aquatic organisms such as fish and shellfish having high concentrations of these toxic compounds (James et al. 2010; Hinder et al. 2011). The living organisms (aquatic or terrestrial) get exposed to these toxins through oral, respiratory, or dermal route. Phycotoxins enter humans mainly by oral route via ingestion of contaminated seafood. These toxins can also enter humans via respiratory system in the form of aerosols. These compounds mainly affect the digestive and nervous system of humans. Intake of shellfish contaminated by phycotoxins causes poisoning in humans and causes syndromes such as paralytic shellfish poisoning, amnesic shellfish poisoning, diarrhetic shellfish poisoning, and neurotoxic shellfish poisoning (Wang 2008).

6.2 Classification of Phycotoxins

Phycotoxins can be classified into different groups according to their chemical structure, mechanism of action, target tissues they affect, and health effects.

Phycotoxins are categorized into two types on the basis of source which they are produced.

1. *Cyanotoxins:* Toxic compounds produced by cyanobacteria blooms (CyanoHABs) are called cyanotoxins. These toxins are present in drinking water, freshwater fish, lakes and cause direct risk to human health.
2. *Marine biotoxins/marine algal toxins:* Toxic compounds produced by dinoflagellates and diatoms are called marine or algal toxins. They get accumulated in tissues of aquatic organisms and from there they enter the food chain.

Phycotoxins have been classified into different types on the basis of their chemical structure and type of effects they produce in humans (Camacho-Muñoz et al. 2021). On the basis of chemical structure phycotoxins are classified as

1. Linear and macrocyclic polyethers such as okadaic acid (OA) and dinophysistoxins (DTXs)
2. Ladder-frame polyethers such as ciguatoxins (CTXs) and brevetoxins (BTXs)
3. Macrocyclic imines such as spirolides (SPXs) and gymnodimines (GYMs)
4. Tetrahydropurines such as saxitoxins (STXs) and gonyautoxins (GTXs)
5. Toxic secondary amines such as domoic acid (DA)

Toxins are classified into eight groups on the basis of their chemical structure (Farabegoli et al. 2018). These include

- Azaspiracids (AZAs)
- Brevetoxins (BTXs)
- Cyclic imines (CIs)
- Domoic acid (DA)
- Okadaic acid (OA)
- Pectenotoxins (PTXs)
- Saxitoxin (STX)
- Yessotoxins (YTXs)

Phycotoxins have also been categorized according to the symptoms produced in humans after the consumption of contaminated seafood (Nicolas et al. 2017).

- Diarrhetic shellfish poisoning (DSP) resulting from the intake of OA, DTXs, pectenotoxins (PTXs), and yessotoxins (YTXs)
- Amnesic shellfish poisoning (ASP) resulting from the intake of DA
- Paralytic shellfish poisoning (PSP) resulting from the intake of STXs and analogues
- Azaspiracid shellfish poisoning (AZP) resulting from the intake of azaspiracids (AZAs)

- Neurotoxic/neurologic shellfish poisoning (NSP) resulting from the intake of BTXs
- Ciguatera fish poisoning (CFP) resulting from the intake of CTX and maitotoxin (MTX)

6.3 Phycotoxins and Their Health Effects

Each type of phycotoxin group produces characteristic clinical symptoms and effects. They also have a different mechanism of action. Most of the phycotoxins cause cellular alteration/damage. Acute level of these toxicants produces neurotoxic affect, disorders of gastrointestinal (GI) tract, cardiovascular, respiratory, and nervous system. Gastrointestinal system gets affected first followed by neurological system. Severity of symptoms increases with time after consumption of a contaminated seafood.

6.3.1 Cyanotoxins

Cyanotoxins prove toxic to living organisms including humans. They are categorized as hepatotoxins, neurotoxins, cytotoxins, dermatoxins, and irritant toxins depending upon the target tissues they affect and biological effect produced by them (Table 6.1). They include chemicals such as cyclic peptides (microcystin and nodularin), alkaloids (anatoxin-a, saxitoxins, cylindrospermopsin, aplysiatoxin, lyngbyatoxin-a), and lipopolysaccharides (LPSs).

Toxicity of these compounds produce gastrointestinal symptoms such as vomiting, nausea, diarrhoea, abdominal pain, cardio-respiratory symptoms such as hypotension, tachycardia, arrhythmia, dyspnoea and neurological symptoms such as perioral burning, paraesthesia, convulsions, and coma. *C*yanobacterial species, namely *Anabaena* spp. and *Microcystis* spp., produce microcystin which causes hepatotoxicity in humans. Anatoxin-a/Homoanatoxin and beta-Methylamino-L-alanine (BMAA) act as neurotoxic cyanotoxins.

6.3.2 Dinoflagellate Phycotoxins

Dinoflagellates produce bioactive compounds, most of which show toxic properties (Garcia-Camacho et al. 2007). These include secondary metabolites of polyketide origin such as polycyclic polyketides (YTXs, CTXs, BTXs, MTXs), palytoxins (PLTXs), macrolides (PTXs and amphidinolides) and linear polyketides (DTXs and OA), cyclic imines (SPXs, GYMs), pinnatoxins (PnTXs), and alkaloids (STX,

Table 6.1 Toxins and their application for different types of drugs

Toxins	Mode of action	Source dinoflagellate	Potential applications
Azaspiracids	Block hERG (human ether-a-go-go related gene) potassium channel by binding to it	*Azadinium spinosum*	Antiarrhythmic drugs
Brevetoxin	Activates voltage-gated sodium channels	*Karenia mikimotoi*	Post-stroke recovery drugs
Cooliatoxin	Blocks unmyelinated nerves	*Coolia* spp.	Unmyelinated nerve fibre-related research, analgesics
Dinophysistoxin	Inhibits protein phosphatases, (serine/threonine phosphatases)	*Prorocentrum* spp.	Tumour promoter
Gonyautoxin	Inhibits sodium channel	*Alexandrium* spp.	Analgesics, anaesthetics, and antidepressant drugs
Haemolysins	Hydrolyze phospholipids in the bilayer Form pores in phospholipid bilayers	*Alexandrium* spp. *Amphidinium carterae* *Cochlodinium polykrikoides* *Gonyaulax* spp. *Heterocapsa circularisquama* *Karenia mikimotoi*	Anticancer drugs Microalgae-harvesting agents
Karlotoxin	Disrupts cell membrane by specific binding to cholesterol	*Karlodinium* spp.	Copepod-killing agent Anticancer drugs Microalgae-harvesting agents
Karmitoxin	Unknown (Ichthyotoxic)	*Karlodinium* spp.	Piscicide Copepod-killing agent
Maitotoxin	Activates calcium channel	*Gambierdiscus toxicus*	As a tool for the analyses of ion channels and insulin secretion Interleukin-1β inducer Sperm acrosome reaction inducer
Neosaxitoxin	Inhibits sodium channel	*Alexandrium* spp.	Analgesics, anaesthetics, and antidepressant drugs Long-acting pain blocker in bladder pain syndrome
Okadaic acid	Inhibits protein phosphatases (serine/threonine phosphatases)	*Dinophysis* spp. *Prorocentrum* spp.	Probe for discovery of neurodegeneration drugs Tumour promoter

(continued)

Table 6.1 (continued)

Toxins	Mode of action	Source dinoflagellate	Potential applications
Palytoxin	Turns Na+/K+ pump into a shape that allows the passive transport of sodium and potassium ions	*Ostreopsis* spp.	Tumour promoter Anticancer drugs
Pectenotoxin	Inhibits protein phosphatases (serine/threonine phosphatases) Depolymerizes actin filaments Disrupts actin cytoskeleton	*Dinophysis* spp.	Anticancer drugs Anti-actin drugs
Prorocentrolides	Acts on nicotinic acetylcholine receptors (nAChRs)	*Prorocentrum* spp.	Anticancer drugs Anticholinergic drugs
Saxitoxin	Sodium channel blocker	*Alexandrium* spp. *Gymnodinium catenatum* *Pyrodinium bahamense*	Analgesics, anaesthetics, and antidepressant drugs Chemical weapon
Yessotoxin	Activates calcium channel Decreases cytosolic 3′,5′-cyclic adenosine monophosphate (cAMP) levels	*Coolia* spp. *Gonyaulax* spp. *Lingulodinium polyedrum* *Protoceratium reticulatum*	Anticancer drugs Antiallergic drugs

GTX, neosaxitoxin (NEO)) (Verma et al. 2019). The phycotoxins cause hydrolysis of phospholipids and lead to pore formation in phospholipid bilayers.

6.4 Major Phycotoxins produced by Dinoflagellates

Alexandrium spp. produce toxins such as saxitoxin (STX), gonyautoxin (GTX), neosaxitoxin (NSTX), and haemolysins (HL). These toxins cause paralysis. They mainly cause modulation of ion channels. STXs specifically act by blocking the sodium channel of cells. They interact with voltage-gated sodium channels and block the receptor, site 1 sodium pump, affecting motor neurons (Louzao et al. 2022). This blockage results in reduction of influx of Na^+ ions into cell thereby affecting the propagation of action potential in excitable membranes. The blockage in conduction of ions affects the generation of action potentials thereby affecting the neuromuscular function causing muscular paralysis. STX and its analogue, NSTX leads to numbness of the lips and gastrointestinal disorders. GTXs bind to site 1 of the α-subunit of the voltage-dependent sodium channel in the postsynaptic membrane of the mammalian nervous system. This affects the synaptic function.

Lipophilic polyether compound namely azaspiracids (AZA) is produced by dinoflagellate Azadinium spinosum. This toxin produces symptoms such as nausea, vomiting, severe diarrhoea, and stomach cramps. High concentrations of this toxin damages intestinal epithelium, lamina propria, liver, and villi. Long-term exposure of AZA causes lung tumours and malignant lymphomas. The toxin interacts with the channel's central pore within the S6 transmembrane domain and blocks potassium-conductance pathway of human ether-a-go-go-related gene (hERG1) channels. Analogues of AZA induce activity of enzyme dehydrogenase which is a part of mitochondrial electron transport chain complex. The activity of this enzyme gets triggered by imbalance of intracellular levels of K^+ and Cl^- ions.

Coolia, a dinoflagellate produces cooliatoxin (CTX) and yessotoxin (YTX). YTX activates nifedipine and the SKF-96365 sensitive calcium channel and decreases cytosolic 3′,5′-cyclic adenosine monophosphate (cAMP) levels. CTX is a monosulfate polyether analogue of YTX and causes depolarization of nerve cell. Binding of the toxin to the site of sodium cell influx results in opening of sodium channels. It leads to blockage of unmyelinated nerves. The symptoms of CTX toxicity include diarrhoea. YTX is also produced by dinoflagellates such as *Gonyaulax spinifera*, *Lingulodinium polyedrum*, and *Protoceratium reticulatum*.

Karlodinium armiger produces karmitoxin (KTX), which is an amine-containing polyhydroxy-polyene toxin and affects membrane permeabilization. *Karlodinium* spp. mainly K. *armiger* and *K. veneficum* produce another compound karlotoxin (KmTx) which causes disruption of the cell membrane by specifically binding to cholesterol.

Dinophysis, a dinoflagellate produces toxins such as okadaic acid (OA), pectenotoxin (PTX), and dinophysistoxin (DPX) (Fu et al. 2019). DPX is an analogue of OA. Both OA and DPXs cause inhibition of serine/threonine (Ser/Thr) phosphatases that induce tumour growth and cause neuronal cell death. PTXs inhibit Ser/Thr phosphatases, depolymerize actin filaments, and disrupt the actin cytoskeleton. DPXs are specifically produced by *D. acuta*, *D. acuminata*, *P. foraminosum*, and *P. lima*, while *D. fortii*, *D. acuta*, *D. acuminata*, and *D. caudata* produce PTXs.

Maitotoxin (MTX) produced by *Gambierdiscus toxicus* affects muscle contraction by activating cellular calcium channels. Dinoflagellates, *namely Karenia brevis* and *K. mikimotoi p*roduce brevetoxin (PbTx) and gymnocin (GC). PbTx activates mammalian voltage-gated sodium channels, thereby causing NSP.

Ostreopsis spp., namely *O. siamensis*, *O. ovata*, and *O. mascarenensis* produce palytoxin (PLTX) and analogues. These compounds affect Na^+/K^+ pump allowing passive transport of sodium and potassium ions. The poisoning symptoms include myalgia, respiratory problems, impairment of the neuromuscular apparatus, and abnormalities in cardiac function.

Prorocentrum spp. produce phycotoxins such as OA, DPX, prorocentrolides (PC), and borbotoxin (BTX) (Hoppenrath et al. 2013; Luo et al. 2017; Pan et al. 2017; Nishimura et al. 2020; Camacho-Muñoz et al. 2021). Some strains also produce DTX1 (Hou et al. 2016; Lee et al. 2016; Hoppenrath, 2017; Pan et al. 2017; Aquino-Cruz et al. 2018; Gu et al. 2019). *Prorocentrum lima* strains produce OA and *Prorocentrum belizeanum, caipirignum, concavum, maculosum*, and

hoffmannianum produce DSP toxins (López-Rosales et al. 2014; Luo et al. 2017; Accoroni et al. 2018; Rodríguez et al. 2018; Lim et al. 2019). Prorocentrolides are cyclic imines and act on muscle and neuronal nicotinic acetylcholine receptors (nAChRs). OA and DTXs act as inhibitors of protein phosphatases 2A (PP2A), 1 (PP1), and 2B (PP2B) which modulate cell signalling pathways (Twiner et al. 2016). They produce gastrointestinal distress and lead to symptoms such as diarrhoea, nausea, vomiting, and abdominal pain (James et al. 2010). DSP toxins inhibit activity of protein phosphatases which leads to diarrhoea. The modulation of neurotransmitters that regulate intestinal mobility, water, and electrolyte secretion leads to DSP syndrome (Louzao et al. 2015). Chronic exposure to OA can induce cytotoxicity, neurotoxicity, immunotoxicity, embryotoxicity, genotoxicity, and cancer (Vilariño et al. 2018). DTXs also show similar effects as OA.

Domoic acid (DA) is a natural neurotoxin produced by red algae and diatom algal species. Some species of red algae and diatoms such as Pseudo-nitzschia, C. armata, Digenea simplex produce DA. DA and analogues bind specific receptors in neurons. They act as excitatory neurotransmitters and cause ASP syndrome. DA affects the central, peripheral, and autonomous nervous system, and skeletal and smooth muscle. The toxin acts as an excitatory amino acid, glutamate agonists which target the kainate receptor present in various vital organs and cause neuronal depolarization. The free radical generation and accumulation induced by DA at the level of the mitochondria leads to the oxidation of vital macromolecules including lipids, proteins, and DNA. This phenomenon can induce apoptosis or necrosis of neurons and glia. Its poisoning results in memory impairment, recurrent seizures, epilepsy, and short-term memory loss. People affected by the poisoning showed gastrointestinal, cardiovascular, and neurologic disorders. Neuronal necrosis affects the physiology of the central nervous system including motor, sensory, and cognitive deficits. Dizziness, nausea, and vomiting are other symptoms, ultimately leading to coma and brain damage or death in the most severe cases. DA poisoning results in three progressive stages. First stage is characterized by the specific appearance of epileptic lesions, second stage includes alterations in physiology and physical damages of the organs; and third (final) stage results in progressive damage with recurrent seizures. DA penetrates the placental membranes and cause detrimental physiological and structural effects on the foetus by producing especially the alterations in the brain development.

Palytoxin (PlTX) is produced by dinoflagellate named Ostreopsis. PlTX and 42-hydroxy-PlTX have also been isolated from marine Trichodesmium spp. cyanobacteria. The compound PlTX includes hydrophilic polyalcohols that bind to extracellular sodium and potassium channels and inhibit active transport of these ions across the membranes. These compounds selectively bind to the Na^+, K^+-ATPase and transform the pump into a channel permeable to monovalent cations. Binding of PITX to the plasma membrane Na^+/K^+-ATPase converts the ion pump into a non-specific ion channel. The channels open permanently and allow uncontrolled transport of ions across the plasma membrane. PlTXs open an H^+ conductive pathway resulting in activation of the Na^+/H^+ exchanger. Intracellular concentration of calcium ion (Ca^{2+}) increases and this stimulates the release of neurotransmitters by

nerve terminals, histamine by mast cells, and vasoactive factors by vascular endothelial cells. It also induces contractions of striated and smooth muscle cells. PTX1 and PTX2 have been described to alter F-actin in several cell types. Rise in Ca levels also lead to activation of phospholipase C and A2. PlTXs induce depolymerization of actin filaments in intestinal and neuroblastoma cells. Excess of intercellular cations causes cell death. PlTX poisoning occurs via oral, cutaneous, inhalational, and ocular exposure routes. The main symptoms of poisoning include nausea, diarrhoea, and vomiting with convulsions, dizziness, numbness, and restlessness in some cases, and subsequently weakness, muscle cramps, myalgia, rhabdomyolysis, bradycardia, tachycardia, respiratory failure and even death in the worst cases. It also acts as a haemolysin which alters the function of excitable cells.

BTXs are lipid-soluble cyclic polyether compounds produced by the dinoflagellates *Karenia brevis, Karenia brevisulcatum, K. mikimotoi, K. selliformis, and K. papilionacea.* BTXs act on the voltage-gated sodium channel (VGSCs) and activate site 5 of this channel. Gymnodimine (GYMs), spirolides (SPXs) and pinnatoxins (PnTXs) block nicotinic acetylcholine receptors which lead to muscular paralysis. Dinophysistoxins (DTXs), a group of DSP toxins inhibit protein phosphatases in vitro.

Yessotoxins (YTXs) are disulfated polycyclic ether compounds. These toxins are produced by dinoflagellates Protoceratium reticulatum, Lingulodinium polyedrum, and Gonyaulax spinifera (Alfonso et al. 2016; Blanco et al. 2020). The toxin and its analogues such as homoyessotoxin and 45-hydroxy-homoyessotoxin mainly affect cardiac muscle cells of heart. The cytotoxicity of YTX occurs via alteration of intracellular levels of calcium and cyclic AMP, cytoskeletal modifications, caspases activation, and the opening of the permeability transition-pore of mitochondria. Theses toxins suggest a possible action of the toxin against VGSC activity. Studies have also shown alteration in cytosolic calcium levels due to YTX. These compounds modulate Ca^{2+} homeostasis in human lymphocytes through activation of phosphodiesterases. They are reported to cause neuronal damage by affecting protein kinase C translocation in primary cortical neurons. An analogue of YTX, i.e., di-desulfo-yessotoxin, induces fatty degeneration in liver and pancreas.

The dinoflagellate Vulcanodinium rugosum produces pinnatoxin (PnTXs), Prorocentrum sp. produces prorocentrolide and the spiro-prorocentrimine. PtTXs show neurotoxic symptoms such as abdominal breathing, respiratory distress, contractions, tremors, and death within 3–20 min after receiving lethal doses of SPXs.

SPXs, GYMs, PnTXs, pteriatoxins (PtTXs), prorocentrolide, portimine, and symbioimine are some of the lipophilic toxins. Spirolides (SPXs) are produced by the dinoflagellates Alexandrium ostenfeldii, Alexandrium peruvianum, gymnodinoid dinoflagellate Gymnodinium cf. mikimotoi. The toxicity of SPXs is due to inhibition of peripheral and central nicotinic acetylcholine receptors (nAChRs) subtypes. The dinoflagellate Karenia selliformis also produces the analogues gymnodimines (GYMs). The neurotoxic symptoms include hyperactivity, paralysis, extension of the hind legs and even death after 15 min of injection. Both SPXs and GYMs inhibit muscarinic (mAChRs) and nicotinic acetylcholine receptors (nAChRs) involved in signal transduction in nervous system. Prorocentrolides act

as chemotaxonomic markers (Li et al. 2020). They depict cytotoxicity against cancer (HCT-116 and Neuro-2a cells) cells and inhibitory activity against human colon adenocarcinoma DLD-1 and human malignant melanoma RP-MI7951 cells under in vitro conditions (Lu et al. 2005; Lee et al. 2019).

Dinophysistoxin-1 (DTX1) was the first toxin of the Okadaic acid (OA) group that was characterized. It was isolated from the dinoflagellate Prorocentrum lima. Dinophysistoxin-2 (DTX2) is another analogue of OA. Besides Prorocentrum species, Dinophysis also produces OA and DTXs. OA and analogues alter the activity of serine/threonine phosphoprotein phosphatases (PPs). OA exerts toxicity by affecting activity of enzymes of digestive system, lipid, amino acid, and sugar metabolism and inducing oxidative stress that interferes with the cell signal transduction in intestinal cells. Toxicity symptoms include diarrhoea, nausea, vomiting, and abdominal pain.

6.5 Various Types of Poisoning Caused by Phycotoxins

6.5.1 Paralytic Shellfish Poisoning (PSP)

Dinoflagellates such as Alexandrium, Gymnodinium, Pyrodinium, and Gymnodinium catenatum produce PSP toxins. PSP toxins are found in certain gastropods, crabs, snails, and bivalve mollusks mainly mussels, clams, oysters, and scallops. These toxins get accumulated in transvectors via feeding in their digestive organs and soft tissues. Paralytic shellfish poisoning is characterized by symptoms such as nausea, vomiting, paralysis, and in severe cases death.

The consumption of contaminated shellfish leads to exposure to saxitoxin (STX) (non-terpene alkaloid) and other analogues such as gonyautoxins (GTXs), neosaxitoxin (NeoSTX), and decarbamoyl-saxitoxin (dc-STX), decarbamoyl-neosaxitoxin (dcneoSTX), decarbamoyl-gonyautoxins (dcGTXs), and the 13-deoxy-decarbamoyl derivatives (doSTX, doGTX) (Manríquez et al. 2015). The STX and analogues bind to VGSCs, inhibit Na^+ influx, and lead to generation and propagation of action potentials in cells. The mild symptoms include numbness around the lips, a prickly sensation in fingertips and toes, headache, dizziness, and nausea. The serious health effects include incoherent speech, stiffness, non-coordination of limbs, difficulty in respiration and rapid pulse. In extreme cases, muscular paralysis leads to difficulty in respiration and choking sensation, while in fatal cases, death occurs due to respiratory paralysis.

Tetrodotoxin (TTX) is an extremely potent toxin. Pufferfish is known source of TTX. Consumption of flesh or organs of the contaminated fish causes poisoning in humans. *Alexandrium tamarense* and *Prorocentrum minimum* (*cordatum*), phytoplankton species are the alternate sources of the toxin. TTX exerts its toxicity by binding to VGSCs followed by blocking Na^+ influx. It induces symptoms such as paralysis of muscles and respiratory failure. In severe cases, cardiac arrhythmias,

muscle paralysis, cranial nerve dysfunction and even death occurs due to respiratory failure.

6.5.2 *Neurotoxic Shellfish Poisoning (NSP)*

Brevetoxins (BTXs) are the toxins responsible for neurotoxic shellfish poisoning (NSP). These toxins are produced by the dinoflagellate Karenia. Algae species such as Chattonella antiqua, Chattonella marina, Fibrocapsa japonica and Heterosigma akashiwo also produce BTX-like compounds. Human exposure occurs through consumption of contaminated clams, oysters and mussels. BTX produces gastrointestinal symptoms such as nausea, vomiting, and diarrhoea and neurological symptoms such as oral paraesthesia, dysarthria, dizziness or ataxia, and walking disorders. The toxins bind to VGSCs in cell membranes and activate them causing depolarization of neuronal and muscle cell membranes. The binding with high affinity to receptor site 5 of the VGSC leads to activation of these channels resulting in uncontrolled Na^{+} influx into cells and depolarization of neuronal and muscle cell membranes. BTXs cause asthma-like symptoms in humans through inhalation exposure.

6.5.3 *Azaspiracid Poisoning (AZP) Syndrome*

Azaspiracids (AZAs) are responsible for the azaspiracid poisoning (AZP) syndrome. These cytotoxic compounds affect cytoskeleton arrangements and inhibit potassium channels (hERG—human ether-à-go-go-related gene) that play an important role in the cardiac action. The poisoning of AZAs leads to symptoms such as cause vomiting, nausea, diarrhoea and stomach cramps in humans. Combination of AZAs and glutaric acid inhibits VGSCs leading to neurotoxicity.

6.6 Applications of Phycotoxins

Besides causing toxicity in humans and other living organism, studies have revealed the potential use of the phycotoxins as therapeutic agents in the area of medicine (Pulz and Gross 2004; Cho et al. 2020; Camacho-Muñoz et al. 2021). Various industrial applications of these toxins have also been reported. Recently their role in production of biofuels, pesticides has also been revealed. The role of phycotoxins in biotechnological applications and drug development has also been explored.

6.6.1 Use in Medicine

Phycotoxins produced by marine dinoflagellates have shown application in therapeutics and drug development (Fig. 6.2). These compounds show specific interactions with their natural targets and can be used to develop new drugs. They act as antimicrobial, anticancer compounds and also find their use in treatment of neurodegenerative disease.

They are also used as biomedical research tools. Phycotoxins inhibit protein phosphatases and enzymes essential in regulation of many cellular processes. These compounds also find their use in cell regulation studies.

Tetrodotoxin (TTX) a neurotoxin isolated from dinoflagellate *Alexandrium tamarense* has shown the capacity to block voltage-gated sodium channels, thus helping in prevention of damage to nerve, muscle function and consequently, the associated pain. Administration of this compound at different doses has shown pain relieving effect. The compound also alleviates neuropathic, inflammatory, and acute pain. It has also shown potential in relieving opioid-resistant pain in cancer patients. Halneuron® (Wex Pharmaceuticals Inc., Vancouver, Canada), a drug developed from TTX used in management of cancer-related pain and chemotherapy-induced neuropathic pain is currently under clinical trials. Since it blocks effect of VGSC, its use as a local anaesthetic has also been explored. TTX has been found potent as it causes minimal toxicity. The toxin has found its application in drug addiction specially heroin.

Saxitoxin (STX) and its analogues have been used as selective sodium channel blockers and found effective as anaesthetic agent (Table 6.1). It has also found its application as an effective pain reliever and modulator without any toxic effect. It

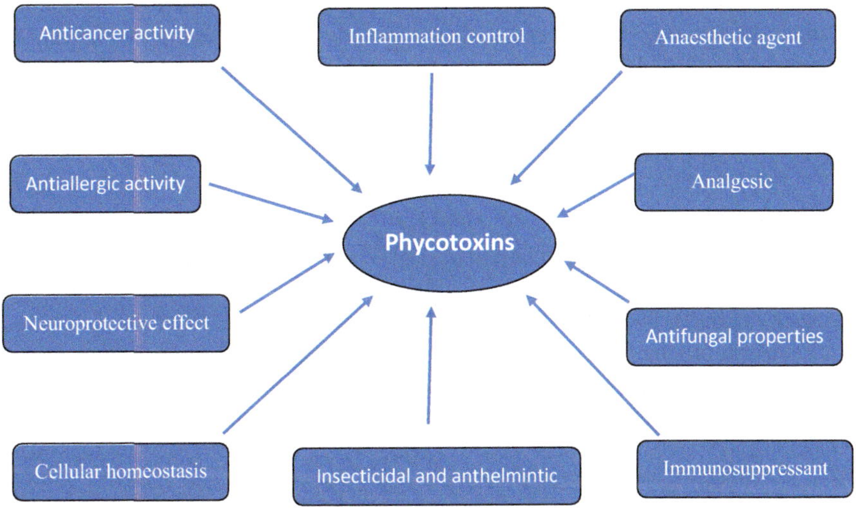

Fig. 6.2 Therapeutic properties of phycotoxins produced by marine dinoflagellates

can minimize or even block pain sensations, muscle spasms, muscle relaxation, and wrinkle reduction. Under in vitro conditions, STXs show antifungal, antibacterial, antialgal, and antiprotozoal activity (Pradhan and Ki 2022). Gonyautoxins (GTXs), analogues of STXs produced by the dinoflagellate *Amphidinium* act as therapeutic agents for acute/chronic anal fissures and are also used as a pain reliever especially in headaches caused due to chronic tension (Garrido et al. 2005).

In vitro and in vivo studies have shown that Okadaic acids (OAs) affect cellular metabolism. They have found their application in study of cellular processes that are regulated by phosphorylation (Pradhan and Ki 2022). OA and its analogues have proven useful in development of neurodegeneration and cancer drugs as they show high activity of serine/threonine protein phosphatases. The compound inhibits protein phosphatases PP1 and PP2A specifically and thus can be used as a probe for study of regulatory mechanisms and neurotoxicity. The compound can also be used in study of neurodegeneration disease such as Alzheimer's disease and as therapeutic for treatment of diseases such as cancer and inflammatory diseases. The toxin also exhibited fungicidal and antimicrobial activity. Inhibition of growth of fungus such as Candida albicans, Aspergillus niger, and Penicillium funiculosum has been noted after application of OA.

Use of Yessotoxin (YTX) as a therapeutic agent for treatment of Alzheimer's disease and many metabolic disorders has been demonstrated using in vitro models. Yessotoxin induces apoptosis in tumoural cells, hence has found its application in antitumour therapies. YTXs showed anticancer and antiallergic activity in a B16F10 melanoma preclinical mouse model (Alfonso et al. 2016). Use of YTX as an anticancer compound in cancer therapy has been proven (Tobío et al. 2016).

Pectenotoxins (PCTs) isolated form *Dinophysis* species have shown anticancer effects. The cytotoxic effect of these toxic compounds against various human cancer cell lines has been demonstrated. PTX-2 has been shown anticancer property against human lung, colon, and breast cancer cells (Espina et al. 2010). Palytoxin (PLTX) isolated from dinoflagellate *Ostreopsis* species has shown immense biological activity (Franchini et al. 2010). Ostreocin, a PLTX-like compound modulates neurotransmitters (acetylcholine and/or norepinephrine) and activates pro-inflammatory signalling cascades such as release of prostaglandin-E2 and histamine. PLTX and ostreocin control changes of cytoskeleton, intestinal and neuroblastoma cells and significantly reduce cytotoxicity. PLTX isolated from *Symbiodinium* has shown a possibility of therapeutic use in the control of lymphoblastic or myelogenous leukaemia. Hence the compounds can be exploited for use in development of anticancer therapeutics (Patocka et al. 2015).

13-desmethylspirolide C acts as an effective therapeutic agent. The compound has shown its efficacy against neurodegenerative diseases under in vitro and in vivo conditions. It has shown to decrease amyloid beta load and Tau hyperphosphorylation in cortical neurons and Alzheimer's disease model under in vivo conditions. In a recent study, the compound has shown neuroprotective effect on a human neuronal stem cell line. Research findings suggested use of SPXs and related compounds can be exploited for the development of therapy against neurodegenerative disorders.

Azaspiracids have shown the action of blocking of hERG potassium channels suggesting their use in development of class III antiarrhythmic drugs. These drugs reduce the risk of arrhythmias by prolonging the action potential, duration, and refractory period of conduction velocity in the myocardium.

Marine toxins have shown the potential to be used as anticarcinogenic compounds. Volume-regulated anion channel (VRAC) inhibitors have shown potential to restrict progression of cancer, preventing the transition of tumoural cells to phase S of the cellular cycle.

Gambierol, toxin produced by dinoflagellate *Gambierdiscus toxicus* and its analogues are compounds that target voltage-gated potassium channels (K_v). These channels play a critical role in regulation of cellular homeostasis and help in controlling diseases such as multiple sclerosis, rheumatoid arthritis, and type I diabetes mellitus by acting as immunosuppressive agents. They also have found effective in treatment of neurodegenerative diseases. Gambierol leads to bidirectional neurite development that benefits patients with brain injury. Studies by Cao et al. (2014) indicated that gambierol induces T-cell proliferation, immunological response, and cytokine production and can be used as a therapeutic agent in T-cell-mediated autoimmune disorders (Cao et al. 2014). Gambierol can be used as an immunosuppressant in diseases involving a malfunctioning immune system such as multiple sclerosis, rheumatoid arthritis, and type 1 diabetes.

Kainoids have shown insecticidal and anthelmintic properties. nAChRs blocking activity of BTX and PCs can be exploited for developing anticholinergic drugs such as nicotinic antagonists. Dinoflagellates species *Karenia* produces brevetoxin (PbTx) and gymnocin (GC). PbTx, an activator of the voltage-gated sodium channel can be used for the development of post-stroke recovery drugs. Brevetoxin in aerosol form affect respiratory function and lead to asthma. PbTx-2 showed capacity to enhance the dendritic arborization and synapse density of the cortical pyramidal neurons in the peri-infarct cortex and improved motor recovery, therefore can be exploited for development of post-stroke recovery drugs. Ciguatoxin (CTX), a toxin procured from *Gambierdiscus toxicus* has shown its therapeutic effect by enhancing contraction of cardiac muscles and increasing discharge of fluid by gastrointestinal cells. This toxin proved useful in studying various human diseases such as cancer, chronic pain, epilepsy, and cardiac arrhythmias. The toxin causes inhibition of unmyelinated nerves and relieves pain, hence has find its application in development of analgesics. The sodium channel blockers, namely GTX and NSTX, have also been considered as potential candidates for the development of analgesic and antidepressant drugs. Use of NSTX in clinical treatment of bladder pain syndrome (BPS) has been demonstrated. NSTX showed the analgesic effect that lasted for 90 days without adverse effects. Toxins such as DPX, OA, and PLTXs showed tumour-promoting activities through different pathways, hence can be used for studying signalling mechanisms involved in carcinogenesis. The potential use of HLs as anticancer agents has been reported.

Spirolides (SPX) are produced by *Karenia selliformis*, *Alexandrium* species. The toxins have been proven to act as a neuroprotector and play a role in providing protection against Alzheimer and other stress related diseases.

Maitotoxin (MTX) showed bioactivity of disruption of cellular calcium homeostasis. *Gambierdiscus toxicus* produces maitotoxin. The compound activates cellular calcium channels (Reyes et al. 2022). It triggers intracellular cascades such as membrane depolarization in excitable cells, neurotransmission, secretion of insulin, phosphoinositide breakdown in cell lipids, cell signalling, programmed cell death and hence can prove a useful tool for cell biology research (Reyes et al. 2022). It is also used in understanding the process of interleukin-1β secretagogue, oncotic death induction, and sperm acrosome reaction induction. Toxin has been used to study innate immune responses and inflammation in vivo studies using mice (Reyes et al. 2022). Since the toxin has shown a role in activation of programmed cell death, cellular regulatory mechanisms, and secretion of neurotransmitter, hence can be exploited for the development of multiple therapies for different diseases.

Ostreopsis spp. produce aerosolized palytoxin (PLTX) along with its analogues, which leads to myalgia, respiratory problems, impairment of the neuromuscular apparatus, and abnormalities in cardiac function. PLTX is considered one of the most lethal marine toxins, and its mode of action is unique wherein it causes the Na^+/K^+ pump to turn into a shape that allows the passive transport of sodium and potassium ions. PLTXs and their analogues can be produced by the dinoflagellates *O. siamensis*, *O. ovata,* and *O. mascarenensis*. PLTX can be used in research for anticancer drugs. PLTX shows toxicity for head and neck carcinoma cells, thus can play a role in tumour promotion and suppression. Phycotoxins including PTX, PCs, and YTXs showed anticancer activity. PTX-2 showed significant cytotoxicity in human cancer cells via downregulation of antiapoptotic Bcl-2, IAP-related proteins, upregulation of Bax protein and ligand (TRAIL)-receptor 1/receptor 2 (DR4/DR5), mitochondrial dysfunction, and apoptosis through suppression of the nuclear factor κB pathway (Kim et al. 2020). Prorocentrolide C derived from dinoflagellate *P. lima* showed ability to induce human cancer cell death via apoptosis and cell cycle arrest (Camacho-Muñoz et al. 2021). Yessotoxins (YTXs) and its analogues have found their use in treatment of Alzheimer's disease via reduction in level of t- and β-amyloid (two compounds formed in the brain) that are responsible for the illness. The compounds are also a proven effective in treatment of lipid, glucose metabolism-related disorders in glioma cells and transcriptional abnormalities of pancreas and liver. The toxin also has some role to play in drugs for treatment of asthma and allergies (Paz et al. 2008).

Prorocentrum spp. produce species-specific diverse phycotoxins, such as OA, DPX, and prorocentrolides (PC) (Pan et al. 1999). PCs act on both muscle and neuronal nicotinic acetylcholine receptors (nAChRs). Brevetoxin (BTX) obtained from the dinoflagellate *Karenia brevis* has shown capacity to reduce symptoms of respiratory infection such as cough, irritability of the nose, bronchoconstriction, congestion, and/or asthma attacks. It triggers immune response in alveolar macrophage cells by boosting cytokines (TNF- and IL-2), reduces phagocytosis and regulates inflammation in pulmonary tissue (Walsh et al. 2008). It also affects cell growth, causes cell death via apoptosis, and has shown genotoxic properties in leukaemic T-cell lines. Properties of BTX-2 such as neuro-activation can improve

neuronal plasticity, hence prove useful in restoring brain function after stroke or other traumatic brain damage.

Karlotoxin (KmTx) a biologically active toxin isolated by *Karlodinium* genus (*K. veneficum*) showed haemolytic, cytotoxic, ichthyotoxic, and antifungal properties. It shows ability to create pores in cholesterol-containing cell membranes and find its use in variety of human diseases, including CHD (coronary heart disease). This toxin can be used as a chemotherapeutic drug used to control cancer including prostate and breast cancer.

Studies proved that Gymnocin-A (GYMA) along with OA can be used as a therapeutic to produce anticancer effects by inducing cell toxicity and acting as chemotherapeutic drug. GYM also can help in reducing amyloid levels and tau phosphorylation, which can treat degenerative illness.

Amphidinolides and colopsinols produced by marine dinoflagellate *Amphidinium* showed anticancer effects via inhibition of cancer cell lines. Amphidinolides showed high cytotoxicity against murine lymphoma L1210 and human epidermoid carcinoma KB cell under in vitro conditions (Kobayashi et al. 2003; Kobayashi and Tsuda 2004).

6.6.1.1 Use in Biofuel Production

The co-culture of dinoflagellate *H. circularisquama* with green microalga *Dunaliella salina* showed effective flocculation activity. This potential of dinoflagellates can be used industrially in biodiesel production (Cho et al. 2020). Haemolysins (HLs) showed bioflocculation properties.

6.6.1.2 Use as Insecticide or Pesticide

KmTx, an allopathic compound isolated from *Karlodinium* spp. exhibits copepod-killing activity. The compound causes cell membrane disruption by binding specifically to cholesterol, hence toxin can be a useful component for anticancer or microalgae-harvesting agents. KTX is also known to affect the copepod *Acartia tonsa*, a natural grazer of *Karlodinium* spp. This toxin showed massive potential as a piscicide and copepod-killing agent because of its ichthyotoxicity and toxicity in copepods.

6.7 Use of Biotechnology in Phycotoxin Production

Many of the phycotoxins show high biological activity, hence these high-value secondary metabolites need to be produced on a large scale for research and commercial applications (Garcia-Camacho et al. 2007). Sustainable phycotoxin production includes chemical synthesis, harvesting of natural blooms, and large-scale

production via in situ, in vitro, and transgenic cultivation. The selection of method depends upon the structural complexity of the phycotoxin, nature of the source and its abundance and pathway of synthesis. Genetic engineering has emerged as a good approach to get enhanced synthesis of these compounds via genetic manipulation and overexpression of specific genes. A thorough study of the biochemical processes occurring within the cells is required to manipulate the genetic framework of dinoflagellates with an aim to increase the production of the metabolites of interest (Verma et al. 2019). Efforts have been made to increase lipid production in non-dinoflagellate algae via technology of genetic engineering (Lin et al. 2019; Park et al. 2019). The complex, large genome size of dinoflagellates, high degree of methylation, introns, redundant repetitive non-coding sequences, and lack of information related to promoter features and transcription factor binding sites restrict the efforts to increase the expression of toxin-related genes and proteins involved in synthesis of some useful phycotoxins (Verma et al. 2019).

6.8 Conclusion

Dinoflagellate-produced phycotoxins are though toxic lethal compounds that can lead to chronic adverse health effects in humans, but these biotoxins have specific modes of action and have shown potential applications in various areas. Some of these toxins can effectively be utilized for producing beneficial materials such as medicines and other therapeutic agents. They also have extensive industrial applications. Biotechnological techniques can be used to increase the production of diverse dinoflagellate phycotoxins based on their intrinsic modes of action. Clinical studies are required to validate the beneficial role of phycotoxins in therapeutics and other fields such as medicine.

References

Accoroni S, Ceci M, Tartaglione L, Romagnoli T, Campanelli A, Marini M et al (2018) Role of temperature and nutrients on the growth and toxin production of *Prorocentrum hoffmannianum* (Dinophyceae) from the Florida keys. Harmful Algae 80:140–148

Ajuzie CC (2007) Palatability and fatality of the dinoflagellate *Prorocentrum lima* to *Artemia salina*. J Appl Phycol 19:513–519

Alfonso A, Vieytes MR, Botana LM (2016) Yessotoxin, a promising therapeutic tool. Mar Drugs 14(2):E30

Anderson DM, Cembella AD, Hallegraeff GM (2012) Progress in understanding harmful algal blooms: paradigm shifts and new technologies for research, monitoring, and management. Annu Rev Mar Sci 4:143–176

Aquino-Cruz A, Purdie DA, Morris S (2018) Effect of increasing sea water temperature on the growth and toxin production of the benthic dinoflagellate *Prorocentrum lima*. Hydrobiologia 813:103–122

Blanco SL, Rodríguez LP, Lago J, Vieites JM, Cabado AG (2020) Potential uses and medical management of neurotoxic phycotoxins. Encyclop Mar Biotechnol 3:1581–1617

Camacho-Muñoz D, Praptiwi RA, Lawton LA, Edwards C (2021) High value phycotoxins from the dinoflagellate *Prorocentrum*. Front Mar Sci 8:638739

Cao Z, Cui Y, Busse E, Mehrotra S, Rainier JD, Murray TF (2014) Gambierol inhibition of voltage-gated potassium channels augments spontaneous Ca^{2+} oscillations in cerebrocortical neurons. J Pharmacol Exp Ther 350:615–623

Cassell RT, Chen W, Thomas S, Liu L, Rein KS (2015) Brevetoxin, the dinoflagellate neurotoxin, localizes to thylakoid membranes and interacts with the light-harvesting complex II (LHCII) of photosystem II. Chembiochem 16:1060–1067

Cho K, Heo J, Han J, Hong HD, Jeon H, Hwang H, Hong C, Kim D, Han JW, Baek K (2020) Industrial applications of dinoflagellate phycotoxins based on their modes of action: a review. Toxins 12(12):805

Cousseau A, Siano R, Probert I, Bach S, Mehiri M (2020) Marine dinoflagellates as a source of new bioactive structures. In: Atta ur R (ed) Studies in natural products chemistry. Elsevier, Amsterdam, pp 125–171. https://doi.org/10.1016/b978-0-12-817905-5.00004-4

Espina B, Louzao MC, Ares IR, Fonfria ES, Vilarino N, Vieytes MR, Yasumoto T, Botana LM (2010) Impact of the pectenotoxin C-43 oxidation degree on its cytotoxic effect on rat hepatocytes. Chem Res Toxicol 23:504–515

Farabegoli F, Lucía B, Rodríguez LP, Vieites JM, Cabado AG (2018) Phycotoxins in marine shellfish: origin, occurrence and effects on humans. Mar Drugs 16(6):188

Franchini A, Malagoli D, Ottaviani E (2010) Targets and effects of yessotoxin, okadaic acid and palytoxin: a differential review. Mar Drugs 8:658–677

Fu LL, Zhao XY, Ji LD, Xu J (2019) Okadaic acid (OA): toxicity, detection and detoxification. Toxicon 160:1–7

Gallardo-Rodríguez J, Sánchez-Mirón A, García-Camacho F, López-Rosales L, Chisti Y, Molina-Grima E (2012) Bioactives from microalgal dinoflagellates. Biotechnol Adv 30:1673–1684

Garcia-Camacho F, Gallardo-Rodríguez J, Sánchez-Mirón A, Cerón-García MC, Belarbi EH, Chisti Y et al (2007) Biotechnological significance of toxic marine dinoflagellates. Biotechnol Adv 25:176–194

Garrido R, Lagos N, Lattes K, Abedrapo M, Bocic G, Cuneo A, Chiong H, Jensen C, Azolas R, Henriquez A (2005) Gonyautoxin: new treatment for healing acute and chronic anal fissures. Dis Colon Rectum 48:335–343

Glibert PM, Wilkerson FP, Dugdale RC, Raven JA, Dupont CL, Leavitt PR et al (2016) Pluses and minuses of ammonium and nitrate uptake and assimilation by phytoplankton and implications for productivity and community composition, with emphasis on nitrogen-enriched conditions. Limnol Oceanogr 61:165–197

Gu S, Xiao SW, Zheng JW, Li HY, Liu JS, Yang WD (2019) ABC transporters in *Prorocentrum lima* and their expression under different environmental conditions including okadaic acid production. Mar Drugs 17:259

Hinder SL, Hays GC, Brooks CJ, Davies AP, Edwards M, Walne AW et al (2011) Toxic marine microalgae and shellfish poisoning in the British Isles: history, review of epidemiology, and future implications. Environ Health 10:54

Hoppenrath M, Chomérat N, Horiguchi T, Schweikert M, Nagahama Y, Murray S (2013) Taxonomy and phylogeny of the benthic *Prorocentrum* species (*Dinophyceae*)—a proposal and review. Harmful Algae 27:1–28

Hoppenrath M (2017) Dinoflagellate taxonomy — a review and proposal of a revised classification. Mar Biodivers 47:381–403

Hou DY, Liang JJ, Zou C, Li HY, Liu JS, Yang WD (2016) MRP functional activity and character in the dinoflagellate *Prorocentrum lima*. J Appl Phycol 28:1667–1676

Ianora A, Bentley MG, Caldwell GS, Casotti R, Cembella AD, Engström-Öst J et al (2011) The relevance of marine chemical ecology to plankton and ecosystem function: an emerging field. Mar Drugs 9:1625–1648

James KJ, Carey B, O'Halloran J, van Pelt FNAM, Škrabáková Z (2010) Shellfish toxicity: human health implications of marine algal toxins. Epidemiol Infect 138:927–940

Karlson B, Andersen P, Arneborg L, Cembella A, Eikrem W, John U, West JJ, Klemm K, Kobos J, Lehtinen S, Lundholm N (2021) Harmful algal blooms and their effects in coastal seas of Northern Europe. Harmful Algae 102:101989

Kim D, Li W, Matsuyama Y, Matsuo A, Yagi M, Cho K, Yamasaki Y, Takeshita S, Yamaguchi K, Oda T (2020) Strain-dependent lethal effects on abalone and haemolytic activities of the dinoflagellate *Karenia mikimotoi*. Aquaculture 520:734953

Kobayashi JI, Tsuda M (2004) Amphidinolides, bioactive macrolides from symbiotic marine dinoflagellates. Nat Prod Rep 21:77–93

Kobayashi JI, Shimbo K, Kubota T, Tsuda M (2003) Bioactive macrolides and polyketides from marine dinoflagellates. Pure Appl Chem 75:337–342

Kohli GS, John U, Van Dolah FM, Murray SA (2016) Evolutionary distinctiveness of fatty acid and polyketide synthesis in eukaryotes. ISME J 10:1877–1890

Lee TCH, Fong FLY, Ho KC, Lee FWF (2016) The mechanism of diarrhetic shellfish poisoning toxin production in *Prorocentrum* spp.: physiological and molecular perspectives. Toxins 8:272

Lee S, Yang AR, Yoo YD, Jeong EJ, Rho JR (2019) Relative configurational assignment of 4-hydroxyprorocentrolide and prorocentrolide C isolated from a benthic dinoflagellate (*Prorocentrum lima*). J Nat Prod 82:1034–1039

Li YY, Tian XQ, Lu YN, Han QH, Ma LY, Fan CQ (2020) Toxins and other chemical constituents from *Prorocentrum lima*. Biochem Syst Ecol 89:104015

Liang Y, Li A, Chen J, Tan Z, Tong M, Liu Z, Qiu J, Yu R (2022) Progress on the investigation and monitoring of marine phycotoxins in China. Harmful Algae 111:102152

Lim ZF, Luo Z, Lee LK, Hii KS, Teng ST, Chan LL et al (2019) Taxonomy and toxicity of *Prorocentrum* from Perhentian Islands (Malaysia), with a description of a non-toxigenic species *Prorocentrum malayense* sp. nov. (*Dinophyceae*). Harmful Algae 83:95–108

Lin WR, Tan SI, Hsiang CC, Sung PK, Ng IS (2019) Challenges and opportunity of recent genome editing and multi-omics in cyanobacteria and microalgae for biorefinery. Bioresour Technol 291:121932

López-Rosales L, Gallardo-Rodríguez JJ, Sánchez-Mirón A, Cerón-García MDC, Belarbi EH, García-Camacho F et al (2014) Simultaneous effect of temperature and irradiance on growth and okadaic acid production from the marine dinoflagellate *Prorocentrum belizeanum*. Toxins 6:229–253

Louzao MC, Fernández DA, Abal P, Fraga M, Vilariño N, Vieytes MR et al (2015) Diarrhetic effect of okadaic acid could be related with its neuronal action: changes in neuropeptide Y. Toxicol Lett 237:151–160

Louzao MC, Vilariño N, Vale C, Costas C, Cao A, Raposo-Garcia S, Vieytes MR, Botana LM (2022) Current trends and new challenges in marine phycotoxins. Mar Drugs 20:198

Lu CK, Chou H-N, Lee CK, Lee TH (2005) Prorocentin, a new polyketide from the marine dinoflagellate *Prorocentrum lima*. Org Lett 7:3893–3896

Luo Z, Zhang H, Krock B, Lu S, Yang W, Gu H (2017) Morphology, molecular phylogeny and okadaic acid production of epibenthic *Prorocentrum* (*Dinophyceae*) species from the northern South China Sea. Algal Res 22:14–30

Manríquez V, Caperan DC, Guzmán R, Naser M, Iglesia V, Lagos N (2015) First evidence of neosaxitoxin as a long-acting pain blocker in bladder pain syndrome. Int Urogynecol J 26:853–858

Nicolas J, Hoogenboom RLAP, Hendriksen PJM, Bodero M, Bovee TFH, Rietjens IM, Gerssen A (2017) Marine biotoxins and associated outbreaks following seafood consumption: prevention and surveillance in the 21st century. Glob Food Secur 15:11–21

Nishimura T, Uchida H, Noguchi R, Oikawa H, Suzuki T, Funaki H et al (2020) Abundance of the benthic dinoflagellate *Prorocentrum* and the diversity, distribution, and diarrhetic shellfish toxin production of *Prorocentrum lima* complex and *P. Caipirignum* in Japan. Harmful Algae 96:101687

Pan Y, Cembella AD, Quilliam MA (1999) Cell cycle and toxin production in the benthic dinoflagellate *Prorocentrum lima*. Mar Biol 134:541–549
Pan L, Chen J, Shen H, He X, Li G, Song X et al (2017) Profiling of extracellular toxins associated with diarrhetic shellfish poison in *Prorocentrum lima* culture medium by high-performance liquid chromatography coupled with mass spectrometry. Toxins 9:308
Park S, Nguyen THT, Jin E (2019) Improving lipid production by strain development in microalgae: strategies, challenges and perspectives. Bioresour Technol 292:121953
Patocka J, Gupta RC, Wu QH, Kuca K (2015) Toxic potential of palytoxin. J Huazhong Univ Sci Technol 35:773–780
Paz B, Daranas AH, Norte M, Riobó P, Franco JM, Fernández JJ (2008) Yessotoxins, a group of marine polyether toxins: an overview. Mar Drugs 6:73–102
Pires E, da Cunha Lana P, Mafra LL Jr (2023) Phycotoxins and marine annelids–a global review. Harmful Algae 122:102373
Pulido OM (2016) Phycotoxins by harmful algal blooms (HABS) and human poisoning: an overview. Int Clin Pathol J 2(6):145–152
Pulz O, Gross W (2004) Valuable products from biotechnology of microalgae. Appl Microbiol Biotechnol 65(6):635–648
Rasmussen SA, Andersen AJC, Andersen NG, Nielsen KF, Hansen PJ, Thomas OL (2016) Rasmussen chemical diversity, origin, and analysis of phycotoxins. J Nat Prod 79:662–673
Reyes JG, Sánchez-Cárdenas C, Acevedo-Castillo W, Leyton P, López-González I, Felix R, Gandini MA, Treviño MB, Treviño CL (2022) Maitotoxin: an enigmatic toxic molecule with useful applications in the biomedical sciences. In: Botana LM (ed) Seafood and freshwater toxins: pharmacology, physiology and detection. CRC Press (Taylor & Francis Group), Boca Raton, pp 677–694
Rodríguez F, Riobó P, Crespín GD, Daranas AH, de Vera CR, Norte M et al (2018) The toxic benthic dinoflagellate *Prorocentrum maculosum* Faust is a synonym of *Prorocentrum hoffmannianum* Faust. Harmful Algae 78:1–8
Rossini GP, Hess P (2010) Phycotoxins: chemistry, mechanisms of action and shellfish poisoning. Int Clin Pathol J 100:65–122
Selander E, Thor P, Toth G, Pavia H (2006) Copepods induce paralytic shellfish toxin production in marine dinoflagellates. Proc Biol Sci 273:1673–1680
Tobío A, Alfonso A, Madera-Salcedo I, Botana L, M., and Blank, U. (2016) Yessotoxin, a marine toxin, exhibits anti-allergic and anti-tumoural activities inhibiting melanoma tumour growth in a preclinical model. PLoS One 11(12):e0167572
Twiner MJ, Doucette GJ, Pang Y, Fang C, Forsyth CJ, Miles CO (2016) Structure–activity relationship studies using natural and synthetic okadaic acid/dinophysistoxin toxins. Mar Drugs 14:207
Verma A, Barua A, Ruvindy R, Savela H, Ajani PA, Murray SA (2019) The genetic basis of toxin biosynthesis in dinoflagellates. Microorganisms 7:222
Vilariño N, Louzao MC, Abal P, Cagide E, Carrera C, Vieytes MR et al (2018) Human poisoning from marine toxins: unknowns for optimal consumer protection. Toxins 10:324
Walsh CJ, Leggett SR, Strohbehn K, Pierce RH, Sleasman JW (2008) Effects of in vitro brevetoxin exposure on apoptosis and cellular metabolism in a leukemic T cell line (Jurkat). Mar Drugs 6:291–307
Wang D (2008) Neurotoxins from marine dinoflagellates: a brief review. Mar Drugs 11:349–371
Zohdi E, Abbaspour M (2019) Harmful algal blooms (red tide): a review of causes, impacts and approaches to monitoring and prediction. Int J Environ Sci Technol 16:1789–1806

Chapter 7
Phytotoxins: Terrestrial Plant Sources

Kuldeep Prasad Agarwala, Sweta Mohanty, and Siddharthan Surveswaran

Abstract Plants produce a wide variety of phytochemicals during biochemical reactions. Plant secondary metabolites (PSMs) are specialised metabolites or phytochemicals that are produced by plants in response to biotic and abiotic stresses or as a by-product of metabolism. PSMs are needed by the cells of the plant to interact with its environment and are produced in very small concentrations. They have various therapeutic effects and are used as medication in various conditions. However, the difference between therapeutic and toxic levels of these compounds is so low that when administered in excess, they can cause adverse conditions. These toxic PSMs are also known as phytotoxins. Phytotoxins are highly versatile in their mode of toxicity, chemical composition, and structure. The major groups of phytotoxins that have been categorised are alkaloids, terpenoids, glycosides, non-protein amino acids, glucosinolates, cardenolides, tannins, phenolics, flavonoids, and phytoecdysteroids. These phytotoxins have several toxic effects like allergenic, pesticidal, hallucinogenic, and allelochemical and may even cause fatalities. The effects of these phytotoxins in humans range from the disturbance in the metabolic pathways taking place in various organs to adverse conditions like cytotoxicity, neurotoxicity, and carcinogenicity. This chapter discusses in detail the various categories of phytotoxic compounds from land plants and their major biological activities.

Keywords Secondary metabolites · Phytochemicals · Phytotoxins · Peptides · Glycosides · Alkaloids · Terpenoids · Tannins

---

K. P. Agarwala · S. Mohanty
Department of Plant Sciences, School of Life Sciences, University of Hyderabad, Hyderabad, Telangana, India
e-mail: sid@uohyd.ac.in

S. Surveswaran (✉)
Department of Plant Sciences, School of Life Sciences, University of Hyderabad, Hyderabad, Telangana, India

Department of Life Sciences, CHRIST (Deemed to be University), Bangalore, India
e-mail: sid@uohyd.ac.in

M. N. V. Prasad et al. (eds.), *Biotoxins*,
https://doi.org/10.1007/978-3-031-75309-1_7

7.1 Introduction

Plants produce a wide variety of phytochemicals during biochemical reactions. These phytochemicals can be classified into primary and secondary metabolites based on the metabolic procedure of their production (Adhikary and Dasgupta 2023). Phenolics, carotenoids, terpenoids, flavonoids, and alkaloids are some of the plant secondary metabolites (PSMs). PSMs are produced in very small concentrations in the plants and their production majorly depends on the developmental and physiological stages in the life cycle of plants (Alugoju and Tencomnao 2023). PSMs play an important role in defence against various predators, by producing phytotoxins and adapting to various environmental conditions and the interaction between the plant and its environment. Phytotoxins or toxic PSMs are very common in the plant kingdom and have several toxic compounds like allergens, pesticides, hallucinogens, allelochemicals, and many fatal compounds as well (Günthardt et al. 2018). The major groups of phytotoxins discussed in this chapter are peptides, glycosides, alkaloids, terpenoids, and peptide.

7.2 Glycosides

Glycosides are secondary metabolites containing aglycone (genin) and glycone (saccharide) linked by a glycosidic bond. They are modified with the addition of a saccharide molecule mediated by glycosyltransferase enzymes. The glycone is mostly a monosaccharide with glucose being the most common. The main groups of glycosides from plants include phenolic glycosides (PHGs), coumarin glycosides (CMGs), flavonoid glycosides (FLGs), saponosides (SPGs), cardiac glycosides (CRGs) and cyanogenic glycosides (CNGs). Then there are anthraquinone glycosides (AQGs), limited to certain families of angiosperms like Fabaceae, Liliaceae, Polygonaceae, Rubiaceae, and Scrophulariaceae, which have been used since prehistoric times because of their laxative effects. Moreover, there are thioglycosides (THGs, also known as glucosinolates). These are the substituted esters of thioamino acids. These are majorly present in the *Brassica* genus of the Brassicaceae family apart from being present in other plant families such as Resedaceae, Capparaceae, and Caricaceae, with sporadic occurrences in Euphorbiaceae, Tropaeolaceae, Tovariaceae, and Moringaceae. THGs have shown anti-cancer effects on breast cancer, rectal cancer, lung, stomach, and colon cancer as well. They are also in the formation of ointments, and liniments, for the treatment of bronchial infections, and detoxification in chronic diseases (Bartnik and Facey 2017).

7.2.1 *Phenolic Glycosides*

Phenols are compounds with a hydroxyl group attached to a benzene ring. PHGs are glycosides of phenols with a benzene unit, one carboxyl group, and one or more hydroxyl groups. They include phenolic acids like hydroxyl derivatives of benzoic or cinnamic acids. They are widely distributed across the plant kingdom such as populin and salicin in the Salicaceae family, arbutin in Ericaceae and Rosaceae family; coniferin in Coniferae, glucovanillin in *Vanilla* sp., and some Gramineae, arbutin, caffeic acid and chlorogenic acid. Moreover, in angiosperms benzoic acid derivatives and their glycosides are universal which includes protocatechuic, p-hydroxybenzoic, vanillic, and syringic acid. Some of the less common ones include rosmarinic acid in the Lamiaceae family, salicylic acid and o-protocatechuic acid in the Ericaceae family, gallic acid and ellagic acid. PHGs have a wide variety of pharmacological significance which includes anti-inflammatory activity, antioxidant activity, antipyretic effect, anti-rheumatic effect, antimicrobial effect, and antidiabetic activity. Some are also used in the treatment of common cold and urinary tract infections, to reduce undesirable skin pigmentation, and treat stomach aches, nausea, and vomiting. They have also found their use in the treatment of mild and moderate hyperlipidaemia, infections of the upper respiratory tract, and superficial wounds. For the above-mentioned purposes, they are mostly administered orally as infusions and decoctions (Bartnik and Facey 2017; Shriti et al. 2023) (Fig. 7.1).

7.2.2 *Coumarin Glycosides*

Coumarins consist of a pyrone ring attached to a benzene ring. They are derived by isomerization of o-coumaric acid side chain followed by lactonization. Based on their structure they are classified as simple coumarins, furanocoumarins, pyranocoumarins, and pyrone-substituted coumarins. The attachment of a sugar moiety to the coumarin nucleus forms the coumarin glycosides. Almost all coumarin glycosides are O-glycosides, with dauroside D being the exception which is a C-glycoside. The most common sugar moieties are D-glucose, D-galactose, L-rhamnose, and D-apiose while sometimes it can be D-xylose and L-arabinose. These glycosides are common in plant families Rutaceae, Rubiaceae, Apiaceae, Asteraceae, Oleaceae, Fabaceae, and Thymelaeaceae. In these plants, their highest levels are present in the fruits followed by stem, roots, and leaves. Some examples of coumarin glycosides include esculin, fraxin, scopolin, esculetin, fraxetin, scopoletin, fraxidin, isofraxidin, cichoriin, melilotoside, herniarin, and umbelliferone with several pharmacological properties. These include anti-inflammatory activity, antipyretic activity, antispasmodic activity, antioxidant and antimicrobial activity. They are also used in the treatment of laryngitis, respiratory problems such as asthma, bronchitis, and gastrointestinal issues like indigestion, and loss of appetite as well as provide relief

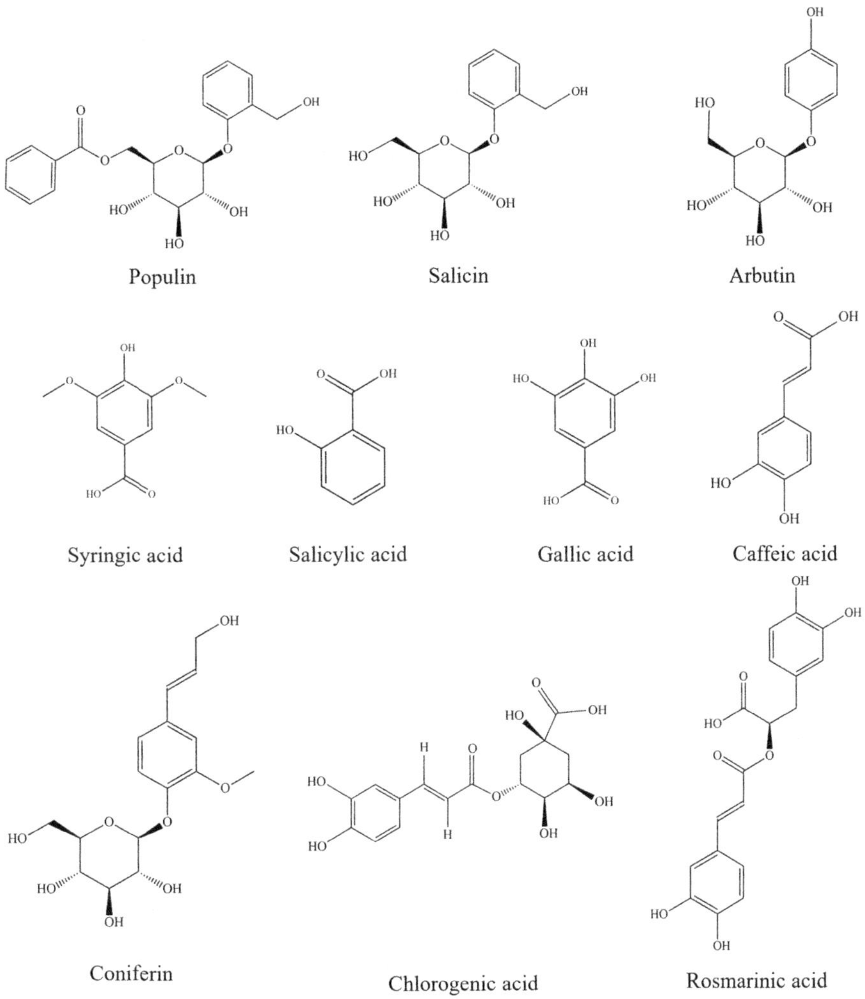

Fig. 7.1 Phenolic glycosides

from menopausal symptoms, enrich blood circulation and in treatment of coronary arteriosclerosis (Penta 2016; Bartnik and Facey 2017) (Fig. 7.2).

7.2.3 *Flavonoid Glycosides*

Flavonoids are polyphenolic compounds synthesised through the phenylpropanoid or acetate-malonate metabolic pathway in plants. Flavonoids are a collection of natural compounds comprised C6-C3-C6 structures forming a 2-phenyl-benzo-γ-pyrane nucleus that consists of two benzene rings (A and B) linked through a

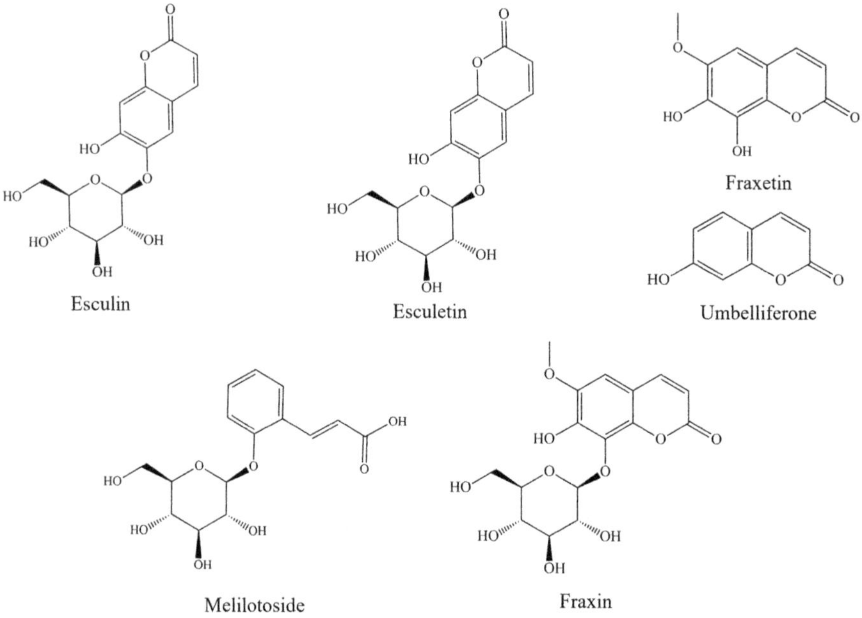

Fig. 7.2 Coumarin glycosides

heterocyclic pyran ring (C). There are various types of flavonoids classified based on the oxidation level of the central pyran C-ring like flavones, flavonols, flavanones, isoflavones, chalcones, aurones, and anthocyanidins. Some of the complex flavonoids also include biflavonoids and flavonolignans. Some others like tri-, tetra-, penta-, or hexa-flavonoids also exist in plants but they are rare. For the formation of flavonoid glycosides, glycosylation may occur by mono-, di-, or trisaccharide. They are widespread in the higher plants. Flavonols and flavones are present in various plants that are used as food while biflavonoids are mainly found in some gymnosperms from Cycadopsida and Coniferopsida classes. Asteraceae family of angiosperms typically have chalcones whereas dihydrochalcones are most common in Ericaceae and Rosaceae. FLGs are almost always water-soluble and are located in vacuoles mostly in the epidermal or mesophyll cells in all parts of a plant and the epidermal cells of flowers. They are responsible for the colour of flowers, fruits and sometimes leaves as well. FLGs like hyperoside, quercetin, isoquercetin, and rutoside have anti-inflammatory, anti-diabetic, anti-haemorrhoid, spasmolytic, and anti-rheumatic activities. These also help treat urinary tract infections, skin problems, and diarrhoea and have vasoprotective activity. Similar activities are also shown by avicularin, astragalin, persicarin, arabinoside, and nicotiflorin. Apart from these, naringenin glycosides have spasmolytic, antioxidant, and antimicrobial activity. Ononin shows anti-diabetic and anti-cancer activities along with its use in the treatment of cardiovascular diseases, obesity, osteoporosis, menopausal syndrome, and anaemia (Bartnik and Facey 2017) (Fig. 7.3).

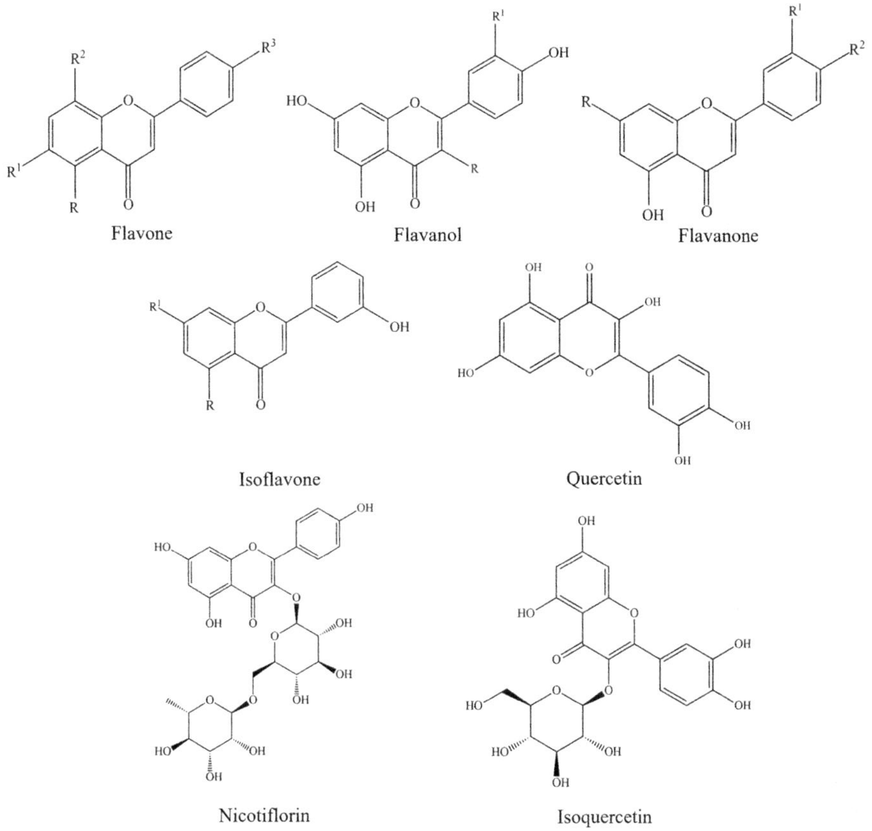

Fig. 7.3 Flavonoid glycosides

7.2.4 Saponosides

Saponins are nonvolatile compounds and are widely distributed in nature. They are named so due to their surfactant properties like the ability to form stable soap-like foams in aqueous solutions. They are amphiphilic in nature, having a hydrophobic aglycone backbone (sapogenin) linked to hydrophilic sugar chain/s. Based on the structure of the sapogenin, there are two classes of saponins—steroidal or triterpenoid. Both are derived from the C30 carbon atoms precursor: 2,3-oxidosqualene. For glycosylation of these saponins, the most common monosaccharides include D-glucose, 3-methyl-D-glucose, D-galactose, L-rhamnose, D-fructose, L-arabinose, D-xylose, D-apiose, and D-chinovose, in addition to D-glucuronic acid and D-galacturonic acid. SPGs are widely distributed in the plant kingdom which includes more than 100 families. Steroidal saponins are distributed in Agavaceae, Alliaceae, Amaryllidaceae, Asparagaceae, Bromeliaceae, Dioscoreaceae, Liliaceae, Palmae, and Scrophulariaceae families. While the families Cucurbitaceae, Fabaceae,

Amaranthaceae, Apiaceae, Aquifoliaceae, Araliaceae, Berberidaceae, Caryophyllaceae, Chenopodiaceae, Myrsinaceae, and Zygophyllaceae have the terpenoid saponins. These have been isolated from different parts of the plants of the mentioned families including leaves, barks, stems, fruits, and seeds. They are part of the plant defence system because of their antimicrobial, fungicidal, allelopathic, insecticidal, molluscicidal, antiparasitic, and antifeeding effects. Moreover, they are also important due to their anti-inflammatory, expectorant, anti-allergic, antiulcer, antiviral, antioxidant, antithrombotic, and anti-haemorrhoidal activities. They are also used due to their hepatoprotective, psychotropic, and anti-diabetic properties, in the treatment of bronchitis, eczema, hypercholesterolaemia, anorexia, and furunculosis. These also have possible applications in the treatment of Alzheimer's disease (Bartnik and Facey 2017).

7.2.5 Cardiac Glycosides

Cardiac glycosides are naturally occurring secondary metabolites with potential cardiovascular action. This class of compound is composed of a lactone ring, steroid ring, and sugar moiety. The structural variability is due to the aglycone steroid moiety. Some of the plant families containing CRGs are Asclepiadaceae, Apocynaceae, Brassicaceae, Celastraceae, Crassulaceae, Liliaceae, Ranunculaceae, and Scrophulariaceae (Bartnik and Facey 2017). Because of the toxic effects and close therapeutic and lethal dose limit, these are not used as crude drugs but their compounds and derivatives find their uses in the pharmaceutical industry. The five-membered and six-membered lactone rings containing cardiac glycosides are called cardenolides and bufadienolides, respectively. Oleandrin, digoxin, and ouabain are well-known cardenolides while bufalin is a bufadienolide. CRGs strengthen the force of the heartbeat by acting as a positive inotrope. CRGs compete with K+ ions by binding to the Na^+-, K^+-ATPase of the cardiomyocyte's sarcolemma. As a result, Ca^{2+} ions concentration increases in cardiac muscle cells causing stronger heart contractions and improving the tone of cardiac muscles (cardiotonic activity). Moreover, CRGs are also useful in the treatment of auricular fibrillation and congestive heart failure. Studies have also shown that breastin, a cold-water extract of oleander leaves, can even induce anti-cancer effects (Bartnik and Facey 2017; Azzalini et al. 2019; Rashan et al. 2023) (Fig. 7.4).

7.2.6 Cyanogenic Glycosides

Cyanogenic glycosides are a specific group of compounds chemically bound to sugars and release cyanide in the form of hydrogen cyanide (HCN, prussic acid) upon hydrolysis. These compounds are stabilised in plants by glycosylation. The most common sugar in CNGs is glucose. It is hypothesised as an evolutionary defence

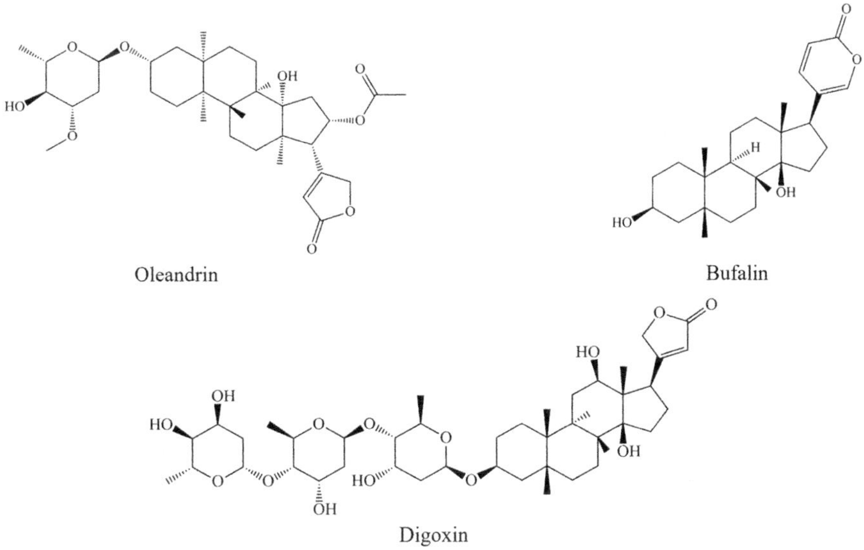

Fig. 7.4 Cardiac glycosides

mechanism in more than 3000 species of plants from Euphorbiaceae, Fabaceae, Poaceae, Linaceae, and Rosaceae families. Some of the most common CNGs are linamarin, lotaustralin, taxiphyllin, neolinustatin, amygdalin, and prunasin. The toxic effects of CNGs are due to the release of hydrogen cyanide causing cyanide poisoning. CNGs like amygdalin and prunasin are considered to be potential anti-cancer drugs. Some CNGs also act as aromatising agents, respiratory stimulants, and antispasmodic agents. Deglycosylation of the components of CNGs releases toxic HCN and the symptoms associated with cyanide poisoning are mental confusion, vomiting, dizziness, rapid respiration, and headache. Severe liver toxicity has been reported due to small intestine bacterial flora having the enzyme to release hydrogen cyanide. The major food plants that do contain cyanogenic glycosides are cassava (*Manihot esculenta*), sorghum (*Sorghum* spp.), and corn (*Zea mays*). The cyanogenic glycoside found at high levels in bitter almonds (*Prunus amygdalus*) is known as amygdalin or laetrile (Bartnik and Facey 2017; Mosayyebi et al. 2020; Crews and Clarke 2014) (Fig. 7.5).

7.3 Terpenoids

Terpenoids are chemicals consisting of isoprene (C5) units and are derived in the mevalonate (MVA) pathway and 2-C-methyl-D-erythriol 4-phosphate (MEP) pathway. It is the most diverse group of natural compounds with more than 50,000

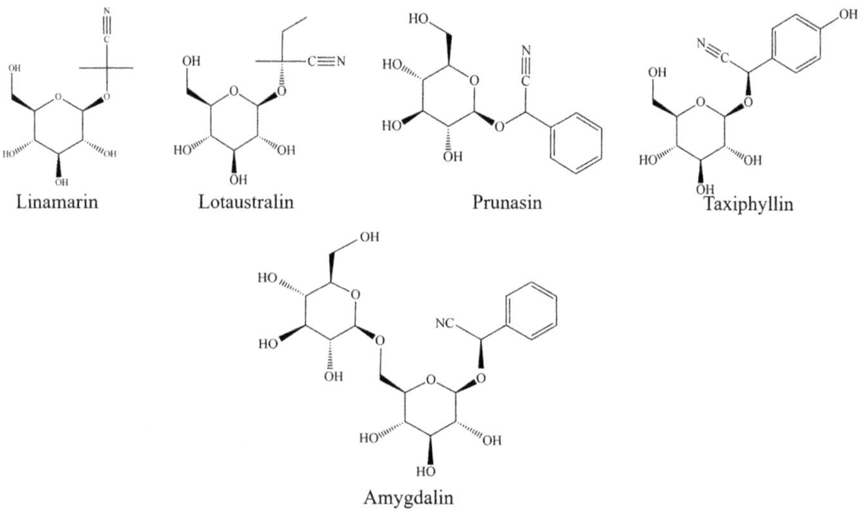

Fig. 7.5 Cyanogenic glycosides

Table 7.1 Types of terpenoids

Terpenoid types	Number of isoprene units
Hemiterpenoids	1
Monoterpenoids	2
Sesquiterpenoids	3
Diterpenoids	4
Sesterterpenoids	5
Triterpenoids	6
Tetraterpenoids	8
Polyterpenoids	> 8

known compounds. Based on the number of isoprene units present, there are various categories of terpenoids (Ludwiczuk et al. 2017) (Table 7.1 and Fig. 7.6).

Monoterpenoids and sesquiterpenoids are the major components of essential oils and are responsible for the characteristic odour of the plants in which they are majorly found. In contrast, the other types of terpenoids are present in resins, waxes, and rubber. Gibberellic acid (a plant hormone) and phytol (side chain of chlorophyll) are universal diterpenoids. Limonin, a triterpenoid, gives bitterness to *Citrus* fruits. Tetraterpenoids, also known as carotenoids, are present in plants as accessory photosynthetic pigment and impart colour to fruits and flowers. Polyterpenoids are found in rubber and as a colloidal suspension in latex. Terpenoids have pharmaceutical as well as several industrial applications ranging from medicines, flavours, and fragrances (Ludwiczuk et al. 2017).

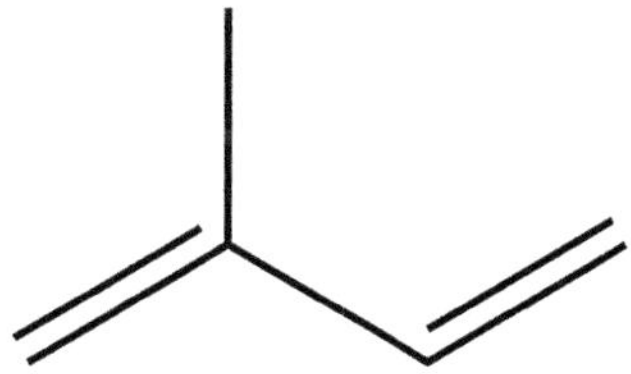

Fig. 7.6 Isoprene

7.3.1 *Monoterpenoids*

Monoterpenoids play a major role in cancer treatment and prevention, antimicrobial, antiviral, and antioxidant properties (Zielińska-Błajet and Feder-Kubis 2020) (Table 7.2).

Other monoterpenoids with antitumour, antimicrobial, antiviral, anti-inflammatory, antiparasitic, antioxidant activities include citral, sabinene, citronellol, pinene, eugenol, borneol, carveol, nerol, verbenol, and pinocarveol, etc. (Carocho et al. 2023) (Fig. 7.7).

7.3.2 *Sesquiterpenoids*

Sesquiterpenoids have several important biological activities such as antiparasitic, antitumour, anti-inflammatory, and analgesic properties (Table 7.3).

Other sesquiterpenoids like helenalin, eupalinin, inuviscolide, angeloylenolin, and nerolidol also show antitumor activity by autophagy in cancer cells and inhibit proliferation of tumour cells by cell apoptosis and arrest of cell cycle. They also possess antiviral, antimicrobial, cytotoxic, and anti-inflammatory activity (Petrović et al. 2019) (Fig. 7.8).

7.3.3 *Diterpenoids*

Diterpenoids have various biological activities like anti-cancer, anti-inflammatory, antimicrobial, analgesic, and antifungal activities. Some of the diterpenoids also possess cardiovascular activity and relaxation of smooth muscles by inhibition of Ca^{2+} channels (Table 7.4).

Other diterpenoids like forskolin, kaurenoic acid, pimaradienoic acid, resiniferatoxin, triterine, triptolidenol, triptonide, and pepulanone also have various biological activities like anti-inflammatory activities, antiviral activities, antispasmodic, and relaxant properties (Salehi et al. 2019; Jin et al. 2021) (Fig. 7.9).

Table 7.2 Types of monoterpenoids

Types	Major pharmacological activities	Occurrence	References
Myrcene	Antiproliferative action against leukaemia, breast cancer and human lung cancer, anti-ageing, antioxidant, antibiotic, anti-mutagenic, and anti-inflammatory properties.	*Cannabaceae*	(Bai and Tang 2020; Surendran et al. 2021)
Limonene	Antitumour, anti-inflammatory, antimicrobial, and anti-diabetic activities. Leads to the apoptosis of different tumours like breast, prostate, colon, pancreas, and liver.	*Rutaceae*	(Mn 2022; da Silva et al. 2021; Carocho et al. 2023)
Camphene	Camphene, found in *Piper cernuum, inhibits* the growth of subcutaneous tumours in humans.	*Piper* sp.	(Carocho et al. 2023)
Carvone	Acts on calcium channels and modulates inflammation mediators. Used in the treatment of inflammation, seizures, pain, infections, obesity, and hypertension.	*Carum carvi, Mentha spicata*	(Pina et al. 2022)
Geraniol	Preventive and therapeutic effects against different types of cancer like lung, prostate, breast, lungs, liver, and pancreatic cancer by inducing cell apoptosis and autophagy.	*Aconitum napellus, Taxus baccata*	(Cho et al. 2016; Yang et al. 2020)
Terpineols	Anti-cancer activity against cancers of the prostate, ovary, breast, lung, and leukaemia, antiulcer, anti-bronchitis, and antioxidant properties.	*Lamiaceae, Myrtaceae*	(Khaleel et al. 2018)
Thymol	Antitumour activity by inducing cytotoxicity, apoptosis, cell cycle arrest, antiproliferative activity and inhibition of signalling pathways.	*Thymus vulgaris, thymus Zygis*	(Sampaio et al. 2021; Islam et al. 2019)
Linalool	Antitumour activities against oral, breast, colorectal, prostate, and liver cancer, anti-inflammatory activities.	*Cinnamomum camphora, Coriandrum sativum*	(Yang et al. 2020; Iwasaki et al. 2016)
Eucalyptol (1,8-cineole)	Treatment of upper respiratory tract disorders, cold. Used as a remedy for influenza, and antispasmodic agent in asthma and bronchitis.	*Eucalyptus globulus, Rosmarinus officinalis*	(Ludwiczuk et al. 2017)
Camphor	Anti-cancer, anti-inflammatory, anti-bacterial, antiviral, and antioxidant activity. Also used as a sedative and counterirritant.	*Cinnamomum camphora*	(Ludwiczuk et al. 2017)

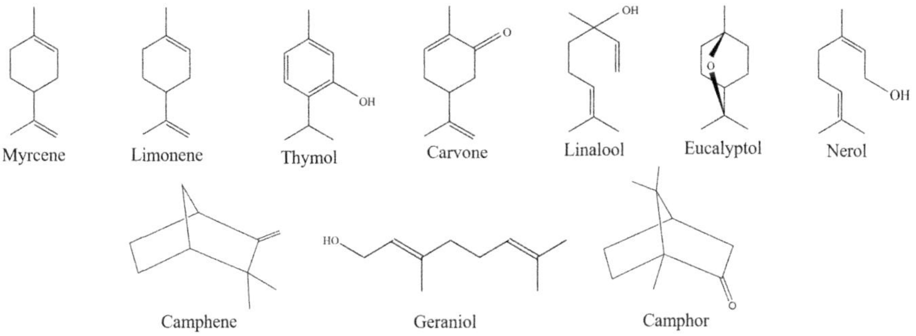

Fig. 7.7 Monoterpenoids

Table 7.3 Types of sesquiterpenoids

Types	Major pharmacological activities	Occurrence	References
β-Caryophyllene	Antimalarial activity, anti-inflammatory, tissue-protective properties, neuroprotective properties (against Parkinson's disease), reduce alcohol-induced liver injury.	*Syzygium aromaticum*	(Kamaraj et al. 2017; Jha et al. 2021; Viveros-Paredes et al. 2017)
Humulene	Anti-inflammatory activity by reducing inflammatory mediators.	*Humulus lupulus*	(Carocho et al. 2023)
Caryophyllene oxide	Anti-inflammatory properties, analgesic activity.		(Carocho et al. 2023)
Patchoulene	Anti-inflammatory, gastroprotective activity like antiulcer agent.		(Zhang et al. 2016; Cox-Georgian et al. 2019)
Parthenolides	Anti-cancer, anti-inflammatory activity.		(Mathema et al. 2012)
Farnesene	Neuroprotective, antitumour, and chemopreventative properties.		(Carocho et al. 2023)
Eudesmol	Antitumour activities, inhibits angiogenesis.		(Srijiwangsa et al. 2018)
Bisabolol	Antitumor activity against pancreatic cancer, therapeutic activity against skin inflammation, antimicrobial properties.		(de Medeiros et al. 2021; Carocho et al. 2023)
Artemisinin	Antimalarial, anti-diabetic.	*Artemisia annua*	(Jiang et al. 2020; Ludwiczuk et al. 2017)
Germacrone	Antiviral activity against influenza virus, porcine virus, feline calicivirus.		(He et al. 2019)

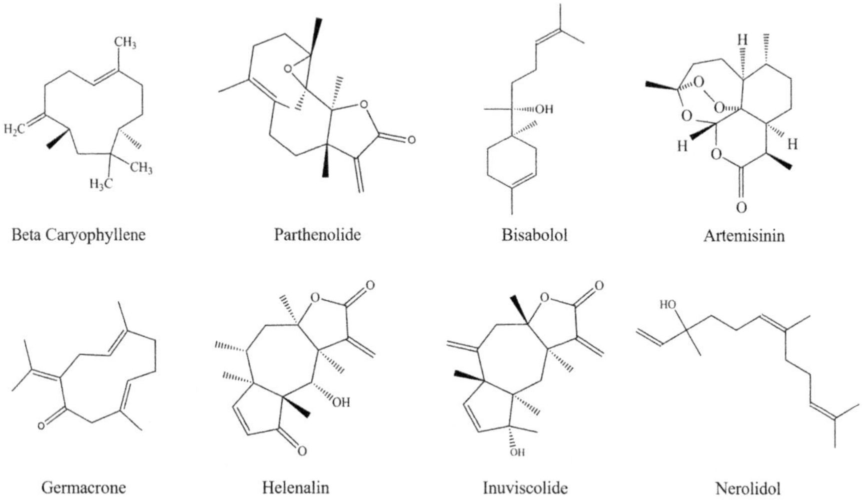

Fig. 7.8 Sesquiterpenoids

Table 7.4 Types of diterpenoids

Types	Major pharmacological activities	Occurrence	References
Tanshinone IIA	Prevents the formation of atherosclerosis, prevents hypertrophy of the heart, and helps vasodilation and blood pressure reduction.	*Salvia miltiorrhiza*	(Wang et al. 2017; Carocho et al. 2023)
Andrographolide	Treats diabetes, has anti-bacterial properties, antiviral activity.	*Andrographis paniculata*	(Zhang et al. 2020)
Stevioside	Possesses an antihyperglycaemic agent and induces blood pressure reduction. Improves cardiometabolic risk in diabetic patients.	*Stevia rebaudiana*	(Gu et al. 2019; Rashad et al. 2019)
Paclitaxel	Anti-cancer activity by acting as antimitotic agent.	*Taxus brevifolia*	(Carocho et al. 2023)
Ferruginol	Antitumour properties in human ovarian cancer, antiparasitic activity and treatment of peptic ulcer disease and oesophagitis.		(González-Cardenete et al. 2021)
Ginkgolides	Has antithrombic properties, improves neurological disorders.	*Ginkgo biloba*	(Cui et al. 2019; Zhang et al. 2021)

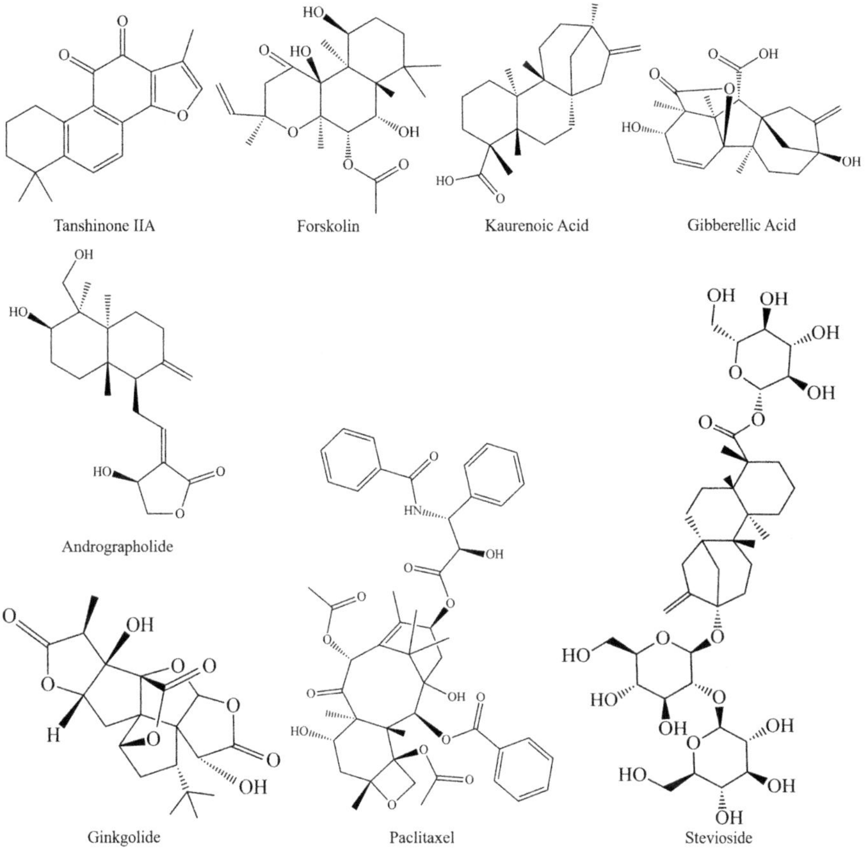

Fig. 7.9 Diterpenoids

7.3.4 Sesterterpenoids

It is a comparatively small group of secondary metabolites. The majority of the sesterterpenoids are cyclic but some are also acyclic. They are widely distributed throughout Kingdom Plantae. They have been isolated from more than 10 families of higher plants including monocotyledons, dicotyledons, ferns, and lichens. The three most important plants containing sesterterpenoids are *Oryza sativa* (rice), *Solanum tuberosum* (potato), and *Triticum aestivum* (wheat). They have many pharmacological activities like anti-cancer, anti-bacterial, antiviral, antifungal, and anti-feedant activity against some pests, and phospholipase inhibition activity. Some of the sesterterpenoids are Leucosceptrine, Heliocide H1, Nitiol, Salvisyriacolide, and salvimirzacolide (Guo et al. 2021) (Fig. 7.10).

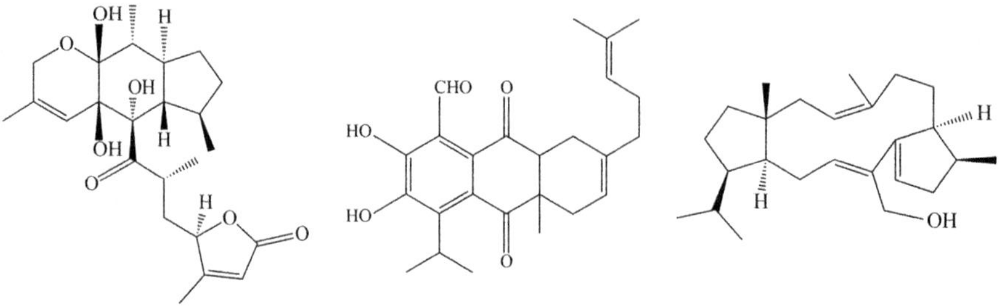

Leucosceptrine Heliocide H1 Nitiol

Fig. 7.10 Sesterterpenoids

7.3.5 Triterpenoids

These are composed of six isoprene units and include steroids and sterols. They are derived from the mevalonate pathway and squalene is the precursor of all triterpenes. Some phytosterols like stigmasterol, campesterol, and sitosterol are found in many higher plants. One of the sterols, α-spinasterol is found in spinach. Most triterpenoids generally occur in the lower members of the plant kingdom like algae and liverworts. The uses of triterpenoids are as well recognised as the other terpenoids. However, the available literature states that they have antioxidant, anti-inflammatory, anti-cancer, anti-diabetic, anti-bacterial, and antiviral activities. They also have other pharmacological effects such as cardiovascular protection, hepatoprotection, and neuroprotection. Moreover, saponins because of their amphipathic nature also act as an excellent detergent in the human digestive system. Apart from these, triterpenoids also include ursolic acid, oleanolic acid, α-amyrin, betulinic acid, cucurbitacin B, aceriphyllic acid A. Azadirachtin from *Azadirachta indica* (Neem tree) is an anti-feedant (Liu et al. 2023a; Cox-Georgian et al. 2019; Aribi et al. 2020; Ludwiczuk et al. 2017) (Fig. 7.11).

7.3.6 Tetraterpenoids

These terpenoids have eight isoprene units. These are fat-soluble pigments also known as carotenoids and impart colours like red, yellow, and orange. Carotenoids, based on their chemical structure, can be classified as hydrocarbons (carotenes) and oxygenated derivatives of hydrocarbons (xanthophylls). Some of the most common examples include lycopene in *Solanum lycopersicum* (Solanaceae), β-carotene in *Daucus carota* (Apiaceae), γ-carotene in *Calendula officinalis* (Asteraceae), lutein in *Spinacia oleracea* (Amaranthaceae), zeaxanthin in *Capsicum annum* (Solanaceae), and crocin in *Crocus sativus* (Iridaceae). The presence of carotenoids like crocetin, crocin, safranal, and picrocrocin in *Crocus sativus* (saffron) gives it anti-inflammatory, antioxidant, antinociceptive, anticonvulsant, antidepressant,

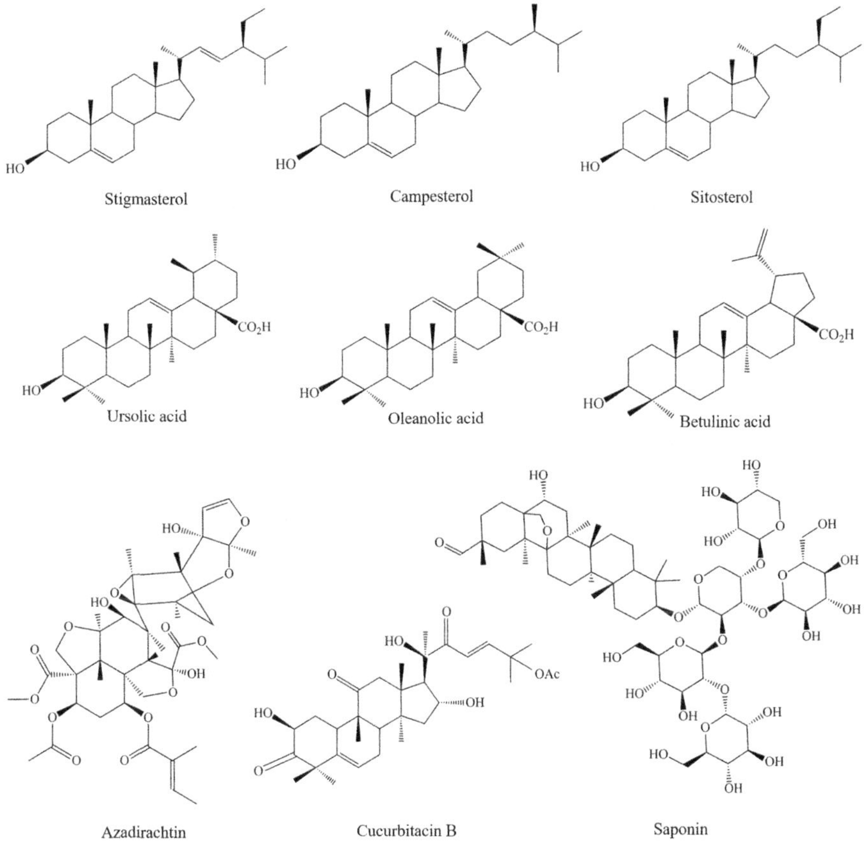

Fig. 7.11 Triterpenoids

antitumorigenic, and immunomodulatory properties. In Ayurveda, it is used in the treatment of asthma, ulcers, arthritis, acne, coronary diseases, and skin diseases. These compounds also have antidepressant properties but only for mild to moderate depression. Moreover, carotenoids like bixin, norbixin, lutein, and zeaxanthin also have anti-diabetic properties. Apart from these, lycopene also has antitumour effects, like in the case of ovarian cancer (Ludwiczuk et al. 2017; da Silveira Vasconcelos et al. 2020; Cox-Georgian et al. 2019) (Fig. 7.12).

7.4 Alkaloids

Alkaloids are a diverse group of naturally occurring metabolites containing nitrogen atom in a heterocyclic ring system, which are mostly derived from amino acids (like lysine, ornithine, tyrosine, tryptophan, phenylalanine, arginine, proline, and

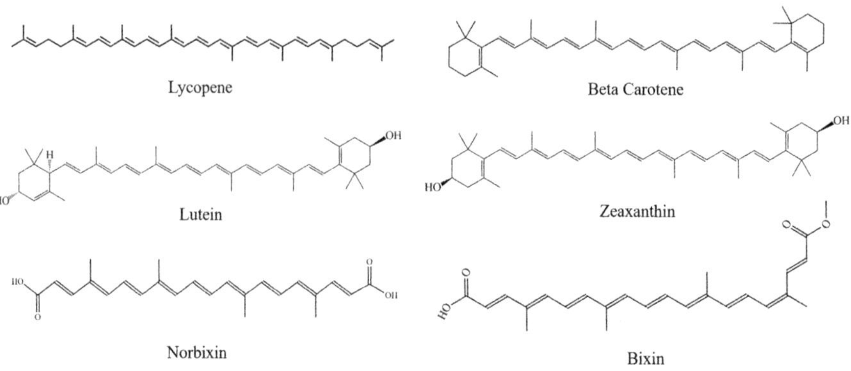

Fig. 7.12 Tetraterpenoids

leucine). More than 25,000 alkaloids have been characterised across 145 families of plants, some of which are Asteraceae, Apocynaceae, Boraginaceae, Berberidaceae, Chenopodiaceae, Buxaceae, Fabaceae, Euphorbiaceae, Lauraceae, Magnoliaceae, Rubiaceae, Papaveraceae, Solanaceae, and Rutaceae. They are either present in the form of water-soluble salts of organic acids (tartaric acid, acetic quinoline acid, citric acid, malic acid, and lactic acid), esters (atropine, scopolamine, cocaine and aconitine), or combined with tannins and sugars. Alkaloids can be classified as heterocyclic and non-heterocyclic alkaloids. Heterocyclic alkaloids are the ones which have a nitrogen atom as part of the heterocyclic ring system and are derived from amino acids. Heterocyclic alkaloids are further categorised based on amino acids of origin. The principal class of alkaloids include isoquinoline alkaloids, tropane alkaloids, pyrrolizidine alkaloids, pyridine alkaloids, indole alkaloids, quinoline alkaloids, and steroidal alkaloids (Kukula-Koch and Widelski 2017; Carocho et al. 2023).

7.4.1 Heterocyclic Alkaloids

7.4.1.1 Isoquinoline Alkaloid

Isoquinoline alkaloids form a large group of alkaloids occurring in plant species of the families Papaveraceae, Berberidaceae, Fumariaceae, Menispermaceae, Rutaceae, Magnoliaeace, Cactaceae, Ranunculaceae, Amaryllidaceae, and Annonaceae. The precursor molecule for this class of alkaloids is dopamine and the basic structural feature is the presence of the isoquinoline nucleus. The clinical usage of isoquinoline alkaloids is extensive due to their diverse structure and functions. This group of alkaloids shows antiviral, anti-cancer, antifungal, cholesterol-lowering, antioxidant, antihyperglycaemic, and anti-inflammatory properties. Morphine and codeine are widely studied isoquinoline alkaloids which are approved as drugs worldwide (Dey et al. 2020). Various functions and sources of isoquinoline alkaloid are given in Table 7.5 (Fig. 7.13).

Table 7.5 Types of isoquinoline alkaloids

Sub-class of isoquinoline alkaloid	Types	Major pharmacological activity	Occurrence	References
Protoberberine	Dehydro-corybulbine	Therapy against malignant melanoma	*Corydalis yanhusuo, Corydalis tuber, Corydalis bulbosa*	(Hu et al. 2019)
		Anti-schizophrenia effects	*Corydalis yanhusuo*	(Wang et al. 2019)
	Berberine	Enhances inflammatory response in the lung tissue, counteracts the cancer-promoting effects on lung tissue cells by adjusting bronchial epithelial cell characteristic	*Berberis vulgaris*	(Yang et al. 2024b)
		Antidepressant effect		(Tang et al. 2024)
Aporphine	Taspine	Potential drug candidate for inflammatory and autoimmune disease therapy	*Caulophyllum* spp.	(Yang et al. 2023)
	Nuciferine	Antioxidant and antiproliferative effect, ingredient for gummy bear supplement	*Nelumbo nucifera*	(Veerichetty et al. 2024)
		Hypoglycaemic and hypolipidaemic activity	*Xylopia aethiopica*	(Liu et al. 2023b)
Benzo-phenanthridine	Sanguinarine	Preclinical therapeutic for regulating microglial function and promote myelin repair	*Sanguinaria canadensis, Argemone mexicana*	(Wang et al. 2024)
	Corynoline	Inhibits oesophageal squamous cell carcinoma progression	*Corydalis bungeana*	(Shi et al. 2024)
		Induces anti-mitosis and polyploidy in cancer cell	*Corydalis longicalcarata*	(Yan et al. 2022)
Protopine		Reduces ovalbumin-induced asthma		(Yang et al. 2024a)

(continued)

Table 7.5 (continued)

Sub-class of isoquinoline alkaloid	Types	Major pharmacological activity	Occurrence	References
Benzylisoquinoline	Liensinine	Inhibiting lung fibroblasts		(Ren et al. 2023)
	Neferine	Reversal of chemo-resistance and increased sensitivity of the cancer cells towards anti-cancer drug	*Nelumbo nucifera*	(Plazas et al. 2022)
	Isoliensinine	Strong effect on several types of tumour cells, induces autophagic death of cervical cancer cell, promotes cisplatin uptake	*Nelumbo nucifera*	(Cheng et al. 2021)
	Tetrandrine			

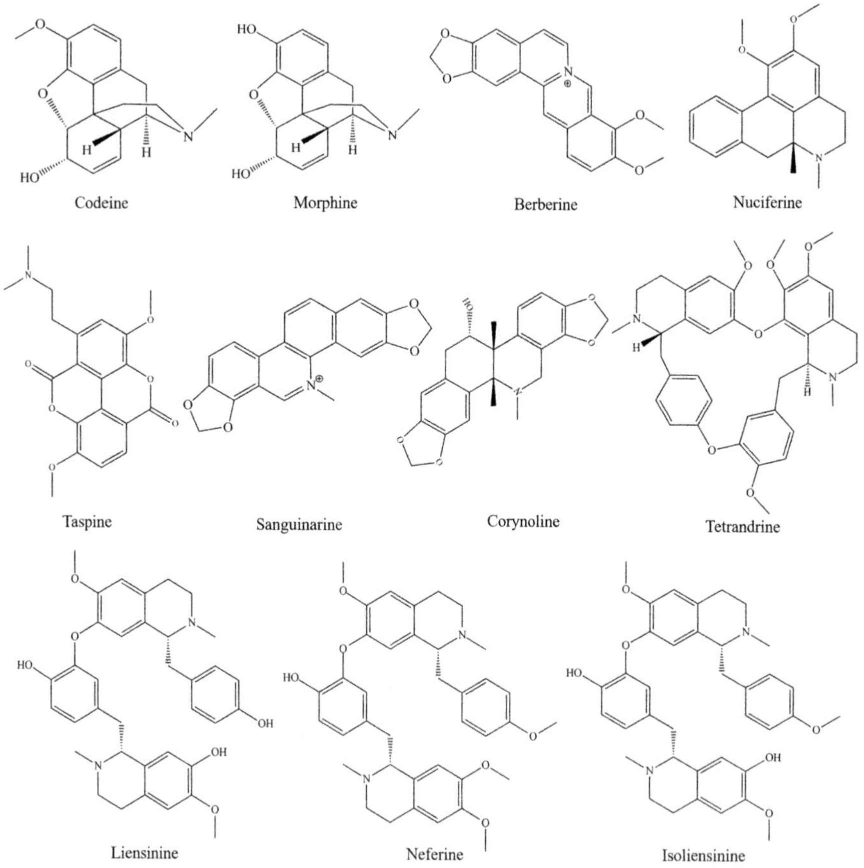

Fig. 7.13 Isoquinoline alkaloids

7.4.1.2 Tropane Alkaloids

Tropane alkaloids have a core tropane skeleton, a two-ringed (pyrrolidine and piperidine) structure with structural variations due to the esterification of tropine with a variety of acids. These are abundant in Solanaceae, Brassicaceae, Convolvulaceae, Euphorbiaceae, and Erythroxylaceae families. Some of the most common examples include atropine, cocaine, hyoscyamine, scopolamine, hyoscine, hygrine, and calystegines. The effects of tropane alkaloids range from fever, nausea, dry mouth, and some skin conditions to very serious respiratory depression, paralysis, and seizures (Torović et al. 2023; Kukula-Koch and Widelski 2017). Tropane alkaloids have various adverse effects including fever, dizziness, tachycardia, tachypnoea, hyperpyrexia, warm dry skin and mucous membranes, mydriasis, hallucinations, and many serious complications like seizures, paralysis, respiratory depression, and cardiac disorders. Apart from these, in various conditions like pregnancy, urinary tract diseases, glaucoma, and prostatic hypertrophy it is suggested that the drugs containing such compounds should be avoided. Apart from their toxic effects, they also have addictive properties (Table 7.6 and Fig. 7.14).

7.4.1.3 Pyrrolizidine Alkaloids

Echimidine are also known as necine bases, which are based on pyrrolizidine structure. It is found in roughly 6000 species of plants which are distributed across more than 12 families of angiosperms like Asteraceae (genera *Senecio*, *Eupatorium*, *Ageratum*), Fabaceae (genera *Crotalaria*, *Chromolaena*, *Lotononis*), Orchidaceae

Table 7.6 Types of tropane alkaloids

Types	Major pharmacological activity	Occurrence	References
Hyoscyamine	Muscarinic receptor blocker and prevent hypervagotonia		(Mai and Kusumoto 2022)
Atropine	Increases systemic blood pressure and pupillary dilation	*Datura stramonium*	(Wasserman et al. 2024)
	Inhibits myopia progression by serotonergic receptor activity		(Thomson et al. 2023)
Scopolamine	Causes neurodegeneration in the brain		(Hawiset et al. 2023)
	Anticholinergic toxicity		(Huerta et al. 2024)
Calystegines	Competitive inhibitors of β-glucosidase and α-galactosidase enzymes	*Ipomoea carnea, ipomoea trifida*	(Salinas et al. 2019)
Anisodamine	Relieves microvascular spasms, reduces stress on endoplasmic reticulum in bleomycin-induced pulmonary fibrosis, reduces oxidative stress, inflammation, and apoptosis	*Anisodus tanguticus*	(Zhang et al. 2023)

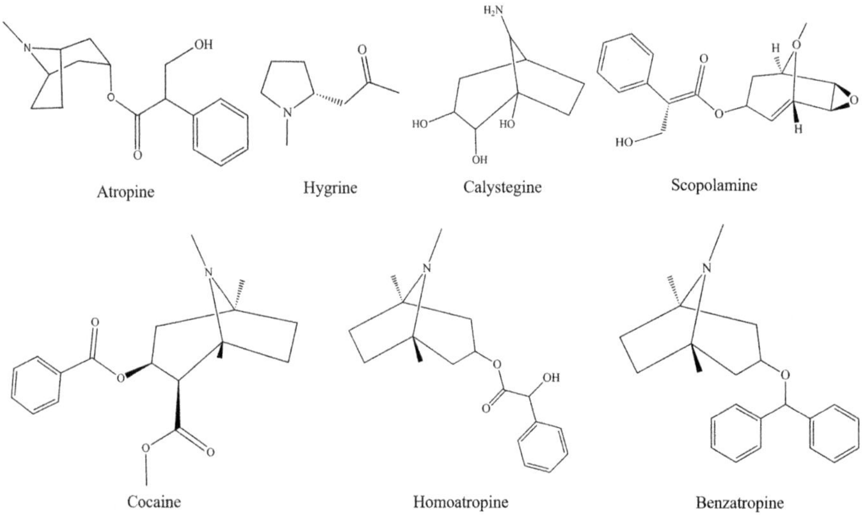

Fig. 7.14 Tropane alkaloids

Table 7.7 Types of pyrrolizidine alkaloids

Types	Major pharmacological activity	Occurrence	References
Echimidine	Impairs cardiac development in zebrafish larvae by disturbing calcium transport	Boraginaceae, Orchidaceae,	(Lin et al. 2023)
Platyphylline	Blindness, weight loss	*Senecio* spp.	
Senecionine	Hepatotoxic		(Fu 2017)

(genus *Liparis*), Boraginaceae, Convolvulaceae, Poaceae, and Lamiaceae to name some of them. More than 650 pyrrolizidine alkaloids have been characterised among known families, more than half of which have hepatotoxic, genotoxic, and carcinogenic effects. The compound becomes toxic when bioactivated by mixing function oxidases to toxic dehydropyrrolizidine alkaloids (pyrroles). Most activation occurs in the liver, which most often results in hepatic damage. The majority of Pyrrolizidine Alkaloids (PA) are considered to be hepatotoxic to mammals, and some are considered carcinogenic. The toxic species damage the adjacent tissues because they are potent electrophiles, and they bind to and cross-link DNA, proteins, amino acids, and glutathione. But acute PA toxicity is highly variable and pyrrole-specific (Stegelmeier 2011; Hardie et al. 2023) (Table 7.7 and Fig. 7.15).

7.4.1.4 Purine Alkaloids

Purine alkaloids are derivatives of xanthines and uric acid such as methylxanthines and methyluric acid. Purine alkaloids provide chemical defence to the plants against insects, vertebrates, fungi, bacteria, and viruses. Leaves and seeds of tea (*Camellia*

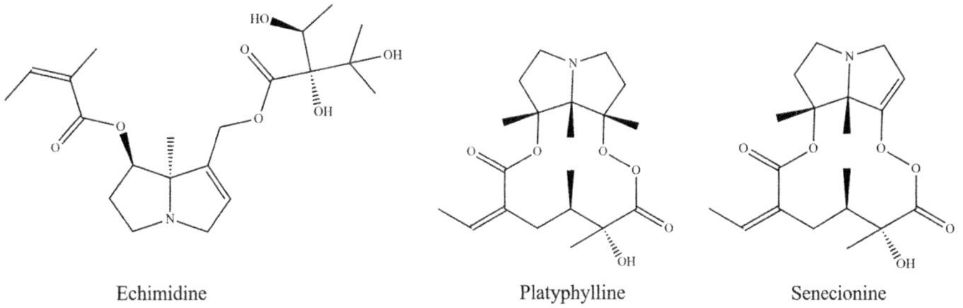

Fig. 7.15 Pyrrolizidine alkaloids

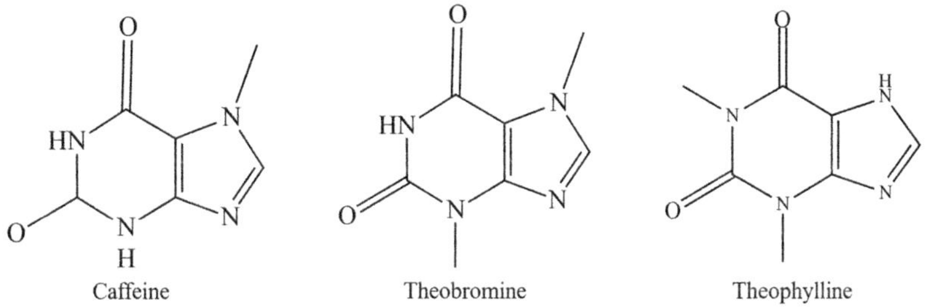

Fig. 7.16 Purine Alkaloids

sinensis) and coffee (*Coffea arabica*) contain caffeine (1,3,7-trimethylxanthine). Caffeine accumulation in young leaves, flowers, and fruits of tea and coffee plants protect young tissues from pathogens (Ashihara et al. 2013). Theacrine is an abundant component in *Camellia assamica* var. kucha which has anti-influenza, anti-inflammation, and ergogenic effects (Lin et al. 2020a). Theobromine (3,7-dimethylxanthine) and theophylline (1,3-dimethylxanthine) are also widely found in purine alkaloids. Theobromine and theophylline have antioxidant properties and help in expansion of the coronary artery, the myocardial excitation, and bronchial smooth muscle relaxation. Recent studies reported better safety and efficacy of Theobromine in contrast with caffeine (Wu et al. 2019; de Veer et al. 2019) (Fig. 7.16).

7.4.1.5 Pyridine Alkaloids

Alkaloids with pyridine form a key component in bioactive compounds. The nitrogen atom of pyridine plays a crucial pharmacological role in many drugs. Cycloalkane pyridines are intermediates of alkaloid biosynthesis. Some of the pyridine alkaloids affecting CNS activity include nicotine, anabasine, cotinine, and anabasamine.

Nicotine was first isolated from dried leaves of *Nicotiana tabacum* from the Solanaceae family and also present in lower amounts in other genera of Solanaceae

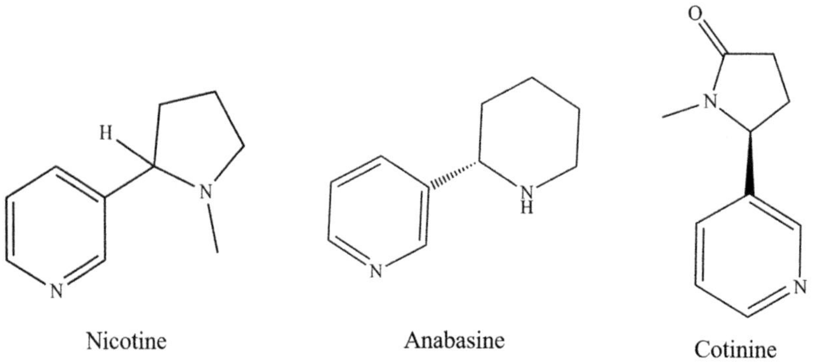

Fig. 7.17 Pyridine Alkaloids

such as tomato, potato, and green peppers. Nicotine is carried to lungs through smoking from where it enters the bloodstream and into the brain. Nicotine has several effects including stress and anxiety reduction, intense arousal, and increased pain thresholds. It also enhances reaction time, attention, and learning skills. These positive reinforcement impacts lead to tobacco addiction. McGrath-Morrow et al. (2020) reported the effect of rapid increased usage of non-combustible nicotine products such as electronic cigarettes (e-cigarettes). The study reported smoking in the first trimester increases birth defects such as limb reduction, gastroschisis, oral clefts and is a risk factor for sudden infant death syndrome (McGrath-Morrow et al. 2020; Lin et al. 2020b).

Anabasine was first isolated from *Anabasis aphylla* and also a CNS stimulant. In lower dosages, it strengthens respiration and raises arterial pressure (Seilkhanov et al. 2016).

Cotinine is a marker for tobacco exposure because of the adequate dose–response relationship of plasma cotinine to daily cigarette consumption. Cotinine is present in *Nicotiana tabacum* and *Duboisia hopwoodii*. Recent studies show cotinine as a potential candidate for treating nicotine addiction (Dominiak et al. 1985; Lin et al. 2020b) (Fig. 7.17).

7.5 Tannins

Tannins are a larger group of plant secondary metabolites which are widely distributed among medicinal plants. Chemically tannins are water-soluble polyphenols which are used in the tanning of animal skins and hides. Tannins range in molecular weights from 300 to 3000 Da. Tannins can form strong complexes with proteins, starches, and other macromolecules. Notable sources for tannins include *Acacia mearnsii, Quercus* spp., *Caesalpinia coriaria, Pinus radiata, and Pinus nigra.*

Tannins are broadly classified into two classes which include hydrolysable tannins and condensed tannins. Hydrolysable tannins include esters of gallic and

ellagic acid with a sugar mainly glucose which are called gallotannins and ellagitannins, respectively. Hydrolysable tannins are not polymeric in nature instead form complex structures where phenolic moieties covalently bonded to the carbohydrates. The relatively simpler gallotannin is the component of commercially available Chinese tannins extracted from *Rhus semialata*. Gallotannins are also found in species of *Acer, Quercus, Rhus* and *Pelargonium*. Ellagitannins are synthesised by oxidative reactions between gallic acid units and are more complex than gallotannins. Castalagin and vescalagin are ellagitannins extracted from chestnut wood (Pizzi 2008; Mueller-Harvey 2001). Red wine storage barrels are made up of heartwoods of *Quercus petraea*, *Quercus robur*, and *Quercus alba* containing ellagitannins which affect sensory profile and colour stability of red wine and also protect against oxidation (Ky et al. 2016).

Condensed tannins also called proanthocyanidins are flavanol polymers. Condensed tannins are the major commercial tannins due to their high reactivity towards aldehyde and other reagents. This chemical nature renders the usage of condensed tannins not only in leather tanning but also used in the preparation of adhesives and resins. Some of the commercial tannins are extracted from *Acacia* (wattle or mimosa bark extract), *Schinopsis* (quebracho wood extract), *Tsuga* (hemlock bark extract), *Rhus* (sumac extract), and *Pinus* (bark extract) species (Fig. 7.18).

The simplest and traditional method of tannin extraction includes solid-liquid extraction. Solvent enters into a tannin rich cell which dissolves and takes out the tannins in extract form. Polar solvents such as water and methanol give high extraction when compared to organic solvents like hexane. Sodium carbonate and sodium bisulfite increase the extraction efficiency by lowering the viscosity of the extract. Alkaline pH increases the yield.

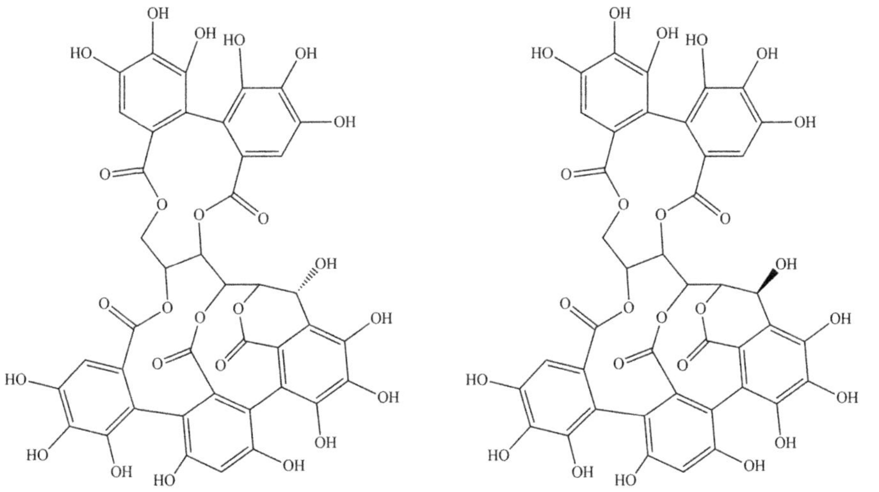

Castalagin Vescalagin

Fig. 7.18 Tannins

Pressurised water extraction uses water at subcritical conditions between atmospheric boiling point (100 °C, 0.1 MPa) and critical point (374 °C, 22.1 MPa). The subcritical conditions increase the transfer of tannins from the sample matrix. This method is different from solid-liquid extraction as the temperature used is above boiling point and pressure applied is also above atmospheric pressure which keeps the water in liquid state. The advantage of pressurised water extraction is extraction of tannins of different polarity by modifying temperature and pressure.

7.6 Peptides as Phytotoxins

Proteins are versatile molecules which perform an array of functions including defence, enzymatic action, as nutrient, as storage, as signalling, and regulatory molecule. Plant peptides act as the first line of defence providing innate immunity. Plant peptides act as nonreceptor ligands and interact with plasma membranes (Salas et al. 2015). The class of toxic plant protein includes lectins, ricin, thionin, cyclotides, urease, arcelins, α-amylase inhibitors, and pore-forming toxins.

Lectins are proteins which selectively bind to glycans which are present either in free form or are part of glycoprotein and glycolipid. Lectins act as a defence molecule by exhibiting diverse structure and substrate specificity. Lectins are produced in high amounts and are constitutively expressed in seeds and storage tissue. Concanavalin A is the first lectin to be isolated from seeds of jack bean (*Canavalia ensiformis*). Majority of plant lectins are made up of one or more lectin domains coupled to unrelated domains. Plant uses these varieties of lectin domains to prevent pathogen attack. Amaranthin is a lectin present in seeds of *Amaranthus caudatus* which provide resistance against aphids when expressed in transgenic tobacco, potato, and cotton. Jacalin-related lectins include T-antigen disaccharide binding lectin domain isolated from jackfruit (*Artocarpus integrifolia*). Jaclin-related lectins provide disease resistance, help in wounding and abiotic stress signalling (Lannoo and Van Damme 2014).

Ricin is a water-soluble glycoprotein obtained from *Ricinus communis* that acts as toxin by inhibiting protein synthesis. Ricin has been exploited as a bioweapon due to easy purification procedures of rough ricin. However, ricin also has a medical application as immunotoxin, eliminating cancer cells. The expression of fusion gene containing sequences encoding anti-CD19-FVS191, cathepsin D proteinase digestion site, and r-Ricin Toxin-A chain results in ricin containing recombinant immunotoxin. Murine monoclonal anti-melanoma antibody coupled to Ricin Toxin-A chain showed promising results against metastatic malignant melanoma in clinical trials (Polito et al. 2019). Ricin has no available antidote, and studies are done for better management of the poisoning (Ataei et al. 2023). Abrin is another similar toxic peptide obtained from *Abrus precatorius*. It can be used as a lethal biotoxic agent as it can stop protein synthesis in cells (Liang et al. 2021).

Plant-produced antimicrobial peptide works by lysing the cell upon interaction with lipid bilayer of plasma membrane. Thionin is one such peptide synthesised

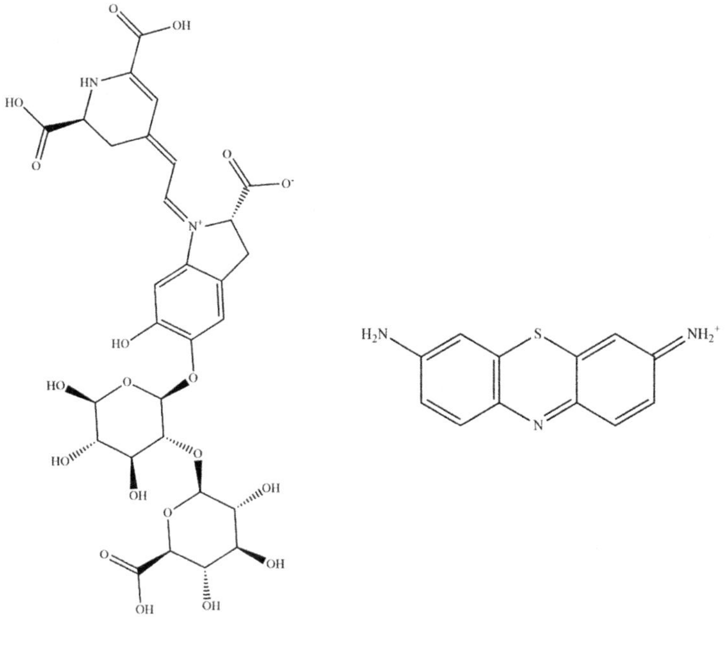

Amaranthin Thionin

Fig. 7.19 Peptide toxins

from *Pyrularia pubera* which causes permeabilization of plasma membrane resulting in immediate death of parasites such as *Leishmania donovani* (Berrocal-Lobo et al. 2009).

Jaburetox is a recombinant insecticidal and membrane-active peptide produced from jack bean (*Canavalia ensiformis*) urease. Recombinant jaburetox disrupts liposomes and has entomotoxic effects (Martinelli et al. 2014).

Cyclotides are macrocyclic peptides produced from plants of Rubiaceae and Violaceae. Cyclotides are inseticidal, anthelmintic, anticancerous, and antimicrobial agents (de Veer et al. 2019). Plan et al. (2008) reported molluscicidal activity of crude cyclotide extracted from *Oldenlandia affinis* and *Viola odorata* (Plan et al. 2008) (Fig. 7.19).

7.7 Conclusion

Thus, compounds of secondary metabolism of plants yield complex chemical compounds which are potential toxins, which at the same time at usually lower doses, have several therapeutic effects. Therefore, plant compounds are double-edged swords and must be put to use wisely for our benefit.

References

Adhikary S, Dasgupta N (2023) Role of secondary metabolites in plant homeostasis during biotic stress. Biocatal Agric Biotechnol 50:102712. https://doi.org/10.1016/j.bcab.2023.102712

Alugoju P, Tencomnao T (2023) Production and role of plants secondary metabolites under various environmental pollution. In: Plants and their interaction to environmental pollution. Elsevier, pp 379–410

Aribi N, Denis B, Kilani-Morakchi S, Joly D (2020) Azadirachtin, a natural pesticide with multiple effects. Med Sci 36:44–49. https://doi.org/10.1051/medsci/2019268

Ashihara H, Yokota T, Crozier A (2013) Biosynthesis and catabolism of purine alkaloids. In: New light on alkaloid biosynthesis and future prospects. Elsevier, pp 111–138

Ataei M, Shirazi FM, Mehrpour O (2023) Ricin and other toxalbumins. In: Reference module in biomedical sciences. Elsevier

Azzalini E, Bernini M, Vezzoli S et al (2019) A fatal case of self-poisoning through the ingestion of oleander leaves. J Forensic Leg Med 65:133–136. https://doi.org/10.1016/j.jflm.2019.05.016

Bai X, Tang J (2020) Myrcene exhibits antitumor activity against lung cancer cells by inducing oxidative stress and apoptosis mechanisms. Nat Prod Commun 15:1934578X2096118. https://doi.org/10.1177/1934578X20961189

Bartnik M, Facey PC (2017) Glycosides. In: Pharmacognosy. Elsevier, pp 101–161

Berrocal-Lobo M, Molina A, Rodríguez-Palenzuela P et al (2009) Leishmania donovani: thionins, plant antimicrobial peptides with leishmanicidal activity. Exp Parasitol 122:247–249. https://doi.org/10.1016/j.exppara.2009.03.019

Carocho M, Heleno SA, Barros L (eds) (2023) Natural secondary metabolites: from nature, through science, to industry. Springer, Cham

Cheng Y, Li H-L, Zhou Z-W et al (2021) Isoliensinine: a natural compound with "drug-like" potential. Front Pharmacol 12:630385. https://doi.org/10.3389/fphar.2021.630385

Cho M, So I, Chun JN, Jeon J-H (2016) The antitumor effects of geraniol: modulation of cancer hallmark pathways (review). Int J Oncol 48:1772–1782. https://doi.org/10.3892/ijo.2016.3427

Cox-Georgian D, Ramadoss N, Dona C, Basu C (2019) Therapeutic and medicinal uses of terpenes. In: Joshee N, Dhekney SA, Parajuli P (eds) Medicinal plants: from farm to pharmacy. Springer, Cham, pp 333–359

Crews C, Clarke D (2014) Natural toxicants: naturally occurring toxins of plant origin. In: Encyclopedia of food safety. Elsevier, pp 261–268

Cui J, Hu L, Shi W et al (2019) Design, synthesis and anti-platelet aggregation activity study of ginkgolide-1,2,3-triazole derivatives. Molecules 24:2156. https://doi.org/10.3390/molecules24112156

da Silva CEH, Gosmann G, de Andrade SF (2021) Limonene and perillyl alcohol derivatives: synthesis and anticancer activity. Mini Rev Med Chem 21:1813–1829. https://doi.org/10.2174/1389557521666210212150504

da Silveira Vasconcelos M, de Oliveira LMN, Nunes-Pinheiro DCS et al (2020) Analysis of tetraterpenes and tetraterpenoids (carotenoids). In: Recent advances in natural products analysis. Elsevier, pp 427–456

de Medeiros CAC, Pinto Â d V, de Oliveira JC et al (2021) Evaluating the antifungal activity of α-bisabolol in association with NaCl on Fusarium oxysporum in maize grains. Curr Microbiol 78:604–610. https://doi.org/10.1007/s00284-020-02313-8

de Veer SJ, Kan M-W, Craik DJ (2019) Cyclotides: from structure to function. Chem Rev 119:12375–12421. https://doi.org/10.1021/acs.chemrev.9b00402

Dey P, Kundu A, Kumar A et al (2020) Analysis of alkaloids (indole alkaloids, isoquinoline alkaloids, tropane alkaloids). In: Recent advances in natural products analysis. Elsevier, pp 505–567

Dominiak P, Fuchs G, von Toth S, Grobecker H (1985) Effects of nicotine and its major metabolites on blood pressure in anaesthetized rats. Klin Wochenschr 63:90–92. https://doi.org/10.1007/BF01733074

Fu PP (2017) Pyrrolizidine alkaloids: metabolic activation pathways leading to liver tumor initiation. Chem Res Toxicol 30:81–93. https://doi.org/10.1021/acs.chemrestox.6b00297

González-Cardenete MA, Rivas F, Basset R et al (2021) Biological profiling of semisynthetic C19-functionalized ferruginol and sugiol analogues. Antibiotics 10:184. https://doi.org/10.3390/antibiotics10020184

Gu W, Rebsdorf A, Anker C et al (2019) Steviol glucuronide, a metabolite of steviol glycosides, potently stimulates insulin secretion from isolated mouse islets: studies in vitro. Endocrinol Diab Metab 2:e00093. https://doi.org/10.1002/edm2.93

Günthardt BF, Hollender J, Hungerbühler K et al (2018) Comprehensive toxic plants-phytotoxins database and its application in assessing aquatic micropollution potential. J Agric Food Chem 66:7577–7588. https://doi.org/10.1021/acs.jafc.8b01639

Guo K, Liu Y, Li S-H (2021) The untapped potential of plant sesterterpenoids: chemistry, biological activities and biosynthesis. Nat Prod Rep 38:2293–2314. https://doi.org/10.1039/d1np00021g

Hardie AG, Olifant K, Smith JFN, Hoffman JE (2023) Scientific evidence of soil transfer of pyrrolizidine alkaloids originating from weed species to rooibos tea. S Afr J Bot 157:159–166. https://doi.org/10.1016/j.sajb.2023.03.053

Hawiset T, Sriraksa N, Kamsrijai U et al (2023) Neuroprotective effect of Tiliacora triandra (Colebr.) Diels leaf extract on scopolamine-induced memory impairment in rats. Heliyon 9:e22545. https://doi.org/10.1016/j.heliyon.2023.e22545

He W, Zhai X, Su J et al (2019) Antiviral activity of germacrone against pseudorabies virus in vitro. Pathogens 8:258. https://doi.org/10.3390/pathogens8040258

Hu H, Dong Z, Wang X et al (2019) Dehydrocorydaline inhibits cell proliferation, migration and invasion via suppressing MEK1/2-ERK1/2 cascade in melanoma. Onco Targets Ther 12:5163–5175. https://doi.org/10.2147/OTT.S183558

Huerta N, Stylli J, Mahar S et al (2024) Supraventricular tachycardia associated with transdermal scopolamine: a case of commonalities leading to an uncommon toxicity. CJC Open 6:569–572. https://doi.org/10.1016/j.cjco.2023.12.004

Islam MT, Khalipha ABR, Bagchi R et al (2019) Anticancer activity of thymol: a literature-based review and docking study with emphasis on its anticancer mechanisms. IUBMB Life 71:9–19. https://doi.org/10.1002/iub.1935

Iwasaki K, Zheng Y-W, Murata S et al (2016) Anticancer effect of linalool via cancer-specific hydroxyl radical generation in human colon cancer. World J Gastroenterol 22:9765–9774. https://doi.org/10.3748/wjg.v22.i44.9765

Jha NK, Sharma C, Hashiesh HM et al (2021) β-caryophyllene, a natural dietary CB2 receptor selective cannabinoid can be a candidate to target the trinity of infection, immunity, and inflammation in COVID-19. Front Pharmacol 12:590201. https://doi.org/10.3389/fphar.2021.590201

Jiang Y-Y, Shui J-C, Zhang B-X et al (2020) The potential roles of artemisinin and its derivatives in the treatment of type 2 diabetes mellitus. Front Pharmacol 11:585487. https://doi.org/10.3389/fphar.2020.585487

Jin J, Zhou M, Wang X et al (2021) Triptolidenol, isolated from Tripterygium wilfordii, disrupted NF-κB/COX-2 pathway by targeting ATP-binding sites of IKKβ in clear cell renal cell carcinoma. Fitoterapia 148:104779. https://doi.org/10.1016/j.fitote.2020.104779

Kamaraj C, Balasubramani G, Siva C et al (2017) Ag nanoparticles synthesized using β-caryophyllene isolated from Murraya koenigii: antimalarial (plasmodium falciparum 3D7) and anticancer activity (A549 and HeLa cell lines). J Clust Sci 28:1667–1684. https://doi.org/10.1007/s10876-017-1180-6

Khaleel C, Tabanca N, Buchbauer G (2018) α-terpineol, a natural monoterpene: a review of its biological properties. Open Chem 16:349–361. https://doi.org/10.1515/chem-2018-0040

Kukula-Koch WA, Widelski J (2017) Alkaloids. In: Pharmacognosy. Elsevier, pp 163–198

Ky I, Le Floch A, Zeng L et al (2016) Tannins. In: Encyclopedia of food and health. Elsevier, pp 247–255

Lannoo N, Van Damme EJM (2014) Lectin domains at the frontiers of plant defense. Front Plant Sci 5:397. https://doi.org/10.3389/fpls.2014.00397

Liang L-H, Yang Y, Geng S et al (2021) Rapid differential detection of abrin isoforms by an acetonitrile- and ultrasound-assisted on-bead trypsin digestion coupled with LC-MS/MS analysis. Toxins 13:358. https://doi.org/10.3390/toxins13050358

Lin P-R, Kuo P-C, Li Y-C et al (2020a) Theacrine and strictinin, two major ingredients for the anti-influenza activity of Yunnan Kucha tea. J Ethnopharmacol 262:113190. https://doi.org/10.1016/j.jep.2020.113190

Lin SX, Curtis MA, Sperry J (2020b) Pyridine alkaloids with activity in the central nervous system. Bioorg Med Chem 28:115820. https://doi.org/10.1016/j.bmc.2020.115820

Lin T, Zhou L, Chen Z et al (2023) Exposure to echimidine impairs the heart development and function of zebrafish larvae. Ecotoxicol Environ Saf 266:115574. https://doi.org/10.1016/j.ecoenv.2023.115574

Liu J, Wang C, Wang Q et al (2023a) Triterpenoids and saponins in drug discovery. In: Privileged scaffolds in drug discovery. Elsevier, pp 613–671

Liu Y, Li Y, Muema FW et al (2023b) Aporphine alkaloids identified from Xylopia aethiopica and their potential hypoglycemic and hypolipidemic activities. J Funct Foods 106:105601. https://doi.org/10.1016/j.jff.2023.105601

Ludwiczuk A, Skalicka-Woźniak K, Georgiev MI (2017) Terpenoids. In: Pharmacognosy. Elsevier, pp 233–266

Mai W, Kusumoto F (2022) Advanced atrioventricular block due to hypervagotonia: treatment with hyoscyamine. HeartRhythm Case Rep 8:343–346. https://doi.org/10.1016/j.hrcr.2022.02.004

Martinelli AHS, Kappaun K, Ligabue-Braun R et al (2014) Structure-function studies on jaburetox, a recombinant insecticidal peptide derived from jack bean (Canavalia ensiformis) urease. Biochim Biophys Acta 1840:935–944. https://doi.org/10.1016/j.bbagen.2013.11.010

Mathema VB, Koh Y-S, Thakuri BC, Sillanpää M (2012) Parthenolide, a sesquiterpene lactone, expresses multiple anti-cancer and anti-inflammatory activities. Inflammation 35:560–565. https://doi.org/10.1007/s10753-011-9346-0

McGrath-Morrow SA, Gorzkowski J, Groner JA et al (2020) The effects of nicotine on development. Pediatrics 145. https://doi.org/10.1542/peds.2019-1346

Mn B (2022) Scientific findings: the amazing use of essential oils and their related terpenes as natural preservatives to improve the ShelfLife of food. FSNT 5:1–12. https://doi.org/10.23880/fsnt-16000215

Mosayyebi B, Imani M, Mohammadi L et al (2020) An update on the toxicity of cyanogenic glycosides bioactive compounds: possible clinical application in targeted cancer therapy. Mater Chem Phys 246:122841. https://doi.org/10.1016/j.matchemphys.2020.122841

Mueller-Harvey I (2001) Analysis of hydrolysable tannins. Anim Feed Sci Technol 91:3–20. https://doi.org/10.1016/S0377-8401(01)00227-9

Penta S (2016) Introduction to coumarin and SAR. In: Advances in structure and activity relationship of coumarin derivatives. Elsevier, pp 1–8

Petrović J, Stojković D, Soković M (2019) Terpene core in selected aromatic and edible plants: natural health improving agents. Adv Food Nutr Res 90:423–451. https://doi.org/10.1016/bs.afnr.2019.02.009

Pina LTS, Serafini MR, Oliveira MA et al (2022) Carvone and its pharmacological activities: a systematic review. Phytochemistry 196:113080. https://doi.org/10.1016/j.phytochem.2021.113080

Pizzi A (2008) Tannins: major sources, properties and applications. In: Monomers, polymers and composites from renewable resources. Elsevier, pp 179–199

Plan MRR, Saska I, Cagauan AG, Craik DJ (2008) Backbone cyclised peptides from plants show molluscicidal activity against the rice pest Pomacea canaliculata (golden apple snail). J Agric Food Chem 56:5237–5241. https://doi.org/10.1021/jf800302f

Plazas E, Avila MMC, Muñoz DR, Cuca S (2022) Natural isoquinoline alkaloids: pharmacological features and multi-target potential for complex diseases. Pharmacol Res 177:106126. https://doi.org/10.1016/j.phrs.2022.106126

Polito L, Bortolotti M, Battelli MG et al (2019) Ricin: an ancient story for a timeless plant toxin. Toxins 11:324. https://doi.org/10.3390/toxins11060324

Rashad NM, Abdelsamad MAE, Amer AM et al (2019) The impact of stevioside supplementation on glycemic control and lipid profile in patients with type 2 diabetes: a controlled clinical trial. Egypt J Intern Med 31:22–30. https://doi.org/10.4103/ejim.ejim_68_18

Rashan LJ, Özenver N, Boulos JC et al (2023) Molecular modes of action of an aqueous nerium oleander extract in cancer cells in vitro and in vivo. Molecules 28:1871. https://doi.org/10.3390/molecules28041871

Ren H-L, Zhang J-H, Xiao J-H (2023) Benzylisoquinoline alkaloids inhibit lung fibroblast activation mainly via inhibiting TGF-β1/Smads and ERK1/2 pathway proteins. Heliyon 9:e16849. https://doi.org/10.1016/j.heliyon.2023.e16849

Salas CE, Badillo-Corona JA, Ramírez-Sotelo G, Oliver-Salvador C (2015) Biologically active and antimicrobial peptides from plants. Biomed Res Int 2015:102129. https://doi.org/10.1155/2015/102129

Salehi B, Staniak M, Czopek K et al (2019) The therapeutic potential of the labdane diterpenoid forskolin. Appl Sci 9:4089. https://doi.org/10.3390/app9194089

Salinas LM, Balseiro A, Jirón W et al (2019) Neurological syndrome in goats associated with Ipomoea trifida and Ipomoea carnea containing calystegines. Toxicon 157:8–11. https://doi.org/10.1016/j.toxicon.2018.11.291

Sampaio LA, Pina LTS, Serafini MR et al (2021) Antitumor effects of carvacrol and thymol: a systematic review. Front Pharmacol 12:702487. https://doi.org/10.3389/fphar.2021.702487

Seilkhanov TM, Nurkenov OA, Isaeva AZ et al (2016) Preparation and NMR spectroscopic studies of the supramolecular inclusion complex of anabasine and β-cyclodextrin. Chem Nat Compd 52:1066–1068. https://doi.org/10.1007/s10600-016-1862-6

Shi Y, Yuan Q, Chen Y et al (2024) Corynoline inhibits esophageal squamous cell carcinoma growth via targeting Pim-3. Phytomedicine 123:155235. https://doi.org/10.1016/j.phymed.2023.155235

Shriti S, Kuldeep Prasad A, Narasimha Sudheer W, Praveen N (2023) Hairy roots as a potential source for the production of rosmarinic acid from genus salvia. Plant Sci Today 10:sp2. https://doi.org/10.14719/pst.2541

Srijiwangsa P, Ponnikorn S, Na-Bangchang K (2018) Effect of β-Eudesmol on NQO1 suppression-enhanced sensitivity of cholangiocarcinoma cells to chemotherapeutic agents. BMC Pharmacol Toxicol 19:32. https://doi.org/10.1186/s40360-018-0223-4

Stegelmeier BL (2011) Pyrrolizidine alkaloid-containing toxic plants (Senecio, Crotalaria, Cynoglossum, Amsinckia, Heliotropium, and Echium spp.). Vet Clin North Am Food Anim Pract 27:419–428. https://doi.org/10.1016/j.cvfa.2011.02.013

Surendran S, Qassadi F, Surendran G et al (2021) Myrcene-what are the potential health benefits of this flavouring and aroma agent? Front Nutr 8:699666. https://doi.org/10.3389/fnut.2021.699666

Tang Y, Su H, Nie K et al (2024) Berberine exerts antidepressant effects in vivo and in vitro through the PI3K/AKT/CREB/BDNF signaling pathway. Biomed Pharmacother 170:116012. https://doi.org/10.1016/j.biopha.2023.116012

Thomson K, Karouta C, Weber D et al (2023) The role of the serotonergic system in atropine's anti-myopic effects. Biomed Pharmacother 167:115542. https://doi.org/10.1016/j.biopha.2023.115542

Torović L, Bursić V, Vuković G (2023) Health concerns associated to tropane alkaloids in maize food products in Serbia. Heliyon 9:e19404. https://doi.org/10.1016/j.heliyon.2023.e19404

Veerichetty V, Saravanabavan I, Pradeep A, Abiraamasundari R (2024) Development of gummy bear supplements and in vitro exploration of antioxidant and antiproliferative potential of Nuciferine. J Ayurveda Integr Med 15:100868. https://doi.org/10.1016/j.jaim.2023.100868

Viveros-Paredes JM, González-Castañeda RE, Gertsch J et al (2017) Neuroprotective effects of β-caryophyllene against dopaminergic neuron injury in a murine model of parkinson's disease induced by MPTP. Pharmaceuticals 10:60. https://doi.org/10.3390/ph10030060

Wang B, Ge Z, Cheng Z, Zhao Z (2017) Tanshinone IIA suppresses the progression of atherosclerosis by inhibiting the apoptosis of vascular smooth muscle cells and the proliferation and

migration of macrophages induced by ox-LDL. Biol Open 6:489–495. https://doi.org/10.1242/bio.024133

Wang L, Zhang Y, Wang C et al (2019) A natural product with high affinity to sigma and 5-HT7 receptors as novel therapeutic drug for negative and cognitive symptoms of schizophrenia. Neurochem Res 44:2536–2545. https://doi.org/10.1007/s11064-019-02873-7

Wang D-J, Wang X, Li S et al (2024) Sanguinarine modulates microglial function via PPARγ activation and protects against CNS demyelination. Int Immunopharmacol 127:111408. https://doi.org/10.1016/j.intimp.2023.111408

Wasserman BN, Massenzio E, Lee K, Plager DA (2024) Systolic hypertension as side effect of topical low dose atropine drops. Am J Ophthalmol Case Rep 33:102004. https://doi.org/10.1016/j.ajoc.2024.102004

Wu F, Liu R, Shen X et al (2019) Study on the interaction and antioxidant activity of theophylline and theobromine with SOD by spectra and calculation. Spectrochim Acta A Mol Biomol Spectrosc 215:354–362. https://doi.org/10.1016/j.saa.2019.03.001

Yan Z, Shi Q, Liu X et al (2022) The phytochemical, corynoline, diminishes Aurora kinase B activity to induce mitotic defect and polyploidy. Biomed Pharmacother 147:112645. https://doi.org/10.1016/j.biopha.2022.112645

Yang W, Chen X, Li Y et al (2020) Advances in pharmacological activities of terpenoids. Nat Prod Commun 15:1934578X2090355. https://doi.org/10.1177/1934578X20903555

Yang J, Pan Y, Zeng X et al (2023) Discovery of novel aporphine alkaloid derivative as potent TLR2 antagonist reversing macrophage polarization and neutrophil infiltration against acute inflammation. Acta Pharm Sin B 13:3782–3801. https://doi.org/10.1016/j.apsb.2023.05.034

Yang J, Zhang M, Luo Y et al (2024a) Protopine ameliorates OVA-induced asthma through modulatingTLR4/MyD88/NF-κB pathway and NLRP3 inflammasome-mediated pyroptosis. Phytomedicine 126:155410. https://doi.org/10.1016/j.phymed.2024.155410

Yang W, Yang T, Huang B et al (2024b) Berberine improved the microbiota in lung tissue of colon cancer and reversed the bronchial epithelial cell changes caused by cancer cells. Heliyon 10:e24405. https://doi.org/10.1016/j.heliyon.2024.e24405

Zhang Z, Chen X, Chen H et al (2016) Anti-inflammatory activity of β-patchoulene isolated from patchouli oil in mice. Eur J Pharmacol 781:229–238. https://doi.org/10.1016/j.ejphar.2016.04.028

Zhang L, Bao M, Liu B et al (2020) Effect of andrographolide and its analogs on bacterial infection: a review. Pharmacology 105:123–134. https://doi.org/10.1159/000503410

Zhang X, Zhong W, Ma X et al (2021) Ginkgolide with intravenous alteplase thrombolysis in acute ischemic stroke improving neurological function: a multicenter, cluster-randomized trial (GIANT). Front Pharmacol 12:792136. https://doi.org/10.3389/fphar.2021.792136

Zhang Y, Zou J, Wan F et al (2023) Update on the sources, pharmacokinetics, pharmacological action, and clinical application of anisodamine. Biomed Pharmacother 161:114522. https://doi.org/10.1016/j.biopha.2023.114522

Zielińska-Błajet M, Feder-Kubis J (2020) Monoterpenes and their derivatives-recent development in biological and medical applications. Int J Mol Sci 21:7078. https://doi.org/10.3390/ijms21197078

Chapter 8
Phytotoxins: Aquatic Plant Sources

Sandeep Kour, Mohd Ali, Deepak Kumar, Roohi Sharma, Parkirti, Vikram, Nitika Kapoor, Harish Changotra, Rajesh Kumari Manhas, Puja Ohri, and Renu Bhardwaj

Abstract Phytotoxins are secondary metabolites that plants produce to defend themselves against diseases, herbivores, insects, biotic and abiotic stress conditions, pests, etc. They exhibit huge diversity in their structures, chemical composition, and toxicity. They have been extricated from both terrestrial and aquatic sources and can be categorized into numerous classes such as terpenes, alkaloids, glycosides, phenolics, cyanotoxins, coumarins, tannins, quinones, and many more. In plant kingdom, the generation of secondary metabolites is primarily an extensive phenomenon of protection from potential predators and unfavourable circumstances. However, aside from shielding plant from various environmental stressors, these plant secondary metabolites have a wide range of applications in various other fields. They can be utilized as biopesticides and herbicides, in medicine and healthcare as anti-inflammatory, antiviral, antimicrobial, and antitumour resources. They have been utilized in vaccines and drug development as well. Aside from these beneficial properties, in terms of mobility, persistence, and toxicity, many of these phytotoxins might have certain characteristics similar to anthropogenic micropollutants. Keeping in view their immense applicability, the present book chapter summarizes the various kinds, aquatic sources, and applications of phytotoxins along with their role as micropollutants.

Keywords Phytotoxins · Secondary metabolites · Alkaloids · Phenolics · Biopesticides · Micropollutants · Antimicrobial · Herbicides

---

S. Kour · M. Ali · D. Kumar · R. Sharma · Parkirti · Vikram · P. Ohri (✉)
Department of Zoology, Guru Nanak Dev University, Amritsar, Punjab, India

N. Kapoor · R. Bhardwaj
Department of Botanical and Environmental Sciences, Guru Nanak Dev University, Amritsar, Punjab, India

H. Changotra
Department of Molecular Biology and Biochemistry, Guru Nanak Dev University, Amritsar, Punjab, India

R. K. Manhas
Department of Microbiology, Guru Nanak Dev University, Amritsar, Punjab, India

M. N. V. Prasad et al. (eds.), *Biotoxins*,
https://doi.org/10.1007/978-3-031-75309-1_8

8.1 Introduction

Plants produce various types of structurally and chemically diverse secondary metabolites. Plant secondary metabolites (PSMs) belong to different classes such as alkaloids, steroids, terpenoids, terpenes, glycosides, phenolics, cyanotoxins, coumarins, tannins, quinones, etc. The concentration of a single PSM ranges from a few g/kg up to more than 200 g/kg dry weight of plant biomass (Poutaraud et al. 2017). Various secondary metabolites produced by plants have beneficial effects such as they help in plant growth, development, ATP synthesis, photosynthesis, attract pollinators, help in seed dispersal, and facilitate mycorrhization (Bucheli 2014). Whilst most of them are beneficial, plants also fabricate some toxic PSMs called as phytotoxins. Phytotoxins can be innately present in different parts of the plants such as leaves, roots, flowers, bark, and fruits (Thakur et al. 2018). They are either deposited on the surface of plants or are accumulated in their tissues. Phytotoxins are synthesized by plants to defend themselves against numerous biotic and abiotic stresses such as pathogens, herbivores, plant parasites, draught, UV rays, temperature, etc. (Bucheli 2014; Chen et al. 2022). These toxic PSMs have diverse modes of actions; for instance, they can be genotoxic, allergenic, hallucinogenic, oestrogenic, or may have fatal effects (Poutaraud et al. 2017).

Owing to their persistence and harmfulness, several anthropogenic chemicals are acknowledged as environmentally significant micropollutants (Gunthardt et al. 2018). Pesticides are frequently condemned and kept under close observation because of their harmful impact on the surface waters (Spycher et al. 2018). Due to their environmental occurrence and toxicity, phytotoxins can also be considered as aquatic pollutants (Bucheli 2014). Although phytotoxins are frequently produced by plants, their potential as micropollutants has gotten very low consideration (Gunthardt et al. 2018). Persistent, mobile, and toxic (PMT) substances have recently come into focus and are of special concern as they can enter the aquatic environment and pose a threat to the water quality by remaining there for a long time (Schulze et al. 2019). Recent in silico predictions of PMT substances have revealed that several phytotoxins are lasting enough and have sufficient mobility to move into the surface waters (Gunthardt et al. 2021). However, not much studies on their monitoring have been conducted and their presence has remained largely unknown. Phytotoxins are mostly considered noxious to humans. But some of these PSMs have pharmacological applications such as antimicrobial and antitumour properties that can be exploited in the field of drug discovery and development (Chen et al. 2022). Moreover, they can help in plant growth promotion by acting as plant growth enhancers and can be utilized as biopesticides, herbicides, etc. (Chen et al. 2022). Keeping into account their immense applicability, the present book chapter highlights various roles played by phytotoxins in both plant and animal kingdoms.

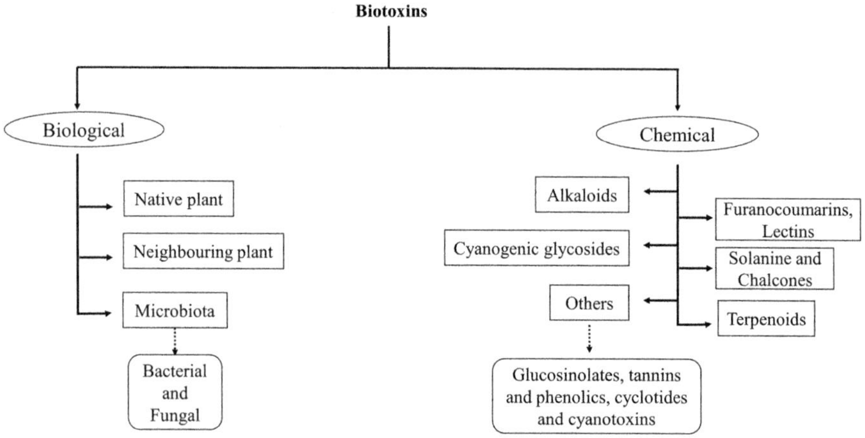

Fig. 8.1 Categorization of phytotoxins

8.2 Categorization of Phytotoxins

Phytotoxins can be categorized on the basis of their occurrence, pharmacological properties, biological properties, chemical properties, taxonomic data, reaction to growth circumstances, or habitat. Nonetheless, by carefully combining some of these elements, other descriptions can be more accurately described. Below are some succinct explanations of potential classification schemes for phytotoxins that are illustrated in Fig. 8.1.

8.2.1 Biological Characteristics

Phytotoxins can be native to the plant or created exogenously by neighbouring plants or connected to the microbiota (bacteria/fungus) that thrives nearby. The biological properties of phytotoxins are not the same. Different traits of phytotoxins may originate from the biological makeup of the accountable endophyte or, if previously investigated, from the bioactivities which the toxins are said to display. Traditional phytotoxin classification methods have classified phytotoxins according to their biological properties. Mycotoxins are fungi-derived secondary metabolites that have detrimental effects on both plants and humans, frequently resulting in fatality, are an excellent illustration of a classification system driven by biological features (Chen et al. 2022). It was necessary to change the classification of mycotoxins along with phytotoxins into two distinct classes. Similar biological activity of toxins, along with other criteria, led to their grouping as endophytic phytotoxins. Additionally, different kinds of bacterial phytotoxins are known to be formed by various families of phytopathogenic bacteria. Most of these toxins interact in a number of ways with the host plant through a clearly characterized mode of action (Chen

et al. 2022). Further, several wild mushrooms contain phytotoxins like muscarine and muscimol. These toxins induce vomiting, dizziness, diarrhoea, headaches, vomiting fits, and salivation when consumed. Delays in the development of symptoms are critical in assessing the effects of poisoning, as they drastically reduce the likelihood of survival due to delayed intervention.

Certain taxonomic characteristics of host plant can also play a major role in classifying toxins conferring to the plants from which they originate. When markers, including race, genus, species, and pathotypes, are employed to ascertain the phylogenetic relationships among endophytes that facilitate the synthesis of certain toxic substances with comparable features, chemotaxonomy has already demonstrated its reliability.

8.2.2 Chemical Characteristics

One of the simplest and highly accessible ways to categorize phytotoxins is through their chemical properties. In addition to other chemical characteristics, this is typically accomplished by looking at chemical structure of the phytotoxins. Chemical structures may be easily categorized based on their biosynthetic processes and the same can be said for phytotoxins. For instance, the chemical structures of phytotoxins produced by *Alternaria* and *Fusarium* infections are comparable.

8.2.2.1 Alkaloids

Alkaloids are a diverse class of chemical molecules that exist naturally and constitute a significant portion of the most extensive group of PSMs. Alkaloids are characterized primarily by the presence of one or more nitrogen atoms in a negative oxidation state at their structural locations, which results in their alkaline qualities and related therapeutic actions (Rasouli et al. 2020). From the perspective of synthesis, a small set of amino acid precursors, including phenylalanine, tyrosine, tryptophan, ornithine, and lysine, are the sources of most alkaloids. Different alkaloids are produced as a result of the conversion of these precursors into adaptable central intermediates. Plant species are regarded as sources of alkaloids when they have an alkaloid content of more than 0.001% (Majnooni et al. 2021).

8.2.2.2 Cyanogenic Glycosides

Cyanogenic glycosides are a class of PSMs that are engaged in a vast variety of processes, including plant growth and development as well as transportation of nitrogen. Although they are a significant and widely distributed class of natural products acquired from plants, cyanogenic glycosides lack the structural diversity of many other natural product classes. In the literature on phytochemistry, 112

naturally occurring cyanogenic glycosides have been described thus far (Yulvianti and Zidorn 2021). As of right now, reports suggest that these distinctive chemicals come from over 2500 different plant species (Thakur et al. 2019). Variations can be detected in the aglycone and sugar portions of natural cyanogenic glycosides. Although glucose is the most common sugar molecule, nature contains a variety of substitution patterns for this molecule. Four main classes—aliphatic, cyclic, aromatic, and heterocyclic aglycones—can be separated in relation to the aglycone molecule (Yulvianti and Zidorn 2021). Amygdalin was the first cyanogenic glycoside to be isolated from a plant source (Rockenbach et al. 1992). Further, plant defence mechanisms rely heavily on cyanogenic glycosides (Adhikary and Dasgupta 2023).

8.2.2.3 Terpenoids

Terpenoids, also referred to as terpenes, are the most chemically, physically, and functionally diversified class of molecules present in living beings, with over 80,000 compounds already recognized to date and many more likely to exist (Zhou and Pichersky 2020). The isoprene unit, a gaseous hydrocarbon with the chemical formula C_5H_8, is a building block of terpenes. They fall into one of the five categories: mono, di, tri, tetra, or sesquiterpenes, based on the number of isoprene units they contain. They typically occur in plants and make up the mainstream of the components in plant-derived essential oils (Cox-Georgian et al. 2019). In many plant-pathogen binomials, one of the defence mechanisms against an attack is the production of terpenes, which function as specific or broad pathogen inhibitors (Toffolatti et al. 2021).

8.2.2.4 Furanocoumarins, Lectins, Solanine, and Chalcones

Furanocoumarins are naturally existing tricyclic aromatic compounds found in plants that are members of the coumarin derivative family and are formed of an α-benzopyrone (coumarin) system bonded to a furan ring (Bruni et al. 2019). The distinction between the two types, i.e., psoralen (linear) and angelicin (angular) varieties is made based on the placement of the substitute. In the first case, the furan ring is joined to the backbone by carbon atoms 6 and 7, but in the latter, the substituent is positioned between C7 and C8. The most prevalent kind in the kingdom of plants is the linear type, particularly among members of the Rutaceae and Apiaceae families. The Apiaceae and Leguminosae (Fabaceae) plant families are only a few that include angular furanocoumarins.

Numerous derivatives can be obtained from angular and linear types. They are created by adding a hydroxyl or alkoxy group to the coumarin ring in place of the hydrogen atom. The compound's physicochemical and biological properties are determined by the kind, position, and quantity of substitutes attached. Plants create

furanocoumarins as a reaction to environmental stress and as a defence against fungal, bacterial, and insect attack (Sumorek-Wiadro et al. 2020).

Further, a growing variety of plant lectins have been reported and studied since ricin was discovered more than 130 years ago. Because they can identify and bind to carbohydrate structures in distinct and reversible ways, all lectins belong to a single protein family; nonetheless, no enzyme activity is assigned to the proteins. As a result, lectins are distinct from catalytically active carbohydrate-active enzymes, like glycosyl hydrolases and transferases. Many of the biological activities of lectins are dependent on their ability to bind to carbohydrates (De Coninck and Van Damme 2021).

Solanine is the glycoalkaloids produced naturally by plants primarily belonging to the Solanaceae family. Aglycone, a steroidal structure containing a nitrogen molecule, and the 3-OH position of a carbohydrate side chain are present. Cholesterol was the source of the steroidal structure. Glucose, galactose, and rhamnose make up the trisaccharide that is the carbohydrate in alpha solanine. They are harmful by nature. The potato plant and its components contain the majority of its source. Due to their frequent usage, potato tubers are said to be a major concern. The plant's self-defence mechanism is attributed to the poisonous nature of these glycoalkaloids. If these substances are consumed in excess of the safety limit, they cause a burning feeling in the tongue, bitterness, and other negative health effects (Kaur and Rani 2022). Further, a chalcone is a chemical made up of an unsaturated α, β-ketone connecting two aromatic rings. Chalcones are a group of bioactive plant metabolites that offer several medical and biological advantages to mankind. Natural chalcones are abundant in plants in the groups Leguminosae, Asteraceae, and Moraceae. A large proportion of the naturally found chalcones is polyhydroxylated aromatic compounds, which are regarded as the bioprecursors of flavonoids, open chain flavonoids, and isoflavonoids. Researchers are interested in studying chalcone-rich plant extracts in the quest for medicinally valuable chemicals since chalcones have the ability to quench radicals due to the inclusion of phenolic groups (Kuttithodi et al. 2022).

8.2.2.5 Tannins and Phenolics

Tannins from plants are polyphenols that are commonly present in both terrestrial and marine plants. Plant tannins have a wide variety of chemical structures. An organized categorization of tannins according to particular structural traits and chemical attributes can offer a useful foundation for associated studies. Condensed tannins and hydrolyzed tannins are the two main categories of plant tannins. Hydrolyzed tannins are composed of polyphenol nuclei with molecular weights ranging from 500 to 3000 Daltons (Da) (13). Catechin, epicatechin, gallocatechin, and epigallocatechin are among the flavan-3-ols that make up the oligomeric or polymeric flavonoids known as condensed tannins. Their molecular weights range from 1000 to 20,000 Da, they are difficult for anaerobic enzymes to break down, and they only depolymerize in the presence of intense oxidation and acid. Tannins

are present in the majority of plant species across the kingdom; they serve to protect the plant from predators and may also aid in growth regulation (Tong et al. 2022). Further, secondary metabolites called phenolic chemicals are found in many species of the plant kingdom and are categorized into two groups: flavonoids and non-flavonoids. The majority of phenolic chemicals are flavonoids. They provide plants their blue, purple, yellow, orange, and red hues, along with carotenoids and chlorophylls. Their structure is C6-C3-C6, and two aromatic rings are joined by a triple carbon bond. The amount, quantity, and location of hydroxyl groups in these substances' chemical structures determine their antioxidant action. Chemicals with simpler and smaller chemical structures than flavonoids fall within this class. However, there are other non-flavonoids with complicated structures and high molar mass. This group is mostly made up of phenolic acids, coumarins, stilbenes, and lignans (Cosme et al. 2020).

8.2.2.6 Other Chemical Categories

In addition to these primary categories of phytotoxins, glucosinolates, tannins, and phenolics, cyclotides and cyanotoxins are some more families of phytotoxins with noteworthy phytotoxicity. Glucosinolates are well-established class of naturally occurring anionic compounds that are present in mustards, cabbages, and other related plants. A variable side chain (R group) formed from an amino acid and a β-thioglucose residue connected by a sulphur atom to a (Z)-N-hydroximinosulfate ester comprises the fundamental structure of all glucosinolates. Plants belonging to the Brassicaceae, Capparaceae, Tovariaceae, and Caricaceae families are reported to produce glucosinolates. About 137 proposed glucosinolates with 88 verified structures consist of various precursor amino acids, varied side-chain lengths due to chain elongation, and substantial side-chain alterations. Due to their putative roles in non-host resistance to bacteria and fungi, insect defence, and health-promoting properties (e.g. anti-carcinogenesis), the significance of these sulfur-containing specialized metabolites has gained attention. Along with other metabolic routes, glucoseolates constitute a network that is crucial to the growth, development, and interactions of plants with their surroundings. Antioxidative defence against oxidative stressors is another function of some glucosinolates (Chhajed et al. 2020). Cyclotides are exceptionally stable mini-proteins generated from plants that exhibit a variety of biological activities, however their primary role is believed to be plant defence. The cyclic cystine knot motif is present in almost all naturally occurring cyclotides. The substantially conserved gap between Cys residues supports a common consensus structure among cyclotides. The circular topology is characterized by a β-hairpin that is linked to a third deformed β-sheet that stabilizes its core. This gives cyclotides their exceptional structural stability against biological degradation and great resistance to heat and chemical denaturation. Because of their stable structure and inherent bioactive properties, cyclotides are useful as therapeutic agents (Ho et al. 2023).

8.3 Phytotoxins from Aquatic Plants

Aquatic plants provide a balance to the aquatic ecosystem. They maintain structure and function of aquatic ecosystem, manage phytochemistry of bottom sediments and water, biotic interactions, and primary production (Nezbrytska et al. 2022). However, the aquatic system is prone to various stresses. There are various methods to control stress in aquatic ecosystem including physical, chemical, and biological methods. Out of these, the biological methods are the foremost in every aspect. It is ecofriendly, cost effective, and more practicable. One such biological method is the use of aquatic plants. Aquatic plants release some phytotoxic chemicals that help in enhancing life quality in aquatic system. For example, *Egeria densa* is a submerged aquatic plant species that releases some allelopathic chemicals such as sulfur compounds, polyacetylenes, oxygenated fatty acids, and polyphenols that help mitigating algal bloom in water bodies (Senavirathna et al. 2021). In a study, the use of aquatic plants, namely, *Crassula helmsii* and *Hydrocotyle ranunculoides*, for allelopathic repression of algal population was done against cyanobacteria *Synechocystis* sp. PCC6803 and algae *Chlorella vulgaris*. The results depicted that the methanolic extract of these plants was effective in controlling the troublesome algae species (Reynolds and Aldridge 2021). In another study, *Spartina alterniflora* that mainly consists of cyclohexane, 2-cyclohexen-1-one, 2,4-Di-tert-butylphenol, heptane, hydrocinnamic, and hexadecanoic acid was effective in inhibiting *Microcystis aeruginosa*. This suggested that by employing rapidly expanding *S. alterniflora*, the *M. aeruginosa* blooms could be controlled (Yuan et al. 2020). Similarly, the water hyacinth (*Eichhornia crassipes*) comprised some chemicals that were effective in inhibiting the growth of *Mimosa pudica*, an extremely invasive herbaceous plant. Their leaf extract mainly consisted of some phenolic compounds that were effective, in a concentration dependent manner, in inhibiting growth, inducing lipid peroxidation, and reducing chlorophyll content of *M. pudica* (Laemoh and Waewthongrak 2022). In yet another study, the inhibitory effect of some macroalga on the growth and photosynthetic activity of cyanobacteria, *Nostoc* sp. and *Aphanizomenon* sp. was observed (Budzałek and Śliwinska-Wilczewska 2021). Similarly, allelopathic activity of *Myriophyllum spicatum* showed inhibitory effect on natural phytoplankton assemblages (Pećić et al. 2018). In another aquatic plant, namely, *M. elatinoides*, the hexadecenoic acid was the main allelochemical effective in inhibiting the growth of two alga, viz., *Selenastrum capricornutum* and *Microcystis aeruginosa* (Bi et al. 2019). Thus, by releasing allelopathic substances with algicidal effect, creating favourable growth conditions for zooplankton and herbivores, reducing light and nutrient availability, the plants control the cyanobacterial growth by various mechanisms that are usually species specific in nature and aid in better control of cyanobacterial population (Nezbrytska et al. 2022). Allelopathic substances are the chemicals that either stimulate or limit the reproduction, development, or survival of the target species and alter the environment. For example, *Lemna minor* L. contains some chemical compounds like phenolic acids that have significant effect on growth, development, and competition among species

that may enhance the allelopathic potential. Its aqueous methanolic extract strongly inhibits growth of cress, rye grass, and lettuce. Thus, the investigation of plants with phytotoxic potential may help in investigating more natural compounds to manage life in aquatic ecosystem. Some aquatic plants with their phytotoxic effects are listed in Table 8.1.

Table 8.1 Phytotoxic effects of aquatic plants

Plant	Phytotoxic compound	Part used	Effect	Reference
Alternanthera philoxeroides	Phenols and alkaloids	Stem	Reduced height, weight, number of branches, and number of leaves in *Eichhornia crassipes*	Bai et al. (2023)
Myriophyllum spicatum	ID52 (melibiose), ID50 (galactinol) and ID55 (α-D-glucopyranoside)	Apical shoots	Cyanobacterial growth was highly inhibited	Jeong et al. (2023)
Spirulina platensis	Palmitic acid, oleic acid, γ-linolenic acid, C_{11} norisoprenoids, and 2-ethyl-3-methylmaleimide	Algal cells	Inhibited the growth of seeds and seedlings of Chinese amaranth	Charoenying et al. (2022)
Ludwigia hexapetala	Linolenic acids, glycosides, polyphenols, quercitrin, and myricitrin	–	Altered water parameters negatively affected the growth of *Utricularia australis*	Pelella et al. (2023)
Najas marina L.	Þ-coumaric, Þ-hydroxybenzoic, syringic acids, ferulic	–	Individual growth dominated; growth of adjoining species was supressed	Puri et al. (2022)
Cerbera manghas L.	Protocatechuic acid and phloridzin	Fruit	Controlled harmful algal bloom	Chen et al. (2021)
Pistia stratiotes L.	Gallic acid, polyphenols, and flavonoids	–	Inhibited the growth of *Microcystis aeruginosa*	Lourenção et al. (2021)
Schumannianthus dichotomus	Methyl syringate and syringic acid	Leaf	Inhibited the growth of Italian ryegrass and cress	Rob et al. (2020)
Ceratophyllum demersum L.	Azelaic acid, alpha linolenic acid, octanedioic acid, phthalic acid, hexanoic acid, pentanedioic acid, butenoic acid, and palmitic acid	–	Potentially inhibited the growth of *Chlorella vulgaris*	Dong et al. (2019)
Marsilea crenata	Isololiolide and Loliolide	Leaves and shoot	Inhibited seedling growth and yield of barnyard grass	Islam et al. (2017)

8.4 Phytotoxins as Environmental Micropollutants

Phytotoxins are substances produced by plants that can have toxic effects on other organisms, including humans (Gunthardt et al. 2018). While the term is commonly associated with plant defence mechanisms against herbivores, it can also be relevant in the context of environmental pollution when these substances enter ecosystems as micropollutants (Bucheli 2014). Phytotoxins, when present in environment, can act as environmental micropollutants, posing a potential risk to ecosystems and organisms due to their toxic effects on plant life and associated ecological balance (Hansen et al. 2021). Micropollutants are pollutants that exist in the environment at very low concentrations, often ranging from micrograms or nanograms per litre. They include a wide range of chemicals, including pharmaceuticals, personal care products, pesticides, and industrial chemicals (Yang et al. 2021). Phytotoxins can become micropollutants when they are introduced into water bodies, soil, or air in small concentrations, potentially impacting the health of ecosystems and organisms.

Phytotoxins, acting as allelochemicals, shape plant interactions by either inhibiting or promoting the growth of neighbouring plants. While this can be a natural form of competition between plant species, the introduction of allelopathic compounds into the environment can disrupt ecosystems and even affect non-target organisms (Kong et al. 2019). When released into the environment, these compounds function as micropollutants, affecting ecosystems even in minute concentrations, as low as a few micrograms per litre. Allelochemicals can volatilize or leach from roots, stem, and leaves which are released as a result of plant injury, defence, or biotic and abiotic stress (Scavo et al. 2018). For instance, black walnut trees produce juglone, a chemical that inhibits the growth of many plants, including strawberry, cucumber, and some ornamental flowers (Medic et al. 2023). Similarly, *Helianthus annuus* is known to release allelopathic compounds that inhibited the germination and growth of neighbouring plants (Rawat et al. 2017). Many other plants such as *Sorghum bicolor, Eucalyptus* spp., *and Brassica* spp. are also known to release the chemicals that affect the agroforestry (Chu et al. 2014; Rehman et al. 2013; Cheema et al. 2007). These allelochemical compounds act as environmental micropollutant in the form of secondary metabolites, can affect invasive species, and also act as biopesticides. Figure 8.2 illustrates the plant phytotoxins that act as environmental micropollutants towards neighbouring plants, terrestrial and aquatic flora and fauna.

A variety of PSMs such as alkaloids, phenolics, and terpenoids are produced by plants which can have toxic effects on other organisms besides their wide range of role in physiological functions. Their functions include facilitating symbiosis with rhizobia and mycorrhiza, supporting fundamental plant metabolism (e.g. photosynthesis, ATP generation) and organ development, and acting as signalling molecules to attract pollinators or seed dispersers. Moreover, they are generated to protect the organism against biotic (such as viruses, bacteria, or fungus) or abiotic (such as UV radiation, temperature, and drought) stress. Furthermore, through a mechanism known as allelopathy, these chemicals may aid the plant in competing with other

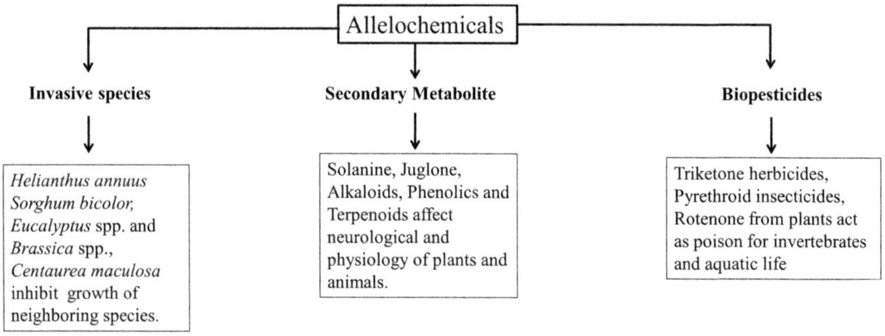

Fig. 8.2 Phytotoxins as environmental micropollutants

plants. When these compounds find their way into the environment, either through natural processes or human activities, they can contribute to the pool of micropollutants. For instance, juglone from walnuts contaminates soil and impacts the health of the nearby plants (Ni et al. 2022). Also, solanine in potatoes consumed by cattle causes neurological, inflammatory, and digestive ailments (Barceloux 2009). Some invasive plant species produce phytotoxins that have a negative influence on the native flora and fauna. When these plants establish themselves in new ecosystems, they may release phytotoxins that act as micropollutants, altering the dynamics of the local environment. For example, invasive species *Centaurea maculosa* replaced native plant species by releasing the 'catechin' phytotoxin from its roots (Bais et al. 2003). Similarly, giant *Solidago gigantea* chemicals were found to invade wheat and scentless mayweed (Ravlić et al. 2015). In a study conducted by Qu et al. (2021), it was observed that the allelopathic compounds released by an invasive species decreased the native plant species' growth and the soil microbial activity, thereby disturbing the microbiota around. Accordingly, leachates from the invasive clonal plant *Wedelia trilobata*, when applied to both itself and its native congener *W. chinensis*, significantly enhanced the growth and competitive ability of *W. trilobata* while suppressing the corresponding traits in *W. chinensis*. These findings suggested that allelochemicals released by *W. trilobata* contributed to its invasion success by promoting its own performance and inhibiting that of the native congener (Ullah et al. 2021). Similarly, azadirachtin application, even in minute concentration of 10–28 μg/L, negatively affected the copepods and zooplanktons of Ontario ponds (Kreutzweiser et al. 2004). Pesticides and herbicides used in agriculture often contain phytotoxic substances (Fu et al. 2020). Phytotoxins can also act as nature's own pesticides. Application of these compounds onto the crops leads to their washing off into nearby water bodies through runoff. As a result, they become micropollutants, affecting aquatic environment and potentially harming non-target organisms (Bucheli 2014). Several highly effective commercial pesticides, such as triketone herbicides derived from leptospermone in *Leptospermum scoparium* and cinmethylin herbicides utilizing 1,4-cineole found in various plant essential oils, have originated from natural toxins (Soltys et al. 2013). Pyrethroid insecticides, exemplified by pyrethrins 1 and 2 from *Chrysanthemum* spp., also serve as successful

commercial products inspired by natural counterparts (Petroski and Stanley 2009). Plants like *Derris* spp., *Lonchocarpus* spp., and *Tephrosia* spp. produce rotenone which is used by tribes as fish poison that eventually poses threat to invertebrates (Burrows and Tyrl 2013). In a study, it was noticed that the treatment of rotenone at 0.15 mg/L concentration caused 21% of aquatic macroinvertebrates to remain missing for 5 years in the Strawberry River, Utah (Ling 2003).

Hence, understanding the presence and effects of phytotoxins as micropollutants is crucial for assessing and mitigating the environmental impact of human activities. Monitoring and research efforts focus on identifying these compounds, determining their sources, and assessing their potential ecological risks. Additionally, developing sustainable agricultural practices and wastewater treatment methods can help minimize the introduction of phytotoxins into the environment as micropollutants (Saravanan et al. 2021). Therefore, adverse effects can be cured by promoting strategies like nature preservation, diverse agriculture and breeding for lower toxins in order to reduce environmental and human exposure proactively.

8.5 Applications of Phytotoxins

The detrimental impacts of phytotoxins on the development and growth of plant have been comprehensively researched. Higher concentrations typically hinder plant development, yet at minimal exposures, growth may paradoxically be enhanced (Weralupitiya et al. 2022). Understanding this dualistic relationship is important for comprehending the nuanced effects of phytotoxins on plant physiology. Hormesis refers to the positive effects of low-dose exposure to phytotoxins, showcasing their stimulatory and advantageous impact (Agathokleous and Calabrese 2019). All-encompassing research has been carried out to elucidate the benefits of these natural herbicidal agents, revealing that various phytotoxins produced by plants possess the ability to improve crop productivity and growth. In agriculture, they are employed as natural herbicides to control weed growth, offering an environmentally friendly alternative to synthetic chemicals (Vurro et al. 2018). Additionally, these compounds serve as valuable tools in scientific research, aiding in the study of plant–pest interactions and providing insights into plant defence mechanisms (Synowiec et al. 2017). The exploration of these compounds has opened avenues for leveraging plant-derived toxins as valuable tools in agriculture as plant stress alleviators, offering insights into sustainable approaches for optimizing crop yields. Some phytotoxins exhibit antimicrobial properties, contributing to their use in medicine for drug development against viral, bacterial, and fungal infections (Wang et al. 2022). These versatile compounds exhibit anti-inflammatory and antitumour properties in medical applications (Miranda et al. 2015). Also, they contribute to drug and vaccine development, showcasing their broad impact in both agriculture and healthcare (Chen et al. 2022). Hence, the multifaceted applications of phytotoxins underscore their potential in diverse fields, offering sustainable solutions and avenues for further exploration. Various applications of phytotoxins are described below (Table 8.2).

Table 8.2 Applications of phytotoxins

Phytotoxin group	Source	Applications	References
Flavonoid	*Combretum leprosum*	Anti-parasitic, cytotoxic, and mutagenic activities	Bantho et al. (2023)
Coumarins	*Umbelliferae*	Improved nutrient uptake and growth in plant	Yang et al. (2023a)
β momorcharin	*Momordica charantia*	Significantly decreased the cancer cell viability without harming the normal cell line	Yang et al. (2023b)
Putrescine		Increased plant growth under salt stress	Ma et al. (2022)
Lupeol	*Zanthoxylum monogynum*	Antiophidic property against venoms of *Bothrops alternatus*	Dos Santos et al. (2021)
Betulinic acid	*Aegiphila integrifolia*	Antibacterial activity against different kinds of bacteria	Nascimento et al. (2021)
Allelopathic hormesis	*Moringa oleifera*	Increased shoot growth	Perveen et al. (2021)
Putrescine		Improved plant growth under salt stress in guava	Ghalati et al. (2020)
Jaburetox	Jack bean	Insecticidal activity	Sá et al. (2020)
Pinocembrin and genistin	Tea, green pepper, and celery leaves	Inhibition of viral replication	Le Lee et al. (2019)
Pokeweed antiviral protein	*Phytolacca americana*	Controls the multiplication of the virus by depurinating its DNA	Neller et al. (2019)
Quercetin-3-*O*-(2-pentosyl)-rhamnoside	*Ziziphus jujuba*	Antioxidant, anti-inflammatory, and antidiabetic properties.	Sakna et al. (2019)
1586 phytotoxins	Different sources	Evaluating environmental micropollution with the use of the hazardous plant-phytotoxin database	Gunthardt et al. (2018)
Artemisinin	*Artemisia annua*	Acts as antimalarial activity	Czechowski et al. (2016)
Phenolic acid	Various sources	Inhibition of viral replication	Medini et al. (2016)
Quercetin-3-O-a-d--arabinopyranoside	Various sources	Prevents viruses from entering cells	Orabi and Orabi (2016)
Oleanolic terpenes	*Camellia japonica*	Prevent viral replication	Yang et al. (2015)

8.5.1 As Plant Stress Alleviators

Abiotic and biotic stresses experienced by plants decrease the agricultural productivity and output (Oshunsanya et al. 2019). Crops are encountered by various forms of environmental stress such as intense salinity, heat, drought, pH, excessive moisture, scarcity of nutrients, competition for space, heavy metal presence, and hazardous radiations (Gull et al. 2019). Additionally, plants can also suffer from biotic stress when they are affected by pathogenic organisms, such as bacteria, fungi, herbivores, nematodes, and various other pests. Plants have evolved natural defence systems to keep their normal metabolic processes and functioning intact in the face of such stress. Plants typically rely on the synthesis of phytotoxins to protect themselves against abiotic stress. A study was conducted to observe the effects of phytotoxins extracted from *P. hysterophorus*, on lettuce plants, under controlled conditions at various temperature ranges (Belz et al. 2011). The results demonstrated that at temperatures ranging from 18 to 24 °C, the phytotoxins generated hormetic stimulation (Belz et al. 2011). Likewise, germination rates were enhanced in radish and mustard and they were able to withstand temperature stress after being treated with phytotoxins extracted from *Ipomoea cairica* L. (Carillo et al. 2011). Similarly, plants can emit phytotoxins that help in increasing the productivity of crops under salinity stress. An essential study of this feature was carried out on rice seeds that were grown with varying NaCl levels, after being primed with different amounts of sunflower extracts (Farooq et al. 2011). Phytotoxins emitted by plants can be used as also growth enhancers. Several techniques, including foliar spraying and seed priming, can be used to provide phytotoxins to target plants, in order to sustain agricultural production, either directly or indirectly. One study examined the foliar application of various sorgaab extracts on maize plants (Jahangeer 2011). Moreover, phytotoxins have the ability to increase gaseous exchange, transpiration, and leaf water content in plants. They are useful in the formation of the cell wall, the synthesis of proteins, and cell division at the cellular level (Chen et al. 2022).

Phytotoxins extracted from plants can be used in the management of various plant diseases as well. Pathogenic bacteria and fungus represent the biggest threat to crop productivity. The harmful metabolites produced by these pathogens cause disease symptoms in the plants. Tissues of such damaged plants can be used to extract phytotoxins. Infection by a phytopathogen leads to impaired functioning of plant tissues. Usually, these bacteria cause poisoning by proliferating in the tissues of plants. The response of host plant tissues to the phytotoxins aids in the quick screening of germplasm to create resistance to plant diseases (Rubiales et al. 2015). Ribosome-inactivating protein (RIPs) are among the vital phytotoxins that plants have employed for number of purposes, including pesticidal, antiviral, antibacterial, and defence against pathogenic and insect attacks (Chen et al. 2022). Several crop plants, such as cowpeas, soybeans, and maize, have disease-resistant lines because of this application (Kankanala et al. 2019).

8.5.2 As Herbicides

Weed management is one among the primary issues faced by the agriculture industry. Weeds significantly reduce agricultural productivity and yield due to their competition with crops plants for nutrients, water, sunlight, and soil (Monteiro and Santos 2022). The development of natural chemicals as herbicides is desperately needed, as modern agriculture mainly relies on the usage of chemical herbicides for controlling weeds (Ofosu et al. 2023). Phytotoxins can be utilized as weed control agents because of their unique features such as low environmental persistence and more environmental safety as compared to synthetic herbicides (Chen et al. 2022). These naturally occurring phytotoxins are fascinating possibilities for the creation of novel natural herbicides with novel modes of action in both natural and controlled contexts due to their vast structural variation and advanced biological activity (Elshafie et al. 2023). Their ability to target novel locations of action contributes to the reduction of herbicide resistance, which is one of their most significant features (Berestetskiy 2023). Synthetic phytotoxins derived from plants have been examined for their potential as herbicide. These herbicides are typically sprayed while the seedlings are just beginning to grow and when they are in weed-sensitive stages. These compounds exhibit toxicity against weeds whilst having no fatal effect on crop plants (Bendejacq-Seychelles et al. 2023). In various studies, aqueous extracts from crops have been efficiently used as weed control agents and to promote crop growth. These plant extracts are sprayed on leaves to suppress weeds. They show hormetic activity (growth stimulation) at low dosages, whereas exposure to higher quantities may have herbicidal impact (Abbas et al. 2017; Makaza et al. 2022; Pouthika and Madhumitha 2023). In a study, it was observed that phytotoxins extracted from *Sphagneticola trilobata* were applied to plant roots in soil at low quantities, including radish, onion, tomato, and cabbage. They generated allelochemicals that stimulated growth of crop plants. An allelopathic effect was also observed among the weed plants (Hernández-Aro et al. 2016). However, the limitations of applying plant extracts to control weeds are known to be dose- and time-dependent and highly specific. Reduced competition for nutrients among weeds and hormetic stimulation are the primary causes of improvements in crop growth (Farooq et al. 2020).

8.5.3 In Medicine and Pharmacology

The term "traditional medicines" or "traditional medicinal system" describes the application of customary wisdom, abilities, and methods to treat and cure illnesses. Almost everybody in the globe still uses and relies on the conventional medical system, despite advancements in contemporary medical systems. Further, the term "use of alternative medicines" refers to procedures including therapeutics and medicinal by-products used for human benefit while being aware of their beneficial,

hazardous, and poisonous effects. The plants that offer positive pharmacological and therapeutic impacts on human health are called medicinal plants. These plants have been employed for millennia and are abundant in bioactive compounds such as volatile oils, flavonoids, sterols, saponins, and alkaloids. Documents from Chinese scriptures and Egyptian papyrus attest to the utilization of herbs as medicine, which was followed by the creation of various conventional medical systems (Aasim et al. 2019). Various studies have reported the production of phytotoxins in aquatic and semiaquatic plants that are used in medicine and pharmacology. The majority of aquatic plants, particularly seaweeds, are utilized extensively as valuable and potential sources of bioactive chemicals in a variety of industries, such as the food, pharmaceutical, cosmetic, nutraceutical, and biomedical. These bioactive substances possess a range of functional biological activities, including antibacterial, antifungal, antiviral, and microbiological capabilities (Pooja et al. 2020). Further, phytotoxins of *Bacopa monnieri*, an aquatic plant, are known to have therapeutic effects. This might be because of the to the existence of Linoleic acid and beta-sitosterol. The former lowers prostate inflammation, while the latter is an anticancer substance. Furthermore, antioxidant chemicals effective against hepatitis C virus, including luteolin, apigenin, and wedelolactone have been identified from an aquatic plant *Eclipta alba* (Manvar et al. 2012). Similarly, numerous studies on the pharmacological characteristics of various *Aponogeton* species have been conducted. The phytotoxins γ-sitosterol, stigmasterol, and cycloartanol having antibacterial, antioxidant, antidiabetic, and anticancer properties have been extracted from *A. appendiculatus* (Chowdhury et al. 2019). Another study employed Prediction of Activity Spectra for Substances (PASS) to evaluate pharmacological importance (antineoplastic, anti-inflammatory, antifungal, and antibacterial activity) of the main constituents of three macrophytes that are common in the Holarctic species of freshwater. The activity spectrum prediction results for the test compounds indicated a strong likelihood of phytotoxic compounds for the presence of anti-inflammatory, antifungal, antitumour, and antibacterial properties (Kurashov et al. 2016).

8.5.4 As Antiophidic Compounds

Snakebite is a complicated issue that impacts individuals, nations, and cultures everywhere in the world. Anti-venom is made using specific substances that are supposed to be antiophidic, in order to prevent snakebites. Numerous plant phytotoxins that may be antiophidic have been investigated. Developing anti-venom drugs from some phytotoxins like triterpenes could be promising. As antiophidic agents, compounds such as camphorans, triterpenes, pterocarpans, and naphthoquinones have been identified and isolated (Fernandes et al. 2014). Terpenes and terpenoids are another phytotoxin class with high antiophidic potential. In a study, it was seen that extracts from *Humirianthera ampla* that included di- and triterpenoids counteracted the venomous effects of *Bothrops jararaca* and *B. jararacussu* (Strauch et al. 2013).

Another study revealed that *Eclipta prostrata* extracts provided defence against the abrasive venom of *Calloselasma rhodostoma*, *Crotalus durissus terrificus*, and *B. jararaca*. The substances stigmasterol, sitosterol, and coumestan (sometimes referred to as wedelolactone) enhance the antiophidic capability. Also, it has been documented that betalactone demonstrates anti-inflammatory properties and counteracts the myotoxic consequences of the crude venom obtained from *B. jararaca* and *B. jararacussu* (Pithayanukul et al. 2004). According to Fernandes et al. (2014), the myotoxicity of the snake venom of *Bothrops* spp. was reduced by *C. leprosum* extract. Lupeol, a different phytotoxin, was also identified from the Indian plant *H. indicus*. It was observed to significantly lessen the impacts of venoms from *Daboia russelii* and *Naja kaouthia*, hence inhibiting the pulmonary abnormalities, cardiovascular disease, and neurotoxicity (Chatterjee et al. 2006). Several substances with strong antiophidic qualities, including steroids, pterocarpans, and naphthoquinones, have also been used to cure snakebites. Extracts of various plants that have been conventionally utilized in folk medicine to treat snakebite should be thoroughly examined in order to make progress in this area. Finding out the pharmacology and toxicology of the necessary phytotoxins requires their isolation and characterization.

8.6 Conclusion

Phytotoxins are very crucial compounds in different areas of the society such as agriculture, health, drug discovery, plant protection, etc. Applications of these secondary metabolites of plants are steadily increasing in various sectors including biotechnology, pharmaceuticals, agroeconomics, medicines, and others. Various studies have investigated their advantages in managing different stresses in plants and combating diseases in humans. However, aquatic plants as sources of phytotoxins have not been explored much and very little information is available in this context. So, more research needs to be done to unravel and utilize phytotoxins released from aquatic plants. This will enable a multidisciplinary contribution in the field of science thus, expanding their application prospects.

References

Aasim M, Khawar KM, Ahmed SI, Karataş M (2019) Multiple uses of some important aquatic and semiaquatic medicinal plants. In: Plant and human health, Vol 2: phytochemistry and molecular aspects, pp 541–577

Abbas T, Nadeem MA, Tanveer A, Chauhan BS (2017) Can hormesis of plant-released phytotoxins be used to boost and sustain crop production? Crop Prot 93:69–76

Adhikary S, Dasgupta N (2023) Role of secondary metabolites in plant homeostasis during biotic stress. Biocatal Agric Biotechnol 50:102712

Agathokleous E, Calabrese EJ (2019) Hormesis can enhance agricultural sustainability in a changing world. Glob Food Sec 20:150–155

Bai SK, Avinash G, Sindhu KK, Geetha KN (2023) Allelopathic potential of Alternanthera philoxeroides (Mart.) Griseb on growth and development of *Eichhornia crassipes* (Mart) Solms. Allelopath J 60(2)

Bais HP, Vepachedu R, Gilroy S, Callaway RM, Vivanco JM (2003) Allelopathy and exotic plant invasion: from molecules and genes to species interactions. Science 301(5638):1377–1380

Bantho S, Naidoo Y, Dewir YH, Singh M, Bantho A (2023) Biological activity of Combretum erythrophyllum: antioxidant, apoptotic, and cytotoxic activity of the leaf and stembark extract. Horticulturae 9(3):305

Barceloux DG (2009) Potatoes, tomatoes, and solanine toxicity (Solanum tuberosum L., Solanum lycopersicum L.). Dis Mon 55(6):391–402

Belz RG, Cedergreen N, Duke SO (2011) Herbicide hormesis–can it be useful in crop production? Weed Res 51(4):321–332

Bendejacq-Seychelles A, Gibot-Leclerc S, Guillemin JP, Mouille G, Steinberg C (2023) Phytotoxic fungal secondary metabolites as herbicides. Pest Manag Sci 80:92. https://doi.org/10.1002/ps.7813

Berestetskiy A (2023) Modern approaches for the development of new herbicides based on natural compounds. Plan Theory 12(2):234

Bi YL, Wu SM, Zhou SN, Wu SH, Su H, Bai ZH, Xu SJ (2019) Allelopathic effects and allelochemicals of *Myriophyllum elatinoides* on *Microcystis aeruginosa* and *Selenastrum capricornutum*. Huan Jing ke Xue= Huanjing Kexue 40(5):2265–2270

Bruni R, Barreca D, Protti M, Brighenti V, Righetti L, Anceschi L, Mercolini L, Benvenuti S, Gattuso G, Pellati F (2019) Botanical sources, chemistry, analysis and biological activity of furanocoumarins of pharmaceutical interest. Molecules 24(11):2163

Bucheli TD (2014) Phytotoxins: environmental micropollutants of concern? Environ Sci Technol 48(22):13027–13033

Budzałek G, Śliwinska-Wilczewska S (2021) Allelopathic effect of *Fucus vesiculosus* (brown alga) and *Coccotylus brodiei* (red alga) on the growth and photosynthesis performance of Baltic cyanobacteria. Ann Univ Paedagog Crac Stud Nat 6

Burrows GE, Tyrl RJ (2013) Toxic plants of North America. John Wiley & Sons, Ames, USA

Carillo P, Annunziata MG, Pontecorvo G, Fuggi A, Woodrow P (2011) Salinity stress and salt tolerance. Abiotic Stress in Plants–Mechanisms and Adaptations 1:21–38

Charoenying P, Laosinwattana C, Chotsaeng N (2022) The allelopathic activity of extracts and isolated from Spirulina platensis. Molecules 27(12):3852

Chatterjee I, Chakravarty AK, Gomes A (2006) Daboia russellii and Naja kaouthia venom neutralization by lupeol acetate isolated from the root extract of Indian sarsaparilla *Hemidesmus indicus* R. Br. J Ethnopharmacol 106(1):38–43

Cheema ZA, Khaliq A, Abbas M, Farooq M (2007) Allelopathic potential of sorghum (Sorghum bicolor L. Moench) cultivars for weed management. Allelopath J 20(1):167

Chen Q, Sun D, Fang T, Zhu B, Liu W, He X, Sun X, Duan S (2021) In vitro allelopathic effects of compounds from *Cerbera manghas* L. on three Dinophyta species responsible for harmful common red tides. Sci Total Environ 754:142253

Chen H, Singh H, Bhardwaj N, Bhardwaj SK, Khatri M, Kim KH, Peng W (2022) An exploration on the toxicity mechanisms of phytotoxins and their potential utilities. Crit Rev Environ Sci Technol 52(3):395–435

Chhajed S, Mostafa I, He Y, Abou-Hashem M, El-Domiaty M, Chen S (2020) Glucosinolate biosynthesis and the glucosinolate–myrosinase system in plant defense. Agronomy 10(11):1786

Chowdhury NS, Islam TB, Farjana F, Jamali S (2019) Pharmacological values and phytochemical analysis of aquatic plant genus aponogeton: a review. Int J Recent Innov Acad Res 3(3):125–141

Chu C, Mortimer PE, Wang H, Wang Y, Liu X, Yu S (2014) Allelopathic effects of Eucalyptus on native and introduced tree species. For Ecol Manag 323:79–84

Cosme P, Rodríguez AB, Espino J, Garrido M (2020) Plant phenolics: bioavailability as a key determinant of their potential health-promoting applications. Antioxidants 9(12):1263
Cox-Georgian D, Ramadoss N, Dona C, Basu C (2019) Therapeutic and medicinal uses of terpenes. In: Medicinal plants: from farm to pharmacy, pp 333–359
Czechowski T, Larson TR, Catania TM, Harvey D, Brown GD, Graham IA (2016) Artemisia annua mutant impaired in artemisinin synthesis demonstrates importance of nonenzymatic conversion in terpenoid metabolism. Proc Natl Acad Sci 113(52):15150–15155
De Coninck T, Van Damme EJ (2021) The multiple roles of plant lectins. Plant Sci 313:111096
Dong J, Chang M, Li C, Dai D, Gao Y (2019) Allelopathic effects and potential active substances of *Ceratophyllum demersum* L. on *Chlorella vulgaris* Beij. Aquat Ecol 53:651–663
Dos Santos BM, Ferreira GM, Tavares MT, De Bona JC, Hirata MH, De Paula VF, Saturnino KC, Soares AM, Mendes MM (2021) Antiophidic activity of the secondary metabolite lupeol isolated from Zanthoxylum monogynum. Toxicon 193:38–47
Elshafie HS, Camele I, Mohamed AA (2023) A comprehensive review on the biological, agricultural and pharmaceutical properties of secondary metabolites based-plant origin. Int J Mol Sci 24(4):3266
Farooq M, Habib M, Rehman A, Wahid A, Munir R (2011) Employing aqueous allelopathic extracts of sunflower in improving salinity tolerance of rice. J Agric Soc Stud 7:75–80
Farooq N, Abbas T, Tanveer A, Jabran K (2020) Allelopathy for weed management. In: Co-evolution of secondary metabolites, pp 505–519
Fernandes FF, Tomaz MA, El-Kik CZ, Monteiro-Machado M, Strauch MA, Cons BL, Tavares-Henriques MS, Cintra AC, Facundo VA, Melo PA (2014) Counteraction of Bothrops snake venoms by *Combretum leprosum* root extract and arjunolic acid. J Ethnopharmacol 155(1):552–562
Fu Y, Bhadha JH, Rott P, Beuzelin JM, Kanissery R (2020) Investigating the use of aquatic weeds as biopesticides towards promoting sustainable agriculture. PLoS One 15(8):e0237258
Ghalati RE, Shamili M, Homaei A (2020) Effect of putrescine on biochemical and physiological characteristics of guava (*Psidium guajava* L.) seedlings under salt stress. Sci Hortic 261:108961
Gull A, Lone AA, Wani NUI (2019) Biotic and abiotic stresses in plants. In: Abiotic and biotic stress in plants, pp 1–19
Gunthardt BF, Hollender J, Hungerbuhler K, Scheringer M, Bucheli TD (2018) Comprehensive toxic plants–phytotoxins database and its application in assessing aquatic micropollution potential. J Agric Food Chem 66(29):7577–7588
Gunthardt BF, Hollender J, Scheringer M, Hungerbühler K, Nanusha MY, Brack W, Bucheli TD (2021) Aquatic occurrence of phytotoxins in small streams triggered by biogeography, vegetation growth stage, and precipitation. Sci Total Environ 798:149128
Hansen HCB, Hilscherova K, Bucheli TD (2021) Natural toxins: environmental contaminants calling for attention. Environ Sci Eur 33(1):1–8
Hernández-Aro M, Hernández-Pérez R, Guillén-Sánchez D, Torres-Garcia S (2016) Allelopathic influence of residues from *Sphagneticola trilobata* on weeds and crops. Planta Daninha 34:81–90
Ho TN, Pham SH, Nguyen LT, Nguyen HT, Nguyen LT, Dang TT (2023) Insights into the synthesis strategies of plant-derived cyclotides. Amino Acids 55:1–17
Islam MS, Iwasaki A, Suenaga K, Kato-Noguchi H (2017) Isolation and identification of two potential phytotoxic substances from the aquatic fern *Marsilea crenata*. J Plant Biol 60:75–81
Jahangeer A (2011) Response of maize (Zea mays L.) to foliar application of three plant water extracts [M.Sc. Hons. thesis]. Department of Agronomy, University of Agriculture
Jeong S, Joo S, Park S (2023) Applying a neural network machine learning model to predict seasonal allelopathic inhibitory effects of *Myriophyllum spicatum* on the growth of Microcystis aeruginosa. Aquat Ecol 58:1–13
Kankanala P, Nandety RS, Mysore KS (2019) Genomics of plant disease resistance in legumes. Front Plant Sci 10:1345
Kaur J, Rani P (2022) Solanine and Chaconine. In: Handbook of plant and animal toxins in food. CRC Press, pp 73–96

Kong CH, Xuan TD, Khanh TD, Tran HD, Trung NT (2019) Allelochemicals and signaling chemicals in plants. Molecules 24(15):2737

Kreutzweiser DP, Sutton TM, Back RC, Pangle KL, Thompson DG (2004) Some ecological implications of a neem (azadirachtin) insecticide disturbance to zooplankton communities in forest pond enclosures. Aquat Toxicol 67(3):239–254

Kurashov EA, Fedorova EV, Krylova JV, Mitrukova GG (2016) Assessment of the potential biological activity of low molecular weight metabolites of freshwater macrophytes with QSAR. Scientifica 2016:1205680

Kuttithodi AM, Nikhitha D, Jacob J, Narayanankutty A, Mathews M, Olatunji OJ, Rajagopal R, Alfarhan A, Barcelo D (2022) Antioxidant, antimicrobial, cytotoxicity and larvicidal activities of selected synthetic bis-chalcones. Molecules 27(23):8209

Laemoh A, Waewthongrak W (2022) Allelopathic effect of *Eichhornia crassipes* aqueous extract against growth of *Mimosa pudica*. Sains Malays 51(10):3153–3162

Le Lee J, Loe MWC, Lee RCH, Chu JJH (2019) Antiviral activity of pinocembrin against Zika virus replication. Antivir Res 167:13–24

Ling N (2003) Rotenoneôa review of its toxicity and use for fisheries management. Sci Conserv 211:1–40

Lourenção A, Mecina GF, Cordeiro-Araújo MK, Bittencourt-Oliveira MC, Chia MA, Bronzel-Júnior JL, Granero FO, Silva LP, da Silva RM (2021) Characterization of allelochemicals from Pistia stratiotes extracts and their effects on the growth and physiology of Microcystis aeruginosa. Environ Sci Pollut Res 28(40):57248–57259

Ma S, Zhou X, Jahan MS, Guo S, Tian M, Zhou R, Liu H, Feng B, Shu S (2022) Putrescine regulates stomatal opening of cucumber leaves under salt stress via the H2O2-mediated signaling pathway. Plant Physiol Biochem 170:87–97

Majnooni MB, Fakhri S, Bahrami G, Naseri M, Farzaei MH, Echeverría J (2021) Alkaloids as potential phytochemicals against SARS-CoV-2: approaches to the associated pivotal mechanisms. Evid Based Complement Alternat Med 2021:1–21

Makaza K, Matigimu M, Sakadzo N (2022) Aqueous leaf extracts of sunflower (Helianthus annuus) for weed management. In: Climate change adaptations in dryland agriculture in semi-arid areas. Singapore, Springer Nature, pp 145–155

Manvar D, Mishra M, Kumar S, Pandey VN (2012) Identification and evaluation of antihepatitis C virus phytochemicals from Eclipta alba. J Ethnopharmacol 144(3):545–554

Medic A, Zamljen T, Slatnar A, Hudina M, Grohar MC, Veberic R (2023) Effect of juglone and other allelochemicals in walnut leaves on yield, quality and metabolites of snack cucumber (Cucumis sativus L.). Food Secur 12(2):371

Medini F, Megdiche W, Mshvildadze V, Pichette A, Legault J, St-Gelais A, Ksouri R (2016) Antiviral-guided fractionation and isolation of phenolic compounds from Limonium densiflorum hydroalcoholic extract. Comptes Rendus Chimie 19(6):726–732

Miranda MA, Varela RM, Torres A, Molinillo JM, Gualtieri SC, Macías FA (2015) Phytotoxins from Tithonia diversifolia. J Nat Prod 78(5):1083–1092

Monteiro A, Santos S (2022) Sustainable approach to weed management: the role of precision weed management. Agronomy 12(1):118

Nascimento LS, Nogueira-Souza PD, Rocha-Junior JR, Monteiro-Machado M, Strauch MA, Prado SA, Melo PA, Veiga-Junior VF (2021) Phytochemical composition, antisnake venom and antibacterial activities of ethanolic extract of *Aegiphila integrifolia* (Jacq) Moldenke leaves. Toxicon 198:121–131

Neller KC, Diaz CA, Platts AE, Hudak KA (2019) De novo assembly of the pokeweed genome provides insight into pokeweed antiviral protein (PAP) gene expression. Front Plant Sci 10:1002

Nezbrytska I, Usenko O, Konovets I, Leontieva T, Abramiuk I, Goncharova M, Bilous O (2022) Potential use of aquatic vascular plants to control cyanobacterial blooms: a review. Water 14(11):1727

Ni ZJ, Zhang YG, Chen SX, Thakur K, Wang S, Zhang JG, Shang YF, Wei ZJ (2022) Exploration of walnut components and their association with health effects. Crit Rev Food Sci Nutr 62(19):5113–5129

Ofosu R, Agyemang ED, Márton A, Pásztor G, Taller J, Kazinczi G (2023) Herbicide resistance: managing weeds in a changing world. Agronomy 13(6):1595

Oshunsanya SO, Nwosu NJ, Li Y (2019) Abiotic stress in agricultural crops under climatic conditions. In: Sustainable agriculture, forest and environmental management, pp 71–100

Orabi MA, Orabi EA (2016) Antiviral and antioxidant activities of flavonoids of Ficus virens: experimental and theoretical investigations. Journal of Pharmacognosy and Phytochemistry 5(3):120

Pećić MN, Predojević DD, Subakov-Simić GV (2018) Allelopathic activity of *Myriophyllum spicatum* L. on natural phytoplankton assemblages. Zbornik Matice srpske za prirodne nauke 134:55–64

Pelella E, Questino B, Ceschin S (2023) Impact of the alien aquatic plant *Ludwigia hexapetala* on the native Utricularia australis: evidence from an indoor experiment. Plan Theory 12(4):811

Perveen S, Mushtaq MN, Yousaf M, Sarwar N (2021) Allelopathic hormesis and potent allelochemicals from multipurpose tree Moringa oleifera leaf extract. Plant Biosyst 155(1):154–158

Petroski RJ, Stanley DW (2009) Natural compounds for pest and weed control. J Agric Food Chem 57(18):8171–8179

Pithayanukul P, Laovachirasuwan S, Bavovada R, Pakmanee N, Suttisri R (2004) Anti-venom potential of butanolic extract of *Eclipta prostrata* against Malayan pit viper venom. J Ethnopharmacol 90(2–3):347–352

Pooja K, Rani S, Rana V, Pal GK (2020) Aquatic plants as a natural source of antimicrobial and functional ingredients. In: Functional and preservative properties of phytochemicals. Academic Press, pp 93–118

Poutaraud A, Michelot-Antalik A, Plantureux S (2017) Grasslands: a source of secondary metabolites for livestock health. J Agric Food Chem 65(31):6535–6553

Pouthika K, Madhumitha G (2023) A review on plant-derived nanomaterials: an effective and innovative insect-resistant strategy for alternate pesticide development. Int J Environ Sci Technol 22:1–24

Puri S, Sidhu MC, Ahluwalia AS (2022) Establishing the dominating behavior of an aquatic plant 'Najas marina'L. Vegetos 35(4):1069–1077

Qu T, Du X, Peng Y, Guo W, Zhao C, Losapio G (2021) Invasive species allelopathy decreases plant growth and soil microbial activity. PLoS One 16(2):e0246685

Rasouli H, Yarani R, Pociot F, Popović-Djordjević J (2020) Anti-diabetic potential of plant alkaloids: revisiting current findings and future perspectives. Pharmacol Res 155:104723

Ravlić M, Baličević R, Peharda A (2015) Allelopathic effect of invasive species giant goldenrod (Solidago gigantea Ait.) on wheat and scentless mayweed. In: Proceedings & abstract of the 8th international scientific/professional conference agriculture in nature and environment protection, Glas Slavonije dd, Osijek, pp 186–190

Rawat LS, Maikhuri RK, Bahuguna YM, Jha NK, Phondani PC (2017) Sunflower allelopathy for weed control in agriculture systems. J Crop Sci Biotechnol 20(1):45–60

Rehman MU, Hussain M, Ali M, Mustafa CB, Shafi J, Iqbal F (2013) Allelopathy of Brassica. A review. Sci Agric 8:222–229

Reynolds SA, Aldridge DC (2021) Embracing the allelopathic potential of invasive aquatic plants to manipulate freshwater ecosystems. Front Environ Sci 8:551803

Rob MM, Hossen K, Iwasaki A, Suenaga K, Kato-Noguchi H (2020) Phytotoxic activity and identification of phytotoxic substances from Schumannianthus dichotomus. Plants 9(1):102

Rockenbach J, Nahrstedt A, Wray V (1992) Cyanogenic glycosides from PS Psydrax and Oxyanthus species [a/t]. Phytochemistry 31(2):567–570

Rubiales D, Fondevilla S, Chen W, Gentzbittel L, Higgins TJ, Castillejo MA, Singh KB, Rispail N (2015) Achievements and challenges in legume breeding for pest and disease resistance. Crit Rev Plant Sci 34(1–3):195–236

Sá CA, Vieira LR, Almeida Filho LCP, Real-Guerra R, Lopes FC, Souza TM, Vasconcelos IM, Staniscuaski F, Carlini CR, Carvalho AFU, Farias DF (2020) Risk assessment of the antifungal and insecticidal peptide Jaburetox and its parental protein the Jack bean (Canavalia ensiformis) urease. Food Chem Toxicol 136:110977

Sakna ST, Mocan A, Sultani HN, El-Fiky NM, Wessjohann LA, Farag MA (2019) Metabolites profiling of Ziziphus leaf taxa via UHPLC/PDA/ESI-MS in relation to their biological activities. Food Chem 293:233–246

Saravanan A, Kumar PS, Jeevanantham S, Karishma S, Tajsabreen B, Yaashikaa PR, Reshma B (2021) Effective water/wastewater treatment methodologies for toxic pollutants removal: processes and applications towards sustainable development. Chemosphere 280:130595

Scavo A, Restuccia A, Mauromicale G (2018) Allelopathy: principles and basic aspects for agroecosystem control. In: Sustainable agriculture reviews 28: ecology for agriculture, pp 47–101

Schulze S, Zahn D, Montes R, Rodil R, Quintana JB, Knepper TP, Reemtsma T, Berger U (2019) Occurrence of emerging persistent and mobile organic contaminants in European water samples. Water Res 153:80–90

Senavirathna MDHJ, Muhetaer G, Atapaththu KSS, Fujino T (2021) *Egeria densa* allelopathy on *microcystis aeruginosa* under different light intensities and preliminary insight into inter-parameter relationships. Water Air Soil Pollut 232:1–13. on red tides. *Science of the Total Environment, 754*, p.142253

Soltys D, Krasuska U, Bogatek R, Gniazdowska A (2013) Allelochemicals as bioherbicides—present and perspectives. In: Herbicides-current research and case studies in use. IntechOpen

Spycher S, Mangold S, Doppler T, Junghans M, Wittmer I, Stamm C, Singer H (2018) Pesticide risks in small streams—how to get as close as possible to the stress imposed on aquatic organisms. Environ Sci Technol 52(8):4526–4535

Strauch MA, Tomaz MA, Monteiro-Machado M, Ricardo HD, Cons BL, Fernandes FF, El-Kik CZ, Azevedo MS, Melo PA (2013) Antiophidic activity of the extract of the Amazon plant *Humirianthera ampla* and constituents. J Ethnopharmacol 145(1):50–58

Sumorek-Wiadro J, Zając A, Maciejczyk A, Jakubowicz-Gil J (2020) Furanocoumarins in anticancer therapy–for and against. Fitoterapia 142:104492

Synowiec A, Kalemba D, Drozdek E, Bocianowski J (2017) Phytotoxic potential of essential oils from temperate climate plants against the germination of selected weeds and crops. J Pest Sci 90:407–419

Thakur A, Sharma V, Thakur A (2018) Phytotoxin: a mini review. J Pharmacogn Phytochem 7(6):2705–2708

Thakur A, Sharma V, Thakur A (2019) An overview of anti-nutritional factors in food. Int J Chem Stud 7(1):2472–2479

Toffolatti SL, Maddalena G, Passera A, Casati P, Bianco PA, Quaglino F (2021) Role of terpenes in plant defense to biotic stress. In: Biocontrol agents and secondary metabolites. Woodhead Publishing, pp 401–417

Tong Z, He W, Fan X, Guo A (2022) Biological function of plant tannin and its application in animal health. Front Vet Sci 8:803657

Ullah MS, Sun J, Rutherford S, Ullah I, Javed Q, Rasool G, Ajmal M, Du D (2021) Evaluation of the allelopathic effects of leachate from an invasive species (Wedelia triobata) on its own growth and performance and those of a native congener (W. chinensis). Biol Invasions 23(10):3135–3149

Vurro M, Boari A, Casella F, Zonno MC (2018) Fungal phytotoxins in sustainable weed management. Curr Med Chem 25(2):268–286

Wang H, Guo Y, Luo Z, Gao L, Li R, Zhang Y, Kalaji HM, Qiang S, Chen S (2022) Recent advances in Alternaria phytotoxins: a review of their occurrence, structure, bioactivity, and biosynthesis. J Fungi 8(2):168

Weralupitiya C, Wanigatunge RP, Gunawardana D, Vithanage M, Magana-Arachchi D (2022) Cyanotoxins uptake and accumulation in crops: Phytotoxicity and implications on human health. Toxicon 211:21–35

Yang JL, Ha TKQ, Dhodary B, Pyo E, Nguyen NH, Cho H, Kim E, Oh WK (2015) Oleanane triterpenes from the flowers of Camellia japonica inhibit porcine epidemic diarrhea virus (PEDV) replication. J Med Chem 58(3):1268–1280

Yang Y, Zhang X, Jiang J, Han J, Li W, Li X, Yee Leung KM, Snyder SA, Alvarez PJ (2021) Which micropollutants in water environments deserve more attention globally? Environ Sci Technol 56(1):13–29

Yang Y, Xu J, Li Y, He Y, Yang Y, Liu D, Wu C (2023a) Effects of coumarin on rhizosphere microbiome and metabolome of Lolium multiflorum. Plan Theory 12(5):1096

Yang D, Peng D, Zhou Y, Qiang Z, Wan L, Fan X, Meng Y, Xu G, Meng Y (2023b) Alpha-Momorcharin, a type I ribosome inactivating protein, induced apoptosis of hepatocellular carcinoma SK-HEP-1 cells through mitochondrial pathway. Nat Prod Res:1–11

Yuan R, Li Y, Li J, Ji S, Wang S, Kong F (2020) The allelopathic effects of aqueous extracts from *Spartina alterniflora* on controlling the *Microcystis aeruginosa* blooms. Sci Total Environ 712:136332

Yulvianti M, Zidorn C (2021) Chemical diversity of plant cyanogenic glycosides: an overview of reported natural products. Molecules 26(3):719

Zhou F, Pichersky E (2020) More is better: the diversity of terpene metabolism in plants. Curr Opin Plant Biol 55:1–10

Chapter 9
Insect Toxins: Selective Interactions in Biological Regime

Manu Rani Paul, Silpa S. Nair, Nelbi P. Thomas, Rabiya Shamshudeen, Sumayya Nizar, and Abin Sebastian

Abstract Insects belong to the largest phylum, namely, Arthropoda in the animal kingdom. Members of this phylum are characterized with a segmented body, three pairs of jointed legs, and an exoskeleton composed of chitin. This phylum came out as the most successful group among animals because of their morphological peculiarities, high adaptability, and a wider range of distribution. However, some insects are harmful to humans as disease vectors while others are capable of human envenomation. Insects belonging to the orders such as Mantodea, Phthiraptera, Hemiptera, Coleoptera, Hymenoptera, Lepidoptera, Diptera, and Siphonaptera are well known to produce toxins. Insect toxins can be peptides, polyamines, lipids, etc., that have different modes of action and varied pathogenic intensities. Insect toxins target nervous system, muscles, and circulatory system of the victims. Aquatic coleopterans like water beetles inject a poison, whereas terrestrial beetles like blister beetles excrete irritants which can cause irritation and blistering upon contact with the human body. Venom of ants, bees, and wasps are also known to cause pain and inflammation in the locally affected areas. Saliva of parasitic insects contains certain proteins which causes allergic reactions in the host. Hence, multiple bites and stings of these insects cause inflammatory reactions such as anaphylactic shock. Also, larvae of a number of species of moths and butterflies have venomous setae or bristles over the body that causes burning sensation and temporary redness as a result of anti-inflammatory reactions. The present chapter discusses diversity of insect toxins and its effect in the biological regime.

Keywords Insect · Venom · Chemical nature · Biological impact · Detoxification

---

M. R. Paul
Department of Zoology, St. George's College Aruvithura, Aruvithura, India

S. S. Nair · N. P. Thomas · R. Shamshudeen · S. Nizar · A. Sebastian (✉)
Department of Botany, St. George's College Aruvithura, Aruvithura, India
e-mail: abin@sgcaruvithura.ac.in

M. N. V. Prasad et al. (eds.), *Biotoxins*,
https://doi.org/10.1007/978-3-031-75309-1_9

9.1 Introduction

Insects are the largest and most diverse group of animals on the Earth. They play pivotal roles in ecosystems such as pollination, decomposition, and serving as a vital food source for various organisms. With over a million described species, they exhibit unparalleled diversity in morphology, behavior, and ecological niches. Insects are essential pollinators, facilitating the reproduction of flowering plants and supporting global food production. Simultaneously, insects aid in nutrient recycling through breaking down organic matter which enriches soil fertility (Sollai and Solari 2022). Despite their ecological importance, some insects pose challenges as pests that threaten crops, livestock, and human health (Skendžić et al. 2021). Balancing the benefits and challenges associated with insects underscores the significance of understanding and conserving insect populations. As stewards of biodiversity, insects exemplify the relationships that sustain life on our planet, making their conservation and study imperative for maintaining ecological balance. The study of insect toxins holds paramount importance due to its far-reaching implications across various scientific disciplines, ranging from ecology to medicine. The co-evolutionary arms race between venomous insects and their prey or predators has profound consequences on biodiversity and ecosystem dynamics.

Insect toxins are often overlooked in comparison to their vertebrate counterparts. But the toxins in insects play a crucial role in the web of biological interactions within ecosystems. Insect toxins serve multifaceted purposes ranging from prey immobilization to defense against predators and competitors (Wang et al. 2000; Yadav and Mishra 2023). Hence, it is imperative to delve into the diverse array of insect toxins, their mechanisms of action, ecological significance, and potential applications in various fields. One of the fundamental functions of insect toxins is predation. Many insects utilize venom as a means to subdue or kill their prey, facilitating a more efficient consumption of resources (Heppner et al. 2008). Venomous insects have evolved an arsenal of toxins that target specific physiological processes in their prey, ranging from neurotoxins disrupting the nervous system to enzymes causing tissue degradation (Fig. 9.1). Notable examples include the venomous cocktail of insects belonging to the order Hymenoptera, Coleoptera, and Hemiptera. Beyond predation, insects deploy toxins as a defense mechanism against potential threats (Oliver and Perlman 2020). This dual role of toxins in predation and defense showcases the adaptability and sophistication of insect biochemical weaponry. Insect toxins often act at the molecular level, targeting specific receptors or enzymes within the prey (Cordero et al. 2015; Heckel 2020). Neurotoxins, a prevalent class of insect toxins, disrupt nerve system function by interfering with ion channels (Monge-Fuentes et al. 2015; Wang et al. 2000). This precise mode of action highlights the intricate biochemical adaptations in insects that have evolved over millions of years.

Insect toxins also have found applications in medical research and therapeutics. Components of bee venoms, for example, have been investigated for their potential as painkillers or anti-rheumatoid and anti-inflammatory properties (Wehbe et al.

Fig. 9.1 Venomous insects. Insects are the largest class in the phylum Arthropoda. The members of this class are characterized by chitinous exoskeleton, segmented body and jointed legs. Many of the members in the class Insecta are characterized by presence of venoms and allergens. Belostoma (**a**), Lepidopteran caterpillar (**b**), and wasp (**c**) are typical examples for insects with venoms and allergens

2019; Monge-Fuentes et al. 2015). So understanding the mechanisms of action of insect toxins not only provides insights into the evolution of biochemical diversity but also offers a valuable resource for developing novel drugs and therapies. In addition to their role in predation and defense, insect toxins contribute significantly to shaping ecological interactions that influence the behavior and evolution of predators. The study of insect toxins thus unveils the interplay between species that reflect the delicate balance of natural communities. Insects, as primary producers or consumers in various food webs, exert a profound influence on ecosystem structure and function. So the ecological impact of insect toxins extends beyond individual species interactions to broader ecosystem dynamics (Skelhorn and Ruxton 2008). The influence of insect toxins on the population dynamics of other organisms shows that resilience and stability of ecosystems largely depend on insects. So the studies on insect toxins will point out intricacies of evolutionary adaptation and ecological interdependence.

Environmental conditions influence venom production in insects (Tobassum et al. 2018). Optimal environmental conditions can enhance the efficiency of venom synthesis and delivery. Many insects are ectothermic, meaning their metabolic activities are temperature-dependent. Additionally, variations in humidity may impact the viscosity and efficacy of the venom (Akpek 2023). Social factors, particularly in species with complex social structures like ants and bees, also affect venom production (Cremer et al. 2007). In these colonies, individuals may release venom in response to intruders, rival colonies, or to defend the nest. The chemical

signals exchanged within the colony serve as cues triggering venom release and point the importance of communication in social insect behavior. Overall, the multifaceted triggers for venom formation in insects underscore the sophisticated adaptations that have evolved over time. This allows insects to efficiently utilize their venomous arsenal in response to ecological challenges. The precise targeting of specific receptors by insect toxins provides a unique advantage in drug development, allowing for tailored interventions with minimal side effects (Cordero et al. 2015). The sustainable utilization of insect toxins as commodities represents a paradigm shift in our approach to healthcare. This evolving understanding of insect toxins reflects the ongoing exploration of nature's biochemical diversity for the betterment of human and environmental health.

Studying insect toxins provides insights into medical research and drug development (Dossey 2010). Investigating the chemical nature and mechanisms of action of insect toxins opens avenues for developing novel therapeutics, contributing to advancements in treating pain, neurological disorders, and other health challenges (Monge-Fuentes et al. 2015). Insect toxins also offer alternative approaches to pest control, aligning with the growing emphasis on sustainable agriculture. Biopesticides based on insect toxins allows for targeted pest management while minimizing the ecological impact associated with traditional chemical pesticides (Kramer and Muthukrishnan 1997). This approach supports pest control strategies that are environment friendly and conducive to maintaining a balanced ecosystem. As key players in ecosystems, insects contribute significantly to biodiversity. Thus, understanding the role of insect toxins in predation, defense, and ecological interactions contributes to the conservation of insect species and their habitats. Hence, unraveling the biochemical nature of insect toxins not only deepens understanding of fundamental ecological processes but also paves the way for innovative solutions in medicine, agriculture, and conservation.

9.2 Diversity of Venomous Insects

Insects are capable of producing human envenomation. Some insect taxa are devoid of toxins or any way of chemical defenses, but orders with high species number especially Hymenoptera, Lepidoptera, etc., are characterized by various ways of chemical defense. There are insects which uses toxins for defense and to kill the prey (Table 9.1). They become harmful to humans by transmitting diseases and causing direct poisoning. Insects belonging to the order Orthoptera (Crickets, grasshoppers, etc.), order Blattodea (Cockroaches), order Hemiptera (Bugs), order Coleoptera (Beetles), order Hymenoptera (Ants, bees, and wasps), order Lepidoptera (Butterflies and moths), order Diptera (True flies and midges), order Siphonaptera (Fleas), etc., can produce toxins. They are equipped with biotoxins that can be poisonous when eaten or harmful upon entry into the blood stream. Insect toxins have different actions and varied pathogenic intensities.

Table 9.1 Venomous and allergic insects

Order	Family	Scientific name	Common name
Orthoptera	Tridactylidae	*Tridactylidae* sp.	Pygmy mole cricket
Blattodea	Blattoidea	*Periplaneta americana*	Cockroaches
Heteroptera	Pentatomidae	*Halyomorpha halys*	Stink bugs
	Belostomatidae	*Lethocerus* sp.	Giant water bugs
	Reduviidae	*Narvesus carolinensis*	Assassin bugs
	Cimicidae	*Cimex lectularius*	Cimex bugs
Coleoptera	Gyrinidae	*Gyrinus natator*	Whirligig beetle
	Silphidae	*Necrophila americana*	Carrion beetles
	Staphylinidae	*Staphylinidae* sp.	Rove beetles
	Lampyridae	*Lampyridae* sp.	Fire fly
	Dermestidae	*Dermestes lardarius*	Larder beetle
	Coccinellidae	*Coccinellidae* sp.	Ladybugs
	Cerambycidae	*Anoplophora glabripennis*	Longhorn beetle
Hymenoptera	Formicidae	*Formica rufa*	Ants
	Vespidae	*Vespula germanica*	Vespids
	Apidae	*Apis mellifera*	Bees
Lepidoptera	Megalopygidae		Moths and butterflies
Diptera	Cecidomyiidae	*Cecidomyiidae* sp.	Gall gnats
	Asilidae	*Zosteria* sp.	Robber flies
	Culicidae	*Anopheles* sp.	Mosquito
	Simuliidae	*Araucnephia* sps.	Black flies
	Ceratopogonidae	*Culicoides impunctatus*	Biting midges
	Hippoboscidae	*Hippobosca equina*	Louse flies
	Tabanidae	*Philoliche* sp.	Horse flies
	Psychodidae	*Phlebotomus* sp.	Sand flies
	Muscidae	*Stomoxys calcitrans*	Stable flies
	Glossinidae	*Glossina* sp.	Tsetse flies
Siphonaptera	Pulicoidea	*Pulex irritans*	Fleas

Aquatic Coleopterans like water beetles inject a poison when it bites. Terrestrial beetles like blister beetle excrete an irritant which can cause irritation and blistering when comes in contact with body. Saliva of a species of water bug Belostoma contains a toxin that causes paralysis in the prey. Venom of ants, bees, wasps, etc., causes pain and inflammation locally in the affected area (Fig. 9.1). Multiple bites and stings of these insects may causes allergic reactions like an anaphylactic shock. Saliva of parasitic insects like flea contains certain proteins that cause allergic reactions in its host. Larvae of a number of species of moths and butterflies have venomous setae or bristles over their body which causes a burning sensation and a temporary redness when comes in contact with human skin. Hairs and setae of some adult moths also cause skin irritation when come in contact with tender skin of human body. Common insects like beetles, wasps, ants, caterpillars of moths, butterflies, etc., are with toxins in them. The members of the order Orthoptera had

chemical defense mechanisms like regurgitation, anal discharges, defecation, glandular defensive secretions, internal toxins, allergens, etc. *Taeniopoda* grasshoppers can be seen in arid and semi-arid brush and grass lands, and it may also live in oak savanna and woodlands. They can discharge their defensive secretions from paired metathoracic spiracles. Pygmy mole crickets (Family: Tridactylidae) which lives in moist sandy soil near water sources such as dam, lakes, streams, etc., can produce chemical repellents.

Among the members of the order Blattodea, cockroaches are nocturnal, during day time they hide under loose bark of trees, rotting logs, etc. They prefer fairly moist habitats. Cockroaches emit defensive secretions from various exocrine glands. According to O'Donel Alexander (1984), they may cause allergic symptoms and bites in humans. The members of the order Heteroptera, stink bugs of the family Pentatomidae cause contact skin injuries. Cockroaches live in cracks or bark of older or dead trees. They can also be seen in walls of buildings, leaf litter, or on vegetation. Insects of the family Belostomatidae can cause painful bites. Members of Belostomatidae (Giant water bugs) are aquatic and are of almost worldwide distribution. It can cause painful bites and their salivary gland secretions are toxic to humans (Haddad Jr et al. 2010). Among the predatory Heteropterans, members of the family Reduviidae possess venoms with non-lethal paralytic effect on the prey (Sahayaraj and Muthukumar 2011). Terrestrial Assassin bugs of the family Reduviidae live outdoors in any terrestrial plants or crops and prey on other insects. They suck blood from humans and transmit diseases. Cimex or bed bugs of family Cimicidae is a hematophagous bug which is an ectoparasite of man and the saliva causes allergic reactions.

Insects in the order Coleoptera include beetles and weevils. Defensive mechanisms of Coleopterans include poisonous bite, defecation, reflex bleeding, etc. Toxins of several species cause skin eruptions in mammals. Defensive chemicals like pederin from the genus *Paederus* of the family Staphylinidae, Cantharidin from the families Meloidae and Oedemeridae, and the serrated larvae of Dermestidae can cause skin eruptions in mammals (O'Donel Alexander 1984). Family Gyrinidae of Coleoptera includes Whirligig beetles that swim on the surface of water. Pygidial secretions of these beetles contain some chemicals which help them in avoiding predation (Miller et al. 1975). Anal and oral secretions of certain genus of Carrion beetles of the family Silphidae contain defensive compounds with antimicrobial activity. Anal secretions of Rove beetles of the family Staphylinidae contain defensive chemicals. Fireflies belong to the family Lampyridae. They are distasteful or unacceptable to a number of predators due to the presence of some defensive compounds (Lloyd 1973; Blum and Sannasi 1974; Sydow and Lloyd 1975; Moosman Jr et al. 2009; Eisner et al. 1978, 1997; De Cock and Matthysen 2001; Tylor 2001). Members of the family Coccinellidae lives mainly in plants. Coccinelline an alkaloid produced by them is the major toxic component of Coccinellidae (Marples 1990). Blister beetles of the family Meloidae distributed mainly in temperate steppes, tropical, and subtropical savannahs and other open habitats (Cardoso et al. 2009). They are phytophagous and feed on flowers and leaves of plants. Its hemolymph contains cantharidin a defensive chemical that is released by a bleeding

reflex when adult beetles are disturbed. About 10–25 mg of cantharidin is enough to kill a human being within 24 h. The long-horned beetle *Onychocerus albitarsis* in the family Cerambycidae which lives in the Amazon and Atlantic forest regions is known to inject a venomous secretion through stingers located at the ends of its antenna (Berkov et al. 2008). This venomous secretion causes skin irritation and subcutaneous inflammation in man (Berkov et al. 2008; Amaral et al. 2018).

Insects of the order Hymenoptera also produce toxins which are harmful to humans. Ants, which belong to the family Formicidae, outnumber all other land animals. Typically, an ant colony comprises workers which are sterile wingless females, soldiers which are modified workers with mouthparts fitted for crushing and fighting, fertile females with a large body, wings, and well-developed reproductive organs, and fertile males with well-developed reproductive organs and male genitals and they live in nest, namely, formicaria. Ant stings are common in humans especially in rural areas. A few species cause serious allergic reactions in humans due to the toxicity of their venom. Ants use their toxins for defense against predators, competitors, and microbial pathogens as well as for social communication. Most venomous ants are the bullet ants found in South and Central America. They live in small groups under fallen tree trunks or rotting log (Sharma and Upadhyay 2021). Fire ants of the species *Solenopsis invicta* are carnivorous and its venom is composed of alkaloids. Fire ants use their powerful jaws to attach to the skin of the prey and they can sting up to 10 times with their abdominal stinger which may cause pain, itchy burning sensation, and urticated papule initially. Multiple stings in children may cause pustulosis and secondary infections (Haddad Junior et al. 2015).

Bees and wasps are social insects that belong to the families Apidae and Vespidae, respectively. In family Vespidae comes hornets and wasps. Their colony consists of fertile queen which is a female and sterile female, they are the workers and males. Wasps make their nests in a variety of places like on tree trunks, in holes or underground, on sun shades of buildings, etc. They use wood fiber for building their nest. Using the saliva queen softens the wood fiber in the form of paper pulp. Parasitoid wasps help to control agriculture pests. They use their venom for paralyzing the prey, which are usually insect pests. So, they are considered beneficial to human beings. Wasp venom is alkaline in nature. The wasp venom causes local irritations and allergic reactions but multiple stings may cause anaphylactic shock. Honey bees of the family Apidae are polymorphic and social insects. Bee venom is acidic in nature. Bee stings cause burning pain and allergic reactions in humans.

Order Lepidoptera includes butterflies and moths. Their larvae cause erucism which is characterized by intense pain associated with mild local inflammation. Larvae of Megalopygidae and Saturniidae cause erucism. Bodies of the caterpillars of these insects are covered by bristles which when contacted with human body causes an intense dermatitis. Insects of the order Diptera show a worldwide distribution. Studies on the larval forms revealed presence of toxic or venomous secretions in a few of members of Diptera. Toxic phenol oxidases were isolated from the salivary glands of the predatory gall midges of the family Cecidomyiidae. These chemicals enable them to paralyze their prey (Mayr 1974). Saliva of the adult Robber flies (Family: Asilidae) contains a venomous secretion which can kill a white mouse or a

locust. Mosquitoes, black flies, biting midges, louse flies, horse flies, sand flies, stable flies, and tsetse flies are hematophagous. They belong to the families Culicidae, Simuliidae, Ceratopogonidae, Hippoboscidae, Tabanidae, Psychodidae, Muscidae, and Glossinidae, respectively. Females of these flies require a blood meal to develop eggs and hence they are hematophagous. Their bites produce irritation and skin lesions in humans. And also, the larval forms of various dipterans may produce cutaneous or furuncular myiasis by invasion of human skin (O'Donel Alexander 1984). Fleas of the order Siphonaptera can produce skin lesions in humans. Secretions of many of them contain anesthetics to reduce the irritations of the host.

9.3 Chemistry of Insect Toxins

Insect toxins are diverse in chemical nature (Kachel et al. 2018; Schwartz et al. 2012). Some common chemical forms of insect toxins are peptides, alkaloids, and organic compounds such as pyrethroids, polyamines, etc. (Fig. 9.2). Each type targets different biological mechanisms in humans. Majority of the toxins targets the nervous system and neuromuscular ion channels. Therefore, insect toxins rapidly affect the behavior of organisms to which the toxins is applied or injected. Other mode of actions of insect toxins also has been identified. For example, apamin from bee venom targets calcium activated potassium channels which can in turn control the release of neuropeptides (Gu et al. 2020).The toxin produced by beetles is diamphotoxin, which is a hemolytic peptide toxin with traditional use as an arrow poison. Philanthotoxins are neurotoxic polyamines found in the venom of certain ant species, particularly in the genus Philanthus. Also, philanthotoxin is used by the digger wasp to paralyze the insect prey which affects the ligand-gated ion channels in the nervous system and neuromuscular junctions (Kachel et al. 2019). So it is clear that this toxin targets the nervous system of the prey affecting motor function and results in paralysis.

Some insect toxins are alkaloids (Arbiser et al. 2007). For example, solenopsins are toxic alkaloid compounds found in the fire ant venom. It cause pain, irritation, and damage upon contact with skin and can trigger allergic reactions in humans. These compounds are part of the defense mechanism of fire ants against predators. Recently, solenopsin analogs have been explored for the potential human psoriasis (Arbiser et al. 2017). Apamin is a peptide neurotoxin found in bee venom. Apamin affects their electrical activity of nerve cells. This can lead to hyper excitability in nerve cells. This compound also regulates gene expression in various signal transduction pathways. Apamin specifically targets Na-K channels, which are mainly found in the brain (Gu et al. 2020). By blocking these channels, apamin inhibits the outflow of potassium ions, leading to increased neuronal excitability and enhanced release of neurotransmitter, respectively. Melittin is another peptide component found in bee venom. This compound is the main component and the major pain producing substance of honey bee venom (Chen et al. 2016). It is also known for

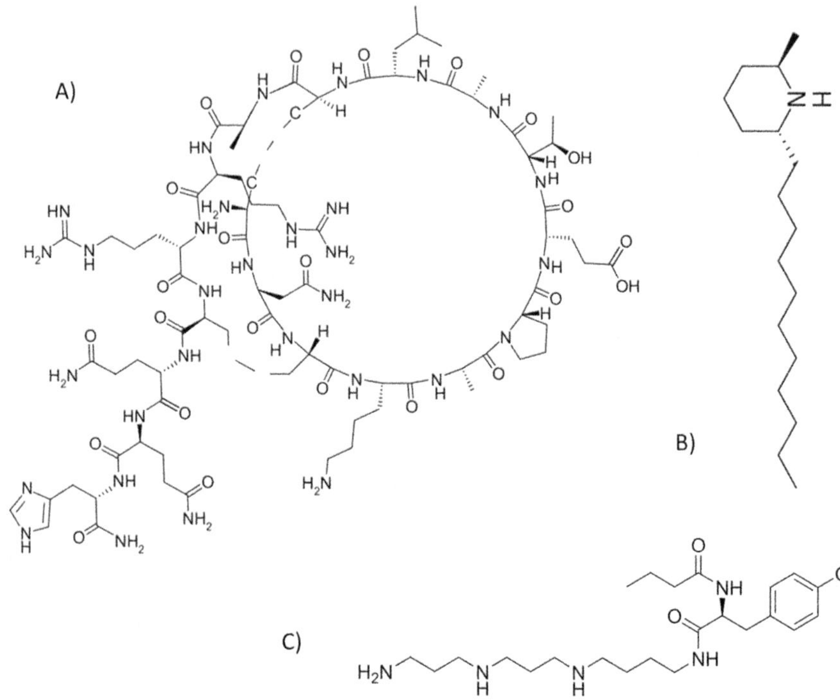

Fig. 9.2 Chemical nature of insect venoms. Insect venom can be peptides or polyamines or enzymes or volatile compounds or bioactive compounds. Most of the insect toxin produces allergic reactions in human body that can be cured with anti-allergic agents. (**a**) Apamin—a neurotoxin in bee venom comprised 18 amino acids, (**b**) Solenopsin—a lipophilic alkaloid in fire ants, and (**c**) Philanthotoxin—a polyamine toxin found in wasps

strong hemolytic and cell membrane destruction properties. This peptide is amphipathic in nature with both hydrophilic and hydrophobic regions. It interacts with cell membranes by forming pores, disrupting their integrity and leading to cell lysis. This ability to permeate cell membranes has drawn attention of melittin in drug delivery research, where it has been explored as a mean to enhance the transport of medications across cell barriers.

Assassin bug produces peptide venom which immobilizes the prey (Walker et al. 2018). This toxin is injected through sharp mouthparts and can cause pain, swelling, and allergic reactions in humans. The saliva of the assassin bug contains the complex cocktail of small and large peptides which plays a role in the immobilization of the prey. However, the severity of the reaction can vary based on the species of assassin bug and individual sensitivity. Diampotoxin is found in the blister beetles of the genus Diamphidia (Whitman et al. 2019). This toxin causes irritation and blistering when in contact with the skin because of hemolytic and cardio toxication. From leaf beetles species of the genus Leptinotarsa, a peptide toxin, leptinotarsin, similar to that of the assassin bug toxin has also been reported (Sablon et al. 2012).

Mastoparan is another peptide derived from wasp venom known for its ability to disrupt cell membranes (de Santana et al. 2022). It is small amphiphilic peptide toxin from wasp venom composed of 14 amino acids. It interacts with cell membrane, disrupting their integrity by forming pores. This led to various cellular effects, including the release of certain molecules and activation of signaling pathways. Mastoparan inhibits all development form of *Trypanosoma cruzi,* the parasite responsible for Chagas disease (Vinhote et al. 2017). Mastoparan is being used in research to study cell signaling and membrane permeability, respectively. It is apparent that some of the insect toxins including philanthotoxins, apamin, and assassin bug toxins are excellent pharmacological tools with selectively to particular ion channel and membrane receptors. So the potential therapeutic roles of insect toxins either in native form or re-engineered form must be explored.

9.4 Physiological Effects and Mode of Action

Insect venoms are the poisonous matter which certain insects secrete and communicate by bites. Insect venoms are also considered as corrosive and irritating chemicals. But majority of insect venoms are ineffectual when they are administrated topically. Insect venoms show their characteristic responses when they are injected to the body or when they come in contact with the cut skin surface of the victims (Müller 2010) (Fig. 9.3). Also, secretions from the glands of insects can be toxic because of the characteristic responses such as inflammation, irritation, and other changes when it comes in contact with sensitive tissues (Müller 2010). The salivary constituents of housefly which irritate and inflame wounded human skin can also be considered as toxin. Irritating hairs and spines of insects also induce chemical responses such as the release of histamine in humans.

The stinging apparatus in insects is more uniform and it is adapted to the single purpose of injecting or discharging venoms (Bissessarsingh and Starr 2021). The apparatus helps in the administration of the toxin to the prey. Most often irritants such as formic acid reach the intra or subdermal tissues only through the wound made by the mandibles of insects. But some insects bite and sting simultaneously where the bite provides a necessary mechanical advantage for inserting the sting. Venom storage vessels in insects permit a ready supply of venom when needed and these vessels are arranged for the efficient ejection of venom (Walker et al. 2017). A forceful injection and a fast diffusion of venom cause rapid action. But the unstable venoms mainly act only on cell surface of neurons and muscles, respectively.

Most of the predatory insects used their saliva to liquefy, degrade, fragment, or lubricate their food to ingestion. These reactions may also occur without immediate toxic actions in the prey. Toxins prevent the functions of animal tissues. For example, larvae of *Galleria mellonella* bitten by Tabanid larvae affect the nerves and muscles where there is no action current resulting in stoppage of all visceral movements followed by rapid dissolution of tissues (Barnoy et al. 2017). It is common that nervous system of the prey is very susceptible to the insect venom. But some

Fig. 9.3 Insects with allergens. (**a**) Honey bee, (**b**) *Cybister* sps., (**c**) Notonecta sps., and (**d**) Lepidopteran caterpillar. Insects are equipped with special appendages that help to inject toxic substances into human body. Insect stings and bites very often end in allergic reactions. Redness, swelling, and itching are the common symptoms that arise due to allergic reactions to insect bites. However, itching, muscle spasms, tingling in the skin, and swollen glands are common symptoms of contact with caterpillars

toxins are transported across the body via fluids. Insect venom injected anywhere in the host is circulated through the blood in animals and it very often affects excitability of muscles (Eitan et al. 1990). Salivary toxins have less fatal effects (Walker et al. 2016). Hypersensitive reactions are the leading cause of death which occurs in higher animals during bee and wasp stings (Hemmer and Wantke 2020a).Venoms of ant, bees, and wasps are considered as allergic and hence venoms from these insects are considered as immediate action toxins. For example, Bee venom contains several enzymes that release histamine involved in allergic reactions (Hossen et al. 2016). Melittin present in the bee venom causes hemolysis of serum free erythrocytes by direct action that makes the cell membrane permeable to hemoglobin (Kwon et al. 2022). The increase in the permeability of skin capillaries results in lowering of blood pressure that may further lead to respiratory paralysis and death of the victim. The venom of the reduviid bug contains a wide range of toxic substances (Walker et al. 2017). The venom contains trypsin like proteolytic fraction and hyaluronidase, respectively. These enzymes act to disrupt the intercellular matrix of animal tissues. It is widely known that allergic reactions are the most serious consequences of insect bites. For example, mosquito bites commonly cause an immediate reaction with the formation of local swelling with a red flare. This disappears and it is replaced by a delayed response in the form of a larger swollen reddened area. These allergic responses are caused by the salivary secretions of

mosquito (Cantillo et al. 2023). But the complication of insect bites and stings are changed according to species, natural susceptibility, difference in wounding mechanism, nature of venom, and the allergic properties (Singh and Mann 2013).

Insect venoms target different signal receptors of cells and cause pains, inflammation, and asphyxiation in some cases (Bohlen and Julius 2012). Venoms of hymenopteran insect wasps contain cytotoxin peptides that disrupt the plasma membrane by interacting with phospholipids which ultimately lead to pore formation and cell lysis (Moreno and Giralt 2015). Neurotoxic venoms of insects target ion channels directly related to pain (Diochot 2021).These venoms are a deadly cocktail of peptides, proteins, and enzymes with diverse bioactivities. Insect neurotoxins mainly react with calcium channels, sodium channels, potassium channels, and acid sensitive ion channels (Diochot 2021). Insect toxins also found to have anticoagulant activity, platelet aggregating and trypsin inhibitory activity. Thus, it is clear that insect venoms are a mixture of numerous bioactive principles that targets intercellular matrix, cell membrane receptors, and membrane channels for predation and defense.

9.5 Detoxification and Therapeutic Properties of Insect Venom

The evolutionary arms race between animals and insects has resulted in the development of diverse defensive traits and counteradaptive features in animals to combat insect toxins (Tarvin et al. 2023). The mechanism of detoxification varies among animals depending on type of insect toxins (Table 9.2). The most important detoxification enzymes of insect toxins include cytochrome P450s monooxygenases, glutathione S-transferases and esterases which are mainly involved in the detoxification of xenobiotics (Koirala et al. 2022). These detoxification enzymes also interact and mediate the transport of toxic compounds across the insect midgut and store them into less toxic forms of excretory products. Animals employ different molecular mechanisms for adaptation against toxic insect venom via rapid excretion with the help of efflux carriers and sequestration. But insects such as monarch butterfly, queen butterfly, and plain tiger butterfly ingest the cardenolides contained within the milkweed plant and sequester them for defense against their predators (Premachandran and Srinivasan 2022). In an experiment, first to second instar stage larvae of the insect, namely, *Olepa* sps. Allowed to feeding on *Calotropis gigantea* leaves (Premachandran and Srinivasan 2022). It is noted that the first and second instar larva avoided the vein region rich in glycosides. However, on reaching the third instar, the larvae started feeding the entire leaf region including veins and fully matured. It is found that carboxyl esterase, glutathione S-transferase, and cytochrome P450s enzymes are the major detoxification enzymes present in the insect body. *Olepa* sps. Larva fed with Calotropis leaves and their excreted fecal waste and fresh Calotropis leaves were subjected to High Resolution Liquid Chromatograph

Table 9.2 Insect venoms, curative measures, and therapeutic application

Name	Toxicant	Effect in human	Curative measure	Therapeutics
Apis mellifera	Aptamin, Mellitin	Localized inflammation	Remove the stinger, wash with soap and water, apply cold pad.	Anti-tumor, anti-parasitic, anti-bacterial, anti-viral, and anti-inflammatory activities.
Polybia sericea	Pompilidotoxins	Irritation of eyes and skin	Clean the area thoroughly with water, place a cool compress on eyelid.	Treatment of pain, multiple sclerosis
Cryptocheilus bicolor	Mutillid venoms	Very sharp and immediate pain on affected area	Clean the area with vinegar and keep the wound clean and dry.	Rheumatoid arthritis, wound healing
Trielis spp.	Kinins	Sometimes residual hard spot in skin, later pain in sting area, redness, swellings	Apply cold pack, anti-inflammatory medications.	Lowering of blood pressure, increase blood flow throughout the body
Dasymutilla occidentalis	Many peptides	Pain in the affected area, burning and corrosive pain	Apply cold pack and anti-inflammatory medicines.	
Epicauta fabricii	Meloe toxin	Skin irritation	Anti-inflammatory drugs.	
Cimex lectularius	Histamine	Irritation in skin and respiratory system	Anti-histamines.	
Photuris lucicrescens	Fatty acid derivatives	Skin irritation	Symptomatic treatment with antidotes.	
Ponerine spp.	Poneratoxin	Nervous system	Supportive care with ant anti-allergic agents.	
Atta geminata	Solenopsin	Skin and respiratory system	Treatment with antidotes.	
Lonomia obliqua	Lonomia venom	Blood and skin	Symptomatic treatment with antidotes.	
Thaumetopoea pityocampa	Thaometopoein	Skin and respiratory system	Anti-inflammatory drugs.	

Mass Spectrometer (HR-LCMS) analysis to confirm the presence of cardenolides. Also, protein-ligand docking was done to predict the interaction between glutathione S-transferase (GST) sigma and delta variant of *Olepa* sps. With GSH-strophanthidin complex. The study also indicated that glutathione S-transferase (GST) enzyme activity was significantly higher in larva fed with *Calotropis* compared to *Ricinus* and banana leaves.

Insect venom induces innate and adaptive immune responses in animals (Avalo et al. 2022). But physiological effects of insect venom in invertebrates are little known. In vertebrates, responses in human are best known. Venomous salivary secretions from insects have common effects such as corrosion of tissues because salivary secretions are digestive in function (Ali 1997). Some insect toxins lead to the production of eicosanoids which is directly related to inflammation, pain, swelling, and redness (Stanley et al. 2012). Allergic reactions against insect toxins will result in a small, red, and itchy bump in the skin (Hemmer and Felix Wantke, 2020b). In insect venom allergies, the immune system attacks certain parts of proteins that enter the body during the insect bites or stings. Sometimes people get stung or bitten over many years without having unusual reaction. Over time though their bodies become more sensitive to the venom in the sting or bite, and they then suddenly have an allergic reaction to it.

Degree of allergic responses varies widely with the sensitivity of individuals with respect to particular insects. Insect bite on mildly sensitive individuals makes them uncomfortable with mild respiratory distress and cramps. Itching and edema are also common. But in severe reactions the insect venom causes cyanosis resulting in diminished circulation and respiration, a flaccid paralysis, and a marked increase in the number of leucocytes. Some insect bites cause localized pains and blisters that result in necrosis of tissues. When the insect saliva comes in contact with sensitive membrane it causes vasodilation and edema, respectively (Smith et al. 2022). A major way our bodies keep away insect toxin is through a built-in mechanism designed to filter out harmful chemicals and eliminate it through skin, lungs, kidney, and colon (Van Ness et al. 2022). Insects are also faced with numerous toxins as they go through life. To survive the natural toxins, insects have evolved various detoxification mechanisms such as reflex bleeding (Knapp et al. 2020). However, this defense differs among different stages, populations, and species of insects.

A curative measure against insect toxins depends on type of the toxin and the severity of the reaction it causes. There are specific antidotes that can be administered against insect toxins. For example, anti-venom is available for certain types of venomous bites or stings such as that of bees (Teixeira-Cruz et al. 2021). Also, antihistamines can help to reduce itching and swelling caused by insect bites or stings. Pain relievers may also be necessary for avoiding discomfort. Lotion made with calamine and hydrocortisone is also promising to reduce symptoms caused by insect bites. Also, anti-inflammatory compounds such as lemon balm essential oils also help to relive toxic effects of insect venom (Ramanauskienė et al. 2015). Glucocorticoids are another candidate compound that help to alleviate insect bite related reactions in the body. Steroid-based medication mainly helps to alleviate allergic reactions during insect bites. For example, the corticosteroids deplete

basophils which in turn decrease leukocyte histamine while concentration of histamine in the single basophil cell remains unchanged (Dunsky et al. 1979). This approach helps to maintain the histamine stores in the skin at a constant rate while reducing allergic reactions.

The bioactive compounds in the insect venoms interact with physiological processes in animals. Hence, these compounds can be used as therapeutic agents. The evolutionary selection over the time makes insect venom a suitable candidate for patho-physiological disorders. The major bioactive compounds in insect venom used for therapeutic purposes are phospholipase A_2, mastoparan, bradykinins, melittin, apamin, hyaluronidase, ectatomin, and mast cell degranulating peptide, respectively (Moran et al. 2022) (Table 9.2). The membrane phospholipids are one of the interactive sites of insect venom which make it a suitable therapeutic agent against cancer. The binding of insect venom causes lysis of cancer cells via formation of membrane pores. Also, the membrane binding property of insect venom helps to formulate target specific drugs. For example, mastoparan targets anionic phosopholipids in cancerous cells and results in the formation of pores in the plasma membrane. Also, mastoparan is found to be effective in release of insulin from pancreatic cell and thereby helps to treat type II diabetes. Similarly, melittin binds with membrane lipids and results in exocytosis. This compound also blocks MAPK signaling cascade via interaction with COX-2 and VGRF-2, respectively. Solenopsin is another potential bioactive compound found to interfere with PI3K signaling cascade and angiogenesis (Arbiser et al. 2007). The bioactive compound from *Scorpio maurus palmatus*, namely, maurotoxin is found to inhibit K_{Ca} channel in human T lymphocyte cells involved in asthma. Above studies indicate that insect toxins are one of the main targets of future drug discovery.

References

Akpek A (2023) Detailed analysis of the effects of viscosity measurement errors caused by heat transfer during continuous viscosity measurements under various temperature changes and the proposed solution of a non-dimensional parameter called the Akpeknumber. Appl Sci 13(19):10684. https://doi.org/10.3390/app131910684

Ali DW (1997) The aminergic and peptidergic innervation of insect salivary glands. J Exp Biol 200(14):1941–1949. https://doi.org/10.1242/jeb.200.14.1941

Amaral A et al (2018) Envenomations in humans caused by the venomous beetle onychocerus albitarsis: observation of two cases in São Paulo state, Brazil. J Clin Toxicol 08(04):1000392. https://doi.org/10.4172/2161-0495.1000392

Arbiser JL et al (2007) Solenopsin, the alkaloidal component of the fire ant (Solenopsis invicta), is a naturally occurring inhibitor of phosphatidylinositol-3-kinase signaling and angiogenesis. Blood 109(2):560–565. https://doi.org/10.1182/blood-2006-06-029934

Arbiser JL et al (2017) Evidence for biochemical barrier restoration: topical solenopsin analogs improve inflammation and acanthosis in the KC-Tie2 mouse model of psoriasis. Sci Rep 7(1):1198. https://doi.org/10.1038/s41598-017-10580-y

Avalo Z et al (2022) Biological effects of animal venoms on the human immune system. Toxins 14(5):344. https://doi.org/10.3390/toxins14050344

Barnoy S et al (2017) The galleria mellonella larvae as an in vivo model for evaluation of Shigella virulence. Gut Microbes 8(4):335–350. https://doi.org/10.1080/19490976.2017.1293225
Berkov A, Rodríguez N, Centeno P (2008) Convergent evolution in the antennae of a cerambycid beetle, Onychocerus albitarsis, and the sting of a scorpion. Sci Nat 95(3):257–261. https://doi.org/10.1007/s00114-007-0316-1
Bissessarsingh M, Starr CK (2021) Comparative morphology of the stinger in social wasps (Hymenoptera: Vespidae). Insects 12(8):729. https://doi.org/10.3390/insects12080729
Blum MS, Sannasi A (1974) Reflex bleeding in the lampyrid Photinus pyralis: defensive function. J Insect Physiol 20(3):451–460. https://doi.org/10.1016/0022-1910(74)90153-x
Bohlen CJ, Julius D (2012) Receptor-targeting mechanisms of pain-causing toxins: how ow? Toxicon 60(3):254–264. https://doi.org/10.1016/j.toxicon.2012.04.336
Cantillo JF et al (2023) Mosquito allergy: immunological aspects and clinical management. Mol Immunol 164:153–158. https://doi.org/10.1016/j.molimm.2023.11.009
Cardoso JL et al (2009) Animaispeçonhentos no Brasil: biologia, clínica E terapêutica. Editora Sarvier, São Paulo
Chen J et al (2016) Melittin, the major pain-producing substance of bee venom. Neurosci Bull 32(3):265–272. https://doi.org/10.1007/s12264-016-0024-y
Cordero M et al (2015) Identifying insect protein receptors using an insecticidal spider toxin. In: Spider venoms. Springer, Dordrecht, pp 1–11
Cremer S, Armitage SAO, Schmid-Hempel P (2007) Social immunity. Curr Biol 17(16):R693–R702. https://doi.org/10.1016/j.cub.2007.06.008
De Cock R, Matthysen E (2001) Do glow-worm larvae (Coleoptera: Lampyridae) use warning coloration? Ethology 107(11):1019–1033. https://doi.org/10.1046/j.1439-0310.2001.00746.x
de Santana CJC et al (2022) Mastoparans: a group of multifunctional α-helical peptides with promising therapeutic properties. Front Mol Biosci 9:824989. https://doi.org/10.3389/fmolb.2022.824989
Diochot S (2021) Pain-related toxins in scorpion and spider venoms: a face to face with ion channels. J Venom Anim Toxins Incl Trop Dis 27:e20210026. https://doi.org/10.1590/1678-9199-jvatitd-2021-0026
Dossey AT (2010) Insects and their chemical weaponry: new potential for drug discovery. Nat Prod Rep 27(12):1737. https://doi.org/10.1039/c005319h
Dunsky E, Zweiman B, Fischler E, Levy D (1979) Early effects of corticosteroids on basophils, leukocyte histamine, and tissue histamine. J Allergy Clin Immunol 63(6):426–432. https://doi.org/10.1016/0091-6749(79)90217-3
Eisner T et al (1978) Lucibufagins: defensive steroids from the fireflies Photinus ignitus and P. marginellus (Coleoptera: Lampyridae). Proc Natl Acad Sci USA 75(2):905–908. https://doi.org/10.1073/pnas.75.2.905
Eisner T et al (1997) Firefly 'femmes fatales' acquire defensive steroids (lucibufagins) from their firefly prey. Proc Natl Acad Sci USA 94(18):9723–9728. https://doi.org/10.1073/pnas.94.18.9723
Eitan M et al (1990) A scorpion venom neurotoxin paralytic to insects that affects sodium current inactivation: purification, primary structure, and mode of action. Biochemistry 29(25):5941–5947. https://doi.org/10.1021/bi00477a009
Gu H, Han SM, Park K-K (2020) Therapeutic effects of apamin as a bee venom component for non-neoplastic disease. Toxins 12(3):195. https://doi.org/10.3390/toxins12030195
Haddad V Jr et al (2010) Bites caused by giant water bugs belonging to Belostomatidae family (Hemiptera, Heteroptera) in humans: a report of seven cases. Wilderness Environ Med 21(2):130–133. https://doi.org/10.1016/j.wem.2010.01.002
Haddad Junior V et al (2015) Venomous and poisonous arthropods: identification, clinical manifestations of envenomation, and treatments used in human injuries. Rev Soc Bras Med Trop 48(6):650–657. https://doi.org/10.1590/0037-8682-0242-2015
Heckel DG (2020) How do toxins from Bacillus thuringiensis kill insects? An evolutionary perspective. Arch Insect Biochem Physiol 104(2):e21673. https://doi.org/10.1002/arch.21673

Hemmer W, Wantke F (2020a) Insect hypersensitivity beyond bee and wasp venom allergy. Allergol Select 4(01):97–104. https://doi.org/10.5414/alx02123e

Hemmer W, Wantke F (2020b) Stichreaktionenabseits von Biene und Wespe. Allergologie 43(02):73–81. https://doi.org/10.5414/alx02123

Heppner JB et al (2008) Venoms and toxins in insects. In: Encyclopedia of entomology. Springer, Dordrecht, pp 4076–4089

Hossen M et al (2016) Impact of bee venom enzymes on diseases and immune responses. Molecules 22(1):25. https://doi.org/10.3390/molecules22010025

Kachel HS, Buckingham SD, Sattelle DB (2018) Insect toxins—selective pharmacological tools and drug/chemical leads. Curr Opin Insect Sci 30:93–98. https://doi.org/10.1016/j.cois.2018.10.001

Kachel HS, Franzyk H, Mellor IR (2019) Philanthotoxin analogues that selectively inhibit ganglionic nicotinic acetylcholine receptors with exceptional potency. J Med Chem 62(13):6214–6222. https://doi.org/10.1021/acs.jmedchem.9b00519

Knapp M, Řeřicha M, Židlická D (2020) Physiological costs of chemical defence: repeated reflex bleeding weakens the immune system and postpones reproduction in a ladybird beetle. Sci Rep 10(1):9266. https://doi.org/10.1038/s41598-020-66157-9

Koirala BK, Moural T, Zhu F (2022) Functional and structural diversity of insect glutathione S-transferases in xenobiotic adaptation. Int J Biol Sci 18(15):5713–5723. https://doi.org/10.7150/ijbs.77141

Kramer KJ, Muthukrishnan S (1997) Insect chitinases: molecular biology and potential use as biopesticides. Insect Biochem Mol Biol 27(11):887–900. https://doi.org/10.1016/s0965-1748(97)00078-7

Kwon N-Y et al (2022) Anticancer activity of bee venom components against breast cancer. Toxins 14(7):460. https://doi.org/10.3390/toxins14070460

Lloyd JE (1973) Firefly parasites and predators. Coleopt Bull 27:91–106

Marples NM (1990) The influence of predation on ladybird colour pattern. University of Wales

Mayr L (1974) UntersuchungenzurFunktion der Speicheldrüsen räuberischer Gallmücken larven (Aphidoletesaphidimyza Rond.)1. Z Angew Entomol 77(1–4):270–273. https://doi.org/10.1111/j.1439-0418.1974.tb03258.x

Miller JR, Hendry LB, Mumma RO (1975) Norsesquiterpenes as defensive toxins of whirligig beetles (Coleoptera: Gyrinidae). J Chem Ecol 1(1):59–82. https://doi.org/10.1007/bf00987720

Monge-Fuentes V et al (2015) Neuroactive compounds obtained from arthropod venoms as new therapeutic platforms for the treatment of neurological disorders. J Venom Anim Toxins Incl Trop Dis 21(1). https://doi.org/10.1186/s40409-015-0031-x

Moosman PR Jr et al (2009) Do courtship flashes of fireflies (Coleoptera: Lampyridae) serve as aposematic signals to insectivorous bats? Anim Behav 78(4):1019–1025. https://doi.org/10.1016/j.anbehav.2009.07.028

Moran D et al (2022) Insect venoms and their bioactive components: a novel therapeutic approach in chronic diseases and cancer. J Cancer Sci Clin Ther 06(04). https://doi.org/10.26502/jcsct.5079176

Moreno M, Giralt E (2015) Three valuable peptides from bee and wasp venoms for therapeutic and biotechnological use: Melittin, apamin and mastoparan. Toxins 7(4):1126–1150. https://doi.org/10.3390/toxins7041126

Müller UR (2010) Insect venoms. In: Anaphylaxis. S. Karger AG, pp 141–156

O'Donel Alexander J (1984) Arthropods and human skin. Springer, London

Oliver KM, Perlman SJ (2020) Toxin-mediated protection against natural enemies by insect defensive symbionts. In: Advances in insect physiology. Elsevier, pp 277–316

Premachandran K, Srinivasan TS (2022) Polyphagous insect Olepa sps. feeding on cardenolide rich Calotropis gigantea (L.) leaves and detoxification mechanism involving GST. Heliyon 8(9):e10596. https://doi.org/10.1016/j.heliyon.2022.e10596

Ramanauskienė K, Stelmakiene A, Majienė D (2015) Assessment of lemon balm (Melissa officinalis L.) hydrogels: quality and bioactivity in skin cells. Evid Based Complement Alternat Med 2015:1–7. https://doi.org/10.1155/2015/635975

Sablon L et al (2012) Chemical ecology of the Colorado potato beetle, Leptinotarsa decemlineata (say) (Coleoptera: Chrysomelidae), and potential for alternative control methods. Insects 4(1):31–54. https://doi.org/10.3390/insects4010031

Sahayaraj K, Muthukumar S (2011) Zootoxic effects of reduviid Rhynocoris marginatus (fab.) (Hemiptera: Reduviidae) venomous saliva on Spodoptera litura (fab.). Toxicon 58(5):415–425. https://doi.org/10.1016/j.toxicon.2011.06.001

Schwartz EF et al (2012) Arthropod venoms: a vast arsenal of insecticidal neuropeptides. Biopolymers 98(4):385–405. https://doi.org/10.1002/bip.22100

Sharma S, Upadhyay RK (2021) Toxic and immune allergic responses of ant venom toxins: a review. Int J Pharm Pharm Sci 13:1–7. https://doi.org/10.22159/ijpps.2021v13i11.42227

Singh S, Mann B (2013) Insect bite reactions. Indian J Dermatol Venereol Leprol 79(2):151. https://doi.org/10.4103/0378-6323.107629

Skelhorn J, Ruxton GD (2008) Ecological factors influencing the evolution of insects' chemical defenses. Behav Ecol 19(1):146–153. https://doi.org/10.1093/beheco/arm115

Skendžić S et al (2021) The impact of climate change on agricultural insect pests. Insects 12(5):440. https://doi.org/10.3390/insects12050440

Smith LB et al (2022) Novel salivary antihemostatic activities of long-form D7 proteins from the malaria vector Anopheles gambiae facilitate hematophagy. J Biol Chem 298(6):101971. https://doi.org/10.1016/j.jbc.2022.101971

Sollai G, Solari P (2022) An overview of 'insect biodiversity. Diversity 14(2):134. https://doi.org/10.3390/d14020134

Stanley D, Haas E, Miller J (2012) Eicosanoids: exploiting insect immunity to improve biological control programs. Insects 3(2):492–510. https://doi.org/10.3390/insects3020492

Sydow SL, Lloyd JE (1975) Distasteful fireflies sometimes emetic, but not lethal. Fla Entomol 58(4):312. https://doi.org/10.2307/3493699

Tarvin RD, Pearson KC, Douglas TE, Ramírez-Castañeda V, Navarrete MJ (2023) The diverse mechanisms that animals use to resist toxins. Annu Rev Ecol Evol Syst 54(1):283–306. https://doi.org/10.1146/annurev-ecolsys-102320-102117

Teixeira-Cruz JM, Strauch MA, Monteiro-Machado M, Tavares-Henriques MS, de Moraes JA, Ribeiro da Cunha LE, Ferreira RS Jr, Barraviera B, Quintas LEM, Melo PA (2021) A novel Apilic antivenom to treat massive, Africanized honeybee attacks: a preclinical study from the lethality to some biochemical and pharmacological activities neutralization. Toxins 13(1):30. https://doi.org/10.3390/toxins13010030

Tobassum S et al (2018) Effect of milking method, diet, and temperature on venom production in scorpions. J Insect Sci 18(4):19. https://doi.org/10.1093/jisesa/iey081

Tylor J (2001) Are glow-worms Lampyris noctiluca (L.) distasteful? vol 9. The Coleopterist

Van Ness KP, Cesar F, Yeung CK, Himmelfarb J, Kelly EJ (2022) Microphysiological systems in absorption, distribution, metabolism, and elimination sciences. Clin Transl Sci 15(1):9–42. https://doi.org/10.1111/cts.13132

Vinhote JFC et al (2017) Trypanocidal activity of mastoparan from Polybia paulista wasp venom by interaction with TcGAPDH. Toxicon 137:168–172. https://doi.org/10.1016/j.toxicon.2017.08.002

Walker A et al (2016) Venoms of heteropteran insects: a treasure trove of diverse pharmacological toolkits. Toxins 8(2):43. https://doi.org/10.3390/toxins8020043

Walker AA et al (2017) Melt with this kiss: paralyzing and liquefying venom of the assassin bug Pristhesancus plagipennis (Hemiptera: Reduviidae). Mol Cell Proteomics 16(4):552–566. https://doi.org/10.1074/mcp.M116.063321

Walker AA et al (2018) Harvesting venom toxins from assassin bugs and other heteropteran insects. J Vis Exp 134:e57729. https://doi.org/10.3791/57729

Wang X-H et al (2000) Discovery and characterization of a family of insecticidal neurotoxins with a rare vicinal disulfide bridge. Nat Struct Biol 7(6):505–513. https://doi.org/10.1038/75921

Wehbe R et al (2019) Bee venom: overview of main compounds and bioactivities for therapeutic interests. Molecules 24(16):2997. https://doi.org/10.3390/molecules24162997

Whitman DW et al (2019) Antiparasitic properties of cantharidin and the blister beetle Berberomeloe majalis (Coleoptera: Meloidae). Toxins 11(4):234. https://doi.org/10.3390/toxins11040234

Yadav T, Mishra G (2023) Insect toxins and their bioprospecting. In: Insects as service providers. Singapore, Springer, pp 131–162

Chapter 10
Natural Toxins in Diverse Foodstuffs and Foodomics

Nalika R. Dayananda

Abstract As opposed to man-made chemicals or environmental pollutants that get into our food supply, toxins can be present due to their natural occurrence in food. Natural toxins found inherently in foods of plant and animal origins serve specific functions in the plant and animal or evolved as chemical defenses against predators, insects, or microorganisms. Natural toxins in foodstuffs have diverse chemical structures and are vastly different in nature and toxicity, including aquatic biotoxins, cyanogenic glycosides, furocoumarins, lectins, mycotoxins, solanines and chaconine, poisonous mushrooms, pyrrolizidine alkaloids, etc., which can be harmful when consumed in sufficient quantities. New nutritional trends are significantly increasing foodborne outbreaks that remain a worldwide problem. "Foodomics"—a discipline that studies the Food and Nutrition domains through the application and integration of advanced -omics technologies to improve consumer's well-being, health, and knowledge—has the potential to comprehensively elucidate what is in food, how this composition differs across the food systems, and how diet composition as an ensemble of foods guides outcomes for nutrition, health, and sustainability. The evolving behaviors of consumers, current production techniques, food processing strategies, shifts in climate, and the escalation of microbial resistance introduce vulnerabilities throughout the food production and distribution chains. These vulnerable points underscore the critical need for implementing foodomics methodologies and advanced protocols as complementary, data-driven approaches aimed at safeguarding food safety across the entire production and distribution lines. By addressing these critical control points, foodomics can minimize contamination risks from natural toxins while fostering advancements in nutritional science and promoting sustainable dietary practices.

Keywords Natural toxins · Food contaminations · Foodomics · Sustainable diets

---

N. R. Dayananda (✉)
Department of Indigenous Medical Resources, Faculty of Indigenous Health Sciences and Technology, Gampaha Wickramarachchi University of Indigenous Medicine, Yakkala, Sri Lanka
e-mail: nalika@gwu.ac.lk

M. N. V. Prasad et al. (eds.), *Biotoxins*,
https://doi.org/10.1007/978-3-031-75309-1_10

10.1 Natural Toxins in Foodstuffs; Classification and Possible Sources

Natural toxins are biochemical materials or secretions that are naturally and originally produced as a defense mechanism of living organisms against predators, competitors, or environmental stressors, possessing a wide range of biological functions and activities that could be detrimental to human and animal health (Bucheli 2014). These bioactive compounds are generally produced by particular organisms to confer competitive advantages in the environment in which they live and enhance survival (Valério et al. 2010). The mechanism of action of these natural toxins ranges from proteolytic, coagulant, hemolytic, to neurotoxic activities. Interestingly, natural toxins could be beneficial to human and animal health as some xenobiotics obtained from these toxins are currently used for therapeutic purposes (Kachel et al. 2018). Natural toxins could be classified based on biological origin, target organ toxicity, or mode of action (Hodgson 2012). Natural toxins can be found in a variety of foods, including plants, fungi, marine organisms, and some animal products Understanding and identifying the classifications and sources of natural toxins is vital for assessing and monitoring their potential risks to human health and implementing interventions to mitigate or manage their behavior in the food supply.

10.1.1 Classification of Natural Toxins

Based on the chemical structures and biological origins of natural toxins, those can be classified into several categories including alkaloids, glycosides, lectins, mycotoxins, and allergenic proteins (Hajšlová et al. 2018). Alkaloids are nitrogen-containing biochemical compounds found especially in plants, known for their diverse pharmacological effects such as nicotine in tobacco, caffeine in coffee and tea, solanine in potatoes, etc. Glycosides are also commonly found in plants, which have the arrangement of a sugar molecule attached to non-sugar constituents. Cyanogenic glycosides including amygdalin in bitter almonds and linamarin in cassava can release cyanide upon hydrolysis, posing a risk to human health if consumed in higher quantities (Bartnik and Facey 2024; Vetter 2000). Some proteins such as lectins can be found in various herbal foods, which tend to bind with carbohydrates which leads to cellular function disruption and gastrointestinal distress (Vijayan and Chandra 1999). Mycotoxins are another important category of natural food toxins and those are secondary metabolites produced by fungal species that contaminate crops directly and indirectly in various stages from growth to storage through harvest. Aflatoxins, produced by *Aspergillus* species, are potent carcinogens commonly found in peanuts, corn, and other commodities such as spices (Alshannaq and Yu 2017). Allergenic proteins, such as gluten in wheat, rye, and barley, can trigger immune responses in susceptible individuals, leading to allergic reactions or intolerances (Lehrer et al. 1996).

Figure 10.1 illustrates the classification of the main natural toxins found in foodstuffs, showing a structured overview of the diverse range of toxic compounds present in different sources of food. Each group represents a distinct source of toxins, reflecting the broad spectrum of natural toxins including plant-origin toxins, fungal toxins (mycotoxins), bacterial toxins, and marine toxins encountered in the food supply chain.

The plant-origin toxins encompass a variety of toxic compounds secreted from plant species. The alkaloids include pyrrolizidine alkaloids, tropane alkaloids, indole alkaloids, and quinolizidine alkaloids which can adversely affect human health, ranging from neurotoxicity to carcinogenicity (Urugo and Tringo 2023). Glycosides are another subgroup of plant-origin toxins which include cyanogenic glycosides, cardiac glycosides, and saponins. Moreover, lectins are protein compounds found as natural toxins in plant foods, such as beans and grains such as ricin and phytohemagglutinin, which can have toxic effects on the gastrointestinal tract and digestive system and interfere with the nutrient absorption process (Vijayan and Chandra 1999). The impacts of natural toxins present in pasture plants are not limited to grazing livestock but can also be carried through the food chain, with some toxins, such as indospicine, being demonstrated to accumulate in the tissues of

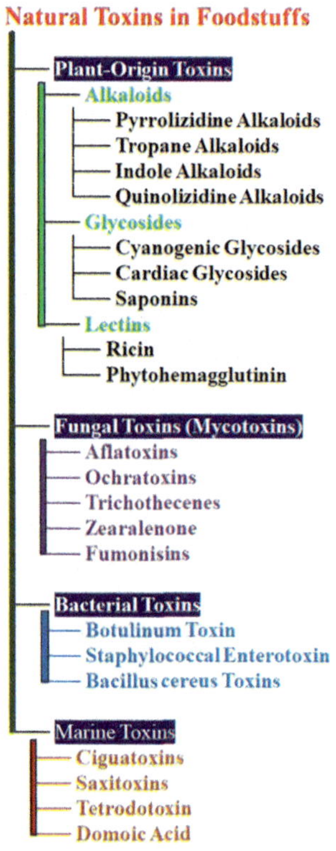

Fig. 10.1 The classification of main natural toxins found in foodstuffs, offering a structured overview of the diverse range of toxic compounds present in different food sources

grazing animals and cause secondary poisoning in animals consuming meat from this livestock (Sultan et al. 2018).

Mycotoxins or fungal toxins consist of toxic metabolites produced by molds and fungi that contaminate food and feed crops with potent carcinogenic aflatoxins (particularly aflatoxin B1 or AFB1) being considered of significant risk to human health (Cao et al. 2022). Once ingested by animals, AFB1 can be carried into milk as the toxic metabolite aflatoxin M1, with the presence of Aflatoxin M1 (AFM1) in milk being of particular concern as young children represent a particularly susceptible age group (Yunus et al. 2019). AFM1 analysis studies have assessed the risk of AFM1 in both liquid and powdered milk marketed in Pakistan, with processed milk shown to have lower AFM1 levels than raw milk (Yunus et al. 2019). Mycotoxin risk is not only limited to the consumption of cereal grains, but can also occur in pollen as used in food supplements (Kostić et al. 2019), and malting and brewing by-products as used in animal feed. Aflatoxins, produced by *Aspergillus* species, are among the most well-known mycotoxins and are associated with hepatotoxicity and carcinogenicity in humans. Ochratoxins, produced by *Aspergillus* and *Penicillium* species, can contaminate grains, coffee, and wine, posing risks to kidney health (Malir et al. 2016). Trichothecenes, zearalenone, and fumonisins are other types of mycotoxins produced by various fungal species (Bennett and Klich 2003), each with its own set of toxic effects on human health, including immunotoxicity, reproductive disorders, and carcinogenesis.

Bacterial toxins are toxic compounds produced by pathogenic bacteria that can contaminate food during processing, storage, or handling. A few examples of bacterial toxins include botulinum toxin, produced by *Clostridium botulinum*, which causes botulism, a severe and potentially fatal form of food poisoning (Hatheway 2018). Staphylococcal enterotoxin, produced by *Staphylococcus aureus*, can cause gastrointestinal illness when present in contaminated foods (Balaban and Rasooly 2000). Bacillus cereus toxins, produced by *Bacillus cereus*, are heat-stable toxins commonly found in rice dishes and can cause vomiting and diarrhea (Granum and Lund 1997). Shiga toxin-producing *Escherichia coli* is a frequent cause of food poisoning with enormous human health and economic impacts, and rapid and accurate identification is imperative for the protection of human health (Armstrong et al. 2018).

The marine toxins include toxic biocompounds produced by marine organisms such as algae, dinoflagellates, bacteria, etc. Ciguatoxins, saxitoxins, tetrodotoxins, and domoic acid are the main marine toxins that can accumulate in seafood (Hall 2009). Seafood toxins, which affect human health adversely if consumed in contaminated fish or shellfish. These toxins can cause various neurological symptoms, including paralysis, respiratory distress, and gastrointestinal disturbances (Blanco et al. 2020).

It is essential to grasp the classification and origins of natural toxins to evaluate their potential hazards to human health and devise strategies to minimize their presence in the food chain of natural ecosystems. Pinpointing significant instances of natural toxins and their origins enables the safeguarding of food safety, security, and public health. Although most natural toxins exist in minimal quantities and pose a

negligible risk to human health, specific toxins can trigger acute or chronic toxicity, resulting in various adverse health outcomes. Therefore, comprehending the health implications and potential risks linked to natural toxins in foodstuffs is imperative for maintaining food safety and safeguarding community health.

10.1.2 Health Effects and Risks Associated with Natural Toxins in Foodstuff

Natural toxins in food can cause both acute (short-term) and chronic health effects (particularly when consumed in high concentrations or over prolonged periods) with a range of clinical symptoms. Acute symptoms range from mild gastrointestinal upset, neurological symptoms, and respiratory paralysis to fatality (Dolan et al. 2010). Depending on the chemical structure and biological activity of the natural toxic compounds, natural toxins can exert their toxic effects through different kinds of mechanisms. Some biotoxins act as enzyme inhibitors, disrupting vital biochemical pathways in the body, while others may interfere with cellular function or induce inflammation and oxidative stress (Janik et al. 2019).

The health effects of natural toxins can vary widely, ranging from mild gastrointestinal disturbances to severe neurological or organ damage. Acute toxicity may manifest as symptoms such as nausea, vomiting, diarrhea, abdominal pain, dizziness, or respiratory distress, while chronic exposure to certain toxins may increase the risk of developing long-term health conditions, including liver disease, kidney dysfunction, neurological disorders, or even cancers (Mohammad et al. 2018). Certain populations, such as infants, pregnant women, the elderly, and individuals with underlying health conditions, may be more vulnerable to the adverse effects of natural toxins due to differences in metabolic capacity, immune function, or physiological susceptibility (Dolan et al. 2010). On the other hand, the consumption of contaminated foods, inadequate food processing or storage practices, food preparation habits, and environmental factors such as climate change or agricultural practices (Sousa 2008) can contribute to the accumulation of toxins in foodstuffs and increase the risk of toxin exposure.

Among the significant health risks and public health concerns due to exposure to natural toxins, aflatoxins are important which produced by molds of the *Aspergillus* genus. Those are potent carcinogens that can contaminate nuts, grains, and legumes, leading to an increased risk of liver cancer and other adverse health outcomes (Pickova et al. 2021). In a variety of plant-borne foods such as cassava, almonds, and stone fruits, Cyanogenic glycosides are present and those can release hydrogen cyanide upon ingestion, causing acute cyanide poisoning and chronic health defects, including thyroid dysfunctions and neurological damage (Cliff et al. 2015). Naturally produced toxins can also bioaccumulate. The marine algal blooms known as "red tides" can result in local filter-feeding organisms such as mussels and oysters becoming toxic (Turner et al. 2021) and coral reef fish can be responsible for the

poisoning known as ciguatera when they accumulate a toxin called ciguatoxin from reef algae (Clausing et al. 2018). Marine natural toxins, such as ciguatoxins and saxitoxins which are secreted by dinoflagellates and cyanobacteria, can bioaccumulate and bioconcentrate in fish and shellfish and go into the marine food chains, leading to ciguatera poisoning and paralytic shellfish poisoning, respectively, resulting in symptoms ranging from gastrointestinal distress to neurological impairment (Holmes et al. 2014).

Therefore, natural toxins present in regular foodstuffs should be considered as a noteworthy public health concern due to their ability to cause acute and chronic toxicity in humans and also due to their potential to transfer through the terrestrial or aquatic food chains. Collaborative efforts among researchers, food producers, regulatory authorities, and public health organizations are needed to assess and monitor natural toxin levels in regular foodstuffs, implement effective mitigatory measures, and build awareness among consumers about safe food handling and processing practices (such as thorough cooking destroys enzyme inhibitors and lectins in beans, soaking in water and boiling also remove some cyanide-generating compounds in the foods concerned, and removal of gonads, skin, and parts of certain fish eliminates toxins concentrated in these tissues) (Gutiérrez-Praena et al. 2013) to minimize the risks of natural toxin exposure and safeguard community health.

10.2 Detection, Analysis, and Regulation of Natural Toxins in Food

Food safety and security remain a global concern and burden. Food regulatory institutions and control agencies work on updating public surveillance and developing procedures while food safety authorities and laboratories are facing the biggest challenges due to practicing conventional methods for foodborne pathogens and biotoxin identification and confirmation no longer providing suitable solutions concerning the accuracy of the ultimate findings. Moreover, more sensitive, robust, and accurate high-throughput technologies and methods gathered under the name foodomics are currently being developed by research laboratories for foodborne pathogen and biotoxin detection (Artavia et al. 2021). Advanced detection and analysis methods play a vital role in identifying and quantifying these toxins accurately and efficiently. Moreover, regulatory agencies establish guidelines and regulations to assess, monitor, and control toxin levels in foodstuffs, ensuring compliance with safety standards/threshold levels (Selak et al. 2023). These advanced analytical and molecular biology methods, including mass spectrometry, immunoassays, biosensors, and other emerging technologies, provide sensitive and specific tools for toxin detection in complex food matrices which are essential for enhancing regulatory frameworks and safeguarding the integrity of the food supply chain.

10.2.1 Advanced Detection and Analysis Methods for Biotoxins Present in Food

Advanced analytical techniques have revolutionized the detection and analysis of natural toxins in foodstuffs, allowing for increased sensitivity, specificity, and efficiency (Rodriguez et al. 2020). Mass spectrometry (MS) is a powerful tool widely used for toxin identification and quantification due to its high sensitivity and ability to provide structural information (Table 10.1). Liquid chromatography coupled with mass spectrometry (LC-MS) and gas chromatography-mass spectrometry (GC-MS) are commonly employed for the analysis of mycotoxins, alkaloids, and other natural toxins in complex food matrices (Rodriguez et al. 2017). Furthermore, high-resolution MS techniques, such as time-of-flight (TOF) and quadrupole-TOF (Q-TOF) MS, offer enhanced resolution and accuracy, enabling the detection of trace levels of toxins (Picardo et al. 2020).

High-performance liquid chromatography (HPLC), GC, and thin-layer chromatography (TLC) are widely used for separating and quantifying natural toxins in complex matrices (Gaillard and Pepin 1999). HPLC coupled with UV, fluorescence, or mass spectrometric detection offers high sensitivity and specificity for analyzing mycotoxins, alkaloids, and other natural toxins in food and feed samples (Rodriguez et al. 2017). For instance, HPLC-MS/MS has been extensively employed for the simultaneous determination of multiple mycotoxins, such as aflatoxins, ochratoxin A, and fumonisins, in a variety of food commodities (Lattanzio et al. 2007). MS tools are for the identification and quantification of natural toxins based on their molecular mass and fragmentation patterns. For example, LC-MS/MS has been

Table 10.1 Analytical and molecular biology methods for detecting the natural toxins in foodstuffs

Detection technique	Instruments used	Analytical or molecular biology principle	Detectable natural toxins in food (Examples)
Chromatography	High-performance liquid chromatography (HPLC), gas chromatography (GC)	Separation based on chemical properties and retention time	Mycotoxins (aflatoxins, ochratoxin A), alkaloids, cyanogenic glycosides
Mass spectrometry	Liquid chromatography-mass spectrometry (LC-MS), gas chromatography-mass spectrometry (GC-MS)	Detection based on molecular mass and fragmentation patterns	Fumonisins, ergot alkaloids, marine biotoxins, tetrodotoxin
Immunoassays	Enzyme-linked immunosorbent assay (ELISA), lateral flow immunoassay (LFIA)	Detection based on antigen–antibody interactions	Mycotoxins (aflatoxins, deoxynivalenol), histamine
Molecular biology	Polymerase chain reaction (PCR), DNA microarray	Detection based on nucleic acid amplification and hybridization	Genes encoding toxins biosynthesis (mycotoxin by synthetic genes)

used for detecting cyanogenic glycosides in cassava products, while GC-MS has been employed for quantifying pyrrolizidine alkaloids in herbal teas and dietary supplements (Kopp et al. 2020).

Similarly to MS, modern spectroscopic techniques such as IR, UV, and NMR are all capable of high sample throughput and automated analysis. Born as tools for structure assignment of compounds in organic chemistry in the first half of the twentieth century, their potentiality was initially either masked by poor knowledge of their theoretical mechanisms or their diffusion limited by the instrumental complexity and cost (Gallo and Ferranti 2016).

Immunoassay techniques, such as enzyme-linked immunosorbent assay (ELISA) and lateral flow immunoassay (LFIA), are commonly used for rapid screening and semi-quantitative analysis of natural toxins in food and feed samples (Bennett 2005). Immunoassays have several benefits such as simplicity, speed, and cost-effectiveness, making them suitable for high-throughput analysis and on-site testing applications. ELISA kits have been developed for detecting mycotoxins, marine biotoxins, and plant toxins in various food matrices, providing rapid results within a few hours (Xie et al. 2018). Biosensors, incorporating biological recognition of elements with transducer systems, offer rapid, real-time, and portable platforms for on-site toxin detection which can be used as effective tools for food safety monitoring in field settings (Eyvazi et al. 2021). Moreover, emerging technologies such as nanotechnology and molecular imprinting have been explored for developing novel innovative sensors with improved sensitivity and selectivity for toxin detection in food (Rajendraprasad et al. 2023).

DNA microarrays offer high-throughput analysis of multiple toxin-producing genes simultaneously, exploring insights into the genetic diversity and toxin profiles of microbial communities (Longueville et al. 2004) in both terrestrial and aquatic ecosystems. Polymerase chain reaction (PCR) and DNA microarray are coming under the molecular biology techniques which have emerged as valuable tools for detecting and identifying natural toxins based on genetic sequences (Rasooly and Herold 2008). These PCR-based procedures enable the amplification and detection of specific toxin-producing genes in microbial or plant samples, allowing for the rapid screening of potential toxin-producing organisms.

10.2.2 Recently Developed More Advanced Methods for Natural Toxin Detection

Table 10.2 showcases innovative approaches and technologies that have emerged in the recent past to address the challenges associated with food safety and quality due to natural toxin contaminations. These methods leverage advancements in biosensing, nanotechnology, microfluidics, aptamer technology, surface plasmon resonance (SPR), electrochemistry, quantum dots, microarray technology, smartphone-based

Table 10.2 Recently developed innovative approaches and technologies to detect and quantify natural toxin contaminations in food

Detection technique	Instrument used	Analytical or molecular biology principle	Detectable natural toxins in food
Biosensors	Optical Biosensors, and Piezoelectric Biosensors	Biorecognition elements (E.g.: antibodies, enzymes)	Mycotoxins (aflatoxins, ochratoxin A), marine biotoxins, histamine
Nanotechnology	Nanomaterial-based sensors and nanoparticles	Nanostructured materials and surface-enhanced Raman spectroscopy	Plant toxins (cyanogenic glycosides, alkaloids), mycotoxins
Microfluidics	Microfluidic chips, lab-on-a-chip devices	Miniaturized fluidic systems and rapid analysis	Marine biotoxins, mycotoxins, cyanogenic glycosides
Aptamer-based assays	Aptamers, aptamer-functionalized sensors	Aptamer-target binding affinity and signal transduction	Marine biotoxins, mycotoxins, alkaloids
Surface plasmon resonance (SPR)	SPR biosensors	Changes in refractive index upon target binding	Marine biotoxins, mycotoxins, plant toxins
Electrochemical sensors	Electrochemical detection systems	Electrochemical reactions and signal amplification	Mycotoxins, marine biotoxins, plant toxins
Microarray technology	DNA microarrays, protein microarrays	Immobilized probes and multiplexed analysis	Toxin-producing genes, mycotoxin biomarkers, allergenic proteins
Smartphone-based assays	Smartphone cameras, mobile apps	Image analysis algorithms and portable devices	Mycotoxins, marine biotoxins, histamine, plant toxins
MALDI-TOF mass spectrometry	Matrix-assisted laser desorption/ionization time-of-flight mass spectrometry	Molecular mass and fragmentation patterns	Mycotoxins, marine biotoxins, alkaloids, cyanogenic glycosides

assays, and mass spectrometry to detect and analyze a wide range of natural toxins present in various food matrices.

Biosensors can be defined as "a compact analytical device or unit incorporating a biological or biologically derived sensitive 'recognition' element integrated or associated with a physio-chemical transducer" (Turner 2000) which represents a promising tool for detecting natural toxins in foodstuffs due to their high sensitivity, specificity, and rapid analysis capabilities. These devices utilize biorecognition elements such as antibodies, enzymes, or aptamers to selectively bind to target toxins, generating a measurable signal that can be detected and quantified (Palchetti and Mascini 2008). Nanotechnology-based sensors and nanoparticles offer novel

platforms for toxin detection, utilizing nanostructured materials and surface-enhanced Raman spectroscopy to enhance sensitivity and selectivity (Rajendraprasad et al. 2023; Bazin et al. 2017).

Microfluidics and lab-on-a-chip devices enable miniaturized fluidic systems for rapid and efficient analysis of natural toxins in food samples (Guo et al. 2015). Various foods are susceptible to contamination and adulteration with mycotoxins, presenting serious health risks to humans. Microfluidic "lab-on-a-chip" devices could integrate and miniaturize versatile functions from sample preparation to detection, showing great potential in rapid, accurate, and high-throughput detection of mycotoxins. These microscale platforms offer advantages such as reduced sample volume, faster analysis times, and multiplexed detection capabilities. Aptamer-based assays utilize synthetic nucleic acid molecules called aptamers, which exhibit high affinity and specificity for target toxins, enabling sensitive and selective detection across a range of food matrices (Bazin et al. 2017).

Surface plasmon resonance (SPR) biosensors detect changes in refractive index upon target binding, offering label-free and real-time analysis of natural toxins in food samples (Hodnik and Anderluh 2009). As an alternative, electrochemical sensor platforms have been developed as powerful quantitative tools for label-free detection of infection-related biomarkers with high sensitivity (Simoska and Stevenson 2019). Electrochemical sensors utilize electrochemical reactions and signal amplification strategies to detect and quantify toxins, providing portable and cost-effective solutions for on-site analysis. Luminescent semiconductor nanocrystals or quantum dots (QDs) are a recently developed class of nanomaterial whose unique photophysical properties are helping to create a new generation of robust fluorescent biosensors. Those quantum dot-based fluorescence assays harness the fluorescent properties of quantum dots for sensitive and multiplexed detection of natural toxins (Sapsford et al. 2006).

Microarray technology enables high-throughput analysis of toxin-producing genes, mycotoxin biomarkers, and allergenic proteins in food samples, offering insights into the composition and safety of food products (Vanhomwegen et al. 2013). Smartphone-based assays leverage image analysis algorithms and portable devices for rapid and accessible detection of natural toxins in foodstuffs, democratizing food safety monitoring (Bui et al. 2023) and surveillance while matrix-assisted laser desorption/ionization time-of-flight mass spectrometry (MALDI-TOF MS) offers high-throughput and high-resolution analysis of natural toxins, based on their molecular mass and fragmentation patterns (Sala-Comorera et al. 2016). This powerful technique enables comprehensive profiling of toxins in complex food matrices, facilitating accurate identification and quantification of contaminants.

10.2.3 Regulatory Frameworks and Guidelines

The regulation of natural toxins in food and feed is governed by national and international regulatory agencies, which establish safety standards, maximum permissible levels, and monitoring programs to ensure consumer protection. Regulatory frameworks for natural toxins vary depending on the type of toxin, its prevalence, toxicity, and potential health risks. Additionally, international organizations, such as the Codex Alimentarius Commission and the European Food Safety Authority (EFSA), provide scientific assessments and recommendations for setting regulatory limits and monitoring programs (Millstone and Van Zwanenberg 2002). When considering the maximum permissible levels (MPLs), they serve as legal standards to ensure that food and feed products are safe for consumption. The European Union (EU) is an example of that and has established maximum levels for mycotoxins, marine biotoxins, and other natural toxins in various food commodities through regulations such as Regulation (EC) No 1881/2006 and Regulation (EU) No 165/2010 (Fürst 2011; van Egmond 2004).

Among those safeguard levels for natural toxins, some regulatory organizations have set up guidelines to control food poisoning through natural toxins. For example, aflatoxins are potent mycotoxins produced by *Aspergillus* fungi, commonly found in crops such as maize, peanuts, and tree nuts. In 2017, the Food and Drug Administration (FDA) issued guidance for the industry on the control of aflatoxin levels in human food, highlighting preventive measures, sampling plans, and analytical methods for aflatoxin analysis (Cai et al. 2020). Also, ciguatera poisoning is caused by the consumption of fish contaminated with ciguatoxins, produced by *Gambierdiscus* algae. The European Food Safety Authority (EFSA) has developed risk assessment methodologies and management strategies for ciguatera toxins in seafood, including monitoring programs and consumer advice (Canals et al. 2021). Furthermore, tetrodotoxin is a potent neurotoxin found in certain species of pufferfish and other marine organisms. The Food and Agriculture Organization (FAO) and World Health Organization (WHO) have established international guidelines for the control of tetrodotoxin in pufferfish and other seafood products, including risk assessment, sampling plans, and regulatory limits (Alkassar et al. 2023).

Moreover, effective regulatory oversight also involves risk assessment and management strategies to prevent and mitigate toxin contamination in foodstuffs. This includes implementing good agricultural practices (GAPs), hazard analysis and critical control points (HACCP), and good manufacturing practices (GMPs) to minimize toxin levels during production, processing, and storage (Babeker et al. 2021). Furthermore, surveillance programs and monitoring systems are established to monitor toxin levels in food samples and identify emerging risks or trends. Rapid alert systems, such as the European Union's Rapid Alert System for Food and Feed (RASFF), facilitate the exchange of information and coordination among regulatory authorities to respond promptly to food safety incidents (Wood 2012).

10.3 What Is Foodomics?

Foodomics is a multidisciplinary field that integrates several "omic" techniques to comprehensively study food components and their effects on human health. These omic techniques include genomics, proteomics, metabolomics, transcriptomics, and microbiomics (Capozzi and Bordoni 2013). Each of these techniques provides unique insights into the composition, structure, and functionality of food components at molecular levels, facilitating a holistic understanding of the complex interactions between diet and health. By combining high-throughput analytical techniques with computational methods, foodomics provides valuable insights into the complex relationships between diet, nutrition, and health, facilitating advancements in food science, nutrition research, and personalized nutrition approaches (Gallo and Ferranti 2016) (Fig. 10.2).

One of the key objectives of foodomics research is to unravel the intricate interactions between food components and biological systems, including the gastrointestinal tract, metabolic pathways, and cellular processes. At its core, foodomics encompasses the comprehensive analysis of food components at the molecular level, including proteins, carbohydrates, lipids, vitamins, minerals, and phytochemicals (Capozzi and Bordoni 2013). These components play essential roles in determining the nutritional value, sensory attributes, and functional properties of foods. By characterizing the molecular profiles of food matrices, foodomics approaches enable researchers to identify and quantify bioactive compounds, assess nutritional quality, detect food contaminants, and trace the origin of food products (Valdés et al. 2017). This information contributes to improving food safety, quality control, and authenticity in the food industry.

Genomics focuses on the study of the complete set of genes (genome) in an organism, providing insights into genetic variations, gene expression patterns, and functional annotations related to food. Researchers can identify genes responsible for desirable traits such as nutritional content, flavor, and disease resistance by analyzing the genetic makeup of food sources. Genomics provides a glimpse into the genetic blueprint of food, while proteomics, metabolomics, and transcriptomics unveil the protein, metabolite, and RNA transcript profiles, respectively, contributing to a holistic understanding of food composition and its health implications (Ibáñez and Cifuentes 2014) while Proteomics involves the study of the complete set of proteins (proteome) expressed in a given sample. It enables the identification, quantification, and characterization of proteins present in foods, including enzymes, allergens, and bioactive peptides. Proteomics can elucidate the impact of processing, storage, and cooking methods on protein structures and functionalities (Valdés et al. 2017; Ibáñez and Cifuentes 2014).

Metabolomics focuses on the study of the complete set of small-molecule metabolites (metabolome) in biological samples, providing insights into metabolic pathways, biochemical reactions, and metabolic fingerprints associated with food intake (Ibáñez and Cifuentes 2014). Metabolomic profiling can reveal biomarkers of dietary exposure, metabolic responses to food components, and nutritional

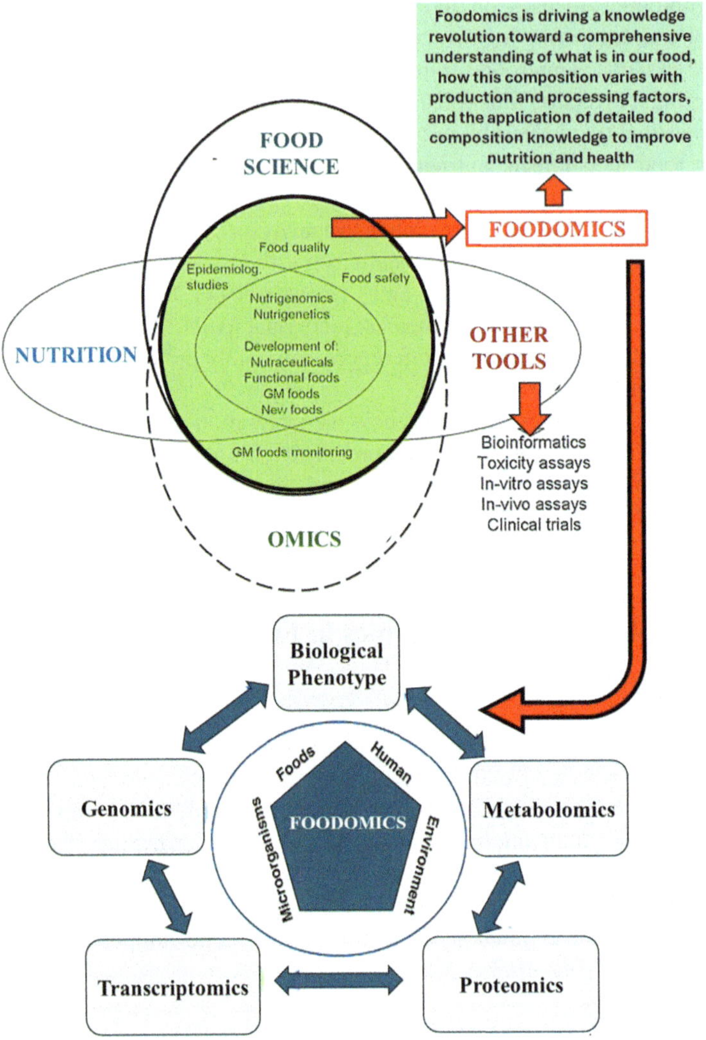

Fig. 10.2 Components of Foodomics; a multidisciplinary field that integrates several "omic" techniques to comprehensively study food components and their effects on human health

imbalances linked to disease risks (Scalbert et al. 2014). Meanwhile, Transcriptomic analysis can uncover gene regulatory networks, transcriptional responses to dietary interventions, and molecular signatures associated with foodborne diseases. Basically, transcriptomics examines the complete set of RNA transcripts (transcriptomes) produced by an organism, offering insights into gene expression patterns, alternative splicing events, and regulatory mechanisms underlying food-related processes (Capozzi and Bordoni 2013).

Microbiomics focuses on the study of the complete set of microorganisms (microbiome) inhabiting a given environment, including the human gut microbiota and food-associated microbial communities (Valdés et al. 2017). It provides insights into microbial diversity, functionality, and interactions within food matrices and the gastrointestinal tract. Microbiomic analysis can elucidate the role of the gut microbiota in food metabolism, nutrient bioavailability, and host–microbe interactions influencing health outcomes (Sharma et al. 2015). By integrating these omic techniques, foodomics enables a comprehensive analysis of the molecular composition, nutritional quality, safety, and authenticity of foods, as well as their impact on human health and disease (Ibáñez and Cifuentes 2014). It facilitates the development of personalized nutrition strategies, functional foods, and dietary interventions tailored to individual needs and preferences, ultimately promoting health and well-being across diverse populations.

In essence, foodomics showcases how the integration of omics technologies and food analysis methods provides a comprehensive framework for studying food and nutrition from a multitude of perspectives and paves the way for informed decisions regarding diet, health, and food safety.

10.4 Applications of Foodomics in Food Quality, Safety, and Authentication

Foodomics is an interdisciplinary field that connects different omics technologies integrated with advanced analytical and molecular biology techniques and bioinformatics tools, has emerged as a powerful approach to study food composition, quality, safety, and authenticity comprehensively. By analyzing the molecular composition, structure, interactions, and functions of foods and other commodities, foodomics enables researchers, policymakers, and the general public to gain valuable insights into the complex relationships between food composition, processing, and health effects (Fig. 10.3).

10.4.1 Applications in Food Quality

One of the primary applications of foodomics is in assessing and enhancing food quality by characterizing the composition, nutritional value, sensory attributes, and functional properties of food products (Balkir et al. 2021). Omics technologies, including genomics, proteomics, metabolomics, and transcriptomics, offer valuable tools for studying the molecular basis of food quality traits and identifying biomarkers associated with desirable attributes. For example, genomics can be used to study the genetic diversity and variation in food crops, livestock, and microorganisms,

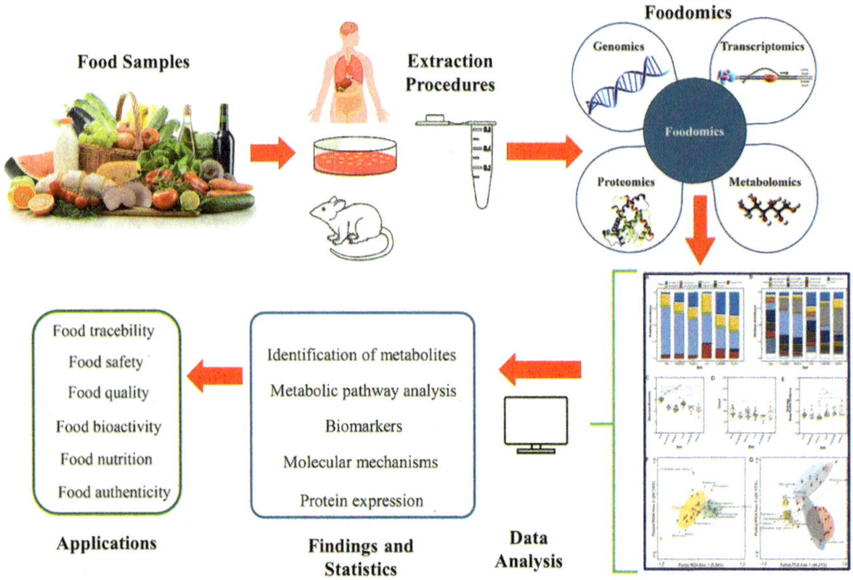

Fig. 10.3 Foodomics as a powerful approach for studying food composition, quality, safety, and authenticity

leading to the development of improved varieties with enhanced nutritional quality, flavor, and shelf life (Eyvazi et al. 2021; Xie et al. 2018). Proteomics enables researchers to identify and quantify proteins in food that contribute to taste, texture, and aroma, allowing for the optimization of food processing and formulation to maintain or enhance sensory attributes (Piras et al. 2016). Metabolomics provides insights into the metabolic pathways and metabolites associated with food flavor, color, and nutritional value, facilitating the selection of ingredients and processing methods to preserve or enhance food quality (Diez-Simon et al. 2019). Transcriptomics allows for the study of gene expression patterns in food organisms and tissues, revealing the molecular mechanisms underlying food quality traits and guiding breeding, cultivation, and processing strategies to produce high-quality food products (Lowe et al. 2017).

10.4.2 Applications in Food Safety

Foodomics is in ensuring food safety by detecting and monitoring food contaminants, toxins, pathogens, and allergens that may pose risks to human health. Omics technologies offer sensitive and specific tools for analyzing food samples and identifying biomarkers of foodborne hazards. For example, genomics can be used to

identify genetic markers associated with foodborne pathogens, allowing for the rapid detection and tracking of foodborne outbreaks and contamination events (Ferone et al. 2020). Proteomics enables the detection and quantification of allergenic proteins in food, helping to assess allergen risks and ensure compliance with labeling regulations. Metabolomics provides insights into the metabolic responses of food organisms to environmental stressors, such as microbial contamination, pesticide residues, and environmental pollutants, facilitating the identification of biomarkers of food safety hazards (Diez-Simon et al. 2019). Transcriptomics allows for the study of gene expression patterns in food organisms in response to food processing, storage, and handling conditions, (Lowe et al. 2017) revealing potential risks and interventions to prevent or mitigate foodborne hazards.

10.4.3 Applications in Food Authentication

Authenticating food products a critical role in foodomics by identifying and quantifying molecular markers and signatures that distinguish genuine products from adulterated or counterfeit ones. Tools in omic technologies help to analyze food samples and detect deviations from expected composition or outcome, origin, and quality. For example, genomics can be used to identify genetic markers specific to certain food species, varieties, or geographical origins, allowing for the authentication of food products based on their DNA fingerprints (Rasooly and Herold 2008; Longueville et al. 2004). Proteomics enables the identification and quantification of protein biomarkers in food that are characteristic of specific species, breeds, or production methods, facilitating the detection of adulteration or substitution (Mamone et al. 2013). Metabolomics investigates insights into the metabolic profiles and chemical fingerprints of food products, allowing for the comparison of authentic samples with adulterated or counterfeit ones (Medina et al. 2019). Transcriptomics allows for the study of gene expression patterns in food organisms in response to environmental factors, such as climate, soil, and cultivation practices, providing additional molecular markers for food authentication (Lowe et al. 2017).

10.4.4 Personalized Nutrition and Disease Prevention with Nutrimetabonomics

One of the most promising applications of foodomics is personalized nutrition, which aims to tailor dietary recommendations and interventions to individual health needs based on genetic, metabolic, and microbial profiles. Precision nutrition approaches leverage omics technologies to stratify individuals into distinct subgroups based on their genetic predispositions, metabolic phenotypes, and gut microbiome composition, enabling targeted interventions to optimize dietary habits and

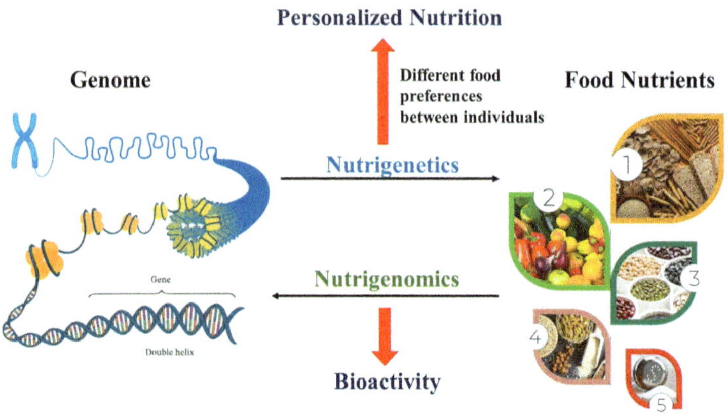

Fig. 10.4 Applications of foodomics in personalized nutrition

prevent chronic diseases (Swann and Claus 2014). For example, individuals carrying genetic variants associated with lactose intolerance may benefit from lactose-free dairy products or lactase enzyme supplementation to alleviate gastrointestinal symptoms and improve lactose tolerance. Likewise, individuals with impaired glucose metabolism may benefit from low-glycemic index diets rich in fiber and complex carbohydrates, which help stabilize blood sugar levels and reduce the risk of type 2 diabetes. By integrating multi-omics data with dietary assessment and clinical outcomes, precision nutrition approaches empower individuals to make informed dietary choices and patterns that support their health goals and improve overall well-being (Vernocchi et al. 2016) (Fig. 10.4).

Furthermore, from assessing food quality traits and nutritional value to detecting foodborne hazards and authenticating food products, foodomics provides valuable insights into the molecular basis of food characteristics and attributes (Capozzi and Bordoni 2013). Continued advancements in omics technologies, analytical methods, and data analysis approaches are essential for furthering our understanding of food composition, safety, and authenticity and addressing emerging challenges in food science and technology.

10.5 Foodomics Techniques for Controlling Natural Toxins in Foodstuffs

These interdisciplinary approaches, including genomics, proteomics, metabolomics, transcriptomics, and microbiomics, provide comprehensive insights into the presence, effects, and mitigation strategies for natural toxins.

Genomics plays a crucial role in identifying genetic markers associated with natural toxin production in food sources. By analyzing the complete set of genes (genome) in organisms, researchers can pinpoint genes responsible for toxin

synthesis in plants, fungi, bacteria, or animals (Valdés et al. 2017). Understanding the genetic basis of toxin production allows for targeted interventions, such as breeding programs to develop toxin-resistant crops or genetic modification to reduce toxin levels in food plants (Capozzi and Bordoni 2013). Proteomics techniques enable the identification and characterization of proteins involved in natural toxin pathways. By studying the complete set of proteins (proteome) expressed in food samples, researchers can identify enzymes responsible for toxin synthesis or detoxification, as well as allergens and other bioactive peptides (Mamone et al. 2009). Proteomic analysis can reveal changes in protein expression patterns in response to environmental factors or processing methods, providing insights into strategies for minimizing toxin contamination during food production and processing.

Metabolomics focuses on studying the complete set of small-molecule metabolites (metabolome) in food samples, offering insights into metabolic pathways and biochemical reactions associated with natural toxins. Metabolomic profiling can identify specific metabolites related to toxin production or detoxification processes, as well as biomarkers of toxin exposure or toxicity. By monitoring changes in metabolite profiles in response to different food processing techniques or storage conditions, metabolomics can help optimize food production practices to minimize toxin contamination and enhance food safety. Transcriptomics examines the complete set of RNA transcripts produced by organisms, providing insights into gene expression patterns and regulatory mechanisms underlying natural toxin production (Lamas et al. 2019). By analyzing gene expression profiles in response to environmental stimuli or genetic modifications, transcriptomics can identify key genes involved in toxin synthesis or regulation. Understanding the molecular mechanisms controlling toxin production allows for the development of targeted interventions to modulate toxin levels in food sources. Microbiomics focuses on studying the complete set of microorganisms (microbiome) inhabiting food matrices, including the human gut microbiota and food-associated microbial communities. Microbiomic analysis can provide insights into microbial diversity, functionality, and interactions within food systems, influencing natural toxin production, degradation, or detoxification (Sharma et al. 2015). By manipulating microbial communities through probiotics, prebiotics, or other interventions, microbiomics offers potential strategies for controlling natural toxins in foodstuffs and improving food safety.

10.6 Conclusion

Foodomics technology presents a groundbreaking opportunity to tackle global issues surrounding food safety, nutrition, sustainability, and security. Through interdisciplinary collaboration and advanced analytical methods, foodomics research can enhance food safety, quality, and sustainability while promoting nutritional health and wellness. By leveraging omics technologies, researchers can develop innovative solutions to address the complex challenges facing the global food system, paving the way for a more resilient and sustainable future. Understanding natural toxins in various food sources is crucial for ensuring food safety and public

health. These toxins, produced by plants, fungi, bacteria, or animals, can pose significant risks when consumed under certain conditions. Foodomics offers a comprehensive approach to studying natural toxins and their impact on food quality, safety, and authenticity. Techniques such as chromatography and mass spectrometry enable the detection and quantification of trace levels of toxins in complex matrices, while regulatory frameworks ensure the safety and quality of food products.

Advancements in foodomics have revolutionized our understanding of natural toxins and their interactions with food components and human physiology. By integrating various omics technologies, researchers can unravel the molecular mechanisms underlying toxin production, toxicity, and metabolism. Integration of these foodomics techniques enables a comprehensive approach to controlling natural toxins in foodstuffs. By combining genomics, proteomics, metabolomics, transcriptomics, and microbiomics data, researchers can gain a deeper understanding of the complex interactions between food components, microorganisms, and environmental factors influencing toxin production and mitigation. This interdisciplinary approach facilitates the development of targeted interventions and management strategies to minimize toxin contamination and ensure the safety and quality of food products for consumers. Through continued research and innovation in foodomics, we can enhance our ability to detect, prevent, and mitigate the risks posed by natural toxins in the food supply, promoting public health and well-being. Moving forward, continued research, collaboration, and innovation in foodomics are essential for addressing emerging challenges and safeguarding public health. By harnessing the power of foodomics, we can mitigate the risks posed by natural toxins and ensure a safer and healthier food supply for all.

References

Alkassar M, Sanchez-Henao A, Reverté J, Barreiro L, Rambla-Alegre M, Leonardo S, Mandalakis M, Peristeraki P, Diogène J, Campàs M (2023) Evaluation of toxicity equivalency factors of tetrodotoxin analogues with a neuro-2a cell-based assay and application to puffer fish from Greece. Mar Drugs 21(8):432

Alshannaq A, Yu JH (2017) Occurrence, toxicity, and analysis of major mycotoxins in food. Int J Environ Res Public Health 14(6):632

Armstrong AK, Krasny ME, Schuldt JP (2018) Communicating climate change: a guide for educators. Cornell University Press, p 144

Artavia G, Cortés-Herrera C, Granados-Chinchilla F (2021) Selected instrumental techniques applied in food and feed: quality, safety and adulteration analysis. Foods 10(5):1081

Babeker AM, Ahmed AI, Ahmed AR, Ebrahiem MA (2021) Assessment of the extent implementing the Good Manufacturing Practices (GMPs) and Hazard Analysis Critical Control Point System (HACCP) in Sudan sugar industry factories. J Agric Sci Food Technol 7(3):368–373

Balaban N, Rasooly A (2000) Staphylococcal enterotoxins. Int J Food Microbiol 61(1):1–10

Balkir P, Kemahlioglu K, Yucel U (2021) Foodomics: a new approach in food quality and safety. Trends Food Sci Technol 108:49–57

Bartnik M, Facey P (2024) Glycosides. In: Pharmacognosy. Academic Press, pp 103–165

Bazin I, Tria SA, Hayat A, Marty JL (2017) New biorecognition molecules in biosensors for the detection of toxins. Biosens Bioelectron 87:285–298

Bennett RW (2005) Staphylococcal enterotoxin and its rapid identification in foods by enzyme-linked immunosorbent assay–based methodology. J Food Prot 68(6):1264–1270

Bennett JW, Klich M (2003) Mycotoxins. Clin Microbiol Rev 16:497–516. https://doi.org/10.1128/CMR.16.3.497-516.2003

Blanco SL, Rodríguez LP, Lago J, Vieites JM, Cabado AG (2020) Potential uses and medical management of neurotoxic phycotoxins. Encycl Mar Biotechnol 3:1581–1617

Bucheli TD (2014) Phytotoxins: environmental micropollutants of concern? Environ Sci Technol 48(22):13027–13033

Bui TH, Thangavel B, Sharipov M, Chen K, Shin JH (2023) Smartphone-based portable biochemical sensors: exploring recent advancements. Chemosensors 11(9):468

Cai Y, McLaughlin M, Zhang K (2020) Advancing the FDA/office of regulatory affairs mycotoxin program: new analytical method approaches to addressing needs and challenges. J AOAC Int 103(3):705–709

Canals A, Martínez CV, Diogène J, Gago-Martínez A, Cebadera-Miranda L, de Vasconcelos FM, Gómez IL, Sánchez EVM, Alférez RC, Núñez D, Friedemann M (2021) Risk characterisation of ciguatera poisoning in Europe. EFSA Support Publications 18(5):6647E

Cao W, Yu P, Yang K, Cao D (2022) Aflatoxin B1: metabolism, toxicology, and its involvement in oxidative stress and cancer development. Toxicol Mech Methods 32(6):395–419

Capozzi F, Bordoni A (2013) Foodomics: a new comprehensive approach to food and nutrition. Genes Nutr 8:1–4

Clausing RJ, Losen B, Oberhaensli FR, Darius HT, Sibat M, Hess P, Swarzenski PW, Chinain M, Bottein MYD (2018) Experimental evidence of dietary ciguatoxin accumulation in an herbivorous coral reef fish. Aquat Toxicol 200:257–265

Cliff J, Nzwalo H, Muquingue H (2015) Cyanide in the production of long-term adverse health effects in humans. In: Toxicology of cyanides and cyanogens: experimental, applied and clinical aspects, pp 98–112

Diez-Simon C, Mumm R, Hall RD (2019) Mass spectrometry-based metabolomics of volatiles as a new tool for understanding aroma and flavour chemistry in processed food products. Metabolomics 15:1–20

Dolan LC, Matulka RA, Burdock GA (2010) Naturally occurring food toxins. Toxins 2(9):2289–2332

Eyvazi S, Baradaran B, Mokhtarzadeh A, de la Guardia M (2021) Recent advances on development of portable biosensors for monitoring of biological contaminants in foods. Trends Food Sci Technol 114:712–721

Ferone M, Gowen A, Fanning S, Scannell AG (2020) Microbial detection and identification methods: bench top assays to omics approaches. Compr Rev Food Sci Food Saf 19(6):3106–3129

Fürst P (2011) European Union Regulations. Mass Spectrometry in Food Safety: Methods and Protocols, pp 1–20

Gaillard Y, Pepin G (1999) Poisoning by plant material: review of human cases and analytical determination of main toxins by high-performance liquid chromatography–(tandem) mass spectrometry. J Chromatogr B: Biomed Sci Appl 733(1–2):181–229

Gallo M, Ferranti P (2016) The evolution of analytical chemistry methods in foodomics. J Chromatogr A 1428:3–15

Granum PE, Lund T (1997) Bacillus cereus and its food poisoning toxins. FEMS Microbiol Lett 157(2):223–228

Guo L, Feng J, Fang Z, Xu J, Lu X (2015) Application of microfluidic "lab-on-a-chip" for the detection of mycotoxins in foods. Trends Food Sci Technol 46(2):252–263

Gutiérrez-Praena D, Jos Á, Pichardo S, Moreno IM, Cameán AM (2013) Presence and bioaccumulation of microcystins and cylindrospermopsin in food and the effectiveness of some cooking techniques at decreasing their concentrations: a review. Food Chem Toxicol 53:139–152

Hajšlová J, Schulzová V, Botek P, Lojza J (2018) Natural toxins in food crops and their changes during processing. Czech J Food Sci 22:S29–S34

Hall S (2009) Seafood toxins. In: Pathogens and toxins in foods: challenges and interventions, pp 233–247

Hatheway CL (2018) Clostridium botulinum and other clostridia that produce botulinum neurotoxin. In: Clostridium botulinum. CRC Press, pp 2–20

Hodgson E (2012) Toxins and venoms. Prog Mol Biol Transl Sci 112:373–415

Hodnik V, Anderluh G (2009) Toxin detection by surface plasmon resonance. Sensors 9(3):1339–1354

Holmes MJ, Brust A, Lewis RJ (2014) Dinoflagellate toxins: an overview. In: Seafood and freshwater toxins: pharmacology, physiology, and detection, 3rd edn, pp 3–38

Ibáñez E, Cifuentes A (2014) Foodomics: food science and nutrition in the postgenomic era. In: Comprehensive analytical chemistry, vol 64. Elsevier, pp 395–440

Janik E, Ceremuga M, Saluk-Bijak J, Bijak M (2019) Biological toxins as the potential tools for bioterrorism. Int J Mol Sci 20(5):1181

Kachel HS, Buckingham SD, Sattelle DB (2018) Insect toxins–selective pharmacological tools and drug/chemical leads. Curr Opin Insect Sci 30:93–98

Kopp T, Abdel-Tawab M, Mizaikoff B (2020) Extracting and analyzing pyrrolizidine alkaloids in medicinal plants: a review. Toxins 12(5):320

Kostić AŽ, Milinčić DD, Petrović TS, Krnjaja VS, Stanojević SP, Barać MB, Tešić ŽL, Pešić MB (2019) Mycotoxins and mycotoxin producing fungi in pollen. Toxins 11(2):64

Lamas A, Regal P, Vázquez B, Miranda JM, Franco CM, Cepeda A (2019) Transcriptomics: a powerful tool to evaluate the behavior of foodborne pathogens in the food production chain. Food Res Int 125:108543

Lattanzio VMT, Solfrizzo M, Powers S, Visconti A (2007) Simultaneous determination of aflatoxins, ochratoxin A and Fusarium toxins in maize by liquid chromatography/tandem mass spectrometry after multitoxin immunoaffinity cleanup. Rapid Commun Mass Spect 21(20):3253–3261

Lehrer SB, Horner WE, Reese G, Taylor S (1996) Why are some proteins allergenic? Implications for biotechnology. Crit Rev Food Sci Nutr 36(6):553–564

Longueville FD, Bertholet V, Remacle J (2004) DNA microarrays as a tool in toxicogenomics. Comb Chem High Throughput Screen 7(3):207–211

Lowe R, Shirley N, Bleackley M, Dolan S, Shafee T (2017) Transcriptomics technologies. PLoS Comput Biol 13(5):e1005457

Malir F, Ostry V, Pfohl-Leszkowicz A, Malir J, Toman J (2016) Ochratoxin A: 50 years of research. Toxins 8(7):191

Mamone G, Picariello G, Caira S, Addeo F, Ferranti P (2009) Analysis of food proteins and peptides by mass spectrometry-based techniques. J Chromatogr A 1216(43):7130–7142

Mamone G, Picariello G, Nitride C, Addeo F, Ferranti P (2013) The role of proteomics in the discovery of marker proteins of food adulteration. In: Proteomics in foods: principles and applications, pp 465–501

Medina S, Pereira JA, Silva P, Perestrelo R, Câmara JS (2019) Food fingerprints–A valuable tool to monitor food authenticity and safety. Food Chem 278:144–162

Millstone E, Van Zwanenberg P (2002) The evolution of food safety policy–making institutions in the UK, EU and codex alimentarius. Soc Policy Adm 36(6):593–609

Mohammad AM, Chowdhury T, Biswas B, Absar N (2018) Food poisoning and intoxication: a global leading concern for human health. In: Food safety and preservation. Academic Press, pp 307–352

Palchetti I, Mascini M (2008) Electroanalytical biosensors and their potential for food pathogen and toxin detection. Anal Bioanal Chem 391:455–471

Picardo M, Sanchís J, Núñez O, Farré M (2020) Suspect screening of natural toxins in surface and drinking water by high performance liquid chromatography and high-resolution mass spectrometry. Chemosphere 261:127888

Pickova D, Ostry V, Malir F (2021) A recent overview of producers and important dietary sources of aflatoxins. Toxins 13(3):186

Piras C, Roncada P, Rodrigues PM, Bonizzi L, Soggiu A (2016) Proteomics in food: quality, safety, microbes, and allergens. Proteomics 16(5):799–815

Rajendraprasad SD, Girigoswami A, Chakraborty S, Girigoswami K (2023) Detection of food contaminants using nanotechnology-a mini review. Curr Nutr Food Sci 19(9):896–905

Rasooly A, Herold KE (2008) Food microbial pathogen detection and analysis using DNA microarray technologies. Foodborne Pathog Dis 5(4):531–550
Rodríguez I, González JM, Botana AM, Sainz MJ, Vieytes MR, Alfonso A, Botana LM (2017) Analysis of natural toxins by liquid chromatography. In: Liquid chromatography. Elsevier, pp 479–514
Rodriguez RS, O'Keefe TL, Froehlich C, Lewis RE, Sheldon TR, Haynes CL (2020) Sensing food contaminants: advances in analytical methods and techniques. Anal Chem 93(1):23–40
Sala-Comorera L, Vilaró C, Galofré B, Blanch AR, García-Aljaro C (2016) Use of matrix-assisted laser desorption/ionization–time of flight (MALDI–TOF) mass spectrometry for bacterial monitoring in routine analysis at a drinking water treatment plant. Int J Hyg Environ Health 219(7):577–584
Sapsford KE, Pons T, Medintz IL, Mattoussi H (2006) Biosensing with luminescent semiconductor quantum dots. Sensors 6(8):925–953
Scalbert A, Brennan L, Manach C, Andres-Lacueva C, Dragsted LO, Draper J, Rappaport SM, Van Der Hooft JJ, Wishart DS (2014) The food metabolome: a window over dietary exposure. Am J Clin Nutr 99(6):1286–1308
Selak A, Reberski JL, Klobučar G (2023) Assessing the persistence, mobility and toxicity of emerging organic contaminants in Croatian karst springs used for drinking water supply. Sci Total Environ 903:166240
Sharma P, Brahma V, Sharma A, Dubey RK, Sidhu GS, Malhotra PK (2015) Microbiomics: an approach to community microbiology. PlantOmics:633–653
Simoska O, Stevenson KJ (2019) Electrochemical sensors for rapid diagnosis of pathogens in real time. Analyst 144(22):6461–6478
Sousa CPD (2008) The impact of food manufacturing practices on food borne diseases. Braz Arch Biol Technol 51:615–623
Sultan I, Rahman S, Jan AT, Siddiqui MT, Mondal AH, Haq QMR (2018) Antibiotics, resistome and resistance mechanisms: a bacterial perspective. Front Microbiol 9:2066
Swann JR, Claus SP (2014) Nutrimetabonomics: nutritional applications of metabolic profiling. Sci Prog 97(1):41–47
Turner AP (2000) Biosensors—sense and sensitivity. Science 290(5495):1315–1317
Turner AD, Lewis AM, Bradley K, Maskrey BH (2021) Marine invertebrate interactions with harmful algal blooms–implications for one health. J Invertebr Pathol 186:107555
Urugo MM, Tringo TT (2023) Naturally occurring plant food toxicants and the role of food processing methods in their detoxification. Int J Food Sci
Valdés A, Cifuentes A, León C (2017) Foodomics evaluation of bioactive compounds in foods. TrAC Trends Anal Chem 96:2–13
Valério E, Chaves S, Tenreiro R (2010) Diversity and impact of prokaryotic toxins on aquatic environments: a review. Toxins 2(10):2359–2410
van Egmond HP (2004) Natural toxins: risks, regulations and the analytical situation in Europe. Anal Bioanal Chem 378:1152–1160
Vanhomwegen J, Berthet N, Mazuet C, Guigon G, Vallaeys T, Stamboliyska R, Dubois P, Kennedy GC, Cole ST, Caro V, Manuguerra JC (2013) Application of high-density DNA resequencing microarray for detection and characterization of botulinum neurotoxin-producing clostridia. PLoS One 8(6):e67510
Vernocchi P, Del Chierico F, Putignani L (2016) Gut microbiota profiling: metabolomics based approach to unravel compounds affecting human health. Front Microbiol 7:1144
Vetter J (2000) Plant cyanogenic glycosides. Toxicon 38(1):11–36
Vijayan M, Chandra N (1999) Lectins. Curr Opin Struct Biol 9(6):707–714
Wood R (2012) Experience of food safety authorities in europe and the rapid alert system. Improving Import Food Safety:133–149
Xie J, Wang L, Wu W, Ou J (2018) Recent advances on determination methods of mycotoxins and marine biotoxins. Med Res 2(3):180011
Yunus AW, Imtiaz N, Khan H, Ibrahim MNM, Zafar Y (2019) Aflatoxin contamination of milk marketed in Pakistan: a longitudinal study. Toxins 11(2):110

Chapter 11
Trace Elements in Healing and Edible Plants as Toxins and Tonics for Human

Majeti Narasimha Vara Prasad

Abstract There is an increased consciousness regionally and globally in production and use of healing plants in addition to edible plants (non-agricultural and non-horticultural plants). These plants are collected wild without knowing the contamination levels in soil and ground water. Agrotechnology is not known for many of these plants. Several of the bioactive molecules and nutraceuticals act as: adaptogen, anticancer, antidiabetic, antidiarrheal, bioavailability enhancers, brain tonics, anti-asthmatic, antihypertensive, antiulcer, antiurolithiatic, hepatoprotective, carminative, stomachic, and purgative. Safety (undisclosed toxic contaminants), efficacy, affordability, availability of bioactive molecules and nutraceuticals as per consumer choice, promising plant products for the treatments for major diseases such as HIV/Malaria/TB/Cancer/Diabetes, etc., are major subjects of investigation. Analytical instrumentation is required for species-selective analysis for trace elements. If phytopharmaceuticals are to be regarded as rational drugs, they need to be standardized and pharmaceutical quality has to be approved by recognized agencies. Therefore, this chapter would focus prominent lacunae in quality control and quality assurance related to trace elements in healing and edible plants. This chapter will shed light upon the criteria of quality assurance and quality control of phytomedicines.

Keywords Deficiency or efficiency of trace elements · Diseases/symptoms · Edible plants · Functional foods · Good manufacturing practices · Medicinal and aromatic plants · Medicinal plants · Medicinally important · Nutraceuticals · Nutritionally important · Peri-urban area · Quality assurance · Quality control · Sophisticated analytical techniques · Trace element contamination in healing plants

M. N. V. Prasad (✉)
Retired, Department of Plant Sciences, School of Life Sciences, University of Hyderabad, Hyderabad, Telangana, India
e-mail: mnvsl@uohyd.ac.in

M. N. V. Prasad et al. (eds.), *Biotoxins*,
https://doi.org/10.1007/978-3-031-75309-1_11

11.1 Introduction

All over the world thousands of firms are manufacturing traditional medicines with or without standardization (Dubey et al. 2004). In the recent past, there has been considerable interest in bioprospecting than at any time in history. About 960 species of medicinal plants are estimated to be in trade both nationally and internationally of which 178 species have annual consumption level exceeding 100 tons per year. It is being observed that there is a global resurgence in the use of medicinal products from natural sources in various ethnic cultures (Table 11.1), traditional and alternative health care systems resulting in growing world herbal trade which stands currently at US$120 billion and is expected to reach US$7 trillion by 2050. In order to attract international and national market, they must have "good quality tag." To implement this Government of India very recently launched the quality control and assurance jointly by the National Medicinal Plants Board (NMPB) and the Quality Council of India (QCI). According to QCI, the scheme has been designed keeping best international practices in view. If phytopharmaceuticals are to be regarded as rational drugs, they need to be standardized and pharmaceutical quality has to be approved.

The scientific leads for the discovery of plant-based pharmaceuticals have been generated through: (a) High-throughput screening and (b) Ethnobotany. Ethnobotany deals with studies among the tribal and rural people for recording their unique knowledge about plant wealth and for search of new resources of herbal drugs, edible plants, and other economic aspects of plants (Fig. 11.1). Table 11.1 details medicinal products from natural sources in various ethnic cultures.

Several of the trace elements serve as essential nutrients and are the constituents of bimolecular as cofactors for various enzymes and in a variety of metabolic functions. A generalized scheme of major contamination pathways is shown in Fig. 11.2. Certain trace elements accumulated in medicinal plants have the healing function to some ailments and disorders; for e.g., zinc—neurochemical transmission; Cr and Mn can be correlated with therapeutic properties against diabetic and cardiovascular diseases. Certain transition group elements regulate hepatic synthesis of cholesterol. Bioorganic farming takes total micronutrient care and is an important

Table 11.1 Medicinal products from natural sources in various ethnic cultures

Complementary medicine	Natural sources	Origin	Trade turnover
Ayurvedic	P, A, M	India	No figure
Chinese	P, A, M	China	More than US$1 billion
Indusyunic	P, A, M	Pakistan	No figure
Islamic	P, A, M	Middle East	No figure
Kampo	P, A, M	Japan	Getting competitive
Korean	P, A, M	Korea	No figure
Other oriental	P, A, M	Other Asian countries	No figure

P medicinal plants, *A* animal sources, *M* mineral sources

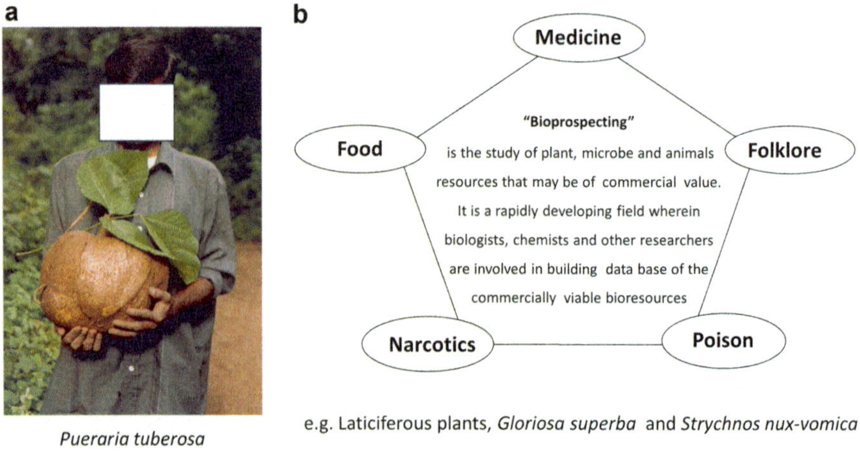

Fig. 11.1 The scientific leads for the discovery of plant-based pharmaceuticals have been generated through: (a) High-throughput screening and (b) Ethnobiology

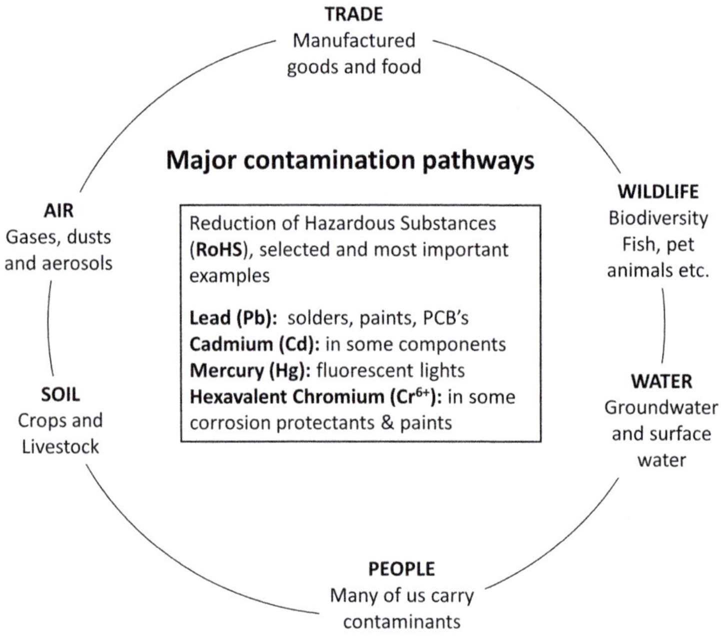

Fig. 11.2 Major contamination pathways

soil management practice for fortification of crop produce with essential nutrients for the health of humans as well as livestock (Raskin et al. 2002).

11.2 About Trace Elements

- Trace elements (= metals, metalloids, and radionuclides) (their percentage of availability is <1 % of the total matter)
- Abundant elements on earth crust: Si, Al, etc.
- Trace elements accumulation in medicinal plants—Toxicity, disease curative and preventive functions are reported and is a major area of research (Hartmann et al. 2003)
- Those essential for the nutrition of higher plants are: B, Cu, Fe, Mn, Mo, and Zn
- Essential for human nutrition are (excess would cause toxicity): Cu, Co, Fe, Mn, Mo, Zn, Cr, Ni, Se, and Sn
- Essential for animal nutrition are: As, Cu, Co, Fe, Mn, Mo, Zn, Cr, Ni, Se, Sn, and V

Cd, Hg, and Pb (Al = abundantly available element) have no metabolic significance, hence are not essential The compliance checking will be done by independent, third party agencies conforming to international standards. Any producer/collector or group of producers/collectors can obtain certification from the designated certification body (CB). The notified CB will undertake surveillance and monitor the herbal materials and products. In the recent past, there has been interest in the development of new commercial drugs from herbal materials and to reestablish traditions of yesterday as drugs of tomorrow (Gurib-Fakim 2006).

A large number of research papers reported trace element contamination in medicinal and edible plants. Reference be made to Table 11.2 (not exhaustive) for some recent observations.

Minerals *sensu lato* (trace elements) are classified in the following three categories and are detailed in Table 11.3.

As with any product from agricultural or wild sources, medicinal plants can be contaminated by organic or inorganic substances of natural or synthetic origin, such as insects, microorganisms, e.g., fungi and their mycotoxins, heavy metals (Chizzola and Franz 1996; Sathiyamoorthy et al. 1997; Wong and Koh 1986; Wong et al. 1993), and radioactive materials (Karavaeva and Molchanova 1998). Glocal demand (global +local = glocal) for plant raw materials (medicinal value) is increasing: Safety (undisclosed toxic contaminants), (Thakral et al. 2024) efficacy, affordability, availability of bioactive molecules and nutraceuticals as per consumer choice, promising plant products for the treatments for major diseases such as HIV/Malaria/TB/Cancer/Diabetes, etc., are gaining momentum. Several of the bioactive molecules and nutraceuticals act as: adaptogen, anticancer, antidiabetic, antidiarrheal, bioavailability enhancers, brain tonics, antiasthmatic, antihypertensive, antiulcer, antiurolithiatic, hepatoprotective, carminative, stomachic, and purgative. However,

Table 11.2 Recent reports of trace element contamination in medicinal and edible plants

	Author(s)	Year
Heavy metal contamination of medicinal plants and potential implications on human health.	Asiminicesei et al.	2020
Toxic metal implications on agricultural soils, plants, animals, aquatic life, and human health.	Okereafor et al.	2020
Heavy metal levels and mineral nutrient status in different parts of various medicinal plants collected from eastern Mediterranean region of Turkey	Karahan et al.	2020
Heavy metals accumulation and translocation in native plants grown on tailing dumps and human health risk.	Pehoiu et al.	2020
Toxic metal implications on agricultural soils, plants, animals, aquatic life, and human health.	Okereafor et al.	2020
Medicinal plants and biogenic metal oxide nanoparticles: a paradigm shift to treat Alzheimer's disease.	Gul et al.	2021
Impacts of heavy metals and medicinal crops on ecological systems, environmental pollution, cultivation, and production processes in China.	Chen et al.	2021
Functional aspects of plant secondary metabolites in metal stress tolerance and their importance in pharmacology.	Anjitha et al.	2021
Heavy metal effects on biodiversity and stress responses of plants inhabiting contaminated soil in Khulais, Saudi Arabia.	Alsherif et al.	2022
An attitude towards the effects of heavy metals in medicinal plants.	Ahandani et al.	2022
Interaction between zinc and selenium biofortification and toxic metals(loid) accumulation in food crops.	Bayanati et al.	2022
Progress in ICP-MS analysis of minerals and heavy metals in traditional medicine	Chen et al.	2022
Cadmium toxicity in medicinal plants: An overview of the tolerance strategies, biotechnological and omics approaches to alleviate metal stress.	Al-Khayri et al.	2023
Toxic elemental impurities in herbal weight loss supplements; a study using ICP-OES microwave-assisted digestion.	Al-Thani et al.	2023
Applied analytical methods for detecting heavy metals in medicinal plants.	Guo et al.	2023
A comparative analysis of heavy metal effects on medicinal plants.	Mukherjee et al.	2023
The content of heavy metals in medicinal plants in various environmental conditions: A review.	Vinogradova et al.	2023
The content of heavy metals in medicinal plants in various environmental conditions:	Vinogradova et al.	2023
Human health risk assessment of trace elements in commonly consumed herbs in Ibadan, Nigeria.	Adeagbo et al.	2024
Quantification of toxic heavy metals, trace elements, and essential minerals contents in traditional herbal medicines commonly utilized in Khyber Pakhtunkhwa, Pakistan.	Anwar et al.	2024
Impact of heavy metal pollution in the environment on the metabolic profile of medicinal plants and their therapeutic potential.	Asiminicesei et al.	2024

(continued)

Table 11.2 (continued)

	Author(s)	Year
Heavy metal accumulation potential of medicinal plants: Phytoremediation perspective.	Awasthi et al.	2024
Determination of heavy metals contamination, risk prediction, and antioxidant properties of anti-malarial herbal mixture sold in Kano state, Nigeria.	Babandi et al.	2024
Phytochemicals involved in mitigating silent toxicity induced by heavy metals.	Ceramella et al.	2024
Assessment of heavy metals and their induced toxicity in medicinal plants: A public health concern.	Chauhan et al.	2024
Cytogenotoxicity evaluation of heavy metals detected in extracts and infusion of *Baccharis trimera*, potential bioaccumulator plant.	da Silva et al.	2024
Pesticides and heavy metal toxicity in fish and possible remediation–a review.	Ghafarifarsani et al.	2024
Phytoremediation potential of senna (*Senna occidentalis*) and neem (*Azadirachta indica*) in detoxification of soils contaminated with Zinc (Zn), Copper (Cu), and Lead (Pb).	Habib	2024
Ensuring the safety of Indian herbal drugs: A comprehensive assessment of heavy metal contaminations with emphasis on lead, arsenic, and cadmium.	Ibrahim et al.	2024
Trace metals contamination of herbal remedies: Health implications and quality control strategies.	Izah et al.	2024
Medicinal, aromatic, and spice plants: biodiversity, phytochemistry, bioactivity, and their processing innovation.	Salamon	2024
Comparison of essential and toxic metals levels in some herbal teas: a systematic review.	Salmani et al.	2024

Table 11.3 Functional basis for the classification of naturally available elements

Medicinally important	Nutritionally important	Toxic elements
Bismuth (Bi)	Chromium (Cr)	Aluminum (Al)
Fluorine (F)	Copper (Cu)	Arsenic (As)
Gallium (Ga)	Iron (Fe)	Cadmium (Cd)
Gold (Au)	Magnesium (Mg)	Cobalt (Co)
Lithium (Li)	Manganese (Mn)	Lead (Pb)
Platinum (Pt)	Selenium (Se)	Mercury—Inorganic (Hg)
	Sulfur (S)	Mercury—Organic ($Hg-CH_3$)
	Zinc (Zn)	Nickel (Ni)
	Silica (Si)	Tin (Sn)
	Iodine (I)	
	Molybdenum (Mo)	

there is a vast gap between demand and supply due to plant material availability becoming rare. Since agro technology is not known for many of the important plants, *in vitro* methods are to be established. Several of these plants are to be collected from nature (Fig. 11.3).

India has a very rich plant biodiversity. In the recent past, there has been considerable interest in drugs obtained from plants than at any time in history. There are

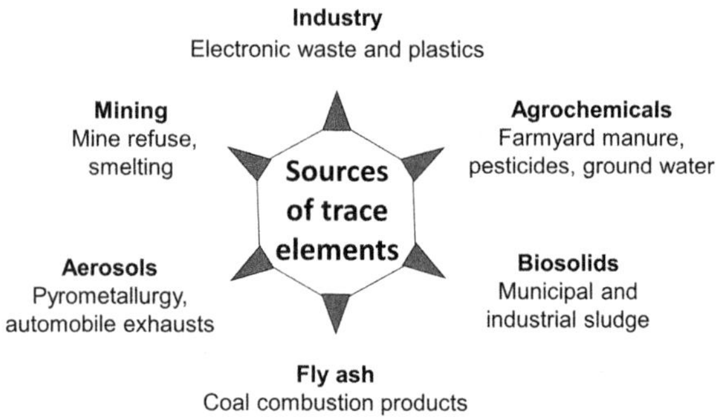

Fig. 11.3 Bioavailability of trace elements to plants (medicinal, aromatic, and edible crops)

15 agro climatic zones and about 18,000 species of flowering plants of which nearly 7000 are believed to have healing properties and are used in Indian System of Medicine, viz., folk medicine, ayurveda, siddha, Unani, and homeopathy including modern medicine. There are several thousands of firms manufacturing traditional medicines with or without standardization (Dubey et al. 2004).

Plants serve as raw materials for food products, agriculture, biotech and pharmaceutical and alternative health and dietary supplements. Therefore, plants are harvested as an alternative source of income. Herbal drugs can be considered as an important export product. However, due to increasing urbanization and uncontrolled expansion of industrial activities during the last decade(s), the environment is often heavily polluted. Concentrations of heavy metals in water, soils exceed permissible limits. When heavy metals are released into the environment, they can accumulate in food crops, medicinal and aromatic plants grown in these areas. Due to the high genetic diversity amongst these plants, there is also a huge potential for trace element accumulation. This could negatively affect human health in case of toxic elements, but it could also be beneficially used in biofortification programs to increase human uptake of essential micronutrients (e.g., Se, Fe, and Zn). Although there is a significant risk for transfer of toxic trace elements to the food chain in the peri-urban Indian areas, relatively little scientific studies focusing on factors affecting trace element uptake by Indian food crops, medicinal and aromatic plants are currently available. Moreover, most of these studies dealt with one particular plant species or only a small selection of species, without comparing the potential for trace element uptake between species.

In the last few decades, there has been probably more interest in drugs obtained from plant sources than at any time in history. About 960 species of medicinal plants are estimated to be in trade both nationally and internationally of which 178 species have annual consumption level exceeding 100 tonnes. It is being observed that there is a global resurgence in traditional and alternative health care systems resulting in growing world herbal trade which stands currently at $120 billion and is expected to reach $7 trillion by 2050 (Fig. 11.4).

In order to attract international and national market they must have "good quality tag." To implement this the Government of India very recently launched the quality control and assurance jointly by the National Medicinal Plants Board (NMPB) and the Quality Council of India (QCI). According to QCI, the scheme has been designed

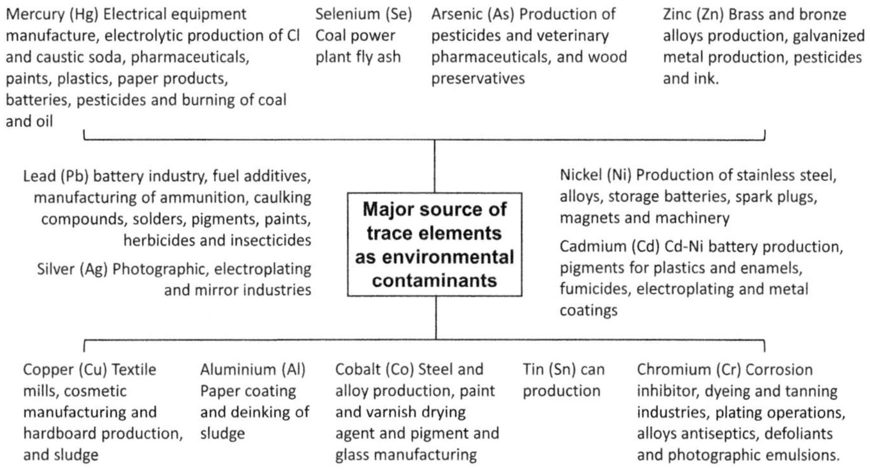

Fig. 11.4 Trace elements as nutrients and contaminants (Prasad 2008; Adeagbo et al. 2024)

keeping best international practices in view—the standards are based on WHO documents which were adopted by NMPB and the compliance checking will be done by independent, third party agencies conforming to international standards. Any producer/collector or group of producers/collectors can obtain certification from the designated certification body (CB). The notified CB will continuously undertake surveillance and monitor the herbal materials and products. In the recent past, there has been a tremendous resurgence of interest in the development of new commercial drugs from herbal materials and to reestablish traditions of yesterday as drugs of tomorrow (Gurib-Fakim 2006).

11.2.1 Dietary Supplement, Health and Education Act

- Phytomedicines: Plant-based pharmaceutical products with proven medical efficacy, 18% of the world's top 150 prescription drugs are from plant sources.
- Herbal remedies: Covering a spectrum of basic formulae made by traditional healers that are sold OTC as capsules, pills, or liquid tonics.
- Herbal teas and infusions: These are hot water infusions made out of herbs and sold in the form of tea bags or in granulated form.
- Homeopathic drugs: They are medicines derived from plant, mineral, and animal sources and used in extremely dilute amounts.
- Ayurvedic drugs: Ayurveda is a South Asian, holistic therapeutic concept, based not only on medical therapy, but including all aspects of human life in its holistic philosophy.
- Dietary supplements: A range of food supplements of plant and animal origin.

- Functional foods: Any food or beverage which has some claim to enhance one's physical or mental health and well-being.
- Nutraceuticals: Hybridized word between nutrition and pharmaceutical. In USA, this term is becoming increasingly popular.
- Cosmeceuticals: Herbal cosmetics enhance the health and beauty of the skin.

Medical geology emphasizes optimal exposure of human being to the essential trace elements in order to maintain or improve health (Fig. 11.5).

The Ayurvedic system of medicine has been in practice for millennia. The ancient seers found that drugs of different origin (herbal, metal, or animal) in addition to codes of conduct and dietary regulations are suitable tools to address several diseases. Use of metallic preparations [(Bhasmas = ash and bhasmikarana (=process of preparation of metallic ash which is zero valent)] in healthcare is unique. Bhasmikarana process with Mercury, Gold, Silver, Lead, Zinc, Copper, etc., was frequently used by seers of the Indian tradition to treat different diseases with great certainty of cure. It is generally claimed that these metals are detoxified during the highly complex bhasmikarana process described in Ayurveda especially Rasashastra manuscripts. Charaka Samhita, one of the scheduled books of Ayurveda also holds ample of references regarding the use of metals for different purposes; selected examples are enumerated in this chapter.

Several of the trace elements serve as essential nutrients and are the constituents of bimolecular as cofactors for various enzymes and in a variety of metabolic functions. Certain trace elements accumulated in medicinal plants have the healing function to some ailments and disorders; for e.g., zinc—neurochemical transmission; Cr

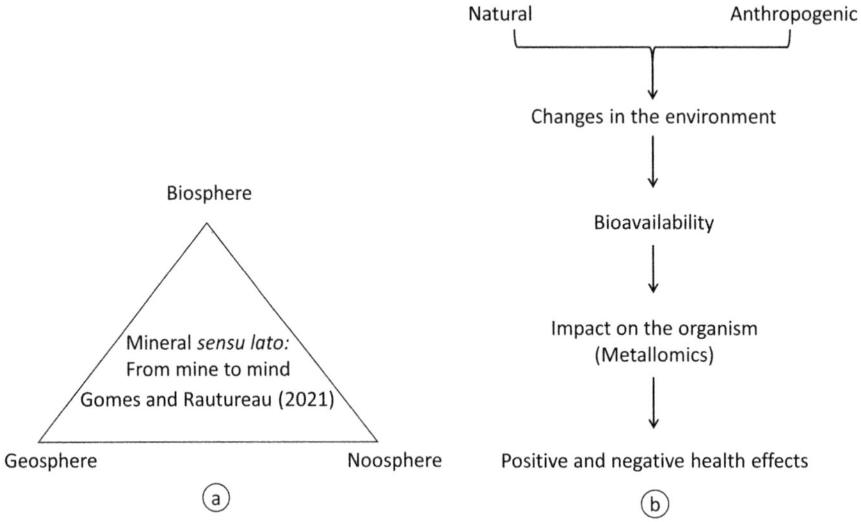

Fig. 11.5 Medical geology and bioavailability of essential trace elements. Pros and cons on human health

and Mn can be correlated with therapeutic properties against diabetic and cardiovascular diseases. Certain transition group elements regulate hepatic synthesis of cholesterol (Curran 1954). Bioorganic farming takes total micronutrient care and is an important soil management practice for fortification of crop produce with essential nutrients for the health of humans as well as livestock.

Nutrigenomics and pharmacogenomics, medical geology (geomedical science), and metallomics are emerging as new areas of research with challenging tasks ahead. The relevance of these cutting-edge fields and their relevance to nutritional security and quality of agricultural produce including healing plants will be discussed with selected examples.

"Metallomics" is the metal-assisted functional biogeochemistry, a new scientific field was first proposed as by Haraguchi in to integrate the research fields related to metal biomolecules and its metals. Metallomics is the scientific field of symbiosis with genomics and proteomics, because syntheses and metabolic functions of genes (DNA and RNA) and proteins cannot function without the coordination of various metal ions and metalloenzymes. In metallomics, metalloproteins, metalloenzymes, and other metal-containing biomolecules are defined as "metallomes," in a similar manner to genomes in genomics as well as proteomes in proteomics. Since the identification of metallomes and the elucidation of their biological or physiological functions in the biological systems is the main research target of metallomics, chemical speciation for specific identification of bioactive metallomes is the crux to establish metallomics as an integrated bio-metal science.

Quality Assurance (QA) is a wide-ranging concept that covers all matters that individually or collectively influence the quality of a drug. It is the total of the organized arrangements made with the objective of ensuring that drugs are of the quality required for their intended use. Quality Assurance therefore incorporates good manufacturing practices, along with other factors.

Quality Control (QC) is the part of good manufacturing practices [GMP] that is concerned with sampling, specifications, testing, documentation, and release procedures. This approach ensures that materials are not released for use, and that drugs (are not) released for sale or supply, until their quality has been deemed satisfactory.

Good manufacturing practices (GMP) are the part of Quality Assurance that ensures that drugs are consistently produced and controlled in such a way as to meet the quality standards appropriate to their intended use, as required by the marketing authorization.

Owing to the great variability in plant chemical composition that results from factors to which plants are exposed during their growth, storage, and the different stages of manipulation, characterization and/or standardization of phytopharmaceuticals are necessary (Marcus and Grollman 2002). Standardization of herbal preparations should allow the knowledge of their composition and prevent, or at least make less likely, the consumption of drugs of questionable quality. The latest Pharmacopeias attempt at standardization, prescribe methods, usually chromatographic, for the analysis and quality control of medicinal products of plant origin (Bauer and Tittel 1996; Gyeresi et al., 1997).

Microbiological contamination and foreign materials are important quality criteria in the testing of medicinal plants. As with any product from agricultural or wild sources, medicinal plants can be contaminated by organic or inorganic substances of natural or synthetic origin, such as insects, microorganisms, e.g., fungi and their mycotoxins (Roy and Chourasia 1990; Kumar and Roy 1993; Chourasia 1995; Efuntoye 1996), heavy metals (Wong and Koh 1986; Wong et al. 1993; Chizzola and Franz 1996; Sathiyamoorthy et al. 1997), and radioactive materials (Graham and Simon 1996; Karavaeva and Molchanova 1998).

Asia is the largest and, with 60% of the world's population also the most populous continent of the world. It is home to the world's oldest documented medical systems, traditional Chinese medicine (TCM) and Ayurveda.

11.3 Medicinal Value of Metals in Ancient Indian System of Medicine (After Charaka Samhita)

The ancient seers found that drugs of different origin (herbal, metal, or animal) in addition to codes of conduct and dietary regulations are suitable tools to address several diseases. Use of metallic preparations [(Bhasmas = ash and bhasmikarana (=process of preparation of metallic ash which is zero valent)] in healthcare is unique. Bhasmikarana process with Mercury, Gold, Silver, Lead, Zinc, Copper, etc., was frequently used by seers of the Indian tradition to treat different diseases with great certainty of cure. It is generally claimed that these metals are detoxified during the highly complex bhasmikarana process described in Ayurveda especially Rasashastra manuscripts. The Ayurvedic system of medicine has been in practice in India for millennia. Charaka Samhita, one of the scheduled books of Ayurveda also holds ample of references regarding the use of metals for different purposes; selected examples are enumerated in this chapter.

Therapeutic potentials of metals in ancient India in Ayurveda and integrative medicine have been well established (Barve et al. 2011). Ancient India demonstrated the use of chemical understanding in technology for manufacture and processing of medicines. Prescription Drugs Different metals (Dhatus) were employed in medicines by ancient Indian chemists to treat or prevent various illnesses (Salunkhe et al. 2023). The three types of Dhatus were Puti Loha (lead; Naga), Misra Loha (alloy; Kamsya, Pitala, Varta), and Sudh Loha (pure metals; gold, silver, copper, and iron). Yasada, or processed zinc, was also utilized in Ayurvedic medicines for conditions like diabetes (Prameha), anemia (Pandu), cough, and eye disorders (Salunkhe et al. 2023).

The Ayurvedic system of medicine dates back to 5000 years B.C. Its Materia Medica contains resources in the form of drugs derived from plant, animal, metal, and mineral sources, the use of which have been advocated in various different pathological manifestations. These drugs have also been converted in to poly-herbal, herbo-mineral, and metallic compound formulations by the seers, who have

documented their clinical experiences and passed on the knowledge to further generations.

Use of certain heavy metals and minerals in Ayurvedic therapeutics increased. Rasashastra, an integral part of Ayurveda, deals with the drugs of mineral origin, and details their varieties, characteristics, processing techniques, properties, therapeutic uses, possibilities of developing adverse effects and their management, etc., in a comprehensive way. Although the roots of this science (Rasa Shastra) exist in the ancient texts of Indian civilization, its development as an independent system of therapy started around the eighth century A.D. Ayurvedic classics written before that time, like Charaka Samhita and Sushruta Samhita, etc., contain descriptions of metals and minerals, their processing techniques and their utilization in therapeutics, etc.

In due course of time, herbo mineral and metallic preparations came to occupy a significant seat in Ayurvedic pharmacopeia and have routinely been used in practice in different parts of India for many centuries. Such preparations are held to be safe, efficacious even in minute doses, and, when manufactured and used following specified classical guidelines, not to lead to any significant untoward effects. The past decade, however, has witnessed concerns by the western scientific community, regarding the safety of Ayurvedic herbal, herbo-mineral, and metallic preparations, which is a major concern for the age-old Ayurvedic heritage Parada (Mercury) Formulations containing mercury are recommended for external application—Kushtha (skin disorders) and to enhance complexion of the skin.

11.3.1 Swarna (Gold) Bhasma

Swarna (Gold) is an important, precious metal known to Indians since antiquity. The "Bhasma" form of Gold is in metallic state. Quantitatively it is a combination of metallic Gold (96.76%), silica (1.14%), ferric oxide (0.14%), phosphates (0.78%), potash (0.16%), salt (0.078%), and traces of copper and magnesium.

In its elemental form, Gold has been employed for centuries as an anti-pruritic agent to relieve itching palms. In 1980, Robert Koch observed that gold inhibits *Mycobacterium tuberculosis* in vitro. This led to trials on arthritis and lupus erythematosus. Also, previous studies carried out in 1973 on gold and its compounds observed beneficial activities at different levels. Gold compounds have the ability to decrease concentrations of rheumatic factors and influence the immunological responses. It has also been established that gold suppresses the anaphylactic release of histamine more effectively than gluco-corticoids. Sodium aurothiomalate (water soluble preparation) was introduced around 25 years ago to treat arthritis and is administered through intramuscular injections. Its effects are probably due to its antimicrobial effects and stimulation of the reticulo endothelial system.

11.3.2 Rajata (Silver) Bhasma

Silver is another precious metal like gold, also attracted the attention of the ancient Acharyas. The use of silver in therapeutics dates back to the period of Charaka and his contemporaries. Though its therapeutic applications are not as extensive as other metals like Tamra or Loha, the ancient classics reveal that silver also enjoyed an important place in Ayurveda therapeutics.

Quantitatively "Rajata Bhasma" is a combination of metallic silver (52–59%), free sulfur (0.675%), ferric oxide (14.33%), calcium (10.769%), silver chloride (0.479%), and traces of sodium, potassium, and aluminum.

Therapeutic uses of Rajata bhasma are the following: Rasayana (Rejuvenators), Pandu Roga (Hematinics), Hicca (hiccup), Swasa (respiratory distress), Tongue scrappers, Nozzle of enema pot, Silver containers, Idol of male gender, Sharp instruments to be placed in labor room, e.g. scissors for cutting umbilical cord.

11.3.3 Tamra (Copper) Bhasma

Copper is another ancient metal known to human civilization. During pre-Vedic times, the metal was part of day-today livelihood functions. Further, it is the earlier known metal for the preparation of the stronger alloy metals such as brass and bronze of which it is a component. Charaka uses the term Arka in a few places which Chakrapani clarifies as synonymous with Tamra. According to the descriptions of Rasa in Vagbhata Ref. there are two forms of Tamra, viz., Nepaliya and Mlechha, only the former being acceptable. Samples with characteristic metallic sheen (Snigdham), soft (Mridulam), bright reddish in color (Shonam), having high tensile strength (Ghanaghata Ksamam), heavy (Guru), and devoid of impurities (Nirvikaram) are identified as best used for medicinal purposes.

Formulations of "Tamra" are useful in a wide range of diseases like Krimi, Sthaulya, Arsha, Ksaya, Pandu, Kusta, Swasa, Kasa, Amlapitta, Sotha, Sula, Yakrit Roga, Grahani dosha, etc. In addition, Charaka advocates the use of Tamra Patra (copper). Therapeutic uses of Tamra bhasma are the following: Rasayana (Rejuvenators), Hicca (hiccup), Swasa (respiratory distress), Visha hara (Anti-poisonous), Kusta (Skin disorders), Granthi (Abscess), Collyrium for Akshi Roga (Eye disorders), Probable translation, Tongue scrapers, Copper containers, Nozzle of enema pot, Aayasa or Loha (Iron).

11.3.4 Mandura (Iron) Bhasma

Mandura, the second form of Iron, has been used for a wide range of therapeutic procedures in classical Ayurveda since antiquity. Purified mandura bhasa, when administered with proper justification is beneficial in inflammations, edematous conditions, jaundice, etc. It is the drug of choice in cases of anemia (Pandu), Chemically, Mandura is a combination of ferric oxide (59.14%), ferrous oxide (26.7%), chlorides (4.4%), magnesium (3.9%), sodium (1.7%), and a few other elements in trace quantities.

Aayasa or Loha is another metal known to ancient civilizations. Iron compounds were particularly employed in diseases such as anemia and other debilitating conditions, where functions of hematopoietic systems are disturbed and the blood has consequently become deficient in iron. Rasashastra classics explain that Loha is par excellence a rejuvenator as it stimulates functional activity of all the organs, promotes life, strength, destroys a number of diseases, and acts as a restorative. The utility of this metal in therapeutics was only identified modern medicine in the first half of the seventeenth century, when its salts were recognized as the best hematinics. Malabsorption syndrome, abdominal lump, abscess of kapha, alopecia, ascitis, disorders of female genital tract, eye, hemorrhoids, anemia, hiccup, respiratory distress, jaundice, leukoderma, obesity, rejuvenators, skin diseases, inflammation of hemorrhoids, anemia, chronic inflammation, provides complexion to psychosis and skin diseases.

11.3.5 Naga/Sisaka (Lead) Bhasma

Quantitatively Naga Bhasma is a combination of lead oxide (75.6%), ferric oxide (7.5%), together with traces of calcium and magnesium chlorides and carbonates. Different formulations of "Naga" are beneficial in diseases such as Hicca (hiccup) respiratory distress, and Kusta (skin diseases).

11.3.6 Vanga/Trapu (Tin) Bhasma

Vanga, one of the Puti Lohas was known to ancient Indian physicians by the name of Trapu. In Charaka Samhita, the metal is categorized under Parthiva Dravyas. According to descriptions in Rasa Vagbhata, there are two varieties of Vanga, viz., Khuraka and Mishraka, the former being acceptable for therapeutic applications. Samples with the characteristics, bright white in color (Dhavalam), soft (Mridulam), shiny, smooth (Snigdham), easily melts (Drutadravam), and heavy (Guru) are identified as Khura Vanga and should be preferred for therapeutic purposes. Quantitatively Vanga Bhasma is a combination of stannic oxide (i.e., of tin) (91.4%), ferric oxide

(2.9%), potassium (2.9%), calcium oxide (2%), aluminum (2%), and magnesium (0.6%) oxides. Formulations of "Vanga" are beneficial in diseases such as skin disease, tongue scrappers, nozzle of enema pot. Singly or in combination with other metals/minerals it is beneficial in disorders of the genitourinary tract.

11.3.7 Misra Loha, i.e., Alloy (Kamsya, Pitala, Varta)

Kamsya is another important Misra Loha, an alloy of Copper and Tin known since the period of Samhita Kala. Charaka used this metal to prepare Vasti netra. It is known as Bell Metal or Bronze. According to the descriptions given in Ayurveda Prakasha, there are two varieties of Kamsya, viz., Pushpa and Tailika, only the former being acceptable for therapeutic applications. Samples giving a sharp sound (Teekshna Shabdam), soft (Mridu), smooth to touch (Snigdha), slightly grayish (Eshat Shyamalam), clear from Pittala (Brass). Pittala is an important Misra Loha, an alloy of Copper and Zinc, known since the period of Samhita Kala. Charaka used this metal to prepare Vasti netra. It is known as Brass. As per the descriptions available in Rasa Ratna Samuchaya, there are two varieties of Pittala, viz., Ritika and Kakatundi. Formulations of "Pittala" are beneficial in diseases like Krimi, Kusta, Pandu, etc. The therapeutic doses given for "Pittala Bhasma" range from 60 mg to 120 mg.

Humanity depended on natural resources (air water soil/sediment containing minerals) for survival. Construction material, viz., aluminum, iron, steel, and cement are produced from mining activity. Several of the excipients are minerals. Minerals in pharmaceuticals fall into one of two main categories: excipient and active substances. The excipients have no intrinsic health benefit on their own; they are used solely as carriers, allowing the intake of minute amount of active substances. pH regulation or adsorbents are the kind of applications for which some minerals are used as active ingredients.

Thus, antacid pills are composed mainly of calcium carbonate, lithium used in anti-depressants is derived from industrial minerals, and many excipients are minerals such as talc, magnesium carbonate, or silica. Many cosmetics incorporate important amount of industrial minerals, such as talc, although others like mica, silica, or borates are utilized for their abrasive, visual, or stabilizing properties (Fig. 11.8). It is necessary to remember that earliest civilizations (e.g., the Romans) already made use of earth pigments for body painting. Industrial minerals are crucial in water management, whether considering drinking water preparation or wastewater treatment. Thus, silica sands are used as filters, perlite, zeolites, or talc as flocculants or adsorbents, bentonites as industrial mineral-based liners and geosynthetic liners, either basal or superficial, are increasingly used to avoid escape of leachate from landfill sites. Air treatment of industrial effluents also largely relies on minerals. Activated carbon is the best-known technique, but other minerals are used as well. For instance, flue gas desulfurization of power station fumes is achieved with calcium carbonate. Industrial minerals such as clays, sand, feldspar, kaolin, and other

minerals are basic to all construction materials, from bricks to tiles and from cement to limes and plastics. Zinc (Yasada) was also used in Ayurvedic medication for eye diseases, diabetes.

11.4 Thus Mineral Geology is Linked to Medicine

Increases in sediment erosion and transport, which can uncover toxic wastes deposited earlier ("chemical time bombs")

- Permafrost and gas-hydrate melting, which can lead to methane degassing and influx of insects and diseases
- Changes in groundwater and surface water quality resulting from accelerated erosion, deforestation, pollution, river diversion, and groundwater mining
- Drop in lake levels and increase in salinity causing loss of water supplies and increase in airborne particulates
- Catastrophic natural events—landslides and avalanches, earthquakes, tsunamis, and volcanic eruptions, with many effects on settlements, water, soils, and air quality

Soil is the soul of life and majority of micro- and macronutrients are available to edible and healing plants (Fig. 11.6). Both plants and humanity depended on natural resources (air, water, and soil/sediment containing minerals) for survival (Figs. 11.7 and 11.8).

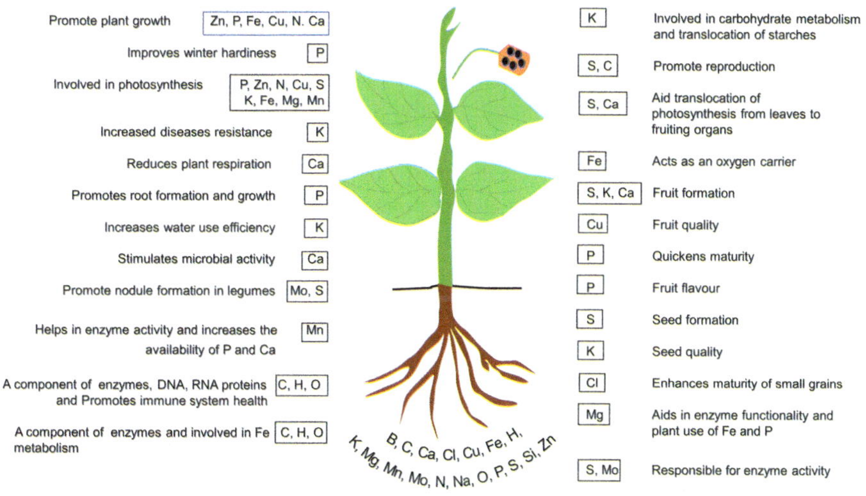

Fig. 11.6 Soil degradation leads to the loss of micro- and macronutrients. Thus, nutrient poor soils are unable to produce healthy food necessary for well-being of human and livestock

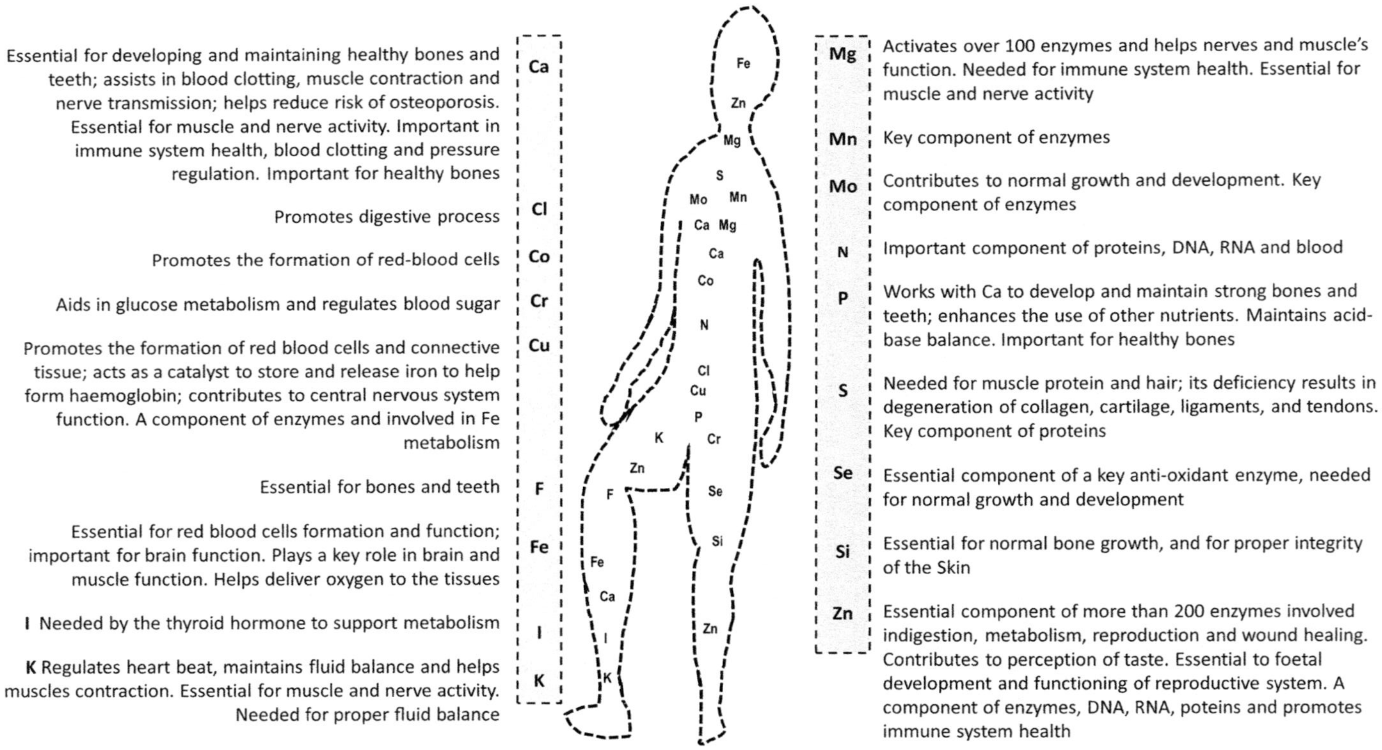

Fig. 11.7 Soil is the foundation of human health via balanced nutrition. Over 2 billion people are estimated to suffer from micronutrient deficiencies. The essential nutrients that are necessary for human health and metabolism are shown in this figure

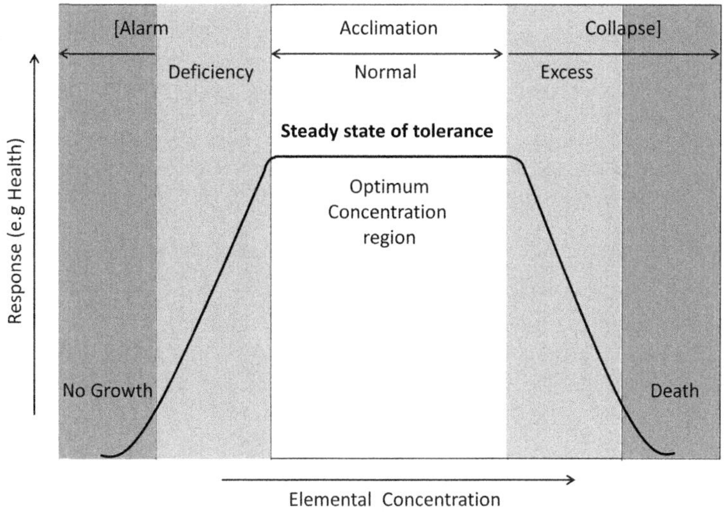

Fig. 11.8 Triphasic kinetics of the general adaptation syndrome (GAS), 1—alarm reaction: this is the somatic response to the first exposure to a localized or general stressor and initially lowers basal state of resistance (Leshem and Kuiper 1996). If the "stressor" is sufficiently strong (e.g., extremes of temperature, salinity, etc.), death may result; 2—stage of resistance or acclimation: if continued exposure to the stressor is compatible with adaptation, resistance mechanisms of either a "syntoxic" (coexistence with stress) or a "catatoxic" (stressor-removal) nature ensues. The cellular manifestations characteristic of the alarm reaction have virtually disappeared or are strictly held in abeyance and resistance rises above normal; 3—state of collapse or exhaustion: following long-term exposure to the stressor, adaptation capability is eventually exhausted. Signs of the alarm reaction reappear and the organ or the whole organism senesces and dies

There are four routes for trace elements intake for humans (Figs. 11.9 and 11.10).Diseases and symptoms associated with deficiency and efficiency of selected trace elements are shown in Table 11.4.

Tables 11.5 depicts the medicinally important elements and functions, Table 11.6 depicts nutritionally important elements and functions, and Table 11.7 depicts toxic elements that do not have metabolic significance in human.

Nutrigenomics and pharmacogenomics, medical geology (geomedical science), and metallomics are emerging as new areas of research with challenging tasks ahead. The relevance of these cutting-edge fields and their relevance to nutritional security and quality of agricultural produce with emphasis on traditional healing plants; selected examples are covered in this chapter (Figs. 11.10 and 11.11; Table 11.8).

The notable medicinal and aromatic plants that have inherent ability of accumulating toxic trace elements are:

a) *Pelargonium*, b) *Chrysopogon zizanioides*, c) *Hemidesmus indicus*, d) *Hydrocotyle umbellata* (*Centella asiatica*), and e) *Alternanthera philoxeroides*.

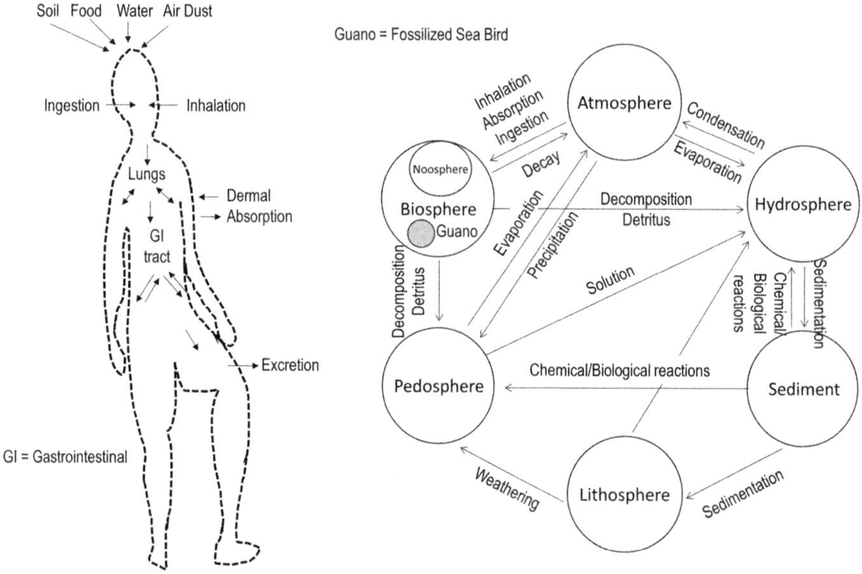

Fig. 11.9 Food, soil, water, and air. The three important pathways for the human intake of minerals: ingestion, inhalation, and skin absorption

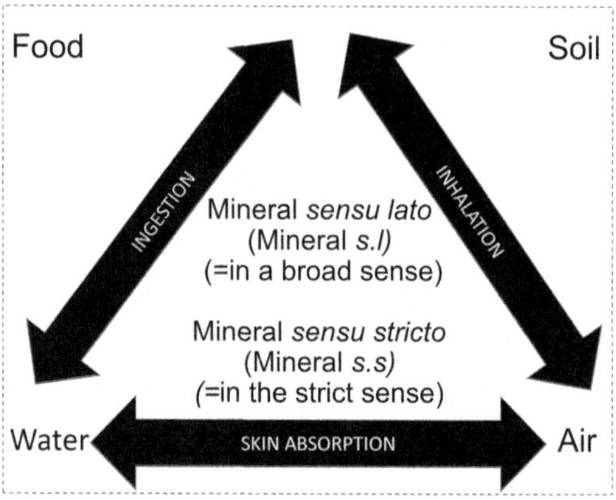

Fig. 11.10 There are four major sources of minerals for human intake, viz., food, soil, water, and air. There are three pathways for the human intake of minerals: ingestion, inhalation, and skin absorption

Amaranthus retroflexus L seed protein is an important animal feed in Eastern Europe and Russia (Fig. 11.11). *A. retroflexus* (pig weed) is notable for stabilization of phosphogypsum dump. Phosphogypsum waste dumps are known to contain radionuclides and toxic elements. The other notable grass native species associated with *A. retroflexus* are *Calamagrostis neglecta, Poa angustifolia, Agropyron cristatum*, and several others (Fig. 11.11).

Aquatic macrophytes, viz., *Ipomea aquatica* (water spinach) and *Alternanthera philoxeroides* (alligator weed) are consumed as leafy vegetable, fodder for animals. Both these plants are known to accumulate high doses of mercury (Göthberg and

Table 11.4 Diseases/symptoms associated with deficiency or efficiency of essential nutrients

Element	Symptoms or disease due to deficiency	Element	Symptom or disease due to excess availability
Selenium	Keshan (China), bone, arthritis, cardiovascular and cancer	Mercury	Minamata disease, neurological disorders
Zinc	Dwarfness (Iran and Egypt) infertility, impaired taste and smell	Cadmium	Itai-Itai, bone crippling, Cancer → prostrate, lung, renal, and heart problems
Copper	Anemia, skeletal defects	Fluorine	Yellowing of teeth, skeletal deformities
Chromium	Glucose metabolism, kidney function, cholesterol disorders	Lead	Lung cancer, neurological disorders
Iodine	Goiter	Manganese	Paraplegia = skeletal deformations → paralysis
Fluorine	Dental caries	Arsenic	Skin, lung cancer
Magnesium	Depression, nervous system. Every organ in the body—especially the heart, muscles, and kidneys need the nutritional element Mg		
Molybdenum	Mouth/esophageal cancer		
Cobalt	Anemia		
Iron	Anemia		
Sodium	Coma		

Greger 2006; Greger M 2008; Zhao et al. 2024). *A. philoxeroides* is an indicator species of polluted sites and proved to be an effective accumulator of heavy metals Fe, Zn, Mn, Pb, and Cd. For its rich and rapid growth, the alligator weed is readily available in the contaminated areas, the local people harvest the leafy shoots and sell in the local markets (Suthari et al. 2017) (Figs. 11.12 and 11.13).

P. stratiotes is a floating, stoloniferous plant found in ponds and streams almost throughout India upto a height of 1000 m. Leaves are green in color, odorless, and bitter in taste. The leaves are approximately 13 cm long and 17 cm wide and of fan-shaped having parallel venation, blunt apex, and entire margin.

A large number of medicinal properties are attributed to the plant, particularly the leaves. The plant is considered antiseptic, antitubercular, and antidysentric. In Gambia, the plant is used as an anodyne for eyewash. Juice of plant is used by Mundas in ear complaints. The ash of plant is applied to the ringworm of the scalp. Leaves are used in eczema, leprosy, ulcers, piles, and syphilis. Juice of leaves boiled in coconut oil is applied externally in chronic skin diseases.

The plant is bitter, pungent flavored having cooling, laxative property. It is useful in "Tridosha," fever, and diseases of blood. The root is laxative, emollient, and diuretic. Leaves infusions have been mentioned in the folklore to be used for dropsy, bladder complaints, kidney afflictions, hematuria, dysentery, and anemia.

A large number of medicinal properties are attributed to *Pistia stratiotes* (water cabbage or water lettuce), particularly the leaves. The plant is considered antiseptic, antitubercular, and antidysentric. In Gambia, the plant is used as an anodyne for

Table 11.5 Medicinally important elements and functions

Element	Functions
Bi Bismuth Atomic number: 83 Atomic mass: 208	Use—Antacids, diarrhea Source—Mining, consumer products Absorption—Intestine Toxicity—Kidney, chronic use results in range of effects Facts—Discovered in 1753, used to treat syphilis and malaria
F Fluorine Atomic number: 9 Atomic mass: 19	Use—Tooth protection Source—Drinking water, food supply Absorption—Intestine Toxicity—Excess causes mottled teeth enamel (fluorosis) Facts—Common water level 0.5–1.5 ppm, 3 ppm effects teeth
Ga Gallium Atomic number: 31 Atomic mass: 69.72	Use—Visualization tool for soft tissues in X-rays Source—Mining, medical injection Absorption—Very poor Toxicity—Kidney Facts—Liquid at room temperature, half-life 4–5 days
Au Gold Atomic number: 79 Atomic mass: 196.97	Use—Treat rheumatoid arthritis, range of industrial uses Source—Mining, medical injection Absorption—Poor Toxicity—Kidney, skin, and mouth lesions Facts—Long half-life
Li Lithium Atomic number: 3 Atomic mass: 6.939	Use—Treat psychiatric disorders Source—Food supply, plants, and meat Absorption—Intestine Toxicity—Wide range, e.g. tremor, seizures, slurred speech, cardiovascular, nausea, vomiting Facts—Daily intake about 2 mg
Pt Platinum Atomic number: 78 Atomic mass: 195.02	Use—Anticancer agent (cisplatin), catalytic converters, metal alloy Source—Mining, road dust Absorption—Poor, as a drug intravenous administration Toxicity—Neuromuscular, kidney Facts—Inhibits cell division, treat ovarian and testicular cancer

eyewash. Juice of plant is used for ear complaints. The ash of plant is applied to the ringworm of the scalp. Leaves are used in eczema, leprosy, ulcers, piles, and syphilis. Juice of leaves boiled in coconut oil is applied externally in chronic skin diseases (Figs. 11.14, 11.15 and 11.16; Table 11.9).

P. stratiotes is medicinally important (Aliotta and Monaco 1991; Achola and Indalo 1997; Tripathi et al. 2010) for its phytochemistry (Fig. 11.8)

Pharmacological Activities
Antifungal activity
Calcium channels involved in calcium oxalate crystal formation in specialized cells

Table 11.6 Nutritionally important elements and functions

Element	Functions
Cr Chromium Atomic number:24 Atomic mass: 52	Use—Essential element, associated with insulin, stainless steel, tanning leather Source—Food supply, inhalation Recommended daily—50–200 μg Absorption—Intestine Toxicity—Acute exposure cause kidney damage, lung cancer Facts—Comes in different oxidized forms—Cr^{3+}, Cr^{6+}
Cu Copper Atomic number: 29 Atomic mass: 63.55	Use—Essential element, widely used Source—Readily available in food Recommended daily—1.5–3.0 mg Absorption—Intestine Toxicity—Copper deficiency—Anemia; excess copper—Rare, Wilson's disease Facts—Excess treated with penicillamine—can be toxic to grazing animals
Fe Iron Atomic number: 26 Atomic mass: 55.85	Use—Oxygen carrying hemoglobin Source—Food Recommended daily–10–15 mg Absorption—Intestine Toxicity—Excess causes bloody feces, bloody vomit, liver damage Facts—3–5 g in the body, 67% associated with hemoglobin
Mg Magnesium Atomic number: 12 Atomic mass: 24.3	Use—Essential nutrient, associated with many enzymes, antacids Recommended daily—280–350 mg Source—Food supply, nuts, cereals, seafood, meats, drinking water Absorption—Small intestine Toxicity—Deficiency—Convulsions; excess—Nervous system Facts—20 g in body
Mn Manganese Atomic number: 25 Atomic mass: 54.94	Use—Trace element, associated with many enzymes Source—Food supply, grains, nuts Recommended daily—2–5 mg Absorption—Intestine poor (5%) Toxicity—Inhalation—Respiratory disease, nervous system, Parkinson's-like syndrome, psychiatric disorders Facts—Half-life 37 days
Se Selenium Atomic number: 34 Atomic mass: 78.96	Use—Essential element, present in most tissue, anticancer, reduces toxicity of metal mercury and cadmium Source—Food supply, shrimp, meat Recommended daily—55–70 μg/day, not to exceed 200 μg/day Absorption—Intestine Toxicity—Deficiency—Heart disorders; excess—"blind staggers," neurological effects
S Sulphur Atomic number: 16 Atomic mass: 32.07	Use—Sulfuric acid, fertilizer, fine chemicals, pesticides, pharmaceuticals, winemaking, and food preservation Source—Sulfur is the fifth most common element on Earth

(continued)

Table 11.6 (continued)

Element	Functions
Zn Zinc Atomic number: 30 Atomic mass: 65.39	Use—Essential element, cofactor with several enzymes, and proteins Source–Food supply, drinking water Recommended daily—12–25 mg Absorption—Intestine Toxicity—Deficiency—Impaired growth, neurological disorders, inhalation can cause metal fume fever
Si Silica Atomic number: 53 Atomic mass: 126.9	Use—Clay minerals and other aluminosilicates Have several medicinal uses Source—Earth crust Silica rich plant leaves are resistant to bacterial and fungal pathogens as they cannot penetrate
I Iodine Atomic number: 53 Atomic mass: 126.9	Use—Chemicals and pharmaceuticals. In chemistry, Iodine compounds are used as catalyst Source—Crustal average abundance is estimated as 0.5 mg/kg
Mo Molybdenum Atomic number: 42 Atomic mass: 95.95	Use—It is the essential component of several enzymes, such as nitrate reductase, xanthine dehydrogenase, aldehyde oxidase, and sulfide oxidase Source—Earth's crust as 1.5 mg/kg

Calcium channel blocking activity and antimicrobial activity
L-Ascorbic acid and L-galactose sources for oxalic acid and Ca-oxalate

Prosopis juliflora tender foliage and pods are important source of animal feed (Fig. 11.17).

Cr is an essential mineral for ruminants in tropical regions. Its pods contain high Cr concentration (up to 150 μg/L). Thus, supplements Cr in fodder for animals (goats) Cr requirement for animals is >0.1 mg/L, while toxic level was 1000 mg/L (Fig. 11.17). *Prosopis juliflora* pods are rich in carbohydrates and minerals. Goats consume the pods to meet their nutritional requirement; for example, dried pods are powdered and used as animal feed.

Agrotechnologies available for several medicinal and aromatic plants, viz., *Artemisia annua, Cymbopogon winterianus* (Citronella), *C. martinii* (Palmarosa), *C. flexuosus* (Lemongrass), *Cassia angustifolia* (Senna), *Chrysanthemum cinerariifolium* (Pyrethrum), *Mentha arvensis* (Menthol mint), *M. citrata* (Bergamot mint), *M. piperata* (Pepper mint), *M. spicata* (Spearmint), *M. cardiaca* (Scotch spearmint), *Ocimum sanctum* (Tulsi), *Pelargonium graveolens* (Geranium), *Chrysopogon zizanioides* (Vetiver), and *Withania somnifera* (Ashwagandha).

Unlike the above-mentioned medicinal and aromatic plants, *Bacopa* is collected from nature as it preferentially grows on wet or marshy lands, on the margins of

Table 11.7 Toxic elements that do not have metabolic significance in human

Element	Functions
Al Aluminum Atomic number: 13 Atomic mass: 26.98	Use—Wide range of consumer products, airplanes to cans Source—Food, drinking water Absorption—Poor Toxicity—Dialysis dementia, possibly neurotoxic Facts—Non-essential, intake 1–10 mg/day
As Arsenic Atomic number: 33 Atomic mass: 74.92	Use—Pesticide and herbicide Source—Food, drinking water Absorption—Intestine Toxicity—Cancer, heart, liver, neurological Facts—Exists in different states—Trivalent (most common), pentavalent, arsenic trioxide, organic and inorganic, etc.
Be Beryllium Atomic number: 4 Atomic mass: 9.01	Use—Metal alloy, nuclear power plants Source—Workplace, coal combustion Absorption—Lung, skin Toxicity—Lung, can be delayed and is progressive, contact dermatitis probable carcinogen Facts—Discovered in 1828, more than 1250 tons from oil and coal combustion
Cd Cadmium Atomic number: 48 Atomic mass: 112.41	Use—Alloy in metal, paint Source—Shellfish, cigarette smoke, workplace—Welding, paints Absorption—Intestine, lungs Toxicity—Lung, emphysema, kidney, calcium metabolism, possible lung carcinogen Facts—"Itai-Itai" is Japanese for "ouch-ouch"—refers to bone pain related to calcium loss
Co Cobalt Atomic number: 27 Atomic mass: 58.93	Use—Component of vitamin B12 Source—Alloy in metals, magnets Recommended daily—None Absorption—Intestine Toxicity—Excessive heart failure, inhalation—"hard metal" lung disease Facts—Once used a foaming agent in beer
Pb Lead Atomic number: 82 Atomic mass: 207.20	Use—Not essential, batteries, old paint and previously gasoline, hobbies Source—Home, paint, dust, kids-hands to mouth, workplace Absorption—Intestine (50% kids, 10% adults) Toxicity—Developmental and nervous system Facts—Developing nervous system very sensitive to low levels of exposure
Hg Mercury Atomic number: 80 Atomic mass: 200	Use—Consumer products, industry, dental amalgams, switches, thermometers Source—Mining, environment Absorption—Inhalation, intestine poor Toxicity—Nervous system toxicant, "Mad Hatters" disease Facts—Liquid silver evaporates at room temperature, bacteria convert to organic methyl mercury

(continued)

Table 11.7 (continued)

Element	Functions
Organic Mercury ($Hg\text{-}CH_3$)	Use—Limited laboratory use—Most common is methyl mercury ($Hg\text{-}CH_3$) Source—contaminated fish (e.g., tuna, shark, pike) Absorption—Intestine very good (90%) Toxicity—Nervous system toxicant, and developmental toxicant Facts—Bacteria convert inorganic mercury to methyl mercury then in to food supply (bioaccumulation)
Ni Nickel Atomic number: 28 Atomic mass: 58.96	Use—Not essential, metal alloy, stainless steel Source—Food supply, jewelry, workplace Absorption—Intestine, skin Toxicity—Carcinogen (lung), contact dermatitis Facts—Discovered in 1751, and 200,000 metric tons used yearly
Sn Tin Atomic number: 50 Atomic mass:118.71	Use—Inorganic—Consumer products; Organic—Fungicide, bactericides Source—Food packaging Absorption—Intestine (low inorganic, high organic) Toxicity—Inorganic—little; organic—Central nervous system Facts—Triethyltin and trimethyltin most toxic

Fig. 11.11 High biomass and seed producing *Amaranthus retroflexus* (pig weed) in green house Kazan [Courtesy Dr Gulshat Gasimova, Kazan (Гульшат Гасимова)]

Table 11.8 Selected examples of trace elements as contaminants in facial cosmetics (Prasad 2023)

Country	Traditional cosmetic	Composition
Saudi Arabia	Henna	Amorphous carbon Cuprite (Cu_2O) Kerdoun et al. 2024 Galena (PbO) Goethite (FeO(OH) Magnetite (Fe_3O_4) Minium (Pb_3O_4) Sassolite (H_3BO_3) Talc ($Mg_3Si_4O.10(OH)_2$) Zincite (ZnO)
Saudi Arabia, India	Kohl	
Middle East	Eyebrow liners	
Morocco, USA, Mauritania, Pakistan, India, Great Britain Saudi Arabia	Kohl	
Bulgaria	Eye shadow, lipstick	
Oman and UAE	Bint al dhahab	
Bahrain	Surma and kohl	
Isreal	Kwali (galena based)	
Nigeria	Kwali eye shadow (graphite based)	

dirty drainage channels, and sewage sludge dumps (Lavu et al. 2013; Sinha 1999; Mondal et al. 2023; Saleem and Urainab 2023) (Figs. 11.18 and 11.19).

Therefore, quality and safety and risks associated with the use of such kind of raw material form an important issue in the science of herbal products management.

11.5 Need for Quality Control and Quality Assurance

Asian medical systems (Indian = Ayurveda, Siddha, Folk, Unani, Tibetan, Homeopathy, and Modern medicinal practices use a variety of plants), Tibetan and Chinese medical systems use a wide range of plant products such as flowers, leaves, roots, rhizomes, whole plant, wood, bark, stem, seeds, fruits, etc., in preparation of herbal formulations. Several of the plants having healing functions have the inherent function of metal accumulation. Thus, the products prepared are often contaminated with toxic levels of heavy metals of natural or anthropogenic Collection of healing plants from trace element contaminated and polluted environment exposed to trace elements and, use of ground water irrigation for cultivation of medicinal plants are the major sources of toxic trace elements. Often the accumulated trace elements pose health risks or may serve as remedy, for which detailed investigations are warranted.

The two commonly used terms related to herbal products quality are: Quality Control (QC) and Quality Assurance (QA). It is a wide-ranging concept that covers all matters that individually or collectively influences the quality of a drug. It is the total of the organized arrangements made with the objective of ensuring that drugs are of the quality required for their intended use (Fig. 11.20).

Fig. 11.12 (**a**) *Alternanthera philoxeroides* (alligator weed) growing luxuriantly on the banks of polluted lakes. (**b**) Plant sold in local market as "Ceylon spinach" at a low price. (**c**) and (**d**) Local vendors and consumers in the local market. (**e**) Large quantities of collected *Alternanthera philoxeroides* (encircled with broken yellow line) along with other leafy vegetables in the market (Suthari et al. 2017)

Fig. 11.13 (**a**)–(**c**) *Hemidesmus indicus*; (**d**, **e**) *Hydrocotyle umbellata* (*Centella asiatica*) well-known medicinal plants and toxic metal accumulators (Sekhar et al. 2005)

Fig. 11.14 *Pistia stratiotes* (water lettuce, water cabbage) growing luxiriantly in polluted water-body. Roots exhibit deposition (plaque) of essential and toxic minerals

Fig. 11.15 Phytoconstituents of healing function are rich in *Pistia stratiotes* (water cabbage or water lettuce) (Sinha and Saxena 2005; Volk 2004)

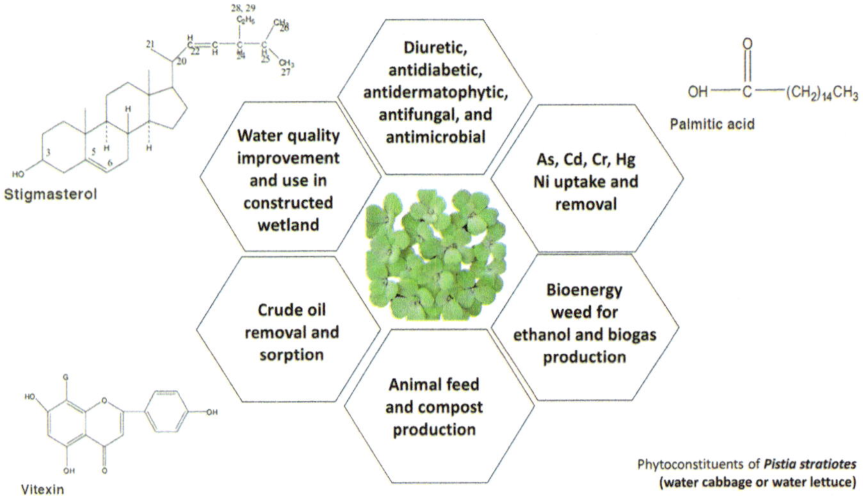

Fig. 11.16 *Pistia stratiotes* (water cabbage or water lettuce) in polluted water

Table 11.9 Nutritional value, pharmacological and metal accumulation significance of *Pistia stratiotes* (water lettuce, water cabbage)

Research topic investigated	Author(s)	Year
Cr (VI) uptake	Sen et al.	1987
As accumulation	Lee et al.	1991
Cr toxicity	Satyakala and Jamil	1992
Ni (II) uptake	Sen and Bhattacharyya	1994
Hg Mercury accumulation	Patra and Panigrahi	1994
Metal accumulation	Cardwell et al.	2002
As reduction	Basu et al.	2003
Cd uptake		
Heavy metal removal	Miretzky et al.	2004
Cd and Cr removal kinetics	Suñe et al.	2007
Hg Mercury uptake	Skinner et al.	2007
Cd removal	Bhakta and Munekage	2008
Phytoremediation	Odjegba and Fasidi,	2007
Concurrent removal and accumulation of heavy metals	Mishra and Tripathi	2008
Crude oil removal	Noyo and George	2008
Cu phytotoxicity	Upadhyay and Panda	2009
Ni and phosphorous sorption efficiencies		
Pb removal kinetics	Espinoza-Quiñones et al.	2009
Nutritional value		
Removal of heavy metals from industrial wastewater	Khan et al.	2009
Cd (II) removal	Miretzky et al.	2010
Cr, Ni, Zn, and phosphorous removal	Mufarrege et al.	2010
Metal and metalloid removal	Marchand et al.	2010
Phytoremediation in engineered wetlands	Zhang et al.	2010
Crude oil sorption	Sanchez-Galvan et al.	2010
Pharmacological activities, phytochemistry and diuretic, antidiabetic, antidermatophytic, antifungal, and antimicrobial properties of *Pistia stratiotes*	Tripathi et al.	2010
As biosorption	Chowdhury and Mulligan	2011
Phytoremediation of arsenic	Rahman and Hasegawa	2011

Fig. 11.17 *Prosopis juliflora* (Sw) DC: Foliage and pods are known to accumulate toxic trace elements, the notable being chromium. (**a**) *P. juliflora* in flowering, (**b**) tender pods, (**c**) mature pods. (**d**) Goats and sheep graze on foliage and pods. (**e**)–(**g**) Mature and dried pods are powdered and used as animal feed (Prasad and Tiwari 2024)

Fig. 11.18 Growth of *Bacopa monnieri* L in natural conditions—Traditional healers and professional plant collectors harvest the material for their livelihood. (**a**) *Bacopa* growing close to sewage sludge and industrial sites, (**b**)–(**d**) In large patches all along the open drainage

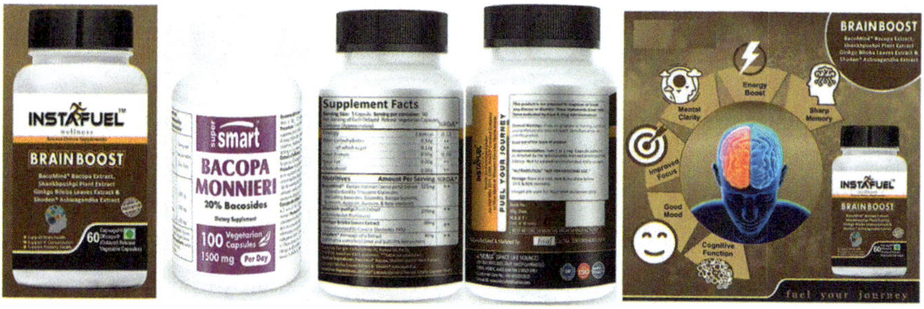

Fig. 11.19 Brain supplements from *Bacopa monnieri*

Fig. 11.20 Routes for accumulation of trace elements in healing plants

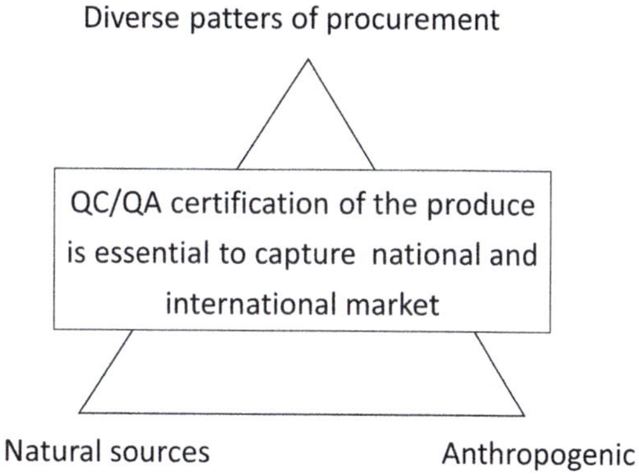

11.6 Some of the Existing Legal Framework for Plant-Derived Ingredients with Medicinal Properties

11.6.1 The World Health Organization

The World Health Organization (WHO) views herbal medicines as herbs, herbal materials, herbal preparations, and finished herbal products that contain active ingredients parts of plants, and other plant materials, or combinations. The WHO recognizes that the traditional use of herbal medicines refers to the long historical use of these medicines and that they may be accepted by national authorities. As a result of this view, the WHO Traditional Medicines Strategy 2001–2005 was developed to review a framework for action for WHO and its partners aimed at enabling TM/CAM to play a far greater role in reducing excess mortality and morbidity especially among impoverished populations.

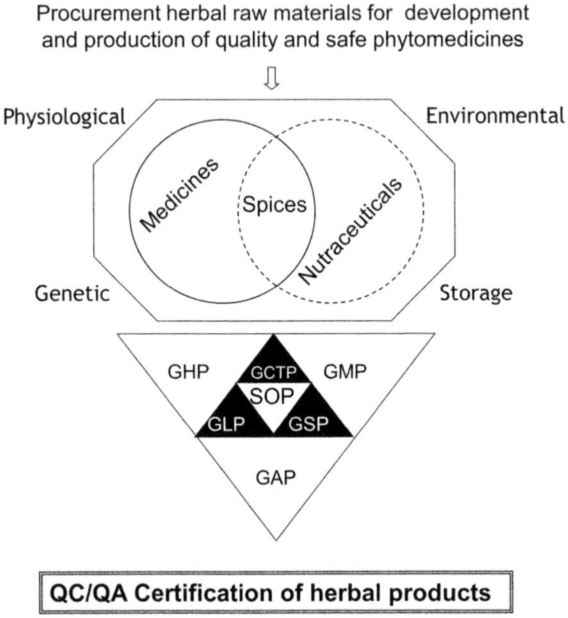

Fig. 11.21 Factors influencing QC/QA of herbal products

11.6.2 European Union

The European Union (EU) directives 2001/83/EC on the Community code relating to medicinal products for human use lay down a general framework for pharmaceutical products requiring pre-marketig approval before gaining access to the market and laying down the requirements for the documentation of quality, safety, and efficacy, the dossier and expert reports. This framework has effectively been in operation. The European Medicines Agency (EMA) is an agency of the European Union (EU) in charge of the evaluation and supervision of pharmaceutical products. Prior to 2004, it was known as the European Agency for the Evaluation of Medicinal Products or European Medicines Evaluation Agency (EMEA). This agency acts as a central agency for single European medicines marketing authorization that operates a Herbal Medicinal Products Working Party (HMPWP). However, individual Member States (UK, Germany, France, Italy, etc.) have taken different approaches in reviewing herbal medicines.

11.6.3 The United States

In the United States (US), the Food and Drug Adminsitration (FDA) has responsibility for food and drug products. Drugs are regarded as products that claim to treat, cure, mitigate, or prevent a disease. Herbal medicines follow the same procedures

as those for a chemical drug. Otherwise natural products are regulated as foods under a requirement for ingredients to be generally recognized as safe (GRAS).

Natural products generally have GRAS status, provided that this is supported by expert consensus. Hence, dietary supplements and herbs are considered to be foods provided that they are generally regarded as safe and do not make medicinal claims. Furthermore, the ingredients and the plants or parts of the plants must be quantified and where, ingredients are listed with a pharmacopeial reference, they must meet the standard laid down in the pharmacopeia. There are also specific requirements for food additives that do need a pre-market approval by the authority (Figs. 11.21, 11.22, 11.23 and 11.24).

References

Achola KJ, Indalo AA (1997) Pharmacologic activities of *Pistia stratiotes*. Pharm Biol 35:329–333

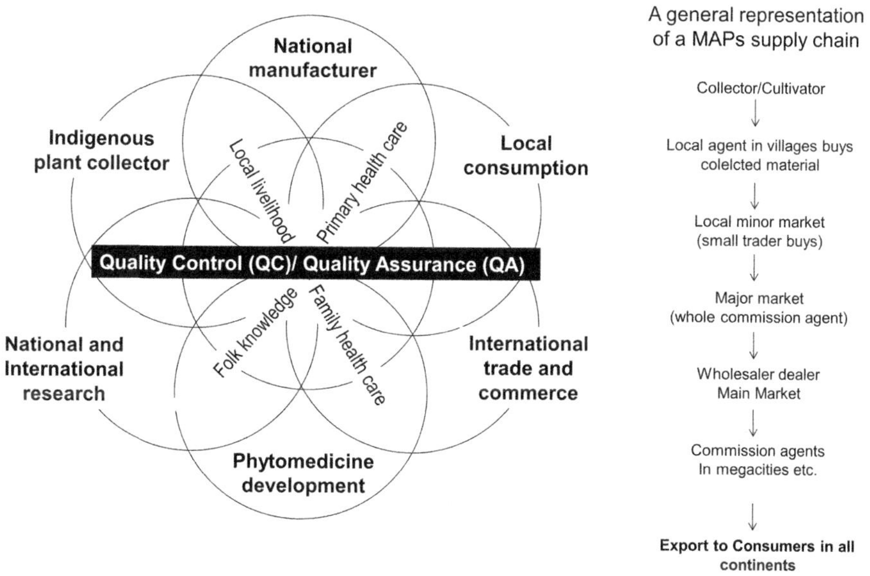

Fig. 11.22 QC and QA of herbal products and supply chain of raw materials for herbal pharma industry

Fig. 11.23 Road map for new phytomedicines Barnes (2003)

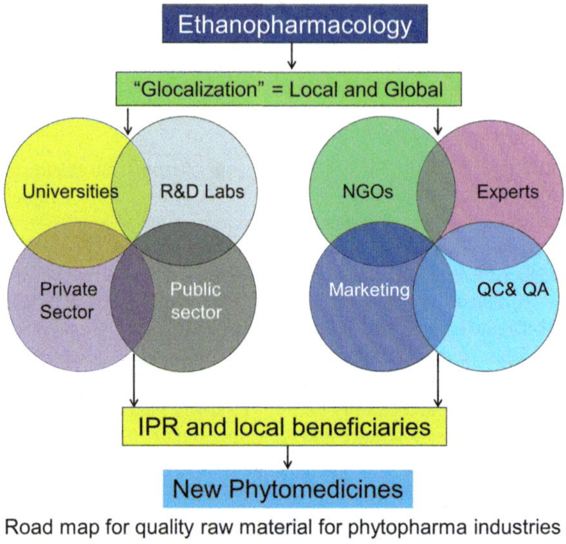

Fig. 11.24 Opportunities and limitations for emerging phytopharma industries

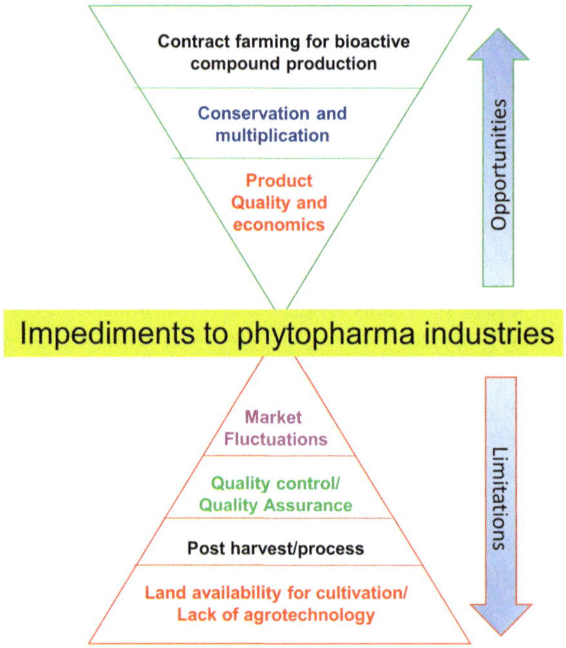

Adeagbo AA, Ogunlaja OO, Johnson HO, Akinseye MO, Oni SO (2024) Human health risk assessment of trace elements in commonly consumed herbs in Ibadan, Nigeria. J Food Compos Anal 126:105838

Ahandani EA, Rafeie F, Alizadeh-Tarpoei Z, Hajipour S, Selamoglu Z, Heidari-Bazardehy SS (2022) An attitude towards the effects of heavy metals in medicinal plants. Bioeng Stud 3(2):1–7

Aliotta G, Monaco P (1991) Acylglycosyl sterols from *Pistia stratiotes*. J Chem Ecol 17:2223–2234

Al-Khayri JM, Banadka A, Nagella P, Alessa FM, Almaghasla MI (2023) Cadmium toxicity in medicinal plants: an overview of the tolerance strategies, biotechnological and omics approaches to alleviate metal stress. Front Plant Sci 13:1047410

Alsherif EA, Al-Shaikh TM, AbdElgawad H (2022) Heavy metal effects on biodiversity and stress responses of plants inhabiting contaminated soil in Khulais, Saudi Arabia. Biology 11(2):164

Al-Thani G, Ibrahim AE, Alomairi M, Salman BI, Hegazy MM, Al-Harrasi A, El Deeb S (2023) Toxic elemental impurities in herbal weight loss supplements; a study using ICP-OES microwave-assisted digestion. Toxics 11(3):272

Anjitha KS, Sameena PP, Puthur JT (2021) Functional aspects of plant secondary metabolites in metal stress tolerance and their importance in pharmacology. Plant Stress 2:100038

Anwar MS, Khan A, Ahmad L, Khan A, Mateen A, Jahan S, Ullah U, AlMasoud N, Alomar TS, Rauf A, Khan FU (2024) Quantification of toxic heavy metals, trace elements and essential minerals contents in traditional herbal medicines commonly utilized in Khyber Pakhtunkhwa, Pakistan. Heliyon 10(3)

Asiminicesei DM, Vasilachi IC, Gavrilescu MARIA (2020) Heavy metal contamination of medicinal plants and potential implications on human health. Rev Chim 71(7):16–36

Awasthi G, Jaiman V, Nama S, Awasthi KK, Pandit PP, Lohar S, Godara V, Kumari P, Singh T, Sankhla MS, Awasthi A (2024) Heavy metal accumulation potential of medicinal plants: phytoremediation perspective. In: Phytoremediation potential of medicinal and aromatic plants. CRC Press, pp 22–34

Asiminicesei DM, Fertu DI, Gavrilescu M (2024) Impact of heavy metal pollution in the environment on the metabolic profile of medicinal plants and their therapeutic potential. Plants 13(6):913

Babandi A, Yussuff YM, Yakasai HM, Ya'u M, Shehu D, Ibrahim S, Abubakar N, Muhammad A, Babagana K, Abubakar SM, Abdullahi SSA (2024) Determination of heavy metals contamination, risk prediction and antioxidant properties of anti-malarial herbal mixture sold in Kano state, Nigeria. Case Stud Chem Environ Eng 9:100576

Barnes J (2003) Pharmacovigilance of herbal medicines: a UK perspective. Drug Saf 26:829–851

Barve M, Mashru M, Jagtap C, Patgiri BJ, Prajapati PK (2011) Therapeutic potentials of metals in ancient India: a review through Charaka Samhita. J Ayurveda Integrative Med 2(2):55–63

Basu A, Kumar S, Mukherjee S (2003) Arsenic reduction from environment by water lettuce (*Pistia stratiotes* L.). Int J Environ Health 45:143–150

Bauer R, Tittel G (1996) Quality assessment of herbal preparations as a precondition of pharmacological and clinical studies. Phytomedicine 2(3):193–198

Bayanati M, Al-Tawaha AR, Al-Taey D, Al-Ghzawi AL, Abu-Zaitoon YM, Shawaqfeh S, Al-Zoubi O, Al-Ramamneh EAD, Alomari L, Al-Tawaha AR, Dey A (2022) Interaction between zinc and selenium bio-fortification and toxic metals (loid) accumulation in food crops. Front Plant Sci 13:1001992

Bhakta JN, Munekage Y (2008) Role of ecosystem components in Cd removal process of aquatic ecosystem. Ecol Eng 32:274–280

Cardwell AJ, Hawker DW, Greenway M (2002) Metal accumulation in aquatic macrophytes from southeast Queensland, Australia. Chemosphere 48:653–663

Ceramella J, De Maio AC, Basile G, Facente A, Scali E, Andreu I, Sinicropi MS, Iacopetta D, Catalano A (2024) Phytochemicals involved in mitigating silent toxicity induced by heavy metals. Foods 13(7):978

Chauhan R, Awasthi S, Kumar S, Singh S, Dwivedi S, Tewari SK (2024) Assessment of Heavy Metals and Their Induced Toxicity in Medicinal Plants: A Public Health Concern. In: Phytoremediation Potential of Medicinal and Aromatic Plants. CRC Press, pp 139–153

Chen W, Yang Y, Fu K, Zhang D, Wang Z (2022) Progress in ICP-MS analysis of minerals and heavy metals in traditional medicine. Front Pharmacol 13:891273

Chen YG, Huang JH, Luo R, Ge HZ, Wołowicz A, Wawrzkiewicz M, Gładysz-Płaska A, Li B, Yu QX, Kołodyńska D, Lv GY (2021) Impacts of heavy metals and medicinal crops on ecological systems, environmental pollution, cultivation, and production processes in China. Ecotoxicol Environ Saf 219:112336

Chizzola R, Franz CH (1996) Metallic trace elements in medicinal and aromatic plants from Austria, P 52-56

Chourasia HK (1995) Mycobiota and mycotoxins in herbal drugs of Indian pharmaceutical industries. Mycol Res 99(6):697–703

Chowdhury MRI, Mulligan CN (2011) Biosorption of arsenic from contaminated water by anaerobic biomass. J Hazard Mater 190:486–492

Curran GL (1954) Effect of certain transition group elements on hepatic synthesis of cholesterol in the rat. J Biol Chem 210(2):765–770

da Silva RMG, de Oliveira Moraes VM, Granero FO, Malaguti Figueiredo CC, Dos Santos VHM, Machado LP, Pereira Silva L (2024) Cytogenotoxicity evaluation of heavy metals detected in extracts and infusion of *Baccharis trimera*, potential bioaccumulator plant. J Toxic Environ Health A 87(3):108–119

Dubey NK, Kumar R, Tripathi P (2004) Global promotion of herbal medicine: India's opportunity. Curr Sci 86(1):37–41

Efuntoye MO (1996) Fungi associated with herbal drug plants during storage. Mycopathologia 136(2):115–118

Espinoza-Quiñones FR, Módenes AN, Costa IL, Palácio SM, Trigueros DEG (2009) Kinetics of lead bioaccumulation from a hydroponic medium by aquatic macrophytes Pistia stratiotes L. Water Air Soil Pollut 203(1–4):29–37

Ghafarifarsani H, Rohani MF, Raeeszadeh M, Ahani S, Yousefi M, Talebi M, Hossain MS (2024) Pesticides and heavy metal toxicity in fish and possible remediation–a review. Ann Anim Sci

Göthberg A, Greger M (2006) Formation of methyl mercury in an aquatic macrophyte. Chemosphere 65:2096–2105

Graham JC, Simon SL (1996) A study of 137Cs in soil profiles from the Marshall Islands. Sci Total Environ 183(3):255–268

Greger M (2008) Trace elements and radionuclides in edible plants. In: Prasad MNV (ed) Trace elements as contaminants and nutrients. Wiley, Hoboken, NJ, pp 121–136

Gul R, Jan H, Lalay G, Andleeb A, Usman H, Zainab R, Qamar Z, Hano C, Abbasi BH (2021) Medicinal plants and biogenic metal oxide nanoparticles: a paradigm shift to treat Alzheimer's disease. Coatings 11(6):717

Guo C, Lv L, Liu Y, Ji M, Zang E, Liu Q, Zhang M, Li M (2023) Applied analytical methods for detecting heavy metals in medicinal plants. Crit Rev Anal Chem 53(2):339–359

Gurib-Fakim A (2006) Medicinal plants: traditions of yesterday and drugs of tomorrow. Mol Asp Med 27(1):1–93

Gyeresi A, Kelemen H, Kata M (1997) Summary of the chromatographic methods recommended by Pharmacopoeias for analysis of medicinal plants and their products. JPC J Planar Chromatogr Modern TLC 10(3):172–177

Habib DW (2024) Phytoremediation potential of Senna (Senna occidentalis) and neem (Azadirachta indica) in detoxification of soils contaminated with Zinc (Zn), Copper (Cu) and Lead (Pb). Journal of Integrated Sciences:23–73

Hartmann AM, Dani'K R, Geist CR (2003) Toxicity of herbal beverages. In: Reviews in Food and Nutrition Toxicity. CRC Press, pp 13–27

Ibrahim M, Detroja A, Bhimani A, Bhatt TC, Koradiya J, Sanghvi G, Bishoyi AK (2024) Ensuring the safety of indian herbal drugs: a comprehensive assessment of heavy metal contaminations with emphasis on lead. Arsenic and Cadmium. https://doi.org/10.21203/rs.3.rs-3950502/v1
Izah SC, Ogidi OI, Iyiola AO, Obasi CP (2024) Trace metals contamination of herbal remedies: health implications and quality control strategies. In: Herbal medicine phytochemistry: applications and trends. Springer International Publishing, Cham, pp 1–38
Karahan F, Ozyigit II, Saracoglu IA, Yalcin IE, Ozyigit AH, Ilcim A (2020) Heavy metal levels and mineral nutrient status in different parts of various medicinal plants collected from eastern Mediterranean region of Turkey. Biol Trace Elem Res 197:316–329
Karavaeva EN, Molchanova IV 1998 Accumulation of radionuclides by medicinal plants in the zone of impact from the Beloyarskaya nuclear power plant, 357–359
Kerdoun MA, Zergui A, Adjaine OEK, Mekhloufi S (2024) Determination of Lead (Pb) in Kohl cosmetics sold in the south of Algeria. J Trace Elem Miner:100170
Khan S, Ahmad I, Shah MT, Rehman S, Khaliq A (2009) Use of constructed wetland for the removal of heavy metals from industrial wastewater. J Environ Manag 90:3451–3457
Kumar S, Roy AK (1993) Occurrence of aflatoxin in some liver curative herbal medicines. Lett Appl Microbiol 17(3):112–114
Lavu RVS, Prasad MNV, Pratti VL, Meißner R, Rinklebe J, Van De Wiele T, Tack F, Du Laing G (2013) Trace metals accumulation in Bacopa monnieri and their bioaccessibility. Planta Med 79(12):1081–1083
Lee CK, Low KS, Hew NS (1991) Accumulation of arsenic by aquatic plants. Sci Total Environ 103:215–227
Leshem YY, Kuiper PJC (1996) Is there a GAS (general adaptation syndrome) response to various types of environmental stress? Biol Plant 38:1–18
Marchand L, Mench M, Jacob DL, Otte ML (2010) Metal and metalloid removal in constructed wetlands, with emphasis on the importance of plants and standardized measurements: a review. Environ Pollut 158:3447–3461
Marcus DM, Grollman AP (2002) Botanical medicines—the need for new regulations. N Engl J Med 347(25):2073–2076
Miretzky P, Munoz C, Carrillo-Chavez A (2010) Cd (II) removal from aqueous solution by *Eleocharis acicularis* biomass, equilibrium and kinetic studies. Bioresour Technol 101:2637–2642
Miretzky P, Saralegui A, Fernandez A (2004) Aquatic macrophytes potential for the simultaneous removal of heavy metals. Chemosphere 57:997–1005
Mishra VK, Tripathi BD (2008) Concurrent removal and accumulation of heavy metals by the three aquatic macrophytes. Bioresour Technol 99:7091–7097
Mondal S, Bhar K, Mondal P, Panigrahi N, Sahoo SK, Swetha P, Chakraborty S, Teja NY, Parveen N (2023) In quest of the mysterious holistic Vedic herb *Bacopa monnieri* (L.) Pennell. Pharm Res 15(3)
Mufarrege MM, Hadad HR, Maine MA (2010) Response of *Pistia stratiotes* to heavy metals (Cr, Ni, and Zn) and phosphorous. Arch Environ Contam Toxicol 58:53–61
Mukherjee S, Chatterjee N, Sircar A, Maikap S, Singh A, Acharyya S, Paul S (2023) A comparative analysis of heavy metal effects on medicinal plants. Appl Biochem Biotechnol 195(4):2483–2518
Noyo EE, George EO (2008) Composition of water soluble fraction (WSF) of amukpe well-head crude oil before and after exposure to *Pistia stratiotes* L. Res J Applied Sci 3:143–146
Odjegba VJ, Fasidi IO (2007) Phytoremediation of heavy metals by *Eichhornia crassipes*. Environmentalist 27:349–355
Okereafor U, Makhatha M, Mekuto L, Uche-Okereafor N, Sebola T, Mavumengwana V (2020) Toxic metal implications on agricultural soils, plants, animals, aquatic life and human health. Int J Environ Res Public Health 17(7):2204

Patra RR, Panigrahi AK (1994) Changes in residual mercury accumulation and pigment contents in some aquatic plants, Pistia and Hydrilla, exposed to solid waste of a chloralkali industry. J Environ Biol 15:299–304

Pehoiu G, Murarescu O, Radulescu C, Dulama ID, Teodorescu S, Stirbescu RM, Bucurica IA, Stanescu SG (2020) Heavy metals accumulation and translocation in native plants grown on tailing dumps and human health risk. Plant Soil 456:405–424

Prasad MNV, Tewari JC (2024) *Prosopis juliflora* (Sw) DC: potential for bioremediation and bio-based economy. In: Bioremediation and bioeconomy. Elsevier, pp 99–118

Prasad MNV (2023) Medical geology: biosphere, geosphere, and noosphere interface. In: Prasad MNV, Vithanage M (eds) Medical geology: en route to one health. Wiley, pp 1–36

Rahman MA, Hasegawa H (2011) Aquatic arsenic: phytoremediation using floating macrophytes. Chemosphere 83:633–646

Raskin I, Ribnicky DM, Komarnytsky S, Ilic N, Poulev A, Borisjuk N, Brinker A, Moreno DA, Ripoll C, Yakoby N, O'Neal JM (2002) Plants and human health in the twenty-first century. Trends Biotechnol 20(12):522–531

Roy AK, Chourasia HK (1990) Mycoflora, mycotoxin producibility and mycotoxins in traditional herbal drugs from India. J Gen Appl Microbiol 36(5):295–302

Salamon I (2024) Medicinal, aromatic, and spice plants: biodiversity, phytochemistry, bioactivity, and their processing innovation. Horticulturae 10:280

Saleem R, Urainab SUMIA (2023) Heavy metal content analysis of herbal plants gathered in various parts of Faisalabad, Pakistan. Int J New Chem 11(2):123–132

Salmani MH, Gholami M, Ranjbar MJ, Mokhberi F (2024) Comparison of essential and toxic metals levels in some herbal teas: a systematic review. Biol Trace Element Res 202(2):615–623

Salunkhe AS, Kolekar GB, Barge JN (2023) The preview of ancient Indian chemistry. Int J Res Anal Rev 10(3):127–131

Sanchez-Galvan G, Mercado FJ, Olguin EJ (2010) Sorption of crude oil by the nonliving biomass of *Pistia stratiotes*. J Biotechnol 150S:S1–S576

Satyakala G, Jamil K (1992) Chromium induced biochemical changes in *Eichornia crassipes* (Mart) Solms and *Pistia stratiotes* L. Bull Environ Contam Toxicol 48:921–992

Sathiyamoorthy P, Van Damme P, Oven M, Golan-Goldhirsh A (1997) Heavy metals in medicinal and fodder plants of the Negev desert. J Environ Sci Health Part A 32(8):2111–2123

Sekhar KC, Kamala CT, Chary NS, Balaram V, Garcia G (2005) Potential of *Hemidesmus* indicus for phytoextraction of lead from industrially contaminated soils. Chemosphere 58(4):507–514

Sen AK, Bhattacharyya M (1994) Studies of uptake and toxic effects of Ni (II) on Salvinia natans. Water Air Soil Pollut 78:141–152

Sen AK, Mondal NG, Mandal S (1987) Studies of uptake and toxic effects of Cr(VI) on *Pistia stratiotes*. Water Sci Technol 19:119–127

Sinha S, Saxena R (2005) Chromium induced lipid peroxidation in the plants of *Pistia stratiotes*: role of antioxidant and antioxidant enzymes. Chemosphere 58:595–604

Sinha S (1999) Accumulation of Cu, Cd, Cr, Mn and Pb from artificially contaminated soil by *Bacopa monnieri*. Environ Monit Assess 57:253–264

Skinner K, Wright N, Porter-Goff E (2007) Mercury uptake and accumulation by four species of aquatic plants. Environ Pollut 145:234–237

Suñe N, Sánchez G, Caffaratti S, Maine M (2007) Cadmium and chromium removal kinetics from solution by two aquatic macrophytes. Environ Pollut 145:467–473

Suthari S, Kiran BR, Prasad MNV (2017) Health risks of leafy vegetable Alternanthera philoxeroides (Alligator weed) rich in phytochemicals and minerals. EuroBiotech J 1(4):293–302

Thakral S, Setia P, Aeron S, Sharma G, Mishra R, Singh J (2024) Unveiling hazards: undisclosed toxic metal arsenic found in herbal medicines used for COVID-19 prevention by family–A case series. Archives of. Legal Med:200527

Tripathi P, Kumar R, Sharma AK, Mishra A, Gupta R (2010) *Pistia stratiotes* (Jalkumbhi). Pharmacognosy Rev 4(8):153–160

Upadhyay RK, Panda SK (2009) Copper-induced growth inhibition, oxidative stress and ultrastructural alterations in freshly grown water lettuce (*Pistia stratiotes* L.). C R Biologies 332:623–632

Vinogradova N, Glukhov A, Chaplygin V, Kumar P, Mandzhieva S, Minkina T, Rajput VD (2023) The content of heavy metals in medicinal plants in various environmental conditions: a review. Horticulturae 9(2):239

Volk GM (2004) Calcium channels involved in Ca-oxalate crystals formation in *P. stratiotes* L. Ann Bot 93:741–753

Wong MK, Koh LL (1986) Mercury, lead, and other heavy metals in Chinese medicines. Biol Trace Elem Res 10:91–97

Wong MK, Tan P, Wee YC (1993) Heavy metals in some Chinese herbal plants. Biol Trace Elem Res 36:135–142

Zhang BY, Zheng JS, Sharp RG (2010) Phytoremediation in engineered wetlands: mechanisms and applications. Proc Environ Sci 2:1315–1325

Zhao Y, Hu C, Wang X, Cheng H, Xing J, Li Y, Wang L, Ge T, Du A, Wang Z (2024) Water spinach (*Ipomoea aquatica* F.) effectively absorbs and accumulates microplastics at the micron level—a study of the Co-exposure to microplastics with varying particle sizes. Agriculture 14(2):301

Chapter 12
Toxic and Antinutritional Substances in Traditional Food Plants

Debabrata Panda, Suraj K. Padhi, Monali Nanda, and Jayanta K. Nayak

Abstract Antinutritional factors are the compounds synthesized naturally in plants and play an important role in determining the use of plants for humans. The antinutrients in plant foods can negatively impact nutrient absorption and reduce their suitability as human food. However, some antinutrients can have positive health effects when present in low amounts. For instance, compounds like tannins, lectins, saponins, amylase inhibitors, phytic acid, and protease inhibitors can decrease nutrient availability and slow down growth. Yet, at lower concentration, these substances can decrease insulin and blood sugar responses to heavy foods, as well as minimize triglyceride and cholesterol levels in the blood. Additionally, phytates, tannins, saponins, protease inhibitors, and oxalates found in certain traditional foods have been linked to a lower risk of cancer. Moreover, a range of indigenous methods and modern technologies are available, which can be used to lower the concentrations of these antinutritional factors. Hence, this chapter offers a comprehensive overview and updated scientific data on the harmful effects and promisive health benefits of key antinutritional factors found in foods from plants. Various traditional processing and indigenous consumption practices are also highlighted for safe use of food plants for humans.

Keywords Antinutrients · Health benefits · Health effects · Millets · Traditional processing · Wild yam

D. Panda (✉) · S. K. Padhi · M. Nanda
Department of Biodiversity and Conservation of Natural Resources, Central University of Odisha, Koraput, Odisha, India

J. K. Nayak
Department of Anthropology, Central University of Odisha, Koraput, Odisha, India

M. N. V. Prasad et al. (eds.), *Biotoxins*,
https://doi.org/10.1007/978-3-031-75309-1_12

12.1 Introduction

In the realm of nutrition, researchers continuously investigate the impact of our dietary choices on our well-being. We usually hear about the good stuff in traditional food that keeps us healthy, but there is also something important that often gets overlooked—antinutrients. These are natural compounds commonly present in many plant-based foods, and they can impact our body's ability to absorb essential nutrients. So, while we focus on the good things in traditional food, it is also essential to understand these antinutrients and how they can impact our health. The term "antinutrients" encompasses a diverse array of compounds, each with its unique chemical properties and physiological effects. For example, phytic acid, which is found in food grains and legumes (Reddy et al. 1982), binds to zinc, calcium, and iron, decreasing their bioavailability and potentially causing mineral deficiencies (Lopez et al. 2002). Lectins, on the other hand, abundant in certain vegetables (Sangronis and Machado 2017), can hinder the absorption of nutrients by binding to the intestinal lining and disrupting gut health (Frohlich and Rogers 2018).

While some antinutrients exhibit antioxidant properties and may confer health benefits, such as certain polyphenols and flavonoids (Frohlich and Rogers 2018), others like oxalates found in certain leafy greens and nuts (Noonan and Savage 1999) can bind to calcium and form insoluble complexes, potentially increasing the risk of kidney stone formation. Tannins, present in various plant foods, can also inhibit digestive enzymes and reduce protein digestibility (Shanmugavadivu and Vinoth 2021).

Understanding the role of antinutrients in traditional food plants is essential for optimizing dietary choices and ensuring adequate nutrient intake. Moreover, exploring strategies to mitigate the adverse effects of antinutrients through traditional food preparation techniques offers valuable insights into promoting overall health and well-being. Fermentation, soaking, sprouting, and cooking are among the methods known to reduce antinutrient levels in foods, thereby enhancing nutrient bioavailability and digestibility (Cummings and Englyst 2009).

In this chapter, we delve into the multifaceted realm of antinutrients in food plants. We examined their sources, mechanisms of action, effects on nutrient absorption, and implications for human health. Furthermore, we explored practical strategies for reducing antinutrient levels in foods (Zhu et al. 2020) and offer recommendations for achieving a balanced and nutritious diet. Through a comprehensive exploration of antinutrients in traditional food plants, this chapter aims to deepen our understanding of the complex interplay between diet and health, empowering individuals to make informed dietary choices and cultivate optimal well-being.

12.2 Antinutrients in Different Plant Foods

Plants inherently possess a range of substances, including some that are classified as antinutrients or toxins. While these compounds serve protective functions for plants, excessive consumption by humans can negatively impact health. The following examples illustrate various antinutrients found in different plant-based foods (Fig. 12.1).

12.2.1 Nitrate

Nitrate is a prominent naturally occurring chemical in vegetables and is crucial in determining their quality. However, excessive concentrations of these anions can be harmful. They have the ability to convert hemoglobin into methemoglobin, which is a non-functional protein unable to transport oxygen like normal hemoglobin, leading to a condition known as hypoxia, specifically termed methemoglobinemia (Natesh et al. 2017). In the gastrointestinal tract, nitrate is converted into nitrite, which is subsequently absorbed into the bloodstream. Upon binding with hemoglobin, nitrite oxidizes ferrous ion to ferric ion, forming methemoglobin. Certain vegetables such as cabbage, lettuce, and spinach have elevated nitrate concentrations, whereas others, including potatoes, carrots, leeks, onions, peas, and beans, contain lower nitrate levels.

12.2.2 Oxalate

Oxalate is a chemical resulting from the combination of oxalic acid with a metal or base. For instance, calcium oxalate, a common form of oxalate found in plants. Some oxalate salts, such as sodium and potassium oxalates, are soluble, but calcium oxalate salts are mostly insoluble.

Calcium oxalate is insoluble and tends to precipitate within the urinary system or kidneys, resulting in the development of sharp-edged crystals at high quantities. When oxalate is expelled in the urine, these crystals aid in the development of kidney stones in the urinary tract (Nachbar et al. 2000). The negative impact on calcium absorption becomes more pronounced when the ratio of oxalate to calcium exceeds 9:4 (Noonan and Savage 1999). Apart from food sources, the human body also produces oxalates as byproducts of ascorbic acid and glyoxylate metabolism. Plants have a variable and uneven distribution of oxalic acid depending on the species. The concentration of oxalate is usually lowest in stems, intermediate in seeds, and highest in leaves.

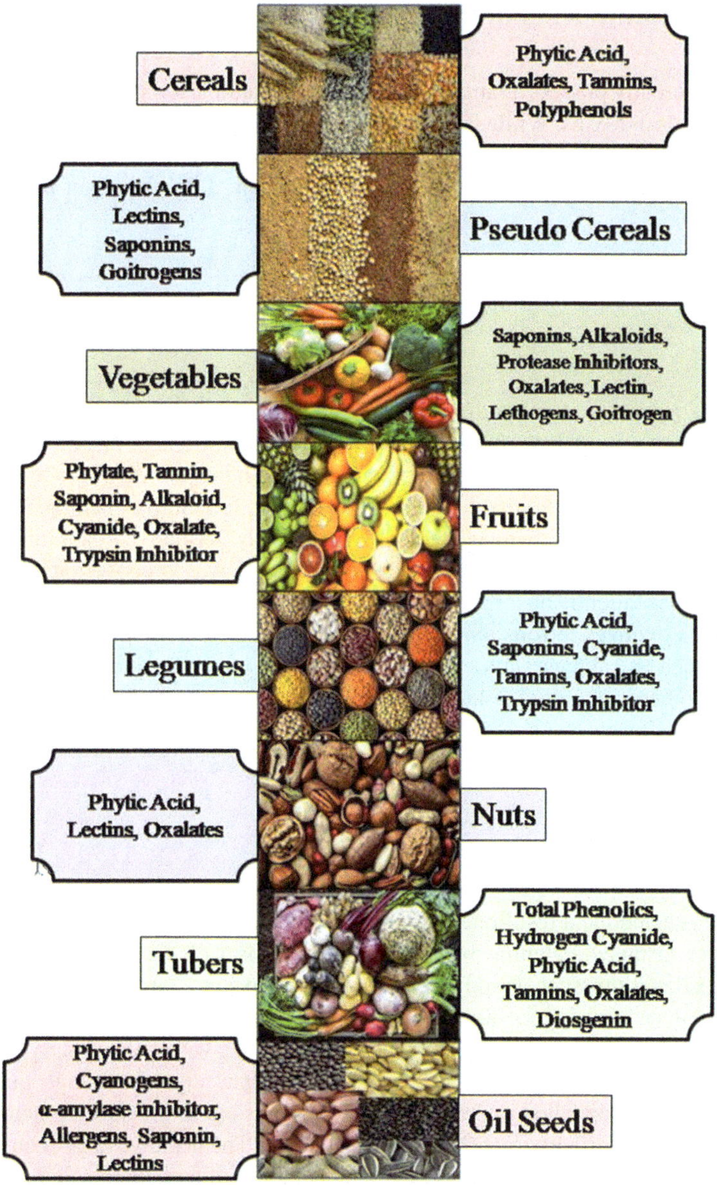

Fig. 12.1 Different types of food plants and their major antinutrients

12.2.3 Phytate

Phytate is a form of stored phosphorus commonly found in various plant tissues, especially seeds and grains. Chemically known as myoinositol-1,2,3,4,5,6-hexakis dihydrogen phosphate, and it is found in foods at different amounts ranging from 0.1 to 6.0% (Gupta et al. 2015). The presence of phytates in the aleurone layer and the germ of the grain makes it easy to separate them during milling in monocotyledons like wheat and rice. Conversely, in dicotyledonous seeds like legumes, nuts, and oilseeds, phytates are tightly bound to proteins and are predominantly concentrated within the protein fraction of these seeds. They act as reservoirs for phosphate and other crucial nutrients such as mineral, required for plant growth, notably during the germination phase (Jansman et al. 1998).

Phytate functions as a highly negatively charged ion throughout a wide pH range. Consequently, its presence in the diet lowers the bioavailability of divalent and trivalent mineral ions, including Ca2+, Cu2+, Zn2+, Fe2+/3+, Mg2+, and Mn2+. It also inhibits the activity of digestive enzymes by chelating mineral cofactors or interaction with proteins. Primarily due to its chelating properties, phytate is considered one of the most potent antinutrients in foods, leading to mineral ion deficiencies in both animal and human nutrition (Bora 2014; Grases et al. 2017).

12.2.4 Saponin

Saponins are widely dispersed throughout plant species and have a variety of biological activities that can be both useful and harmful. These natural compounds are characterized by their ability to create lather or foam when mixed with water. With a high molecular weight, saponins are glycosides that find diverse applications in beverages, confectionery, cosmetics, and pharmaceuticals. Saponins have an amphiphilic character due to their structure, which consists of a lipid-soluble aglycone and one or more water-soluble sugar chains. As a result, they are surface-active chemicals with detergent, emulsifying, wetting, and foaming activities (Shanthakumari et al. 2008).

In dietary plants, high level of saponins contributes to astringency and a bitter taste. The primary limitation to their utilization is their strong bitter taste. Historically, saponins have been identified as antinutritional factors because of their negative consequences, such as limiting growth and lowering food intake due to their animosity and potential to cause throat irritation (Liener 2003). Although minimal amount of saponins in legumes might not provide substantial health hazards, higher dietary concentrations can potentially be toxic (Jansman et al. 1998). Saponins can decrease cholesterol levels by binding to cholesterol and preventing its absorption. Furthermore, they have been shown to reduce blood glucose and insulin responses to starchy meals (Gemede and Ratta 2014). Additionally, saponins can cause hemolysis of red blood cells and exhibit toxicity in rats (Johnson et al. 1986).

12.2.5 Tannin

Tannin, a plant-derived polyphenolic compound, is characterized by its astringent and bitter qualities. Tannin has the ability to adhere or precipitate proteins and other organic molecules such as amino acids and alkaloids (Redden et al. 2005). Tannins are known for their heat stability and their ability to decrease protein digestion in both human and animal. This effect is likely caused by the partial unavailability of proteins or the inhibition of digestive enzymes, leading to an increase in fecal nitrogen. Tannins are found in many different foods and are well known for their ability to block the actions of enzymes including trypsin, chymotrypsin, lipase, and amylase. Additionally, dietary iron absorption can be impeded by tannins, which can also lower the level of protein of meals (Felix and Mello 2000). Excessive levels of tannins in the diet can impede microbial enzyme activities, particularly those involved in cellulose digestion, consequently reducing intestinal digestion (Aletor 2005).

12.2.6 Lectins and Hemagglutinins

Lectins have the capability to bind to carbohydrates. In modern terminology, proteins capable of agglutinating red blood cells and exhibiting specific sugar-binding properties are named as "lectins" (Fereidoon 2014). The term "hemagglutinins" is used when the specific sugar-binding properties of a substance are not yet identified. These carbohydrate-binding proteins are predominantly found in plants, particularly in the seeds of certain cereals and legumes, as well as in potato tubers. Lectins attach to intestinal epithelial cells and obstruct the absorption of nutrients. Additionally, they can cause damage to the intestinal tract, creating conditions that facilitate bacterial contact with the bloodstream (Muramoto 2017).

12.2.7 Alkaloids

Alkaloids are a major class of chemical compounds that are produced by plants. They are often found as salts of plant acids such as citric, tartaric, malic, or oxalic acid. Alkaloids are widely distributed, present in 15–20% of vascular plants. These tiny chemical compounds are classed as antinutritional substances due to their impact on the neurological system, where they can disrupt or excessively enhance electrochemical transmission. For instance, ingesting high amounts of tropane alkaloids can result in symptoms such as rapid heartbeat, paralysis, and, in extreme cases, death (Aletor 1993). Ingesting an unintended high dose of tryptamine alkaloids may result in symptoms like an unsteady gait and, in extreme instances, death. Alkaloids have notable physiological impacts on humans, and these compounds can

also be toxic, including causing disruption of the cell walls within the alimentary canal (Fernando and Pinto 2012).

12.2.8 *Enzyme Inhibitors*

Enzyme inhibitors such as protease inhibitors are commonly found in cultivated legumes and they may disrupt proteolytic enzyme activity in animal digestive tracts. While heat treatment can degrade these inhibitors, some residual activity may persist in the final products. The antinutrient effects of protease inhibitors are associated with growth retardation and pancreatic hypertrophy. Triticale, barley, and rye have elevated levels of protease (trypsin) inhibitor activity. Trypsin inhibitors operate by inhibiting protease enzymes in the digestive system, resulting in unpalatable complexes with food proteins that remain resistant to digestion despite high levels of digestive enzymes. Protease inhibitors limit trypsin activity and, to a lesser extent, chymotrypsin activity, reducing protein digestion in some juvenile ruminants and monogastric animals (Friedman et al. 2003). Alpha-amylase inhibitors, on the other hand, target alpha-amylase, the enzyme responsible for breaking down complex carbohydrates into simpler sugars (Choi et al. 2019).

12.2.9 *Cyanogenic Glycosides*

Cyanogenic glycosides (CN) are secondary metabolites generated from amino acids and are present in more than 2500 plant species (Vetter 2000). These compounds have attracted considerable research attention due to the well-known toxic nature of hydrogen cyanide (HCN) associated with them. According to Ferreira et al. (1995), bamboo shoots have cyanide contents ranging from 1000 to 8000 mg/kg of hydrogen cyanide. The essential enzyme cytochrome-C oxidase, which is part of the electron transport pathway of cellular respiration, is rendered ineffective by cyanide. This inhibition disrupts cellular oxygen utilization, leading to cellular hypoxia. Furthermore, it can cause a rapid and life-threatening reduction in oxygen delivery to tissues, potentially resulting in cardiac arrest. In severe cases, cyanide poisoning can lead to failure of multiple organ systems, including the cardiovascular, respiratory, and nervous systems (Fortin et al. 2010).

12.2.10 *Oligosaccharides*

Oligosaccharides are short-chain carbohydrates comprised of three to ten simple sugars (monosaccharides) linked together. Common oligosaccharides include raffinose and stachyose, which are found in various legumes, beans, and certain

vegetables. Oligosaccharides can be classified as antinutrients because humans lack the enzymes required to fully digest them. When these oligosaccharides reach the colon undigested, they can undergo fermentation by gut bacteria, leading to gas production and discomfort. Additionally, they may disrupt the absorption of specific minerals (Bora 2014).

12.2.11 Isoflavonoids

Isoflavonoids, a subgroup of flavonoids, are polyphenolic compounds derived from plants and display a diverse array of biological activities. They are commonly found in legumes, particularly soybeans, making soy a valuable source of isoflavones such as daidzein and genistein. Despite their health-promoting properties, isoflavonoids can also be considered antinutrients in certain scenarios. Research has shown that they inhibit enzymes responsible for protein digestion, which may lead to reduced protein bioavailability. However, the extent of this effect may vary depending on individual dietary patterns and the overall nutritional context (Abbas and Ahmad 2018).

12.2.12 Lupin Alkaloids

Lupin alkaloids refer to a collection of substances that are naturally present in different species of lupin plants. Certain lupin alkaloids, particularly quinolizidine alkaloids like lupanine and sparteine, can be toxic when present in high concentrations. Consumption of lupins containing elevated levels of these alkaloids may result in adverse effects, including gastrointestinal issues, neurological symptoms, potential liver damage, and respiratory problems (Múzquiz et al. 1989). To minimize alkaloid amount in lupin seeds and lupin-derived products and improve their safety, different techniques for processing such as fermentation, boiling, and soaking are employed (Mohammed et al. 2017).

12.2.13 Non-Protein Amino Acids

Nature encompasses numerous non-protein amino acids, each serving diverse functions within living organisms. While some non-protein amino acids add to the nutritional profile and flavor of certain foods, others may possess antinutrient properties. Examples such as lathyrogens and cycasin have been linked to neurodegenerative diseases in animals (Ottesen and Magnuson 2009; D'Mello 2015). Additionally, non-protein amino acids like mimosine and canavanine are considered toxic (Akande and Fabiyi 2010). Non-protein amino acids can disrupt the balance of

essential amino acids, potentially leading to insufficient protein synthesis and compromising the overall nutritional quality of the diet.

12.2.14 Phenolic Compounds

Secondary metabolites such as phenolic compounds and polyphenols are frequently found in a variety of plant-based meals, such as grains, legumes, fruits, vegetables, and drinks like tea, cocoa, and wine. These substances have the ability to inhibit the working efficiency of digestive enzymes like amylase and protease, which are crucial for the breakdown of carbohydrates and proteins during digestion. Excessively high levels of phenolic antioxidants may interfere with enzymatic reactions and metabolic processes in the body, potentially affecting nutrient utilization (Patterson et al. 2017).

Gossypol, a phenolic chemical was initially identified in 1899, is produced by pigment glands located in different parts of the cotton plant, including flower buds, stems, seeds, and leaves, with the highest concentration found in the seeds. Acute toxicity resulting from gossypol exposure can present as respiratory distress, compromised body weight gain, lethargy, weakness, loss of appetite, and, in severe cases, fatalities occurring several days later (Rogers et al. 2002).

12.2.15 Alpha-Galactosides

Alpha-galactosides, such as raffinose and stachyose, are regarded as antinutrients in certain food plants. These compounds are oligosaccharides that pose a challenge for the human digestive system to break down. Consequently, alpha-galactosides can cause flatulence and discomfort in some individuals as they reach the colon undigested, where they are fermented by gut bacteria, leading to gas production. Foods like legumes and cruciferous vegetables are known to contain higher levels of alpha-galactosides. However, soaking, cooking, and fermenting these foods can help to reduce the levels of these antinutrients, thereby enhancing their digestibility (Khokhar and Chauhan 1986).

12.2.16 Anti-vitamin Factors

These compounds, found in certain food plants, act as antinutrients by interfering with the utilization, absorption, or activity of essential vitamins. For example, an anti-vitamin E component found in uncooked kidney beans can lead to muscular degeneration and liver necrosis. Furthermore, linseed contains a factor that inhibits vitamin B6 (pyridoxine) activity, which can suppress growth. This component is

related to L-amino acids, more precisely D-proline, which is found in association with glutamic acid as the peptide linatine. L-amino-D-proline exhibits approximately four times the activity of linatine (Salunke 2006). The effect of this component can be mitigated by treating the meal with water.

12.2.17 Glycoalkaloids

In plants, glycoalkaloids serve as a defense mechanism against herbivores by deterring them from consuming the plant. However, these protective compounds can pose risks to humans when consumed in excessive amounts. The peel contains three to ten times more glycoalkaloids than the flesh of the potato. Alkaloid concentrations in bitter varieties can reach up to 80 mg/100 g throughout the tuber and 150–220 mg/100 g in the peel. Solanine in potatoes and tomatine in tomatoes can be toxic when consumed in elevated amount, potentially causing symptoms like abdominal pain, vomiting, nausea, and, in severe cases, neurological effects. However, due to the higher temperature required for their decomposition (243 °C), they remain unaffected during normal cooking processes (Sinha and Khare 2017).

12.2.18 Goitrogens

Substances that interfere with thyroid gland function and contribute to the development of goiter are known as goitrogens. The term "goitrogen" is derived from "goitre," which signifies the hypertrophy or enlargement of the thyroid gland. When the thyroid hormone synthesis is impaired, the gland compensates for the inadequate hormone levels by enlarging. There are primarily two groups of vegetables known to interfere with the production of thyroid hormones in humans—foods associated with soybeans and vegetables belonging to the cruciferous family (Sinha and Khare 2017).

12.2.19 Diosgenin

Diosgenin is a compound that occurs naturally and present in specific plants, particularly in the roots of species within the *Dioscorea* genus. High doses or concentrated forms of diosgenin are linked to potential adverse effects (Padhan et al. 2020). Although rare, allergic reactions to diosgenin or plants containing diosgenin may occur in sensitive individuals. Symptoms can include skin rashes, itching, or swelling.

12.3 Positive Effects of Antinutrients on Human Health

While antinutrients are generally considered detrimental to human health due to their interference with nutrient absorption or utilization, some antinutrients may have positive effects on human health under certain conditions. These potential positive effects are often context-dependent and may include:

12.3.1 Antioxidant Properties

Certain antinutrients, such as tannins and flavonoids, have antioxidant properties which aid the body's defense against free radicals, which are associated with various chronic diseases and the aging process (Olawoye and Gbadamosi 2017).

12.3.2 Anticancer Properties

Some antinutrients, such as lectins and protease inhibitors, have garnered attention for their potential anticancer effects, as they might contribute to preventing the development and proliferation of cancer cells. Nath et al. (2022) highlighted the anticancer properties of phytic acid.

12.3.3 Immune System Modulation

Certain antinutrients, like lectins and saponins, have been demonstrated to influence the immune system. While excessive consumption may present risks, controlled intake could potentially assist in regulating immune responses (Chukwuebuka and Chinenye 2015).

12.3.4 Cholesterol-Lowering Effects

Phytosterols, a group of antinutrients found in plants, have been associated with lowering blood cholesterol levels. These compounds can compete with cholesterol for absorption in the digestive tract, potentially resulting in reduced cholesterol levels. Additionally, phytic acid has also been observed to assist in bringing down blood cholesterol levels (Parca et al. 2018).

12.3.5 Blood Sugar Regulation

Some antinutrients, such as alpha-amylase inhibitors found in certain legumes, have been associated with the regulation of blood sugar levels. These inhibitors function by causing a delay in the release of glucose into the bloodstream via inhibiting the digestion of carbohydrates (Parca et al. 2018).

12.3.6 Bone Health

While excessive intake of oxalates can contribute to kidney stone formation, moderate consumption may have a protective effect on bone health. Oxalates may form insoluble crystals in the kidneys, potentially reducing the risk of calcium crystal formation in bones (Thakur and Sharma 2019).

Furthermore, antinutrients like oligosaccharides in legumes serve as prebiotics, fostering the growth of beneficial gut bacteria. This supports enhanced digestive health and nutrient absorption. It is crucial to highlight that the beneficial effects of antinutrients are typically linked to controlled and moderate consumption. Excessive intake or consumption of foods rich in antinutrients in inappropriate ways may result in adverse effects. Moreover, the impact of antinutrients, whether positive or negative, can vary from person to person depending on things like general eating habits, health, and genetic susceptibility.

12.4 Adverse Effects of Antinutrients

Antinutrients found in plant foods can have adverse effects on health when consumed in large quantities. For example, phytates, oxalates, and tannins can bind to essential minerals like iron, zinc, and calcium, inhibiting their absorption and potentially leading to mineral deficiencies (Reddy and Pierson 1994; Noonan and Savage 1999). Lectins and saponins, other antinutrients, may cause digestive discomfort such as bloating and diarrhea, especially when consumed in high amounts (Muramoto 2017). Additionally, certain antinutrients like cyanogenic glycosides and protease inhibitors can be toxic in large quantities, leading to symptoms of poisoning (Vetter 2000; Reddy et al. 1982). Protease inhibitors found in soybeans can interfere with protein digestion by inhibiting protease enzymes, potentially leading to poor protein utilization (Reddy et al. 1982). Goitrogens present in foods like cruciferous vegetables can impair thyroid function by inhibiting iodine uptake, which is essential for thyroid hormone production (Zhu et al. 2020). Moreover, lectins and other antinutrients may trigger allergic reactions in susceptible individuals, resulting in symptoms such as hives or swelling (Peumans and Van Damme 1996). Although antinutrients can have negative effects, they are typically not

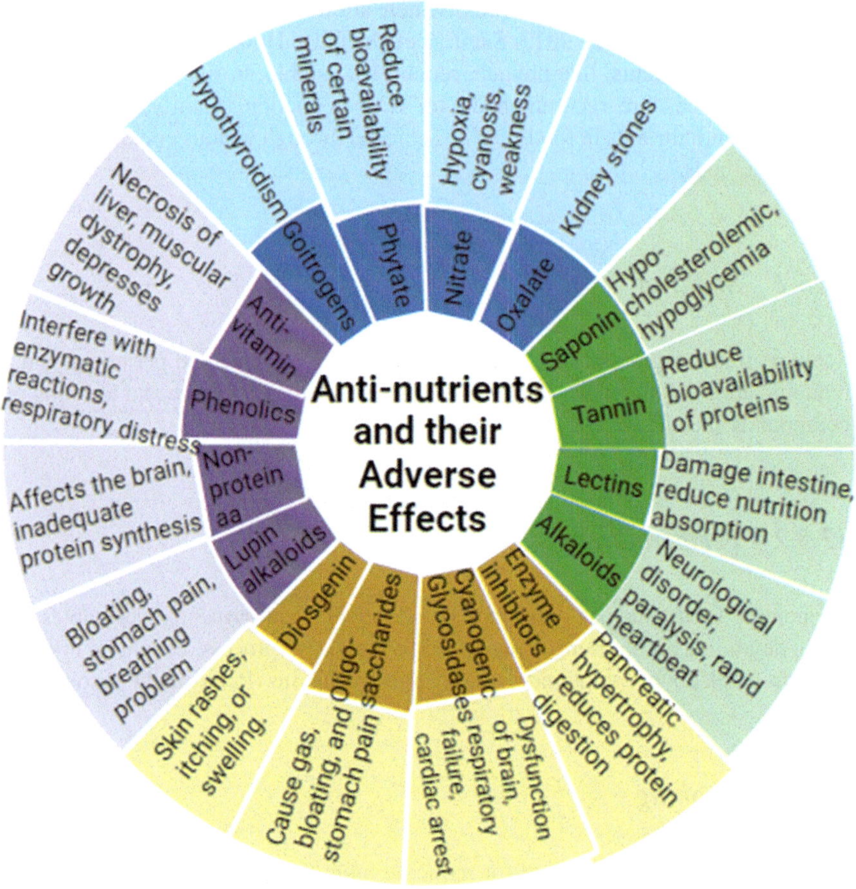

Fig. 12.2 Various types of antinutrients and their detrimental effects on human health

harmful when ingested in an adequate manner as part of a balanced diet. In fact, some may even provide health benefits when consumed in appropriate amounts (Samtiya et al. 2020). Some common antinutrients and their negative impact on health of human being have been described in Fig. 12.2.

12.5 Mitigation of Antinutrients

The most appropriate approach for lowering the concentration of antinutrients depends on their types, structures, and chemical properties, as emphasized by Vikram et al. (2020). Generally, heat treatments, including autoclaving and microwave-heating, are widely regarded as highly effective in decreasing antinutrient levels, especially in leafy vegetables. This effectiveness is due to the capacity of

heat to break down plant cell walls, which allows for the extraction of soluble antinutrients (Natesh et al. 2017; Samtiya et al. 2020). However, it is crucial to note that some antinutrients, like phytate, remain heat-stable. In such cases, alternative physical methods are recommended for their reduction (Samtiya et al. 2020). Mitigating antinutrients in foods can be achieved through various cooking and preparation methods.

12.5.1 Soaking

Soaking grains, legumes, nuts, and seeds in water for a specified period can aid in reducing antinutrient levels. This process assists in breaking down and leaching out compounds such as phytates (Samtiya et al. 2020).

12.5.2 Fermentation

Fermenting foods, such as sourdough bread, yogurt, and fermented vegetables, can aid in degrading antinutrients. During the fermentation process, beneficial microorganisms break down compounds like phytates and lectins (Reddy and Pierson 1994).

12.5.3 Cooking

Cooking is an effective method for reducing antinutrient levels, as heat can break down or deactivate various antinutrients. Common cooking techniques, such as roasting, boiling, and steaming, are known to reduce antinutrient levels (Yadav and Sehgal 2003).

12.5.4 Sprouting

Sprouting entails germinating seeds, grains, or legumes, which can result in a decrease in antinutrient content. This process activates enzymes that break down compounds like phytates (Mbithi et al. 2001). A longer sprouting period notably increases ascorbic acid levels in chickpea, millets, mung bean, and cowpea (Panda et al. 2020).

12.5.5 Blanching

Blanching, a process that involves briefly immersing foods in boiling water followed by rapid cooling, can help decrease antinutrient levels, particularly in vegetables (Mosha et al. 1995).

In addition to these methods, combining foods can also mitigate the effects of certain antinutrients. It is important to recognize that while these approaches can reduce antinutrient levels; complete elimination may not be achievable. However, integrating these techniques into food preparation can contribute to improved nutritional quality in diets. Additionally, individuals with particular dietary needs or health conditions should consult healthcare professionals or registered dietitians for personalized recommendations. Consuming a varied and balanced diet can help minimize the impact of antinutrients. Eating a diverse range of foods ensures that even if some foods contain antinutrients, overall nutrient intake remains adequate.

12.6 Acceptable Daily Intake of Antinutrients

The presence and impact of antinutrients can vary among individuals based on factors such as overall diet, health status, and genetic differences. In many cases, the potential adverse effects of antinutrients are associated with extremely high or unbalanced intakes, and most people can tolerate moderate amounts of these compounds without experiencing negative effects. It is important to note that many foods containing antinutrients are also excellent sources for crucial nutrients, and their ingestion is an integral part of a balanced diet. The antinutrients present in different plant foods and their Acceptable Daily Intake (ADI) were highlighted in Table 12.1.

12.7 Overview and Conclusion

Individuals with particular dietary concerns or health conditions are advised to consult healthcare professionals or registered dietitians for personalized guidance. Consuming a varied and balanced diet can help minimize the impact of antinutrients. Eating diverse foods ensures that even if some contain antinutrients, overall nutrient intake remains adequate. Balanced and varied diets that include a wide range of foods typically provide a mix of nutrients and antinutrients, contributing to overall health and well-being. For issues related to specific dietary components, it is prudent to obtain individualized advice from a healthcare professional or registered dietitian. In conclusion, the presence of antinutrients and toxins in traditional food plants highlights the intricate relationship between dietary components and human health. While antinutrients can present challenges by interfering with nutrient

Table 12.1 Antinutrients in different types of food plants and their acceptable daily intake (ADI)

Antinutrients	Food plant types					ADI (mg/kg bw)/ (mg/day)	Reference (s)
	Grains	Legumes	Nuts and seeds	Tubers and roots	Fruits and vegetables		
Nitrate	NA	Lentils, peas, and beans	NA	Beets, carrots, and radishes	Spinach, lettuce, arugula, kale, and Swiss chard	3.7 mg/kg bw	Jonvik et al. (2016); Bahadoran et al. (2016)
Oxalate	NA	Beans, lentils, and peas	Almonds, cashews, and sesame seeds	Beets, Sweet potato, and yams	Spinach, rhubarb, Swiss chard, and cocoa	50 mg/day	Bargagli et al. (2020); Radek and Savage (2008); Santamaria et al. (1999); Noonan and Savage (1999)
Phytate	Wheat, rice, oats, barley, and corn	Beans and lentils	Almonds, walnuts, sunflower, and sesame seed	NA	Spinach, beets, carrots, apples, and berries	629.0 mg/kg bw	Amirabdollahian and Ash (2010); Reddy (2001)
Saponin	Oats	Chickpeas, lentils, and soybeans	Quinoa seeds	Sweet potato	Alfalfa, ginseng, and red beans	NA	Wu et al. (2001)
Tannin	NA	Lentils and chickpeas.	walnuts and almonds, coffee beans	NA	Grapes, pomegranate, blueberries, blackberries, and cranberries	2.5 g/day	Sharma et al. (2019); Montes-Ávila et al. (2017); Mohammed (2008)
Lectins	Wheat	Peas, beans, and lentils	Peanuts and sunflower	Potatoes	Tomatoes, eggplants, and peppers	27 mg/kg bw	Peumans and Van Damme (1996); Roh (2008)
Alkaloids	Rye	Beans and peas	Poppy seeds	Potatoes	Chili peppers, eggplants, and tomatoes	0.125 mg/kg bw	Koleva et al. (2012); Srinivasan (2013)
Enzyme inhibitors	NA	Soybean	NA	Potatoes	NA	5 mg/kg bw	Zu et al. (2017); Lajolo and Genovese (2002)
Cyanogenic glycosidases	Sorghumand maize	beans,	Almonds and flaxseeds	Cassava	Apricots, cherries, and apples, spinach, and bamboo shoots	0.075 mg/kg bw	Abraham et al. (2016); McMahon et al. (1995)

Oligosaccharides	Wheat, barley, and rye	Beans, lentils, chickpeas	Almonds Cashews Chia seeds Flaxseeds Sesame Seed	Onions and garlic	Apples and watermelon, leeks, and asparagus	13,000 mg/day	Delzenne (2003); Jovanovic-Malinovska et al. (2014)
Isoflavonoids	NA	Soybeans	NA	Not significant sources of isoflavonoids	NF	50 mg/day	Kim et al. (2022); Barnes (2010)
Lupin alkaloids	NF	Lupin seeds	NF	NF	NF	0.125 mg/kg bw	Koleva et al. (2012)
Non-protein amino acids	Maize and rice	Lentils and soybeans	Almonds and sunflower seeds	Sweet potatoes, and cassava	Tomato, banana, spinach, and broccoli	240 mg/kg bw	Elango (2023)
Phenolic compounds	Wheat, rice, barley, and oats	Peas, beans, and lentils	Almonds, walnuts, flaxseeds, chia seeds, and sunflower seeds	Potatoes, ginger and turmeric, Onions, and garlic	Berries, apples, oranges, lemons, red grapes, spinach, kale, and tomatoes	0.25 mg/kg bw	Shahidi and Ambigaipalan (2015); Määttä-Riihinen et al. (2004)
Alpha-galactosidases	Wheat, barley, and rye	Lentils, chickpeas, black beans, Kidney beans, and navy beans	Almonds and flaxseeds	Potatoes and sweet potatoes	Berries, onions, broccoli, and Brussels sprouts	NA	Guo et al. (2017)
Anti-vitamin factor	Wheat, rice, and oats	Soybeans	NA	NA	Spinach and rhubarb	NA	Thakur and Sharma (2019)
Glycoalkaloids	NA	NA	NA	Potatoes	Tomatoes, eggplants, bell peppers, and chili peppers	0.5 mg/kg bw	Schrenk et al. (2020); Friedman (2002)
Goitrogens	Millet and sorghum	Soybeans and peanuts	Flaxseeds and pine nuts	Cassava	Strawberries, peaches, broccoli, cauliflower, cabbage, and Brussels sprouts	NA	Sood and Bandral (2019)
Diosgenin	NA	NA	NA	Yams	NA	300 mg/kg	Wang and Wang (2022); Padhan and Panda (2020)

NA not abundant, *NF* not found, *BW* body weight

absorption, it is crucial to acknowledge that their levels can often be reduced through various cooking and preparation methods (Yadav and Sehgal, 2003). Moreover, when consumed in moderation, some antinutrients may offer health benefits such as antioxidant properties and potential anticancer effects. Conversely, toxins found in certain traditional food plants can adversely affect health, ranging from mild discomfort to severe illness. Therefore, carefully selecting, preparing, and consuming traditional foods are essential for balancing nutritional benefits and potential risks (Liener 1994). Highlighting the importance of diverse and balanced diets while considering individual health conditions will contribute to optimizing the nutritional value of traditional diets while minimizing potential drawbacks. Regular monitoring, research, and awareness are crucial for advancing our understanding of the dynamic interplay between conventional food plants, antinutrients, toxins, and human health.

References

Abbas Y, Ahmad A (2018) Impact of processing on nutritional and antinutritional factors of legumes: A review. Ann Food Sci Technol 19(2):199–215

Abraham K, Buhrke T, Lampen A (2016) Bioavailability of cyanide after consumption of a single meal of foods containing high levels of cyanogenic glycosides: A crossover study in humans. Arch Toxicol 90(3):559–574

Akande KE, Fabiyi EF (2010) Effect of processing methods on some antinutritional factors in legume seeds for poultry feeding. Int J Poult Sci 9(10):996–1001

Aletor VA (1993) Allelochemicals in plant foods and feeding stuffs. Part I. Nutritional, and physiopathological aspects in animal production. Vet Hum Toxicol 35(1):57–67

Aletor VA (2005) Anti-nutritional factors as nature's paradox in food and nutrition securities. Inaugural lecture series 15, delivered at The Federal University of Technology, Akure (FUTA)

Amirabdollahian F, Ash R (2010) An estimate of phytate intake and molar ratio of phytate to zinc in the diet of the people in the United Kingdom. Public Health Nutr 13(9):1380–1388

Bahadoran Z, Mirmiran P, Jeddi S, Azizi F, Ghasemi A, Hadaegh F (2016) Nitrate and nitrite content of vegetables, fruits, grains, legumes, dairy products, meats and processed meats. J Food Compos Anal 51:93–105

Bargagli M, Tio MC, Waikar SS, Ferraro PM (2020) Dietary oxalate intake and kidney outcomes. Nutrients 12(9):2673

Barnes S (2010) The biochemistry, chemistry and physiology of the isoflavones in soybeans and their food products. Lymphat Res Biol 8(1):89–98

Bora P (2014) Antinutritional factors in foods and their effects. J Acad ind res 3(6):285–290

Choi WC, Parr T, Lim YS (2019) The impact of four processing methods on trypsin-, chymotrypsin- and alpha-amylase inhibitors present in underutilized legumes. J Food Sci Technol 56:281–289

Chukwuebuka E, Chinenye IJ (2015) Biological functions and antinutritional effects of phytochemicals in living systems. J Pharm Biol Sci 10(2):10–19

Cummings JH, Englyst HN (2009) Fermentation in the human large intestine and the available substrates. Am J Clin Nutr 59(5):777S–784S

Delzenne NM (2003) Oligosaccharides: state of the art. Proc Nutr Soc 62(1):177–182

D'Mello JPF (2015) Toxicology of non-protein amino acids. In: Amino acids in higher plants. CAB, Wallingford UK, pp 507–537

Elango R (2023) Tolerable upper intake level (UL) for individual amino acids in humans: A narrative review of recent clinical studies. Adv Nutr 14(4):885–894
Felix JP, Mello D (2000) Farm animal metabolism and nutrition. CABI, Wallingford
Fereidoon S (2014) Beneficial health effects and drawbacks of antinutrients and phytochemicals in foods. Appl Microbiol Biotechnol 97:45–55
Fernando RP, Pinto MDP (2012) Goitrogenic food and prevalence of goitre in Sri Lanka. J Food Sci 41:1076–1081
Ferreira VLP, Yotsuyanagi K, Carvalho CRL (1995) Elimination of cyanogenic compounds from bamboo shoots Dendrocalamus giganteus Munro. Trop Sci 35:342–346
Fortin JL, Desmettre T, Manzon C, Judic-Peureux V, Peugeot-Mortier C, Giocanti JP, Capellier G (2010) Cyanide poisoning and cardiac disorders: 161 cases. J Emerg Med 38(4):467–476
Friedman M (2002) Tomato glycoalkaloids: role in the plant and in the diet. J Agric Food Chem 50(21):5751–5780
Friedman M, Henika PR, Mackey BE (2003) Effect of feeding solanidine, solasodine and tamatidine to non-pregnant and pregnant mice. Food Chem Toxicol 41:61–71
Frohlich MW, Rogers QR (2018) An overview of the health considerations of plant-based diets. Am J Clin Nutr 70(3):539S–550S
Gemede HF, Ratta N (2014) Antinutritional factors in plant foods: Potential health benefits and adverse effects. Int J Food Sci Nutr 3(4):284–289
Grases F, Prieto RM, Costa-Bauza A (2017) Dietary phytate and interactions with mineral nutrients. In: Gutiérrez OM, Kalantar-Zadeh K, Mehrotra R (eds) Clinical aspects of natural and added phosphorus in foods. Springer, New York, pp 175–183
Guo L, Wang P, Gu Z, Jin X, Yang R (2017) Proteomic analysis of broccoli sprouts by iTRAQ in response to jasmonic acid. J Plant Physiol 218:16–25
Gupta RK, Gangoliya SS, Singh NK (2015) Reduction of phytic acid and enhancement of bioavailable micronutrients in food grains. J Food Sci Technol 52(2):676–684
Jansman AJ, Hill GD, Huisman J, Vander Poel AF (1998) Recent advances of research in antinutritional factors in legumes seeds. Wageningen Pers, Wageningen, The Netherlands
Johnson IT, Gee JM, Price K, Curl C, Fenwick GR (1986) Influence of saponin on gut permeability and active nutrient transport in vitro. J Nutr 116:2270–2277
Jonvik KL, Nyakayiru J, Pinckaers PJ, Senden JM, van Loon LJ, Verdijk LB (2016) Nitrate rich vegetables increase plasma nitrate and nitrite concentrations and lower blood pressure in healthy adults. J Nutr 146(5):986–993
Jovanovic-Malinovska R, Kuzmanova S, Winkelhausen E (2014) Oligosaccharide profile in fruits and vegetables as sources of prebiotics and functional foods. Int J Food Prop 17(5):949–965
Khokhar S, Chauhan BM (1986) Anti-nutritional factors in moth beans (*Vigna aconitifolia*): Varietal difference and effects of methods of domestic processing and cooking. J Food Sci 51(3):591–594
Kim Y, Kim DW, Kim K, Choe JS, Lee HJ (2022) Usual intake of dietary isoflavone and its major food sources in Koreans: Korea National Health and Nutrition Examination Survey 2016-2018 data. Nutr Res Pract 16(1):134–146
Koleva II, van Beek TA, Soffers AE, Dusemund B, Rietjens IM (2012) Alkaloids in the human food chain-natural occurrence and possible adverse effects. Mol Nutr Food Res 56(1):30–52
Lajolo FM, Genovese MI (2002) Nutritional significance of lectins and enzyme inhibitors from legumes. J Agric Food Chem 50(22):6592–6598
Liener IE (1994) Implications of antinutritional components in soybean foods. Crit Rev Food Sci Nutr 34(1):31–67
Liener IE (2003) Phytohemagglutinins: Their nutritional significance. J Agric Food Chem 22:17
Lopez HW, Leenhardt F, Coudray C, Remesy C (2002) Minerals and phytic acid interactions: Is it a real problem for human nutrition? Int J Food Sci Technol 37(7):727–739
Määttä-Riihinen KR, Kamal-Eldin A, Törrönen AR (2004) Identification and quantification of phenolic compounds in berries of Fragaria and Rubus species (family *Rosaceae*). J Agric Food Chem 52(20):6178–6187

Mbithi S, Van Camp J, Rodriguez R, Huyghebaert A (2001) Effects of sprouting on nutrient and antinutrient composition of kidney beans (*Phaseolus vulgaris* var. Rose coco). Eur Food Res Technol 212:188–191

McMahon JM, White WL, Sayre RT (1995) Cyanogenesis in cassava (*Manihot esculenta* Crantz). J Exp Bot 46(7):731–741

Mohammed HS (2008) Inhibitory effect of tannic acid extracted from grape seeds and pomegranate peels on some microorganisms. Mesop J Agric 36(1):12–18

Mohammed MA, Mohamed EA, Yagoub AEA, Mohamed AR, Babiker EE (2017) Effect of processing methods on alkaloids, phytate, phenolics, antioxidants activity and minerals of newly developed lupin (*Lupinus albus* L.) cultivar. J Food Process Preserv 41(1):12960

Montes-Ávila J, López-Angulo G, Delgado-Vargas F (2017) Tannins in fruits and vegetables: Chemistry and biological functions. In: Yahia EM (ed) Fruit and vegetable phytochemicals: chemistry and human health, 2nd edn. Springer, New York, pp 221–268

Mosha TC, Gaga HE, Pace RD, Laswai HS, Mtebe K (1995) Effect of blanching on the content of antinutritional factors in selected vegetables. Plant Foods Hum Nutr 47:361–367

Muramoto K (2017) Lectins as bioactive proteins in foods and feeds. Food Sci Technol Res 23:487–494

Múzquiz M, Burbano C, Gorospe MJ, Ródenas I (1989) A chemical study of Lupinus hispanicus seed-toxic and antinutritional components. J Sci Food Agric 47:205–214

Nachbar MS, Oppenheim JD, Thomas JO (2000) Lectins in the US diet: Isolation and characterization of a lectin from the tomato (Lycopersicon). J Biol Chem 255:2056

Natesh HN, Abbey L, Asiedu SK (2017) An overview of nutritional and antinutritional factors in green leafy vegetables. Hortic Int J 1:00011

Nath H, Samtiya M, Dhewa T (2022) Beneficial attributes and adverse effects of major plant based foods anti-nutrients on health: A review. Hum Nutr Metab 28:200147

Noonan SC, Savage GP (1999) Oxalate content of foods and its effect on humans. Asia Pac J Clin Nutr 8:64–74

Olawoye BT, Gbadamosi SO (2017) Effect of different treatments on in vitro protein digestibility, antinutrients, antioxidant properties and mineral composition of *Amaranthus viridis* seed. Cogent Food Agric 3:1296402

Ottesen AR, Magnuson BA (2009) Naturally occurring toxins in plants. In: Juneja VK, Sofos JN (eds) Pathogens and toxins in foods: challenges and interventions. ASM, Washington, DC, pp 301–313

Padhan B, Panda D (2020) Potential of Neglected and Underutilized Yams (*Dioscorea spp.*) for Improving Nutritional Security and Health Benefits. Front Pharmacol 11:496

Padhan B, Biswas M, Panda D (2020) Nutritional, anti-nutritional and physico-functional properties of wild edible yam (*Dioscorea spp.*) tubers from Koraput, India. Food Biosci 34:100527

Panda D, Sailaja NH, Padhan B, Lenka K (2020) Sprouting-associated changes in nutritional and physico-functional properties of indigenous millets from Koraput, India. Proc Natl Acad Sci India Sect B Biol Sci 90:79–86

Parca F, Koca YO, Aydın UNAY (2018) Nutritional and antinutritional factors of some pulses seed and their effects on human health. Int J Second Metab 5:331–342

Patterson CA, Curran J, Der T (2017) Effect of processing on antinutrient compounds in pulses. Cereal Chem 94:2–10

Peumans WJ, Van Damme EJM (1996) Prevalence, biological activity and genetic manipulation of lectins in foods. Trends Food Sci Technol 7:132–138

Radek M, Savage GP (2008) Oxalates in some Indian green leafy vegetables. Int J Food Sci Nutr 59:246–260

Redden RJ, Chen W, Sharma B (2005) Chickpea breeding and management. CABI, Wallingford

Reddy NR (2001) Occurrence, distribution, content, and dietary intake of phytate. In: Reddy NR, Sathe SK (eds) Food phytates. CRC, Boca Raton, pp 41–68

Reddy NR, Pierson MD (1994) Reduction in antinutritional and toxic components in plant foods by fermentation. Food Res Int 27(3):281–290

Reddy NR, Sathe SK, Salunkhe DK (1982) Phytates in legumes and cereals. Adv Food Res 28:1–92
Rogers GM, Poore MH, Paschal JC (2002) Feeding cotton products to cattle. Vet Clin North Am Food Anim 18(2):267–294
Roh KS (2008) Biochemical properties of eggplant fruit lectin. J Life Sci 18(3):350–356
Salunke BK (2006) Anti-nutritional constituents of different grain legumes grown in North Maharashtra. J Food Sci 43(5):519–521
Samtiya M, Aluko RE, Dhewa T (2020) Plant food anti-nutritional factors and their reduction strategies: an overview. Food Prod Process Nutr 2:1–14
Sangronis E, Machado CJ (2017) Lectins and their significance in the diet. Nutr Hosp 34(3):712–716
Santamaria P, Elia A, Serio F, Todaro E (1999) A survey of nitrate and oxalate content in fresh vegetables. J Sci Food Agric 79(13):1882–1888
Schrenk D, Bignami M, Bodin L, Chipman JK, del Mazo J, Grasl-Kraupp B (2020) Risk assessment of glycoalkaloids in feed and food, in particular in potatoes and potato derived products. EFSA J 18(8):e06222
Shahidi F, Ambigaipalan P (2015) Phenolics and polyphenolics in foods, beverages and spices: antioxidant activity and health effects—a review. J Funct Foods 18:820–897
Shanmugavadivu M, Vinoth A (2021) Tannins in food plants: A review of their chemistry, distribution, and potential health effects. Crit Rev Food Sci Nutr 61(5):742–755
Shanthakumari S, Mohan V, Britto J (2008) Nutritional evaluation and elimination of toxic principles in wild yam (*Dioscorea spp.*). Trop Subtrop Agroecosyst 8(3):319–225
Sharma K, Kumar V, Kaur J, Tanwar B, Goyal A, Sharma RK, Gat Y, Kumar A (2019) Health effects, sources, utilization and safety of tannins: a critical review. Toxin Rev 40(3):432–444
Sinha K, Khare V (2017) Review on: antinutritional factors in vegetable crops. J Pharm Innov 6(12):353–358
Sood M, Bandral JD (2019) Composition, quality and uses. In: Sharma RM, Yamdagni R, Dubey AK, Pandey V (eds) Strawberries: production, postharvest management and protection. CRC, Boca Raton, p 23
Srinivasan K (2013) In: Ramawat K, Mérillon JM (eds) Biological activities of pepper alkaloids, vol 45. Natural products, Springer, Berlin, pp 1397–1437
Thakur A, Sharma V (2019) An overview of anti-nutritional factors in food. Int J Chem Stud 7(1):2472–2479
Vetter J (2000) Plant cyanogenic glycosides. Toxicon 38(1):11–36
Vikram N, Katiyar SK, Singh CB, Husain R, Gangwar LK (2020) A review on anti nutritional factors. Int J Curr Microbiol Appl Sci 9(5):1128–1137
Wang D, Wang X (2022) Diosgenin and its analogs: potential protective agents against atherosclerosis. Drug Des Dev Ther 16:2305–2323
Wu J, Lin L, Chau FT (2001) Ultrasound-assisted extraction of ginseng saponins from ginseng roots and cultured ginseng cells. Ultrason Sonochem 8(4):347–352
Yadav SK, Sehgal S (2003) Effect of domestic processing and cooking on selected antinutrient contents of some green leafy vegetables. Plant Foods Hum Nutr 58:1–11
Zhu Y, Sang S, Chu Y (2020) Antinutrients in plant-based foods: A review. Trends Food Sci Technol 100:58–69
Zu K, Pizzurro DM, Lewandowski TA, Goodman JE (2017) Pharmacokinetic data reduce uncertainty in the acceptable daily intake for benzoic acid and its salts. Regul Toxicol Pharmacol 89:83–94

Part III
Biotoxins Classified by their Mode of Action

Chapter 13
Biotoxins and Their Conjugates for Use in Targeted Medicinal Applications with Focus on Eye Diseases

Pavithra Arumugam, Sukanya, Sukanya Madhan Mohan, Prema Padmanaban, Narayanan Janakiraman, and Sailaja V. Elchuri

Abstract Toxins produced in biological organisms are poisons that have evolved over time in animals/plants and microbes. They are toxic to predating organisms including humans. Biotoxins give adaptive advantage to the producing organisms against predators and to combat altering climatic conditions. Recently, due to anthropogenic activities and warming of Oceans, Eutrophication alterations in pH of waters have increased the biotoxins produced in the organisms. Additionally, biotoxins can synergistically act on contaminants of emerging concern to attenuate the toxic effect on organism at ecosystem level in addition to affecting human health. Several classes of biotoxins are present and the identification of new compounds is ongoing. Over the preceding five decades, much research has gone into investigating their mode of action, how they contribute to pathogenicity, and developing antidotes that negate their influence. Further, advances in detection methods including chromatography, mass spectrometry, vibrational spectroscopy, nanotechnology have led to detect novel toxins and their mode of actions. From the regulatory perspective, there are also new demands set by global regulators like Food and Drug Administration, European Union, Therapeutic Goods Administration, and Ministry of Health and Labour Welfare, and the detection methods/production needs standardization from the industries. These are vital for food safety, bioterrorism vigilance, and environmental protection. DNA affecting property of the toxins can have therapeutic potential. Interestingly, several research groups have concentrated on the therapeutic potential of biotoxins which can be used to impede vital biological functions and/or damage the integrity of their target cells. A large-scale

P. Arumugam · Sukanya · S. Madhan Mohan · N. Janakiraman · S. V. Elchuri (✉)
Department of Nanobiotechnology, Vision Research Foundation, Sankara Nethralaya, Nungambakkam, Chennai, Tamil Nadu, India

P. Padmanaban
Department of Cornea, Medical Research Foundation, Sankara Nethralaya, Nungambakkam, Chennai, Tamil Nadu, India

M. N. V. Prasad et al. (eds.), *Biotoxins*,
https://doi.org/10.1007/978-3-031-75309-1_13

production of biotoxins is explored using synthetic biology and biotechnological approaches. The therapeutic potential of not only biotoxins but their conjugates with aptamers, antibodies, affibodies, and nanobodies is thrust area of research. The conjugates are characterized efficiently using physical and chemical properties in the laboratories. The mechanism of action of biotoxins and their conjugates is studied using Omic approaches such as transcriptomics, proteomics, and metabolomics in preclinical evaluations for several diseases including eye diseases. The chapter focuses on emergent biotoxins and their conjugates as potential drugs in the preclinical and clinical evaluations.

Keywords Biotoxin conjugates · Omic studies · Mechanism of action · Therapy · Synthetic biotoxins · Clinical studies

13.1 Introduction

Biotoxins are biological in origin and may come from microbial, plant, or animal species (Fig. 13.1). These toxins are hazardous to other organisms yet benign in the organisms that make them. Furthermore, they provide a selection advantage for the producing organism to survive in the environment. The biotoxins from various sources will be discussed further. This review focuses on biotoxins' therapeutic potential for a variety of human diseases. Plant toxins can cause various side effects including, slight itching, vomiting, nausea, psychological issues, paralysis, abnormalities in embryonic development, and irregular heart rhythms (Kocyigit et al. 2023). Animal venom toxins trigger the onset of inflammation, the production of pro-inflammatory substances (cytokines), reactive oxygen species (ROS), proteolytic enzymes, and stimulate the movement of white blood cells and macrophages and triggering several signaling pathways (Matkivska et al. 2024).

Plants inherently contain toxins as a defense mechanism against animals, producing natural products as a means of protection. One distinctive trait of higher plants is their ability to produce a vast array of organic compounds known as

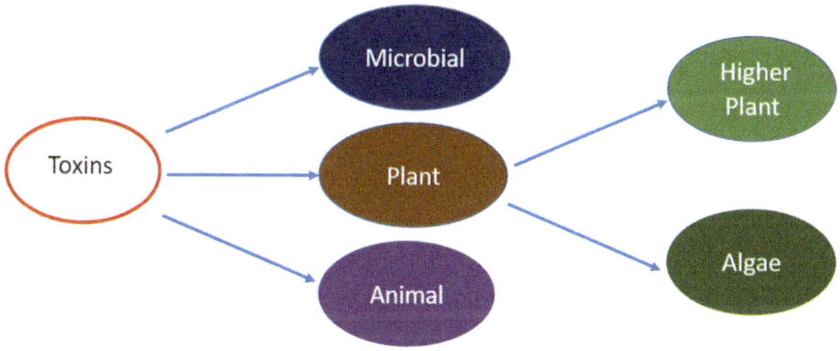

Fig. 13.1 A general understanding of biotoxin and its origin

secondary metabolites. These toxins, often secondary metabolites, are present in common food sources like fruits and vegetables. Their production is an evolutionary response by plants to safeguard themselves from threats such as bacteria, fungi, insects, and predators (Chandra Sekhar et al. 2012). Most plant species worldwide are not suitable for consumption primarily due to the presence of toxins they generate. The resistance of plants against pathogens and herbivores is significantly influenced by chemical protection. Secondary metabolites, a distinct feature of plants, play a crucial role in safeguarding them against a broad spectrum of microorganisms (such as viruses, bacteria, fungi) and herbivores (including arthropods and vertebrates) (Wink 1988). The development of thorns, spikes, trichomes, glandular hairs, and bark has been understood as a defensive mechanism against predators (Deverall 1977).

13.2 Plant Toxins as Therapeutics

Alkaloids, terpenes, steroids, and phenolic compounds are plant secondary metabolites that display clear toxicity to both humans and animals. These substances have served as the foundation for the development of drugs or as templates for designing pharmaceuticals (Rates et al. 2015). Traditional healing systems have utilized toxins for various purposes since ancient times, the contemporary transformation of toxins into medicines has commenced. Toxins derived from plants, specifically categorized as "ribosome inactivating proteins" (RIPs), have been employed in the development of immunotoxins. The frequently utilized toxins include ricin, saporin, and pokeweed antiviral protein (PAP), originating from *Ricinus communis*, *Saponaria officinalis*, and *Phytolacca americana*, respectively. These toxins are highly potent and exert their toxic effects by enzymatically disrupting the protein synthesis process in eukaryotic cells (Shapira and Benhar 2010). Several recognized categories of toxic proteins comprise lectins, ribosome-inactivating proteins, protease inhibitors, α-amylase inhibitors, ureases, arcelins, antimicrobial peptides, and pore-forming toxins. Most of these proteins have a tendency to amass in the susceptible regions of the plant, notably in seeds and vegetative storage tissues. Notably, the initial proteins identified as ribosome-inactivating proteins (RIPs), including arcelins, canatoxins, and lectins, all have their origins in seeds, where these proteins are notably abundant (Carlini and Guimarães 1981).

13.3 Ribosome-Inactivating Proteins

Ribosome-inactivating proteins (RIPs) are toxins with the capacity to selectively and irreversibly halt protein translation (Olsnes and Pihl 1973). Plants employ various RIPs (ribosome-inactivating proteins) such as ricin, abrin, and saporins as a defensive mechanism to counteract potential threats from viruses or parasites (Endo

et al. 1987). For over a century, the existence of two powerful toxins, ricin from the seeds of *Ricinus communis* and abrin from the seeds of *Abrus precatorius*, was known. However, the term "RIPs" (ribosome-inactivating proteins) was introduced much later (Stirpe 1982) after it was revealed that these toxins inflict damage on ribosomes. Saporin damaging the function of ribosomes is illustrated in Fig. 13.2. RIPs derived from plants are categorized into three primary types: Type I comprises a solitary polypeptide chain of around 30 kDa; Type II is a heterodimer, comprising an A chain that performs a function equivalent to the Type I polypeptide, linked to a B subunit with lectin-binding capabilities; meanwhile, Type III is produced as inactive precursors (ProRIPs) that necessitate proteolytic processing for the formation of an active RIP (Zhu et al. 2018). These exhibit diverse activities such as antitumor, anticancer, antiviral, glycosidase, and neurotoxic effects. Several RIPs, including trichosanthin (TCS), α-momorcharin, MAP30, abrin, and ricin, have demonstrated the ability to induce apoptosis in cancer cells (Zeng et al. 2015). Figure 13.2 gives a basic understanding of how saporin, ricin, and abrin toxins affect the central dogma of the cell. Consequently, they inhibit the growth of tumor cells in various types of cancer, including breast cancer, leukemia/lymphoma, and hepatoma (Kocyigit et al. 2023). The RIPs from many plants have shown antitumor activity both in vitro and in vivo (Fan et al. 2008). RIPs have been employed in the creation of immunotoxins through antibody conjugation, presenting significant potential as therapeutic agents. The study of MAP30 has shed light on the biomedical uses of RIPs, demonstrating their effectiveness as antitumor and antiviral agents, particularly in combating HIV. This suggests a promising avenue for the use of RIPs as an alternative treatment for AIDS (Puri et al. 2012). PAP Pokeweed Antiviral Protein (PAP) has demonstrated efficacy against a wide range of viruses, encompassing HIV, herpes simplex virus, cytomegalovirus, influenza virus, polio virus, hepatitis B virus, and DNA viruses (He et al. 2008).There is a suggested hypothesis that ribosome-inactivating proteins (RIPs) might exert their antiviral influence through signaling

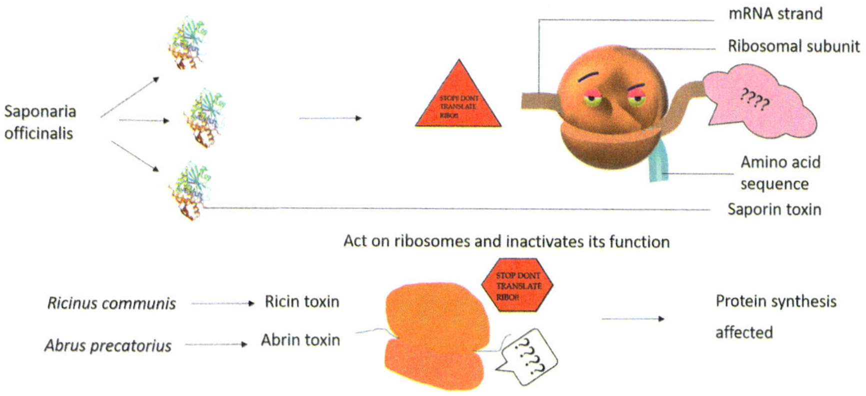

Fig. 13.2 Effect of toxin derived from plants, ribosome-inactivating proteins (RIPs), RCSB PDB: 3HIT Crystal structure Saporin-L1 from *Saponaria officinalis*. Similar mechanism of action is seen in ricin toxin and abrin toxin. This image is made using Microsoft Office applications

pathways. In the context of viral infection, RIPs have been observed to enhance the activity of p53 and c-Jun N-terminal kinase (JNK), while concurrently inhibiting the activation of KF-B, p38MAPK, and Bcl-2. The modulation of these pathways is thought to induce the demise of infected cells, thereby impeding the propagation of the virus (Akkouh et al. 2015). Plant RIPs are also employed as traditional natural antibiotics due to their diverse antibacterial, antifungal, and insecticidal properties (Wang et al. 2016). Certain RIPs possess DNA-nicking, DNase, or RNAse activities which can disrupt various stages of viral processes including replication, transcription, translation, and assembly (Citores et al. 2021). Consequently, these RIPs exhibit an antiviral effect by impeding protein synthesis in cells infected by the virus (Lu et al. 2020). Heterophyllin, a novel plant toxic lectin classified under the type 2 RIP family, demonstrated the capability to induce apoptosis, as evidenced by increased Annexin V positivity and minimal necrosis. This feature is advantageous in therapeutic applications compared to other cytotoxic agents, as apoptosis induction avoids the inflammatory responses associated with necrosis activation (Bortolotti et al. 2023).

13.4 Lectins in Cancer Therapy

The biological functions of lectins, such as immune responses, interactions between cells, signaling pathways, and cell growth, are dictated by their abilities to bind to carbohydrates (Ghazarian et al. 2011).The present approaches to cancer treatment and its after-effects lead to various harmful and life-threatening conditions; therefore, there is a demand for an alternative, effective, and target-specific potential drug therapeutics against cancer which would exert minimum side effects. In order to counteract these harmful conditions, certain specific structures are required to be targeted, which would provide a novel target for cancer therapeutics (Bhutia et al. 2019). As there are aberrant expression and secretion of glycoconjugates with an unusual glycan structure in the cancer cells, it is possible to detect these glycoconjugates easily by using lectins, the plant proteins which bind specifically to sugar molecules (Mody et al. 1995). Plant lectins have the capability to trigger mechanisms that inhibit cancer. Additionally, plant lectins play a role in regulating cell death through processes such as apoptosis and autophagy, suggesting their potential involvement of immune cells to provide defense against cancer (Narimatsu et al. 2010). Research conducted both in laboratory settings (in vitro) and within living organisms (in vivo) has shown that lectins possess the ability to inhibit the proliferation of cancer cells. They achieve this by acting as agents that counteract angiogenesis, metastasis, and cell proliferation (Gautam et al. 2020). This suggests that lectins could be valuable in the treatment of cancer. Although there have been limited clinical studies exploring the potential use of lectins as anticancer medications, their impact on human cancer cell lines appears to be associated with their ability to induce apoptosis and autophagy by regulating the synthesis of proteins (Chan et al. 2015).

13.5 Lectins in Ocular Biology

Galectins, a family of proteins, that bind to carbohydrates have demonstrated the ability to interact with various receptors containing β-galactosides. This interaction plays a crucial role in the regulation of immunopathological processes linked to viral and bacterial infections, transplantation, wound healing, dry eye, angiogenesis, and lymphangiogenesis (AbuSamra and Argüeso 2018).The existence of a galectin family member within the cornea and its possible role in promoting bacterial infection is well studied. Using antibody targeting galectin-3 could block the adhesion of *Pseudomonas aeruginosa* to corneal epithelial cells (Gupta et al. 1997). Dectin-1 a C-type lectin receptor is essential in the cornea as it regulates infections caused by *Candida albicans* and *Fusarium solani* (Kolar et al. 2017). Galectin-3 interacts with clusterin, a highly expressed protein in the human corneal epithelium (Bauskar et al. 2015). Maintaining the integrity of this interaction could offer therapeutic benefits for various dry conditions affecting the surface of the eye (Fini et al. 2016). Galectin-3 serves a pro-angiogenic function in the cornea promoting corneal angiogenesis (Markowska et al. 2010). Hence, regulation of galectin-3 can have therapeutic applications in reducing ocular fibrosis and neovascularization (Chen et al. 2017). Sialic acids are commonly found as end sugars on glycans of glycoproteins and glycolipids in the plasma membranes of eukaryotic cells. They play a crucial role in immune surveillance and maintaining homeostasis by presenting self-associated molecular patterns (Varki 2011). A family of receptors known as Siglecs (sialic acid-binding immunoglobulin-type lectins), particularly CD33-related Siglecs (CD33rSiglecs), exhibit specific recognition of sialic acids on the surfaces of certain immune cells (Lübbers et al. 2018).

Age-related macular degeneration (AMD) is a prominent contributor to global visual impairment and dysfunction. Revitalizing glyco-immune checkpoint regulation using a sialic acid that targets microglial/macrophage Siglecs represents a promising therapeutic approach in visual function across all stages of macular degeneration (Tolentino et al. 2023). Sialic acid polymers, specifically polysialic acid with an average degree of polymerization 20 (polySia avDP20), hinder the generation of reactive oxygen species (ROS) of human mononuclear phagocytes through the inhibitory receptor known as sialic acid-binding immunoglobulin-like lectin-11 (SIGLEC11). When administered through low-dose intravitreal injection in humanized transgenic mice expressing SIGLEC11, polySia avDP20 mitigates vascular leakage. Furthermore, higher doses of this polysialic acid not only impede the expression of tumor necrosis factor-α (TNF-α/TNFSF2) but also prevent the production of superoxide (Karlstetter et al. 2017; Shahraz et al. 2015).

13.6 Ricin

Ricin, an extremely toxic protein found in castor beans, hinders protein synthesis even at minuscule concentrations (Olsnes and Pihl 1973). The seeds of the castor bean plant (*Ricinus communis*) contain ricin, and these seeds have been widely utilized in traditional medicine to address various ailments. The catalytically active chain-A of ricin toxin (RTA, approximately 30 kDa) is connected to the ricin toxin chain-B (RTB, approximately 32 kDa) through a single disulfide bridge (Olsnes 2004; Lord 1985). The RTB functions as a lectin, attaching to glycolipids and glycoproteins present on the cell surface (Moya et al. 1985).The therapeutic potential lies in the utilization of ricin's capacity to deactivate ribosomes and induce the death of eukaryotic cells, which is employed for the treatment of cancer (Mosinger 1951). Immunotoxins based on the ricin A-chain exhibited considerable selectivity but limited effectiveness against cancer cells (Wawrzynczak et al. 1991). The majority of immunotoxins derived from ricin were quickly broken down within lysosomes, leading to a reduction in the quantity of ricin molecules translocating to the cytosol to carry out their cytotoxic impact (Antignani and FitzGerald 2013). Immunotoxins that undergo routing through the Golgi complex demonstrate higher efficacy. Utilizing pre-embedding immunoperoxidase methods, the investigation into the internalization mechanism of unlabeled immunotoxins with peroxidase-labeled antibodies and permeabilized cells revealed the presence of ricin chains in the Golgi complex. Additionally, a significant portion of the conjugate managed to evade lysosomal degradation (Timar et al. 1991). Ricin, a plant-derived cytotoxic protein, possesses anti-proliferative properties and holds potential as an effective anticancer agent. Nevertheless, similar to other therapeutic agents, it comes with several limitations, including non-specificity, immunogenicity, rapid degradation in the bloodstream, and lysosome degradation within cells (Tyagi et al. 2015).

13.7 Saporin

Saporin is another plant-based toxin that enters mammalian cells via the endocytic route and travels to the endoplasmic reticulum and Golgi apparatus. Saporin is a ribosomal inactivating protein of type 1 with N glycosidase activity. The toxicity of the toxin was increased by adding *Gypsophila* saponin along with saporin. This increased the toxicity and half-life of the toxin within the cell and enabled the toxin to escape the process of endocytosis. To investigate this, His-tagged saporin was tritium-labeled, and its cytotoxic effects were examined both independently and in conjunction with Gypsophila saponins. The cellular transport and degradation processes of His-tagged saporin were comparable between cells treated with Gypsophila saponins and control cells. However, following ultracentrifugation of a post-nuclear supernatant, the concentration of cytosolic His-tagged saporin was notably higher in saponin-treated cells compared to cells exposed solely to His-tagged saporin. This

suggests that saponins facilitate the escape of saporin from endosomes (Weng et al. 2009). Saporin toxicity differs from that of ricin in that it follows a Golgi-independent route and colocalizes with a late endocytic/lysosome signal. Saporin, a plant toxin that exploits endocytic machinery, is being tested for targeted tumor therapy (Polito et al. 2013). The targeting efficacy of saporin conjugate has been demonstrated to be superior to that of other toxin conjugates, particularly in retinoblastoma (Athyala et al. 2019). The cancer cell surface overexpressed protein mediated uptake of Saporin, property was utilized to conjugate this toxin to antibody that specifically targets overexpressed protein, to avoid off-targeting. Epithelial cell adhesion molecule (EpCAM) exhibits elevated expression in epithelial tumors (Maetzel et al. 2009). The antibody to EpCAM is used for targeted delivery. The effectiveness of saporin–EpCAM (SAP–EpAB) conjugates was evaluated against MCF-7 (breast cancer) cells (Lund et al. 2014) and WERI-Rb1 (retinoblastoma) cells. The study demonstrated that the SAP-EpAB conjugate selectively targets retinoblastoma and breast cancer cells expressing EpCAM. The research aimed to identify the altered cellular pathway in treated cells, specifically focusing on the significant reduction in cancer cell proliferation through Wnt pathway (Veeck and Dahl 2012). The conjugate exhibited a pronounced impact on the Wnt pathway, leading to the suppression of cancer cell proliferation.

To explore the potential role of melanopsin cells as conduits for photic information detected by rods and/or cones, researchers devised a targeted immunotoxin. This immunotoxin comprised an anti-melanopsin antibody linked to the ribosome-inactivating protein saporin. Through intravitreal injection, they observed that the selective elimination of these cells in the adult mouse retina resulted in modifications to the light-induced effects (Göz et al. 2008). Ricin-mAb35 is an immunotoxin that combines ricin, a ribosomal toxin, with a monoclonal antibody (mAb35) that targets acetylcholine receptors found on skeletal muscle. This unique composition allows Ricin-mAb35 to exert targeted cytotoxic effects within skeletal muscle cells, including those in the extraocular muscles (Christiansen et al. 2000). Research findings indicate that the injection of ricin-mAb35 into rabbit extraocular muscle leads to enduring muscle loss. This has been suggested as a potential therapeutic approach for strabismus. The significant reduction in extraocular muscle force generation persists for at least 12 weeks post-injection (Christiansen et al. 2003).

13.8 Animal Toxins

Animal toxins, like plant toxins, have a variety of therapeutic functions that have been thoroughly documented in the literature. Toxins produced by the animal system typically target human cell types, with potential applications in tailored therapies. The lethal effectiveness has been exploited on multiple occasions by conjugating with a suitable targeted ligand. Natural peptides found in the venoms of animals like snakes, scorpions, bees, and spiders, for example, have been shown to reduce inflammation, regulate glutamate release, modify neurotransmitter levels, block ion

channel activation, decrease the number of protein aggregates, and increase neuroprotective factor levels. As a result, these venom components may be employed as therapeutic strategies to slow or even stop neurodegeneration. However, significant technological hurdles must be overcome because venom peptides are difficult to collect and characterize, and the amount collected from natural sources is insufficient to undertake all necessary experiments and testing. Fortunately, improvements in heterologous protein expression and peptide chemical synthesis will assist in producing enough quantities and allowing chemical and pharmacological alterations of these proteins. Toxins derived from snake venom have been extensively researched for potential therapeutic applications in Alzheimer's disease and other diseases. These venoms are mostly classified as neurotoxins or dendrotoxins. The dendrotoxins were isolated from African mambas (*Dendroaspis* spp.) and these toxins possessed the ability to inhibit potassium channels (Fig. 13.3) (Koh et al. 2006). The fasciculins are a class of toxins secreted by snake species especially African green Mamba snake, are known to inhibit acetylcholine esterase activity (AChE) in neuromuscular junction by binding to a peripheral location of this enzyme, potentiating acetylcholine action, and causing generalized muscular fasciculation (Fig. 13.4a). They are substantially less poisonous than the other components of *Dendroaspis angusticeps* (green mamba) venom (Mebs 1989) The venom of the Indian snake *Daboia russelli* contains several PLA2 isoforms, procoagulant enzymes (factor X and V activators), hemorragins, nucleases, proteases, hyaluronidases, and other chemicals (Tsai 1997). Figure 13.3 simply shows the mechanism of action of Dendrotoxin on K+ channels in neurons. Scorpion toxins are another class of animal toxins with potential therapeutic applications. *Buthus martensii karsch* (Bmk) scorpion venom has been utilized in Chinese medicine for thousands of years to treat nervous system diseases (Wang et al. 2020). Increased neurogenesis, neuron maturation, and BDNF expression have been reported following Scorpion venom heat resistant peptide (SVHRP) treatment of neural stem cell cultures (Wang et al. 2014).

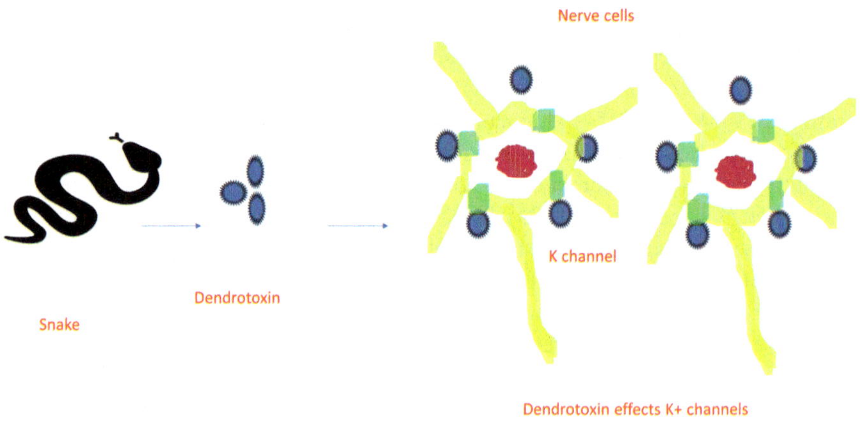

Fig. 13.3 Dendrotoxin a snake toxin affects potassium channels

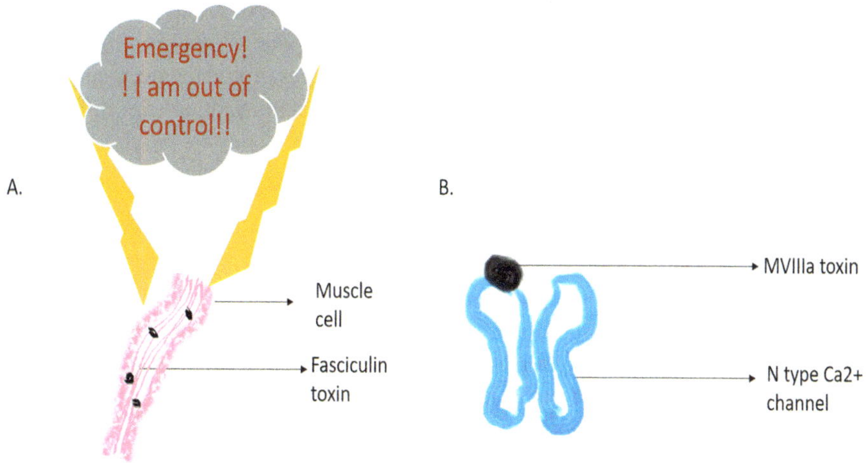

Fig. 13.4 Toxins binds to cells and cell surface proteins. (**a**) Fasciculin binds to muscle cells. (**b**) The conotoxin MVIIIa binds on N type calcium channels. The binding on muscle cells effects function of muscles and binding to Calcium channels may block the uptake of Calcium ions

Furthermore, anti-inflammatory properties of Bmk extracts in human chondrocyte and macrophage cultures have been documented (Kim et al. 2005). These findings point to the potential usefulness of Bmk venom components like SVHRP in treating other neurological illnesses including Alzheimer's. Many studies have been undertaken on spiders, wasps, and bee venom. With approximately 37,000 species, spiders are one of the most diverse families of venomous organisms, although only a small percentage of spider toxins have been found (Escoubas and Bosmans 2007). Most of the research on spider venom toxins has focused on them as prospective therapeutic agents for cancer, cardiovascular disease, and analgesic and antinociceptive effects (Pineda et al. 2014). However, there has been little research into the possible usefulness of spider poisons in neurodegenerative diseases. Another conotoxin that has been intensively researched for its neuroprotective qualities is MVIIA, an N-type Ca2+ channel blocker (Fig. 13.4b). Notably, the creation of a synthetic form of MVIIA, the SNX-111, encouraged more research (Olivera et al. 1987). The earliest experiments demonstrated that SNX-111 can prevent neuronal damage following acute global ischemia in rats, even when provided up to 24 h after the ischemic insult (Valentino et al. 1993). SNX-111 significantly ameliorates brain damage and lowers infarct size in a transient focal ischemia rat model subjected to 2 h MCA occlusion (Yenari et al. 1996). In addition to MVIIC and MVIIA, another conopeptide termed GVIA from the snail *Conus geographus* has been studied for its neuroprotective qualities. GVIA prevents excessive glutamate release during ischemia by blocking N-type Ca2+ channels, and this inhibition has a considerable protective impact in neurons (Madden et al. 1990). Spider venom toxins are proven to block ion channels. For example, the venom of the spider *Phoneutria nigriventer* has a combination of toxins that impact ion channels and have been studied for the

therapy of neurodegenerative disorders. PhTx3 toxin from *Phoneutria nigriventer* venom is a broad-spectrum neuronal Ca2+ channel blocker that also inhibits glutamate absorption (Gomez et al. 2002; Reis et al. 1999). The neuroprotective effects of PhTx3, GVIA, and MVIIC were investigated using hippocampal slices and a mouse cholinergic septal neuronal cell line (SN56 cells) subjected to ischemia caused by oxygen deprivation and low glucose insult. According to the findings of this investigation, PhTx3 totally recovers neuronal cell death, whereas both snail toxins only provide partial cell protection (Pinheiro et al. 2006). Several additional animal toxins have also been found to reduce ischemia injury. The most predominant poisonous component in the crude venom of the Chinese bird spider, *Ornithoctonus huwena* is HWTX-I, an N-type Ca^{2+} channel blocker (Liang et al. 1993). HWTX-I alters the expression and activity of various apoptotic cascade components, including SOD, glutathione peroxidase, Fas, TNF, caspase-8, and caspase-3, in a rat model of global cerebral ischemia-reperfusion injury (Wang et al. 2007)). *Bothrops atrox* and *Bothrops brazili* have venoms that contain serine proteinases capable of inducing angiogenesis via the PI3K/Akt pathway (Bhat et al. 2016). Recent research has shown that spider venom chemicals have the potential to treat glaucoma, which is an eye disorder where intraocular pressure is increased due to several reasons like genetic, lifestyle, etc. Intraocular pressure built can have the potential to damage the retinal cells which are important for processing vision. It has been demonstrated, for example, that components of the venom of the Brazilian spider *Parawixia bistriata* are neuroprotective. Using a rat glaucoma paradigm based on retinal ischemia/reperfusion, it was demonstrated that the compound FrPbAII improved neuroprotection of the retinal cell layers while also crossing the blood brain barrier (BBB) (Liberato et al. 2006). Like FrPbAII, another chemical isolated from the venom of *Parawixia bistriata* displays neuroprotective benefits in retina ischemia/reperfusion glaucoma models (Fachim et al. 2015). Spider venom can be used in treating the diseased eye. PbTx1.2.3, produced from *Parawixia bistriata* venom, has been demonstrated to protect neurons against degeneration in an ischemia/reperfusion retina model, most likely by anti-excitotoxic effect, which refers to the glutamate receptor, which is implicated in neuroexcitation. Excitotoxicity can cause neuronal death, which can be avoided by the venom of *Parawixia bistriata* (Fontana et al. 2003). These findings strongly imply that *Parawixia bistriata* venom components could be used as medicinal agents.

13.9 Conjugates Using Toxins

Toxins have been employed in the development of conjugates for a variety of diseases, including cancer. Conjugates are created using various linkers or ligands. They could be antibodies, affibodies, aptamers, nanobodies, or anything else. These linkers are created by targeting a certain disease marker protein, which is predominantly produced on the cell surface via several pathways. Before toxin conjugations, these surface indicators must be identified for disease specificity. Figure 13.5 gives a visualization of a toxin conjugate.

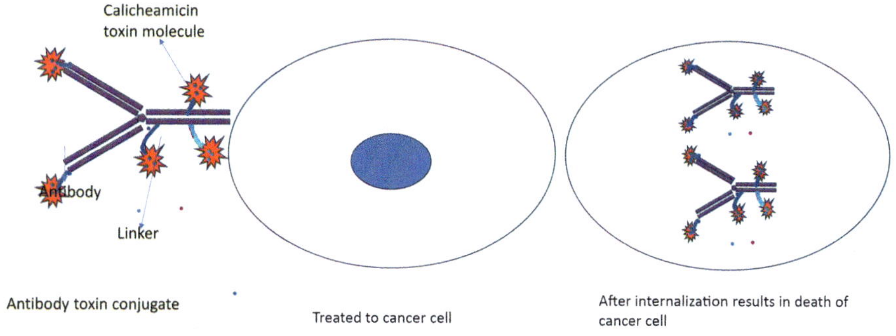

Fig. 13.5 Toxin conjugate used in therapy, featuring an antibody linked to a toxin molecule like calicheamicin targeting a cancer cell

13.10 Antibody Conjugates for Therapy

Antibody drug conjugates (ADCs) are precise therapeutic agents compared to traditional chemotherapy that may harm healthy cells (Fig. 13.5). They transport chemotherapy agents to cancer cells through a linker connected to a monoclonal antibody designed to attach to a particular target present on cancer cells. Once bound to the target, such as a cancer protein or receptor, the ADC releases a cytotoxic drug within the cancer cell (Peters and Brown 2015). Antibody drug conjugates (ADCs) merge the targeted precision of antibodies with the powerful cell-killing capabilities of drugs (Lambert and Berkenblit 2018). Like cancer, numerous ocular proliferative disorders may be addressed by utilizing antibodies linked to toxins. The application of ADC agents for actively targeting and suppressing the proliferation of epithelial and endothelial cells in the eye has undergone testing in both preclinical and clinical settings (Shen and Attar 2015). In the development of ADCs, toxins typically incorporate reactive groups like amines or thiols, which are employed for binding to antibodies through specific linkers (Hoffmann et al. 2020).The initial therapeutic monoclonal antibody, muromonab, received FDA approval in 1986 for use as an immunosuppressant in organ transplants. It took two decades, until 2006, to witness the first approval of a monoclonal Fab in ophthalmology—specifically, ranibizumab (Lucentis®). The first FDA-approved ADC in 2001 was gemtuzumab ozogamicin (Mylotarg®) for acute myelogenous leukemia (Bross et al. 2001), but it was later withdrawn due to safety concerns related to ADC stability. Currently, only two ADCs, brentuximab vedotin (Adcetris®) and trastuzumab emtansine (Kadcyla), are on the market although numerous other ADC candidates are undergoing clinical testing.

13.10.1 Aptamer Conjugates for Targeted Therapy

Aptamers are short oligonucleotide (DNA/RNA) or amino acid stretches that develop secondary structures in order to fit perfectly onto target protein structures (Egli and Manoharan 2023). Aptamers are ligands, unlike antibodies; they are less immunogenic and have a minuscule molecular size. They can be generated in vitro and chemically modified to boost their stability. Aptamers can conjugate not just the drug molecules but also siRNA or imaging agents, making it more appealing therapeutic agents due to the above distinctive features of aptamers (Zhao et al. 2023). Aptamer drug conjugates are used to target a variety of cancers, including lymph node malignancies via the CCL21 receptor (Che et al. 2023). Not only can the conjugates target tumor cells, but adding an imaging moiety allows them to be used for both targeting and imaging the tumors. Recently, zirconium-based metallic frames have been employed for theranostics, in which an aptamer is coupled with a zirconium metallic frame for imaging and doxorubicin for therapeutic reasons, and it has been verified on breast cancer cells (Feng et al. 2023). DNA and RNA aptamers have been investigated for specific targeting of tumor cells for therapy. Triple negative breast cancer has been targeted using DNA aptamer targeting cMYC successfully. The aptamer is a nucleic acid-based aptamer that naturally targets cMYC proteins and is further conjugated with pomalidomide, an analog of thalidomide that limits triple negative breast cancer cell proliferation (Li et al. 2023). The epithelial cell adhesion molecule (EpCAM) protein binding aptamer has delivered conjugated Doxorubicin in mice model (Mashreghi et al. 2021). DNA and RNA aptamers being relatively small in molecular size, their structure needs to be studied biophysically with its target before they are tried for experimental conjugation and efficacy, as the binding depends on the secondary structure it forms (Athyala et al. 2016). Aptamer conjugates for ocular drug delivery using aptamers that can selectively bind to proteins overexpressed on cell surface (EpCAM protein) are used as selective targeting molecules. The EpCAM protein is overexpressed in several cancers including breast, pancreatic, lung, retinoblastoma and is also considered as cancer stem cell marker. The cells overexpressing this protein are usually considered drug resistant and targeted therapies that can selectively kill cancer cells sparing healthy cells are in research recently. The anticancer drug doxorubicin was conjugated to EpCAM and this conjugate can selectively kill retinoblastoma cells compared to normal cells. Neocarzinostatin obtained from *Streptomyces carzinostaticus* is a peptide toxin and was tried as drug for cancer. However, the drug failed phase 2 clinical trials due to adverse reactions caused by the drug when it was used on cancer patients. The off-target effects of the toxin were reduced by conjugating the toxin to the cell surface molecule aptamer EpCAM. The side effects were reduced, and the toxin aptamer conjugate could selectively kill retinoblastoma cells. The mechanism of action was studied using global transcription by DNA microarrays. The toxin conjugate elicited cancer cell death by modulating cell cycle through elongation factors and TGF-β signaling (Athyala et al. 2017). The Mucin protein is expressed on corneal epithelial cells and this protein was targeted using an aptamer functionalized

with liposome. This drug was reported recently for treating dry eyes syndrome (Wong et al. 2023). The first licensed aptamer medicine (pegaptanib) has been used to control angiogenesis in age-related macular degeneration (AMD) and diabetic retinopathy (DR), with clinical trials demonstrating its efficacy and safety. Aptamers targeting angiogenesis-related markers, such as E10030, RBM-007, AS1411, and complement factor have also been discovered and tested in clinical trials (Cao et al. 2023). There are two aptamer drugs that are approved for geographical atrophy (GA), that is usually caused after the dry AMD advancement. These are lesions caused on the macula of the retina without rod cells, retinal pigment epithelial cells and choriocapillaris (Fleckenstein et al. 2018; Wong et al. 2014). Avacincaptad pegol is a pegylated RNA aptamer that targets complement factor 5 (C5) and was found efficient in regulating the GA. In a phase three clinical trial efficacy was observed (Patel et al. 2023). Later FDA approved this drug for second line of therapy. The siRNA are short chain RNA molecules that can inhibit translation of protein. Several proteins are overexpressed in cancers and the use of siRNAs against the overexpressed proteins can downregulate the overexpressed protein enabling selective killing of cancer cells compared to normal cells. Stathmin, survivin siRNAs are conjugated with RNA aptamer using different chemistry such as amide bond formation, di sulfide bond formation (Subramanian et al. 2015). These chimeras could effectively kill cancer cells. In addition to cytoplasmic or membrane expressed proteins as targets, nuclear proteins can be specifically targeted as potential drugs for cancer treatment. Nucleolin, for example, is a nuclear protein that regulates transcription and rRNA transport and has been shown to be overexpressed on the cell surface in a few epithelial cancer cells. Nucleolin aptamer NCL-APT (AS1411) has shown efficacy in treating cancer in mouse models of retinoblastoma (Subramanian et al. 2016).

13.11 Bacterial Toxins and Eye Research Models

Bacterial toxins are of two types, exotoxins and endotoxins. Endotoxins are lipopolysaccharides (LPS) present in the cell wall. Exotoxins are peptides formed inside the cell wall which are released. LPS consist of O-antigen, polysaccharide, and lipid A molecules; these are secreted by *Escherichia coli*, *Pseudomonas*, and *Klebsiella*. These cause septic shock with multiple organ failure. The potent toxins have no adaptive immune response. Therefore, the toxins can cause septic shock repeatedly (Sheehan et al. 2022). There are toxin producing microbes such as *Staphylococcus aureus* or *Streptococcus pyogenes* which are considered opportunistic pathogens which are present as normal microbiota in the human skin, mucous membrane. These produce exotoxins which activate the polyclonal lymphocyte activation. This activation leads to rapid shock, multiple organ failure, and death. This toxin is called as toxin shock syndrome toxin-1 (TSST-1). Enterotoxin from *S. aureus* (SpE), A, B, and C from *S. pyogenes* and superantigen A are together called super antigenic exotoxins. The pathogenicity of the TSS through super antigenic exotoxin induces unconventional T cell activation using antigen presenting cells. Super antigen in

turn activates NF-κβ leading to release of inflammatory cytokines. The immediate action of these superantigens, if understood better right from toxicity level, binding site, mode of action, will have a potential for anesthetic application. Interestingly, apart from protein synthesis inhibiting antibiotics (clindamycin and linezolid), polyspecific intravenous immunoglobulins G (IVIG) and two probiotics (*Lactobacillus acidophilus* and *Lacticaseibacillus rhamnosus*) interfere with *S. aureus* growth and TSST-1 production and these therapy options are under phase II clinical trials (Atchade et al. 2024). Another rare toxin which comes from the bacterial toxin–antitoxin (TA) systems is from Chlamydia-related *Waddlia chondrophila*, which is type II TA system. Bacterial toxin–antitoxin systems are small genetic elements (bacterial genome) encoding stable toxin leading to poisoning by inhibiting translation and an unstable antitoxin or substances which acts like antagonist to the toxin system. It is called a system and is classified as type I to type VI toxins based on the activity or the interaction of the toxin and antitoxin systems.

Type I: These antitoxins are small regulatory RNAs which inhibit toxin's translation of mRNA to protein/toxin process. They are generally found in plasmid for maintenance during phage infection by creating toxin RNA/antitoxin RNA duplex.
Type II: These TA systems have toxin and antitoxin interacting in protein–protein level as either lock key or induced fit type.
Type III: These TA systems have antitoxin which is RNAs interacting with the protein and inhibits the toxin.
Type IV: These TA systems have antitoxin inhibiting the toxin effect by binding to its target molecule.
Type V: Here the antitoxin is a protein which cleaves the mRNA of toxin during the protein synthesis thereby inhibiting the toxin's production.
Type VI: These TA systems have an antitoxin which does not interact with toxins synthesis but target it to undergo proteolysis.

The *Waddlia chondrophila* TA system is considered rare because this bacterium is an intracellular bacterium and especially in members of Chlamydiales disease condition. The TA system is induced by iron starvation. These TA systems are useful to study stress conditions.

Some of the application for biotoxins could be using conjugated cholera toxin A subunit (SP-CTA) which acts through neurokinin receptors to alter non-rapid eye movement (NREM) sleep slow wave activity thereby activating cells expressing SP receptors in sleep active neuronal areas in brain. This attempt could pave a strategic therapy related to insomnia and other sleep deprived condition (Zielinski and Gerashchenko 2017). Another application of biotoxins which is extensively used is to use it for creating model systems to study disease conditions. Disease models incorporate biomarkers of disease conditions progression, severity, and clinical outcomes. It is important to study the molecular mechanisms involved in the disease condition in a controlled environment to first study the molecular interactions. A lot of diseases are challenging to treat, and clinicians lack the in vitro and in vivo disease models to study disease progression and therapy options. Even if disease models are available it is challenging to study the drug treatment on disease progression. The use of toxins for glaucoma research is presented in Fig. 13.6. A fluorescent

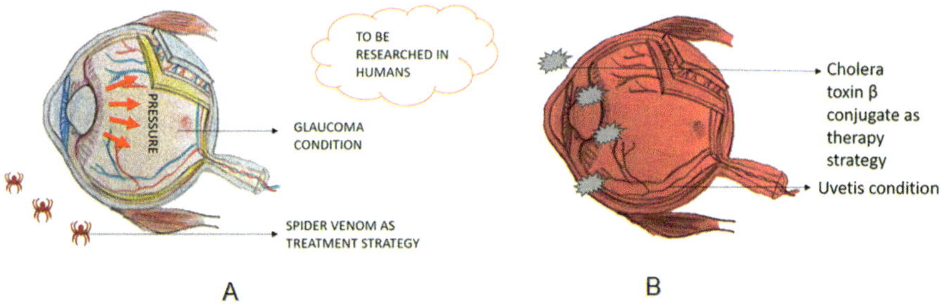

Fig. 13.6 (**a**) How spider venom toxin can be used in glaucoma eye condition (hand drawn). (**b**) Cholera toxin conjugated with DHEA nanoparticle used as delivery vehicle

cholera toxin subunit β (CTB) as intravitreal injections on rats was used along with phosphate buffered saline to assess transport. This toxin treatment increased the ganglion cells degeneration and reduction of immune labeled ganglion cell bodies. The researchers used this approach to see the effect of Brimonidine on the reduction of retinal ganglion cells which is crucial for understanding disease conditions. The glaucoma disease progressed with increase in age, by creating an increased intraocular hypertension in rats. Immunolabeled cholera toxin β (CTB) was used to study the retinal ganglion cells that are usually affected in Glaucoma (Angelucci et al. 1996). This application uses 0.5 mg of cholera toxin conjugated with Alexa 594 or 488 (Lambert et al. 2011). Apart from being used for immunohistochemical applications for disease models, biotoxins can also be used to study immune mechanisms and disease pathology. For example, Scott W Mc Pherson et al. used diphtheria toxin to study the involvement of $CD8^+$ T regulatory cells in uveoretinitis, a condition known as posterior uveitis, is an inflammatory condition that affects the uvea and retina of the eye. In an autoimmune mouse model (C5TBL/6) which has uveoretinitis induction, using diphtheria toxin (DTX) and its receptor they showed local action of this toxin in presenting toxin sensitive regulatory T cells (β gal specific CD8+ T cells) in disease induced eye compared to untreated disease-free another eye. This shows diphtheria toxin's role in regulating the immune response (CD8 T cells). The concentration used here is 25 ηg/μl DTX, but 1ηg of DTX is sufficient to eliminate the dendritic cell population in retinal parenchyma (McPherson et al. 2012).

Behcet's Disease (BD) is an autoimmune disorder which is characterized by severe uveitis along with systemic symptoms such as ulcers in the oral and genital regions, cutaneous, vascular, joint and neurological problems (Pickering and Haskard 2000). This is a significant cause of blindness in Mediterranean and Eastern countries. The microbial heat shock protein HSP65 from *Streptococcus sanguinis* variants and human analog HSP 60 were implicated in the BD disease due to their cross immune reactions (Lehner et al. 1991). An epitope for T cells that is from microbe that cause BD disease was identified (Hasan et al. 1996). This peptide caused uveitis in Rats and it was conjugated to cholera toxin B subunit. The toxin conjugate could efficiently reduce uveitis in Phase 1 clinical trial in humans (Stanford et al. 2004). Delivering drugs to brain and retina is difficult due to the presence of blood brain barrier (BBB) and blood retinal barrier (BRB), respectively.

Therefore, several strategies are used to deliver drugs through these barriers. The transmucosal carrier cholera toxin B (CTB) subunit was expressed in chloroplasts and then purified to have a plant coating on it. This was fused with myelin basic protein and then targeted to brain and retina in transgenic Alzheimer's disease mice models. The toxin conjugate when given orally could effectively penetrate BBB and BRB and deliver drug to brain and retina reducing amyloid plaque formation in brain and retinal ganglion cell death, respectively. These observations have implications for Glaucoma (Figure 13.6.) (Kohli et al. 2014). Recently, CTB was conjugated to hsCD80 which is known to modulate immune response in cancer. The conjugate was specifically expressed in *L. lactis* bacteria. These organisms were tested in mouse model of colorectal cancer. This probiotic served as antitumor immunotherapy. This was more efficient in reducing tumor burden compared to hsCD80 expressing *L. lactis* (Lin et al. 2023). A nanoparticle toxin conjugate was attached to FDA-approved drug for sustainable release and targeted delivery to retinal ganglions that are affected in Glaucoma. The CTB was used as targeted molecule to deliver drugs to retinal ganglion. The drug dehydroepiandrosterone (DHEA) was delivered using polymeric nanoparticles. The nanocomposite could effectively stop degeneration of retinal ganglions. If nanoparticle-drug complex was used without the CTB toxin as targeted delivery molecule, the drug was unable to reach retinal ganglion cells and stop the degradation indicating the role of toxin molecule in targeted delivery of drugs (Zhao et al. 2017). However, several nanoformulations are studied as potential therapy molecules for eye diseases (Mahaling et al. 2023). The CTB toxin could improve corneal graft formation when the toxin was mixed with corneal cells and injected into mice (Ma et al. 1998). Figure 13.6b shows how the DHEA nanoparticle interacts with the diseased eye.

13.11.1 Bacterial Toxins Causing Eye Diseases

Endotoxins are early biomarkers for age-related macular degeneration (Larsen et al. 2023). LPS and cytokines such as IFN-γ affect microglial cells and are known to cause inflammation and uveitis. Research is in progress to identify natural compounds that can reduce inflammation. Apigenin, a plant compound could reduce the inflammation in mouse models (Shu et al. 2023). Endotoxins can cause bacterial keratitis (Liang et al. 2007) affecting the Cornea. Figure 13.7 shows how keratitis eye looks compared to normal eye and how endotoxins affect the normal eye inducing keratitis. Corneal inflammation was observed which was mediated by LPS through Toll Like receptor (TLR) signaling. Exotoxins are peptides that are secreted inside bacteria. The common eye pathogen is *Staphylococcus aureus*, which infects tear ducts, cornea, conjunctiva, anterior chamber, and posterior chamber. The organism is known to secrete α-toxin, β-toxin, γ-toxin, and Panton-Valentine leukocidin (PVL). This organism infections were difficult to treat lately due to the phenomenon of multidrug resistance (Lee et al. 2018). The important strains are Methicillin resistant Staphylococcus aureus (MRSA) pathogens that have become difficult to treat in hospital-based settings. The identification of Staphylococcus species is a lengthy

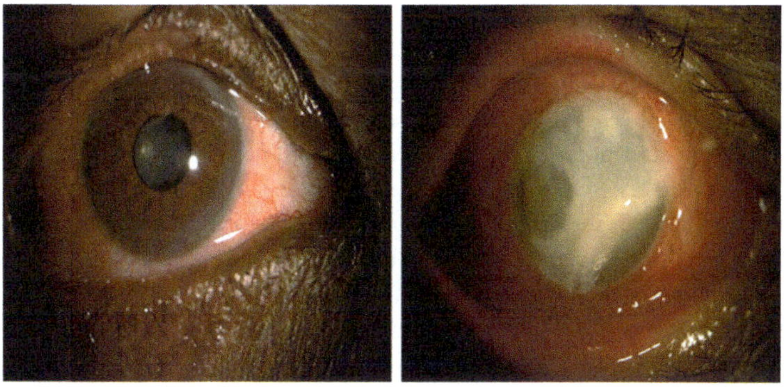

Fig. 13.7 (**a**) Normal right eye and (**b**) Keratitis eye. Pictures are given by The Cornea Department of Sankara Nethralaya after patient informed consent

process involving culturing the organism. Therefore, research is in progress to identify the causative organism quickly. A micropore technology was used along with machine learning techniques to identify various species of *Staphylococcus* (Morimura et al. 2024). Antibiotic-resistant *Staphylococcus* sp. and other microbes causing ocular diseases are reported in ocular diseases like keratitis and conjunctivitis (Petrillo et al. 2021; Asbell et al. 2020; André et al. 2023). The MDR phenomenon is thought to be due to horizontal transfer of resistant genes from organism to organism. Usually, transfer of genetic material is from parent to offspring. The horizontal transfer of genes gives fast adaptation to changing environment and gives survival advantage to the organism and this becomes a clinical problem in a community and in the world due to increased travel in the age of globalization (Prasad and Elchuri 2022). The toxins secreted from *Staphylococcus* such as α-toxin, β-toxin, γ-toxin, and Panton-Valentine leukocidin (PVL) were implicated in causing eye diseases (Astley et al. 2019). The α-hemolysin is the name given to α-toxin as the first identification of this toxin was by lysis of Red blood cells (RBCs) (Glenny and Stevens 1935). This toxin was known to cause severe infections systemically in addition to eye diseases such as keratitis and endophthalmitis. This toxin is a pore-forming toxin that behaves like an ionic channel and regulates the passage of ions into the cell. This in turn is regulated by integrin molecules and metalloproteinases. The toxicity to the cell is observed in severe infections and in mild conditions. The toxin can reduce adherens junctions function enabling the loss of barrier function in cornea and iris causing inflammation (Dal Peraro and van der Goot 2016; Li et al. 2021; Ulhuq and Mariano 2022). The role of toxins in causing eye diseases was demonstrated in animal models of inflammation. Additionally, mutated toxins were used compared to wild type alpha- toxins; the muted toxins could not cause the eye disease in the animal models whereas the wild type toxin caused disease indicating that the toxins present in the microbes are the main causative agents of eye infections rather than entire organism (Dempsey and Conrady 2023). The γ-toxin is a pore-forming toxin and has two secretory proteins. These are Hlg B and either Hlg C or HlgA. This toxin was tested on rabbit cornea keratitis eye model and was found to cause the disease (Bierdeman et al.

2017). β-toxin, secreted by *Staphylococcus* is a sphingomyelinase that lyses the lipids in the epithelial cell membranes and is known to cause keratitis (Dajcs et al. 2002). The PVL toxin is present in MRSA strains of *Staphylococcus*. Additionally, the bacterium produces fibronectin binding protein, collagen-adhesion binding protein, Staphopain A, and extracellular adherence protein (Eap), which mediate bacteria–host cell interactions. The other bacterium that causes keratitis is *Pseudomonas aeruginosa*. It secretes several toxins and virulence factors that mediate corneal keratitis. These are: *Pseudomonas* protease IV (PIV), exotoxins such as *exoS*, *exoU*, and *exoT*, *P. aeruginosa* small protease (PASP), elastases, and extracellular factors such as lipopolysaccharides (LPS). Several of these toxins are demonstrated as causative agents for keratitis leading to blindness if proper therapy is not given timely (Shah and Wozniak 2023; Dempsey and Conrady 2023). Several emerging therapy options are investigated recently as MDR strains are more prevalent. Several enterotoxins as causative agents of keratitis are proven recently (Johnson et al. 2023)

13.12 Adverse Ocular Events in Clinical Trials

Kadcyla (formerly ado-trastuzumab emtansine (T-DM1) from Genentech is a compound that combines the human epidermal growth factor receptor 2 (HER2)-targeting antibody trastuzumab (marketed as Herceptin® by Genentech) with DM1, a maytansinoid (Pegram and Slamon 2000). During a phase I clinical trial involving dose escalation, individuals diagnosed with metastatic HER2-positive breast cancer were administered T-DM1 intravenously every 3 weeks, with doses ranging from 0.3 to 4.8 mg/kg (Krop et al. 2010). Although most adverse events (AEs) were unrelated to the eyes, instances of conjunctivitis, photophobia, and swollen tear duct, severity, were seldom documented (Eaton et al. 2015). In the second phase of the study, ocular adverse events (AEs) were observed in 31.3% of participants, encompassing issues such as dry eye, heightened lacrimation, vision blurring or impairment, and conjunctivitis at the maximum tolerated dose (MTD) of 3.6 mg/kg (Burris et al. 2011). Mylotarg (Pfizer), also known as gemtuzumab ozogamicin (GO), is a compound composed of a CD33-binding antibody and a calicheamicin cytotoxin (Lapusan et al. 2012). In a clinical trial, nine elderly patients diagnosed with untreated or relapsed acute myeloid leukemia (AML) received a combination treatment of GO along with cytarabine. One participant encountered grade 4 ocular bleeding (Piccaluga et al. 2004). Lorvotuzumab mertansine, also known as IMGN901, is a compound consisting of a CD56-targeting antibody conjugated with $N^{2'}$-deacetyl-$N^{2'}$-(3-mercapto-1-oxopropyl)-maytansine (DM1) (Woll et al. 2010). The DM1 is a synthetic derivative of naturally occurring maytansinoid toxins that are more potent than natural analogs. The toxins are derived from plants, mosses, and several microbes (Kirschning et al. 2008). In a phase I clinical trial involving dose escalation, IMGN901 was assessed in patients with solid tumors expressing CD56. While ocular adverse events were infrequent, three patients reported eye redness at a severity level of grade 1-2. BT-062 is a compound that combines monoclonal antibody targeting CD138 molecule to maytansinoid toxins (Ikeda et al. 2009).

In a phase I dose-escalation trial involving individuals with relapsed or refractory multiple myeloma, a female participant encountered dry eyes and blurred vision, while a male participant developed corneal epithelial damage and crystal corneal inclusions. These adverse events were classified as grade 2 and considered medically significant in both cases (Chanan-Khan et al. 2009). Several antibody drug conjugates (ADCS) are in clinical trials (Maecker et al. 2023; Zhao et al. 2020; Sun et al. 2023). Some ADCs in various phases of clinical trials are given in Table 13.1.

Table 13.1 ADCs in clinical trials

ADC	Target antigen	Toxin	Status/ phase	Clinical trial reg. number
Gemtuzumab ozogamicin	CD33	Calicheamicin derivative	Approved	NCT01548911
Brentuximab vedotin	CD30	Monomethyl auristatin E	Approved	NCT04685616
Trastuzumab deruxtecan	HER2	Deruxtecan	Approved	NCT03734029
Loncastuximab tesirine	CD19	Pyrrolobenzodiazepine dimer	Approved	NCT03589469
Inotuzumab ozogamicin	CD22	*N*-acetyl-gamma-calicheamicin dimethylhydrazide	Approved	NCT01371630
Trastuzumab emtansine	HER2	DM1 (a maytansine derivative)	Approved	NCT02658734
Polatuzumab vedotin-piiq	CD79b	Monomethyl auristatin E	Approved	NCT04231877
Sacituzumab govitecan	Trop-2	SN-38	Approved	NCT05089734
Belantamab mafodotin	BCMA	Monomethyl auristatin-F	III	NCT06160609
Tusamitamab ravtansine	CEACAM5	Maytansinoid	III	NCT04394624
Rovalpituzumab tesirine	DLL3	Pyrrolobenzodiazepine dimer	III	NCT01901653
Telisotuzumab vedotin	c-Met	Monomethyl auristatin E	III	NCT06093503
Datopotamab deruxtecan	Trop-2	Topoisomerase I inhibitor	III	NCT06176261
Naratuximab emtansine	CD37	Maytansine-derived microtubule disruptor, DM1	II	NCT01702571
Lorvotuzumab mertansine	CD56	Tubulin-binding maytansinoid DM1.	II	NCT02452554
Coltuximab ravtansine	CD19	Maytansine	II	NCT 01472887
Glembatumumab vedotin	gpNMB	Monomethyl auristatin E	II	NCT02487979
Indatuximab ravtansine	CD138	Pomalidomide and dexamethasone	II	NCT01638936

13.13 Algal Toxins Effect on Eye and Therapeutic Potential of the Algal Toxins

Toxins from Cyanobacteria are secreted by harmful algal blooms. Climate change and nutrient availability are thought be the lead factors that increased toxin production. Exposure to blue-green alga *Lyngbya majuscula* in the marine waters causes corneal abrasion and conjunctivitis. The toxin substances that were observed in the algae were Aplysiatoxin, debromoaplysiatoxin, and oscillatoxin A. Of these, debromoaplysiatoxin was known to cause severe dermatitis in humans (Werner et al. 2012). However, *Lyngbya* sp. are a source of several secondary metabolites that have potential as antibacterial, antifungal, and antiparasitic activity (Youssef et al. 2022). There are several drugs derived from Cyanobacteria or shellfish that harbor them are FDA approved for various diseases. One such example is Polatuzumab vedotin marketed by Genentech/Roche. This drug is approved for treating Non-Hodgkin lymphoma, chronic lymphocytic leukemia, and B-cell lymphoma. The drug Enfortumab Vedotin marketed by Astellas Pharma and Seagen is approved for treating metastatic urothelial cancer. Tisotumab vedotin after phase 3 clinical trial (Coleman et al. 2021) is approved for metastatic cervical cancer. The drug consists of Algal/shellfish toxin monomethyl auristatin conjugated to antibody that targets tissue factor protein overexpressed on cervical cancer cells. There are 14 drugs isolated from Cyanobacteria/shellfish in Phase I clinical trial. Several drugs target various solid tumors such as breast, lung, prostrate, urothelial, and ovarian. Thirteen drugs are in Phase 2 clinical trials for solid tumors and a single drug ARX-788 is in phase 3 clinical trial for breast cancer and gastric cancer (https://www.marinepharmacology.org/ accessed on 28-03-2024). The algal toxins are conjugated to antibodies to specifically target cancer cells and spare normal cells. Additionally, several marine products were approved by FDA for various therapies. The nucleosides from sponges' target DNA polymerases were approved for Leukemia and Viruses mitigation. Alkaloids, peptides from sponges and Tunicates were approved as cancer drugs. Several marine products are in clinical trials for cancer, neurological problems, and as antimicrobial compounds.

13.14 Bioprospecting of Marine drugs

The biodiversity is explored for new resources which have social and economic value. Pharmaceutical industry is well known for utilizing these resources for bringing new drugs into the market which are of commercial values. There is a criticism against them as they sometimes exploit indigenous knowledge for drug preparations without acknowledging or sharing profits from them. Therefore, several international agencies come together to frame laws so that indigenous tribes could be benefited and there could be empowerment of modern knowledge in the host countries (Beattie et al. 2011). The Millennium Ecosystem Assessment (MEA) gives us a

chance to find products for bioprospecting (http://www.millenniumassessment.org). The MEA is sponsored by UNEP, UN-FAO World bank, etc. Bioresources are used in several areas of agriculture, manufacturing, bioremediation, and many industries. Bioprospecting is the backbone of several industries.

In conclusion, toxins cause several eye diseases. Several toxins are used for investigations in making in vivo and in vitro eye disease models. Additionally, toxin conjugate drugs are approved, and some are in clinical trials for various diseases. These toxin conjugate drugs after completion of clinical trials and FDA approval, in future can overcome presently unmet needs for curing eye diseases. There are several adverse events on eye during the use of toxin drugs. Future drug combinations can lower these effects. Several novel therapeutic candidates can be researched using bioprospecting of drugs from biological sources. However, indigenous populations and countries where drugs are available should be given adequate importance by international community.

References

AbuSamra DB, Argüeso P (2018) Lectin-glycan interactions in corneal infection and inflammation. Front Immunol 9:2338. https://doi.org/10.3389/fimmu.2018.02338

Akkouh O, Ng TB, Cheung RCF, Wong JH, Pan W, Ng CCW, Sha O, Shaw PC, Chan WY (2015) Biological activities of ribosome-inactivating proteins and their possible applications as antimicrobial, anticancer, and anti-pest agents and in neuroscience research. Appl Microbiol Biotechnol 99:9847–9863. https://doi.org/10.1007/s00253-015-6941-2

André C, Lebreton F, Van Tyne D, Cadorette J, Boody R, Gilmore MS, Bispo PJM (2023) Microbiology of eye infections at the massachusetts eye and ear: an 8-year retrospective review combined with genomic epidemiology. Am J Ophthalmol 255:43–56. https://doi.org/10.1016/j.ajo.2023.06.016

Angelucci A, Clascá F, Sur M (1996) Anterograde axonal tracing with the subunit B of cholera toxin: a highly sensitive immunohistochemical protocol for revealing fine axonal morphology in adult and neonatal brains. J Neurosci Methods 65:101–112

Antignani A, FitzGerald D (2013) Immunotoxins: the role of the toxin. Toxins (Basel). https://doi.org/10.3390/toxins5081486

Asbell PA, Sanfilippo CM, Sahm DF, DeCory HH (2020) Trends in antibiotic resistance among ocular microorganisms in the United States from 2009 to 2018. JAMA Ophthalmol 138:439–450. https://doi.org/10.1001/jamaophthalmol.2020.0155

Astley R, Miller FC, Mursalin MH, Coburn PS, Callegan MC (2019) An eye on Staphylococcus aureus toxins: roles in ocular damage and inflammation. Toxins (Basel) 11. https://doi.org/10.3390/toxins11060356

Atchade E, De Tymowski C, Grall N, Tanaka S, Montravers P (2024) Toxic shock syndrome: a literature review. Antibiotics. https://doi.org/10.3390/antibiotics13010096

Athyala PK, Kanwar JR, Alameen M, Kanwar RK, Krishnakumar S, Watson J, Vetrivel U, Narayanan J (2016) Probing the biophysical interaction between neocarzinostatin toxin and EpCAM RNA aptamer. Biochem Biophys Res Commun 469:257–262. https://doi.org/10.1016/j.bbrc.2015.11.109

Athyala PK, Kanwar JR, Chitipothu S, Kanwar RK, Krishnakumar S, Watson JP, Narayanan J (2017) Neocarzinostatin, aptamer conjugates for targeting EpCAM-positive tumor cells. Anticancer Res 37:3615–3629. https://doi.org/10.21873/anticanres.11732

Athyala PK, Chitipothu S, Kanwar JR, Krishnakumar S, Narayanan J (2019) Synthesis of saporin-antibody conjugates for targeting EpCAM positive tumour cells. IET Nanobiotechnol 13:90–99. https://doi.org/10.1049/iet-nbt.2018.5222

Bauskar A, Mack WJ, Mauris J, Argüeso P, Heur M, Nagel BA, Kolar GR, Gleave ME, Nakamura T, Kinoshita S, Moradian-Oldak J, Panjwani N, Pflugfelder SC, Wilson MR, Fini ME, Jeong S (2015) Clusterin seals the ocular surface barrier in mouse dry eye. PLoS One 10:e0138958. https://doi.org/10.1371/journal.pone.0138958

Beattie AJ, Hay M, Magnusson B, de Nys R, Smeathers J, Vincent JFV (2011) Ecology and bioprospecting. Austral Ecol 36:341–356. https://doi.org/10.1111/j.1442-9993.2010.02170.x

Bhat SK, Joshi MB, Ullah A, Masood R, Biligiri SG, Arni RK, Satyamoorthy K (2016) Serine proteinases from Bothrops snake venom activates PI3K/Akt mediated angiogenesis. Toxicon 124:63–72

Bhutia SK, Panda PK, Sinha N, Praharaj PP, Bhol CS, Panigrahi DP, Mahapatra KK, Saha S, Patra S, Mishra SR, Behera BP, Patil S, Maiti TK (2019) Plant lectins in cancer therapeutics: Targeting apoptosis and autophagy-dependent cell death. Pharmacol Res 144:8–18. https://doi.org/10.1016/j.phrs.2019.04.001

Bierdeman MA, Torres AM, Caballero AR, Tang A, O'Callaghan RJ (2017) Reactions with antisera and pathological effects of Staphylococcus aureus gamma-toxin in the cornea. Curr Eye Res 42:1100–1107. https://doi.org/10.1080/02713683.2017.1279636

Bortolotti M, Biscotti F, Zanello A, Polito L, Bolognesi A (2023) Heterophyllin: a new adenia toxic lectin with peculiar biological properties. Toxins (Basel) 16. https://doi.org/10.3390/toxins16010001

Bross PF, Beitz J, Chen G, Chen XH, Duffy E, Kieffer L, Roy S, Sridhara R, Rahman A, Williams G, Pazdur R (2001) Approval summary: gemtuzumab ozogamicin in relapsed acute myeloid leukemia. Clin Cancer Res 7:1490–1496

Burris HA 3rd, Rugo HS, Vukelja SJ, Vogel CL, Borson RA, Limentani S, Tan-Chiu E, Krop IE, Michaelson RA, Girish S, Amler L, Zheng M, Chu Y-W, Klencke B, O'Shaughnessy JA (2011) Phase II study of the antibody drug conjugate trastuzumab-DM1 for the treatment of human epidermal growth factor receptor 2 (HER2)-positive breast cancer after prior HER2-directed therapy. J Clin Oncol 29:398–405. https://doi.org/10.1200/JCO.2010.29.5865

Cao J, Zhang F, Xiong W (2023) Discovery of aptamers and the acceleration of the development of targeting research in ophthalmology. Int J Nanomed 18:4421–4430. https://doi.org/10.2147/IJN.S418115

Carlini CR, Guimarães JA (1981) Isolation and characterization of a toxic protein from Canavalia ensiformis (jack bean) seeds, distinct from concanavalin A. Toxicon 19:667–675. https://doi.org/10.1016/0041-0101(81)90104-5

Chan YS, Yu H, Xia L, Ng TB (2015) Lectin from green speckled lentil seeds (Lens culinaris) triggered apoptosis in nasopharyngeal carcinoma cell lines. Chin Med 10:25. https://doi.org/10.1186/s13020-015-0057-6

Chanan-Khan AA, Jagannath S, Heffner LT, Avigan D, Lee KP, Lutz RJ, Haeder T, Ruehle M, Uherek C, Wartenberg-Demand AB (2009) Phase I study of BT062 given as repeated single dose once every 3 weeks in patients with relapsed or relapsed/refractory multiple myeloma. Blood 114(22):1862

Chandra Sekhar J, Sandhya S, Ravindran V, Banji D, Kommu S, Rsnakk C (2012) Plant toxins-useful and harmful effects. Hygeia J Drugs Med 44:79–90

Che Y-J, Ren X-H, Wang Z-W, Wu Q, Xing K, Zhang M, Xu C, Han D, Yuan S, Zheng S-H, Chen Y-Y, Liao X-R, Shi F, Zhong X-H, Cai X, Cheng S-X (2023) Lymph-node-targeted drug delivery for effective immunomodulation to prolong the long-term survival after heart transplantation. Adv Mater 35:e2207227. https://doi.org/10.1002/adma.202207227

Chen W-S, Cao Z, Leffler H, Nilsson UJ, Panjwani N (2017) Galectin-3 inhibition by a small-molecule inhibitor reduces both pathological corneal neovascularization and fibrosis. Invest Ophthalmol Vis Sci 58:9–20. https://doi.org/10.1167/iovs.16-20009

Christiansen SP, Sandnas A, Prill R, Youle RJ, McLoon LK (2000) Acute effects of the skeletal muscle-specific immunotoxin ricin-mAb 35 on extraocular muscles of rabbits. Invest Ophthalmol Vis Sci 41:3402–3409

Christiansen SP, Becker BA, Iaizzo PA, McLoon LK (2003) Extraocular muscle force generation after ricin-mAb35 injection: implications for strabismus treatment. J. AAPOS Off. Publ Am Assoc Pediatr Ophthalmol Strabismus 7:1–6. https://doi.org/10.1067/mpa.2003.S1091853103000569

Citores L, Iglesias R, Ferreras JM (2021) Antiviral activity of ribosome-inactivating proteins. Toxins (Basel) 13. https://doi.org/10.3390/toxins13020080

Coleman RL, Lorusso D, Gennigens C, González-Martín A, Randall L, Cibula D, Lund B, Woelber L, Pignata S, Forget F, Redondo A, Vindeløv SD, Chen M, Harris JR, Smith M, Nicacio LV, Teng MSL, Laenen A, Rangwala R, Manso L, Mirza M, Monk BJ, Vergote I, Raspagliesi F, Melichar B, Gaba Garcia L, Jackson A, Henry S, Kral Z, Harter P, De Giorgi U, Bjurberg M, Gold M, O'Malley D, Honhon B, Vulsteke C, De Cuypere E, Denys H, Baurain J-F, Zamagni C, Tenney M, Gordinier M, Bradley W, Schlumbrecht M, Spirtos N, Concin N, Mahner S, Scambia G, Leath C, Farias-Eisner R, Cohen J, Muller C, Bhatia S (2021) Efficacy and safety of tisotumab vedotin in previously treated recurrent or metastatic cervical cancer (innovaTV 204/GOG-3023/ENGOT-cx6): a multicentre, open-label, single-arm, phase 2 study. Lancet Oncol 22:609–619. https://doi.org/10.1016/S1470-2045(21)00056-5

Dajcs JJ, Thibodeaux BA, Girgis DO, O'Callaghan RJ (2002) Corneal virulence of Staphylococcus aureus in an experimental model of keratitis. DNA Cell Biol 21:375–382. https://doi.org/10.1089/10445490260099656

Dal Peraro M, van der Goot FG (2016) Pore-forming toxins: ancient, but never really out of fashion. Nat Rev Microbiol 14:77–92. https://doi.org/10.1038/nrmicro.2015.3

Dempsey MP, Conrady CD (2023) The host–pathogen interplay: a tale of two stories within the cornea and posterior segment. Microorganisms. https://doi.org/10.3390/microorganisms11082074

Deverall BJ (1977) Defence mechanisms of plants, Cambridge monographs in experimental biology. Cambridge University Press, Cambridge

Eaton JS, Miller PE, Mannis MJ, Murphy CJ (2015) Ocular adverse events associated with antibody-drug conjugates in human clinical trials. J Ocul Pharmacol Ther 31:589–604. https://doi.org/10.1089/jop.2015.0064

Egli M, Manoharan M (2023) Chemistry, structure and function of approved oligonucleotide therapeutics. Nucleic Acids Res 51:2529–2573. https://doi.org/10.1093/nar/gkad067

Endo Y, Mitsui K, Motizuki M, Tsurugi K (1987) The mechanism of action of ricin and related toxic lectins on eukaryotic ribosomes. The site and the characteristics of the modification in 28 S ribosomal RNA caused by the toxins. J Biol Chem 262:5908–5912

Escoubas P, Bosmans F (2007) Spider peptide toxins as leads for drug development. Expert Opin Drug Discov 2:823–835. https://doi.org/10.1517/17460441.2.6.823

Fachim HA, Mortari MR, Gobbo-Netto L, Dos Santos WF (2015) Neuroprotective activity of parawixin 10, a compound isolated from Parawixia bistriata spider venom (Araneidae: Araneae) in rats undergoing intrahippocampal NMDA microinjection. Pharmacogn Mag 11:579

Fan J-M, Luo J, Xu J, Zhu S, Zhang Q, Gao D-F, Xu Y-B, Zhang G-P (2008) Effects of recombinant MAP30 on cell proliferation and apoptosis of human colorectal carcinoma LoVo cells. Mol Biotechnol 39:79–86. https://doi.org/10.1007/s12033-008-9034-y

Feng H, Zhao L, Bai Z, Xin Z, Wang C, Liu L, Song J, Zhang H, Bai Y, Feng F (2023) Aptamer modified Zr-based porphyrinic nanoscale metal-organic frameworks for active-targeted chemo-photodynamic therapy of tumors. RSC Adv 13:11215–11224. https://doi.org/10.1039/d3ra00753g

Fini ME, Bauskar A, Jeong S, Wilson MR (2016) Clusterin in the eye: An old dog with new tricks at the ocular surface. Exp Eye Res 147:57–71. https://doi.org/10.1016/j.exer.2016.04.019

Fleckenstein M, Mitchell P, Freund KB, Sadda S, Holz FG, Brittain C, Henry EC, Ferrara D (2018) The progression of geographic atrophy secondary to age-related macular degeneration. Ophthalmology 125:369–390. https://doi.org/10.1016/j.ophtha.2017.08.038

Fontana ACK, Guizzo R, Beleboni RDO, Meirelles e Silva AR, Coimbra NC, Amara SG, dos Santos WF, Coutinho-Netto J (2003) Purification of a neuroprotective component of Parawixia bistriata spider venom that enhances glutamate uptake. Br J Pharmacol 139:1297–1309

Gautam AK, Sharma D, Sharma J, Saini KC (2020) Legume lectins: Potential use as a diagnostics and therapeutics against the cancer. Int J Biol Macromol 142:474–483. https://doi.org/10.1016/j.ijbiomac.2019.09.119

Ghazarian H, Idoni B, Oppenheimer SB (2011) A glycobiology review: carbohydrates, lectins and implications in cancer therapeutics. Acta Histochem 113:236–247. https://doi.org/10.1016/j.acthis.2010.02.004

Glenny AT, Stevens MF (1935) Staphylococcus toxins and antitoxins. J Pathol Bacteriol 40:201–210. https://doi.org/10.1002/path.1700400202

Gomez MV, Kalapothakis E, Guatimosim C, Prado MAM (2002) Phoneutria nigriventer venom: a cocktail of toxins that affect ion channels. Cell Mol Neurobiol 22:579–588

Göz D, Studholme K, Lappi DA, Rollag MD, Provencio I, Morin LP (2008) Targeted destruction of photosensitive retinal ganglion cells with a saporin conjugate alters the effects of light on mouse circadian rhythms. PLoS One 3:e3153. https://doi.org/10.1371/journal.pone.0003153

Gupta SK, Masinick S, Garrett M, Hazlett LD (1997) Pseudomonas aeruginosa lipopolysaccharide binds galectin-3 and other human corneal epithelial proteins. Infect Immun 65:2747–2753. https://doi.org/10.1128/iai.65.7.2747-2753.1997

Hasan A, Fortune F, Wilson A, Warr K, Lehner T, Sanderson J, Shinnick T, Mizushima Y, van der Zee R, Stanford M (1996) Role of γδ T cells in pathogenesis and diagnosis of Behçet's disease. Lancet 347:789–794. https://doi.org/10.1016/S0140-6736(96)90868-5

He Y-W, Guo C-X, Pan Y-F, Peng C, Weng Z-H (2008) Inhibition of hepatitis B virus replication by pokeweed antiviral protein in vitro. World J Gastroenterol 14:1592–1597. https://doi.org/10.3748/wjg.14.1592

Hoffmann RM, Mele S, Cheung A, Larcombe-Young D, Bucaite G, Sachouli E, Zlatareva I, Morad HOJ, Marlow R, McDonnell JM, Figini M, Lacy KE, Tutt AJN, Spicer JF, Thurston DE, Karagiannis SN, Crescioli S (2020) Rapid conjugation of antibodies to toxins to select candidates for the development of anticancer antibody-drug conjugates (ADCs). Sci Rep 10:8869. https://doi.org/10.1038/s41598-020-65860-x

Ikeda H, Hideshima T, Fulciniti M, Lutz RJ, Yasui H, Okawa Y, Kiziltepe T, Vallet S, Pozzi S, Santo L, Perrone G, Tai Y-T, Cirstea D, Raje NS, Uherek C, Dälken B, Aigner S, Osterroth F, Munshi N, Richardson P, Anderson KC (2009) The monoclonal antibody nBT062 conjugated to cytotoxic Maytansinoids has selective cytotoxicity against CD138-positive multiple myeloma cells in vitro and in vivo. Clin. cancer Res. an Off. J Am Assoc Cancer Res 15:4028–4037. https://doi.org/10.1158/1078-0432.CCR-08-2867

Johnson WL, Sohn M, Woeller CF, Wozniak RAF (2023) Staphylococcal enterotoxins promote virulence in bacterial keratitis. Invest Ophthalmol Vis Sci 64:5. https://doi.org/10.1167/iovs.64.5.5

Karlstetter M, Kopatz J, Aslanidis A, Shahraz A, Caramoy A, Linnartz-Gerlach B, Lin Y, Lückoff A, Fauser S, Düker K, Claude J, Wang Y, Ackermann J, Schmidt T, Hornung V, Skerka C, Langmann T, Neumann H (2017) Polysialic acid blocks mononuclear phagocyte reactivity, inhibits complement activation, and protects from vascular damage in the retina. EMBO Mol Med 9:154–166. https://doi.org/10.15252/emmm.201606627

Kim K-S, Cho H-S, Lee S-D, Kim K-H, Cho J-Y, Chung K-H, Lee Y-C, Moon S-K, Kim C-H (2005) Inhibitory effect of Buthus martensi Karsch extracts on interleukin-1β-induced expression of nitric oxide (NO) synthase and production of NO in human chondrocytes and LPS-induced NO and prostaglandin E2 production in mouse peritoneal macrophages. Toxicol In Vitro 19:757–769. https://doi.org/10.1016/j.tiv.2005.04.008

Kirschning A, Harmrolfs K, Knobloch T (2008) The chemistry and biology of the maytansinoid antitumor agents. Comptes Rendus Chim 11:1523–1543. https://doi.org/10.1016/j.crci.2008.02.006

Kocyigit E, Kocaadam-Bozkurt B, Bozkurt O, Ağagündüz D, Capasso R (2023) Plant toxic proteins: their biological activities, mechanism of action and removal strategies. Toxins (Basel) 15. https://doi.org/10.3390/toxins15060356

Koh DCI, Armugam A, Jeyaseelan K (2006) Snake venom components and their applications in biomedicine. Cell Mol Life Sci 63:3030–3041. https://doi.org/10.1007/s00018-006-6315-0

Kohli N, Westerveld DR, Ayache AC, Verma A, Shil P, Prasad T, Zhu P, Chan SL, Li Q, Daniell H (2014) Oral delivery of bioencapsulated proteins across blood-brain and blood-retinal barriers. Mol Ther 22:535–546. https://doi.org/10.1038/mt.2013.273

Kolar SS, Baidouri H, McDermott AM (2017) Role of pattern recognition receptors in the modulation of antimicrobial peptide expression in the corneal epithelial innate response to F. solani. Invest Ophthalmol Vis Sci 58:2463–2472. https://doi.org/10.1167/iovs.16-20658

Krop IE, Beeram M, Modi S, Jones SF, Holden SN, Yu W, Girish S, Tibbitts J, Yi J-H, Sliwkowski MX, Jacobson F, Lutzker SG, Burris HA (2010) Phase I study of trastuzumab-DM1, an HER2 antibody-drug conjugate, given every 3 weeks to patients with HER2-positive metastatic breast cancer. J Clin Oncol 28:2698–2704. https://doi.org/10.1200/JCO.2009.26.2071

Lambert JM, Berkenblit A (2018) Antibody-drug conjugates for cancer treatment. Annu Rev Med 69:191–207. https://doi.org/10.1146/annurev-med-061516-121357

Lambert WS, Ruiz L, Crish SD, Wheeler LA, Calkins DJ (2011) Brimonidine prevents axonal and somatic degeneration of retinal ganglion cell neurons. Mol Neurodegener 6:4. https://doi.org/10.1186/1750-1326-6-4

Lapusan S, Vidriales MB, Thomas X, de Botton S, Vekhoff A, Tang R, Dumontet C, Morariu-Zamfir R, Lambert JM, Ozoux M-L, Poncelet P, San Miguel JF, Legrand O, DeAngelo DJ, Giles FJ, Marie J-P (2012) Phase I studies of AVE9633, an anti-CD33 antibody-maytansinoid conjugate, in adult patients with relapsed/refractory acute myeloid leukemia. Invest New Drugs 30:1121–1131. https://doi.org/10.1007/s10637-011-9670-0

Larsen PP, Féart C, Pais de Barros J-P, Merle BMJ, Gayraud L, Delyfer M-N, Korobelnik J-F, Delcourt C (2023) Association of age-related macular degeneration with a blood biomarker of lipopolysaccharide, a gut bacterial proinflammatory toxin. Invest Ophthalmol Vis Sci 64:47. https://doi.org/10.1167/iovs.64.14.47

Lee AS, de Lencastre H, Garau J, Kluytmans J, Malhotra-Kumar S, Peschel A, Harbarth S (2018) Methicillin-resistant Staphylococcus aureus. Nat Rev Dis Prim 4:18033. https://doi.org/10.1038/nrdp.2018.33

Lehner T, Lavery E, Smith R, van der Zee R, Mizushima Y, Shinnick T (1991) Association between the 65-kilodalton heat shock protein, Streptococcus sanguis, and the corresponding antibodies in Behçet's syndrome. Infect Immun 59:1434–1441. https://doi.org/10.1128/iai.59.4.1434-1441.1991

Li Y, Li Y, Mengist HM, Shi C, Zhang C, Wang B, Li T, Huang Y, Xu Y, Jin T (2021) Structural basis of the pore-forming toxin/membrane interaction. Toxins (Basel) 13. https://doi.org/10.3390/toxins13020128

Li X, Zhang Z, Gao F, Ma Y, Wei D, Lu Z, Chen S, Wang M, Wang Y, Xu K, Wang R, Xu F, Chen J-Y, Zhu C, Li Z, Yu H, Guan X (2023) c-Myc-targeting PROTAC based on a TNA-DNA bivalent binder for combination therapy of triple-negative breast cancer. J Am Chem Soc 145:9334–9342. https://doi.org/10.1021/jacs.3c02619

Liang S, Zhang D, Pan X, Chen Q, Zhou P (1993) Properties and amino acid sequence of huwentoxin-I, a neurotoxin purified from the venom of the Chinese bird spider Selenocosmia Huwena. Toxicon 31:969–978. https://doi.org/10.1016/0041-0101(93)90256-I

Liang H, Brignole-Baudouin F, Labbé A, Pauly A, Warnet J-M, Baudouin C (2007) LPS-stimulated inflammation and apoptosis in corneal injury models. Mol Vis 13:1169–1180

Liberato JL, Cunha AOS, Mortari MR, Gelfuso ÉA, de Oliveira Beleboni R, Coutinho-Netto J, dos Santos WF (2006) Anticonvulsant and anxiolytic activity of FrPbAII, a novel GABA uptake inhibitor isolated from the venom of the social spider Parawixia bistriata (Araneidae: Araneae). Brain Res 1124:19–27

Lin Z, Tang Y, Chen Z, Li S, Xu X, Hou X, Chen Z, Wen J, Zeng W, Meng X, Fan H (2023) Soluble CD80 oral delivery by recombinant Lactococcus suppresses tumor growth by enhancing antitumor immunity. Bioeng Transl Med 8:e10533. https://doi.org/10.1002/btm2.10533

Lord JM (1985) Precursors of ricin and Ricinus communis agglutinin. Glycosylation and processing during synthesis and intracellular transport. Eur J Biochem 146:411–416. https://doi.org/10.1111/j.1432-1033.1985.tb08667.x

Lu J-Q, Zhu Z-N, Zheng Y-T, Shaw P-C (2020) Engineering of ribosome-inactivating proteins for improving pharmacological properties. Toxins (Basel) 12. https://doi.org/10.3390/toxins12030167

Lübbers J, Rodríguez E, van Kooyk Y (2018) Modulation of immune tolerance via siglec-sialic acid interactions. Front Immunol 9:2807. https://doi.org/10.3389/fimmu.2018.02807

Lund K, Bostad M, Skarpen E, Braunagel M, Kiprijanov S, Krauss S, Duncan A, Høgset A, Selbo PK (2014) The novel EpCAM-targeting monoclonal antibody 3-17I linked to saporin is highly cytotoxic after photochemical internalization in breast, pancreas and colon cancer cell lines. MAbs 6:1038–1050. https://doi.org/10.4161/mabs.28207

Ma D, Mellon J, Niederkorn JY (1998) Conditions affecting enhanced corneal allograft survival by oral immunization. Invest Ophthalmol Vis Sci 39:1835–1846

Madden KP, Clark WM, Marcoux FW, Probert AW Jr, Weber ML, Rivier J, Zivin JA (1990) Treatment with conotoxin, an 'N-type' calcium channel blocker, in neuronal hypoxic-ischemic injury. Brain Res 537:256–262

Maecker H, Jonnalagadda V, Bhakta S, Jammalamadaka V, Junutula JR (2023) Exploration of the antibody-drug conjugate clinical landscape. MAbs 15:2229101. https://doi.org/10.1080/19420862.2023.2229101

Maetzel D, Denzel S, Mack B, Canis M, Went P, Benk M, Kieu C, Papior P, Baeuerle PA, Munz M, Gires O (2009) Nuclear signalling by tumour-associated antigen EpCAM. Nat Cell Biol 11:162–171. https://doi.org/10.1038/ncb1824

Mahaling B, Low SWY, Ch S, Addi UR, Ahmad B, Connor TB, Mohan RR, Biswas S, Chaurasia SS (2023) Next-generation nanomedicine approaches for the management of retinal diseases. Pharmaceutics 15. https://doi.org/10.3390/pharmaceutics15072005

Markowska AI, Liu F-T, Panjwani N (2010) Galectin-3 is an important mediator of VEGF- and bFGF-mediated angiogenic response. J Exp Med 207:1981–1993. https://doi.org/10.1084/jem.20090121

Mashreghi M, Zamani P, Karimi M, Mehrabian A, Arabsalmani M, Zarqi J, Moosavian SA, Jaafari MR (2021) Anti-epithelial cell adhesion molecule RNA aptamer-conjugated liposomal doxorubicin as an efficient targeted therapy in mice bearing colon carcinoma tumor model. Biotechnol Prog 37:e3116. https://doi.org/10.1002/btpr.3116

Matkivska R, Samborska I, Maievskyi O (2024) Effect of animal venom toxins on the main links of the homeostasis of mammals (Review). Biomed Rep 20:16. https://doi.org/10.3892/br.2023.1704

McPherson SW, Heuss ND, Gregerson DS (2012) Regulation of CD8+ T cell responses to retinal antigen by local FoxP3+ regulatory T cells. Front Immunol 3:26171

Mebs D (1989) Snake venoms: toolbox of the neurobiologist. Endeavour 13:157–161. https://doi.org/10.1016/s0160-9327(89)80003-1

Mody R, Joshi SH, Chaney W (1995) Use of lectins as diagnostic and therapeutic tools for cancer. J Pharmacol Toxicol Methods 33:1–10. https://doi.org/10.1016/1056-8719(94)00052-6

Morimura A, Taniguchi M, Takei H, Sakamoto O, Naono N, Akeda Y, Onozuka D, Yoshimura J, Tomono K, Kutsuna S, Hamaguchi S (2024) Using novel micropore technology combined with artificial intelligence to differentiate Staphylococcus aureus and Staphylococcus epidermidis. Sci Rep 14:6994. https://doi.org/10.1038/s41598-024-55773-4

Mosinger M (1951) [Necrosing or clastic effects of ricin on different organs and on experimental sarcomas]. C R Seances Soc Biol Fil 145:412–415

Moya M, Dautry-Varsat A, Goud B, Louvard D, Boquet P (1985) Inhibition of coated pit formation in Hep2 cells blocks the cytotoxicity of diphtheria toxin but not that of ricin toxin. J Cell Biol 101:548–559. https://doi.org/10.1083/jcb.101.2.548

Narimatsu H, Sawaki H, Kuno A, Kaji H, Ito H, Ikehara Y (2010) A strategy for discovery of cancer glyco-biomarkers in serum using newly developed technologies for glycoproteomics. FEBS J 277:95–105. https://doi.org/10.1111/j.1742-4658.2009.07430.x

Olivera BM, Cruz LJ, de Santos V, LeCheminant GW, Griffin D, Zeikus R, McIntosh JM, Galyean R, Varga J, Gray WR (1987) Neuronal calcium channel antagonists. Discrimination between calcium channel subtypes using omega-conotoxin from Conus magus venom. Biochemistry 26:2086–2090. https://doi.org/10.1021/bi00382a004

Olsnes S (2004) The history of ricin, abrin and related toxins. Toxicon 44:361–370. https://doi.org/10.1016/j.toxicon.2004.05.003

Olsnes S, Pihl A (1973) Different biological properties of the two constituent peptide chains of ricin, a toxic protein inhibiting protein synthesis. Biochemistry 12:3121–3126. https://doi.org/10.1021/bi00740a028

Patel SS, Lally DR, Hsu J, Wykoff CC, Eichenbaum D, Heier JS, Jaffe GJ, Westby K, Desai D, Zhu L, Khanani AM (2023) Avacincaptad pegol for geographic atrophy secondary to age-related macular degeneration: 18-month findings from the GATHER1 trial. Eye 37:3551–3557. https://doi.org/10.1038/s41433-023-02497-w

Pegram M, Slamon D (2000) Biological rationale for HER2/neu (c-erbB2) as a target for monoclonal antibody therapy. Semin Oncol 27:13–19

Peters C, Brown S (2015) Antibody-drug conjugates as novel anti-cancer chemotherapeutics. Biosci Rep 35. https://doi.org/10.1042/BSR20150089

Petrillo F, Pignataro D, Di Lella FM, Reibaldi M, Fallico M, Castellino N, Parisi G, Trotta MC, D'Amico M, Santella B, Folliero V, Della Rocca MT, Rinaldi M, Franci G, Avitabile T, Galdiero M, Boccia G (2021) Antimicrobial susceptibility patterns and resistance trends of Staphylococcus aureus and coagulase-negative staphylococci strains isolated from ocular infections. Antibiotics (Basel, Switzerland) 10. https://doi.org/10.3390/antibiotics10050527

Piccaluga PP, Martinelli G, Rondoni M, Malagola M, Gaitani S, Visani G, Baccarani M (2004) First experience with gemtuzumab ozogamicin plus cytarabine as continuous infusion for elderly acute myeloid leukaemia patients. Leuk Res 28:987–990. https://doi.org/10.1016/j.leukres.2004.01.009

Pickering MC, Haskard DO (2000) Behçet's syndrome. J R Coll Physicians Lond 34:169–177

Pineda SS, Sollod BL, Wilson D, Darling A, Sunagar K, Undheim EAB, Kely L, Antunes A, Fry BG, King GF (2014) Diversification of a single ancestral gene into a successful toxin superfamily in highly venomous Australian funnel-web spiders. BMC Genomics 15:1–16

Pinheiro ACN, Gomez RS, Massensini AR, Cordeiro MN, Richardson M, Romano-Silva MA, Prado MAM, De Marco L, Gomez MV (2006) Neuroprotective effect on brain injury by neurotoxins from the spider Phoneutria nigriventer. Neurochem Int 49:543–547. https://doi.org/10.1016/j.neuint.2006.04.009

Polito L, Bortolotti M, Mercatelli D, Battelli MG, Bolognesi A (2013) Saporin-S6: a useful tool in cancer therapy. Toxins (Basel) 5:1698–1722. https://doi.org/10.3390/toxins5101698

Prasad MNV, Elchuri SV (2022) Pharmaceuticals and personal care products in the environment with emphasis on horizontal transfer of antibiotic resistance genes. Chemistry-Didactics-Ecology-Metrology 27:35–51. https://doi.org/10.2478/cdem-2022-0005

Puri M, Kaur I, Perugini MA, Gupta RC (2012) Ribosome-inactivating proteins: current status and biomedical applications. Drug Discov Today 17:774–783. https://doi.org/10.1016/j.drudis.2012.03.007

Rates SMK, Betti AH, Müller LG, Nunes JDM (2015) Plant toxins as sources of drugs. In: Gopalakrishnakone P, Carlini CR, Ligabue-Braun R (eds) Plant toxins. Springer, Dordrecht, pp 1–21. https://doi.org/10.1007/978-94-007-6728-7_5-1

Reis HJ, Prado MA, Kalapothakis E, Cordeiro MN, Diniz CR, De Marco LA, Gomez MV, Romano-Silva MA (1999) Inhibition of glutamate uptake by a polypeptide toxin (phoneutria-toxin 3-4) from the spider Phoneutria nigriventer. Biochem J 343(Pt 2):413–418
Shah S, Wozniak RAF (2023) Staphylococcus aureus and Pseudomonas aeruginosa infectious keratitis: key bacterial mechanisms that mediate pathogenesis and emerging therapeutics. Front Cell Infect Microbiol 13:1250257. https://doi.org/10.3389/fcimb.2023.1250257
Shahraz A, Kopatz J, Mathy R, Kappler J, Winter D, Kapoor S, Schütza V, Scheper T, Gieselmann V, Neumann H (2015) Anti-inflammatory activity of low molecular weight polysialic acid on human macrophages. Sci Rep 5:16800. https://doi.org/10.1038/srep16800
Shapira A, Benhar I (2010) Toxin-based therapeutic approaches. Toxins (Basel) 2:2519–2583. https://doi.org/10.3390/toxins2112519
Sheehan JR, Sadlier C, O'Brien B (2022) Bacterial endotoxins and exotoxins in intensive care medicine. BJA Educ 22:224–230. https://doi.org/10.1016/j.bjae.2022.01.003
Shen J, Attar M (2015) Antibody-drug conjugate (ADC) research in ophthalmology—a review. Pharm Res 32:3572–3576. https://doi.org/10.1007/s11095-015-1728-9
Shu N, Zhang Z, Wang X, Li R, Li W, Liu X, Zhang Q, Jiang Z, Tao L, Zhang L, Hou S (2023) Apigenin alleviates autoimmune uveitis by inhibiting microglia M1 pro-inflammatory polarization. Invest Ophthalmol Vis Sci 64:21. https://doi.org/10.1167/iovs.64.5.21
Stanford M, Whittall T, Bergmeier LA, Lindblad M, Lundin S, Shinnick T, Mizushima Y, Holmgren J, Lehner T (2004) Oral tolerization with peptide 336-351 linked to cholera toxin B subunit in preventing relapses of uveitis in Behcet's disease. Clin Exp Immunol 137:201–208. https://doi.org/10.1111/j.1365-2249.2004.02520.x
Stirpe F (1982) On the action of ribosome-inactivating proteins: are plant ribosomes species-specific? Biochem J. https://doi.org/10.1042/bj2020279
Subramanian N, Kanwar JR, Kanwar RK, Krishnakumar S (2015) Targeting cancer cells using LNA-modified aptamer-siRNA chimeras. Nucleic Acid Ther 25:317–322. https://doi.org/10.1089/nat.2015.0550
Subramanian N, Srimany A, Kanwar JR, Kanwar RK, Akilandeswari B, Rishi P, Khetan V, Vasudevan M, Pradeep T, Krishnakumar S (2016) Nucleolin-aptamer therapy in retinoblastoma: molecular changes and mass spectrometry-based imaging. Mol Ther Nucleic Acids 5:e358. https://doi.org/10.1038/mtna.2016.70
Sun T, Niu X, He Q, Liu M, Qiao S, Qi R-Q (2023) Development, efficacy and side effects of antibody-drug conjugates for cancer therapy (Review). Mol Clin Oncol 18:47. https://doi.org/10.3892/mco.2023.2643
Timar J, McIntosh DP, Henry R, Cumber AJ, Parnell GD, Davies AJ (1991) The effect of ricin B chain on the intracellular trafficking of an A chain immunotoxin. Br J Cancer 64:655–662. https://doi.org/10.1038/bjc.1991.377
Tolentino MJ, Tolentino AJ, Tolentino EM, Krishnan A, Genead MA (2023) Sialic acid mimetic microglial sialic acid-binding immunoglobulin-like lectin agonism: potential to restore retinal homeostasis and regain visual function in age-related macular degeneration. Pharmaceuticals (Basel) 16. https://doi.org/10.3390/ph16121735
Tsai I-H (1997) Phospholipases A2 of Asian snake venoms. J Toxicol Toxin Rev 16:79–113. https://doi.org/10.3109/15569549709016450
Tyagi N, Tyagi M, Pachauri M, Ghosh PC (2015) Potential therapeutic applications of plant toxin-ricin in cancer: challenges and advances. Tumour Biol 36:8239–8246. https://doi.org/10.1007/s13277-015-4028-4
Ulhuq FR, Mariano G (2022) Bacterial pore-forming toxins. Microbiology 168. https://doi.org/10.1099/mic.0.001154
Valentino K, Newcomb R, Gadbois T, Singh T, Bowersox S, Bitner S, Justice A, Yamashiro D, Hoffman BB, Ciaranello R (1993) A selective N-type calcium channel antagonist protects against neuronal loss after global cerebral ischemia. Proc Natl Acad Sci USA 90:7894–7897

Varki A (2011) Since there are PAMPs and DAMPs, there must be SAMPs? Glycan "self-associated molecular patterns" dampen innate immunity, but pathogens can mimic them. Glycobiology. https://doi.org/10.1093/glycob/cwr087

Veeck J, Dahl E (2012) Targeting the Wnt pathway in cancer: the emerging role of Dickkopf-3. Biochim Biophys Acta 1825:18–28. https://doi.org/10.1016/j.bbcan.2011.09.003

Wang Y-R, Liu R-Y, Wang L-C, Mao H-F, Chen J-Q (2007) Effect of Huwentoxin-I on the Fas and TNF apoptosis pathway in the hippocampus of rat with global cerebral ischemia. Toxicon 50:1085–1094

Wang T, Wang S-W, Zhang Y, Wu X-F, Peng Y, Cao Z, Ge B-Y, Wang X, Wu Q, Lin J-T, Zhang W-Q, Li S, Zhao J (2014) Scorpion venom heat-resistant peptide (SVHRP) enhances neurogenesis and neurite outgrowth of immature neurons in adult mice by up-regulating brain-derived neurotrophic factor (BDNF). PLoS One 9:e109977. https://doi.org/10.1371/journal.pone.0109977

Wang S, Li Z, Li S, Di R, Ho C-T, Yang G (2016) Ribosome-inactivating proteins (RIPs) and their important health promoting property. RSC Adv 6:46794–46805

Wang X, Zhang S, Zhu Y, Zhang Z, Sun M, Cheng J, Xiao Q, Li G, Tao J (2020) Scorpion toxins from Buthus martensii Karsch (BmK) as potential therapeutic agents for neurological disorders: state of the art and beyond. Med Toxicol. https://doi.org/10.5772/intechopen.90889

Wawrzynczak EJ, Watson GJ, Cumber AJ, Henry RV, Parnell GD, Rieber EP, Thorpe PE (1991) Blocked and non-blocked ricin immunotoxins against the CD4 antigen exhibit higher cytotoxic potency than a ricin A chain immunotoxin potentiated with ricin B chain or with a ricin B chain immunotoxin. Cancer Immunol Immunother 32:289–295. https://doi.org/10.1007/BF01789046

Weng A, Bachran C, Fuchs H, Krause E, Stephanowitz H, Melzig MF (2009) Enhancement of saporin cytotoxicity by Gypsophila saponins—more than stimulation of endocytosis. Chem Biol Interact 181:424–429. https://doi.org/10.1016/j.cbi.2009.07.007

Werner KA, Marquart L, Norton SA (2012) Lyngbya dermatitis (toxic seaweed dermatitis). Int J Dermatol 51:59–62. https://doi.org/10.1111/j.1365-4632.2011.05042.x

Wink M (1988) Plant breeding: importance of plant secondary metabolites for protection against pathogens and herbivores. Theor Appl Genet 75:225–233. https://doi.org/10.1007/BF00303957

Woll PJ, O'Brien M, Fossella F, Shah MH, Clinch Y, O'Keeffe J, Qin A, O'Leary J, Lorigan P (2010) Phase I study of lorvotuzumab mertansine (Imgn901) in patients with Cd56-positive solid tumours. J Clin Oncol 21:viii175–viii176

Wong WL, Su X, Li X, Cheung CMG, Klein R, Cheng C-Y, Wong TY (2014) Global prevalence of age-related macular degeneration and disease burden projection for 2020 and 2040: a systematic review and meta-analysis. Lancet Glob Health 2:e106-16. https://doi.org/10.1016/S2214-109X(13)70145-1

Wong KY, Liu Y, Zhou L, Wong M-S, Liu J (2023) Mucin-targeting-aptamer functionalized liposomes for delivery of cyclosporin A for dry eye diseases. J Mater Chem B 11. https://doi.org/10.1039/D3TB00598D

Yenari MA, Palmer JT, Guo HS, de Crespigny A, Moseley ME, Steinberg GK (1996) Time-course and treatment response with SNX-111, an N-type calcium channel blocker, in a rodent model of focal cerebral ischemia using diffusion-weighted MRI. Brain Res 739:36–45. https://doi.org/10.1016/S0006-8993(96)00808-6

Youssef DTA, Mufti SJ, Badiab AA, Shaala LA (2022) Anti-infective secondary metabolites of the marine Cyanobacterium Lyngbya morphotype between 1979 and 2022. Mar Drugs. https://doi.org/10.3390/md20120768

Zeng M, Zheng M, Lu D, Wang J, Jiang W, Sha O (2015) Anti-tumor activities and apoptotic mechanism of ribosome-inactivating proteins. Chin J Cancer 34:325–334. https://doi.org/10.1186/s40880-015-0030-x

Zhao L, Chen G, Li J, Fu Y, Mavlyutov TA, Yao A, Nickells RW, Gong S, Guo L-W (2017) An intraocular drug delivery system using targeted nanocarriers attenuates retinal ganglion

cell degeneration. J. Control. release Off. J Control Release Soc 247:153–166. https://doi.org/10.1016/j.jconrel.2016.12.038
Zhao P, Zhang Y, Li W, Jeanty C, Xiang G, Dong Y (2020) Recent advances of antibody drug conjugates for clinical applications. Acta Pharm Sin B 10:1589–1600. https://doi.org/10.1016/j.apsb.2020.04.012
Zhao X, Wu X, Wang H, Lai S, Wang J (2023) Targeted therapy for cisplatin-resistant lung cancer via aptamer-guided nano-zinc carriers containing USP14 siRNA. MedComm 4:e237. https://doi.org/10.1002/mco2.237
Zhu F, Zhou Y-K, Ji Z-L, Chen X-R (2018) The plant ribosome-inactivating proteins play important roles in defense against pathogens and insect pest attacks. Front Plant Sci 9:146. https://doi.org/10.3389/fpls.2018.00146
Zielinski MR, Gerashchenko D (2017) Sleep-inducing effect of substance P-cholera toxin A subunit in mice. Neurosci Lett 659:44–47

Chapter 14
Neurotoxins of Natural Origin

Isha Madaan, Jaspreet Kour, Shamsher Singh, Akanksha Sharma, Simranpreet Kaur, Manjot Kaur, Jahanvi, Jetina, Nivedita Sharma, Tarandeep Kaur, Renu Bhardwaj, and Geetika Sirhindi

Abstract Neurotoxins of natural origin wield formidable potential, showcasing a diverse array of compounds derived from plants, animals, and microorganisms that possess the ability to perturb normal neural function. This book chapter endeavors to explore the wide spectrum of neurotoxins found in nature, elucidating their intricate mechanisms of action and consequential effects on the nervous system. The chapter begins by offering an overview of neurotoxins' classification based on their sources, encompassing venoms, poisons, and other biological materials. It delves into the multifaceted chemical structures and intricate pathways through which these toxins exert their influence, disrupting neurotransmission, ion channels, and synaptic processes. Special emphasis is placed on their diverse physiological effects, including paralysis, convulsions, and neuromuscular blockade, highlighting the critical role these toxins play in predator–prey interactions, defense mechanisms, and pharmaceutical research.

Furthermore, this chapter explores case studies of prominent neurotoxins, such as botulinum toxin, tetanus toxin, tetrodotoxin, and α-latrotoxin, elucidating their specific mechanisms of action and potential therapeutic applications. The evolving landscape of research into harnessing these toxins for medicinal purposes and drug development is also discussed, shedding light on their promising roles in neurology,

I. Madaan
Department of Botany, Punjabi University, Patiala, Punjab, India

Science Department, Government College of Education, Jalandhar, Punjab, India

J. Kour · R. Bhardwaj
Department of Botanical & Environmental Sciences, Guru Nanak Dev University, Amritsar, Punjab, India

S. Singh
Government Brijindra College, Faridkot, Punjab, India

A. Sharma · S. Kaur · M. Kaur · Jahanvi · Jetina · N. Sharma · T. Kaur
Science Department, Government College of Education, Jalandhar, Punjab, India

G. Sirhindi (✉)
Department of Botany, Punjabi University, Patiala, Punjab, India
e-mail: geetika@pbi.ac.in

M. N. V. Prasad et al. (eds.), *Biotoxins*,
https://doi.org/10.1007/978-3-031-75309-1_14

pain management, and neuromodulation therapies. The comprehensive analysis of neurotoxins of natural origin provided in this chapter not only underscores their profound impact on the nervous system but also highlights their potential as valuable tools in understanding neural pathways and developing novel therapeutic interventions.

Keywords Neurotoxins · Neurology · Pharmaceutical · Plants · Animals · Microorganisms

14.1 Introduction

Neurotoxins, derived from natural sources, encompass a vast spectrum of compounds with the capacity to disrupt neural function and elicit physiological alterations within the nervous system. Their origins span the botanical, animal, and microbial realms, each offering distinct chemical compositions and modes of action that intricately interplay with the complex machinery of the nervous system. In this chapter, we embark on a journey through the multifaceted landscape of natural neurotoxins, beginning with an exploration of their classification into three primary categories: plant neurotoxins, animal neurotoxins, and microbial neurotoxins. The realm of plant neurotoxins unveils an extensive repertoire of botanical sources harboring potent neurotoxic compounds. From the lush foliage of rainforests to the arid deserts, plants have evolved an arsenal of chemical defenses to deter herbivores and fend off pathogens, many of which exhibit neurotoxic properties (Schiavo et al. 2000). Within this domain, we delve into the biochemical intricacies of these compounds, elucidating their effects on neural signaling pathways and their diverse applications in traditional medicine and modern pharmacology. From the iconic ricin derived from the castor bean plant to the intricate array of alkaloids found in deadly nightshade, the plant kingdom offers a rich tapestry of neurotoxic molecules ripe for exploration. Transitioning from plants to animals, we venture into the realm of animal neurotoxins, where venomous creatures wield an array of potent toxins as weapons for predation and defense (Polito et al. 2019). From the venoms of snakes and spiders to the toxins secreted by marine organisms such as cone snails and jellyfish, animal neurotoxins represent a fascinating convergence of evolutionary adaptations honed through millennia of ecological interactions. Here, we unravel the complex biochemical machinery underlying venom production and delivery, shedding light on the mechanisms by which these toxins target specific components of the nervous system, often with remarkable precision. Through this exploration, we gain insights into the diversity of animal neurotoxins and their implications for both basic research and therapeutic innovation.

Further the chapter delves into the realm of microbial neurotoxins, where bacteria, fungi, and other microorganisms produce an array of toxic compounds capable of perturbing neural function. From the infamous botulinum toxin produced by *Clostridium botulinum* to the neurotoxic peptides secreted by cyanobacteria,

microbial neurotoxins showcase the remarkable biochemical diversity harnessed by microorganisms to manipulate host physiology (Schantz and Johnson 1992). Here, we unravel the intricate mechanisms by which these toxins interact with neural receptors and ion channels, disrupting synaptic transmission and eliciting deleterious effects on nervous system function. Through comprehensive analysis, we aim to illuminate the role of microbial neurotoxins in disease pathogenesis and their potential as targets for therapeutic intervention.

Beyond classification, this chapter delves into the intricate mechanisms by which neurotoxins exert their effects on neural pathways, elucidating their biochemical and physiological impacts on nervous system function. By unraveling the molecular underpinnings of neurotoxin action, we lay the foundation for understanding their pharmacological and therapeutic applications (Gao et al. 2017). From the management of movement disorders with botulinum toxin injections to the treatment of chronic pain with conotoxins, natural neurotoxins offer promising avenues for therapeutic innovation, albeit with challenges and considerations for future research. In the realm of public health, the impact of neurotoxins on human health cannot be overstated. From foodborne illnesses caused by microbial toxins to envenomations resulting from animal bites and stings, neurotoxic exposures pose significant risks to human well-being. Here, we explore the implications of neurotoxin exposure for public health and elucidate strategies for their management, treatment, and prevention. Through interdisciplinary collaboration and ongoing research efforts, we endeavor to mitigate the adverse effects of neurotoxins and safeguard human health in an ever-evolving landscape of natural hazards.

In summation, this chapter serves as a comprehensive guide to the world of natural neurotoxins, offering insights into their classification, mode of action, pharmacological applications, and impact on human health. Through meticulous exploration and analysis, we aim to deepen our understanding of these fascinating compounds and their roles in shaping the intricate tapestry of life on Earth. Neurotoxins are the toxins which hinder the transmission of nerve impulse from one place to another. They act as both postsynaptically and presynaptically.

14.2 Classification of Neurotoxins

Neurotoxins, originating from a diverse array of natural sources, represent a complex class of compounds with the remarkable ability to disrupt neural function and induce physiological alterations within the intricate network of the nervous system. As agents of toxicity, these compounds have garnered significant attention from researchers, clinicians, and public health officials due to their potential impacts on human health and the environment. Understanding the classification of neurotoxins provides a foundational framework for elucidating their mechanisms of action, pharmacological applications, and implications for human health and well-being. Neurotoxins, originating from a variety of natural sources, can be classified into distinct categories based on their origins and biological properties, i.e., plant,

animal, and microbial neurotoxins. This classification serves as a framework for understanding the diverse array of compounds that impact neural function.

14.2.1 Plant Neurotoxins

Plant toxins are compounds derived from plants, which can be either naturally occurring low-molecular-weight toxins or by-products of secondary metabolism. These secondary metabolites are specific to every plant species and contribute to their distinctive traits, such as pigmentation, flavors, and protective compounds. In general, these secondary metabolites are not directly involved in growth of plant but produced in response to environmental stress. But, there is also another group of plant chemicals that are important for moving resources around the plant when the environment changes. These can include chemicals that were previously thought of as either primary or secondary metabolites. Nervous system affecting compounds belong to this mixed group and this includes a wide range of substances like indoleacetic acid, caffeine, epinephrine, acetylcholine, opiates, nicotine, levodopa, serotonin, melatonin, hyperforin, and others. Melatonin, a common neurotransmitter in humans (scientifically called *N*-acetyl-5-methoxytryptamine), is found in many organisms, including plants, and helps control daily rhythms or body clocks. Other substances like hyperforin, found in different plants, may work in human brains and possibly in plants too, by either blocking the movement of serotonin (a mood-related chemical) or by triggering certain ion channels, which are like gates for ions in cells. Recently researchers have begun to hypothesize a role for neurological signaling molecules in cellular communications in plants. Neurotransmitters that are produced by plants include acetylcholine, γ-aminobutyric acid (GABA), catecholamines (Świędrych et al. 2004), indoleamines, and auxin (Murch et al. 2000). In animals, these compounds are stored in neurons and released into the synaptic space, in response to a stimulus. In plant cells, acetylcholine is produced in the stems, leaves, nodes, and roots in response to heat shock (Momonoki and Momonoki 1991). The neurotransmitter GABA accumulates in plants following stress and is involved in pH regulation, nitrogen storage, plant development and defense against insects (Shelp et al. 2003). Catecholamines are also found at elevated levels in plant tissues exposed to stress, but the biochemical mechanisms associated with these observations have yet to be elucidated (Świędrych et al. 2004). There are currently more than 12,000 papers on the presence of melatonin in diverse species ranging from mammals, insects, planarians, mollusks, dinoflagellates, and algae to grains, vegetables, and medical plants. The first identifications of melatonin in plant tissues were made in 1995 (Dubbels et al. 1995) and the discovery of melatonin in intact plants was fairly recent (Murch et al. 1997, 2000). Early research questioned whether melatonin found in plants was actually produced by the plant tissues or was a by-product of plant–microbe interactions. In studies with axenic cultures of plant cells and sterile cultures of plantlets, the first evidence of a putative biosynthetic pathway for melatonin in higher plants has been

provided (Murch et al. 2000). Till date, none of the enzymes as melatonin biosynthesis have been identified in plants, no genes have been cloned, and no receptors for indoleamines have been isolated. It has been hypothesized that the role of melatonin in plants may be analogous to its function in mammals as a chemical signal of photoperiod as a calmodulin binding factor or as an antioxidant (Balzer and Hardeland 1996). Preliminary evidence has indicated that the rate of melatonin biosynthesis may indeed follow a circadian rhythm in plants, with the highest level in dark periods and the lowest level during daylight (Balzer and Hardeland 1996). In the dynoflagellate, *Gonyaulax polyedra*, melatonin mimicked the classic short-day response that is manifested in the formation of asexual cysts (Hardeland 1993). A similar rhythmic fluctuation in the content of melatonin in the short-day plant *Chenopodium rubrum* has been reported (Kolář et al. 1997). Neuroregulators are compounds that alter the function of neurons through either stimulation or inhibition of response. Neuroregulation can occur in two ways, viz., (a) alteration of synaptic activity by altering the effectiveness of a neurotransmitter or (b) alteration of neurotransmitter concentrations by alteration of synthesis, release, receptor interactions, reuptake, or metabolism. One of the most commonly known neuroregulatory compounds from plants is caffeine. It is thought that caffeine is produced by plants to act as a chemical defense against predators such as insect larvae (Nathanson 1984). Caffeine may be released from seed coats into soil to inhibit germination of other seeds (Suzuki and Waller 1987). These data provide evidence that neurologically active compounds, like caffeine, produced by plant cells may serve important functions for the plants as well as being useful for humans. Within the class of neuroregulatory molecules, there are a number of compounds that overstimulate human neurons, eventually resulting in neuronal cell death. Neurotoxins in plants fall into several complex categories that include fairly common compounds like auxins and the more exotic compounds like djenkolic acid and mimosine from Mimosa pudica (Murch et al. 2004). Recent research has begun to investigate the role of neurotoxic amino acids in higher plants. BOAA (Beta-N-OxalylAmino-L-Alanine) accumulates in plant tissues and seeds under water stress (Zhang et al. 2005). BMAA (β-MethylAmino-L-alanine) is an effective agonist of glutamate receptors in plant cells and has been used to generate mutants of Arabidopsis with differing glutamate receptor sensitivity (Brenner et al. 2000). Arabidopsis seedlings grown on a medium supplemented with BMAA have an etiolated phenotype. Hypocotyls were elongated in the light-grown Arabidopsis seedlings (Brenner et al. 2000). The hypocotyl elongation and etiolated phenotype could partially be reversed by supplementation of the BMAA medium with glutamate. These data provide the first evidence of interactions between the neurotoxic amino acid and receptor functions in higher plants.

14.2.2 Animal Neurotoxins

Animal neurotoxins are venomous compounds produced by a variety of animals, including snakes, spiders, scorpions, and marine organisms. These toxins are primarily used for predation, defense, and competition within ecological niches. Animal neurotoxins target specific components of the nervous system, such as ion channels, neurotransmitter receptors, and synaptic proteins, leading to disruption of neural signaling and paralysis in prey or predators (Postma 2009). Examples of animal neurotoxins include snake venom neurotoxins, such as α-neurotoxins and β-neurotoxins, which target acetylcholine receptors, and conotoxins from cone snails, which modulate ion channel function (Table 14.1). The study of animal neurotoxins not only provides insights into predator–prey interactions but also offers opportunities for the development of novel therapeutics and pharmacological tools. TTX is a non-protein, low-molecular-weight neurotoxin. **TTX** was initially thought to exist only in ***Tetraodontidae pufferfish***, and there is a debate as to whether **TTX** in fish is endogenous (produced by the pufferfish itself) or exogenous (taken from the outside and accumulated). The use of **TTX** preserves its mechanism of action primarily by blocking voltage-gated sodium channel (VGSCs), and thus altering neuronal activity. Blockade of VGSCs converts these toxic substances into effective therapeutic agents, especially for pain. **Tetrodotoxin** is a poisonous substance that reaches the land and water surfaces and affects many taxa, from bacteria to

Table 14.1 Venomous creatures and their neurotoxic components and their effects

Name of venomous creatures	Neurotoxic components	Their effects
1. Snake venoms	• Hyaluronidase (present in all snakes) • Elapids (present in all snakes) • Beta-bungarotoxin • Cardiotoxins	• Helps in distribution of other venom components throughout tissues of prey • Causes death of individual • Causes neuromuscular block • Augmentation of myocardial contraction at low concentration and systolic arrest at high concentration
2. Spider venoms	• Alpha-latrotoxin (found in black widow spider) • Agatoxins (found in funnel web spider)	• Causes muscle spasms, respiratory embarrassment, vomiting, and hypertension • Causes numbness, nausea, vomiting, abdominal colic, sweating, and salivation
3. Scorpion venoms	• Alpha scorpion toxin • Beta scorpion toxin	• Delays inactivation of sodium channel and thus prolongs action potential, autonomic nervous system excitation such as dilatation of pupil • Affect activation in addition to slowing inactivation of sodium channel, vomiting, diarrhea
4. Sea snail: cone shell	• Alpha, mu, omega conotoxin	• Causes respiratory paralysis in humans
5. Corals: blue jelly fish	• Discharge stinging capsules nematocysts	• Causes cardiorespiratory failure

vertebrates (Lago et al. 2015). **Saxitoxin** (STX) is a guanidinium salt shellfish toxin which is similar to TTX. It exhibits good inhibitory activity on TTX-sensitive VGSC; STX has more than 50 natural analogs that exhibit VGSC inhibitory similar to STX. Like zetekitoxin AB also inhibits TTX-resistant VGSCs as well as cardiac isoform NaV1.5. Tetrodotoxin has evaluated at preclinical and clinical levels for the treatment of neuropathy or cancer, showing good efficacy and safety. Effects of TTX in cancer pain is tested in only one study or TTX has not been evaluated in some models of neuropathy such as diabetic neuropathy (González-Cano et al. 2021).

14.2.3 Microbial Neurotoxins

Various microbial neurotoxins are being used to understand the physiology of the nervous system and may have the potential value in the treatment of certain types of muscular disorders through the modification of nervous stimulation of muscle activity. Neurotoxins have highly specific actions on molecular targets and therefore offer an effective means of characterizing the growing number of identified and channels and receptors in nervous system.

Clostridium botulinum and *Clostridium tetani* secrete very potent neurotoxins that lead to several neurological disorders in humans. Botulinum neurotoxins (BoNTs) are produced by anaerobic, spore-forming bacteria ***Clostridium botulinum***, ***Clostridium butyricum***, ***Clostridium baratii***, and ***Clostridium argentinensis***, as well as some other species (Aguilar 2018). Although all BoNTs consist of two peptide chains linked by disulphide bonds, the amino acid sequence of various peptide chains found in each BoNT subtype is different. The first botulinum toxin approved for clinical use in humans was licensed in United States in 1989. There are currently six US FDA-approved models on the market. There are some minor differences in toxicological properties of the preparation and it is best not to use them interchangeably (Choudhury et al. 2021). BoNT-A is effective in primary and secondary dystonia. But the latter will require special treatment to prevent complication such as death and Wilson's disease (Tyślerowicz et al. 2020). Originally, in the 1970s, BoNT was used to treat strabismus, but today BoNT has many medical uses such as treatment of muscle hyperactivity including tics, essential tremor, cerebral palsy, secretory disorders, and pain syndromes like chronic migraine. Today, advances in molecular biology have enabled the development of BoNTs that have potential to treat many diseases. BoNTs form botulinum complexes, also known as progenitor toxin complexes (PTCs), through non-covalent binding to various non-toxin proteins. Another important thing in future is optogenetic technology for local manipulation of photoactivatable BoNTs, indicating controlled neurotransmission (Rasetti-Escargueil and Popoff 2020). Although BoNT can be life-threatening, it is a versatile therapeutic agent. However, the most potent chemical, BoNT is a drug of choice in United States, placing *C. botulinum* research under strict government regulation. BoNT's toxicity, impact on public safety, and many medical uses will soon require the development of safe solutions to expand *C. botulinum* research

(Nowakowska et al. 2022). There are thousands of species capable of producing toxins, most of which are neurotoxins. After millions of years of evolution, biotoxins have selectivity to interfere with physiological function and discompose synaptic transmission (Zhou et al. 2024). Cervical dystonia (CD) is the most difficult to treat with botulinum neurtoxin (BoNT). However, for many patients, this recovery is often negative and unsatisfactory. Since the beginning of BoNT treatment, injections conducted taking into consideration anatomical landmarks (Sławek and Jost 2021). Botulinum neurotoxin type A (BoNT-A) has been the standard treatment for hypertonia in cerebral palsy for more than 20 years. Currently, evidence has developed to support its use in ambulant and non-ambulant children with cerebral palsy (Paget et al. 2018).

Tetanus is a disease caused by **toxins** produced by ***Clostridium tetani*** bacteria. This potent **neurotoxin** blocks central synaptic transmission, causing muscle stiffness and spasm and in severe cases, instability of the cardiac system (Lu et al. 2022). Additionally, the use of **TeNTs** (Tetanus toxin) may be useful for research to better understand the effects associated with skeletal muscle contractions and lateral/collateral alternative body movement. Labeling with new bright fluorescent probes using together with **TeNT** derivatives may help identify and characterize the spinal cord and brainstem neuronal circuits involved in neurophysiology and pathogenesis of tetanus (Megighian et al. 2021). ***C. tetani*** is a homogeneous bacterial species which produce only one type of tetanus toxins whereas botulinum neurotoxin producing strains are heterogeneous. *C. botulinum* is divided into four groups which on the basis of phenotypic and genotypic parameters, corresponds to different species. In addition, some strains of other species such as *Clostridium butyricum* and *Clostridium baratii* can produce related botulinum neurotoxin type E and F, respectively. This toxin has been found to be a typical protein exhibiting no unique composition or physicochemical properties and is apparently devoid of a prosthetic group. Botulinum toxin exerts its action on a muscle indirectly by blocking the release of the neurotransmitter acetylcholine at the nerve ending, resulting in reduced muscle activity or paralysis. *Streptococcus pneumoniae* (*or Pneumococcus*) is the most common cause of bacterial meningitis, which is characterized by neuronal cell death in the hippocampus as well as by an inflammatory response. Pneumolysin (PLY) produced by ***S. pneumoniae***, is one of the virulence factors leading to neuronal cell damages and which also causes injury to pulmonary alveolar epithelial cells

Staphylococci produce a whole range of toxins and extracellular enzymes, such as proteases, a hyaluronidase, a lipase, and a nuclease that facilitate tissue destruction and spreading, membrane-damaging toxins that cause cytolytic effects on host cells and tissue damage. They are very potent and neurospecific, they have a limited diffusion when locally injected, and their action is reversible with time. All types of neurotoxins are synthesized as single chain protein molecule of about 1,50,000 molecular weight with the low toxicity (Ek et al. 2012). *Clostridium botulinum* strains were cleaved by extracellular proteases into two chain molecules consisting of heavy subunit of about 1,00,000 molecular weight and light subunit of about 50,000 molecular weight. Tetanus toxins are produced

by an anaerobic spore-forming rod bacteria that has similar morphology to *Clostridium W botulinum*. Unlike botulinum, Tetanus toxin can enter into central nervous system by retrograde intra-axonal transport through motor nerves (Rasetti-Escargueil and Popoff 2020). Seven serologically distinct toxins have been identified (types A–G) and, of these, botulinum toxin A has been approved for clinical use to treat blepharospasm, strabismus, cervical dystonia, primary axillary hyperhidrosis, and moderate-to-severe glabellar frown lines. Botulinum toxin B has been approved for the treatment of cervical dystonia. Botulinum toxins may also be employed as biological weapons. Botulism and tetanus are responsible for severe neurological diseases, i.e. while BoNTs cause flaccid paralysis, TeNT causes spastic paralysis (Cai et al. 2021).

14.3 Mode of Action of Neurotoxins

Neurotoxins are substances that can cause damage to the nervous system, leading to adverse effects on neural pathways and overall neurological functions (Cao et al. 2021). The mode of action of neurotoxins can vary depending on the specific toxin, but they generally interfere with normal neural signaling and can lead to a range of symptoms and health issues (Zhou et al. 2024).

14.3.1 Pathway of Neurotoxins

The pathway of neurotoxins involves the entry of these toxic substances into the body, their distribution through various biological systems (Bellinger 2008), and their specific interactions with components of the nervous system (National Research Council (US) Committee 1992). Here is a general outline of the pathway of neurotoxins (Fig. 14.1):

(i) **Entry into the body:** Neurotoxins can enter the body through multiple routes, including ingestion, inhalation, and skin absorption (Boyes and van Thriel 2020). Contaminated food, water, air, or exposure to toxic substances in the environment are common sources of neurotoxin entry (Cardenas-Iniguez et al. 2022).

(ii) **Absorption and distribution:** Once neurotoxins enter the body, they may be absorbed into the bloodstream (Popoff and Poulain 2010). The mode of absorption depends on the chemical properties of the neurotoxin. Some neurotoxins are absorbed through the gastrointestinal tract, while others may enter through the respiratory system or skin. Neurotoxins are then distributed throughout the body via the bloodstream. Circulation carries these toxic substances to various organs and tissues, including the nervous system (Spencer and Lein 2014).

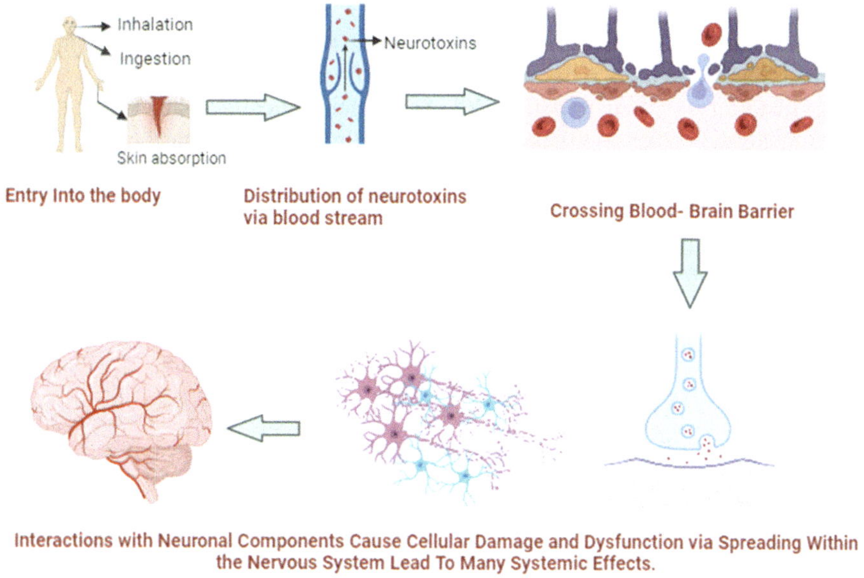

Fig. 14.1 A general outline of the pathway of neurotoxins

(iii) **Blood-brain barrier crossing:** To affect the nervous system, neurotoxins often need to cross the blood-brain barrier (Wu et al. 2021), a protective barrier that limits the passage of substances from the bloodstream into the brain (Furtado et al. 2018). Some neurotoxins have specific mechanisms that allow them to cross the blood-brain barrier and directly impact neural tissues (Chiodo et al. 2007).

(iv) **Localization in neural tissues:** Once inside the brain or peripheral nerves, neurotoxins may accumulate in specific neural tissues (Calne 1991). The distribution pattern can vary depending on the chemical properties of the neurotoxin and its affinity for certain cell types or structures within the nervous system (Abbott et al. 2010).

(v) **Interactions with neuronal components:** Neurotoxins exert their toxic effects by interacting with various components of neurons (Popoff and Poulain 2010). This can include interference with neurotransmitter release, modulation of ion channels, inhibition of enzymes involved in neural function, disruption of mitochondrial activity, and other mechanisms outlined in the previous responses (Cao et al. 2021).

(vi) **Cellular damage and dysfunction:** The interactions between neurotoxins and neuronal components lead to cellular damage and dysfunction. Depending on the specific mechanisms involved, this damage can manifest as impaired signaling, cell death, or structural damage to neural tissues (Zhao et al. 2019).

(vii) **Spread within the nervous system:** Some neurotoxins may spread within the nervous system, affecting multiple areas and pathways. This spread can

contribute to the widespread impact of neurotoxic exposure (Spencer and Lein 2014).

(viii) **Systemic effects:** In addition to their specific effects on the nervous system, neurotoxins can also have systemic effects on other organs and tissues. This can lead to secondary health issues beyond neurological symptoms (Ge et al. 2005; Saubaméa et al. 2012).

Understanding the pathway of neurotoxins is essential for developing strategies to prevent exposure, diagnose toxicity, and design targeted interventions to mitigate their effects on the nervous system (Lanphear et al. 2000). It also emphasizes the importance of environmental monitoring and regulatory measures to minimize the risk of neurotoxin exposure (Cao et al. 2021).

14.3.2 Biochemical and Physiological Impacts on Nervous System Functions

Here is a detailed elucidation of how neurotoxins affect neural pathways and the biochemical and physiological impact on nervous system functions:

(a) **Neural signal transmission:** Neurotransmitters play a crucial role in transmitting signals between nerve cells (neurons) at synapses. Neurotoxins can disrupt this process by affecting neurotransmitter release, binding, or reuptake (Zhou et al. 2022). Some neurotoxins block the release of neurotransmitters from the presynaptic neuron, preventing the transmission of signals to the postsynaptic neuron. Neurotoxins generally inhibit or promote the release of neurotransmitters or bind to receptors that are located in the pre- or post-synaptic membranes, thereby affecting physiological functions of synapses and affecting biological processes (Pirazzini et al. 2017).

(b) **Blood-brain barrier disruption:** Some neurotoxins can compromise the integrity of the blood-brain barrier, allowing harmful substances to enter the brain (Wu et al. 2021). This breach can lead to inflammation and further damage to neural tissues (Romero et al. 2023)

(c) **Neuroinflammation:** Neurotoxins can trigger inflammatory responses in the nervous system, contributing to additional damage and dysfunction (Cao et al. 2021).

(d) **Ion channel modulation:** Neurotoxins may target ion channels, which are crucial for the generation and propagation of action potentials along the neuron (Matsubara et al. 1993; Bao et al. 2009). Toxins can block ion channels, disrupting the normal flow of ions across the cell membrane and affecting the generation and propagation of action potentials (Cao et al. 2021).

(e) **Enzyme inhibition:** Certain neurotoxins act as enzyme inhibitors, disrupting key enzymes involved in neurotransmitter synthesis or degradation (Searle et al. 2014). Inhibition of enzymes can lead to the accumulation or depletion of

neurotransmitters, disrupting the delicate balance necessary for normal neural function (Cao et al. 2021).

(f) **Excitotoxicity:** Some neurotoxins induce excitotoxicity by overstimulating neurons (Wang and Qin 2010), leading to excessive release of neurotransmitters such as glutamate (Connolly et al. 2016). The excess glutamate can trigger a cascade of events, including increased intracellular calcium levels, activation of destructive enzymes, and ultimately cell death (Cabeza et al. 2021).

(g) **Mitochondrial dysfunction:** Neurotoxins may interfere with mitochondrial function, disrupting energy production and leading to oxidative stress. Mitochondrial dysfunction can contribute to neuronal damage and death (Zhao et al. 2019).

(h) **Axonal transport disruption:** Neurotoxins can interfere with the transport of cellular components along the axon, disrupting the normal functioning of neurons. This disruption can lead to the degeneration of axons and the overall impairment of neural communication (Cao et al. 2021).

(i) **Apoptosis and cell death:** Prolonged exposure to neurotoxins may induce apoptosis (programmed cell death) in neurons (Kobayashi et al. 2006). Cell death can lead to the loss of specific neural pathways, resulting in functional deficits in the affected areas of the nervous system (Lanphear et al. 2019).

Overall, neurotoxins exert their effects by disrupting various aspects of neural function, including neurotransmission, ion channel activity, enzyme function, mitochondrial health, and overall cellular integrity (Banks et al. 1997). The consequences of neurotoxin exposure can range from temporary impairment to permanent damage, depending on the specific toxin and the extent of exposure (Cao et al. 2021). Understanding the mode of action of neurotoxins is crucial for developing therapeutic interventions and preventive measures (Segura-Aguilar 2018)

14.4 Pharmacological and Therapeutic Applications

Toxins which have specific neuronal action have become topics of interest, due to their effects on fundamental processes with the capability for targeting or delivery of biocargo and therapeutic vectors to neurons. Various neurotoxins having pharmacological and therapeutic applications are Botulinum toxin, Tetrodotoxin, alpha-latrotoxin, tetanus toxin, dendrotoxin, conotoxin , and beta-bungarotoxin described as follows (Fig. 14.2).

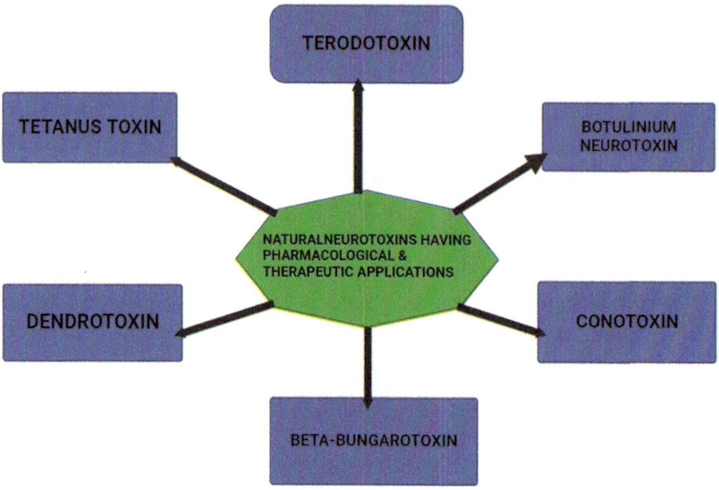

Fig. 14.2 Some of natural neurotoxins

14.4.1 Tetrodotoxin

Tetrodotoxin (TTX) is a toxic substance found in pufferfish, named after its family Tetraodontidae (Tahara and Hirata 1909). It has reached both terrestrial and aquatic environments also in different taxonomic groups from bacteria to vertebrates. TTX causes blockage of voltage-gated sodium channels (VGSCs). Moreover, it also causes blockage of L-type Ca2+ current in canine cardiac cells (Lago et al. 2015). Therefore, TTX is a promising tool as therapeutic drug especially used as analgesic for treatment of different types of pains (Lago et al. 2015), and researches are going on for its therapeutic use to treat cancer pains (Bane et al. 2014; Lago et al. 2015). Indeed, TTX does not pose any genotoxic risk to patients (Guzmán et al. 2007; Lago et al. 2015). In ancient times, globe fish (containing TTX) was used as Japanese traditional medicine to treat leprosy, since globe fish flesh alleviated neuralgia of patients affected with leprosy (Lago et al. 2015). Though, widespread usage of TTX in pharmacology still much of its potential and the antidote of TXX toxicity is yet to be discovered .

14.4.2 Botulinum Neurotoxins

Botulinum toxin also called "miracle poison" is one of the most poisonous biological compounds known, produced by the bacterium ***Clostridium botulinum***. Its mechanism of action is the blockage of neurotransmission by preventing the release of neurotransmitter; acetylcholine at neuromuscular junction that leads to muscle paralysis. Initially, Scott A.B. in 1981 demonstrated the effectiveness of botulinum

toxin type A for control of strabismus in humans. Later, FDA approved its usage for cosmetic purposes temporarily to reduce forehead frown lines. Currently, it is used in every sub-specialty of medicine. Botulinum neurotoxin plays a vital role in management of strabismus, hemifacial spasms, focal dystonia and different spastic movement disorders. Moreover, it is also used as an alternative to sphincterotomy in patients with chronic anal fissures (Maria et al. 1998) and achalasia. Not only this cosmetic use of BTX has widened its scope from correction of forehead lines to creases, wrinkling all over the face, neck, chest, and many more, to dermatological applications such as localized axillary or palmar hyperhidrosis. Injections of BTX cause mild pain, transient numbness, headache, malaise, or mild nausea. Its effect decreases with the increase in distance from the site of injection, the most serious side effect of it is temporary paralysis of nearby musculature caused by BTX. In addition to this cosmetic usage, toxins too have side effects as dysphagia because of diffusion of BTX into oropharynx (when injected in or near neck), ptosis that usually lasts for 2–6 weeks. Type A botulinum toxin has wide scope in medicinal chemistry but risk of developing antibodies against it decreases the repeated and high doses of it. Research is yet to be conducted related to the combination of type A and type F BTX that have been suggested to minimize total units and overall antigenic doses of it (Mahrhold et al. 2006).

14.4.3 Beta-Bungarotoxin

Animal venoms attracted researchers all around the world because of their richness, specialization, and effective components; most of these components are peptides having high selectivity and affinity for huge amount of targets such as ion channels, hemostatic pathways, enzymes, or various membrane receptors (Mohamed Abd El-Aziz et al. 2019). Various venomous organisms in the Animal kingdoms are snakes, scorpions, spiders, amphibians, cone snails and their venoms may be useful in biotechnological or pharmacological applications (Mohamed Abd El-Aziz et al. 2019). However, venomous snakes are rich sources of toxins among other venomous species (Tu 1996). In Ayurveda, cobra venom is used to treat joint pain, arthritis (Gomes et al. 2010), whereas in Chinese medicine it is used to treat opiate addiction. With the advancement in biotechnology and medicine, Italian naturalist Felice Fontana described the influence of snake venom on blood coagulation (Mohamed Abd El-Aziz et al. 2019) In blood coagulation process factors V, IX, and X along with Ca2+ ions, phospholipids activate prothrombin and induce fibrin clot formation . Therefore, the venom of Russell's viper (***Daboia russelli***) contains coagulation activators for factor V and factor X but not for factor IX (Mukherjee 2014), venom of ***Bothrops atrox*** (Hofmann and Bon 1987), king cobra (***Ophiophagus hannah***), banded krait (***Bungarus fasciatus***) contains coagulation activators for factor X that can be used as anticoagulants (Mohamed Abd El-Aziz et al. 2019). In

addition to this, cytotoxin present in snake venom interacts with lipoproteins present in the plasma membrane of cells to cause shrinkage of the cell ultimately leading to its death (Chakrabarty and Sarkar 2017; Kerkkamp et al. 2018). This cytotoxic effect can be used to destroy or degrade tumor cells; for instance, Elapid venoms have cytotoxicity toward both B16F10 melanoma and chondrosarcoma. Venoms of snakes are also used for its antimicrobial activity. FDA approved the following snake venom based drug for therapeutic applications; captopril, integrilin, and Aggrastat (Mohamed Abd El-Aziz et al. 2019). Snake venoms can be considered as mini-drug libraries in which each drug has pharmacological and therapeutic applications (Hardy et al. 2014). However, less than 0.01% of these toxins have been identified and characterized (Mohamed Abd El-Aziz et al. 2019) and a lot of work is yet to be done in this regard.

14.4.4 Dendrotoxins

Dendrotoxins are non-enzymatic venom neurotoxins extracted from ***Dendroaspis angusticeps*** (green mamba snake) and venoms of other ***Dendroaspis*** sp. (Laustsen et al. 2018). Dendrotoxins are made up of the sequence of 57–60 amino acids that are cross linked by three disulphide bonds (Harvey 1997). Due to their high efficiency and selectivity for neuronal potassium channels they are therapeutic agents (Harvey 2001). It has potential to block K+ channel and enhance acetylcholine release hence can be used to treat Alzheimer's disease induced by acetylcholine deficit (Chen et al. 2018).

14.4.5 Tetanus Toxin

It is a deadly poison produced by anaerobic, gram-positive bacteria ***Clostridium tetani*** (Ovsepian et al. 2019). It blocks release of neurotransmitter leading to spastic paralysis (Bayart et al. 2022) .Tetanus toxin is treated with formaldehyde to produce tetanus toxoid (TTd) which plays an important role in pharmaceutical world, especially in vaccines (Bayart et al. 2022). Tetanus toxin fragment C (TTFC) which is derived from a part of tetanus toxin is now used as both neuroprotective agent for CNS disorders and also as a carrier protein in vaccines and is a better alternative for tetanus toxoid owing to its non-toxic and neuronal binding properties (Bayart et al. 2022).

14.4.6 Conotoxin

Conotoxin is produced by the Conoidea superfamily (e.g., cone snails, etc.) (Chen et al. 2018). Conotoxins are 10–30 amino acids in length and rarely more than 60 having high affinity and specificity to their targets. Its mechanism of action is targeting the ion channels or receptors in the nervous systems (Chen et al. 2018).Various conotoxins are natural ligand to many ion channels (e.g., Na+ channel, K+ channel, Ca2+ channel) and receptors (such as nicotinic acetylcholine receptor, serotonin receptor). U.S Food and Drug Administration (FDA) approved Ziconotide, which displays remarkable analgesic activity in treating chronic pain via its potent inhibitory effect on N type Ca2+ channels (Rigo et al. 2013; Chen et al. 2018).

14.5 Impact on Human Health

Neurotoxins can have adverse effects on human health, targeting the nervous system and potentially leading to a range of neurological disorders and impairments. These substances interfere with the normal functioning of neurons, neurotransmitters, and other components of the nervous system, causing symptoms such as cognitive deficits, motor impairments, sensory disturbances, and behavioral changes. Neurotoxins such as paralytic shellfish toxins (PST), ciguatoxins, and tetrodotoxin can cause various health issues ranging from mild symptoms to severe illness or even death. Neurotoxins, produced by cyanobacteria like *Anabaena* and *Microcystis*, cause rapid paralysis of muscles affecting the nervous system. Anatoxin-A mimics acetylcholine, leading to muscle paralysis and suffocation. These toxins pose severe health risks including convulsions and death. Neurotoxins, such as those found in chemotherapy drugs like taxanes, platinum drugs, and vinca alkaloids, as well as proteasome inhibitors like bortezomib, also have significant adverse effects on human health (Guillotin and Delcourt (2021)). These drugs affect sensory nerves, leading to slow action potential, considerable pain, functional loss, and ultimately chemotherapy withdrawal. Chemotherapy-induced peripheral neurotoxicity (CIPN) is characterized by symptoms such as pain, tingling, numbness, and impaired sensory function in the hands and feet. In some cases, motor nerves and the autonomic nervous system may also be affected, depending on the specific drugs used. Neurotoxins, such as lead, methylmercury, arsenic, polychlorinated biphenyls (PCBs), and toluene, have detrimental effects on human health, particularly on neurodevelopment. These toxins, widely present in the environment, pose significant risks, especially during critical developmental stages such as fetal development and early childhood.

Raising public awareness about the presence and effects of neurotoxins is essential for preventing and minimizing their impact on human health. Educational campaigns can inform individuals about common sources of neurotoxins, including certain chemicals, pesticides, heavy metals, and environmental pollutants. Public

awareness efforts should also highlight the symptoms and health risks associated with neurotoxin exposure, encouraging people to take proactive measures to protect themselves and their families.

Public awareness regarding the potential neurotoxic effects of chemotherapy drugs is crucial for early detection and management of CIPN. Healthcare professionals should educate patients undergoing chemotherapy about the symptoms of neurotoxicity and monitor them closely during treatment. Management strategies for CIPN include dose reduction, cessation of chemotherapy, and alternative or novel approaches to treatment. Understanding the link between environmental exposures and neurodevelopmental disorders can empower individuals and communities to advocate for stricter regulations, safer practices, and preventive measures. Educating the public about recognizing symptoms of neurotoxin poisoning can facilitate early intervention and treatment. Providing clear information about the sources of neurotoxins, such as contaminated seafood or harmful algal blooms, can help individuals make informed decisions. Effective management of neurotoxin exposure requires a multifaceted approach involving regulation, monitoring, and intervention. Government agencies and regulatory bodies play a crucial role in setting standards and guidelines for the use, handling, and disposal of neurotoxic substances in various industries and products. Monitoring programs can track levels of neurotoxins in the environment, food, water, and occupational settings, helping to identify sources of contamination and assess potential health risks. Early detection and intervention are essential for managing neurodevelopmental disorders. Medical treatments and therapies can help mitigate symptoms and improve quality of life for affected individuals. Additionally, healthcare professionals need to be equipped with the knowledge and resources to diagnose and treat neurotoxin-related health conditions promptly.

Treatment strategies for neurotoxin-related health issues may vary depending on the specific substance involved, the severity of symptoms, and individual factors such as age and overall health. Medical interventions may include pharmacological treatments to alleviate symptoms, supportive therapies (e.g., EDTA chelation therapy) to manage complications, and rehabilitation programs to improve neurological function and quality of life for affected individuals. Early detection and intervention are crucial for maximizing treatment effectiveness and minimizing long-term damage. Utilizing advanced toxicological methods, such as in vitro testing and computational modeling, enables the identification and assessment of neurotoxicity more efficiently and ethically. These methods accelerate the discovery of potential neurotoxicants and facilitate risk assessment. Various strategies have been proposed for the treatment of CIPN, such as dose optimization (e.g., OPTIMOX stop-and-go) and the use of complementary and alternative medicine (CAM) methods. However, dose reduction or cessation of chemotherapy can increase cancer-related morbidity and mortality, thus requiring the consideration of alternative approaches. CAM methods, such as herbal medicines and acupuncture, show promising effects in alleviating CIPN symptoms and may offer viable options for prevention and treatment.

Preventing neurotoxin exposure is paramount for safeguarding public health. Prevention efforts may include implementing occupational safety measures to reduce workplace hazards, promoting environmental policies that limit the release

of neurotoxins into the air, water, and soil, and advocating for safer alternatives to neurotoxic substances in consumer products and industrial processes. Individuals can also take steps to minimize their exposure by adopting healthy lifestyle habits, such as eating a balanced diet, avoiding smoking and excessive alcohol consumption, and using protective equipment when working with potentially hazardous materials. Strategies to prevent exposure to neurotoxins include promoting safe handling and disposal of hazardous chemicals, encouraging breastfeeding to minimize infant exposure, and advocating for clean air and water regulations. Implementing regulations to monitor and control the use of neurotoxic chemicals in industries, consumer products, and agriculture can also mitigate exposure risks. Collaboration among government agencies, healthcare providers, researchers, and community organizations is vital for effective management and prevention of neurotoxin-related health risks.

14.6 Conclusion and Future Perspectives

In conclusion, the exploration of neurotoxins of natural origin has revealed a diverse array of compounds with profound impacts on neural function, human health, and environmental ecosystems. Through classification into plant, animal, and microbial categories, valuable insights into their mechanisms of action have been gained, illuminating the intricate ways in which these compounds disrupt neural pathways. The examination of their pharmacological and therapeutic applications underscores their potential as tools in medical research and therapeutics, despite challenges. Looking ahead, future perspectives hold promise for advancing our understanding of neural function and disease pathogenesis. Continued research efforts, interdisciplinary collaborations, and effective strategies for mitigating neurotoxin exposure's impact are vital for progress in this field. Overall, the study of natural neurotoxins represents a dynamic and evolving discipline with far-reaching implications for biomedical research, medicine, and public health, offering opportunities for innovative discoveries and interventions to benefit humanity and safeguard our planet's well-being.

References

Abbott NJ, Patabendige AA, Dolman DE, Yusof SR, Begley DJ (2010) Structure and function of the blood–brain barrier. Neurobiol Dis 37(1):13–25

Aguilar SJ (2018) Learning analytics: at the nexus of big data, digital innovation, and social justice in education. TechTrends 62:37–45

Balzer I, Hardeland R (1996) Melatonin in algae and higher plants—possible new roles as a phytohormone and antioxidant. Botanica Acta 109(3):180–183

Bane V, Lehane M, Dikshit M, O'Riordan A, Furey A (2014) Tetrodotoxin: chemistry, toxicity, source, distribution and detection. Toxins 6(2):693–755

Banks EC, Ferretti LE, Shucard DW (1997) Effects of low level lead exposure on cognitive function in children: a review of behavioral, neuropsychological and biological evidence. Neurotoxicology 18(1):237–281

Bao QS, Lu CY, Song H, Wang M, Ling W, Chen WQ, Deng XQ, Hao YT, Rao S (2009) Behavioural development of school-aged children who live around a multi-metal sulphide mine in Guangdong province, China: a cross-sectional study. BMC Public Health 9:1–8

Bayart C, Mularoni A, Hemmani N, Kerachni S, Jose J, Gouet P, Paladino J, Le Borgne M (2022) Tetanus toxin fragment C: structure, drug discovery research and production. Pharmaceuticals 15(6):756

Bellinger DC (2008) Very low lead exposures and children's neurodevelopment. Curr Opin Pediatr 20(2):172–177

Boyes WK, van Thriel C (2020) Neurotoxicology of nanomaterials. Chem Res Toxicol 33(5):1121–1144

Brenner ED, Martinez-Barboza N, Clark AP, Liang QS, Stevenson DW, Coruzzi GM (2000) Arabidopsis mutants resistant to S (+)-β-methyl-α, β-diaminopropionic acid, a cycad-derived glutamate receptor agonist. Plant Physiol 124(4):1615–1624

Cabeza OA, Zhang Z, Khoury ES, Sheldon RA, Sharma A, Zhang F, Slusher BS, Kannan RM, Kannan S, Ferriero DM (2021) Neuroprotective effects of a dendrimer-based glutamate carboxypeptidase inhibitor on superoxide dismutase transgenic mice after neonatal hypoxic-ischemic brain injury. Neurobiol Dis 148:105201

Cai S, Kumar R, Singh BR (2021) Clostridial neurotoxins: Structure, function and implications to other bacterial toxins. Microorganisms 9(11):2206

Calne DB (1991) Neurotoxins and degeneration in the central nervous system. Neurotoxicology 12(3):335–339

Cao Y, Li B, Ismail N, Smith K, Li T, Dai R, Deng Y (2021) Neurotoxicity and underlying mechanisms of endogenous neurotoxins. Int J Mol Sci 22(23):12805

Cardenas-Iniguez C, Burnor E, Herting MM (2022) Neurotoxicants, the developing brain, and mental health. Biol Psychiatry Global Open Sci 2(3):223–232

Chakrabarty D, Sarkar A (2017) Cytotoxic effects of snake venoms. In: Inagaki H, Vogel CW, Mukherjee A, Rahmy T (eds) Snake venoms. Springer, Dordrecht, pp 1–16

Chen N, Xu S, Zhang Y, Wang F (2018) Animal protein toxins: origins and therapeutic applications. Biophys Rep 4(5):233–242

Chiodo LM, Covington C, Sokol RJ, Hannigan JH, Jannise J, Ager J, Greenwald M, Delaney-Black V (2007) Blood lead levels and specific attention effects in young children. Neurotoxicol Teratol 29(5):538–546

Choudhury S, Baker MR, Chatterjee S, Kumar H (2021) Botulinum toxin: an update on pharmacology and newer products in development. Toxins 13(1):58

Connolly NM, D'Orsi B, Monsefi N, Huber HJ, Prehn JH (2016) Computational analysis of AMPK-mediated neuroprotection suggests acute excitotoxic bioenergetics and glucose dynamics are regulated by a minimal set of critical reactions. PLoS One 11(2):e0148326

Dubbels R, Reiter RJ, Klenke E, Goebel A, Schnakenberg E, Ehlers C, Schiwara HW, Schloot W (1995) Melatonin in edible plants identified by radioimmunoassay and by high performance liquid chromatography-mass spectrometry. J Pineal Res 18(1):28–31

Ek CJ, Dziegielewska KM, Habgood MD, Saunders NR (2012) Barriers in the developing brain and neurotoxicology. Neurotoxicology 33(3):586–604

Furtado D, Björnmalm M, Ayton S, Bush AI, Kempe K, Caruso F (2018) Overcoming the blood–brain barrier: the role of nanomaterials in treating neurological diseases. Adv Mater 30(46):1801362

Gao B, Peng C, Yang J, Yi Y, Zhang J, Shi Q (2017) Cone snails: A big store of conotoxins for novel drug discovery. Toxins 9(12):397

Gomes A, Bhattacharjee P, Mishra R, Biswas AK, Dasgupta SC, Giri B, Debnath A, Gupta SD, Das T, Gomes A (2010) Anticancer potential of animal venoms and toxins. Indian J Exp Biol 48(2):93–103

González-Cano R, Ruiz-Cantero MC, Santos-Caballero M, Gómez-Navas C, Tejada MÁ, Nieto FR (2021) Tetrodotoxin, a potential drug for neuropathic and cancer pain relief? Toxins 13(7):483
Guillotin S, Delcourt N (2021) Marine neurotoxins' effects on environmental and human health: an OMICS overview. Mar Drugs 20(1):18
Guzmán A, de Henestrosa ARF, Marín AP, Ho A, Borroto JIG, Carasa I, Pritchard L (2007) Evaluation of the genotoxic potential of the natural neurotoxin Tetrodotoxin (TTX) in a battery of in vitro and in vivo genotoxicity assays. Mutat Res 634(1–2):14–24
Hardeland R (1993) The presence and function of melatonin and structurally related indoleamines in a dinoflagellate, and a hypothesis on the evolutionary significance of these tryptophan metabolites in unicellulars. Experientia 49:614–622
Hardy MC, Cochrane J, Allavena RE (2014) Venomous and poisonous Australian animals of veterinary importance: A rich source of novel therapeutics. BioMed Res Int. https://doi.org/10.1155/2014/671041
Harvey AL (1997) Recent studies on dendrotoxins and potassium ion channels. Gen Pharmacol 28(1):7–12
Harvey AL (2001) Twenty years of dendrotoxins. Toxicon 39(1):15–26
Hofmann H, Bon C (1987) Blood coagulation induced by the venom of Bothrops atrox. 1. Identification, purification, and properties of a prothrombin activator. Biochemistry 26(3):772–780
Kerkkamp H, Bagowski C, Kool J, van Soolingen B, Vonk FJ, Vlecken D (2018) Whole snake venoms: Cytotoxic, anti-metastatic and antiangiogenic properties. Toxicon 150:39–49
Kobayashi H, Oikawa S, Kawanishi S (2006) Mechanism of DNA damage and apoptosis induced by tetrahydropapaveroline, a metabolite of dopamine. Neurochem Res 31:523–532
Kolář J, Macháčková I, Eder J, Prinsen E, Van Dongen W, Van Onckelen H, Illnerová H (1997) Melatonin: occurrence and daily rhythm in Chenopodium rubrum. Phytochemistry 44(8):1407–1413
Lago J, Rodríguez LP, Blanco L, Vieites JM, Cabado AG (2015) Tetrodotoxin, an extremely potent marine neurotoxin: Distribution, toxicity, origin and therapeutical uses. Mar Drugs 13(10):6384–6406
Lanphear BP, Dietrich K, Auinger P, Cox C (2000) Cognitive deficits associated with blood lead concentrations< 10 microg/dL in US children and adolescents. Public Health Rep 115(6):521
Lanphear BP, Hornung R, Khoury J, Yolton K, Baghurst P, Bellinger DC, Canfield RL, Dietrich KN, Bornschein R, Greene T, Rothenberg SJ (2019) Erratum: "Low-level environmental lead exposure and children's intellectual function: an international pooled analysis". Environ Health Perspect 127(9):099001
Laustsen AH, Gutiérrez JM, Knudsen C, Johansen KH, Bermúdez-Méndez E, Cerni FA, Jürgensen JA, Ledsgaard L, Martos-Esteban A, Øhlenschlæger M, Pus U (2018) Pros and cons of different therapeutic antibody formats for recombinant antivenom development. Toxicon 146:151–175
Lu P, Ghiasi S, Hagenah J, Hai HB, Hao NV, Khanh PNQ, Khoa LDV, Consortium VITAL, Thwaites L, Clifton DA, Zhu T (2022) Classification of tetanus severity in intensive-care settings for low-income countries using wearable sensing. Sensors 22(17):6554
Mahrhold S, Rummel A, Bigalke H, Davletov B, Binz T (2006) The synaptic vesicle protein 2C mediates the uptake of botulinum neurotoxin A into phrenic nerves. FEBS Lett 580(8):2011–2014
Maria G, Brisinda G, Bentivoglio AR, Cassetta E, Gui D, Albanese A (1998) Botulinum toxin injections in the internal anal sphincter for the treatment of chronic anal fissure: long-term results after two different dosage regimens. Ann Surg 228(5):664–669
Matsubara K, Collins MA, Akane A, Ikebuchi J, Neafsey EJ, Kagawa M, Shiono H (1993) Potential bioactivated neurotoxicants, N-methylated β-carbolinium ions, are present in human brain. Brain Res 610(1):90–96
Megighian A, Pirazzini M, Fabris F, Rossetto O, Montecucco C (2021) Tetanus and tetanus neurotoxin: From peripheral uptake to central nervous tissue targets. J Neurochem 158(6):1244–1253

Mohamed Abd El-Aziz T, Soares AG, Stockand JD (2019) Snake venoms in drug discovery: Valuable therapeutic tools for life saving. Toxins 11(10):564

Momonoki YS, Momonoki T (1991) Changes in acetylcholine levels following leaf wilting and leaf recovery by heat stress in plant cultivars. Jpn J Crop Sci 60(2):283–290

Mukherjee AK (2014) A major phospholipase A2 from Daboia russelii russelii venom shows potent anticoagulant action via thrombin inhibition and binding with plasma phospholipids. Biochimie 99:153–161

Murch SJ, Simmons CB, Saxena PK (1997) Melatonin in feverfew and other medicinal plants. Lancet 350(9091):1598–1599

Murch SJ, KrishnaRaj S, Saxena PK (2000) Tryptophan is a precursor for melatonin and serotonin biosynthesis in in vitro regenerated St. John's wort (Hypericum perforatum L. cv. Anthos) plants. Plant Cell Rep 19:698–704

Murch SJ, Cox PA, Banack SA (2004) A mechanism for slow release of biomagnified cyanobacterial neurotoxins and neurodegenerative disease in Guam. Proc Natl Acad Sci USA 101(33):12228–12231

Nathanson JA (1984) Caffeine and related methylxanthines: possible naturally occurring pesticides. Science 226(4671):184–187

National Research Council, Division on Earth, Life Studies, Commission on Life Sciences, Board on Environmental Studies, Committee on Neurotoxicology and Models for Assessing Risk (1992) Environmental neurotoxicology

Nowakowska MB, Selby K, Przykopanski A, Krüger M, Krez N, Dorner BG, Dorner MB, Jin R, Minton NP, Rummel A, Lindström M (2022) Construction and validation of safe Clostridium botulinum Group II surrogate strain producing inactive botulinum neurotoxin type E toxoid. Sci Rep 12(1):1790

Ovsepian SV, O'Leary VB, Zaborszky L, Ntziachristos V, Dolly JO (2019) Amyloid plaques of Alzheimer's disease as hotspots of glutamatergic activity. Neuroscientist 25(4):288–297

Paget SP, Swinney CM, Burton KL, Bau K, O'Flaherty SJ (2018) Systemic adverse events after botulinum neurotoxin A injections in children with cerebral palsy. Dev Med Child Neurol 60(11):1172–1177

Pirazzini M, Rossetto O, Eleopra R, Montecucco C (2017) Botulinum neurotoxins: biology, pharmacology, and toxicology. Pharmacol Rev 69(2):200–235

Polito L, Bortolotti M, Battelli MG, Calafato G, Bolognesi A (2019) Ricin: An ancient story for a timeless plant toxin. Toxins 11(6):324

Popoff MR, Poulain B (2010) Bacterial toxins and the nervous system: neurotoxins and multipotential toxins interacting with neuronal cells. Toxins 2(4):683–737

Postma TL (2009) Neurotoxic animal poisons and venoms. Clin Neurotoxicol 55(2–3):463–489

Rasetti-Escargueil C, Popoff MR (2020) Engineering botulinum neurotoxins for enhanced therapeutic applications and vaccine development. Toxins 13(1):1

Rigo FK, Dalmolin GD, Trevisan G, Tonello R, Silva MA, Rossato MF, Klafke JZ, Cordeiro MDN, Junior CJC, Montijo D, Gomez MV (2013) Effect of ω-conotoxin MVIIA and Phα1β on paclitaxel-induced acute and chronic pain. Pharmacol Biochem Behav 114:16–22

Romero IA, Abbott NJ, Bradbury MW (2023) The blood-brain barrier in normal CNS and in metal-induced neurotoxicity. In: Toxicology of metals, vol I. CRC, Boca Raton, pp 561–585

Saubaméa B, Cochois-Guégan V, Cisternino S, Scherrmann JM (2012) Heterogeneity in the rat brain vasculature revealed by quantitative confocal analysis of endothelial barrier antigen and P-glycoprotein expression. J Cereb Blood Flow Metab 32(1):81–92

Schantz EJ, Johnson EA (1992) Properties and use of botulinum toxin and other microbial neurotoxins in medicine. Microbiol Rev 56(1):80–99

Schiavo G, Matteoli M, Montecucco C (2000) Neurotoxins affecting neuroexocytosis. Physiol Rev 80(2):717–766

Searle AK, Baghurst PA, Van Hooff M, Sawyer MG, Sim MR, Galletly C, Clark LS, McFarlane AC (2014) Tracing the long-term legacy of childhood lead exposure: a review of three decades of the port Pirie cohort study. Neurotoxicology 43:46–56

Segura-Aguilar J (2018) Neurotoxins as preclinical models for Parkinson's disease. Neurotox Res 34(4):870–877

Shelp BJ, Van Cauwenberghe OR, Bown AW (2003) Gamma aminobutyrate: from intellectual curiosity to practical pest control. Can J Bot 81(11):1045–1048

Sławek J, Jost WH (2021) Botulinum neurotoxin in cervical dystonia revisited—Recent advances and unanswered questions. Neurol Neurochir Pol 55(2):125–132

Spencer PS, Lein PJ (2014) Neurotoxicity. In: Wexler P (ed) Encyclopedia of toxicology, 3rd edn. Academic, London

Suzuki T, Waller GR (1987) Allelopathy due to purine alkaloids in tea seeds during germination. Plant Soil 98:131–136

Świędrych A, Lorenc-Kukuła K, Skirycz A, Szopa J (2004) The catecholamine biosynthesis route in potato is affected by stress. Plant Physiol Biochem 42(7–8):593–600

Tahara Y, Hirata Y (1909) Studies on the puffer fish toxin. J Pharm Soc Jpn 29(587625):2

Tu AT (1996) Overview of snake venom chemistry. In: Singh IBR, Tu AT (eds) Natural toxins 2: structure, mechanism of action, and detection. Springer, New York, pp 37–62

Tyślerowicz M, Kiedrzyńska W, Adamkiewicz B, Jost WH, Sławek J (2020) Cervical dystonia—improving the effectiveness of botulinum toxin therapy. Neurol Neurochir Pol 54(3):232–242

Wang Y, Qin ZH (2010) Molecular and cellular mechanisms of excitotoxic neuronal death. Apoptosis 15:1382–1402

Wu YC, Sonninen TM, Peltonen S, Koistinaho J, Lehtonen Š (2021) Blood–brain barrier and neurodegenerative diseases—modeling with iPSC-derived brain cells. Int J Mol Sci 22(14):7710

Zhang D, Xing G, Xu H, Yan Z, Wang C, Wang Y, Li Z (2005) Relationship between oxalic acid and the metabolism of β-N-oxalyl-α, β-diaminopropionic acid (ODAP) in grass pea (Lathyrus sativus L.). Isr J Plant Sci 53(2):89–96

Zhao M, Chen J, Mao K, She H, Ren Y, Gui C, Wu X, Zou F, Li W (2019) Mitochondrial calcium dysfunction contributes to autophagic cell death induced by MPP+ via AMPK pathway. Biochem Biophys Res Commun 509(2):390–394

Zhou K, Luo W, Liu T, Ni Y, Qin Z (2022) Neurotoxins acting at synaptic sites: A brief review on mechanisms and clinical applications. Toxins 15(1):18

Zhou CY, Pan CG, Peng FJ, Zhu RG, Hu JJ, Yu K (2024) Simultaneous determination of trace marine lipophilic and hydrophilic phycotoxins in various environmental and biota matrices. Mar Pollut Bull 203:116444

Chapter 15
Toxic Effects of Oestrogenic Biomolecules

Catherine Bennetau-Pelissero

Abstract Some biomolecules essentially from plants or fungi have been shown to exhibit oestrogenic properties. These are prenyl-flavanones, coumestans, resorcylic acid lactones, isoflavones and lignans. Their oestrogenic characteristics have been exploited in medicinal preparations for specific physio-pathological situations like menopause, contraception or osteoporosis. However, the oestrogenic substances are not required for the general populations. Indeed, generally speaking, in animals and humans, oestrogenic effects can be assimilated to endocrine disruptions and should be avoided. In this chapter are related the toxicological effects of plant biomolecules with oestrogenic effects. This is performed by first describing the physiological roles of oestrogens in healthy and sick subjects. Then the actual exposure to oestrogenic biomolecules will be estimated with references to the historical situation. This will lead to proposals to reduce such an exposure. Afterward, the toxicological information actually recorded on animal models by international organizations will be exposed before relating the data presently available in humans. Finally, it will be shown that oestrogenic biomolecules can also exhibit other deleterious effects, reinforcing the need to reduce the animal and human exposures.

Keywords Phytoestrogens · Mycotoxins · Rodents · Cattle · Human · Exposure · Toxic effects · Reference doses · Endocrine disruptors · Reduction

15.1 Introduction

This chapter is dedicated to oestrogenic natural substances and to their deleterious effects in animals and humans. Such substances act as endocrine disruptors and can be included in large cocktails that encompass anthropogenic molecules with similar

C. Bennetau-Pelissero (✉)
Department of Feed & Food, Bordeaux Sciences Agro, Gradignan, France

University of Bordeaux, Talence, France
e-mail: catherine.bennetau@agro-bordeaux.fr

M. N. V. Prasad et al. (eds.), *Biotoxins*,
https://doi.org/10.1007/978-3-031-75309-1_15

mode of actions. This makes the oestrogenic biotoxins particularly undesirable nowadays. Here are mentioned plant molecules whose oestrogenic activities are known for ages (Bennetts et al. 1946). Some of them have been used as antifertility agents in many previous civilizations (Farnsworth et al. 1975). They include plants containing 8-prenylnaringenin (8-PN) like hop or citrus and other medicinal plants (Pohjanvirta and Nasri 2022). They comprise plants containing oestrogenic coumestans like coumestrol (COUM) (Tu et al. 2021) or oestrogenic isoflavones like genistein (GEN), daidzein (DAI), glycitein (GLY), formononetin, or biochanin A (Bennetau-Pelissero 2018). These plants are, for instance, alfalfa, clover, kudzu, soybean, black beans, brooms, heath or gorse (Farnsworth et al. 1975). Beside molecules that are directly produced by plants, some other natural oestrogens are produced by fungi or by the gut microflora. These are, respectively, zearalenone (ZEN) and zearalenol (ZEL) from several fusarium species growing and enterolactone (ENL) and enterodiol (END) biosynthesized in animal gut from the lignans secoisolariciresinol diglucoside, matairesinol, lariciresinol, pinoresinol or syringaresinol (Canivenc-Lavier and Bennetau-Pelissero 2023). As will be seen in this chapter if the most potent oestrogens are monitored and present at low levels in human and animal food, this is not the case for isoflavones from soybean since this pulse is increasingly used in animal feed and also for human food.

15.2 The Target Tissues of Oestrogens and Their Physiological Consequences

15.2.1 Reproductive Physiology

Oestrogens are molecules whose activities evoke those of oestradiol (E_2). E_2 is the main steroid produced by animals controlling physiological situations. Beside are other oestrogens: oestrone (E_1), oestriol (E_3) and oestetrol (E_4) which interact with the oestrogen receptors (ERs) with lower affinity (Alderman 3rd and Taylor 2021). Basically, E_2 is produced by the ovary, brain, adrenal glands, adipose tissues and acts on multiple target organs from skin and bone to gut but also all reproductive organs including uterus, vagina, mammary gland, foetal tissues and so on (Prior 2020). E_2 is also required for a complete male reproductive efficiency and ERs are necessary in the testis, prostate and other tissues of the male reproductive tract (Cooke et al. 2017). Reproduction in vertebrates is cyclically controlled by the pituitary hormones FSH and LH whose secretions are regulated by E_2 and GnRH (Goodman et al. 2022). The latter is produced by hypothalamic neurones. GnRH secretion is pulsatile during the reproductive period and is also controlled by ovarian glycoproteins like inhibin A and B and neuropeptides like kisspeptin (Goodman et al. 2022). Kisspeptin is retro-controlled by E_2 which induces or represses its secretions and the availability of its receptor depending on its own circulating level and the hypothalamus area considered (Starrett and Moenter 2023). Indeed,

kisspeptin is produced by neurones identified in the arcuate nucleus and the antero-ventral periventricular area. These neurones seem to be oppositely regulated by E_2 (Starrett and Moenter 2023). Basically, low E_2 levels decrease GnRH production while high levels present at the end of the follicular phase increase it. GnRH increases ovarian and testicular activity and when it is not released, the gamete production is reduced. E_2 analogues like ethynyl-oestradiol are active substances of contraceptive pills. Their mode of action passes via the kisspeptin and GnRH inhibition (Starrett and Moenter 2023).

E_2 in females is mainly synthesized in the maturing ovary by the De Graaf follicle. E_2 produced by this follicle induces the production of vaginal mucus, the oestrous behaviour and the maturation of the uterine endometrium (Prior 2020). These phenomena cooperate to welcome the male semen and then the embryos insuring their best survival in the female genital tract. The last development stage of the follicle is oocyte expulsion induced by a LH surge which is also induced by high E_2 plasma levels (Starrett and Moenter 2023). Then, the De Graaf follicle turns into a corpus luteum synthesizing progesterone (P). The main role of this steroid is to protect gestation and embryos' development. For instance, P is involved in the regulation of oviduct secretions allowing the young embryos survival before implantation (Prior 2020). During gestation, P and E_2 are synthesized at very high levels: 50–100 times higher than those present during menstrual cycles. The main roles of E_2 are to increase food metabolism efficiency and induce vasodilatation allowing a better transfer of oxygen and food metabolites to the foetuses. High levels of P and E_2 also reduce the immune defences of the mother in order to increase her tolerance to her foetuses (Tal and Taylor 2021).

E_2 is also required for a proper sperm production and when E_2 receptors are deficient in males the reproduction is impaired. This phenomenon is linked to a defect in the hypothalamus–pituitary axis but also to an alteration of testis development and secretions (Cooke et al. 2017).

The rapid description of these reproductive steps points out the crucial role of E_2 and also of potential xenoestrogens on animal and human reproductive physiology.

15.2.2 Developmental Physiology

During foetal development, E_2 and other oestrogens are synthesized by the placenta or derive from the mother's pool and there is no difference of foetus impregnation according to sex (Alderman 3rd and Taylor 2021). At birth, a dramatic drop occurs in neonates. However, E_2 rises again in infants during minipuberty.

During pregnancy, the foetus' liver produces a specific protein called α-fetoprotein which has a great affinity for the maternal E_2 and several fatty acids (Mizejewski 2004). This α-fetoprotein would protect the foetus from oestrogenization. This protein is also present in children, and in rodent, it was shown to decrease free E_2 in plasma and delay sexual maturity when injected to off-springs (Mizejewski 2004).

The foetal sexualization occurs in utero. This process has been studied in animal models and in humans based on heterozygous or homozygous mutations of crucial genes. From these, it appears that autosomal and gonosomal genes are involved in sexualization. Their expression may be transient and essentially governed by transcriptional factors under epigenetic control (Rey et al. 2020). The basic sex is the female one and without inputs from factors encoded by genes present on the Y chromosomes, like SRY, the foetus evolves towards a female phenotype with ovaries, female genital tract and vagina.

The female genital tract derives from Müllerian ducts which initially appear into the mesonephros close to the Wolffian ducts. The early development of the female genital tract is independent of E_2 (Alderman 3rd and Taylor 2021). E_2 can only be active when the foetal organs and cells synthesize its receptors. In human female foetuses, ERβ appears at 8 weeks post-conception in the uterovaginal canal and ERα at 12 weeks in the uterovaginal canal and the uterus (Cunha et al. 2021). These receptors play different roles, ERβ being mainly involved in differentiation and ERα in proliferation processes.

When the foetus has a Y chromosome, its Sertoli cells synthesize an anti-Müllerian hormone (AMH) from the 8th week post-conception in human, which progressively destroys the Müllerian ducts. AMH synthesis exists also in females but is lower and only appears from the 24th week of gestation. At that time the Mullerian ducts are no longer sensitive to AMH. Indeed, the timing of secretion is sexually different and crucial for proper sex differentiation. In males, the genital tract derives from the Wolffian ducts (Rey et al. 2020). The male genital tract develops under partial and transient E_2 control since both ERβ and aromatase are expressed in Sertoli, Leydig, and germ cells from 13 to 24 weeks after conception (Rochira and Carani 2023). This E_2 secretion has been shown to play a role in the development of rete testis, efferent ducts, epididymis and vas deferens. T is involved in this process and E_2 or xenoestrogens may disturb it as will be seen later. By the end of the male foetal development, testis which are initially within the peritoneal cavity move to the scrotum. This is required to ensure an adequate temperature for adult sperm production.

During the first months of life (4–12 months) human infants undergo a minipuberty characterized by elevated levels of LH in boys and FSH in girls (Pasterski et al. 2015). In infant boys, quite high levels of T and E_2 are recorded, the second being produced from the former. In girls E_2 is produced but at levels 10–50 times lower than in boys (Kuiri-Hänninen et al. 2014). Such a sex-differentiated steroid exposure is believed to be involved in a late gender differentiation of adult behaviour (Pasterski et al. 2015). In parallel, in rodents, it was shown that the early E_2 secretion in male off-springs is responsible for an enlargement of the Sexually Dimorphic Nucleus of the Preoptic Area (SDN-POA). This enlargement is controlled by the neonatal secretion of GnRH and LH inducing T secretion by the testis. T masculinizes the SDN-POA when it is aromatized locally into E_2. Finally, it is E_2 which is responsible for the masculinization process acting through nuclear and membrane E_2 receptors (Dovey and Vasudevan 2020). In females, the E_2 levels are too low to induce such enlargement. In monkeys, it was shown that when the

neonatal LH synthesis is blocked, this has consequences on the male behaviour in adulthood (Wallen et al. 1995). In human, the role of oestrogens during human minipuberty still needs to be deciphered.

15.2.3 Oestrogen-Dependent Pathologies

Some pathologies are obviously linked to oestrogens including E_2. These are oestrogens and also some androgens-dependent diseases. Indeed, androgens can be converted into oestrogens by the endogenous enzyme aromatase and some androgen-dependent tissues express both E_2 receptors and aromatase. In that case, E_2 locally synthesized from T can act together with androgens in the development of the pathology. This is observed in human prostate which is under androgen control but which also exhibit ERα and/or ERβ together with CYP19A1 (Rochira and Carani 2023).

The main pathologies that depend on E_2 are oestrogen-dependent cancers like breast, vaginal or uterine cancers. In these pathologies, tumour cells express the E_2 receptors, namely, ERα, ERβ or GPER and can respond to E_2 stimulation by a proliferation event. ERα is the receptor mainly involved in proliferation. It is also the main receptor in the mammary gland while vagina and uterus equally express ERβ and also GPER (Luo and Liu 2020). The proliferative effect of E_2 on breast and uterine cancer cells has long been studied in vivo and in vitro. It is used as control molecule when detecting environmental oestrogenic chemicals in toxicological test such as the E-screen test on MCF-7 cells (Soto et al. 1995). Cells that are negative for ERs including GPER generally do not grow under E_2 stimulation (Girgert et al. 2017).

Some troubles or pathologies are related to oestrogen deficiency. These are menopausal symptoms and osteoporosis for instance. In these cases, E_2 treatments are known to attenuate pathophysiological issues and hormone replacement therapies (HRT) were developed based on oestrogen mimicking chemical substances (Patel and Bihani 2018). E_2 action goes through both membrane and nuclear receptors. The latter require precise interactions with their ligand to be able to recrute transcription co-activators. Therefore, the chemical structure of the E_2 substitutes is crucial. Indeed, the allostery of any xenoestrogen bound to the ligand binding domain or the ER induces a specific terciary structure of the receptor. According to this structure, the dimeric ER will be able or not to recruit nuclear transcription co-activator and RNA polymerase. Because these activators may be specific of one target cell or another, it appears possible to elicit E_2 substitutes with specific actions on specific target cells avoiding cancer proliferative action. Such chemicals able to exhibit only parts of E_2 action have been called SERMs for Selective Oestrogen Receptors Modulators (Patel and Bihani 2018).

Other pathologies linked to immunity are apparently influenced by E_2. These are some autoimmune diseases like multiple sclerosis (MS) and systemic lupus erythematosus (SLE). Both are more frequently observed in premenopausal women (Khan and Ahmed 2016). E_2 has been shown to play a role in the innate and adaptative

immune system. In addition, E_2 can stimulate several pathways of the immune system and ERs are expressed in immune cells. Indeed, ERα has been found in haematopoietic cells, bone marrow cells, thymus stromal cells, in thymocytes and in peritoneal macrophages, while ERβ is preferentially expressed in thymus and spleen of human mid-gestational foetus and lymphocytes in human lymph nodes (Khan and Ahmed 2016). T lymphocytes CD4+ cells, have higher ERα levels than ERβ whereas CD8+ cells only weakly express both receptors. In B cells, the ERβ expression is higher than ERα. Even though both ERs have comparable affinity to oestrogen and recognize the same ERE (Oestrogen Responsive Element) on the double-stranded DNA, they may have distinct, non-overlapping or even antagonist effects. This gives more complexity to the E_2 action even if it is likely that E_2 can act on immune cells which possess ERs. Physiologically, E_2 is usually associated with progesterone (P) and unless considering pure E_2 treatments, it is hard to decipher in humans the proper action of each hormone. Therefore, the physiological, epigenetic and local contexts including tissue, inflammation state, microbiome, environment, can influence the immune response to E_2.

Considering autoimmune diseases influenced by E_2, MS occurs predominantly in women and E_2 seems to protect against the inflammation and the demyelination of neurones associated with this pathology (Khan and Ahmed 2016). E_2 acts mainly through ERα-dependent pathways reducing autoantigen-specific proinflammatory biomolecules like IFNγ, TNFα, IL-17 or iNOS. In parallel, DPN, the ERβ agonist, protects oligodendrocytes and increases endogenous myelination. Contrarily, in SLE, E_2 seems to be deleterious and E_2 increases the severity of flares of disease and the production of anti-double-stranded DNA auto-antibodies, IgG, IgM in serum from patients with SLE. In addition, the female : male susceptibility ratio of SLE is 9–20 : 1 (Khan and Ahmed 2016). In their review, the authors mentioned that E_2 activates the immune system towards both exogenous antigens and endogenous autoantigens. E_2 also promotes systemic inflammation (induction of B cell activating factor and IFN signature genes). More than 50% of the genes involved in menstrual cycle are also altered in SLE patients and during menstrual cycles, in SLE women, the TNF Receptor SF14 is reduced resulting in an overall immune enhancement (Khan and Ahmed 2016).

Finally, E_2 interactions with the immune system are far from being simple to summarize.

15.3 The Exposure of Consumers to Natural Oestrogenic Substances

15.3.1 Actual Exposure by Compounds

8-PN is mainly consumed through hop flavoured beers and therefore essentially in Western countries. Hops are added to beer as extracts obtained from female hop cones which are rich in prenylated chalcones and prenylated flavanones including

8-PN (Pohjanvirta and Nasri 2022). The proportions of each substance vary during beer brewing but the concentration of Xanthohumol (XN) may amount to 1% of final product while 8-PN is more than 10 times less abundant. XN is a metabolic precursor of isoxanthohumol (IXN) which in turn can be converted into 8-PN. According to Pohjanvirta and Nasri (2022), the exposure of moderate beer lovers to 8-PN and its potential metabolic precursor IXN varies from 1 to 2 mg/day. The exposure to 8-PN through food supplement varies with their quality from undetectable to 180 μg/day. In some studies, the pharmacokinetic of 8-PN was established with huge pharmacological doses but the ingested amount modifies the pharmacokinetic parameters (van Breemen et al. 2014). Nevertheless, when the intake level is high, only 1.5% of the ingested dose is retrieved in urine. According to Van Breemen et al. (2014), serum concentrations of 8-PN in post-menopausal women after standardized food supplement intake were in the nM range (2–6 ng/ml) which is far less than the concentrations tested in vitro.

COUM which is present in alfalfa and clover after fungi attacks is usually controlled as other oestrogenic and genotoxic agents. COUM may be present in food supplements containing alfalfa at concentrations lower than 100 mg/kg (Gaweł 2012). In the USA, Reger et al. (2018) recorded exposures up to 1.2 mg/day. In China, the COUM exposure is low and serum concentrations were hardly detectable and considered as traces (Hu et al. 2014). Finally, no data are available on COUM human pharmacokinetic because this compound is toxic.

Oestrogenic mycotoxins from fusarium: ZEN and ZELα and β are carefully monitored in human food. In Spain, ZEN was found in food and in human urine (Carballo et al. 2021). The authors estimated that the probable daily intake was sometimes higher than the tolerable daily intake (25 μg/kg/day). ZEN was found in 10% of the urine samples analysed. Again, ZEN and ZEL being toxic, their bioavailability has only been reported in rats (Shin et al. 2009) with a modellization for humans.

In the human diet, isoflavones are mainly present in soy and soy-based foodstuffs but can also be found at high concentrations in legumes like red beans. In soy, three isoflavones are present which exhibit oestrogenic activities. Table 15.1 presents means and range of variation of measurements obtained on different soy-based food from the European market.

In Asia, the exposure is considered to be higher than in the Western countries because of traditional soy consumption. Indeed, in China, the isoflavone intake was recently evaluated on more than 53,000 people to range from 0.8 to 78.0 mg/day (Zhu et al. 2021), while in Japan, the isoflavones exposure estimated on more than 30,000 people, ranged from 14 to 75 mg/day (Murai et al. 2022). In both cases the intake estimations were based on isoflavone measurements performed on soyfood that were bought in food-shops and which were prepared following industrial processes. However, for Asian people, soy is traditional and is quite often prepared at home following family recipes. Besides, in Western countries, the isoflavones intake is of a few mg/day in the general population. However, some subpopulations like vegan consumers or the Adventists of the 7th day in the USA are much more exposed, with doses reaching 17.9 mg/day as a median (Jacobsen et al. 2014).

Table 15.1 Mean isoflavone content of different soy-based foods. Values are sums of GEN and DAI

Type of product	Number of measurements	Portion size (g)	Mean dose of isoflavones/portion (mg)	Range of values mg/portion
Soy beans raw	5	100	94.8	80.6–120.6
Toasted soy beans	5	100	111.7	71.8–251.9
Edamame	4	120	10.4	29.2–64.3
Soybean flour	5	100	42.5	32.9–42.9
Protein isolate (sports)	3	50	28.9	14.2–64.5
Soy-based drinks	12	1 mug (250 ml)	32.0	7.1–44.2
Soy-based desert cream	15	150	16.2	1.7–22.3
Soy-based yogurt	14	125	8.9	4.1–25.4
Soy-based cream	12	50 ml	4.5	3.1–6.7
Soy-based Vegan steak	21	100	23.0	4.3–51.2
Soy-based raw pasta	2	100	23.5	22.1–25.3
Soy-based cooked pasta	2	100	8.9	8.2–9.3
Tempeh	6	100	24.4	13.8–36.4
Tofu	12	100	44.7	22.1–79.8
Soy-based cheese	8	50	23.8	23.5–36.2
Miso	5	10	1.24	12.2–12.6
Soy sauce	9	30 ml	0.2	0.1–0.3

Recently, the French population exposure to soy isoflavones was estimated to be of 1.9 mg/day of isoflavones in the non-soy eaters while the mean exposure among soy consumers was about 6.9 mg/day (Lee et al. 2022). However, the variability is huge around these figures since in the same study, soy isoflavone intake varied from 0 to 213 mg/day (Lee et al. 2022). In plasma, the total isoflavones including the conjugated forms considered to be inactive can reach 10 μM (Mathey et al. 2006). However, this is usually obtained using daily intake of food supplements. The food intake usually leads to plasma concentrations of 2.5 μM when GEN, DAI, GLY and their inactive conjugates are summed. The free forms of all three isoflavones rarely exceed 1 μM at the C_{max} of pharmacokinetic. The kinetics of absorption and elimination of isoflavones are known in human for all three isoflavones as well as for the metabolite equol (Shinkaruk et al. 2012).

Enterolignans are produced by the gut microflora from lignan precursors (Canivenc-Lavier and Bennetau-Pelissero 2023). The efficiency of the enterolignans production is variable among people. However, the exposure in European countries was estimated at 1.23 (0.89–1.73) mg/day (Molina-Montes et al. 2016).

Nevertheless, although the lignan exposure is low compared to isoflavones, in Western studies, when isoflavones and lignans are measured in urine samples, enterolignans tend to be more concentrated than isoflavones (Bensaada et al. 2023). This is not the case in plasma. The pharmacokinetics of ENL and END were studied by Kuijsten et al. (2005).

15.3.2 Historical Exposure

In the past, mycotoxins could sometimes appear in culture episodically (Ramos et al. 2011). They were initially thought to develop during storage before being detected directly in the field (Richard 2007). Although plant protection products did not exist, in the past, farmers used to grow their crops at low density and to organize culture rotations and associations that were beneficial for the plants. Nowadays, the increasing yield of plant culture is obtained by plant selection and concentration which are propitious to dissemination of plant diseases with high financial impacts. Fortunately, detection tools now exist and help in mycotoxin management even if losses can be significant (Shabeer et al. 2022).

ZEN and ZEL being mycotoxins their occurrence is tightly monitored and can lead to crop destruction. The oestrogenic effect of ZEN was first reported in 1958 in swine fed moulded corn and then in rat via the uterotrophy test (Stob et al. 1962).

Besides, COUM isolated in 1957 (Bickoff et al. 1957) was considered to be an oestrogenic phytoalexin. Therefore, its occurrence in feed and food is linked to pest attacks especially *Pseudopeziza medicaginis* in alfalfa.

In parallel, reports of the oestrogenic effect of isoflavones date from 1946 (Bennetts et al. 1946) and were first observed in ewes grazing Australian clover pastures. The effects were similar to those observed on swine contaminated by ZEN (Stob et al. 1962). Equol was the first oestrogenic isoflavone discovered in human urine and related to soy intake (Axelson et al. 1984). This discovery led to isoflavone quantification in the urine of Asian people known to eat soy-based food fodder (Adlercreutz et al. 1991). The first measurements were performed using gas chromatography and urine samples collected from Japanese people from a small rural village. These people made their food following traditional recipes and their mean isoflavone urinary excretion was 5.4 μM/day. Ten years later, Arai et al. (2000) measured the isoflavone excretion in the urine of Tokyo citizens. They found a mean excretion level as high as 63.2 μM/day. One of the main difference between rural and urban Japaneses was the origin of their soyfood. In countrysides people had "home made" soyfood while in cities they tend to have industrial soyfood bought in food-shops. Besides, Liu reported a daily soy consumption at 38 g/day in rural Chinese women and an isoflavone exposure mainly below 20 mg/day (Liu et al. 2004). Looking at Table 15.1, it can be seen that the portions of modern soy-food are greater and that the daily exposure to isoflavones can be much higher than via the traditional one. Additionally, Bensaada et al. (2023), measured a large variability of urine excretion levels in French consumers, with values ranging from 0 to 1728 μM, a mean at 33.6 μM and a median at 0.56 μM.

Lignans being present in fruits and vegetables they have always been consumed by human being and vegan people are the most exposed especially if they have grains regularly. Nevertheless, considering the data available so far, the exposure remains low compared to isoflavones and usually in the range of 1–5 mg/day (Canivenc-Lavier and Bennetau-Pelissero 2023). At these doses no effect were ever reported on human fertility.

15.3.3 Reduction Opportunities

As mentioned, COUM, ZEN and ZEL are carefully monitored and their occurrences in human food are reduced to the minimum. Considering 8-PN, the exposure can only induce physiological effects in heavy beer drinkers and in people taking hop-based food-supplements. Because the enterolignans usually express beneficial effects at the doses usually consumed by humans, the only substances which have to be reduced to limit their potential adverse effects on health are isoflavones. Indeed, isoflavones can be used in women to face oestrogen deficiency and in that case their effects are beneficial (Canivenc-Lavier and Bennetau-Pelissero 2023). However, such effects are not required for all people and oestrogenic effects may be deleterious in children, pregnant women, premenopausal women, men and persons with oestrogen-dependent diseases.

Looking at the past soy consumption in rural regions of China it can be seen that isoflavone exposures were limited to a few mg/day (Liu et al. 2004). At these doses the urinary excretion was low and no adverse effect was recorded (Adlercreutz et al. 1991). As mentioned previously, in these rural regions, people were used to cook soybean in a traditional fashion. When traditional recipes were analysed, it could be noticed that dried beans were first rinsed, soaked and simmered in water (Fernandez-Lopez et al. 2016). Indeed, in soybeans and all derived products, isoflavones are essentially under glucoside forms and heating processes can transform them into acetyl or malonyl forms. All conjugated forms are water soluble as can be seen in Table 15.2 which gives Log P values of monoconjugates. When the isoflavones are conjugated with two glycoside-residues their water solubility is even higher. Indeed, the lower the LogP, the higher the water solubility. The aglycone forms in soy generally represent less than 5% of the total isoflavone forms.

Table 15.2 LogP values of the main oestrogenic isoflavones from soybean

Log P substances	Aglycone	Glucoside	Acetyl	Malonyl
Genistein	2.7	0.9	0.9	0.6
Daidzein	2.5	0.7	0.7	0.4
Glycitein	2.4	0.6	0.7	0.4

Data from https://pubchem.ncbi.nlm.nih.gov

Therefore, water treatments allow isoflavones to leak from the soybean matrix (Fernandez-Lopez et al. 2016) and the resulting food is low in isoflavones. This can be observed in Fig. 15.1.

Figure 15.1 shows that the final concentrations of isoflavones obtained in soy-based food following traditional recipes involving water treatments, can be ten times lower than those recorded in commercial foodstuffs industrially produced. This shows that the traditional exposure of Asian people to isoflavones was much probably lower in the past than what is observed nowadays.

Indeed, prolonged soaking, boiling and simmering in renewed water was, in the past, performed to clean soybeans that were stored in rough conditions (Fernandez-Lopez et al. 2016). This legume contains several antinutritional factors, including phytoestrogens and it is not attacked by rodents or birds. In the past, soybeans could be kept in cellars or in farmyards between two cultures. To reduce the antinutritional factors which prevent degradations by pest, the first soy recipes involved water boiling and fermentation steps. The latter required both hydrated and sterilized soy-seeds to avoid spoiling the final product with undesired moults (Liu et al. 2022). This is why all traditional recipes included rinsing, soaking, prolonged boiling or simmering in water. Nowadays, such treatments are difficult to perform at industrial scale since they require tons of water. This is probably why the soy-based commercialized products are so rich in isoflavones. However, as mentioned by Bensaada et al. (2023), water can be recycled after cleaning. This operation allows to recollect the isoflavones which can then be used for specific purposes in specific populations. Finally, Fig. 15.1 shows that, isoflavones can be dramatically reduced in soy-based products following traditional recipes and using domestic water treatments during food processing. Such treatments that significantly reduce human exposure to oestrogenic and anti-thyroid substances, when required, could allow a greater consumption of soy by all human subpopulations.

15.4 The Reported Physiological Consequences of the Exposure to Natural Oestrogenic Substances

15.4.1 Reproductive Impairments in Animal Models and Toxicological Values Derived

As previously mentioned, oestrogens are involved in many mammalian and human physiological processes. Therefore, xenoestrogens can affect human and animal physiology at many levels and endocrine disruptors are studied at different stages of development including foetal, infancy, childhood, adolescence, adulthood and old age. Toxicological studies performed on animal models are usually prolonged through multiple generations including some which are not exposed directly but exposed in utero or are off-springs from animals exposed before mating (NTP 2008). Table 15.3 shows data collected on the toxic effects of the oestrogenic

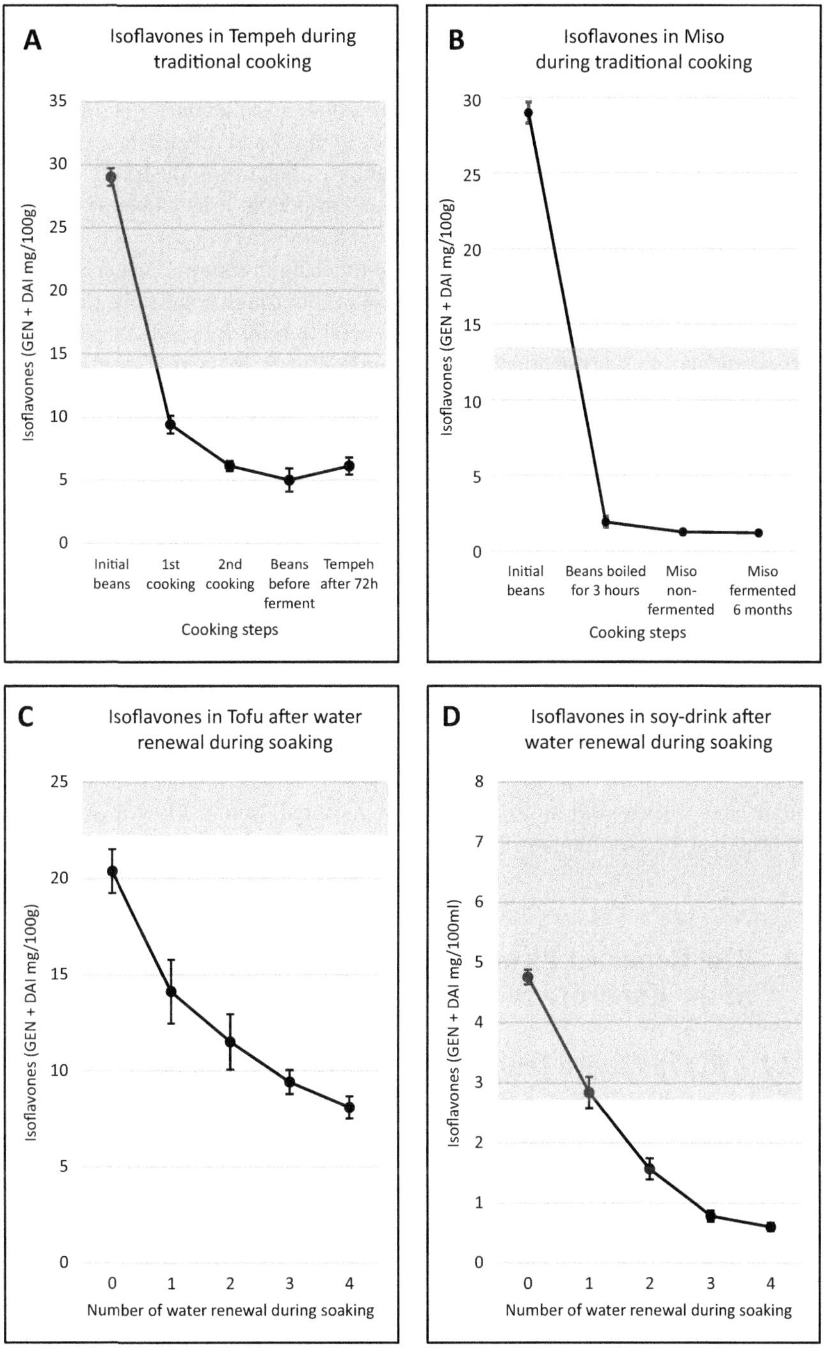

Fig. 15.1 Evolution of isoflavone concentrations in different soy-based food prepared according to traditional recipes. (**a**) tempeh boiled and then soaked twice for 20 min according to an Indonesian recipe, (**b**) miso prepared from soybeans boiled for 3 h in water renewed twice, (**c**) tofu made from soybeans soaked for 9 h in tap water renewed several times, (**d**) soy-milk prepared from soybeans soaked for 9 h in tap water renewed several times. The grey zone covers the lowest ranges of values observed for commercial equivalent foodstuff (Bensaada et al., 2024)

Table 15.3 Reference doses and potential intakes in France for the main natural oestrogenic substances

Compounds	Chemical family	NOAEL or LOAEL[a] in animal	Model species	Theoretical reference dose for human[b] (RfD)	Potential intake in France	References
Diethylstilbestrol	E_2 analogue	NOAEL 5 mg/kg/day	Rat	0.05 mg/kg bw/day	Drug forbidden	NIH/PubChem (2023)
8-PN	Phytoestrogen	–	Human	400 μg/day[a]	No data available	Arrete Royal (1997)
Coumestrol	Phytoestrogen	LOAEL 100 mg/kg/day	Mouse	0.33 mg/kg bw/day	0.016 μg/kg bw/day	Pocock et al. (2002), Anses (2011)
Zearalenone	Mycotoxin	NOAEL 40 μg/kg/day	Pig	0.4 μg/kg bw/day	0.042 μg/kg bw/day	WHO-JECFA (1993), Anses (2011)
Genistein	Phytoestrogen	LOAEL 35 mg/kg/day	Rat	0.12 mg/kg bw/day	0–1.5 mg/kg bw/day	NTP (2008), Lee et al. (2022)
Daidzein	Phytoestrogen	NOAEL 50 mg/kg/day	Hen	0.5 mg/kg bw/day	0–0.8 mg/kg bw/day	Shi et al. (2013), Lee et al. (2022)
Biochanin A	Phytoestrogen	LOAEL 25 mg/kg/day	Rat	0.083 mg/kg bw/day	0.0003 mg/kg bw/day	Soujanya et al. (2016), Anses (2011)
Formononetin	Phytoestrogen	NOAEL 5 mg/kg/day	Mouse	0.05 mg/kg bw/day	0.0013 mg/kg bw/day	Parandin and Mohammadi (2019), Anses (2011)

[a]The value is the limit fixed for human adults by the Belgian health authorities
[b]Safety factor from LOAEL to NOAEL = 3 in this table

compounds evoked in this chapter. They include No Observable Adverse Effect Levels (NOAEL) or Lowest Observable Adverse Effect Levels (LOAEL) defined experimentally on animal models and theoretical Reference doses for humans obtained by dividing NOAEL by 100 and LOAEL by 300. Beside are given examples of potential intakes in France.

As seen in this table, the only compounds that can potentially overcome their theoretical reference values are soy isoflavones, *i.e.* GEN and DAI, because their concentrations in modern soy-food are high and that these soy-food are diverse and attractive. Consequently they can be associated in the same day or even in the same meal.

As mentioned previously, in Australian ewes grazing subterranean clover, dietary oestrogenic isoflavones and coumestrol were shown to induce persistent oestrus, vaginal secretion, infertility, stillbirths, dystocia and prolapse of the uterus. In rams,

reduced sperm count, prostate and bulbourethral gland pathologies were observed. Additionally, abnormal secretions of mammary fluid were also shown in immature females and in castrated males (Bennetts et al. 1946; Farnsworth et al. 1975). Within 4 years, the proportions of females with lambs were reduced from 80% to less than 10% and rearing activity was dropped out by many Australian farmers. In other cattle species like cow or swine, fertility problems were also recorded (see Bennetau-Pelissero 2018 for a review). Looking at the effects observed in rats, dogs or monkeys, the natural oestrogenic substances can reduce age at vaginal opening in females, induce abnormal cycles and mammary development (NTP 2008). They can also decrease sperm production in males (Perez-Rivero et al. 2009). These compounds were also shown by OECD to induce uterotrophy at dietary levels (Owens et al. 2003) and the use of soy-containing diet was discouraged especially when endocrine disruption was studied. Rat exposure, in early life, can disrupt the mating behaviour of females and males (Patisaul 2017). Natural oestrogenic substances were shown to reduce litter size and the anogenital distance in pups (NTP 2008). Phytoestrogens were also shown to decrease the reproductive life of female rats by inducing premature ovarian apoptosis (Moon et al. 2009).

15.4.2 Reproductive Impairments In Humans

The review by Canivenc-Lavier and Bennetau-Pelissero (2023) made a valuable point on the reproductive disorders recorded nowadays in humans exposed to effective doses of phytoestrogens. From this review, it appears that soy isoflavones with oestrogenic and anti-thyroid effects are active when ingested at doses over 20 mg/day for an adult of 60 kg body-weight. Such a dose should be declined accordingly to body-weight in children but probably not in infants experiencing minipuberty. At this specific physiological stage, endocrine disruptors being man-made or natural should be avoided. As shown in Table 15.3, the only substances which can nowadays overcome their active doses in humans are the soy isoflavones because of their high concentrations in industrial foodstuffs (Table 15.1).

Recently, it was shown that infants exposed during minipuberty to high amounts of isoflavones (*i.e.* 4–11 mg/kg/day) through soy-based infant formulas are more frequently affected by uterine fibrosis or endometriosis (Upson et al. 2016). These women also report irregular menstrual cycles and pains, and epigenetics marks were found in the vaginal tissues of young girls neonatally exposed to soy-based infant formula (see Canivenc-Lavier and Bennetau-Pelissero 2023 and references). In 4-months-old boys fed soy from birth or from the third week of life a reduction of the testicular size measured by ultrasounds was noticed (Gilchrist et al. 2010). Gender play behaviour was also affected in 42-months-old girls fed soy-based infant formula during their first months of life (Adgent et al. 2011). Although the steroid hormones in plasma were not found so far to be affected in infants fed soy-based formulas, these data strongly suggest that GnRH, FSH and LH secretions could be altered by high isoflavone exposure during minipuberty. This was reported

in marmoset neonatally fed either human cow-milk-based formulas or soy-based infant formulas (Sharpe et al. 2002). Interestingly, the results obtained on human (Adgent et al. 2011) and in non human primate (Sharpe et al., 2002) are the only allowing to determine specifically the effect of xenoestrogens in the neonatal period. This is because it is always ethically disputable to obtain scientific data exposing human newborns. Indeed, when other endocrine disruptors are considered the exposure usually starts to occur in utero like for DES or other human-made disruptors. Therefore, it is impossible to sort the effect of prenatal and postnatal exposure. In the case of soy-based infant formulas, the isoflavone concentrations are so huge in soy-based infant formula, that their effects can be distinguished from the prenatal period when isoflavones only reach foetus via their mother tissues. According to Setchell (1998), the isoflavone plasma levels of infants fed soy-based formulas is 1,000,000 times higher than those observed in breast-fed infants. Adgent et al. (2011) also suggest that, like in rodents or in non-human primates, an exposure to xenoestrogens during minipuberty may, at least transiently, affect gender-behaviour in humans.

In their review, Canivenc-Lavier and Bennetau-Pelissero (2023) mentioned that 8-PN and soy isoflavones were found to disturb women's menstrual cycles. Soy isoflavones can be active at doses of 45 mg/day. Chronic isoflavones exposures were responsible for a delayed LH surge occurring at mid cycle, resulting in a 2 days increase of menstrual cycle duration. This dose seems also responsible for luteal deficiency in American premenopausal women. An increased risk of never being pregnant was observed in American women from the Adventist of the 7th day community who have a mean isoflavone intake of 50 mg/day (Jacobsen et al. 2014). This threshold dose is also mentioned to induce proliferation of normal breast cells, in premenopausal women (McMichael-Phillips et al. 1998). Additionally, in adults, reproductive anomalies were observed when isoflavones were over-consumed for long periods of time. Indeed, Chandrareddy et al. (2008) reported cases of women with endometriosis, fibroids and whose contraceptive treatment was impaired by daily and large soy intakes. In parallel, five population studies linked high isoflavones levels in men's biological fluids and reduced sperm quality and count (see Canivenc-Lavier and Bennetau-Pelissero 2023 for details). Finally, Imai et al. (2022) reported the case of a man consuming 1.4 l of soy-drinks per day for 4 years. He presented an idiopathic hypogonadism, gynaecomastia and altered libido. Soy arrest allowed restoring gonadotrophins and testosterone plasma levels within 1 month.

15.4.3 Other Consequences on Human Pathologies

As mentioned previously there are concerns regarding the role of natural oestrogens in oestrogen-related pathologies. When affections are due to oestrogen deficiencies, natural oestrogens can be beneficial (Canivenc-Lavier and Bennetau-Pelissero 2023). However, when oestrogens induce pathologies, controversies emerged. This

includes breast, uterus or prostate cancers as well as immune diseases. In addition, for isoflavones and coumestrol, thyroid-dependent diseases should also be considered.

Several meta-analyses suggest that natural oestrogens can relief from menopausal symptoms (Sarri et al. 2017). This was tested with 8-PN from hops, lignans from linseed and isoflavones from soybean, clover, alfalfa or kudzu. However, coumestrol that could sometimes be present in alfalfa extracts without being analysed may be responsible for some effects and could partly explain discrepancies observed between studies (Franco et al. 2016). Nevertheless, studies were not all convergent mainly due to large variations in the quality of plant resources. The better characterized the source, the more convincing the beneficial effects (Canivenc-Lavier and Bennetau-Pelissero 2023).

Osteoporosis occurs mainly with oestrogen deficiencies although other factors like lacks of calcium and vitamin D, low level of physical activity and ageing play a role in the occurrence of this pathology. Osteoporosis is classically delayed using HRTs but these treatments are less prescribed since the results of the Women's Health Initiative were largely diffused. Since then, there was an increasing interest in alternative therapies for pathologies linked to oestrogen deficiency and experiments and meta-analyses tended to decipher the role of phytoestrogens on bone health. The compounds most studied were soy isoflavones and recently a positive effect of isoflavones on bone mineral density (BMD) in post-menopausal women was reported (Inpan et al. 2024). The improvement was significant at the lumbar spine, the femoral neck and distal radius. However, this required treatments with at least 50 mg/day of GEN for more than 12 months. In soybean extracts, DAI usually represents half the amount of GEN. This means that the effective dose of isoflavones is about 80 mg/day. Such results sustain the oestrogenic potency of soybean isoflavones in humans.

While there are no significant link between phytoestrogens including ZEN and ZEL and endometrial cancers, discrepancies exist about the links between natural oestrogenic substances and breast cancers. At the levels authorized neither ZEN, ZEL, COUM or 8-PN showed any positive correlations with breast cancers in women populations. Additionally, ENL and END which are ingested through vegetables, fruits and grains are associated with reduced risk of breast cancers (Canivenc-Lavier and Bennetau-Pelissero 2023). However, data obtained in vitro on oestrogen-responsive human breast cancer cells showed that all phytoestrogens except ENL and END exhibited a proliferative activity at relevant doses (Canivenc-Lavier and Bennetau-Pelissero 2023). In athymic nude mice, oestrogenic isoflavones and not ENL and END enhanced the proliferation of human breast cancer cells implanted, despite the presence of letrozole or tamoxifen (Andrade et al. 2015 and references included). In healthy women, soy-based supplements containing GEN and DAI increase normal breast cell proliferation (McMichael-Phillips et al. 1998) and in women with oestrogen-dependent breast cancers a soy-based supplementation for duration <30 days increases the expression of tumour genes involved in cell proliferation (Shike et al. 2014). Despite all these converging data showing an oestrogenic effect of isoflavones in vitro and in vivo on oestrogen-dependent

breast cells, population studies tend to show that soy is protective against breast cancer. This has been shown at least in Asian populations exposed to isoflavones from childhood in an Asian context (Chen et al. 2014). The beneficial association is not so clear in Western countries and this could be linked to genetic and/or environmental factors. The current hypothesis is that GEN and possibly other soy-compounds interact epigenetically with DNA preventing the transcription of pro-oncogenes. This hypothesis would explain why an early exposure to significant doses of isoflavones, like in Asian countries, could prevent breast cancer events occurring later in life. Additionally, this hypothesis would combine both genetic and environmental parameters. Because in Western countries, the exposure to isoflavones is incidental and much lower than in Asia, the protection is generally not observed.

Prostate cancers (PCs) are under T control but ERs and aromatase being present in prostate cells, the action of E_2 on some PC is possible. When soy or isoflavones are tested on human PCs, a reduction of the pathology can sometimes be observed. However, this is controversial and is probably due to the type of tumour considered (Bennetau-Pelissero 2018).

Finally, isoflavones not only interact with the oestrogen pathways but also disturb the thyroid function. In vitro, isoflavones can reduce TPO (Thyroid PerOxidase) activity and the production rates of thyroid hormone (Sosvorová et al. 2012) and can bind to the thyroid hormone transporter transthyretin reducing the bioavailability of T_3 and T_4 (Radović et al. 2006). Additionally, isoflavones can bind to the thyroid receptors TRα and TRβ inducing a competitive inhibition on the action of thyroid hormones (Ariyani et al. 2018). Resulting effects can be seen at systemic level especially in hypothyroid patients on Levothyroxine (Conrad et al. 2004). Cases of soy impacts on hypothyroid patients are exposed in Canivenc-Lavier and Bennetau-Pelissero (2023). Additionally, Sathyapalan et al. (2011) showed that 20 mg of soy isoflavones can aggravate the status of women patients with mild hypothyroidism and Tonstad et al. (2016) also observed that soy intake was positively correlated with TSH levels only in American women from the Adventist of the 7th day. Finally, COUM was associated with the occurrence of thyroid cancer in humans (Wang et al. 2020).

Considering phytoestrogens and immunity, few data are presently available. COUM and soybean isoflavones could increase the deleterious symptoms of SLE in animal models but there are no convincing data in human so far (Canivenc-Lavier and Bennetau-Pelissero 2023).

15.5 Conclusions

Natural oestrogenic substances are present in plants and fungi. They were initially identified via their deleterious effects in cattle, and several plants containing these substances were used as antifertility agents in traditional medicines. Nowadays, isoflavones from legumes are the oestrogenic substances that are the most frequent

in the human environment and which deserve most concerns. The exposure remains low in the general population except for regular modern soybean eaters for whom the regular daily intake can overcome the theoretical reference doses obtained from toxicological models. This is the case in a significant proportion of Chinese, Japanese, Korean, Vietnamese and Taiwanese people and also in vegan or vegetarian people in the Western countries. Physiological effects of these substances were demonstrated in humans at different physiological stages including infancy, childhood, adulthood and menopause in women. Other effects are still controversial or doubtful. These effects were either beneficial or adverse depending on the level of exposure and the physiological stage of the consumers. An active dose could be identified in adults and derived for children. Thus, modern soy-food intended for the general population would better contain low isoflavone concentrations and these oestrogenic substances would better be reserved for specific applications in specific populations. Finally and because soybean is a very interesting pulse with the best and largest proportion of edible proteins, fibres and poly-unsaturated fatty acids, its consumption by human population should increase in the future. However, the exposure to oestrogenic isoflavones should be reduced, in a context where many other endocrine disruptors are present in the environment. Solutions exist implying water treatments that were previously used in traditional cooking practices. Thus, efforts should be made by soy-food producers to reduce isoflavones concentrations in their foodstuffs. This would allow soy-based products to be more largely consumed by all categories of human populations.

References

Adgent MA, Daniels JL, Edwards LJ, Siega-Riz AM, Rogan WJ (2011) Early-life soy exposure and gender-role play behavior in children. Environ Health Perspect 119(12):1811–1816

Adlercreutz H, Honjo H, Higashi A, Fotsis T, Hämäläinen E, Hasegawa T, Okada H (1991) Urinary excretion of lignans and isoflavonoid phytoestrogens in Japanese men and women consuming a traditional Japanese diet. Am J Clin Nutr 54(6):1093–1100

Alderman MH 3rd, Taylor HS (2021) Molecular mechanisms of estrogen action in female genital tract development. Differentiation 118:34–40

Andrade JE, Ju YH, Baker C, Doerge DR, Helferich WG (2015) Long-term exposure to dietary sources of genistein induces estrogen-independence in the human breast cancer (MCF-7) xenograft model. Mol Nutr Food Res 59(3):413–423

Anses (2011) Étude de l'alimentation totale (EAT 2) Contaminants inorganiques, minéraux, polluants organiques persistants, mycotoxines, phyto-estrogènes. Avis de l'Anses, Rapport d'expertise. ISBN: 978-2-11-128112-7

Arai Y, Uehara M, Sato Y, Kimira M, Eboshida A, Adlercreutz H, Watanabe S (2000) Comparison of isoflavones among dietary intake, plasma concentration and urinary excretion for accurate estimation of phytoestrogen intake. J Epidemiol 10(2):127–135

Ariyani W, Iwasaki T, Miyazaki W, Yu L, Takeda S, Koibuchi N (2018) A possible novel mechanism of action of genistein and daidzein for activating thyroid hormone receptor-mediated transcription. Toxicol Sci 164(2):417–427

Arrete Royal du 29 Aout 1997. Relatif à la fabrication et au commerce de denrées alimentaires composées ou contenant des plantes ou préparations de plantes (M.B. 21.XI.1997)

Axelson M, Sjövall J, Gustafsson BE, Setchell KD (1984) Soya—a dietary source of the non-steroidal oestrogen equol in man and animals. J Endocrinol 102(1):49–56
Bennetau-Pelissero C (2018) Natural estrogenic substances, origins, and effects. In: Mérillon J-M, Ramawat KG (eds) Bioactive molecules in food, Reference series in phytochemistry. Springer, Cham. 70 p
Bennetts HW, Underwood EJ, Shier FL (1946) A specific breeding problem of sheep on subterranean clover pastures in Western Australia. Aust Vet J 22(1):2–12
Bensaada S, Raymond I, Pellegrin I, Viallard JF, Bennetau-Pelissero C (2023) Validation of ELISAs for isoflavones and enterolactone for phytoestrogen intake assessment in the French population. Nutrients 15(4):967–985
Bensaada S, Peruzzi G, Cubizolles L, Denayrolles M, Bennetau-Pelissero C (2024) Traditional and domestic cooking dramatically reduce estrogenic isoflavones in soy foods. Foods 13(7):999–1018
Bickoff EM, Booth AN, Lyman RL, Livingston AL, Thompson CR, Deeds F (1957) Coumestrol, a new estrogen isolated from forage crops. Science 126(3280):969–970
Canivenc-Lavier M-C, Bennetau-Pelissero C (2023) Phytoestrogens and health effects. Nutrients 15:317–361
Carballo D, Pallarés N, Ferrer E, Barba FJ, Berrada H (2021) Assessment of human exposure to deoxynivalenol, ochratoxin A, zearalenone and their metabolites biomarker in urine samples using LC-ESI-qTOF. Toxins (Basel) 13(8):530
Chandrareddy A, Muneyyirci-Delale O, McFarlane SI, Murad OM (2008) Adverse effects of phytoestrogens on reproductive health: a report of three cases. Complement Ther Clin Pract 14(2):132–135
Chen M, Rao Y, Zheng Y, Wei S, Li Y, Guo T, Yin P (2014) Association between soy isoflavone intake and breast cancer risk for pre- and post-menopausal women: a meta-analysis of epidemiological studies. PLoS One 9(2):e89288–e89298
Conrad SC, Chiu H, Silverman BL (2004) Soy formula complicates management of congenital hypothyroidism. Arch Dis Child 89(1):37–40
Cooke PS, Nanjappa MK, Ko C, Prins GS, Hess RA (2017) Estrogens in male physiology. Physiol Rev 97(3):995–1043
Cunha GR, Li Y, Mei C, Derpinghaus A, Baskin LS (2021) Ontogeny of estrogen receptors in human male and female fetal reproductive tracts. Differentiation 118:107–131
Dovey JL, Vasudevan N (2020) Does GPER1 play a role in sexual dimorphism? Front Endocrinol (Lausanne) 11:595895–595903
Farnsworth NR, Bingel AS, Cordell GA, Crane FA, Fong HS (1975) Potential value of plants as sources of new antifertility agents II. J Pharm Sci 64(5):717–754
Fernandez-Lopez A, Lamothe V, Delample M, Denayrolles M, Bennetau-Pelissero C (2016) Removing isoflavones from modern soyfood: Why and how? Food Chem 210:286–294
Franco OH, Chowdhury R, Troup J, Voortman T, Kunutsor S, Kavousi M, Oliver-Williams C, Muka T (2016) Use of plant-based therapies and menopausal symptoms: a systematic review and meta-analysis. JAMA 315(23):2554–2563
Gaweł E (2012) Chemical composition of lucerne leaf extract (EFL) and its applications as a phytobiotic in human nutrition. Acta Sci Pol Technol Aliment 11(3):303–310
Gilchrist JM, Moore MB, Andres A, Estroff JA, Badger TM (2010) Ultrasonographic patterns of reproductive organs in infants fed soy formula: comparisons to infants fed breast milk and milk formula. J Pediatr 156(2):215–220
Girgert R, Emons G, Gründker C (2017) 17β-estradiol-induced growth of triple-negative breast cancer cells is prevented by the reduction of GPER expression after treatment with gefitinib. Oncol Rep 37:1212–1218
Goodman RL, Herbison AE, Lehman MN, Navarro VM (2022) Neuroendocrine control of gonadotropin-releasing hormone: Pulsatile and surge modes of secretion. J Neuroendocrinol 5:e13094

Hu XJ, Song WR, Gao LY, Nie SP, Eisenbrand G, Xie MY (2014) Assessment of dietary phytoestrogen intake via plant-derived foods in China. Food Addit Contam Part A Chem Anal Control Expo Risk Assess 31(8):1325–3135

Imai H, Nishikawa H, Suzuki A, Kodama E, Iida T, Mikura K, Hashizume M, Kigawa Y, Tadokoro R, Sugisawa C, Endo K, Iizaka T, Otsuka F, Nagasaka S (2022) Secondary hypogonadism due to excessive ingestion of isoflavone in a man. Intern Med 61(19):2899–2903

Inpan R, Na Takuathung M, Sakuludomkan W, Dukaew N, Teekachunhatean S, Koonrungsesomboon N (2024) Isoflavone intervention and its impact on bone mineral density in postmenopausal women: a systematic review and meta-analysis of randomized controlled trials. Osteoporos Int 35(3):413–430

Jacobsen BK, Jaceldo-Siegl K, Knutsen SF, Fan J, Oda K, Fraser GE (2014) Soy isoflavone intake and the likelihood of ever becoming a mother: the Adventist Health Study-2. Int J Womens Health 6:377–384

Khan D, Ahmed AS (2016) The immune system is a natural target for estrogen action: opposing effects of estrogen in two prototypical autoimmune diseases. Front Immunol 6:635–643

Kuijsten A, Arts ICW, Vree TB, Hollman PCH (2005) Pharmacokinetics of enterolignans in healthy men and women consuming a single dose of secoisolariciresinol diglucoside. J Nutr 135:795–801

Kuiri-Hänninen T, Sankilampi U, Dunkel L (2014) Activation of the hypothalamic-pituitary-gonadal axis in infancy: minipuberty. Horm Res Paediatr 82(2):73–80

Lee A, Bensaada S, Lamothe V, Lacoste M, Bennetau-Pelissero C (2022) Endocrine disruptors on and in fruits and vegetables: Estimation of the potential exposure of the French population. Food Chem 373(Pt B):131513

Liu Z, Li W, Sun J, Liu C, Zeng Q, Huang J, Yu B, Huo J (2004) Intake of soy foods and soy isoflavones by rural adult women in China. Asia Pac J Clin Nutr 13(2):204–209

Liu L, Chen X, Hao L, Zhang G, Jin Z, Li C, Yang Y, Rao J, Chen B (2022) Traditional fermented soybean products: processing, flavor formation, nutritional and biological activities. Crit Rev Food Sci Nutr 62(7):1971–1989

Luo J, Liu D (2020) Does GPER really function as a G protein-coupled estrogen receptor in vivo? Front Endocrinol 11:148–161

Mathey J, Lamothe V, Coxam V, Potier M, Sauvant P, Bennetau-Pelissero C (2006) Concentrations of isoflavones in plasma and urine of post-menopausal women chronically ingesting high quantities of soy isoflavones. J Pharm Biomed Anal 41(3):957–965

McMichael-Phillips DF, Harding C, Morton M, Roberts SA, Howell A, Potten CS, Bundred NJ (1998) Effects of soy-protein supplementation on epithelial proliferation in the histologically normal human breast. Am J Clin Nutr 68(6 Suppl):1431S–1435S

Mizejewski GJ (2004) Biological roles of alpha-fetoprotein during pregnancy and perinatal development. Exp Biol Med (Maywood) 229(6):439–463

Molina-Montes E, Sánchez MJ, Zamora-Ros R, Bueno-de-Mesquita HB, Wark PA, Obon-Santacana M, Kühn T, Katzke V, Travis RC, Ye W, Sund M, Naccarati A, Mattiello A, Krogh V, Martorana C, Masala G, Amiano P, Huerta JM, Barricarte A, Quirós JR, Weiderpass E, Angell Åsli L, Skeie G, Ericson U, Sonestedt E, Peeters PH, Romieu I, Scalbert A, Overvad K, Clemens M, Boeing H, Trichopoulou A, Peppa E, Vidalis P, Khaw KT, Wareham N, Olsen A, Tjønneland A, Boutroun-Rualt MC, Clavel-Chapelon F, Cross AJ, Lu Y, Riboli E, Duell EJ (2016) Flavonoid and lignan intake and pancreatic cancer risk in the European prospective investigation into cancer and nutrition cohort. Int J Cancer 139(7):1480–1492

Moon HJ, Seok JH, Kim SS, Rhee GS, Lee RD, Yang JY, Chae SY, Kim SH, Kim JY, Chung JY, Kim JM, Chung SY (2009) Lactational coumestrol exposure increases ovarian apoptosis in adult rats. Arch Toxicol 83(6):601–608

Murai U, Sawada N, Charvat H, Inoue M, Yasuda N, Yamagishi K, Tsugane S (2022) JPHC Study Group. Soy product intake and risk of incident disabling dementia: the JPHC Disabling Dementia Study. Eur J Nutr 61:4045–4057

NIH/PubChem (2023) Diethylstilbestrol. https://pubchem.ncbi.nlm.nih.gov/compound/Diethylstilbestrol

NTP (2008) Multigenerational reproductive study of genistein (Cas No. 446-72-0) in Sprague-Dawley rats (feed study). Natl Toxicol Program Tech Rep Ser 539:1–266
Owens W, Ashby J, Odum J, Onyon L (2003) The OECD program to validate the rat uterotrophic bioassay. Phase 2: dietary phytoestrogen analyses. Environ Health Perspect 111(12):1559–1567
Parandin R, Mohammadi L (2019) The effects of formononetin derived from red clover during pregnancy on puberty, some reproductive parameters and lordosis behavior of female mice. Armaghane Danesh 24(2–133):199–213
Pasterski V, Acerini CL, Dunger DB, Ong KK, Hughes IA, Thankamony A, Hines M (2015) Postnatal penile growth concurrent with mini-puberty predicts later sex-typed play behavior: Evidence for neurobehavioral effects of the postnatal androgen surge in typically developing boys. Horm Behav 69:98–105
Patel HK, Bihani T (2018) Selective estrogen receptor modulators (SERMs) and selective estrogen receptor degraders (SERDs) in cancer treatment. Pharmacol Ther 186:1–24
Patisaul HB (2017) Endocrine disruption by dietary phyto-oestrogens: impact on dimorphic sexual systems and behaviours. Proc Nutr Soc 76(2):130–144
Perez-Rivero JJ, Martinez-Maya JJ, Perez-Martinez M, Aguilar-Setien A, Garcia-Suarez MD, Serrano H (2009) Phytoestrogen treatment induces testis alterations in dogs. Potential use in population control. Vet Res Commun 33(1):87–95
Pocock VJ, Sales GD, Milligan SR (2002) Comparison of the oestrogenic effects of infant milk formulae, oestradiol and the phytoestrogen coumestrol delivered continuously in the drinking water to ovariectomised mice. Food Chem Toxicol 40:643–651
Pohjanvirta R, Nasri A (2022) The potent phytoestrogen 8-prenylnaringenin: a friend or a foe? Int J Mol Sci 23(6):3168–3198
Prior JC (2020) Women's reproductive system as balanced estradiol and progesterone actions—A revolutionary, paradigm-shifting concept in women's health. Drug Discov Today Dis Models Womens Health 32:31–40
Radović B, Mentrup B, Köhrle J (2006) Genistein and other soya isoflavones are potent ligands for transthyretin in serum and cerebrospinal fluid. Br J Nutr 95(6):1171–1176
Ramos A, Sanchis V, Marín S (2011) The prehistory of mycotoxins: related cases from ancient times to the discovery of aflatoxins. World Mycotox J 4(2):101–112
Reger MK, Zollinger TW, Liu Z, Jones JF, Zhang J (2018) Dietary intake of isoflavones and coumestrol and the risk of prostate cancer in the prostate, lung, colorectal and ovarian cancer screening trial. Int J Cancer 142(4):719–728
Rey R, Josso N, Racine C (2020) Sexual differentiation. In: Feingold KR, Anawalt B, Blackman MR et al (eds) Endotext [Internet]. MDText.com, Inc., South Dartmouth, MA. Available from https://www.ncbi.nlm.nih.gov/books/NBK279001/. Updated 27 May 2020
Richard JL (2007) Some major mycotoxins and their mycotoxicoses—an overview. Int J Food Microbiol 119(1–2):3–10
Rochira V, Carani C (2023) Estrogens, male reproduction and beyond. In: Feingold KR, Anawalt B, Blackman MR et al (eds) Endotext [Internet]. MDText.com, Inc., South Dartmouth, MA. Updated 21 Mar 2023. Available from https://www.ncbi.nlm.nih.gov/sites/books/NBK278933/
Sarri G, Pedder H, Dias S, Guo Y, Lumsden MA (2017) Vasomotor symptoms resulting from natural menopause: a systematic review and network meta-analysis of treatment effects from the National Institute for Health and Care Excellence guideline on menopause. BJOG 124(10):1514–1523
Sathyapalan T, Manuchehri AM, Thatcher NJ, Rigby AS, Chapman T, Kilpatrick ES, Atkin SL (2011) The effect of soy phytoestrogen supplementation on thyroid status and cardiovascular risk markers in patients with subclinical hypothyroidism: a randomized, double-blind, cross-over study. J Clin Endocrinol Metab 96(5):1442–1449
Setchell KD (1998) Phytoestrogens: the biochemistry, physiology, and implications for human health of soy isoflavones. Am J Clin Nutr 68(6 Suppl):1333S–1346S
Shabeer S, Asad S, Jamal A, Ali A (2022) Aflatoxin contamination, its impact and management strategies: an updated review. Toxins (Basel) 14(5):307

Sharpe RM, Martin B, Morris K, Greig I, McKinnell C, McNeilly AS, Walker M (2002) Infant feeding with soy formula milk: effects on the testis and on blood testosterone levels in marmoset monkeys during the period of neonatal testicular activity. Hum Reprod 17(7):1692–1703

Shi SR, Gua H, Chang LL, Wang ZY, Tong HB, Zou JM (2013) Safety evaluation of daidzein in laying hens: Part I. Effects on laying performance, clinical blood parameters, and organs development. Food Chem Toxicol 55:684–688

Shike M, Doane AS, Russo L, Cabal R, Reis-Filho JS, Gerald W, Cody H, Khanin R, Bromberg J, Norton L (2014) The effects of soy supplementation on gene expression in breast cancer: a randomized placebo-controlled study. J Natl Cancer Inst 106(9):189–201

Shin BS, Hong SH, Bulitta JB, Lee JB, Hwang SW, Kim HJ, Yang SD, Yoon HS, Kim DJ, Lee BM, Yoo SD (2009) Physiologically based pharmacokinetics of zearalenone. J Toxicol Environ Health A 72(21-22):1395–1405

Shinkaruk S, Durand M, Lamothe V, Carpaye A, Martinet A, Chantre P, Vergne S, Nogues X, Moore N, Bennetau-Pelissero C (2012) Bioavailability of glycitein relatively to other soy isoflavones in healthy young Caucasian men. Food Chem 135(3):1104–1111

Sosvorová L, Mikšátková P, Bičíková M, Kaňová N, Lapčík O (2012) The presence of monoiodinated derivates of daidzein and genistein in human urine and its effect on thyroid gland function. Food Chem Toxicol 50(8):2774–2779

Soto AM, Sonnenschein C, Chung KL, Fernandez MF, Olea N, Serrano FO (1995) The E-SCREEN assay as a tool to identify estrogens: an update on estrogenic environmental pollutants. Environ Health Perspect 103(Suppl 7):113–122

Soujanya MGS, Prathap Reddy K, Sridevi V, Ramachandra Reddy P, Sreenivasula RP (2016) In utero exposure of biochanin-A alters female reproduction in rat. J Clin Mol Endocrinol 1(2):08

Starrett JR, Moenter SM (2023) Hypothalamic kisspeptin neurons as potential mediators of estradiol negative and positive feedback. Peptides 163:170963–170973

Stob M, Baldwin RS, Tuite J, Andrews FN, Gillette KG (1962) Isolation of an anabolic, uterotrophic compound from corn infected with Gibberella zeae. Nature 196:1318

Tal R, Taylor HS (2021) Endocrinology of pregnancy in comprehensive free online endocrinology book. In: Feingold KR, Anawalt B, Blackman MR et al (eds) Endotext [Internet]. MDText.com, South Dartmouth, MA

Tonstad S, Jaceldo-Siegl K, Messina M, Haddad E, Fraser GE (2016) The association between soya consumption and serum thyroid-stimulating hormone concentrations in the Adventist Health Study-2. Public Health Nutr 19(8):1464–1470

Tu Y, Yang Y, Li Y, He C (2021) Naturally occurring coumestans from plants, their biological activities and therapeutic effects on human diseases. Pharmacol Res 169:105615–105639

Upson K, Harmon QE, Baird DD (2016) Soy-based infant formula feeding and ultrasound-detected uterine fibroids among young African-American women with no prior clinical diagnosis of fibroids. Environ Health Perspect 124(6):769–776

van Breemen RB, Yuan Y, Banuvar S, Shulman LP, Qiu X, Alvarenga RF, Chen SN, Dietz BM, Bolton JL, Pauli GF, Krause E, Viana M, Nikolic D (2014) Pharmacokinetics of prenylated hop phenols in women following oral administration of a standardized extract of hops. Mol Nutr Food Res 58(10):1962–1969

Wallen K, Maestripieri D, Mann DR (1995) Effects of neonatal testicular suppression with a GnRH antagonist on social behavior in group-living juvenile rhesus monkeys. Horm Behav 29(3):322–337

Wang Q, Huang H, Zhao N, Ni X, Udelsman R, Zhang Y (2020) Phytoestrogens and thyroid cancer risk: a population-based case-control study in Connecticut. Cancer Epidemiol Biomarkers Prev 29(2):500–508

WHO-JECFA (1993) Zearalenone. Evaluations of the Joint FAO/WHO Expert Committee on Food Additives (JECFA). TRS 896-JECFA 53/93

Zhu J, Qi Zhao Q, Qiu Y, Zhang Y, Cui S, Yu Y, Chen B, Zhu M, Wang N, Liu X, Jiang Y, Xu W, Zhao G (2021) Soy isoflavones intake and obesity in Chinese adults: a cross-sectional study in Shanghai, China. Nutrients 13(8):2715–2729

Part IV
New Applications of Biotoxins

Chapter 16
Therapeutic Potential of Biotoxins and Biotechnological Innovations

Prince Kumar Singh, Gereraj Sen Gupta, Indrajeet Kumar, Umesh Kumar, Parvati Madheshiya, Akanksha Dwivedi, Priyanka Singh, Ashish Kumar Mishra, Saumya Mishra, Shivani Gupta, Rajesh Kumar Sharma, and Supriya Tiwari

Abstract The captivating world of biotoxins, often perceived as nature's formidable weapons, presents an intricate and promising landscape for scientific exploration and innovation. Biotoxins, originating from a vast spectrum of life forms such as bacteria, fungi, plants, and animals, exhibit a remarkable diversity of bioactive compounds. Their structural diversity, coupled with potent biological activities, positions them as valuable resources for drug discovery, agricultural and environmental management. This chapter highlights into specific case studies where biotoxins have been harnessed to develop antivenom, anti-cancer agents, analgesics, other therapeutic marvels, etc. The comprehensive understanding of biotoxin mechanisms, bioprospecting, and biotechnological tools, including genetic engineering and synthetic biology, offers exciting avenues for the sustainable extraction, production, and modification of biotoxins. The biotechnological harnessing of biotoxins emphasizes the significant progress and potential future applications of biotoxins reliant on collaboration among researchers, clinicians, and regulators to unlock their full capacity.

Keywords Biotoxins · Bioactive compound · Antivenom · Anti-cancerous · Environmental management

16.1 Introduction to Biotoxins: Nature's Powerful Agents

Toxins include both biological and chemical compounds derived from living and non-living beings, although biotoxins are unique among toxic substances as they come solely from living organisms (Lecointre et al. 2023). Biotoxins, pervasive

P. K. Singh · G. S. Gupta · I. Kumar · U. Kumar · P. Madheshiya · A. Dwivedi · P. Singh · A. K. Mishra · S. Mishra · S. Gupta · R. K. Sharma · S. Tiwari (✉)
Laboratory of Ecotoxicology, Department of Botany, Institute of Science, Banaras Hindu University, Varanasi, Uttar Pradesh, India

M. N. V. Prasad et al. (eds.), *Biotoxins*,
https://doi.org/10.1007/978-3-031-75309-1_16

throughout nature, have long fascinated scientists and elicited admiration for their formidable potency and adaptability. These compounds, sourced from plants, animals, and microorganisms, form a diverse array of chemical defenses finely tuned for different ecological functions, such as attacking, defending, and hunting for prey (Parak et al. 2024). The toxicity of biotoxins is contingent upon both dosage and host susceptibility, with each toxin possessing a defined safe threshold beyond which adverse effects manifest (Edo et al. 2024). Generally, biotoxins exhibit a remarkable capacity to selectively target specific molecular pathways and biological processes with unparalleled precision (Wang et al. 2023a, b). Their structural diversity further amplifies their potential as reservoirs of chemical scaffolds for drug discovery, presenting an invaluable resource in the pursuit of novel therapeutics (Ouassaf et al. 2023). Through elucidating the intricate molecular mechanisms dictating their actions, scientists can harness the therapeutic potential of biotoxins to engineer precise interventions for a spectrum of human diseases. However, utilization of nature's resources for human well-being necessitates thoughtful conservation of our ecological resources and equitable access to scientific advancements.

Over the past few years, the field of biotechnology has experienced a significant evolution, enabling researchers to explore the immense therapeutic possibilities of biotoxins (Bouley et al. 2023). Nature, within the realm of biotechnology, emerges as an infinite reservoir of inspiration for groundbreaking innovation and exploration. Utilizing the principles of evolutionary design and molecular engineering, researchers seek to personalize biotoxins for targeted biomedical applications and therapeutic outcomes (Wang et al. 2023a, b). Through deliberate genetic manipulation and evolution, scientists manipulate the genetic blueprint of biotoxins to amplify their potency, specificity, and pharmacokinetic profiles. The pivotal role of biotoxins in addressing drug discovery, antitoxin, and cancer therapy underscores the extraordinary capacity of nature's pharmacopeia in combating human maladies (Linton 2005). Through the integration of advanced research with clinical perspectives, establish intricate molecular pathways and cellular targets through which biotoxins influence pain perception, and integrating advanced research and clinical insights help to elucidate the molecular pathways and cellular targets through which biotoxins modulate pain perception and tumor progression (Pereira et al. 2023).

Transition from nature's laboratory to clinical application presents a multitude of challenges and despite their vast potential in healthcare, biotoxins pose significant hurdles in drug development due to their intricate structures and complex biological activities (Javaid et al. 2021). Interdisciplinary collaboration and innovative approaches are imperative for mitigating challenges such as resource scarcity, poor absorption kinetics, and off-target effects, thereby facilitating a thorough exploration of biotoxins' biotechnological advancements and therapeutic prospects (Vincy et al. 2022). By deciphering the mysterious properties of nature's biotoxin resources, new pathways are uncovered for combating diseases and improving human health in the foreseeable future (Pradhan and Ki 2022). Furthermore, effective management of biotoxin risks necessitates concerted efforts and partnerships among governmental agencies, non-governmental organizations, and stakeholders across various sectors. Through regulatory oversight, public education, and outreach

initiatives, governments play a significant role in safeguarding public health and preserving environmental integrity in the face of biotoxin-related threats.

16.2 Exploring the Diverse Sources of Biotoxins: Understanding Their Origins

Biotoxins, originating from a diverse range of organisms, i.e., microbes, algae, fungi, plant, and animal species (Fig. 16.1), intricately influence ecological dynamics and human health outcomes, emphasizing the necessity for comprehensive strategies to address their broad implications (Pavaux et al. 2020). Identifying the primary reservoirs of toxins enables us to delineate the intricate routes through which these compounds infiltrate ecosystems and ascend through food chains, empowering us to enhance the precision of our predictive frameworks and anticipate the ecological and societal impacts more effectively (Edo et al. 2024). Understanding the fundamental origins of biotoxins facilitates the evaluation of their spatiotemporal dynamics, intricately influenced by environmental factors such as temperature fluctuations, pH variations, and human-induced perturbations, thereby empowering us to refine our predictive models and mitigate potential ecological impacts effectively (Free et al. 2022). This understanding permits anticipation of fluctuations in toxin levels and their ramifications for ecosystem health. Moreover, delving into biotoxin origins provides a comprehensive grasp of their

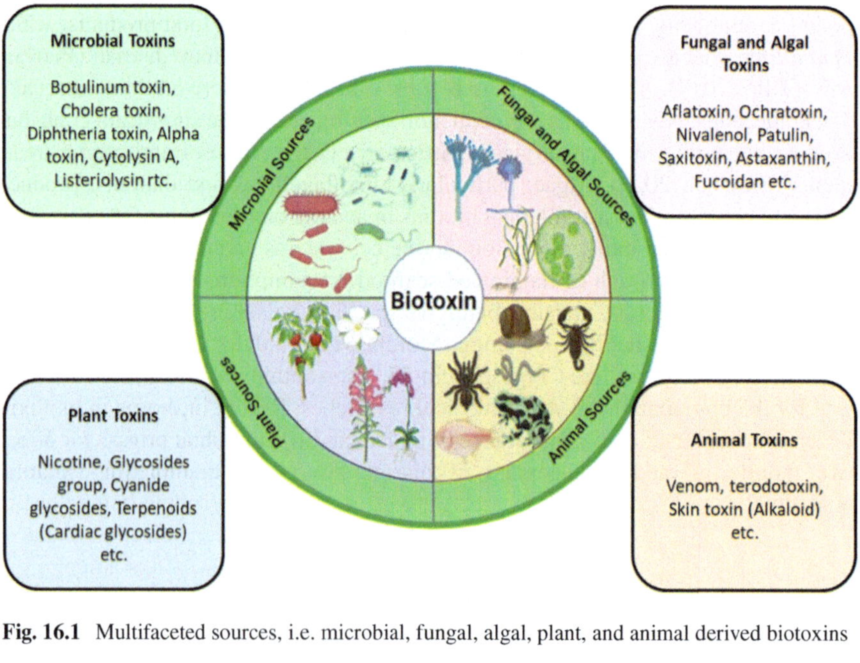

Fig. 16.1 Multifaceted sources, i.e. microbial, fungal, algal, plant, and animal derived biotoxins

diverse effects on both ecosystems and human health, enhancing our ability to manage and mitigate their impacts (Yang et al. 2024).

16.2.1 Microbial, Fungal, and Algal Origins

Microbial sources are key contributors to the diverse array of biotoxins present in ecosystems. Within the microbial realm, bacteria, fungi, and algae exhibit remarkable diversity in their ability to produce potent toxins, impacting both environmental and human health (Fan et al. 2022). Bacteria, exemplified by *Clostridium botulinum* and *Vibrio cholerae*, are renowned for synthesizing biotoxins with severe consequences for human health. Botulinum toxin, produced by *C. botulinum*, is recognized as one of the most potent neurotoxins, leading to botulism—a severe paralytic illness even in minute doses (Thakur et al. 2014). Similarly, *V. cholerae* produces cholera toxin, responsible for the severe gastrointestinal illness cholera (Clark et al. 2019). Moreover, bacterial-derived biotoxins like Colibactin, Cytolethal distending toxin, Diphtheria toxin, *Pseudomonas aeruginosa* exotoxin A, Alpha-toxin, Cytolysin A, and Listeriolysin O, etc., play important roles in various medical and clinical fields (Lucas et al. 2020). Understanding the microbial origins of these toxins is vital for effectively managing outbreaks and implementing preventive measures. Fungi, such as *Aspergillus* and *Penicillium* species, are prolific producers of mycotoxins, toxic secondary metabolites that contaminate food and feed supplies. These mycotoxins pose significant risks to human and animal health, ranging from acute poisoning to chronic diseases. *Aspergillus* species produce aflatoxins, potent carcinogens, and hepatotoxins found in contaminated food products, while *Penicillium* species produce ochratoxin A, implicated in kidney disease (Perrone and Gallo 2017). Furthermore, numerous other fungal-derived biotoxins like Trichothecenes, Nivalenol, Deoxynivalenol, Patulin, etc., make significant contributions to bolstering medical services, particularly in the realm of anti-cancer treatments (Strachan 2023). Algae, particularly dinoflagellates and diatoms, produce toxins responsible for harmful algal blooms in aquatic environments. These marine biotoxins have devastating effects on marine ecosystems and pose significant risks to human health through contaminated seafood consumption. A number of examples include saxitoxins, causing paralytic shellfish poisoning, and domoic acid, leading to amnesic shellfish poisoning (Morabito et al. 2018). Furthermore, several algal-derived biotoxins like Cryptophycin-55, Astaxanthin, Fucoidan, etc., are integral for the upgradation of cancer medical researches. Recent increases in biotoxin occurrence underscore the importance of understanding microbial origins for effective monitoring and management of marine biotoxin contamination (Sharma et al. 2022).

16.2.2 Plant Origins

Plant sources include a wide range of biotoxins, a secondary metabolite mainly to impede herbivores and defend against diseases. These chemicals play crucial roles in plant defense processes, serving as powerful deterrents to herbivores and microbial invaders (De Filippis 2016). Gaining knowledge about the characteristics and impacts of plant-derived toxins is crucial for effectively controlling agricultural methods, conserving the variety of living organisms, and ensuring the well-being of both humans and animals. Alkaloids are a significant group of plant metabolites that are distinguished by their nitrogen-containing structures and a wide range of physiological impacts. Nicotine, present in tobacco plants, is a prominent alkaloid renowned for its powerful neurotoxic properties that interfere with the neurological systems of herbivores, discouraging them from eating the plant. The usage of nicotine in humans can result in addiction and a range of detrimental health consequences, which are of great concern to public health authorities (Kennedy and Wightman 2011). Glycosides form another crucial group of plant toxins, often comprising a sugar molecule bound to a non-sugar moiety (aglycone). Cyanogenic glycosides, such as those present in *Cassava*, release cyanide upon hydrolysis (Chen et al. 2022). Cyanide poisoning can result in respiratory failure and death, posing serious risks to humans and animals consuming inadequately processed cassava products (Uchechukwu-Agua et al. 2015). Terpenoids, including cardiac glycosides found in plants like foxglove (*Digitalis purpurea*), are characterized by their isoprene-derived structures and diverse biological activities (Mitra et al. 2023). Cardiac glycosides interfere with the normal functioning of the heart muscle, leading to cardiac arrhythmias and, in severe cases, cardiac arrest (Roberts et al. 2016). Consumption of plants containing these biotoxins can cause poisoning in both humans and animals, with symptoms ranging from mild gastrointestinal discomfort to severe neurological or cardiac effects, depending on the toxin and the amount ingested (Janik et al. 2019). Recognizing and understanding the effects of plant toxins are imperative for preventing and managing cases of plant poisoning, both in agricultural contexts and in natural environments where humans and animals may encounter toxic plants (Shahid et al. 2017).

16.2.3 Animal Origins

Animal sources present a varied and intriguing assortment of biotoxins, synthesized by diverse organisms as part of their defensive or predatory strategies. These toxins serve critical roles in predator–prey interactions, chemical defenses, and symbiotic relationships with microorganisms (Zupo et al. 2019). Understanding the nature and effects of animal toxins is crucial for managing human–animal interactions, developing treatments for envenomation, and gaining insights into evolutionary adaptations (Duda et al. 2021). Venomous animals, such as snakes, spiders, scorpions, and

cone snails, are well-known producers of biotoxins, utilizing venom as a potent weapon for subduing prey or defending against predators (Bordon et al. 2020). Injecting toxins into their prey or attackers, these creatures can disrupt physiological processes, leading to paralysis, tissue damage, or death (Pagán 2018). Comprehending the composition and mechanisms of action of animal venoms is pivotal for developing antivenom treatments and mitigating the risks associated with venomous encounters (Bolon et al. 2023).

Marine organisms also contribute to the biotoxin repertoire, producing a wide range of toxins as chemical defenses against predators. For instance, pufferfish contain tetrodotoxin, a potent neurotoxin capable of inducing paralysis and death if ingested in sufficient quantities (Butzke and Luch 2010). Certain frog species produce skin toxins, known as alkaloids, deterring predators with their bitter taste or toxic effects (Santos et al. 2016). Understanding the ecological roles of these toxins is essential for conservation efforts and managing human interactions with marine and terrestrial ecosystems. Additionally, animals may harbor symbiotic bacteria producing toxins aiding in digestion or defense against competitors (Alava et al. 2017). Termites, for example, host gut bacteria producing enzymes capable of breaking down cellulose, facilitating their digestion of cellulose-rich diets (Li et al. 2022). Similarly, certain marine animals host symbiotic bacteria producing toxins, which deter potential predators or competitors, making understanding that animal toxins are vital for managing human–animal interactions and advancing scientific knowledge and medical treatments (Bolon et al. 2023). By studying the composition, function, and evolutionary origins of animal toxins, scientists can develop novel therapies for conditions such as chronic pain, hypertension, and neurodegenerative diseases (De Souza et al. 2018). Furthermore, insights gained from studying animal toxins can inform conservation efforts and help protect vulnerable species from anthropogenic threats.

16.3 Empowering Biotoxins: Custom Solutions for Enhanced Welfare

Biotoxins, once dreaded for their toxicological implications, are now undergoing a paradigm shift in perception—a recognition of their latent potential to empower humanity. Rather than being solely regarded as hazards to health and safety, biotoxins are now being appreciated as valuable resources capable of addressing a diverse array of societal benefits (Fig. 16.2).

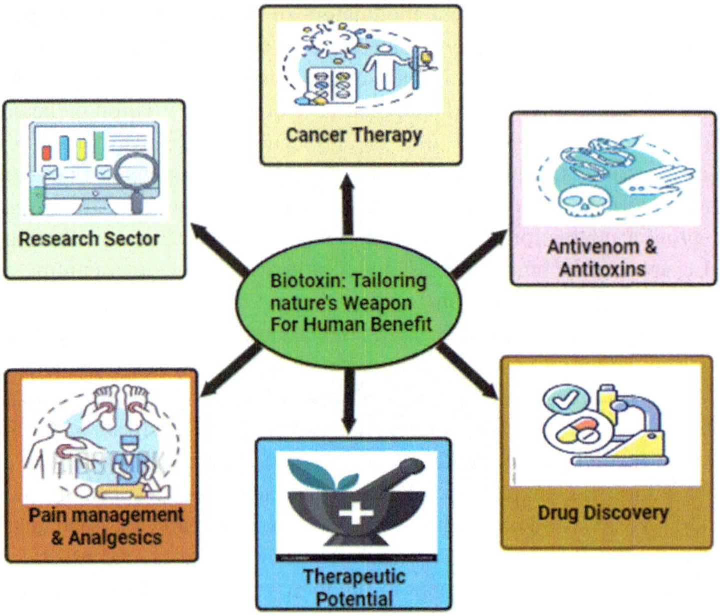

Fig. 16.2 Diverse biotoxin applications across multiple sectors of human welfare

16.3.1 *Biotoxin Generation for Research Purposes*

Biotoxins, synthesized by a range of organisms, hold significant potential for both research and medical purposes. Understanding their biosynthesis pathways, molecular structures, and physiological effects is crucial for leveraging them effectively in drug development, diagnostics, and therapeutics. This comprehension is pivotal for harnessing the therapeutic capabilities of biotoxins in diverse medical applications, from drug discovery to treatment interventions (Li and Wang 2018). With their varied structural and functional properties, biotoxins serve as valuable molecular tools for deciphering complex cellular processes. These natural compounds, originating from various organisms, provide researchers with unique insights into the molecular mechanisms underlying cellular physiology and pathology (Assunção et al. 2017). Biotoxins play diverse roles that include targeted binding, modulation of signaling pathways, exploration of protein–protein interactions, and regulation of gene expression, all of which are essential for advancing our understanding of cellular biology and facilitating drug discovery endeavors. Central to their effectiveness is the ability of biotoxins to selectively bind to specific cellular targets. Through precise molecular recognition, biotoxins can engage with receptors, ion channels, enzymes, and other biomolecules with high affinity and specificity. This targeted binding enables researchers to investigate the function and regulation of these critical cellular components, offering insights into their roles in both health and disease (Zeng et al. 2023).

Biotoxins also serve as potent modulators of cellular signaling pathways. By interfering with intracellular signaling cascades such as MAPK, PI3K/AKT, and NF-κB pathways, biotoxins can influence fundamental cellular processes including proliferation, differentiation, and apoptosis (Dai et al. 2022). Through these interactions, researchers gain insights into the molecular mechanisms driving cellular responses to various stimuli, paving the way for the development of targeted therapeutic interventions. Biotoxins serve as invaluable metabolites for investigating protein–protein interactions, which form the foundation of nearly all cellular processes. Leveraging the precise specificity of biotoxin–protein interactions, scientists can delineate intricate protein interaction networks, unravel signaling pathways, and unveil novel regulatory mechanisms. Techniques such as affinity chromatography, co-immunoprecipitation, and fluorescence resonance energy transfer (FRET) facilitate the visualization and quantification of protein–protein interactions within live cells, yielding comprehensive insights into cellular organization and function. Beyond their involvement in protein-level interactions, biotoxins also play a crucial role in governing gene expression. Through interactions with transcription factors, chromatin modifiers, and other regulatory proteins, biotoxins can modulate the transcriptional activity of target genes. Transcriptomic analyses, combined with biotoxin treatments, enable researchers to pinpoint genes and pathways that undergo dysregulation in various disease states, thus presenting potential targets for therapeutic intervention.

Immunotoxins are composed of an antibody combined with a toxin protein sourced from plants or bacteria, such as *Corynebacterium diphtheriae* (diphtheria toxin). Diphtheria toxin (DT) operates through well-understood mechanisms that alter mammalian elongation factor 2 (EF-2), thus hindering protein translation and inducing cell death. The antibody segment of immunotoxins targets receptors that are abundantly present on cancer cells (the specific receptor varies based on the immunotoxin and cancer type). Upon receptor binding, the toxin gains entry into the cell, ultimately leading to cancer cell death as illustrated in Fig. 16.3 (Shafiee et al. 2019).

Cholera toxin exerts a tumor-inhibitory effect in human bladder TCC cells by specifically inhibiting cell proliferation, inducing G1 phase arrest, and orchestrating PKA-dependent inactivation of the c-Raf/Mek/Erk pathway, coupled with a significant reduction in cyclin D1, Cdk4, and Cdk6 expression (Zheng et al. 2014). Applications of biotoxins extend beyond basic research to drug discovery and development. By serving as screening tools for identifying novel drug targets and lead compounds, biotoxins facilitate the development of therapeutics for a wide range of diseases, including cancer, neurodegenerative disorders, and infectious diseases. Moreover, biotoxin-inspired molecules hold promise as innovative therapeutics with enhanced specificity and efficacy compared to conventional drugs. Despite their immense potential, the use of biotoxins as molecular probes poses certain challenges, including issues related to dosage, delivery, and toxicity. Addressing these challenges requires careful consideration of experimental design and optimization of biotoxin-based approaches to ensure their safe and effective use in research and clinical settings (Das et al. 2023).

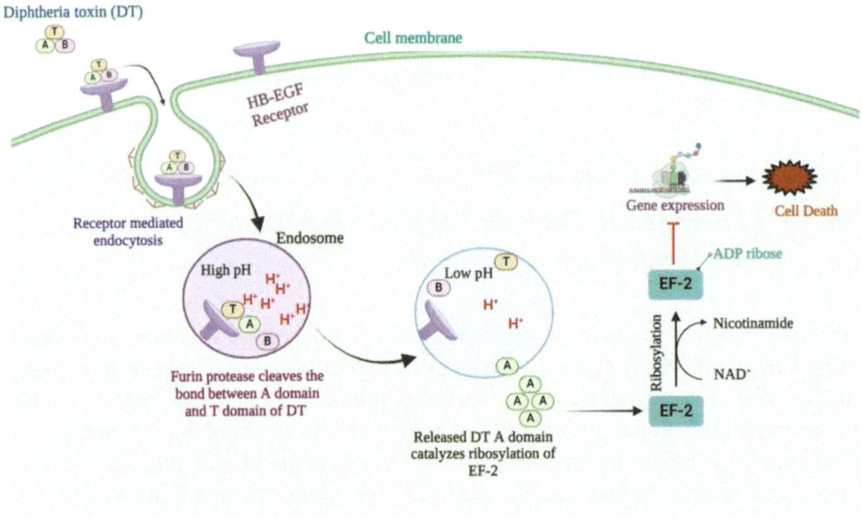

Fig. 16.3 Diagrammatic representation of the mechanism of action of diphtheria toxin (DT) induced cell death in carcinoma research

16.3.2 Use of Biotoxins in Drug Discovery

Despite their usual role as harmful agents in predator–prey interactions, however these chemicals have attracted considerable attention in the field of drug discovery and development because of their potential pharmacological effects. Various poisonous chemicals derived from plants are classified into several categories, such as alkaloids, glycosides, flavonoids, proteins, and saponins (Ma et al. 2015). Alkaloids exhibit remarkable structural diversity and possess a broad range of pharmacological effects, including muscle relaxant properties as well as cardiovascular and respiratory activities (Botha and Penrith 2008). Flavonoids, characterized by their diverse structures, exhibit a wide array of actions.

However, they are infrequently utilized as a primary cause of toxicity in mammals, with the exception of certain compounds with unique structures like rotenoids, which are utilized for inducing fish poisoning. Approximately 80% of poisons consist of cardioactive substances, primarily cardiac glycosides derived from the *Acokanthera*, *Parquetina*, and *Strophanthus* genera (such as ouabain, which causes acute cardiac insufficiency). Additional genera containing a substantial number of toxic species include *Rauwolfia serpentina*, *Catharanthus roseus*, *Digitalis purpurea*, *Taxus brevifolia*, *Colchicum autumnale*, *Camptotheca acuminata*, *Papaver somniferum*, *Erythroxylum coca*, *Atropa belladonna*, *Datura stramonium*, and *Hyoscyamus niger* (Anywar 2020). Several other plant poisons exhibit diverse activities; however, their poisonous effects are predominantly chronic rather than immediate. Examples include sesquiterpene lactones, iridoids, pyrrolizidine alkaloids, and tannins. Alkaloid-bearing plants commonly used as potent toxins include

Strychnos spp. (specifically strychnine, a compound of interest in neuropharmacological research), *Boophone disticha*, *Crinum* spp., *Triclisia dictyophylla*, *Nicotiana* spp., *Physostigma venenosum*, *Sarcocephalus latifolius*, and *Erythrophleum* spp. (Panter et al. 2013).

16.3.3 Therapeutic Potential: Harnessing Biotoxins for Medical Breakthroughs

Biotoxins, known as natural toxins, despite their toxicological nature, are increasingly being explored for their potential as therapeutic agents. There has been a significant rise in the demand of newly developed biocompounds and their derivatives in recent years, showing significant potential in the fields of biotechnology and pharmacology (Glaser and Mayer 2009). Researchers are investigating ways to harness these toxins for therapeutic purposes, leveraging their unique biochemical properties. This novel approach holds to combat several pathophysiological problems such as pain management in several tissues, tumor, cancer, and neurological disorders (Yarsan and Aktaş 2017). Based on their origin, the natural toxins can be divided into many categories: Mycotoxins: toxins produced by fungi, such as ochratoxins, zearalenone, fumonisins from trichothecenes, and aflatoxins. Most are heat-stable and have relatively small molecular weights; Bacterial toxins: toxins produced by bacteria such as botulin and *Bacillus cereus* enterotoxin, etc. Many bacterial toxins are proteins that are not heat-stable (van Egmond 2004). A variety of bacterial toxins have been chemically or genetically connected to ligands that adhere to cancer cells Phytotoxins: toxins produced by edible plant species such as pyrrolizidine alkaloids; Zootoxins: Animal-produced toxins like bufotoxin.

Drug addicts in the Netherlands have a deadly habit of licking the head of toad, resulting in hallucinations (van Egmond 2004) and Phycotoxins: Toxins produced by algae are small to medium-sized naturally occurring substances classified into a wide variety of chemical categories (Rossini and Hess 2010). Taxonomic Reference List of Toxic Plankton Algae of the Intergovernmental Oceanographic Commission (IOC) indicates that within the 179 species of marine microalgae, at least 95 dinoflagellates are known to produce phycotoxins (Moestrup, 2009). These bio compounds constitute a variety of compounds with different functional compositions and consequently, they are capable of the possibility of having a significant impact on many different biological receptors and metabolic pathways (Cembella 2003). The mechanism of their detrimental impact can vary significantly among various substances. Because of their diverse pharmacological actions, which could include analgesic, anti-cancer, anticholesterol, cytotoxic, anti-infective, immunosuppressive, and neurological disease treatments, they could find application to both human and veterinary medicine. A different therapeutic method, hirudotherapy, or medicinal leech therapy, that was authorized by the FDA (Food and Drug Administration) in 2004. Leeches have been utilized for medical purposes since the beginning of

civilization (Abdualkader et al. 2013). Leeches are hematophagous organisms and their saliva contains around 100 biologically active compounds, the majority of which have anticoagulant, anti-inflammatory, and analgesic properties (Singh 2010) also applied in arthritic conditions (Cooper and Mologne 2017), cardiovascular diseases and diabetes mellitus (Abdualkader et al. 2013).

16.3.4 Biotoxins Insights: From Venom to Medicine

Antivenoms play a crucial role in treating envenomations, which are the result of venomous bites or stings from animals such as snakes, spiders, scorpions, and certain marine creatures. These antivenoms are developed by injecting animals with small, controlled doses of venom to stimulate an immune response. The antibodies produced by the animals are then harvested and purified to create antivenom. Developed toward the latter part of the nineteenth century, antivenoms persist as the primary therapeutic intervention for envenomations. These formulations consist predominantly of antibodies, chiefly immunoglobulins G (IgG), which are generated within animals following immunization with venom-derived antigens (Chippaux and Goyffon 1998). Immunization involves the repeated administration of minute doses of venom, sometimes modified or combined with adjuvants. Upon envenomation, patients receive antivenom therapy to counteract and eliminate the venom's effects. Despite the early recognition of their clinical efficacy through empirical and experimental evidence, systematic investigation of antivenoms in clinical trials or pharmacological studies did not commence until after 1950 (Chippaux et al. 2010). The mechanism of antivenoms in treating envenomations involves several steps (Silva and Isbister 2020):

(A) **Antibody binding**: Antivenoms contain antibodies that specifically bind to the toxins present in the venom (Ledsgaard et al. 2018). These antibodies are raised in animals (such as horses or sheep) through controlled exposure to venom. When administered to a patient, the antibodies in the antivenom recognize and bind to the venom toxins.

(B) **Toxin neutralization**: Once the antibodies bind to the venom toxins, they neutralize their effects. This neutralization can occur through various mechanisms, including blocking the toxins' ability to bind to cellular receptors (Fig. 16.4), inhibiting their enzymatic activity, or facilitating their clearance from the body through immune system processes (Tan et al. 2016; Wong et al. 2016).

(C) **Prevention of further damage**: By neutralizing the venom toxins, antivenoms prevent further damage to tissues and organs. This includes mitigating effects such as tissue necrosis, hemorrhage, coagulopathy, neurotoxicity, and systemic effects like shock or organ failure.

(D) **Alleviation of symptoms**: Antivenoms not only neutralize venom toxins but also help alleviate symptoms associated with envenomation. These may include

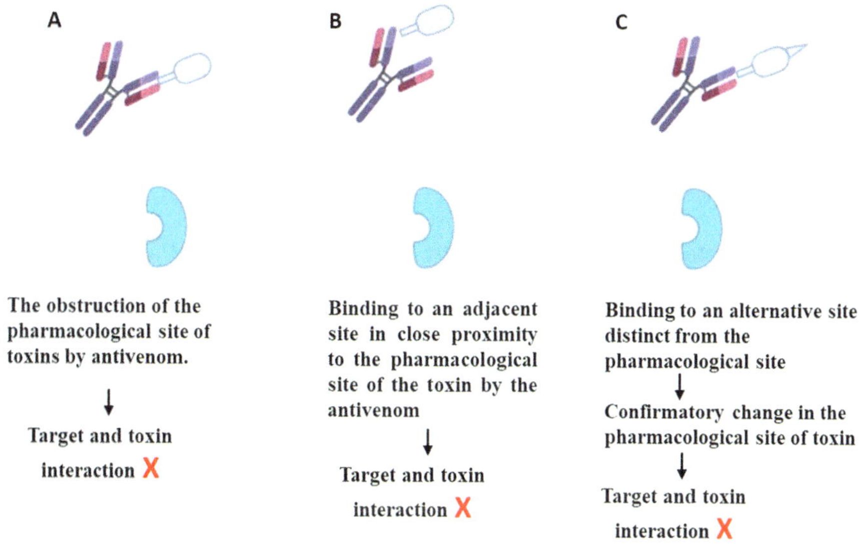

Fig. 16.4 Antivenom neutralizes toxins through alternate mechanisms (**a**, **b**, and **c**), hindering their binding to the pharmacologically active site on target tissues

pain, swelling, inflammation, and systemic symptoms such as nausea, vomiting, sweating, and difficulty breathing (Boyer et al. 2015).

(E) **Enhancement of endogenous defenses**: Antivenoms can also enhance the body's natural defenses against venom toxins (Bermúdez-Méndez et al. 2018). By neutralizing a significant portion of the venom, antivenoms allow the patient's immune system to more effectively combat the remaining toxins and clear them from the body.

Critical Antivenom Therapies: Saving Lives

Life-saving antivenom treatments are those that effectively counteract the toxic effects of venom from various venomous animals, preventing severe morbidity and mortality in envenomed individuals. Here are several examples of life-saving antivenom treatments:

1. **Snake antivenom**

   (a) ***Polyvalent antivenom***: These antivenoms are designed to treat envenomations caused by multiple snake species within a particular geographic region. They are crucial for regions with diverse snake fauna, such as sub-Saharan Africa, Southeast Asia, and South America. The Fav-Afrique antivenom, produced by Sanofi Pasteur (France), is considered safe and effective. It is polyvalent, targeting most of the medically important snake species in sub-Saharan Africa. In particular, it is highly effective in treating envenoming by *Echis ocellatus*, the West African saw-scaled viper (Chippaux et al. 1999; Wolf et al. 2011) that causes great morbidity and mortality

throughout the West and Central African savannah. The venom of *E. ocellatus* may induce systemic hemorrhage, coagulopathy, and shock, as well as extensive local tissue damage. In the absence of treatment, the case fatality rate is 10–20% (Pugh and Theakston 1980). In India, polyvalent anti-snake venom (ASV) raised in horses using venoms of four important venomous snakes in India, viz., Indian cobra (*Naja naja*), Indian krait (*Bungarus caeruleus*), Russell's viper (*Daboia russelii*), and saw-scaled viper (*Echis carinatus*), is widely used in the management of snake bites (Pore et al. 2015).

(b) ***Monovalent antivenoms***: These antivenoms target specific species of venomous snakes. Examples include antivenoms for bites from the Indian cobra (*Naja naja*), Russell's viper (*Daboia russelii*), common krait (*Bungarus caeruleus*), and saw-scaled viper (*Echis carinatus*). Examples include *Trimeresurus albolabris* Monovalent Antivenom (TaMAV) and *Gloydius brevicaudus* Monovalent Antivenom (GbMAV) (Yong et al. 2021).

2. **Scorpion antivenom**: Scorpion antivenom is essential for treating envenomations caused by scorpion stings, which can lead to severe neurotoxicity and cardiovascular effects in some cases (Feola et al. 2020). Scorpion antivenoms are used in regions where scorpions, such as the *Leiurus quinquestriatus* (deathstalker) or *Centruroides* spp., are prevalent.
3. **Spider antivenom**: Antivenoms for spider envenomations are available for species with medically significant venom, such as the black widow spider (*Latrodectus* spp.) (Isbister and White 2004) and the Brazilian wandering spider (*Phoneutria* spp.) (Bucaretchi et al. 2000). These antivenoms are crucial for managing symptoms like severe pain, muscle rigidity, and systemic effects.
4. **Marine animal antivenom**: Antivenoms are available for envenomations caused by various marine animals, including CSL-antivenom for box jellyfish (*Chironex fleckeri*) and certain species of sea snakes (Fernandez et al. 2011). These antivenoms are essential for counteracting the potent toxins injected during marine animal encounters, preventing cardiovascular collapse, respiratory failure, and other severe complications (Endean and Sizemore 1988; Winkel et al. 2003; Hughes et al. 2012).
5. **Recombinant antivenom**: In the realm of cutting-edge antivenom research, the utilization of phage display technology alongside recent breakthroughs in toxicovenomics has empowered investigators to explore novel therapeutic avenues aimed at enhancing the efficacy of envenomation treatments (Laustsen et al. 2016). Recombinant antivenoms represent a promising advancement in antivenom therapy. Manufactured through recombinant DNA techniques, these antivenoms present benefits such as diminished likelihood of adverse effects and enhanced uniformity in effectiveness. Although they remain in the developmental stage, recombinant antivenoms exhibit promise in addressing envenomations caused by diverse venomous species (Laustsen et al. 2018; Jenkins and Laustsen 2020).

16.3.5 Delving into Biotoxins: Pain Management and Analgesics

Pain management remains a critical aspect of healthcare, with millions worldwide facing acute and chronic pain challenges (Tompkins et al. 2017). Traditional approach for mitigating pain frequently revolves around pharmaceutical drugs, which pose potential hazards including dependency, adverse effects, and the propensity for tolerance development with prolonged usage (Yaksh and Wallace 2011). In recent times, there has been a growing interest in exploring alternative avenues for pain management, with an emphasis on the potential utilization of biotoxins. Through their ability to selectively target distinct pain pathways or nociceptive processes, biotoxins present an avenue for physicians to deliver more individualized pain treatment. Tailoring biotoxin-based treatments to fit the individual pain profiles of patients allows physicians to pursue superior results while minimizing the risk of adverse effects (Ovsepian et al. 2019). One of the most well-known examples is botulinum toxin, frequently utilized in aesthetic procedures for its ability to relax muscles (Witmanowski and Błochowiak 2020). Nevertheless, surpassing its cosmetic role, botulinum toxin has showcased efficacy in managing chronic pain disorders like migraines, neuropathic pain, and muscle spasms. Its mechanism of action revolves around impeding the release of neurotransmitters associated with pain signaling, thereby diminishing the sensation of pain and facilitating muscle relaxation (Wanat et al. 2022).

Alternative biotoxins, including conotoxins derived from cone snails, represent potent analgesic properties by precisely targeting specific receptors within the nervous system (Mackieh et al. 2021). These peptides have demonstrated encouraging results in preclinical research for managing neuropathic pain and could potentially offer a safer alternative to conventional opioid medications (Starnowska-Sokół and Przewłocka 2020). Furthermore, venom-derived peptides from spiders, scorpions, and venomous snakes have been explored for their potential analgesic effects, with certain compounds demonstrating selective inhibition of pain receptors or modulation of neurotransmitter release. By precisely targeting particular receptors or ion channels crucial in the pain signaling pathway, biotoxins leverage their diverse mechanisms of action and potential analgesic benefits to open up new possibilities for confronting the intricacies of pain management and enhancing patient outcomes (Díaz-Gómez et al. 2024). However, further study is needed to improve our understanding of their mechanisms of action, optimize their analgesic benefits, and ensure their safety for deployment in clinical environments.

16.3.6 Biotoxins: Revolutionizing Cancer Therapy Approaches

On a global scale, cancer remains a significant concern for public health, being recognized as the second leading cause of death in developed nations (Shakeri et al. 2020; Kashyap et al. 2022). Every year, tens of millions of individuals are diagnosed with various cancer forms, and more than half ultimately die due to cancer growth. Despite technological progress, recurrence, uncontrolled cell division, resistance to apoptosis, impaired immune surveillance, and metastasis remain prominent factors in mortality. As per WHO (World Health Organization) reports that cancer is the leading cause of death globally, with an estimated ten million deaths projected by 2020. In terms of new cancer cases in 2020, breast, lung, colon and rectum, prostate, skin (non-melanoma), and stomach cancers were the most common. Additionally, the primary causes of cancer-related deaths in 2020 were lung (1.80 million deaths), colon and rectum (0.93 million deaths), liver (0.83 million deaths), stomach (0.76 million deaths), and breast cancers (0.68 million deaths) (Al-Yozbaki et al. 2021; WHO 2021). Several therapies such as radiotherapy, surgical therapy, cytotoxic chemotherapy, oncolytic virotherapy, hormonal replacement therapy, and immunotherapy are currently used for cancer treatment and are associated with several side effects. Among different chemotherapeutic options, the use of biotoxins with recent advancement becomes the choice of therapy against cancer cells with improved efficacy of current therapies and benefit of minimized side effects (Table 16.1). One of the most remarkable aspects of biotoxins in cancer therapy is their ability to selectively target cancer cells while sparing healthy tissues (Citrin 2017). Biotoxin selectivity is derived from the distinct molecular features of cancer cells, including altered surface receptor expression, disrupted signaling pathways, and increased metabolic activity (Pandey et al. 2023). Utilizing these specific attributes, biotoxins can effectively target tumor tissues, offering a precise method for cancer therapy while minimizing toxicity to healthy cells.

At the molecular level, the mechanism of action of biotoxins for cancer therapy involves intricate interactions between the toxin molecules and specific cellular components within cancer cells. Biotoxins typically exploit molecular features unique to cancer cells, such as altered surface receptor expression or dysregulated intracellular signaling pathways, to exert their cytotoxic effects selectively (Lahiani et al. 2017). One fundamental aspect of biotoxin action is their ability to recognize and bind to molecular targets on the surface of cancer cells.

This recognition typically emerges from high-affinity interactions between specific regions of the biotoxin molecule and complementary binding sites present on cell surface receptors (Zhao et al. 2018). These receptors may be overexpressed or aberrantly activated in cancer cells, providing an opportunity for selective targeting. For instance, certain biotoxins may target receptors such as EGFR, HER2, or CD22, which are frequently overexpressed in various cancer types (Trivanović et al. 2021). Following receptor binding, biotoxins can undergo internalization into the cancer cell through endocytic processes. Once inside the cell, the biotoxin may undergo intracellular trafficking to reach its intended target or site of action. This may involve

Table 16.1 Some crucial biotoxins and their sources with anti-cancer potential

Biotoxin sources	Biotoxin name	Anti-cancerous efficacy	References
Flora			
Stellera chamaejasme L	Chamaejasmenin B Neochamaejasmin C	Cancer therapy	Yu et al. (2022)
Persea americana	Persin	Induces Bim-dependent apoptosis in human breast cancer cells	Zhang and Kanakkanthara (2020)
Stellera chamaejasme L	Gnidimacrin	Antitumor activities against mouse leukemia P-388 and L-1210	Otsuki et al. (2019)
Ricinus communis	Ricin toxin	Vaccine adjuvant/carrier to design cancer vaccines	Polito et al. (2019)
Saponaria officinalis L	Saporin	Hematological tumors	Giansanti et al. (2018)
Abrus precatorius	Abrin	Against colon cancer	Yu et al. (2016)
Fauna			
Bee	Melittin	Inhibit prostate cancer	Pandey et al. (2023)
Japanese "fugu" or puffer fish	Tetrodotoxin	Anti-cancer drug	González-Cano et al. (2021)
Radianthus macrodactylus	Actinoporin	Potential cancer-preventive activity through the induction of p53-independent apoptosis and inhibition of the oncogenic AP-1 and NF-κB nuclear factors activity	Kvetkina et al. (2020)
Sea sponges	Halitoxin	Toxic against KB line of cancer cells	Shady et al. (2020)
Ocean sunfish, salamandrine	Maculotoxin	Anti-cancer	Ramachandiran et al. (2020)
Snake	Disintegrins (Salmosin, Contortrostatin) Aggretin	Potent inhibitors of both platelet aggregation and integrin-dependent cell adhesion	Schönthal et al. (2020)
Scorpaenidae	Stonefish toxin	Antiproliferative, cytotoxic, and apoptogenic activity against human leukemic cells	Ramachandran et al. (2019)
Blarina brevicauda	Soricidin	Inhibits the transient receptor potential of vallinoid type 6 (TRPV6) calcium channels that are up-regulated in a number of cancers	Anand et al. (2019)
Toad	Bufalin	Endometrial and ovarian cancers	Lan et al. (2019)

(continued)

Table 16.1 (continued)

Biotoxin sources	Biotoxin name	Anti-cancerous efficacy	References
Bufo bufo gargarizans	Cinobufacini	Suppresses the side effects of chemotherapy	Chen et al. (2018)
Sea cucumber	Holotoxin	Induced apoptosis in human leukemia cells	Yun et al. (2018)
Frogs and octopus	Chiriguitoxin	Anti-cancer	Agrawal et al. (2013)
Microbes			
Dinoflagellates (*Alexandrium* sp., *Gymnodinium* sp., *Pyrodinium* sp.)	Saxitoxin	Anti-cancer and antimicrobial compound	Montuori et al. (2023)
Gambierdiscus toxicus	Gambierol	Modulatory action upon voltage gated K^+ channels	Raposo-Garcia et al. (2023)
Prorocentrum	Okadaic acid	Potent tumor promoter	Fujiki et al. (2023)
Clostridium botulinum	Botulinum toxin a	Treatment against breast cancer	Grenda et al. (2022)
Escherichia coli	Shiga toxin 1 (Stx1)	Targets human cancer cells	Robert and Wiels (2021)
Karenia brevis	Brevetoxin	Powerful neurotoxic	Colon et al. (2021)
Ostreopsis siamensis	Palytoxin	A wide spectrum of biological effects	Kraus (2020)
Omphalotus illudens	Illudin S	Anti-cancer drugs	Uto et al. (2019)
Staphylococcus	Staphylococcal enterotoxin C2 (SEC2)	Against bladder cancer	Liu et al. (2017)
Amphidinium	Karlotoxin	Blood cancer	Cai et al. (2016)
Gonyaulax spinifera	Yessotoxin	Anti-cancer activity	Alfonso et al. (2016)

trafficking through endosomal compartments or direct translocation across cellular membranes, depending on the specific properties of the biotoxin and its cellular target (Gupta et al. 2017).

Upon reaching their intracellular targets, biotoxins can disrupt key molecular pathways essential for cancer cell survival and proliferation. For example, some biotoxins inhibit protein synthesis by targeting ribosomes or elongation factors, thereby preventing the production of essential proteins required for cell growth and function (Márquez-López and Fanarraga 2023). On the other hand, biotoxins may interfere with DNA replication or repair mechanisms, leading to DNA damage and activation of apoptotic or necrotic cell death pathways (Alfonso et al. 2015). In addition to direct cytotoxic effects, biotoxins can also modulate intracellular signaling cascades involved in cell survival, apoptosis, and metastasis. By targeting key signaling molecules or downstream effectors within these pathways, biotoxins can disrupt cancer cell signaling networks and induce cell cycle arrest, apoptosis, or

senescence (Li et al. 2023). Moreover, biotoxins may exert anti-angiogenic effects by inhibiting the formation of new blood vessels within tumors, thereby depriving cancer cells of oxygen and nutrients essential for their growth and survival (Khalily and Soydan 2023). Efforts to enhance the tumor-targeting properties of biotoxins, optimize their pharmacokinetic properties, and minimize off-target effects on normal tissues are ongoing areas of research in the development of biotoxin-based cancer therapies.

16.4 Exploring New Frontiers: Artificial Intelligence (AI) Applications in Biotoxin-Based Treatments

Over the past few years, the integration of emerging trends and pioneering technologies, including artificial intelligence (AI), has spurred remarkable advancements in biotoxin-based therapies. Nevertheless, the traditional approaches to biotoxin discovery, optimization, and distribution have been characterized by their laborious and time-consuming nature (Josić et al. 2017). The rise of AI has fundamentally altered these methodologies, presenting unparalleled chances to hasten drug discovery, fine-tune treatment methodologies, and elevate patient outcomes in biotoxin therapy (Perez Santin et al. 2021). In the scientific domain, machine learning algorithms systematically analyze comprehensive datasets containing molecular structures and biological activity profiles to anticipate the efficacy, selectivity, and safety of potential biotoxin candidates. Utilizing AI-driven virtual screening and computational modeling, researchers can efficiently identify promising leads for further experimental validation, significantly expediting the drug discovery process (Sadybekov and Katritch 2023). Moreover, AI allows for the investigation of vast chemical space and the generation of novel biotoxin variants with optimized pharmacological characteristics, paving the way for the advancement of next-generation therapeutics. AI-driven target identification and pathway analysis plays a significant role in biotoxin-based therapy. Through revealing the molecular targets and signaling pathways responsible for biotoxin-induced therapeutic effects, AI algorithms facilitate the strategic optimization of biotoxins with elevated specificity and potency (Wang et al. 2022). Utilizing computational modeling and network analysis, scientists are able to identify critical receptors, ion channels, or signaling molecules implicated in the pathogenesis of diseases targeted by biotoxin therapies (Odoemelam 2023). Furthermore, AI-driven platforms promote the innovation of novel drug delivery technologies, like nanoparticles, liposomes, and micelles, which can deliver biotoxins precisely to target tissues or cells (Belgamwar et al. 2023). As AI continues to progress and develop, it is positioned to reshape biotoxin-based therapies, introducing new treatment paradigms and elevating outcomes for patients affected by a wide array of medical conditions.

16.5 Role of Government Organization (GO) and Non-government Organization (NGO) in Management of Biotoxin

Coordinated efforts between governmental organization (GO) and non-governmental organizations (NGO) are indispensable for the effective management of biotoxins, which derive from living organisms and present substantial risks to human health and the environment (Jewett et al. 2008). In the realm of biotoxin management, GO and NGO each contributes specialized yet complementary roles, capitalizing on their unique capabilities and resources to mitigate hazards and protect public health. The fundamental role of GO lies in their mandate for regulatory oversight and policy implementation, essential for establishing and maintaining standards (Table 16.2) concerning biotoxin management (Dean et al. 2020). These organizations, spanning environmental protection agencies, health sectors, and regulatory entities, assume the task of monitoring biotoxin levels in diverse environments and setting standards for allowable exposure thresholds. GO, through the enactment of laws and regulations, provides the structural foundation for biotoxin management strategies, ensuring adherence to safety protocols and enhancing public understanding of the associated risks (Zhang et al. 2018). GO contributes essential financial resources to research endeavors focused on elucidating biotoxin dynamics and formulating progressive mitigation strategies. Through the strategic use of their authority and resources, these entities play a critical role in safeguarding public health and environmental harmony (Lazonick and Tulum 2011).

NGO contributes significantly to biotoxin management through their multifaceted approaches, including advocacy efforts, educational campaigns, and community outreach endeavors (Duong 2017). Comprising environmental advocacy groups, public health organizations, and scientific research institutions, these entities serve as indispensable stakeholders in raising consciousness about biotoxin-related matters and advocating for policy transformations to address underlying environmental and public health concerns. NGO frequently lead research efforts aimed at deepening our understanding of the ecological effects of biotoxins and their potential long-term implications for ecosystems and biodiversity (Reid et al. 2019). Additionally, NGO may initiate public outreach endeavors aimed at informing individuals about the risks linked with biotoxin exposure and encouraging the adoption of sustainable practices to reduce environmental contamination (Krishna et al. 2017). Through synergistic collaboration between GO and NGO, both entities can collectively address the complex challenges associated with biotoxin management, thereby safeguarding public health and environmental well-being for current and future generations.

Table 16.2 Global and national legislation established for the various classes of biotoxins

Country	Biotoxin	Allowable limit
Japan[a]		
	PSP	800 μg STX eq./kg
	DSP	160 μg OA eq./kg
	AZAs	160 μg OA eq./kg
	DA	20 mg/kg
	NSP	20 MU/100 g
	TTX	10 MU/g. Restricted capture and sale
	CTX	0.025 MU/g. Banned import and sale of ciguateric fish
USA[b]		
	PSP	800 μg STX eq./kg
	DA	20 mg/kg
	DSP	160 μg OA eq./kg
	AZAs	160 μg AZA-1 eq./kg
	NSP	800 μg BTX-2 eq./kg
	CTXs	10–100 ng/kg P-CTX-1 eq.
Australia[c]		
	PSP	800 μg STX eq./kg
	DA	20 mg/kg
	NSP	200 MU/kg (≈ 800 μg BTX-2 eq./kg)
	CTX	
Canada[d]		
	PSP	800 μg STX eq./kg
	DA	20 mg/kg
	OA	200 μg OA eq./kg
	DTXs	200 μg OA eq./kg
	PTXs	200 μg OA eq./kg
European Union[e]		
	PSP	800 μg STX eq./kg
	DA	20 mg/kg
	ASP	20 mg/kg
	DSP	160 μg OA eq./kg
	PTX	160 μg PTX2-eq/kg
	YTX	3.75 mg/kg
	AZP	160 μg AZA1-eq/kg
	AZAs	160 μg OA eq./kg
Mexico[f]		
	STX	800 μg STX eq./kg
	DA	20 mg/kg
	BTX	20 MU/100 g
	OA	160 μg OA eq./kg
	DTXs	160 μg OA eq./kg
	CTX	2.5 MU/100 g

(continued)

Table 16.2 (continued)

Country	Biotoxin	Allowable limit
New Zealand[g]		
	PSP	800 µg STX eq./kg
	DA	20 mg/kg
	OA	160 µg OA eq./kg
	DTXs	160 µg OA eq./kg
	PTXs	160 µg OA eq./kg
	AZAs	160 µg OA eq./kg
	BTX	20 MU/100 g (≈ 800 µg BTX-2 eq./kg)
Chile[h]		
	STX	800 µg STX eq./kg
	DA	20 mg/kg
	OA	160 µg OA eq./kg
	DTXs	160 µg OA eq./kg
	PTXs	160 µg OA eq./kg
	YTXs	3.75 mg/kg
	AZAs	160 µg OA eq./kg
CODEX[i]		
	STX	800 µg STX eq./kg
	DA	20 mg/kg
	OA	160 µg OA eq./kg
	DTXs	160 µg OA eq./kg
	AZAs	160 µg OA eq./kg
	BTX	200 MU/kg
India[j]		
	AFs	10–15 µg/kg
	AF M1	0.5 µg/kg
	OTA	20 µg/kg
	PAT	50 µg/kg
	DON	1000 µg/kg

AZA azaspiracid; *BTX* brevetoxin; *DA* domoic acid; *PTX* pectenotoxin; *OA* okadaic acid; *STX* saxitoxin; *YTX* yessotoxin; *CTXs* ciguatoxins; *TTX* tetrodotoxin; *PLTX* palytoxins; *DTX* dinophysistoxins; *AFs* aflatoxin; *OT* ochratoxins; *PAT* patulin; *NSP* neurologic shellfish poisoning; *CEP* Ciguatera fish poisoning; *ASP* azaspiracid poisoning; *ASP* amnesic shellfish poisoning; *DSP* diarrhetic shellfish poisoning; *PSP* paralytic shellfish poisoning

[a]Vilariño et al. (2020)
[b]Anon (2011a)
[c, g]Anon (2011b)
[a, d]Vilariño et al. (2020)
[e, f, h]Alexander et al. (2009, 2010)
[i]Anon (2008)
[j]Chatterjee et al. (2023)

16.6 Challenges and Future Prospects

Navigating the complexities of nature's biotoxins presents both challenges and promising avenues in biotechnology and therapeutics. The intricate molecular configurations and functional mechanisms inherent in biotoxins, derived from a spectrum of organisms like bacteria, marine life, and venomous animals, represent a considerable challenge. Moreover, the vast diversity of biotoxins adds to the complexity of identifying and characterizing their therapeutic potential. Despite the widespread research and medical application of certain biotoxins, many others remain undiscovered or unexplored, indicating significant potential for therapeutic advancement. Furthermore, the progress and production of biotoxin-based therapies encounter significant technical and regulatory barriers. The procedures involved in isolating, purifying, and synthesizing biotoxins can be technically intricate and resource-demanding. In addition, ensuring the safety and efficacy of biotoxin therapies mandates thorough preclinical and clinical evaluations, as well as adherence to regulatory protocols. Regulatory approval processes for biotoxin-based therapies may also present challenges, particularly in comprehending the unique attributes of these intricate biomolecules and their potential for off-target effects. Conquering these hurdles will necessitate interdisciplinary cooperation among scientists, clinicians, regulatory agencies, and industry associates to establish resilient. Despite facing these challenges, the outlook for harnessing nature's biotoxins is optimistic. Biotoxins embody a varied reservoir of therapeutic promise, exhibiting multifaceted pharmacological attributes such as analgesic effects, muscle relaxant properties as well as in cancer research. Progress in biotechnology, genomics, and bioinformatics is advancing the exploration, characterization, and engineering of biotoxins for therapeutic utility. Bioinformatics applications and computational modeling are projecting the structure and actions of emerging biotoxins, expediting drug discovery. Furthermore, biotechnological strategies such as genetic modification and synthetic biology are empowering the generation of recombinant biotoxins with personalized attributes and augmented therapeutic efficacy.

16.7 Conclusion

Exploring biotoxins for biotechnological progress and healthcare reveals an intriguing account of utilizing natural resources for the betterment of humanity. Addressing the intricate mechanisms and diverse applications of biotoxins elucidates their significant potential in diverse fields as in research, medicine, and industry. Biotoxin's complex nature yields a multitude of opportunities for scientific exploration and innovation. From serving as potent pharmacological agents to being indispensable in biotechnological processes like drug discovery and bioengineering, biotoxins emerge as invaluable resources for tackling critical healthcare challenges. Moreover, exploration highlights the necessity of responsible stewardship and ethical

discernment in the prudent utilization of biotoxins. Although their different medicinal potency is considerable, it is crucial to proceed carefully to minimize any potential risks to both human health and the environment. Coordination between GO and NGO facilitates a comprehensive strategy in biotoxin management, addressing the varied challenges posed by these compounds while ensuring the safety of the public. This study delves into the intricate operations of biotoxins charting a path toward a future where this natural agent become vital tools for enhancing human health and well-being.

References

Abdualkader AM, Ghawi AM, Alaama M, Awang M, Merzouk A (2013) Leech therapeutic applications. Indian J Pharm Sci 75(2):127

Agrawal A, Gopal K, Agrawal A, Gopal K (2013) Fate of biotoxins in the environment and its health implications. Biomonit Water Waste Water 109–120

Alava JJ, Cheung WW, Ross PS, Sumaila UR (2017) Climate change—contaminant interactions in marine food webs: Toward a conceptual framework. Glob Chang Biol 23(10):3984–4001

Alexander J, Benford D, Cockburn A, Cravedi JP, Dogliotti E, Di Domenico A, Fernández-Cruz ML, Fürst P, Fink-Gremmels J, Galli C, Grandjean P (2009) Marine biotoxins in shellfish-saxitoxin group scientific opinion of the Panel on Contaminants in the Food Chain. EFSA J 1019:1–76. Adopted on 25 Mar 2009

Alexander J, Benford D, Cockburn A, Cradevi J-P, Dogliotti E, Domenico AD, Fernandez-Cruz ML, Fink-Gremmels J, Furst P, Galli C, Grandjean P, Heinemeyer G, Johansson N, Mutti A, Schlatter J, van Leeuwen R (2010) Scientific opinion on marine biotoxins in shellfish—emerging toxins: ciguatoxin group. Eur Food Saf Auth J 8(6):1–38

Alfonso A, Fernández-Araujo A, Vieytes MR (2015) Marine toxins as modulators of apoptosis. In: Botana LM, Alfonso A (eds) Phycotoxins: chemistry and biochemistry. Wiley, Hoboken, pp 203–224

Alfonso A, Vieytes MR, Botana LM (2016) Yessotoxin, a promising therapeutic tool. Mar Drugs 14(2):30

Al-Yozbaki M, Wilkin PJ, Gupta GK, Wilson CM (2021) Therapeutic potential of natural compounds in lung cancer. Curr Med Chem 28(39):7988–8002

Anand P, Filipenko P, Huaman J, Lyudmer M, Hossain M, Santamaria C, Huang K, Ogunwobi OO, Holford M (2019) Antitumor effects of Tv1 venom peptide in liver cancer. BioRxiv. https://doi.org/10.1101/518340

Anon (2008) Standard for live and raw bivalve molluscs. CODEX STAN 292–2008

Anon (2011a) Chapter 6: Natural toxins (a chemical hazard). In: Fish and fisheries products hazards and control guidance. Food and Drug Administration

Anon (2011b) Contaminants and natural toxicants. Australia New Zealand Food Standards Code Standard 1.4.1

Anywar G (2020) Historical use of toxic plants. In: Mtewa AG et al (eds) Poisonous plants and phytochemicals in drug discovery. Wiley, Hoboken, pp 1–17

Assunção J, Guedes AC, Malcata FX (2017) Biotechnological and pharmacological applications of biotoxins and other bioactive molecules from dinoflagellates. Mar Drugs 15(12):393

Belgamwar A, Khan S, Rai M (2023) Drug targeting: nanotechnology principles, future perspectives, and challenges. Nanotechnology principles in drug targeting and diagnosis. Elsevier, Amsterdam, pp 3–17

Bermúdez-Méndez E, Fuglsang-Madsen A, Føns S, Lomonte B, Gutiérrez JM, Laustsen AH (2018) Innovative immunization strategies for antivenom development. Toxins 10(11):452

Bolon B, Heinz-Taheny K, Yeung KA, Oguni J, Erickson TB, Chai PR, Goldfine CE (2023) Animal toxins. In: Haschek and Rousseaux's handbook of toxicologic pathology, vol 3. Academic, New York, pp 547–628

Bordon KDCF, Cologna CT, Fornari-Baldo EC, Pinheiro-Júnior EL, Cerni FA, Amorim FG, Anjolette FAP, Cordeiro FA, Wiezel GA, Cardoso IA, Ferreira IG (2020) From animal poisons and venoms to medicines: achievements, challenges and perspectives in drug discovery. Front Pharmacol 11:1132

Botha CJ, Penrith ML (2008) Poisonous plants of veterinary and human importance in southern Africa. J Ethnopharmacol 119(3):549–558

Bouley TA, Machalaba C, Keast J, Gerwick WH, Fleming LE (2023) Marine biotechnology: A One Health approach to linking life on land to life underwater. In: Fleming LE, Alcantara-Creencia LB, Gerwick WH, Goh HC, Gribble MO, Maycock B, Solo-Gabriele H (eds) Oceans and human health. Academic, San Diego, pp 149–180

Boyer L, Alagón A, Fry BG, Jackson TNW, Sunagar K, Chippaux JP (2015) Signs, symptoms and treatment of envenomation. In: Fry BG (ed) Venomous reptiles and their toxins: evolution, pathophysiology and biodiscovery. Oxford University Press, New York, pp 32–60

Bucaretchi F, Deus Reinaldo CRD, Hyslop S, Madureira PR, De Capitani EM, Vieira RJ (2000) A clinico-epidemiological study of bites by spiders of the genus Phoneutria. Rev Inst Med Trop Sao Paulo 42:17–21

Butzke D, Luch A (2010) High-molecular weight protein toxins of marine invertebrates and their elaborate modes of action. EXS 100:213–232

Cai P, He S, Zhou C, Place AR, Haq S, Ding L, Chen H, Jiang Y, Guo C, Xu Y, Zhang J (2016) Two new karlotoxins found in *Karlodinium veneficum* (strain GM2) from the East China Sea. Harmful Algae 58:66–73

Cembella AD (2003) Chemical ecology of eukaryotic microalgae in marine ecosystems. Phycologia 42(4):420–447

Chatterjee S, Dhole A, Krishnan AA, Banerjee K (2023) Mycotoxin monitoring, regulation and analysis in India: a success story. Foods 12(4):705

Chen D, Chen J, Guo Y, Li Y (2018) Cinobufacini promotes apoptosis of bladder cancer cells by influencing the expression of autophagy-related genes. Oncol Lett 15(5):7104–7110

Chen H, Singh H, Bhardwaj N, Bhardwaj SK, Khatri M, Kim KH, Peng W (2022) An exploration on the toxicity mechanisms of phytotoxins and their potential utilities. Crit Rev Environ Sci Technol 52(3):395–435

Chippaux JP, Goyffon M (1998) Venoms, antivenoms and immunotherapy. Toxicon 36(6):823–846

Chippaux JP, Lang J, Amadi-Eddine S, Fagot P, Le Mener V (1999) Treatment of snake envenomations by a new polyvalent antivenom composed of highly purified F (ab) 2: results of a clinical trial in northern Cameroon. Am J Trop Med Hyg 61(6):1017–1018

Chippaux JP, Stock RP, Massougbodji A (2010) Methodology of clinical studies dealing with the treatment of envenomation. Toxicon 55(7):1195–1212

Citrin DE (2017) Recent developments in radiotherapy. N Engl J Med 377(11):1065–1075

Clark GC, Casewell NR, Elliott CT, Harvey AL, Jamieson AG, Strong PN, Turner AD (2019) Friends or foes? Emerging impacts of biological toxins. Trends Biochem Sci 44(4):365–379

Colon R, Wheater M, Joyce EJ, Ste. Marie EJ, Hondal RJ, Rein KS (2021) The marine neurotoxin brevetoxin (PbTx-2) inhibits Karenia brevis and mammalian thioredoxin reductases by targeting different residues. J Nat Prod 84(11):2961–2970

Cooper EL, Mologne N (2017) Exploiting leech saliva to treat osteoarthritis: A provocative perspective. J Tradit Complement Med 7(3):367–369

Dai C, Tian E, Hao Z, Tang S, Wang Z, Sharma G, Jiang H, Shen J (2022) Aflatoxin B1 toxicity and protective effects of curcumin: Molecular mechanisms and clinical implications. Antioxidants 11(10):2031

Das S, Singh S, Chawla V, Chawla PA, Bhatia R (2023) Surface plasmon resonance as a fascinating approach in target-based drug discovery and development. TrAC Trends Anal Chem 171:117501

Dean WS, Adejumo HA, Caiati A, Garay PM, Harmata AS, Li L, Rodriguez EE, Sundar S (2020) A framework for regulation of new and existing PFAS by EPA. J Sci Policy Governance 16(1):1–14
De Filippis LF (2016) Plant secondary metabolites: From molecular biology to health products. In: Azooz MM, Ahmad P (eds) Plant-environment interaction: responses and approaches to mitigate stress. Wiley Blackwell, Hoboken, pp 263–299
De Souza JM, Goncalves BD, Gomez MV, Vieira LB, Ribeiro FM (2018) Animal toxins as therapeutic tools to treat neurodegenerative diseases. Front Pharmacol 9:145
Díaz-Gómez JL, Martín-Estal I, Rivera-Aboytes E, Gaxiola-Muñíz RA, Puente-Garza CA, García-Lara S, Castorena-Torres F (2024) Biomedical applications of synthetic peptides derived from venom of animal origin: A systematic review. Biomed Pharmacother 170:116015
Duda R, Monteiro WM, Giles-Vernick T (2021) Integrating lay knowledge and practice into snakebite prevention and care in central Africa, a hotspot for envenomation. Toxicon X 11:100077
Duong HT (2017) Fourth generation NGOs: Communication strategies in social campaigning and resource mobilization. J Nonprofit Public Sector Market 29(2):119–147
Edo GI, Samuel PO, Oloni GO, Ezekiel GO, Ikpekoro VO, Obasohan P, Ongulu J, Otunuya CF, Opiti AR, Ajakaye RS, Essaghah AEA (2024) Environmental persistence, bioaccumulation, and ecotoxicology of heavy metals. Chem Ecol 1–28
Endean R, Sizemore DJ (1988) The effectiveness of antivenom in countering the actions of box-jellyfish (Chironex fleckeri) nematocyst toxins in mice. Toxicon 26(5):425–431
Fan P, Yu H, Xi B, Tan W (2022) A review on the occurrence and influence of biodegradable microplastics in soil ecosystems: are biodegradable plastics substitute or threat? Environ Int 163:107244
Feola A, Perrone MA, Piscopo A, Casella F, Della Pietra B, Di Mizio G (2020) Autopsy findings in case of fatal scorpion sting: a systematic review of the literature. Healthcare 8(3):325. MDPI
Fernandez I, Valladolid G, Varon J, Sternbach G (2011) Encounters with venomous sea-life. J Emerg Med 40(1):103–112
Free CM, Moore SK, Trainer VL (2022) The value of monitoring in efficiently and adaptively managing biotoxin contamination in marine fisheries. Harmful Algae 114:102226
Fujiki H, Sueoka E, Watanabe T, Komori A, Suganuma M (2023) Cancer progression by the okadaic acid class of tumor promoters and endogenous protein inhibitors of PP2A, SET and CIP2A. J Cancer Res Clin Oncol 1–9
Giansanti F, Flavell DJ, Angelucci F, Fabbrini MS, Ippoliti R (2018) Strategies to improve the clinical utility of saporin-based targeted toxins. Toxins 10(2):82
Glaser KB, Mayer AM (2009) A renaissance in marine pharmacology: from preclinical curiosity to clinical reality. Biochem Pharmacol 78(5):440–448
González-Cano R, Ruiz-Cantero MC, Santos-Caballero M, Gómez-Navas C, Tejada MÁ, Nieto FR (2021) Tetrodotoxin, a potential drug for neuropathic and cancer pain relief? Toxins 13(7):483
Grenda T, Grenda A, Krawczyk P, Kwiatek K (2022) Botulinum toxin in cancer therapy—current perspectives and limitations. Appl Microbiol Biotechnol 106(2):485–495
Gupta N, Noël R, Goudet A, Hinsinger K, Michau A, Pons V, Abdelkafi H, Secher T, Shima A, Shtanko O, Sakurai Y (2017) Inhibitors of retrograde trafficking active against ricin and Shiga toxins also protect cells from several viruses, Leishmania and Chlamydiales. Chem Biol Interact 267:96–103
Hughes RJ, Angus JA, Winkel KD, Wright CE (2012) A pharmacological investigation of the venom extract of the Australian box jellyfish, Chironex fleckeri, in cardiac and vascular tissues. Toxicol Lett 209(1):11–20
Isbister GK, White J (2004) Clinical consequences of spider bites: recent advances in our understanding. Toxicon 43(5):477–492
Janik E, Ceremuga M, Saluk-Bijak J, Bijak M (2019) Biological toxins as the potential tools for bioterrorism. Int J Mol Sci 20(5):1181
Javaid M, Haleem A, Singh RP, Rab S, Suman R (2021) Exploring the potential of nanosensors: A brief overview. Sens Int 2:100130

Jenkins TP, Laustsen AH (2020) Cost of manufacturing for recombinant snakebite antivenoms. Front Bioeng Biotechnol 8:703
Jewett EB, Lopez CB, Dortch Q, Etheridge SM, Backer LC (2008) Harmful algal bloom management and response: assessment and plan
Josić D, Rešetar D, Peršurić Ž, Martinović T, Pavelić SK (2017) Detection of microbial toxins by-omics methods: A growing role of proteomics. In: Colgrave ML (ed) Proteomics in food science. Academic, London, pp 485–506
Kashyap S, Pal S, Chandan G, Saini V, Chakrabarti S, Saini NK, Mittal A, Thakur VK, Saini AK, Saini RV (2022) November. Understanding the cross-talk between human microbiota and gastrointestinal cancer for developing potential diagnostic and prognostic biomarkers. Semin Cancer Biol 86:643–651. Academic
Kennedy DO, Wightman EL (2011) Herbal extracts and phytochemicals: plant secondary metabolites and the enhancement of human brain function. Adv Nutr 2(1):32–50
Khalily MP, Soydan M (2023) Peptide-based diagnostic and therapeutic agents: Where we are and where we are heading? Chem Biol Drug Des 101(3):772–793
Kraus JL (2020) Artificial intelligence applied to the production of high-added-value dinoflagellates toxins. AI Soc 35:851–855
Krishna IM, Manickam V, Shah A, Davergave N (2017) Environmental management: science and engineering for industry. Butterworth-Heinemann, Oxford
Kvetkina A, Malyarenko O, Pavlenko A, Dyshlovoy S, von Amsberg G, Ermakova S, Leychenko E (2020) Sea anemone Heteractis crispa actinoporin demonstrates in vitro anticancer activities and prevents HT-29 colorectal cancer cell migration. Molecules 25(24):5979
Lahiani A, Yavin E, Lazarovici P (2017) The molecular basis of toxins' interactions with intracellular signaling via discrete portals. Toxins 9(3):107
Lan YL, Lou JC, Jiang XW, Wang X, Xing JS, Li S, Zhang B (2019) A research update on the anticancer effects of bufalin and its derivatives. Oncol Lett 17(4):3635–3640
Laustsen AH, Lohse B, Gutiérrez JM, Lomonte B, McCafferty J, Rasmussen AR (2016) Recombinant antivenoms. University of Copenhagen, Copenhagen
Laustsen AH, Gutiérrez JM, Knudsen C, Johansen KH, Bermúdez-Méndez E, Cerni FA, Jürgensen JA, Ledsgaard L, Martos-Esteban A, Øhlenschlæger M, Pus U (2018) Pros and cons of different therapeutic antibody formats for recombinant antivenom development. Toxicon 146:151–175
Lazonick W, Tulum Ö (2011) US biopharmaceutical finance and the sustainability of the biotech business model. Res Policy 40(9):1170–1187
Lecointre G, Aish A, Améziane N, Chekchak T, Goupil C, Grandcolas P, Vincent JF, Sun JS (2023) Revisiting nature's "unifying patterns": a biological appraisal. Biomimetics 8(4):362
Ledsgaard L, Jenkins TP, Davidsen K, Krause KE, Martos-Esteban A, Engmark M, Rørdam Andersen M, Lund O, Laustsen AH (2018) Antibody cross-reactivity in antivenom research. Toxins 10(10):393
Li J, Wang H (2018) Research progress of biotoxins. J Food Saf Qual 9(13):3359–3367
Li J, Wang S, Zhao J, Dong Z, Shao T (2022) Gut microbiota of Ostrinia nubilalis larvae degrade maize cellulose. Front Microbiol 13:816954
Li S, Wang A, Wu Y, He S, Shuai W, Zhao M, Zhu Y, Hu X, Luo Y, Wang G (2023) Targeted therapy for non-small-cell lung cancer: New insights into regulated cell death combined with immunotherapy. Immunol Rev 321(1):300–334
Linton DS (2005) Emil von Behring: Infectious disease, immunology, serum therapy, vol 255. American Philosophical Society, Philadelphia
Liu T, Li L, Yin L, Yu H, Jing H, Liu Y, Kong C, Xu M (2017) *Superantigen staphylococcal* enterotoxin C1 inhibits the growth of bladder cancer. Biosci Biotechnol Biochem 81(9):1741–1746
Lucas R, Hadizamani Y, Gonzales J, Gorshkov B, Bodmer T, Berthiaume Y, Moehrlen U, Lode H, Huwer H, Hudel M, Mraheil MA (2020) Impact of bacterial toxins in the lungs. Toxins 12(4):223
Ma L, Gu R, Tang L, Chen ZE, Di R, Long C (2015) Important poisonous plants in Tibetan ethnomedicine. Toxins 7(1):138–155

Mackieh R, Abou-Nader R, Wehbe R, Mattei C, Legros C, Fajloun Z, Sabatier JM (2021) Voltage-gated sodium channels: a prominent target of marine toxins. Mar Drugs 19(10):562
Márquez-López A, Fanarraga ML (2023) AB toxins as high-affinity ligands for cell targeting in cancer therapy. Int J Mol Sci 24(13):11227
Mitra R, Ghosh S, Mukherjee G, Acharya Chowdhury A (2023) Secondary metabolites: treasure trove for future medicine. In: Plant specialized metabolites: phytochemistry, ecology and biotechnology. Cham, Springer Nature, pp 1–45
Moestrup O (2009) IOC-UNESCO taxonomic reference list of harmful micro algae. http://www.marinespecies.org/HAB
Montuori E, De Luca D, Penna A, Stalberga D, Lauritano C (2023) Alexandrium spp.: from toxicity to potential biotechnological benefits. Mar Drugs 22(1):31
Morabito S, Silvestro S, Faggio C (2018) How the marine biotoxins affect human health. Nat Prod Res 32(6):621–631
Odoemelam CS (2023) Computational modelling of supramolecular human and animal structures: applications to enzymes relevant in comparative physiological studies. Doctoral dissertation, Nottingham Trent University
Otsuki K, Li W, Asada Y, Chen CH, Lee KH, Koike K (2019) Daphneodorins A–C, anti-HIV gnidimacrin related macrocyclic daphnane orthoesters from Daphne odora. Org Lett 22(1):11–15
Ouassaf M, Bourougaa L, Al-Mijalli SH, Abdallah EM, Bhat AR, Kawsar A, S.M. (2023) Marine-Derived Compounds as Potential Inhibitors of Hsp90 for Anticancer and Antimicrobial Drug Development: A Comprehensive In Silico Study. Molecules 28(24):8074
Ovsepian SV, O'Leary VB, Ayvazyan NM, Al-Sabi A, Ntziachristos V, Dolly JO (2019) Neurobiology and therapeutic applications of neurotoxins targeting transmitter release. Pharmacol Ther 193:135–155
Pagán OR (2018) Strange survivors: how organisms attack and defend in the game of life. BenBella Books, Dallas
Pandey P, Khan F, Khan MA, Kumar R, Upadhyay TK (2023) An updated review summarizing the anticancer efficacy of melittin from bee venom in several models of human cancers. Nutrients 15(14):3111
Panter KE, Welch KD, Gardner DR, Green BT (2013) Poisonous plants: effects on embryo and fetal development. Birth Defects Res C Embryo Today Rev 99(4):223–234
Parak M, Asgari A, Nourian YH, Ghanei M (2024) A review of poisoning with various types of biotoxins and its common clinical symptoms. Toxicon 240:107629
Pavaux AS, Berdalet E, Lemée R (2020) Chemical ecology of the benthic dinoflagellate genus Ostreopsis: review of progress and future directions. Front Mar Sci 7:498
Pereira AFM, Cavalcante JS, Angstmam DG, Almeida C, Soares GS, Pucca MB, Ferreira Junior RS (2023) Unveiling the pain relief potential: harnessing analgesic peptides from animal venoms. Pharmaceutics 15(12):2766
Perez Santin E, Rodríguez Solana R, González García M, García Suárez MDM, Blanco Díaz GD, Cima Cabal MD, Moreno Rojas JM, Lopez Sanchez JI (2021) Toxicity prediction based on artificial intelligence: A multidisciplinary overview. Wiley Interdiscip Rev Comput Mol Sci 11(5):e1516
Perrone G, Gallo A (2017) Aspergillus species and their associated mycotoxins. Mycotoxigenic fungi. Methods Protoc 1542:33–49
Polito L, Bortolotti M, Battelli MG, Calafato G, Bolognesi A (2019) Ricin: An ancient story for a timeless plant toxin. Toxins 11(6):324
Pore SM, Ramanand SJ, Patil PT, Gore AD, Pawar MP, Gaidhankar SL, Ghanghas RR (2015) A retrospective study of use of polyvalent anti-snake venom and risk factors for mortality from snake bite in a tertiary care setting. Indian J Pharm 47(3):270
Pradhan B, Ki JS (2022) Phytoplankton toxins and their potential therapeutic applications: A journey toward the quest for potent pharmaceuticals. Mar Drugs 20(4):271
Pugh RNH, Theakston RDG (1980) Incidence and mortality of snake bite in savanna Nigeria. Lancet 316(8205):1181–1183

Ramachandiran S, Satpaty S, Shankar K, Kalimuthu K, Muthuvel A (2020) Anti-cancer properties of protein hydrolysate from the posterior salivary gland of Amphioctopus membranaceus (Quoy & Gaimard, 1832). Int J Pept Res Ther 26:1429–1436

Ramachandran S, Rajagopal S, Ramachandran S, Rajagopal S (2019) Biomedical importance of marine peptides/toxins. In: Zebrafish: a model for marine peptide based drug screening. Springer, Singapore, pp 1–14

Raposo-Garcia S, Costas C, Louzao MC, Vieytes MR, Vale C, Botana LM (2023) Synergistic effect of brevetoxin BTX-3 and ciguatoxin CTX3C in human voltage-gated Nav1. 6 sodium channels. Chem Res Toxicol 36(12):1990–2000

Reid AJ, Carlson AK, Creed IF, Eliason EJ, Gell PA, Johnson PT, Kidd KA, MacCormack TJ, Olden JD, Ormerod SJ, Smol JP (2019) Emerging threats and persistent conservation challenges for freshwater biodiversity. Biol Rev 94(3):849–873

Robert A, Wiels J (2021) Shiga toxins as antitumor tools. Toxins 13(10):690

Roberts DM, Gallapatthy G, Dunuwille A, Chan BS (2016) Pharmacological treatment of cardiac glycoside poisoning. Br J Clin Pharmacol 81(3):488–495

Rossini GP, Hess P (2010) Phycotoxins: chemistry, mechanisms of action and shellfish poisoning. In: Luch A (ed) Molecular, clinical and environmental toxicology: Volume 2: Clinical toxicology. Birkhäuser, Basel, pp 65–122

Sadybekov AV, Katritch V (2023) Computational approaches streamlining drug discovery. Nature 616(7958):673–685

Santos JC, Tarvin RD, O'Connell LA (2016) A review of chemical defense in poison frogs (Dendrobatidae): ecology, pharmacokinetics, and autoresistance. In: Schulte BA, Goodwin TE, Ferkin MH (eds) Chemical signals in vertebrates, vol 13. Springer, Cham, pp 305–337

Schönthal AH, Swenson SD, Chen TC, Markland FS (2020) Preclinical studies of a novel snake venom-derived recombinant disintegrin with antitumor activity: A review. Biochem Pharmacol 181:114149

Shady NH, Khattab AR, Ahmed S, Liu M, Quinn RJ, Fouad MA, Kamel MS, Muhsinah AB, Krischke M, Mueller MJ, Abdelmohsen UR (2020) Hepatitis c virus ns3 protease and helicase inhibitors from red sea sponge (Amphimedon) species in green synthesized silver nanoparticles assisted by in silico modeling and metabolic profiling. Int J Nanomed 3377–3389

Shafiee F, Aucoin MG, Jahanian-Najafabadi A (2019) Targeted diphtheria toxin-based therapy: a review article. Front Microbiol 10:2340

Shahid M, Dumat C, Khalid S, Schreck E, Xiong T, Niazi NK (2017) Foliar heavy metal uptake, toxicity and detoxification in plants: A comparison of foliar and root metal uptake. J Hazard Mater 325:36–58

Shakeri S, Ashrafizadeh M, Zarrabi A, Roghanian R, Afshar EG, Pardakhty A, Mohammadinejad R, Kumar A, Thakur VK (2020) Multifunctional polymeric nanoplatforms for brain diseases diagnosis, therapy and theranostics. Biomedicines 8(1):13

Sharma PC, Sharma D, Sharma A, Bhagat M, Ola M, Thakur VK, Bhardwaj JK, Goyal RK (2022) Recent advances in microbial toxin-related strategies to combat cancer. Semin Cancer Biol 86:753–768. Academic

Silva A, Isbister GK (2020) Current research into snake antivenoms, their mechanisms of action and applications. Biochem Soc Trans 48(2):537–546

Singh AP (2010) Medicinal leech therapy (hirudotherapy): a brief overview. Complement Ther Clin Pract 16(4):213–215

Starnowska-Sokół J, Przewłocka B (2020) Multifunctional opioid-derived hybrids in neuropathic pain: Preclinical evidence, ideas and challenges. Molecules 25(23):5520

Strachan CR (2023) Mycotoxin and their impact on animal health. In: World Mycotoxin Forum, October 2023, vol 21, p 4

Tan KY, Tan CH, Fung SY, Tan NH (2016) Neutralization of the principal toxins from the venoms of Thai Naja kaouthia and Malaysian Hydrophis schistosus: insights into toxin-specific neutralization by two different antivenoms. Toxins 8(4):86

Thakur G, Raghu HV, Tehri N, Kumar N, Yadav A, Malik RV (2014) Biochip based detection-An emerging tool for ensuring safe milk: A review. J Innovat Biol 3:147–154

Tompkins DA, Hobelmann JG, Compton P (2017) Providing chronic pain management in the "Fifth Vital Sign" Era: Historical and treatment perspectives on a modern-day medical dilemma. Drug Alcohol Depend 173:S11–S21

Trivanović D, Pavelić K, Peršurić Ž (2021) Fighting cancer with bacteria and their toxins. Int J Mol Sci 22(23):12980

Uchechukwu-Agua AD, Caleb OJ, Opara UL (2015) Postharvest handling and storage of fresh cassava root and products: a review. Food Bioprocess Technol 8:729–748

Uto Y, Sasaki K, Takahashi M, Morimoto K, Inoue K (2019) Application of high-speed countercurrent chromatography for the purification of high-purity illudin S from Omphalotus japonicus. Anal Sci 35(7):789–792

van Egmond HP (2004) Natural toxins: risks, regulations and the analytical situation in Europe. Anal Bioanal Chem 378:1152–1160

Vilariño N, Louzao MC, Fraga M, Botana LM (2020) 15 From science to policy: dynamic adaptation of legal regulations on aquatic biotoxins. In: Botana LM, Louzao MC, Vilariño N (eds) Climate change and marine and freshwater toxins. De Gruyter, Berlin, pp 607–654

Vincy A, Mazumder S, Banerjee I, Hwang KC, Vankayala R (2022) Recent progress in red blood cells-derived particles as novel bioinspired drug delivery systems: challenges and strategies for clinical translation. Front Chem 10:905256

Wanat M, Nowak B, Świątko A, Mirkowski K, Domański J, Dąbrowski P, Domagała Z (2022) Trigger points in medical practice-current therapeutic directions. Med J Cell Biol 10(4):129–137

Wang Q, Yang Q, Wu W (2022) Ensuring seafood safe to spoon: a brief review of biosensors for marine biotoxin monitoring. Crit Rev Food Sci Nutr 62(9):2495–2507

Wang S, Yang Z, Ma X, Zhou Z, Ma Z, Chen J, Ren X (2023a) Emerging synchrotron radiation X-ray-sensitive probes for in situ bioimaging at the nanoscale. TrAC Trends Anal Chem 170:117453

Wang T, Wu M, Cao L, Liu B (2023b) Organic functional substance engineered living materials for biomedical applications. Biomaterials 301:122248

Winkel KD, Hawdon GM, Fenner PJ, Gershwin LA, Collins AG, Tibballs J (2003) Jellyfish antivenoms: past, present, and future. J Toxicol Toxin Rev 22(1):115–127

Witmanowski H, Błochowiak K (2020) The whole truth about botulinum toxin—a review. Adv Dermatol Allergol/Postępy Dermatol Alergol 37(6):853–861

Wolf A, Mazenot C, Spadoni S, Calvet F, Demoncheaux JP (2011) [FAV-Africa: a polyvalent antivenom serum used in Africa and Europe]. Med Trop (Mars) 71:537–540

Wong KY, Tan CH, Tan NH (2016) Venom and purified toxins of the spectacled cobra (Naja naja) from Pakistan: insights into toxicity and antivenom neutralization. Am J Trop Med Hyg 94(6):1392

World Health Organization (WHO) (2021) https://www.who.int/news-room/fact-sheets/detail/cancer

Yaksh TL, Wallace MS (2011) Opioids, analgesia, and pain management. In: Brunton LL, Chabner BA, Knollmann BC (eds) Goodman and Gilman's the pharmacological basis of therapeutics. McGraw-Hill, New York, pp 481–526

Yang H, Feng Q, Zhu J, Liu G, Dai Y, Zhou Q, Xia S, Wu Z, Zhang Y (2024) Towards sustainable futures: A review of sediment remediation and resource valorization techniques. J Clean Prod 435:140529

Yarsan E, Aktaş İ (2017) Usage of biotoxins as analgesic. Turk J Occup/Environ Med Saf 2(1 (1)):257–257

Yong MY, Tan KY, Tan CH (2021) Potential para-specific and geographical utility of Thai Green Pit Viper (Trimeresurus albolabris) Monovalent Antivenom: Neutralization of procoagulant and hemorrhagic activities of diverse *Trimeresurus* pit viper venoms. Toxicon 203:85–92

Yu Y, Yang R, Zhao X, Qin D, Liu Z, Liu F, Song X, Li L, Feng R, Gao N (2016) Abrin P2 suppresses proliferation and induces apoptosis of colon cancer cells via mitochondrial membrane depolarization and caspase activation. Acta Biochim Biophys Sin 48(5):420–429

Yu XQ, Yan QL, Shen S, Cheng ZY, Huang XX, Yao GD, Song SJ (2022) Chamaejasmenin E from Stellera chamaejasme induces apoptosis of hepatocellular carcinoma cells by targeting c-Met in vitro and in vivo. Bioorg Chem 119:105509

Yun SH, Sim EH, Han SH, Han JY, Kim SH, Silchenko AS, Stonik VA, Park JI (2018) Holotoxin A1 induces apoptosis by activating acid sphingomyelinase and neutral sphingomyelinase in K562 and human primary leukemia cells. Mar Drugs 16(4):123

Zeng S, Liu X, Kafuti YS, Kim H, Wang J, Peng X, Li H, Yoon J (2023) Fluorescent dyes based on rhodamine derivatives for bioimaging and therapeutics: recent progress, challenges, and prospects. Chem Soc Rev 52(16):5607–5651

Zhang D, Kanakkanthara A (2020) Beyond the paclitaxel and vinca alkaloids: Next generation of plant-derived microtubule-targeting agents with potential anticancer activity. Cancers 12(7):1721

Zhang Z, Godefroy SB, Lyu H, Sun B, Fan Y (2018) Transformation of China's food safety standard setting system—review of 50 years of change, opportunities and challenges ahead. Food Control 93:106–111

Zhao L, Huang Y, Dong Y, Han X, Wang S, Liang X (2018) Aptamers and aptasensors for highly specific recognition and sensitive detection of marine biotoxins: recent advances and perspectives. Toxins 10(11):427

Zheng X, Ou Y, Shu M, Wang Y, Zhou Y, Su X, Zhu W, Yin W, Li S, Qiu P, Yan G (2014) Cholera toxin, a typical protein kinase A activator, induces G1 phase growth arrest in human bladder transitional cell carcinoma cells via inhibiting the c-Raf/MEK/ERK signaling pathway. Mol Med Rep 9(5):1773–1779

Zupo V, Mutalipassi M, Ruocco N, Glaviano F, Pollio A, Langellotti AL, Romano G, Costantini M (2019) Distribution of toxigenic Halomicronema spp. in adjacent environments on the Island of Ischia: Comparison of strains from thermal waters and free living in Posidonia oceanica meadows. Toxins 11(2):99

Chapter 17
The Potential of Biotoxins for Military Use

Monika Gałwa-Widera

Abstract The use of biological agents for military or terrorist objectives, with the aim of destroying or reducing the size of an enemy army, to assault a city, or another location, is what biological threats are all about. The most common definition of biological hazards is the mass killing of humans by microorganisms such as viruses, bacteria, or poisons. Herds of farm animals or microorganisms that cause crop plant diseases, such as some fungi, can also be considered biological hazards.

Biological hazards are not a recent development. Numerous attempts to employ infectious microbes for military objectives have been documented in historical texts. Using the corpses of those who perished from the plague as disease carriers is one example.

During the fourteenth-century Battle of Caffa, the Tatars sent to the invading Mongol army to cause an epidemic. They perished in large numbers, negotiated with Italians, and left their city.

The British army's "gifting" of blankets used by smallpox victims to Native Americans in North America is another instance of the deployment of biological threats in the eighteenth century. Native Americans who had never been exposed to this illness before were totally susceptible to it and perished in large numbers.

Today, the Geneva Convention forbids the use, storage, and investigation of biological dangers. In spite of this, numerous nations have carried out or are currently carrying out covert studies.

Large anthrax bacilli were thrown into the atmosphere at Sverdlovsk (Yekaterinburg, modern-day Russia) in 1972 due to an accident that claimed 66 lives. During Operation Desert Storm in 1990–1991 in Iraq, significant amounts of readily usable botulinum toxin were found, which is among the most lethal poisons known to man.

M. Gałwa-Widera (✉)
Faculty of Infrastructure and Environment, Czestochowa University of Technology, Czestochowa, Poland
e-mail: monika.galwa-widera@pcz.pl

M. N. V. Prasad et al. (eds.), *Biotoxins*,
https://doi.org/10.1007/978-3-031-75309-1_17

This chapter offers an overview of the threat and historical use of biological threats, and it underscores the critical need for stringent international regulation and cooperation to prevent their development and deployment.

Keywords Biological threats · Toxins · War · Terrorism · Microbiology · Microorganisms · Pandemic · Plague

17.1 Introduction

Biological threats remain one of the most terrifying threats of war created by man. Although its use is widely condemned, it cannot be ruled out that it will be used for large-scale damage to people during current and possible future with incompatible diverse people. The real possibilities of its use by people necessitate prompt analysis of the biological agents as threats.

The literature presents many definitions in the field of biological threats. In the simplest terms, biological threats, causemass destruction through pathogenic microorganisms, and their toxins. These include mainly viruses, bacteria, or protozoa and their toxins responsible for disease epidemics among people and their surroundings (Michaliuk 2016). It is therefore a dangerous threat with a massive impact on people by contaminating animals, water, and food with pathogenic microorganisms and toxins produced by them.

17.2 History

According to historical sources, biological threats have been used since ancient times. However, there are no clear indications when it was first used. It is unknown whether these were intentional attacks or of an intuitive nature, as these were not supported by scientific knowledge at that time leading to speculations by present-day scientists and historians.

In the fourth century BC, Scythian archers used arrows dipped in decaying corpses and excrement. They were famous for their very cruel fighting methods.

Famous leaders in ancient times also used biological threats. Alexander of Macedon considered a brilliant strategist and politician, ordered corpses to be left unburied when retreating after a lost battle to cause a plague. Hannibal used another method, using animal venom as a poison. During naval battles against King Eumenes II of Pergamon, he used clay projectiles filled with poisonous snakes as catapults.

Coating arrows with poison, poisoning wells, or throwing carcasses beyond city walls were everyday battles between Germanic tribes and the Romans. The Romans, although they officially condemned the use of poisons, used these methods during battles in Anatolia.

South American Indians eagerly used toxins to poison the arrows of their traditional threats—blowpipes. For this purpose, the venom of large spiders and the secretion of small tree frogs were used. In the Middle Ages, biological threats were used many times as a means to facilitate the capture of besieged cities. The most famous biological attack of the Middle Ages was when the Tatars catapulted the bodies of those who died of the plague to the Caffa fortress (Janeczek et al. 2010).

Cases of dead people or dead animals being abandoned near water intakes are also described. The probable assumption behind such actions was an attempt to spread a plague that would decimate the enemy. Another example is the use by the Assyrians in the sixth century B.C. of poisonous red cardi mushrooms (*Claviceps purpurea*) to poison the wells of enemies. The alkaloids contained in this parasitic fungus have a very wide-ranging effect on the human body, from hallucinations to death.

In ancient times, a successful case of a biological attack using the bacillus *Francisella tularensis* was the expansion of the Hittites into the Middle East. For this reason, tularemia is still called Hittite disease. It devastated the areas of the Middle East at the end of the fourteenth century BC. The Hittites sent animals infected with the bacillus, to the enemy camp, which led to the spread of the disease among them. As a result of these actions, Uhhazitis the ruler of Arzawa fell ill, and the Hittite kings, Suppilium I and Arnuwand II also fell victim to this disease.

The secret bacteriological center in Suzdal investigated the possibility of using many infectious agents, including tularemia, as well as plague, typhus, and Q fever, as biological threats before the outbreak of World War II. The research that began then contributed to the use of this microorganism on the World War II war front. Soviet soldiers in 1942–1943 infected German soldiers with the tularemia bacterium during the siege of Stalingrad, which resulted in numerous illnesses on both fighting sides (Binczycka-Anholcer and Imiołek 2011).

Simultaneously with the US entry into World War II, the Americans began a research program on bacteriological threats. For this purpose, Camp Detrick and Camp Frederick were built, which conducted experiments in field conditions. Initially, the research was conducted only on anthrax and bacteria from the Brucella family. In a short time, the research program was expanded to include other biological pathogens for humans, primarily tularemia and plague. Additionally, the pathogens affecting animals and crops were investigated too.

The research on the use of the tularemia bacillus in the USA was reduced to a laboratory scale, solely for research purposes after the end of World War II. In the 1960s, the USA conducted many experiments on the use of biological threats. Until the research on biological threats was discontinued in the late 1960s, the United States possessed seven types of standardized biological threats. These include, among others: anthrax and tularemia bacteria, SEB toxins, and botulinum. Furthermore, organisms and factors that cause brucellosis, Q fever, and Venezuelan equine encephalitis (VEE) were researched. The inability by humans to fight these diseases on a large scale makes them biological threats (Mierzyńska 2002).

The first records of mass plague cases date back to ancient Rome. In the second century AD, Roman plans included expanding the borders of the empire to include

lands belonging to Germanic peoples from what is now Czechia and Moravia and creating a new province named Marcomannia. However, in 165 AD, the legionnaires were attacked by an unknown disease termed plague. Cassius Dio, a Roman historian of Greek origin, recorded that in the very center of the empire, approximately 2000 people died of this disease in one day. Overall, it was estimated that the number of people who died from the plague was at least 5 million inhabitants of the entire Roman Empire; including two emperors: Lucius Verus (169 AD) and Marcus Aurelius (180 AD). The weakening of the legionary army due to disease, and not the strength of the Germanic peoples, stopped Rome's expansion in the north and thwarted the creation of a new province, Marcomannia.

Epidemics also took a toll on soldiers during the wars of Crusades. The army stationed near Antioch decreased by 94% due to disease in the First Crusade war in 1096–1099). Another Crusade under Baldwin's leadership did not reach Jerusalem due to the raging plague. For over 2000 years, 41 cases of the "Black Death" epidemic have been recorded. The plague was a biological threat used by the conflicting parties in their battles. The fear of the possible return of the plague also had a different face. It became the reason for closing baths in cities and popularizing the "cult of dirt." The plague pathogen was used in the Middle Ages to kill enemy troops. In the Battle of Carolstein (1422), the bodies of dead soldiers with the disease were catapulted onto enemy troops to reduce their population (Hawley and Etizen Jr. 2001).

In 1710, during the Russo-Swedish war, the plague pathogen was used, just as in the Middle Ages, during the Battle of Carolstein. The Russians, unable to capture the city effectively defended by the Swedes, transferred the bodies of those who died of the plague to Reval (now Tallinn), leading to the fall of the city.

In the early 1940s, Japan dropped ceramic bombs containing fleas infected with the plague bacteria over Manchuria and nearby regions. The force of the explosion scattered the grain, attracting rats, which were then infected with fleas. These, in turn, spread the disease further into the society. Chinese cities were repeatedly attacked by Japan with various types of biological agents contaminating food and water using agrotechnical sprays. Japan's defeat in World War II thwarted plans to use bacteriological threats against the Allies. After the capitulation, Japanese commanders ordered the disbandment of units located in Manchuria, the destruction of all documents related to bacteriological threats research, and the demolition of research laboratories. The research program in the USSR on plague bacilli (but also on anthrax) was carried out not only by the plants in Sverdlovsk but also in Stepnogorsk in Kazakhstan. To conceal the fact of research on biological threats, efforts were made to camouflage external installations that could be associated with the production of bacteriological threats. Russian biologists probably conducted research on the use of disease-causing pathogens as threats until at least the 1980s. Ken Alibek, a military doctor who was employed as the head of the Stepnogorsk plant and "Biopreparat," claims that research activities on biological warfare agents have not been interrupted despite Yeltsin's ban in this field, and may continue to this day (Dzwolak and Dzwolak 2009).

The first mentions of anthrax date back to the twelfth century and refer to the Greek word anthracite, that is, coal, associated with black skin lesions occurring on the body of infected people and were described in "Bartholomaeus Anglicus' De Proprietatibus Rerum."

Before the outbreak of World War I, Dr. Anton Dilger, a US citizen of German origin and a surgeon employed at Johns Hopkins University in Baltimore, set up a laboratory in his private home in which he multiplied strains of anthrax and glanders. He received them from the German government and gave them to workers of the Baltimore shipyard who sympathized with the Germans. The workers' task was to infect 4000 horses, mules, and cattle intended for transport to the European continent as an aid to the Allies. The Germans were not only the creators but also the victims of sabotage biological attacks, including German intelligence discovered that the French staying in Switzerland were spreading the Burkholderia mallei bacterium among horses intended for transport to Germany. In the year of the outbreak of World War I, German pilots dropped fruit, children's toys, and sweets infected with anthrax and glanders over Romania. It was only in the interwar period, in 1924, that an international commission blamed Germany for the use of biological agents in Italy, Russia, and Romania; however, the allegations were not proven.

During World War II, Great Britain conducted research on the possibility of using disease-causing pathogens as threats to protect against a possible attack by Germany using biological threats. At this time, the Third Reich was also researching anthrax bacteria for use as a biological threat. In the 1940s, anthrax bombs were dropped on the uninhabited island of Griunard near Scotland. As a result of this experiment, all animals on the island died, and the activity of Bacillus anthracis spores was observed for another half a century after the experiment. The disinfection of the island with liquid formaldehyde was undertaken after 1980. (Riedel 2004).

Also during World War II, the Polish Home Army tried to use letters containing anthrax in its operations, but without much success. In the spring of 1943, a soap factory in London produced 5 million portions of feed injected with an anthrax suspension. It was a deliberate retaliatory action in the event of German biological aggression against Great Britain, which included scattering infected fodder over the enemy's meadows and fields. The goal of the plan was to infect beef cattle with anthrax.

In 1997, the terrorist organization Counter Holocaust Lobbyists of Hitler secretly delivered signed anthrax and plague tiles to public buildings. Shortly after the incident, the situation turned out to be a false alarm, resulting in huge costs associated with neutralizing the biological threat. In the late 1990s, the alarm was raised in abortion hospitals where letters containing anthrax spores were supposed to arrive. After the attack on the World Trade Center in 2001, an attempt was made to cause an epidemic of this disease using letters containing anthrax. Contaminated letters were sent to the headquarters of mass media (TV stations, New Jersey Post) and American Media (on the Florida peninsula). Politicians from the Democratic Party were to receive further letters containing anthrax spores. The biological terrorist attack, which occurred less than a few days after the most tragic attack on the World Trade Center in the history of the United States, killed five people.

A very dangerous infectious viral disease with an acute course and high mortality is smallpox (Variola vera—smallpox). History indicates that its first fatality was Pharaoh Ramesses V (1160 BC), while the first smallpox epidemic dates back to the Peloponnesian War between the Athenians and Spartans (431–404 BC). Hundreds of years later, the Spanish conquistadors were victorious not only because they had better threats (cannons and muskets), but it was the biological threats that they unknowingly carried with them that caused more damage and losses than the threats. Herman Cortes sailed to the then-Aztec lands at the beginning of the sixteenth century. He was seen by the natives as a god, but over time this perception changed to a more negative one, rebellions were started and the Spanish colonizers were forced to leave Aztec territory. During the fighting, one of the soldiers died of smallpox. The Aztecs, while searching the bodies of the fallen, unknowingly became infected with the smallpox virus. The disease slowly but systematically destroyed the Aztec community—from ordinary workers, through military units, to the emperor himself. The inability to bury a large number of victims resulted in the houses in which the dead were buried being collapsed to become their graves. The conquistadors reached as far as Tenochtitlan, where Cortes said that at that point smallpox had taken more Aztecs than Spanish cannons. Smallpox quickly spread throughout the Incas, killing the emperor, his heir, and important military commanders in the empire within a few months. Shortly thereafter, the Inca Empire was attacked by Francisco Pizarro. However, in this case, there is no clarity. Researchers analyzed whether the inhabitants of the Old Continent resorted to the use of biological threats to conquer the lands of America before 1700 and whether the natural spread of diseases became the cause of the death of many natives who had no immunity to the ailments of Europeans, which largely facilitated the colonization of the new continent. Probably since the first contact of the locals with Europeans in the early sixteenth century, approximately 95% of people have died as a result of various epidemics and diseases transported across the ocean.

There is no confirmation of intentional action in the previously described case; however, records confirm the intentional use of the smallpox pathogen in French warfare against Indians in the years 1754–1760. Colonel Henry Bouquet and Gen. Jeffrey Amherst gave rugs and clothing contaminated with the smallpox virus to the natives loyal to the French. Although the unfair actions had negligible effects, they nevertheless became a turning point in biological warfare. Fifteen years after this incident, the American Revolution broke out, in which commander Benedict Arnold led his troops to conquer Canadian lands. The attempt to capture Quebec failed, forcing the retreat of Arnold's army. Information quickly spread among the ranks that the British were spreading the smallpox virus among them thanks to people in contact with the ground forces. The views were later confirmed by Thomas Jefferson himself, accusing the leader of the British in Quebec of actions aimed at infecting American troops. During the Civil War, the English general Charles Cornwallis proposed to send 700 dark-skinned men infected with smallpox to the plantations supporting the rebellion. However, the effects of this action are unknown.

The examples discussed show that microorganisms considered as potential biological threats have been known to humanity for millennia. Even though they occur

naturally in nature, their incidence is relatively low. However, the work of microbiologists allows us to increase their activity. Information about pathogens and microorganisms that are less typical for a given climate zone and that have contributed in the past and currently affect global health issues is very limited. It is highly probable that over time, due to natural reasons such as the adaptation of pathogens to new environmental conditions, as well as due to the uncontrolled manipulation of the genetic material of pathogens, natural environmental barriers (mountains, seas, ambient temperature, and intermediate hosts) will be broken down and they spread around the world. An example of such a disease is malaria.

It is estimated that the losses of British troops resulting from malaria during the Napoleonic conquests in 1793–1815 amounted to over 200,000 soldiers and half a century later (during the Civil War of 1861–1865), over 1.3 million incidents of febrile malaria were recorded, of which, as many as 10,000 cases were fatal.

The prevalence of Malaria disease on the Macedonian war front during World War I, held off French, English, and German troops for approximately 36 months, forcing over 80% of the population to undergo hospital treatment. However, during World War II, more people died from malaria in the Pacific Ocean than from battles fought there. According to some researchers, 50 percent of people who have ever lived have died from this disease.

Until the nineteenth century, the use of biological threats was an occasional use of a primitive method of infecting an opponent with a disease naturally occurring at that time. The birth of modern microbiology, and the discovery of bacteria and microorganisms and their properties, turned out to be a real breakthrough. Engaging the potential of scientific achievements has taken the use of biological threats to a new level. When the discoveries of Louis Pasteur and Robert Koch were presented, no one probably expected how sinister their use could be. Research conducted for strictly peaceful purposes, aimed at combating the epidemic, was used in the twentieth century to improve methods of killing.

17.3 Classification and Characteristics of Agents as Biological Threats

Biological threats are divided into three categories A, B, and C, taking into account the pathogenicity of the agent, the difficulty of identifying it and recognizing the disease it causes, as well as the ease of its spread. Group A includes the most dangerous factors due to their high virulence, causing disease with high mortality, and the need to use specialized methods for their identification. These include bacteria such as *Francisella tularensis* (tularemia), *Yersinia pestis* (plague), *Bacillus anthracis* (anthrax), viruses such as *Poxvirus variola* (smallpox), and *Clostridium botulinum* toxins (botulinum). Filoviridae (e.g., Ebola) and Arenaviridae (Lassa fever) consist of several members that belong to category A viruses.

Tularemia is an acute bacterial infectious disease whose etiological factor is the pleomorphic, Gram-negative bacillus *Francisella tularensis*, which occurs in humans and animals. A typical symptom of the disease is swelling of the affected lymph nodes, and sepsis may also occur. In natural conditions, the disease occurs as a result of human bites by insects carrying bacteria in their saliva. In the event of a potential biological attack, as a result of bacteria being sprayed into the air, the so-called typhoid forms (Weiner and Kubajka 2015). Pharyngitis, often accompanied by pneumonia, may occur in up to 80 percent of people (Figs. 17.1 and 17.2).

Plague (Pestis) is a bacterial infectious disease of humans. The etiological agent of plague is the non-spore-producing bacterium *Yersinia pestis*. Plague occurs in several clinical forms—bubonic, septicemic, and pneumonic (primary and secondary). Without treatment, bubonic disease is fatal in 50 percent of cases. In the pulmonary form (pestis pneumonica), there are symptoms of severe, exudative pneumonia, with hemoptysis, shortness of breath, and cyanosis, and the mortality rate reaches 100%. According to WHO data, in the years 2010–2015, 3248 cases and 584 deaths were recorded. The plague spreads primarily by rats but can also spread by other small rodents such as squirrels, wild rabbits, voles, ground squirrels, and dormice. Fleas transmit germs from infected animals to humans. Only the pulmonary form of the plague can be transmitted through droplets. When germs enter the body, they first attack the lymph nodes and then spread through the blood throughout the body (Figs. 17.3 and 17.4).

Anthrax (Latin: anthrax) is an infectious and contagious disease caused by a Gram-positive bacterium, the anthrax bacillus (*Bacillus anthracis*). Anthrax most often occurs in cattle, horses, sheep, and goats, while birds are immune to natural infection (WHO 2008). Each year, approximately 2000 people are infected with anthrax (Inglessby and Tonat 2002). The cutaneous form of anthrax affects 95% cases which, if untreated, 24 percent end in death. In the remaining 5% of infected cases, in the intestinal and pulmonary forms, despite the treatment, the mortality

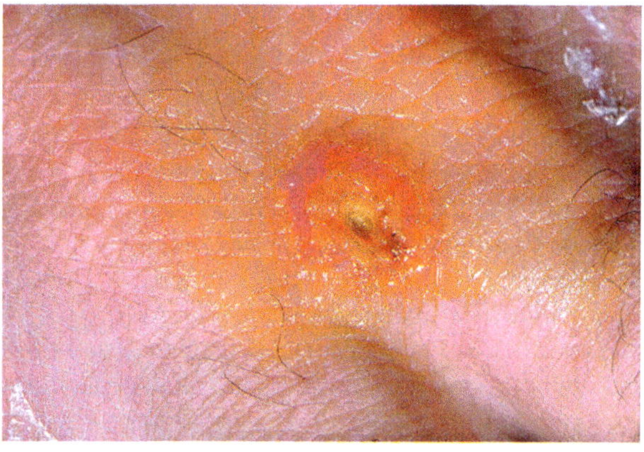

Fig. 17.1 Hand skin wound infected with Francisella tularensis (Dennis et al. 2001)

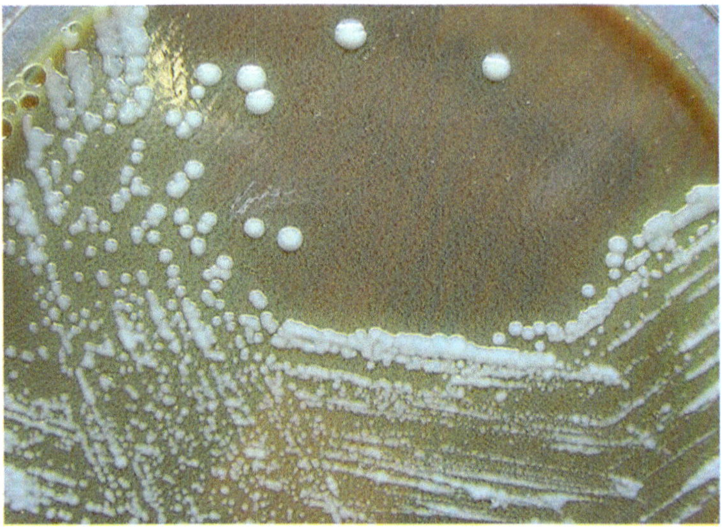

Fig. 17.2 Bacterial colonies of Francisella tularensis on agar medium (CDC/Larry Stauffer, Oregon Public Health Laboratory (PHIL nr 1910), 2002

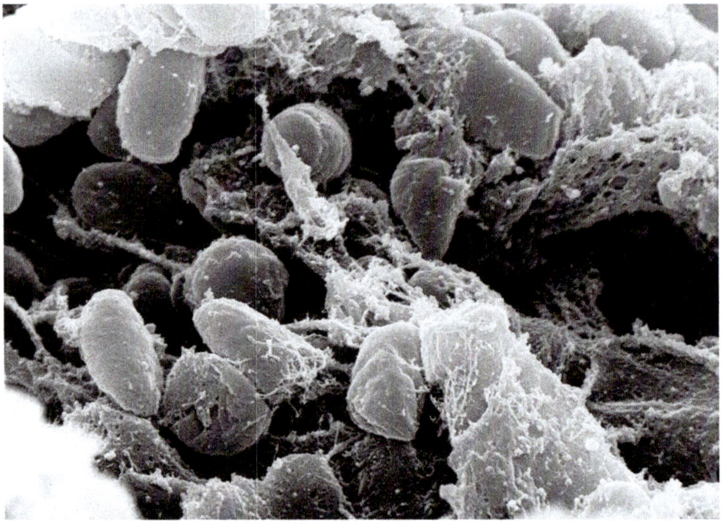

Fig. 17.3 Yersinia pestis bacterium, image from a scanning electron microscope (public domain, Rocky Mountain Laboratories, NIAID, NIH http://www3.niaid.nih.gov/biodefense/Public/Images.htm

rate reaches 75 and 85%, respectively. The spore forms produced by anthrax are exceptionally resistant to all disinfectants, acids, and bases, ionizing radiation, and heating. They can only be eliminated by fire or steam sterilization, at 100 °C, under high pressure. Spore forms can survive in soil or water for up to several dozen years

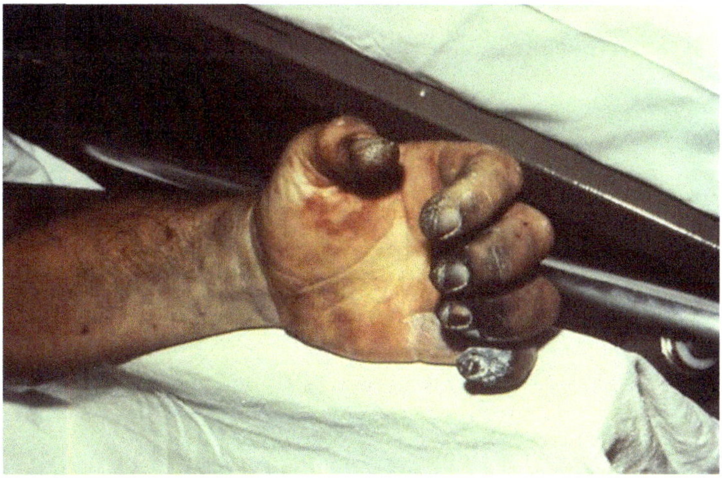

Fig. 17.4 Fingers of a patient diagnosed with septic plague (public domain, this media comes from the Centers for Disease Control and Prevention's Public Health Image Library (PHIL), with identification number #4137)

while retaining their ability to infect. Anthrax bacilli produce a toxin that is a combination of three specific proteins that alone are completely harmless, but only this combination gives them their lethal power. People are most often infected with anthrax by introducing the germ through the skin, which leads to the development of the cutaneous form. Inhalation of dust containing anthrax germs or their spores causes the pulmonary form while eating infected meat and its products leads to the intestinal form. There was no human-to-human transmission (Figs. 17.5 and 17.6).

Smallpox is a severe, viral infectious disease with a high mortality rate. A typical symptom is a characteristic rash that appears at the same time all over the body, after about 2–4 days of fever. In the final stage of the disease, the pustules become covered with scabs, which fall off and leave permanent, unsightly scars. After organizing an anti-epidemic campaign, in 1966, an intensive campaign to eliminate this dangerous disease took place around the world. The eradication of smallpox was announced by the World Health Organization (WHO) in 1980. This means that the environment is free from the smallpox virus, but collections of it are kept in selected laboratories around the world (Fig. 17.7).

Ebola hemorrhagic fever is a severe often fatal infectious disease caused by the Ebola virus. According to the International Statistical Classification of Diseases and Related Health Problems, it is included in the three-character category A98: "Other viral hemorrhagic fevers, not elsewhere classified."

Symptoms of the disease appear two days to three weeks after infection with the virus, most often the incubation period is from 8 to 10 days. The disease begins with fever (over 38.5 °C), headache, muscle and throat pain, and significant weakness. The next stage of the disease includes vomiting, diarrhea, rash, kidney and liver dysfunction, and in some cases, both internal and external bleeding (Fig. 17.8).

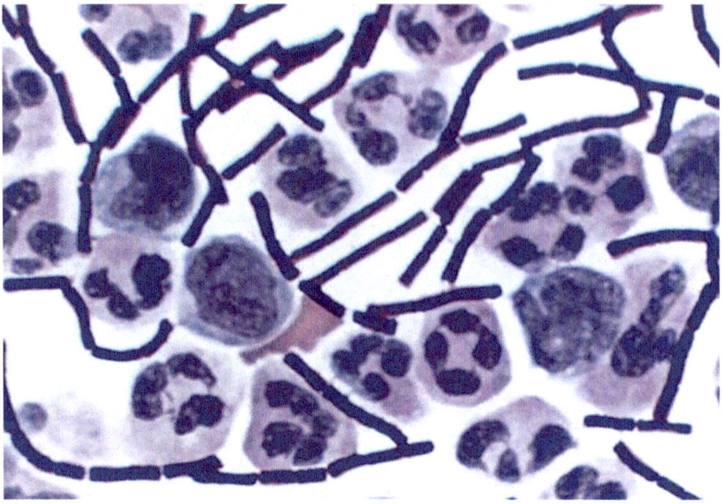

Fig. 17.5 Anthrax bacteria (bacilli) stained with the Gram method, visible spherical forms—neutrophils (public domain, https://en.m.wikipedia.org/wiki/File:Gram_Stain_Anthrax.jpg)

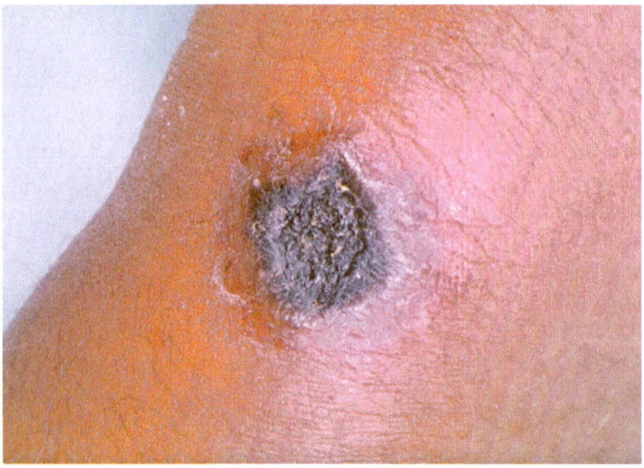

Fig. 17.6 Skin lesion caused by anthrax (public domain, this media comes from the Centers for Disease Control and Prevention's Public Health Image Library (PHIL), ID #2033)

Group B included factors with lower pathogenicity and mortality, easier to identify. These include bacteria such as Brucella spp. (brucellosis), Burkholderia maleic and pseudomaleii (glanders), Chlamydia psittaci (ornithosis), Rickettsia prowazekii (scratch fever), *Salmonella* (salmonellosis), *Shigella* (dysentery), *Vibro cholera* (cholera), the protozoan *Cryptosporidium parvum* (diseases of the digestive system), a group of Alphavirus viruses causing encephalitis, e.g., *Venezuelan encephalitis* virus and toxins such as ricin, epsilon toxin produced by *Clostridium perfringens*,

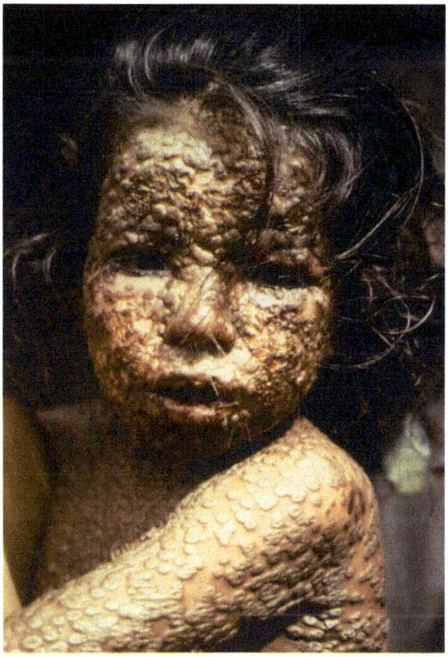

Fig. 17.7 Characteristic facial rash caused by smallpox (public domain, this is a file from Wikimedia Commons)

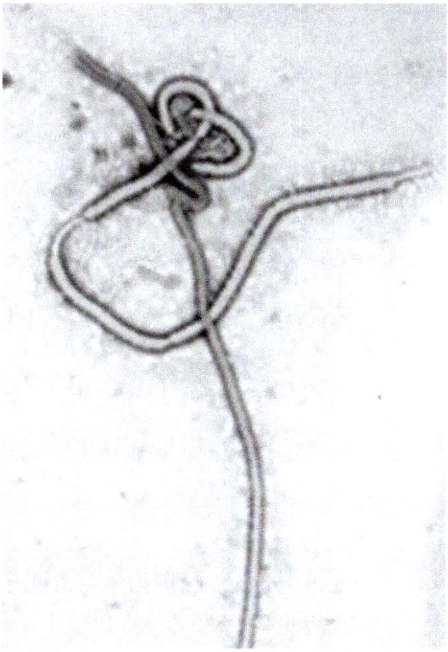

Fig. 17.8 Ebola virus (public domain, this photo is the work of the Centers for Disease Control and Prevention, part of the U.S. Department of Health and Human Services, taken or performed as part of the employee's official duties. As a work of the United States federal government, the image is in the public domain)

Table 17.1 Time of development of symptoms and methods of treatment (WIHE)

Name	Type	Incubation time	Damage level and healing
Anthrax	Bacteria	3–5 days	Fatal if untreated/penicillin treatment before symptoms appear
Vibrio cholera	Bacteria	1–3 days	50% mortal
Bakteria Yersinia pestis	Bacteria	2–5 days	Fatal if untreated/curable with antibiotics
Tularaemia	Bacteria	2–4 days	Fatal if untreated/curable with antibiotics
Typhus abdominals	Bacteria	10–14 days	50% mortal/curable
Febris Flava	Virus	3–6 days	10% fatal if untreated/curable
Variola vera	Virus	10–14 days	60–80% fatal if untreated/ curable
Botulus	Toxin	18–36 hours	65% fatal if untreated/curable with toxoid
Ricinus communis	Toxin	2 hours	High mortality

or staphylococcal enterotoxin B (*Staphylococcus aureus*). Some of these factors may be used to poison food and water (Magdzik and Czarkowski 1995).

Group C includes factors that may be used in the future. These include mainly viruses such as hantaviruses, yellow fever virus, tick-borne hemorrhagic fever virus, and tick-borne encephalitis virus. *Mycobacterium tuberculosis*, a tuberculosis bacillus resistant to antibiotics belongs to this class of organisms (Tylak 2012) (Table 17.1).

Whether a specific pathogen can be used as a biological threat is influenced by many of its characteristics. From the point of view of bioterrorism, an advantageous feature is the ease of obtaining it. Moreover, compared to other threats of mass destruction, e.g., the atomic bomb, it is cheap, both in production and use. The production of "cargo" does not require excessively complicated technologies. Both the manufacturing site and the final product can be easily hidden. Depending on the purpose and needs, it may be used, for example, by an airplane, a balloon, a skyscraper, and wind or air movements, which will spread the bacteria over large areas. In the case of a smaller-scale attack, e.g., a ventilation system in the building or access to a water intake will suffice. Germs can also be transmitted in a seemingly natural way by their normal carriers, e.g., lice, fleas, mice, or rats. Biological threats can attack humans directly or indirectly by attacking animals or plants. Many of the pathogens are infectious to both humans and animals, causing the so-called zoonoses.

It is also important that biological threats work insidiously. It is not visible, has no taste or smell, and its use is not accompanied by noise, flash, or any other effects such as fire or building collapse.

It is impossible to ignore such a feature of biological threats as their effectiveness. It usually causes death or illness of the recipient, not always immediately. This phenomenon hinders and delays effective treatment and influences the development of the epidemic. An important feature of this threat is the lack of quick and effective ways to defend against it. Collective and individual protective equipment can provide limited protection, but only after recognizing the threat and identifying the

infectious agent, similar to antibiotic treatment or treatment with antisera. In the case of anthrax, the antibiotic is only effective when administered quickly, within 24 to 48 hours of infection. Protective vaccinations will also not work, because to be effective they must be performed within a certain time before the attack. It is not possible to vaccinate everyone as a preventive measure, because many factors constitute potential biological threats, and there are not always effective vaccines against them. Modern science, especially in the field of biotechnology and genetic engineering, also poses a threat to biological threats. This is the department responsible for the appearance of microorganisms against which we have no threats, e.g., resistant to all known antibiotics, disinfectants, or available vaccines. Bacillus anthracis—the anthrax bacillus—is still very popular among those interested in the use of biological threats. Anthrax bacteria are relatively easy to cultivate, and their spores have a very long survival period. They can live for decades without losing their virulence and can be delivered simply by mail or dispersed in the air, depending on needs. Anthrax acquired through the respiratory or oral route is severe and has a poor prognosis. Black pustule, a cutaneous form of anthrax, has a slightly milder course.

There is a huge risk of using the smallpox virus in a bioterrorist attack. In the world, most of the population has no immunity to this disease. After the elimination of this virus in the natural environment, human vaccinations were discontinued, and the immunity of people vaccinated in the distant past was almost extinguished. Calculations indicate that the mortality rate if it were used would be up to 30%.

Although after signing the UN Biological Threats Convention in 1972, the signatory countries officially withdrew from research and stockpiling of biological threats, unofficially such research may be still ongoing, and it is doubtful that stocks of these deadly threats will be eliminated. There was a fear that some of them could have fallen into the wrong hands, including terrorists (Szot 2012).

It should be noted that the use of biological agents in combat operations does not remain the exclusive domain of states. Terrorist attacks using biological threats carried out by extremist groups were recorded in the twentieth century. These groups saw a chance to gain publicity through such acts. In 1984 in Oregon (USA), the Rajneesh sect, trying to disrupt local elections, used *Salmonella typhimurium* bacteria to infect salad bars, which caused serious food poisoning in 751 people. In the 1990s, the Japanese sect Aum Shinrikyō conducted research on the use of anthrax, and in 1995 it used sarin gas to attack the Tokyo subway. In 2001, after the attack on the World Trade Center, bioterrorist attacks took place in the United States using mail containing anthrax bacilli. Anthrax letters were sent by unknown perpetrators to government institutions and the media, resulting in the death of 5 people and 32,000 had to undergo prophylactic ciprofloxacin therapy. These attacks caused enormous panic and once again drew public attention to the real threat of using biological agents.

17.4 Biological Threats: Legal Regulations

The first legal regulations regarding the production and use of biological threats appeared at the beginning of the twentieth century. The Fourth Hague Convention of 1907, concerning the laws and customs of land warfare, stated that "the parties do not have an unlimited right to choose the means of harming the enemy" (Article 22 of the annex to the convention—Regulations). Moreover, it is forbidden to use poison or poisoned threats there. The interpretation of these provisions should not raise any doubts, but the experience of World War I showed that such regulations were insufficient. Then, in the Geneva Protocol of 1925, the list of banned substances was expanded to include biological agents. This happened at the request of Polish General Kazimierz Sosnkowski, who called for "in the name of civilization to prevent the use of bacteriology for barbaric purposes." Before the outbreak of World War II, only a few countries managed to ratify the protocol. The countries that did not sign the agreement at that time included: the Soviet Union, Japan, Great Britain, the United States, and France (Kocik et al. 2014).

The lack of provisions prohibiting research on this type of threat was a serious problem. Some countries also had reservations about the right to retaliate if an opponent used such measures.

In 1969, the United States, which had a developed arsenal of advanced biological threats, made a declaration to destroy its military potential and to discontinue further research. Great Britain, Canada, and Sweden also joined this declaration. As a result of this declaration, a renewed attempt was made by the UN to normatively and finally exclude biological threats from use and further research. In 1972, the above-mentioned Biological Threats Convention entered into force. The signatory countries have declared their determination to eliminate the use of biological agents or toxins. In the act, they also indicated not to conduct research on, produce, or acquire bacteriological threats. This is in the interest of all humanity. This convention is the first to commit to complete disarmament, unlike other disarmament agreements, which only aim to stop quantitative or qualitative progress in armaments (Michaliuk and Malicki 2003).

However, the effectiveness of this regulation was undermined by cases of violation of its provisions and the lack of specific control mechanisms.

The most famous and obvious example of non-compliance with the convention was the 1979 disaster in Sverdlovsk (today Yekaterinburg). As a result of a laboratory failure, anthrax bacilli were released into the atmosphere, causing a local epidemic and the death of 64 people. The USSR authorities tried to cover up the real causes of the tragedy, but research into biological threats was reported by scientists who had fled to the West, and in 1992, President Boris Yeltsin admitted that the USSR had repeatedly violated the provisions of the convention (Alibek and Handelman 2000).

In 2004, the UN made another attempt to limit the multiplication of microbial cells that constitute threats of mass destruction in the face of the growing terrorist threat.

UN Security Council Resolution 1540 of 2004 made commitments to prevent non-state actors from gaining access to such threats and related materials. This act raised the issue of controlling the so-called dual-use materials and technologies that are used for peaceful purposes but could be used to produce threats of mass destruction. However, the resolution is based only on states' declarations, and its adoption did not result in the creation of effective verification and control tools.

An interesting international mechanism aimed at controlling the flow of dual-use materials is the Australian Group. This informal organization of states, established in 1985, monitors transfers of substances and technologies that can be used to obtain chemical and biological threats. Currently, it has 41 members, including the European Commission. As part of the jointly developed checklists, several biological factors (viruses, bacteria, fungi, and toxins) pathogenic to humans, animals, and plants were verified, and their genetic material was also controlled. Some potentially dangerous technologies related to research and development in the field of microbiology and aerosol dispersion were also subject to surveillance procedures. A significant disadvantage of this project is its limited scope. The international initiatives mentioned above are based on mutual trust between partners. In the absence of effective control mechanisms, this approach seems very idealistic. It is also impossible to expect that terrorists will comply with international conventions and resolutions.

17.5 Effectiveness of Biological Threats

Analyzing the history of modern armed conflicts, it should be noted that mainly the losses of the human factor determine the outcome of each of them. Eliminating human resources, both military and civilian, permanently or temporarily of the enemy allows you to gain an advantage over him or directly determines victory. The decision to use violence always entails consent to kill, and combat technologies and the strategies and tactics of action adapted in parallel to their development often remain in opposition to generally accepted principles of conduct.

Biological threats are extremely dangerous and unfair. It is characterized by high effectiveness and relatively low cost of acquisition and use. The cost of the previously mentioned 1984 Salmonella typhimurium attack in Oregon, which sickened 751 people, was approximately $100. This means, according to expert estimates, that the cost of causing human losses in an area of 1 km is only $1. The cost for a similar area in the case of chemical threats will be $800, and for traditional threats about $2000. Biological threats are sometimes referred to as "the poor man's threats of mass destruction" precisely because they are easy to obtain and conceal. To transmit bacteriological agents of destruction, sophisticated and advanced technical means are not needed, natural carriers such as insects or rodents are sufficient. Contamination of water supplies, crops, or ready-made food products, as well as leaving contaminated items in places where there are a lot of people, is not a problem. Any ecosystem element can be the subject of a biological attack: people,

Table 17.2 Effectiveness of threats of mass destruction (AAP-6 2013)

Head type	Killed	Injured
Conventional	6	13
Chemical (sarin)	200–300	200–300
Nuclear	40,000	40,000
Biological (anthrax)	20,000–80,000	–

animals, water, soil, and air. This circumstance makes it even more difficult to detect contamination and effectively protect against it. In addition, a huge range of microorganisms and toxins can cause population damage on various scales, from incapacitating to fatal, and environmental contamination lasting from several hours to several weeks or even years. Bacteriological threats cannot be seen with the naked eye, which makes recognition even more difficult, and there is also the possibility of having a delayed effect, increasing the surprise. The death of an infected person does not have to occur immediately, and in milder cases, the symptoms of the disease may not appear immediately, which further complicates treatment and effective protection against the spread of bacteria (Stojek 2008).

The psychological effect of the threat causing a wave of panic in society in attacks using biological threats is also important. In the twenty-first century, this was visible, for example, after the anthrax attacks in the United States. Aroused by terrorists and their actions, and sometimes even just by the threat of an attack, fear and a sense of threat remain one of the main elements characterizing contemporary terrorism. The threat of using biological agents should be considered an extremely strong bargaining chip in the hands of terrorists, especially since from the point of view of fanatics blinded by ideology, the choice of combat tools is of no importance, as the most important goal remains the pursuit of their interests.

Spectacularity, cruelty, and, above all, unpredictability and relatively low costs make biological threats seem to be an ideal tool for carrying out a terrorist attack on basically every continent (Table 17.2).

17.6 How to Recognize a Biological Threats Attack and Its Effects

Various facts may suggest that a terrorist attack using biological threats may have occurred. What may be suspicious is the appearance of many cases of the disease in areas previously free from it or at unusual times of the year, in an unusual age group. Unusual routes of infection may also raise suspicions, as may an unusual course of the disease and difficulties in treatment. Another reason to suspect that we are dealing with a bioterrorist attack may be the outbreak of an epidemic of a zoonotic disease without previous episodes of occurrence, the appearance of several epidemics of the same disease in remote areas, or, for example, the discovery of illegal transports or warehouses of such threats. Biological threats may be used with or without warning of their use.

An actual attack with a warning is easier to control and may have less impact than an attack without a warning. In the case of a warning, the alarm may be false, but the threat itself fulfills its role, with virtually no financial outlay. It causes fear, panic, and disorganization of life. It may also cause financial losses due to, for example, putting appropriate services on alert or implementing specialized tests. Acacia from 1997 was such an example. During the operation, plates were planted with the suggestion that they contained plague and anthrax germs. The cost of the terrorist action was $1, while the cost of evacuation, decontamination, and quarantine of several hundred people, as well as microbiological tests, was very high, at several thousand dollars. In the event of an attack without warning, the first victims are not always different from those infected naturally. This leads to an epidemic in which the aggressor is no longer involved. This delays the determination of the actual epidemiological situation. The consequence of such action is a delay in preventive reactions (Kaznowski et al. 2003).

17.7 Protecting the Country against Biological Attack

It is impossible to eliminate biological factors dangerous to humans and animals from our environment. However, it is possible to limit or avoid them. Each state, when creating public health policy to maintain the well-being of its citizens, introduces normative acts that regulate the principles of risk prevention. The issue of biological protection requires the creation of an organizational, legal, and institutional system that protects populations, crops, and breeding. To secure these areas, regulations regarding the storage, use, and transport of potentially hazardous substances are necessary.

The Act on Preventing and Combating Infections and Infectious Diseases in Humans defines the principles and procedures for preventing infectious diseases and the principles for monitoring the epidemiological situation. The legislator detailed the obligations of administrative bodies in the field of prevention and protection against infectious diseases. The obligations of entities providing health care services and the manner of dealing with persons staying on the territory of Poland in the event of epidemiological threats are also specified.

The State Sanitary Inspection is responsible for maintaining the country's sanitary and epidemiological barrier. Border sanitary control is intended to prevent the introduction into the territory of a given country of food and materials and products intended for contact with it, the contamination of which with biological agents would pose a threat to human health and life. The development of international transport (mainly air and road), as well as the threat of using infectious agents in terrorist acts, have resulted in a real threat to public health on a transnational scale. Therefore, they react and assess threats and the epidemiological situation in the border area (Otłowski 2009).

Other elements serving the country's biological defense include the Act on the Protection of Animal Health and Combating Infectious Animal Diseases, which

regulates issues related to combating and monitoring zoonotic diseases and zoonotic agents, as well as the Act on Plant Protection, which introduces mechanisms to protect crops against harmful factors. The protection of crops and livestock remains extremely important for every country, not only due to the need to protect animal welfare and the safety of the entire ecosystem but also due to the need to secure the economic potential of the country, as agriculture is one of the key branches of the economy of virtually every country.

Legislators also regulate issues related to research, genetic engineering, and the safety of genetically modified crops. Contained use of GMOs requires the consent of the Minister of the Environment, and there is also a special committee working under the Minister's office to review research directions and issue permits. In addition to representatives of science, it includes representatives of the ministries of health, agriculture, and national defense.

The threat of biological agents may also result from inappropriate, unintentional behavior, e.g., during transport, storage, or testing. The transport of goods considered potentially dangerous in our country is regulated by international regulations (Stępień and Tołwiński 2015).

Infectious biological agents, including waste, require special care in terms of transport and packaging and must be transported using appropriate means of transport and properly labeled. These issues are regulated by the Act on the Transport of Dangerous Goods. The Act and implementing acts specify detailed requirements that must be met during transport operations, introduce the obligation to prepare detailed instructions in the event of possible packaging leakage, specify persons responsible for the proper course of transport, as well as control and supervision bodies. Carriers, loaders, unloaders, packers, and pourers (e.g., tankers) are obliged to submit detailed reports on the hazardous substances transported every year. Accidents and breakdowns during transport are also recorded. Based on the data from the reports, information is collected about entities involved in the transport, loading, and unloading of dangerous goods. They do not serve only statistical purposes but allow monitoring and control of the flow of these materials (Zacher 2012).

Active threats to public health of a cross-border nature imply the need for continuous state cooperation with the World Health Organization (WHO), the European Commission, the European Center for Disease Prevention and Control (ECDC), and the European Food Safety Authority in both warning and response procedures. However, at the national and ministerial level, cooperation between the Ministry of Interior, National Defense, and Health with the Government Center for Security and national experts and consultants in specific fields is necessary.

Another area of risk are production and research processes carried out within the country. Supervision of the safety of use in Geneva on September 30, 1957. The provisions of the ADR agreement are amended every two years.

It remains necessary to physically secure facilities, technologies, and the so-called dual-use materials against uncontrolled takeover by unauthorized persons and undesirable use and trade. Authorization of access and protection of the conditions of storage and use of potentially hazardous biological agents for the safety of personnel and the environment are regulated in occupational health and safety

(OSH) regulations. Entrepreneurs using genetically modified biological agents are obliged to inform the sanitary inspection about this. A certain difficulty remains the fact that, unlike most chemical and physical factors, there are currently neither clear criteria for assessing exposure to biological factors nor generally recognized normative (reference) values, and no appropriate methodological recommendations have been developed.

17.8 Summary

Nowadays, biological threats are the first among the factors that may threaten the security of countries around the world. There is a lack of effective supervision over the spread of biological threats, which necessitates strengthening international arms control. Biological threats are ideal for psychological warfare and causing panic in society. This is often used by groups that do not have high technological capabilities or classic means of supporting and securing combat operations. The weaker sides of conflicts often use unsophisticated, cheap-to-use methods that involve significant suffering for the population, which allow them to influence the will to fight the opponent who has a technological advantage on the battlefield (Pochylska 2017).

Combating the effects of the use of biological threats is a difficult task. It is not difficult to imagine a situation in which artificially induced epidemics of farm animal diseases are used to destabilize the economy, or crops are infected to destroy the harvest. There is even a term for agroterrorism, i.e., attacks aimed at the agricultural economy (Langbein et al. 2003).

Biological threats may significantly affect the economic situation of the country: first, losses in crops and breeding result in price changes and the need to import from abroad, second, there is a severe loss of markets and reputation for producers exporting their products (WHO 2010, 2017).

The use of new or appropriately modified bacteria may have consequences not only for agriculture but also for the entire ecosystem. One of the new dangerous forms of terrorism is the phenomenon of the so-called food terrorism. This term describes the intentional use of food products to spread pathogenic biological (or chemical) agents to commit terrorist attacks. Such acts have already appeared in highly developed countries, although they were usually motivated by the desire to extort ransom from food-producing companies (Dzwolak and Dzwolak 2009).

Effective control of emerging threats requires obtaining an appropriate level of knowledge about them. This is the only way to protect yourself against them. Genetic engineering, biotechnology, and synthetic biology are developing very dynamically. However, another threat appears here: although scientific research is intended to serve humanity and peaceful purposes, it can also be used contrary to these intentions, to the detriment of other people. For this reason, the publication of research results on the so-called factors is controversial.

In 2010, the World Health Organization drew attention to the issue of researchers' liability. New technologies and the development of research, on the one hand,

create opportunities for the development of medicine, but on the other hand, they carry the risk of their inappropriate use, e.g., to create new biological threats. The World Health Organization therefore calls for work to raise awareness of the risk of proliferation of biological threats and to introduce supervision over research, as well as to create a code of ethics for researchers.

Another important problem is the appropriate security of trade in dual-use materials and technologies. The level of this security seems to be insufficient at the moment. Entities that would like to plan a bioattack have many opportunities to obtain pathogens. For example, they can use natural sources to grow dangerous bacteria themselves—some can be obtained even at home or in an averagely equipped laboratory. They can also steal such substances from often poorly protected research laboratories or hospitals. From time to time, the media reports cases of the theft of dangerous bacteria from research centers in former Soviet republics. It is also possible to obtain biological agents from a rogue, weak, or decaying state, or a state sponsor.

International regulations require continuous improvement in the context of reducing the threat of attacks using biological threats. The lack of real control mechanisms means that conventions and resolutions often remain at the declarative level. The progressive increase in the activity of terrorist organizations and the degeneration of state structures in many countries, especially in the North Africa and Middle East regions, mean that the threat of bioterrorism in the broad sense cannot be underestimated today. The fear of the consequences of using biological threats has so far effectively stopped aggressors, but it is impossible to predict how long this state of affairs will last, which is why bioterrorism and the threat of biological threats are an ongoing issue. Our imagination should therefore precede the ideas of those who want to threaten our security.

Finally, it seems necessary to develop existing non-proliferation regimes while improving the potential of defense units. Given the contemporary threats of using biological agents, both military and civilian structures should be prepared for a possible attack in the event of military operations other than war in the classic sense of the word.

References

AAP-6 (2013) NATO dictionary of terms and definitions containing military terms and their definitions used in NATO (in Polish)

Alibek K, Handelman S (2000) Biohazard, Warszawa

Binczycka-Anholcer M, Imiołek A (2011) Bioterrorism as one of the forms of modern terrorism. Hygeia Public Health 46(3):329. 326–333 (in Polish)

Dennis DT, Inglesby TV, Henderson DA, Bartlett JG, Ascher MS, Eitzen E, Fine AD, Friedlander AM, Hauer J, Layton M, Lillibridge SR (2001) Tularemia as a biological weapon: medical and public health management. JAMA 285(21):2763–2773. https://doi.org/10.1001/jama.285.21.2763

Dzwolak W, Dzwolak W (2009) Food terrorism - threat factors. Przemysł Spożywczy 63(9):43–45

Hawley RJ, Etizen EM Jr (2001) Biological threats a primer for microbiologists. Ann Rev Microbiol 55:235–253
Inglessby TV, Tonat K (2002) Anthrax as a biological threats. JAMA, pp 2236–2252
Janeczek M, Chrószcz A, Ożóg T (2010) Attempts to use biological and chemical weapons in ancient times, epidemics and their impact on history. Veterinary Life 85(3):230–233. (In Polish)
Kaznowski A, Mokracka J, Koczura R (2003) Tasks of microbiological laboratories in the event of a terrorist attack. Post Microbiol 42:319–343
Kocik J, Niemcewicz M, Bielecka-Oder A (2014) Bioterrorism. In: Heczko PB, Wróblewska M, Pietrzyk A (eds) Medical microbiology. National Institute of Medical Publications, Warsaw, p 652
Langbein K, Skalnik C, Smolek I (2003) Bioterrorism. Muza SA, Warsaw
Magdzik W, Czarkowski MP (1995) Infectious diseases in Poland in 1993. Przegl Epidemiol 49:77–80
Michaliuk B (2016) Biological weapons and bioterrorism. Zeszyty Naukowe AON 102(1):17
Michaliuk B, Malicki B (2003) Biological weapons - selected problems. Terrorism and weapons of mass destruction: diagnoses, views, conclusions, No. 1 (50), 137–141
Mierzyńska D (2002) Hermanowska – Szpakowicz ttularemia as a potential weapon of bioterrorists. Med Pr 53(3):279–281
Otłowski T (2009) Psychological aspects of terrorism - outline of the problem
Pochylska A (2017) "Humanization of war"? Limiting the possibility of using certain weapons in armed conflict in the context of ethical norms. Sci J Soc Doc Stud Jagiellonian Univ. Nauki Społeczne 18
Riedel S (2004) Biological warfare and bioterrorism: a historical review. Proc (Bayl Univ Med Cent) 17(4):400–406. https://doi.org/10.1080/08998280.2004.11928002
Stępień M, Tołwiński M (2015) Effectiveness of international regimes for the non-proliferation of biological and chemical weapons. Polityka i Społeczeństwo 4(13):115
Stojek NM (2008) Contemporary threats with biological weapons - bioterrorism. Medycyna Ogólna 14(XLIII):176
Szot W (2012) Bioterrorism and security in the era of globalization. Sec Theory Pract 2(VII):37
Tylak T (2012) Bioterrorism, e-Terroryzm.pl. Online Bulletin of the Center for the Study of Terrorism and the quarterly e-Studia nad Bezpiecześć i Terroryzmem
Weiner M, Kubajka M (2015) Tularemia – serious zoonotic disease. Health Probl Civiliz 9(1):39–46. https://doi.org/10.5114/hpc.2015.57104
Wojskowy Instytut Higieny i Epidemiologii (WIHE) (n.d.) Military Institute of Hygiene and Epidemiology, presentation. http://www.wihe.waw.pl/wihe/o-nas
World Health Organization (2010) Responsible life sciences research for global health security: a guidance document, Geneva
World Health Organization (2017) World malaria report
Zacher LW (2012) Individual and social contexts and challenges for security. In: Sienkiewicz P, Marszałek M, Świeboda H (eds) Methodology of national security research, vol 3. Warsaw Management University, Warsaw, pp 88–104

Chapter 18
Microbial Biotoxins in Biotechnological Applications for Plant Disease and Pest Management

Chitranjan Kumar, Vashista Kotra, Ajay Tomar, and Majeti Narasimha Vara Prasad

Abstract Microbial biotoxins, produced by bacteria, fungi, and algae, have gained attention for their potential applications in biotechnology, especially in managing plant diseases and pests. These biotoxins have a range of bioactive properties, including antimicrobial, insecticidal, nematicidal, and herbicidal activities, making them promising candidates for sustainable and environmentally friendly solutions in agriculture. This review explores the diversity of microbial biotoxins and their mechanisms of action against plant pathogens and pests, as well as the biotechnological approaches used to harness these biotoxins for plant protection, including the development of biopesticides and genetically modified crops. The integration of microbial biotoxins into pest and disease management strategies has the potential to reduce reliance on chemical pesticides, minimize environmental impacts, and enhance crop yield and quality. The use of microbial biotoxins for pest control and their impact on agroecosystems has been documented. Biocontrol agents, such as predators and parasitoids, derived from bacteria, fungi, viruses, algae, protozoa, and entomopathogenic nematodes, are increasingly being used to develop microbial biopesticides as an alternative to chemical pesticides. As biotechnological advances continue to unravel their full potential, their role in sustainable agriculture is likely to expand, contributing to a more resilient and ecologically balanced food production system. The review highlights the diverse sources of microbial biotoxins, emphasizing their natural occurrence in the environment. Regulatory frameworks must ensure the safety of biotechnological products for both human health and the environment. Developing resistance in target pests and potential impacts on nontarget organisms must be carefully monitored and managed. Microbial biotoxins represent a valuable resource in the quest for sustainable and eco-friendly solutions to plant disease and pest management.

C. Kumar (✉) · V. Kotra · A. Tomar
Amity Institute of Organic Agriculture, Amity University Uttar Pradesh, Noida, India
e-mail: ckumar@amity.edu

M. N. V. Prasad
University of Hyderabad, Hyderabad, India

M. N. V. Prasad et al. (eds.), *Biotoxins*,
https://doi.org/10.1007/978-3-031-75309-1_18

Keywords Ecological balance · Microbial biotoxins · Molecular mechanisms · Plant protection · Sustainable agriculture · Target pests

18.1 Introduction

Plant disease and pest management are crucial for food security and economic stability in agriculture. Conventional methods, such as chemical pesticides, have been criticized for their environmental and health impacts. As a result, there has been a growing focus on sustainable and environmentally friendly alternatives (Olson 2015; Adejumo and Vögele 2021; Hernández-Melchor et al. 2023). Biotechnological solutions have emerged as promising alternatives, offering effective and eco-friendly means of safeguarding plant health. Plant diseases and pest infestations pose significant challenges to crop yields and quality, especially in a world with a rising global population and increasing demand for food production. To meet this demand, it is essential to ensure crops remain healthy and free from harmful pathogens and pests. A pathogenic bacterium, *Pseudomonas gingeri,* secretes a biotoxin called monoacetylphloroglucinol (MAPG). The application of MAPG at 750 μg/ml caused mycelial death in *Agaricus bisporus*, which reduced the severity of ginger blotch disease in mushrooms (Huang et al. 2023). In recent decades, there has been a shift toward sustainability in agricultural practices. Traditional chemical-based pest control methods have raised concerns about soil and water pollution, harm to nontarget organisms, and the development of pesticide-resistant pests. This has led to the development and adoption of biotechnological solutions.

Plants and microbes have coexisted for millions of years, forming mutually beneficial relationships. Microbes, including bacteria and fungi, colonize the root zones of plants and contribute to plant health and growth. In return, plants provide nutrients and a suitable environment for these microbes to thrive. Plant growth-promoting rhizobacteria (PGPRs) and other soil microbes have emerged as key players in promoting plant health (Kumar et al. 2023). Understanding the role of these microbes is crucial to harnessing their potential in sustainable agriculture. This chapter explores the utilization of microbial biotoxins in plant disease and pest management, focusing on their mechanisms, applications, and future prospects for safeguarding plant health and promoting sustainable agriculture. Case studies, scientific research, and real-world applications will be used to illustrate the significance of microbial biotoxins in the biotechnological arsenal for plant protection.

18.2 Plant Growth-Promoting Rhizobacteria (PGPRs)

Plant growth-promoting rhizobacteria (PGPRs) are soil bacteria that colonize plant roots, providing numerous benefits to their host plants. They are crucial in agriculture for enhancing plant growth, health, and protection against stressors such as

diseases and pests. This section provides a detailed understanding of PGPRs, their definition, characteristics, ecological significance, mechanisms of action, and intriguing microbial biotoxins they produce.

18.2.1 PGPRs: A Brief Overview

- **Definition and Characteristics of PGPRs**

PGPRs are rhizosphere-associated bacteria that promote plant growth and health. They inhabit the root zone and form a symbiotic relationship with plants. PGPRs can be diverse, including Pseudomonas, Bacillus, and Rhizobium, each with unique attributes. They can colonize and persist in the rhizosphere, engaging in beneficial interactions with plants.

- **The Diversity of PGPRs and Their Ecological Significance**

The diversity of PGPRs, with varying growth-promoting mechanisms and ecological roles, allows for the selection of PGPRs tailored to specific agricultural and ecological contexts. These PGPRs play a crucial role in maintaining soil health and fertility by contributing to nutrient cycling, suppressing pathogens, and mitigating environmental stressors.

18.2.2 Mechanisms of PGPR Action: Unveiling Their Plant-Boosting Powers

Plant growth-promoting rhizobacteria (PGPRs) are champions of plant health and vitality, armed with an array of mechanisms that directly benefit their botanical hosts. Among these mechanisms, nutrient cycling, improved nutrient uptake, and induced systemic resistance (ISR) take center stage, showcasing the remarkable ways in which PGPRs foster plant growth, resilience, and protection against biotic and abiotic stressors (Kumar et al. 2023).

18.2.2.1 Nutrient Cycling and Improved Nutrient Uptake

PGPRs are effective in optimizing nutrient availability for plants through two methods: solubilization and nitrogen fixation. Solubilization involves the secretion of organic acids, enzymes, and chelating compounds that break down insoluble minerals in soil, providing a nutrient buffet for the plant's roots. Nitrogen fixation, on the other hand, converts atmospheric nitrogen into ammonia or other plant-usable forms, boosting nitrogen supply without the need for synthetic fertilizers. PGPRs act as natural nitrogen factories, enhancing plant growth and vitality by

transforming nutrient-poor soils into thriving ecosystems. They are indispensable partners in the plant's quest for growth and vitality.

18.2.2.2 Induced Systemic Resistance (ISR) and Its Role in Pest and Pathogen Management

In the ongoing battle against pathogens and pests, PGPRs support plants in inducing systemic resistance (ISR). For example, *Bacillus* species induce a favorable environment for nitrogen fixation, hormone production, mineralization, solubilization of plant nutrients, siderophore production, and the induction of systemic resistance in plants (Saxena et al. 2020) (Fig. 18.1). This remarkable defense mechanism empowers plants to ward off attackers with finesse for the following functions:

- **Activation of Plant Immune Responses**
- PGPRs initiate a molecular dialogue with plants, activating signaling pathways that reinforce their immune system. This proactive response involves producing antimicrobial compounds, activating defense-related genes, and synthesizing pathogenesis-related proteins (Jalmi and Sinha 2022).
- **Broad-Spectrum Protection**
- ISR is a holistic approach to plant protection that provides plants with a versatile defense system against various pathogens, reducing their vulnerability to diseases and pests, regardless of the plant species (Heil and Bostock 2002).
- **Reduced Reliance on Chemical Pesticides**
- ISR reduces plant reliance on chemical pesticides, reducing agriculture's environmental footprint and contributing to sustainable pest management practices. PGPRs are now integrated into IPM strategies, promoting a harmonious coexistence between agriculture and nature.

PGPRs play a crucial role in soil microorganisms and plant roots, promoting nutrient efficiency, resilience, and defense. They enhance crop yields and plant health, contributing to sustainable agriculture and ecosystem balance (Mohanty et al. 2021). Understanding these mechanisms allows us to use PGPRs effectively, reducing our reliance on synthetic chemicals and fostering a healthier farming future.

18.2.3 Microbial Biotoxins Produced by PGPRs

18.2.3.1 Types of Biotoxins Produced by PGPRs

Plant growth-promoting rhizobacteria (PGPRs) are essential for enhancing plant growth and protecting them from stressors like pathogens. They produce a variety of microbial biotoxins, each with unique properties and mechanisms. This

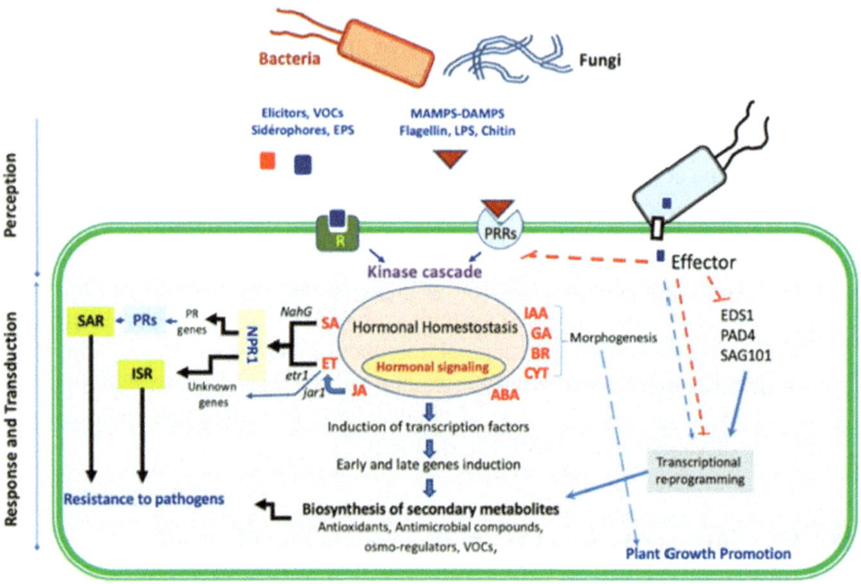

Fig. 18.1 Microbial signaling response and transduction: Microbial interactions produce MAMPs or DAMPs, such as flagellin, LPS, and chitin, which are perceived by PRRs, and other elicitors, such as VOCs, and siderophores. The active receptors trigger various signaling cascades, which biosynthesize phytohormones, which activate defensive pathways. Abbreviations: abscisic acid (ABA); brassinosteroid (BR); cytokinin (CYT); damage-associated molecular patterns (DAMPs); enhanced disease susceptibility (EDS); ethylene (ET); exopolysaccharides (EPS); gibberellic acid (GA); indole acetic acid (IAA); induced systemic resistance (ISR), jasmonic acid (JA); lipopolysaccharides (LPS); microbe-associated molecular patterns (MAMPs), nonexpressor of pathogenesis-related genes (NPR); pathogenesis-related protein (PR); pattern-recognition receptors (PRRs), peptidyl arginine deiminase (PAD), salicylic acid (SA); senescence-associated gene (SAG), systemic acquired resistance (SAR); volatile organic compounds (VOCs) (MDPI Source: Lahlali et al. 2022)

exploration will examine the types of biotoxins produced by PGPRs, their classification, biosynthesis, and ecological significance.

- **Antimicrobial Peptides**
- PGPRs are prolific producers of antimicrobial peptides, which are small proteins with potent antimicrobial properties. These peptides, such as bacteriocins, directly inhibit the growth of plant pathogens by disrupting their cell membranes. Bacteriocins, in particular, are specialized toxins that target closely related bacterial species, making them valuable weapons in the competitive microbial world (Simons et al. 2020).
- **Secondary Metabolites**
- PGPRs secrete secondary metabolites such as antibiotics and siderophores, which are crucial tools for outcompeting pathogenic microorganisms. Antibiotics disrupt the metabolic processes of pathogens, creating an unfavorable

environment for their growth. Siderophores, which are iron-chelating molecules, compete with pathogens for essential iron, thereby limiting its availability to potential invaders. Secondary metabolites can be biotoxins, biopesticides, pheromones, enzyme inhibitors, anti-tumor agents, immune-modulators, and growth promoters, imparting several biological functions in plants/animals (Mehta et al. 2018; Verma et al. 2024).

- **Volatile Organic Compounds (VOCs)**
- PGPRs emit volatile organic compounds (VOCs) that significantly impact plant health. These compounds can inhibit pathogen growth by disrupting cellular processes and can induce induced systemic resistance (ISR) in plants. ISR occurs when plants are exposed to beneficial microbes or their products, preparing them to respond more effectively to future pathogens or pests. VOCs act as signaling molecules, triggering a cascade of plant defense responses that enhance resistance to diseases and pests (Zhu et al. 2022).

18.2.3.2 Modes of Action of PGPR Biotoxins in Plant Protection

The biotoxins produced by PGPRs exhibit a range of mechanisms to protect plants. Understanding these modes of action is crucial for harnessing the full potential of PGPRs in sustainable agriculture and environmental management (Fig. 18.2).

- **Antimicrobial Peptides**
- Bacteriocins are antimicrobial peptides that directly kill or inhibit the growth of pathogens by disrupting their cell membranes, leading to cell lysis and death. They target closely related bacterial species, leaving nontarget organisms unaffected (Simons et al. 2020).
- **Secondary Metabolites**
- Secondary metabolites, such as antibiotics and siderophores, protect plants by disrupting the metabolic processes of pathogens and limiting their growth and virulence. Antibiotics disrupt replication and thrive in the rhizosphere, while siderophores sequester essential iron, making it less available to invaders, thus limiting the resources (Saeed et al. 2021).

PGPRs play a protective role by engaging in interactions with plants and the rhizosphere, such as nutrient mobilization, competition for root colonization sites, and production of growth-promoting substances. This comprehensive examination of PGPR biotoxins reveals their intricate strategies for protecting plants from various threats. Utilizing PGPRs and their biotoxins in sustainable agriculture, bioremediation, and environmental management can enhance crop yield and resilience while minimizing chemical interventions. This can lead to a more sustainable approach to crop management, ultimately enhancing crop resilience (Backer et al. 2018).

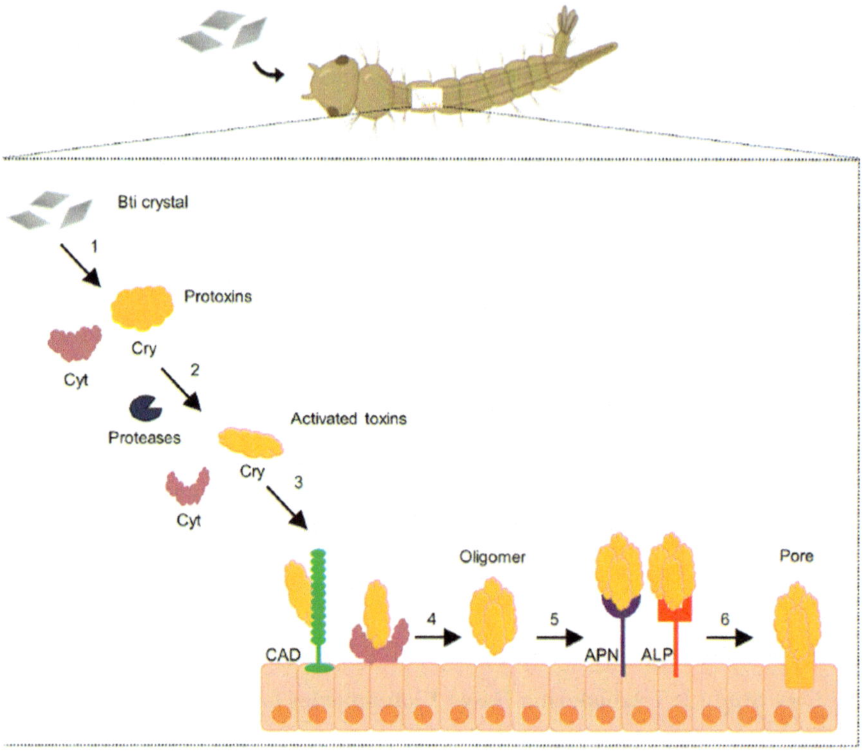

Fig. 18.2 Mode of action of Cyt and Cry toxins, along with the synergy of Cyt1Aa after Bti ingestion, osmotic cell lysis in mosquito larvae (Dipteran pests). (1) The crystal is first solubilized, and (2) proteolyzed and activated by insect gut proteases. Cadherin or Cyt1Aa protein functions as a receptor for (3) Cry4Ba and (4) Cry11Aa, promoting Cry oligomerization. Finally, biotoxins are inserted into the membrane to form pores through (5) APN (aminopeptidases) and (6) ALN (alkaline phosphatases) leading to osmotic cell lysis. Sp: spore, C: Crystal. (MDPI Source: Valtierra-de-Luis et al. 2020; Silva-Filha et al. 2021)

18.2.4 Case Studies: Real-World Examples of PGPRs Utilizing Microbial Biotoxins for Pest and Disease Control

In this section, we delve into compelling case studies that exemplify how plant growth-promoting rhizobacteria (PGPRs) harness microbial biotoxins to combat plant pests and diseases effectively. These case studies not only provide concrete instances of the practical application of PGPR-based biotechnological solutions in agriculture but also offer in-depth insights into their mechanisms, effectiveness, sustainability, and potential for broader adoption in modern farming practices.

- **Case Study 1: Bacillus-Based Biopesticides for Insect Control**

Bacillus-based PGPRs, such as *Bacillus thuringiensis* (Bt), are used to manage insect pests. Bt maize, engineered to express Bt crystal proteins, has been

successfully integrated into agricultural systems, reducing the need for chemical insecticides, minimizing environmental contamination, and increasing farmers' profitability. Studies demonstrate how PGPRs can contribute to sustainable pest management by reducing the ecological impact of pest control measures, highlighting the potential of PGPRs, such as Bacillus spp. in pest control. Bt Cry3A toxins effectively reduce the damage caused by the larvae of the beetle *Crioceris quatuordecimpunctata,* to agricultural crops (Gao et al. 2012). Bt cry proteins are considered biotoxins, which are utilized against several crop pests, and their mechanisms have been revealed by ELISA, electrochemistry, and fluorescence (Gu et al. 2021). Pest larvae ingest toxins into their midgut, which are solubilized in alkaline conditions, forming a toxic substance (Nascimento et al. 2022). The toxic fragments can form transmembrane pores, causing osmotic imbalance and cell rupture, ultimately leading to the death of insects by bacteremia (Bravo et al. 2013). As a mode of action, inactive Bt toxins are sprayed on crops, consumed by pests, and then activated by the alkaline pH of the midgut. The activated toxins cause swelling and death of pests (Peng et al. 2019). Two biotoxins that have been commercialized for cotton production are Cry1Ac and Cry2Ac, under the name of Bollgard II, to protect them against lepidopterans (Liu et al. 2022a, b). CryIAc proteins disrupt the peritrophic membrane of cabbage looper larvae by inhibiting the binding of mucin protein with chitin calcofluor, resulting in increased mortality of the *Trichoplusia ni* (Guo et al. 2019).

- **Case Study 2: Trichoderma-Based Biofungicides for Suppressing Diseases**

Trichoderma spp., a group of soil-borne PGPRs, is used in cropping systems to combat root diseases. They produce secondary metabolites such as chitinases and antifungal compounds (Fig. 18.3), which suppress soil-borne pathogens like *Rhizoctonia.* These biofungicides, when applied as seed treatments or soil inoculants, enhance plant health and reduce crop losses caused by soil-borne pathogens (Abbas et al. 2022; Yao et al. 2023). This case study highlights the role of PGPRs in sustainable disease management. *Trichoderma* produces several metabolites, such as trichomycin, chlorotrichomycin, gelatinomycin, and antimicrobial peptides (Maruyama et al. 2020). Some commonly applied species of *Trichoderma* in biocontrol are *T. harzianum, T. koningii, T. asperellum, T. hamatum, T. longibrachiatum, T. polysporum,* and *T. viride* (Di Marco et al. 2022).

- **Case Study 3: Induced Systemic Resistance (ISR) with *Pseudomonas* spp.**

Pseudomonas spp. are PGPRs that can induce systemic resistance in plants by triggering their natural defense mechanisms. *Pseudomonas entomophila* possesses a biotoxin secretion system for a process mediated by biotoxin (Vodovar et al. 2006). Farmers have successfully integrated these PGPRs into their disease management strategies, enhancing the plant's innate immunity and reducing the need for chemical fungicides. This approach minimizes the risk of pathogen resistance development and contributes to sustainable disease control while preserving the ecological balance of agroecosystems. *Pseudomonas* sp. (QJX-1) exerts algicidal action on *Microcystis aeruginosa* by secreting 2,4-di-tert-butylphenol to damage

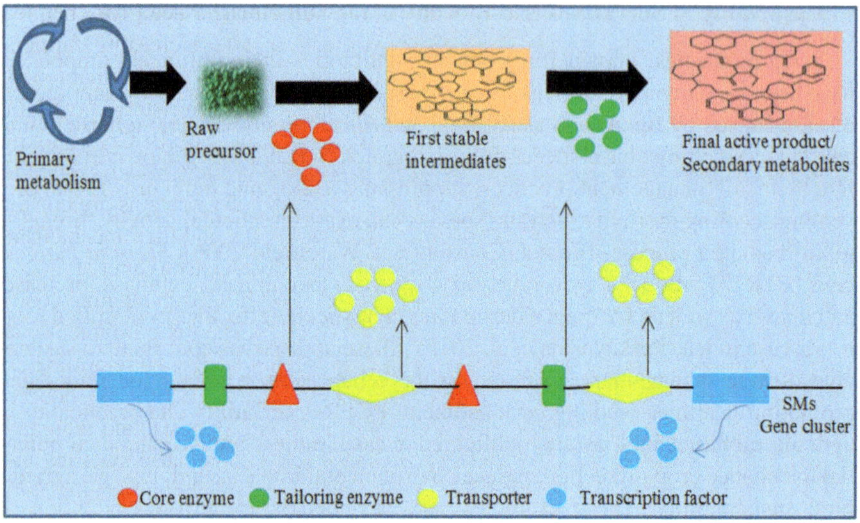

Fig. 18.3 Biosynthetic pathways of secondary metabolites and biotoxins and their regulation, in *Trichoderma* spp. (MDPI Source: Khan et al. 2020)

the photosynthetic structure in algae (Qi et al. 2021). Similarly, *Pseudomonas aeruginosa* secretes 2-heptyl-4-quinolone which binds to cytochrome (B6F), impeding the rate of photosynthesis in *Phaeodactylum tricornutum.* This leads to the biocontrol of algal blooms in water bodies (Dow et al. 2020). These case studies highlight the practicality and effectiveness of PGPRs in pest and disease control in agriculture, highlighting the potential of PGPR-based biotechnological solutions to reduce the environmental footprint, enhance crop health, and contribute to sustainable and resilient agricultural systems. A Gram-negative *Pseudomonas* symbiont can produce a defensive biotoxin pederin in *Paederus fuscipes.* The concentration of pederin up to 1312 and 596 ppm was lethal to two stored product pests, *Ephestia kuehniella* Zeller and *Tribolium confusum* Jacquelin, respectively, showing a unique mode of action for pest control (Amirmohammadi et al. 2019).

- **Case Study 4: Rhizobium-Based Nitrogen Fixation and Pest Resistance**

Rhizobium species, known for fixing atmospheric nitrogen, can also enhance pest resistance in certain crops. In soybean cultivation, certain strains induce systemic resistance against chewing herbivores through secondary metabolites and signaling molecules. Farmers who adopt this approach report reduced damage from pests like caterpillars and increased soybean yields without excessive use of pesticides (Igiehon et al. 2019). This case study highlights the multifaceted benefits of PGPRs in promoting crop health and reducing chemical inputs, emphasizing the potential of Rhizobium species in enhancing crop health.

- **Case Study 5: Serratia-Based Biocontrol for Soil-Borne Pathogens**

Serratia species, known for producing antibiotics and antifungal compounds, have been used in biocontrol against soil-borne pathogens, particularly damping-off disease caused by fungi such as *Pythium* and *Rhizoctonia*. Farmers in horticultural and nursery settings have successfully integrated Serratia-based biocontrol agents into their soil management practices, suppressing pathogenic fungi growth and preventing seedling mortality. Serratia species can exert insecticidal activity through a toxin-mediated process (Ffrench-Constant and Waterfield 2006). *Serratia marcescens* (ETR17) produces hydrolytic enzymes, such as chitinase, cellulase, protease, and lipase, as well as IAA and siderophores. The species effectively controlled root-rot disease in tea (Purkayastha et al. 2018). These approaches contribute to sustainable disease management by reducing the reliance on synthetic fungicides and promoting healthier seedling establishment. PGPRs' versatility and practicality in agricultural biotechnology are highlighted in case studies, emphasizing their potential to enhance crop protection, reduce environmental impacts, and improve agricultural sustainability. These examples serve as models for broader adoption of PGPR-based solutions in modern farming practices.

- **Case Study 6: Mycorrhizal Fungi Enhancing Nutrient Uptake and Disease Resistance**

Mycorrhizal fungi, specifically arbuscular mycorrhizal (AM) fungi, form symbiotic relationships with plant roots, enhancing nutrient uptake and water absorption. They also activate the plant's immune response, making it more resistant to soil-borne pathogens like *Fusarium* and *Verticillium*. Farmers who adopt mycorrhizal fungi have reported improved tomato yields and disease resistance. This case study highlights the potential of fungal-based PGPRs to enhance nutrient acquisition and plant defenses against diseases. This highlights the potential of utilizing beneficial fungi in sustainable agriculture by reducing the reliance on synthetic fertilizers and chemical fungicides. This case study emphasizes the potential of beneficial fungi in agriculture. A biotoxin crude destruxin derived from an entomopathogenic fungus, *Metarhizium anisopliae* (Metch.) caused 90% mortality in the larvae of *Spodoptera litura* (Fab.) causing oxidative stress and alteration in superoxide dismutase, catalase, and peroxidase enzymes (Sree and Padmaja 2008). BjαIT, a toxin gene isolated from *Buthotus judaicus*, was expressed in the fungus *Metarhizium acridum*, enhancing its virulence against *Locusta migratoria* by reducing the infection period (Peng and Xia 2015).

- **Case Study 7: Trichoderma-Based Biopesticides for Soil Health and Pest Control**

This case study explores the use of *Trichoderma* species as biofungicides in agriculture. Trichoderma-based PGPRs are effective against plant pathogens and nematode pests, promoting soil health and controlling nematode populations in agriculture (Yao et al. 2023; Verma et al. 2024). Farmers in regions with nematode infestations have successfully integrated Trichoderma-based biopesticides into their cropping

systems for effective pest control. *Trichoderma* species produce enzymes that break down nematode eggshells, immobilizing these pests in agriculture. This holistic approach to PGPRs addresses multiple agricultural challenges, including nutrient management, disease control, and pest resistance in modern farming. The case studies highlight the versatility and efficacy of PGPRs, including bacteria and fungi, in enhancing crop health and sustainability in modern farming.

- **Case Study 8: Beneficial Endophytic Fungi for Enhanced Crop Resilience**

Endophytic fungi, which live within plant tissues, have been found to improve plant resilience, particularly in rice crops in agriculture. These fungi colonize the rice plant's vascular system, aiding in water and nutrient uptake. They trigger stress responses, leading to improved drought tolerance and resistance against fungal pathogens such as *Magnaporthe oryzae*, the cause of rice blast disease (Ganie et al. 2022). This case study demonstrates how endophytic fungi, a subgroup of PGPRs, contribute to crop resilience in the face of multiple stressors. This offers a sustainable approach to crop management and reduces the need for excessive irrigation and chemical fungicides.

- **Case Study 9: Beneficial Yeasts for Postharvest Disease Control**

Yeast-based PGPRs have found a niche in agriculture, especially in mitigating postharvest diseases in fruits. *Candida oleophila*, a yeast species with biocontrol properties against pathogenic fungi, colonizes the fruit's surface, outcompeting and suppressing their growth. This case study demonstrates the potential of yeast-based PGPRs in sustainable agriculture, particularly in reducing food waste (Díaz et al. 2020). By offering a natural and effective means of postharvest disease control, these yeasts contribute to increased shelf life and improved fruit quality. The integration of beneficial endophytic fungi and yeast-based PGPRs highlights the broad spectrum of microorganisms that can be leveraged to address agricultural challenges, such as drought resistance and postharvest disease control (Table 18.1).

18.3 Soil Microbes

Soil microbes are essential for maintaining soil health and fertility, which in turn affect plant health and productivity. This section delves into the diverse world of soil microbes, their biotoxins, and potential in pest management. It also addresses challenges in harnessing their power for sustainable agriculture, highlighting the importance of understanding their role in ecosystems.

Table 18.1 Microbial biotoxins and beneficial microorganisms in agriculture: case studies in sustainable crop management

Case study	Microorganism	Target pest/ disease	Mechanism	Benefits	References
1	*Bacillus* spp.	Insect pests (e.g., European corn borer)	Production of insecticidal crystal proteins (Bt toxins)	This leads to reduced reliance on chemical insecticides, decreased environmental contamination, and enhanced profitability.	Gu et al. (2021)
2	*Trichoderma* spp.	Soil-borne pathogens (e.g., Rhizoctonia, Fusarium)	Production of chitinases and other antifungal compounds.	Improved plant health, reduced crop losses, decreased reliance on chemical fungicides, and enhanced soil health.	Di Marco et al. (2022), Yao et al. (2023)
3	*Pseudomonas* spp.	Various plant diseases	Induction of systemic resistance (ISR)	Enhanced disease resistance, reduced need for chemical fungicides, and minimized risk of pathogen resistance.	Qi et al. (2021)
4	*Rhizobium* spp.	Chewing herbivores (e.g., caterpillars)	Induction of systemic resistance.	Reduced pest damage and increased crop yields without excessive pesticide use.	Igiehon et al. (2019)

(continued)

Table 18.1 (continued)

Case study	Microorganism	Target pest/ disease	Mechanism	Benefits	References
5	*Serratia* spp.	Soil-borne pathogens (e.g., Pythium, Rhizoctonia)	Production of antibiotics and other antifungal compounds.	Reduced seedling mortality, sustainable disease management, and decreased reliance on synthetic fungicides.	Purkayastha et al. (2018)
6	Mycorrhizal fungi	Soil-borne pathogens (e.g., Fusarium, Verticillium)	Enhanced nutrient uptake, activation of plant immune response	Improved crop yields, enhanced disease resistance, reduced need for synthetic fertilizers, and fungicides.	Peng and Xia (2015)
7	*Trichoderma* spp.	Nematode pests	Production of enzymes that break down nematode eggshells.	Nematode population suppression, improved soil health, and reduced reliance on synthetic nematicides.	Yao et al. (2023)
8	Endophytic fungi	Drought tolerance, fungal pathogens (e.g., Magnaporthe oryzae)	Colonization of plant vascular system, triggering plant stress responses.	Improved drought tolerance, and enhanced resistance against fungal pathogens.	Ganie et al. (2022)
9	Beneficial yeasts	Postharvest diseases (e.g., Botrytis cinerea)	Colonization of fruit surfaces, outcompeting pathogenic fungi	Reduced postharvest losses, extended shelf life, reduced reliance on chemical fungicides.	Díaz et al. (2020)

(continued)

Table 18.1 (continued)

Case study	Microorganism	Target pest/ disease	Mechanism	Benefits	References
10	Entomopathogenic fungi	Insect pests (e.g., aphids, whiteflies)	Production of toxins that penetrate insect cuticles.	Precise and environmentally friendly pest control, reduced reliance on chemical insecticides, and nanopesticidal properties.	Bihal et al. (2023)
11	Mycoviruses	Fungal pathogens (e.g., rice blast)	Infection and weakening of pathogenic fungi.	Novel genetic-level approach to disease management in crops.	Moriyama et al. (2018)
12	*Streptomyces* spp.	Soil-borne diseases (e.g., Phytophthora)	Production of antibiotics.	Sustainable disease management, and enhanced soil health.	Khan et al. (2023)
13	Bacteriophages	Bacterial plant pathogens (e.g., Xanthomonas)	Production of lytic enzymes.	Reduced bacterial diseases and decreased use of chemicals.	Vu and Oh (2020)
14	*Penicillium* spp. (non-virulent)	Postharvest molds (e.g., Virulent *Penicillium expansum*)	Production of antifungal compounds.	Reduced postharvest losses, and improved fruit quality.	Llobregat et al. (2024)
15	*Frankia* spp.	Nitrogen fixation in non-leguminous plants.	Nitrogen fixation.	Enhanced soil fertility, and reduced use of synthetic nitrogen fertilizers.	Santi et al. (2013)
16	*Azospirillum* spp.	Nitrogenfixation in cereals and grasses	Biological nitrogen fixation.	Improved nutrient availability, increased crop yields.	Zeffa et al. (2019)

(continued)

Table 18.1 (continued)

Case study	Microorganism	Target pest/ disease	Mechanism	Benefits	References
17	*Burkholderia* spp.	Plant growth promotion and disease suppression.	Production of siderophores and antibiotics.	Enhanced plant growth, reduced soil-borne diseases, and induction of immune resistance in wheat.	An et al. (2022)
18	Actinomycetes spp.	Suppression of root-knot nematodes.	Production of nematode-trapping networks.	Reduction of nematode populations and improved root health.	Ruanpanun and Chamswarng (2016)
19	Arbuscular mycorrhizal fungi	Nutrient uptake and drought tolerance.	Mycorrhizal association.	Enhanced nutrient acquisition, and increased drought resistance.	Begum et al. (2019)
20	*Streptomyces* spp.	Induction of systemic resistance (ISR).	Activation of plant defense genes.	Enhanced disease resistance, and reduced use of chemical fungicides.	Hata et al. (2021)
21	*Agrobacterium tumefaciens*	Crown gall disease control.	Inhibition of *Agrobacterium tumefaciens*.	Reduced tumor formation and healthier plants.	Lee et al. (2009)
22	*Paenibacillus polymyxa*	Soil health enhancement.	Production of plant growth-promoting compounds	Improved soil structure, nutrient availability, and plant growth.	Weselowski et al. (2016)
23	*Xylariaceae* spp. (as endophytes)	Defense against herbivores.	Production of defensive secondary metabolites.	Enhanced herbivore resistance in plants, and reduced herbivore damage to plants.	Alam et al. (2021)
24	Chitosan oligosaccharides	Various plant diseases.	Activation of plant defense responses.	Enhanced disease resistance, and reduced reliance on chemical fungicides.	Guarnizo et al. (2020)

(continued)

Table 18.1 (continued)

Case study	Microorganism	Target pest/ disease	Mechanism	Benefits	References
25	*Enterobacter* spp.	Plant growth promotion.	Production of plant growth hormones.	Improved plant vigor, and increased crop yields.	Ngalimat et al. (2021)
26	*Rhizophagus irregularis*	Phosphorus uptake enhancement.	Mycorrhizal association.	Increased phosphorus acquisition, and improved nutrient utilization.	Le Pioufle et al. (2019)
27	Lactic acid bacteria	Postharvest preservation.	Production of antimicrobial compounds.	Extending the shelf life of fruits and vegetables can reduce food waste.	Zapaśnik et al. (2022)
28	*Lactobacillus* spp.	Mycotoxin degradation.	Production of enzymes to detoxify mycotoxins.	Improved food safety, and reduced mycotoxin contamination.	Muhialdin et al. (2020)
29	*Roseobacter* spp.	Algal blooms in aquaculture.	Algicidal activity.	Control of harmful algal blooms, and improved water quality in aquaculture systems.	Coyne et al. (2022)
30	Bacterial consortia.	Soil health and nutrient cycling.	Synergistic microbial interactions.	Enhanced nutrient availability and improved soil fertility.	Vishwakarma et al. (2020)
31	*Acetobacter diazotrophicus*	Nitrogen fixation in sugarcane.	Biological nitrogen fixation in sugarcane.	Reduced use of synthetic nitrogen fertilizer, and increased sugarcane yield.	Carvalho et al. (2022)
32	Myxobacteria	Control of fungal plant pathogens.	Production of antibacterial compounds.	Biocontrol of bacterial diseases in plants, and reduced chemical use.	Bhat et al. (2021)

(continued)

Table 18.1 (continued)

Case study	Microorganism	Target pest/ disease	Mechanism	Benefits	References
33	Lichen Symbiosis	Environmental stress tolerance.	Mutualistic association between fungi and algae.	Improved plant resilience to harsh environmental conditions.	Gasulla et al. (2021)
34	Nitrifying bacteria	Nitrification in soils.	Conversion of ammonium to nitrate.	Enhanced nitrogen availability for plants and reducing nitrogen leaching.	Norton and Ouyang (2019)
35	Actinobacteria	Suppression of damping-off disease.	Production of antifungal metabolites.	Improved seedling survival, and sustainable disease management.	Ebrahimi-Zarandi et al. (2022)
36	Diazotrophic iron-reducing bacteria.	Nitrogen fixation in paddy fields.	Biological nitrogen fixation.	Reduced synthetic nitrogen fertilizer use, and increased rice yield.	Masuda et al. (2023)
37	*Clostridium* spp.	Soil organic matter decomposition.	Anaerobic breakdown of organic materials.	Enhanced soil carbon sequestration and improved soil structure.	Anderson et al. (2018)
38	Algal Biostimulants	Plant growth promotion	Production of plant growth hormones, and beneficial metabolites.	Improved plant vigor, increased crop yields, and reduced chemical inputs.	Ali et al. (2021)
39	*Rhizobium rhizogenes*	Root-knot nematodes affect tomatoes.	Genetic transformation can provide nematode resistance.	Reduced nematode damage, and improved tomato yield.	Zhang et al. (2023)
40	*Streptomyces scabies*	Potato scab disease.	Production of thaxtomin toxins.	Decreased scab incidence, and improved potato quality.	Ismail et al. (2020)

(continued)

Table 18.1 (continued)

Case study	Microorganism	Target pest/ disease	Mechanism	Benefits	References
41	*Verticillium lecanii*	Aphids in greenhouse crops.	Insect parasitism.	Effective control of aphid populations, and reduced virus transmission.	Aqueel and Leather (2013)
42	*Nitrosomonas europaea*	Ammonia-oxidizing bacteria.	Conversion of ammonia to nitrite.	Improved nitrogen cycling, and reduced ammonia volatilization.	Laanbroek et al. (2002)
43	*Photorhabdus luminescens*	Soil-dwelling insect larvae.	Production of toxins, and mutualistic associations with nematodes.	Biocontrol of insect pests, and reduced larval damage.	Regaiolo et al. (2020)
44	*Bradyrhizobium japonicum*	Soybean nodulation.	Biological nitrogen fixation.	Enhanced nitrogen availability for soybeans, and increased yields.	Ulzen et al. (2016)
45	*Lactobacillus* spp.	Silage preservation	Fermentation and lactic acid production.	Improved silage quality, and reduced spoilage.	Guo et al. (2023)
46	*Streptomyces avermitilis*	Control of plant-parasitic nematodes.	Production of avermectin toxins.	Biocontrol of nematode pests in crops.	Meidani et al. (2020)
47	*Clostridium beijerinckii*	Biofuel production from lignocellulosic biomass.	Enzymatic breakdown of biomass.	Enhanced biofuel production, and sustainable energy source.	Olorunsogbon et al. (2022)
48	*Methylobacterium extorquens*	Plant methanol utilization.	Methanol assimilation, and plant growth promotion.	Improved nutrient availability for plants, and enhanced plant growth.	Krug et al. (2020)

18.3.1 Soil Microbes: A Diverse Community

Soil, considered Earth's living skin, is home to a diverse ecosystem of microorganisms such as bacteria, archaea, fungi, protozoa, and viruses. Each microbe plays a unique role in shaping the soil microbiome, contributing to its complexity and resilience. These microbes play a crucial role in ecosystems and agriculture.

18.3.1.1 Overview of Soil Microbial Diversity

Soil microbial diversity is a key indicator of the balance within the soil ecosystem, with millions to billions of cells in a single gram representing thousands of species. This diversity is influenced by the diverse niches and microhabitats provided by soil, each promoting specific microbial communities, which in turn exhibit diverse metabolic capabilities. High microbial diversity enriches soil ecological functioning processes including nutrient cycling and rhizodeposition, in the rhizosphere (Maron et al. 2018).

18.3.1.2 The Interplay Between Soil Microbes and Plant Health

Soil microbes play a crucial role in plant health, facilitating nutrient cycling and mycorrhizal associations. These microbes break down organic matter, making essential nutrients accessible to plants. Mycorrhizal associations between fungi and plants enhance plant health and improve access to resources. Soil microbes produce signaling molecules that influence plant growth and defense responses, ultimately affecting plant health, productivity, and resilience in terrestrial ecosystems. Biological nitrogen fixation, root exudation, and triggering of specific biochemical pathways, are valuable ecological services in the rhizosphere (Jacoby et al. 2017).

18.3.2 Biotoxins from Soil Microbes

18.3.2.1 Discussion of Microbial Biotoxins Produced by Various Soil Microbes

Soil microbes produce a variety of biotoxins, including antibiotics, antifungal compounds, bacteriocins, and secondary metabolites, which are crucial for survival, resource competition, and protection against predators or pathogens. These biotoxins can kill or inhibit other competing microorganisms (Mullis et al. 2019). Understanding the diversity and roles of these biotoxins is essential for comprehending the complexities of soil microbial communities.

18.3.2.2 The Ecological Roles of These Biotoxins

Microbial biotoxins are crucial for soil ecosystems, as they inhibit the growth of competing microorganisms, allowing biotoxin-producing species to access resources. Bacteriocins, produced by bacteria, kill or inhibit closely related strains, enabling niche specialization. They are successfully being used as biopreservatives and in veterinary medicines. Similarly, thuricin, a subclass of bacteriocin, produced by Bt NEB17, assists crops in signaling against abiotic stress (climate resilience)

(Nazari and Smith 2020). Secondary metabolites, acting as signaling molecules or toxins, influence the behavior and survival of other microorganisms, contributing to the complex competition and cooperation within soil microbial communities.

18.3.3 Soil Microbes in Pest Management

Soil microbes, microorganisms living beneath the Earth's surface, offer sustainable alternatives to chemical pesticides in pest management. These microbes, including bacteria and fungi, interact with plants and pests, offering potential for natural pest control. Recent research shows that soil microbes can suppress populations and enhance plant defenses. Emerging technologies are promising an environmentally friendly, economically viable, and sustainable future for pest management in agriculture, transforming agricultural practices and providing greener, more effective solutions.

18.3.3.1 How Soil Microbes Contribute to Natural Pest Suppression

Soil microorganisms armed with biotoxins, which are toxic substances, play a crucial role in naturally suppressing pests. These biotoxins reduce insect pest populations and create an inhospitable environment for beneficial soil microbes, providing a sustainable and eco-friendly pest control method (Chalivendra 2021). The market for biopesticides is continuously increasing, with global commercial acceptability in modern agricultural systems (Kumar et al. 2021).

18.3.3.2 The Potential of Harnessing Soil Microbes for Biotechnological Pest Management

Soil microbes and their biotoxins offer potential for biotechnological pest management in sustainable agriculture. The aggregation of microbes develops soil microbiome, which is an emerging, potent, and vital topic of research in sustainable agriculture (Suman et al. 2022). Researchers and practitioners are exploring strategies to increase soil microbe abundance and activity, such as targeted inoculation or optimizing soil management practices. This approach aims to reduce reliance on chemical pesticides, thereby reducing their environmental and health effects and promoting sustainable agriculture.

18.3.4 Applications and Challenges

Soil microbes, rich in bacteria, fungi, and other microorganisms, produce biotoxins that have diverse applications in agriculture, medicine, and environmental science. Biotoxins are used on a commercial scale for the preparation of biofungicides, biopesticides, and biofertilizers (Verma et al. 2024). These naturally occurring compounds have varying properties and effects, offering potential solutions to pressing challenges. However, harnessing these biotoxins for practical applications faces hurdles such as isolation, purification, safety concerns, and regulatory issues. This exploration of biotoxins from soil microbes will explore their diverse applications, from pest control and bioremediation to drug development and beyond. The wide-reaching impact of these microbial weapons will be examined, while addressing the challenges faced by researchers and industries in utilizing soil microbial biotoxins for societal and environmental improvement.

18.3.4.1 Case Studies Showcasing the Use of Soil Microbes and their Biotoxins in Agriculture

Soil microbe-based strategies are being used by agricultural practitioners worldwide to tackle pest issues. These strategies are adaptable and effective, reducing the need for synthetic pesticides in organic farming and disrupting pest life cycles in large-scale monoculture farming. These real-world applications highlight the ecological benefits of using soil microbes for pest management. A Gram-negative proteobacterium, Photorhabdus luminescens, forms a symbiotic relationship with nematodes (Heterorhabditis). These nematodes are used as biocontrol agents for crop pests. Photorhabdus species secrete toxins and secondary metabolites that exhibit insecticidal properties (Kushwah and Somvanshi 2016).

18.3.4.2 Challenges and Limitations in Implementing Soil Microbe-Based Pest Management Strategies

Soil microbe-based pest management faces challenges such as variability in effectiveness across soil types and environmental conditions, concerns about off-target effects on non-pest organisms, and potential shifts in microbial community dynamics. The complexity of soil ecosystems necessitates a deeper understanding of microbial interactions, as unintended consequences may emerge from microbial interventions. Soil microbe-based strategies hold great promise for sustainable pest management, but further investigation is needed to determine their scalability and economic feasibility on a large scale. Factors such as production costs, application logistics, and regulatory considerations will influence the practicality of these approaches in mainstream agriculture.

Soil microbes and their biotoxins offer ecological knowledge and sustainable pest management solutions, combating pests, reducing reliance on chemical pesticides, and promoting environmentally friendly agricultural practices. The interplay between soil microbes, their biotoxins, and plant health exemplifies the intricate relationships that govern agricultural ecosystems.

Despite challenges and limitations, the continued exploration of soil microbe-based strategies offers a promising path toward a more resilient and sustainable agriculture. Research, innovation, and collaboration among scientists, farmers, and policymakers will be crucial in harnessing the full potential of soil microbes in pest management, ultimately leading to healthier crops, reduced environmental impact, and a more sustainable future for agriculture.

18.4 Future Prospects

As we peer into the future of agricultural innovation, the utilization of microbial biotoxins in plant disease and pest management stands as a promising frontier. In this section, we explore the potential advancements, sustainability implications, regulatory considerations, and the transformative role that microbial biotoxins may play in shaping the agriculture of tomorrow.

18.4.1 Advancements in Biotechnological Research: Exploring Microbial Biotoxins

18.4.1.1 Emerging Technologies and Tools for Studying Microbial Biotoxins

The world of biotechnology is constantly evolving, and the study of microbial biotoxins is no exception. Advancements in analytical techniques, such as mass spectrometry and next-generation sequencing, enable scientists to identify, characterize, and understand biotoxins with unprecedented precision. These technological leaps have ushered in a new era in our ability to unravel the mysteries of microbial biotoxins, offering exciting possibilities in various fields.

- **Mass Spectrometry (MS) in Biotoxin Analysis:**

- Mass spectrometry is a significant advancement in biotoxin research due to its high sensitivity and accuracy, enabling scientists to detect and quantify biotoxins in complex environmental samples, aiding in monitoring and managing biotoxin-related issues.
- **Next-Generation Sequencing (NGS) for Biotoxin-Producing Microorganisms:**

- NGS technologies have revolutionized microbial community study, enabling the identification of biotoxin-producing organisms in diverse environments. Metagenomics, a subset of NGS, helps researchers explore genetic diversity.
- **Omics Approaches for Deeper Insights:**

- Metabolomics, proteomics, and transcriptomics are enhancing our understanding of biotoxin production and its regulation. By analyzing metabolites, proteins, and gene expression profiles, scientists can uncover intricate molecular mechanisms, aiding in the development of targeted biotoxin management interventions.

18.4.1.2 Genomic and Metagenomic Approaches to Enhance Our Understanding

Genomics and metagenomics are opening new avenues to explore the genetic foundations of microbial biotoxin production. These cutting-edge techniques are instrumental in advancing our understanding of the genetic basis of biotoxin synthesis and its ecological significance.

- **Genomic Insights into Biotoxin Synthesis:**

- By sequencing the genomes of biotoxin-producing microorganisms, researchers can identify the genes responsible for biotoxin synthesis. This knowledge not only aids in understanding the biochemical pathways involved but also provides potential targets for genetic manipulation to control biotoxin production. For instance, in the context of agriculture, this approach holds promise for developing crops with inherent resistance to biotoxin-producing pests or pathogens (Liu et al., 2022).
- **Metagenomics Unraveling Complex Microbiome Dynamics:**
- Metagenomics allows scientists to examine entire microbial communities in the context of the production of biotoxins. In soil ecosystems, where diverse microbial populations interact, metagenomics provides insights into the complex dynamics between producers of biotoxins, potential antagonists, and their roles in shaping soil health. Understanding these dynamics is crucial for sustainable strategies for managing pests and diseases that harness the power of beneficial microorganisms.
- **Bioprospecting for Novel Biotoxin-Producing Microbes:**

- Genomic and metagenomic approaches aid in bioprospecting, uncovering unknown producers of biotoxins, leading to the identification of novel biotoxins with unique properties or applications in various fields.

The integration of advanced technologies with genomic and metagenomic methods is transforming research on microbial biotoxins, enhancing our understanding of biotoxins, and paving the way for innovative applications in agriculture and

environmental management. These advancements aim to create a sustainable future by harnessing the power of microbial biotoxins for the greater good.

18.4.2 Sustainable Agriculture: A Paradigm Shift with Microbial Biotoxins

Sustainable agriculture aims to balance environmental stewardship, economic viability, and social equity. Microbial biotoxins can help address the heavy reliance on chemical pesticides, reshaping the agricultural landscape and addressing the cornerstone of a greener, more equitable future.

18.4.2.1 The Role of Microbial Biotoxins in Achieving Sustainable Agriculture Goals

Sustainable agriculture aims to meet present needs without compromising future generations' ability to meet their own. It minimizes environmental impacts, ensures economic viability, and promotes social fairness in farming practices. Microbial biotoxins, as natural tools, align with sustainable agriculture principles and can significantly contribute to achieving sustainability goals:

- **Minimizing Environmental Impact:**
- Chemical pesticides, which have long been regarded as indispensable for pest and disease management, have raised concerns about their impact on the environment. Runoff from agricultural fields can contaminate water bodies, leading to adverse effects on aquatic ecosystems and the quality of drinking water. Chemical residues can persist in the environment, affecting nontarget organisms, including beneficial insects and pollinators. In contrast, microbial biotoxins offer a way to significantly reduce these negative environmental consequences.
  - Preservation of Biodiversity: Microbial biotoxin-based strategies are selective, targeting specific pests while sparing nontarget organisms. This selectivity reduces collateral damage to beneficial insects and predators, contributing to the preservation of biodiversity in agricultural landscapes.
  - Soil Health: The health of the soil, a cornerstone of sustainable agriculture, is intricately linked to the use of chemical pesticides. Microbial biotoxins, applied judiciously, can foster a healthier soil ecosystem by minimizing the disruption of soil microorganisms and microbial diversity. Healthy soils lead to improved nutrient cycling, enhanced water retention, and overall resilience.
  - Water Quality: By reducing the reliance on chemical pesticides, microbial biotoxins can mitigate the risk of pesticide runoff into water bodies. This benefits aquatic ecosystems and drinking water quality, aligning with sustainable agriculture's commitment to safeguarding natural resources.

- **Promoting Economic Viability:**
- Sustainable agriculture emphasizes economic sustainability for farmers, ensuring that farming practices remain economically viable. Microbial biotoxins offer economic advantages that extend beyond their reduced environmental impact:
  - Lower Input Costs: Chemical pesticides often constitute a significant portion of production costs for farmers. Microbial biotoxins, once established and optimized, can offer a cost-effective alternative, reducing the financial burden on farmers.
  - Decreased reliance on synthetic chemicals: Overreliance on synthetic chemicals can lead to the development of pesticide-resistant pests, necessitating increased pesticide use. Microbial biotoxin-based strategies provide a sustainable approach by minimizing resistance development, ultimately safeguarding crop yields and economic returns.
- **Ensuring social equity:**
- Sustainable agriculture places great emphasis on social fairness, seeking to ensure that farming practices benefit not only the producers but also the broader community. Microbial biotoxin adoption contributes to social equity in several ways:
  - Health and Safety: Chemical pesticides can pose health risks to farmers and agricultural workers. The shift toward microbial biotoxins reduces exposure to hazardous chemicals, enhancing the safety and well-being of those working in agriculture.
  - Community and Consumer Trust: Consumers increasingly seek food produced with minimal chemical inputs, aligning with the ethos of sustainable agriculture. The adoption of microbial biotoxin-based strategies enhances consumer trust in the safety and sustainability of agricultural products.

18.4.2.2 Reducing Chemical Pesticide Dependence through Biotechnological Solutions

Chemical pesticides have been a cornerstone of modern agriculture for decades, providing effective means to combat pests and diseases that threaten crop yields. However, the adverse environmental and health effects associated with their widespread use have spurred a global call for alternative approaches. In this context, microbial biotoxin-based solutions offer precisely the alternative that agriculture needs: effective, selective, and environmentally friendly methods for pest and disease management.

1. Precision Pest Control: Microbial biotoxins have the unique ability to target specific pests while sparing nontarget organisms. This precision allows farmers to effectively manage pest populations while minimizing harm to beneficial insects, pollinators, and other wildlife.

2. Reduced Environmental Impact: The reduction in chemical pesticide usage translates into a significant decrease in the environmental footprint of agriculture. By minimizing pesticide runoff and contamination of water bodies, microbial biotoxin-based strategies contribute to cleaner ecosystems and healthier environments.
3. Preservation of Soil Health: Healthy soils are the foundation of sustainable agriculture, and chemical pesticides can disrupt soil ecosystems. Microbial biotoxins, with their selectivity and minimal impact on nontarget soil organisms, promote soil health and resilience.
4. Meeting Consumer Preferences: Consumer preferences are evolving toward food products produced with minimal chemical inputs. Microbial biotoxin-based strategies align with these preferences, enhancing the marketability of agricultural products and ensuring the long-term viability of farming practices.
5. Mitigating Pesticide Resistance: The overuse of chemical pesticides has led to the development of pesticide-resistant pests, posing significant challenges to crop protection. Microbial biotoxin-based solutions offer a sustainable approach by reducing the selective pressure for resistance development.
6. Enhancing Farm Worker Safety: Farm worker safety is a paramount concern in agriculture. The shift away from chemical pesticides toward microbial biotoxins reduces the exposure of farmers and agricultural workers to hazardous chemicals, promoting a safer working environment.

18.4.3 Regulatory and Ethical Considerations

As the promising realm of microbial biotoxins unfolds in the realm of agriculture, it brings with it a host of regulatory and ethical considerations. The responsible and safe application of these biotechnological solutions necessitates a structured framework that addresses potential risks, ensures compliance with established guidelines, and considers the ethical implications of their use. In this section, we delve into the intricate web of regulatory and ethical considerations that will shape the future landscape of microbial biotoxin utilization in agriculture.

18.4.3.1 The Importance of Regulatory Frameworks for Biotoxin-Based Products

As microbial biotoxins are increasingly used in agricultural pest and disease management, regulatory frameworks must evolve to ensure their safe and responsible use. Regulatory bodies worldwide will need to establish guidelines for the development, testing, and deployment of biotoxin-based products. These frameworks will evaluate the effectiveness, environmental impact, and human safety of biotoxin-based products. Robust regulations will instill confidence among farmers,

consumers, and stakeholders, ensuring that biotoxin-based solutions meet stringent quality and safety standards.

18.4.3.2 Ethical Implications and Public Perception of Microbial Biotoxin Use

The ethical dimensions of using microbial biotoxins in agriculture merit careful consideration. Transparent communication and education about the benefits and potential risks of biotoxin-based strategies are essential. Ethical concerns may arise regarding unintended ecological consequences or the manipulation of microbial communities. Public perception and acceptance of these approaches are pivotal, and proactive engagement with stakeholders can help address concerns and foster acceptance. Ethical stewardship ensures that the adoption of biotoxin-based solutions aligns with broader societal values and aspirations for sustainable agriculture. To assess the health risks associated with biotoxins, genomic analysis of secondary metabolites and biotoxins is becoming vital (Deising et al. 2017).

18.5 Conclusion

In agriculture, microbial biotoxins represent a promising paradigm shift toward sustainability, precision, and environmental stewardship. As we reflect on the multifaceted journey through this chapter, several key takeaways emerge. First and foremost, microbial biotoxins offer a compelling alternative to conventional chemical pesticides. Their precise targeting of pests, minimal impact on nontarget organisms, and reduced environmental footprint align seamlessly with the principles of sustainable agriculture. These biotechnological solutions pave the way for preserving biodiversity, soil health, and water quality, marking significant progress toward a more harmonious coexistence with our planet.

Yet, the integration of microbial biotoxins into agricultural practices is not without complexity. Regulatory frameworks must evolve to ensure responsible and safe use of biotoxin-based products. Ethical considerations, including transparency, safety, and public perception, require careful navigation. Engaging stakeholders, from farmers to policymakers and the broader public, is essential to building trust and fostering acceptance of these innovative approaches.

Microbial biotoxins offer a compelling narrative of innovation, sustainability, and ecological harmony in agriculture. With continued research, responsible implementation, and a commitment to ethical stewardship, we have the opportunity to cultivate a resilient and sustainable agricultural future. In this future, the power of microbes will enrich our harvests while preserving the delicate balance of our planet, ensuring a bountiful harvest not just for today but for generations to come.

References

Abbas A, Mubeen M, Zheng H, Sohail MA, Shakeel Q, Solanki MK et al (2022) *Trichoderma* spp. genes involved in the biocontrol activity against *Rhizoctonia solani*. Front Microbiol 13:884469. https://doi.org/10.3389/fmicb.2022.884469

Adejumo TO, Vögele RT (2021) Introduction: potential of botanicals and microorganisms as biopesticides. In: Biopesticides. Botanicals and microorganisms for improving agriculture and human health. Logos Verlag, Berlin, pp 17–34

Alam B, Lǐ J, Gě Q, Khan MA, Gōng J, Mehmood S, Yuán Y, Gǒng W (2021 Dec) Endophytic fungi: from symbiosis to secondary metabolite communications or vice versa? Front Plant Sci 17(12):791033. https://doi.org/10.3389/fpls.2021.791033

Ali O, Ramsubhag A, Jayaraman J (2021) Biostimulant properties of seaweed extracts in plants: implications towards sustainable crop production. Plants (Basel). 10(3):531. https://doi.org/10.3390/plants10030531

Amirmohammadi F, Bandani A, Amirheidari B, Sabahi G (2019) Effect of pederin on activity of digestive and detoxifying enzymes of *Ephestia kuehniella* (Lep: Pyralidae) and *Tribolium confusum* (Col: Tenebrionidae). J Crop Protect 8(1):45–55

An C, Ma S, Liu C, Ding H, Xue W (2022) Burkholderia ambifaria XN08: a plant growth-promoting endophytic bacterium with biocontrol potential against sharp eyespot in wheat. Front Microbiol 13:906724. https://doi.org/10.3389/fmicb.2022.906724

Anderson CR, Peterson ME, Frampton RA, Bulman SR, Keenan S, Curtin D (2018) Rapid increases in soil pH solubilise organic matter, dramatically increase denitrification potential and strongly stimulate microorganisms from the Firmicutes phylum. PeerJ:e6090. https://doi.org/10.7717/peerj.6090

Aqueel MA, Leather SR (2013) Virulence of Verticillium lecanii (Z.) against cereal aphids; does timing of infection affect the performance of parasitoids and predators? Pest Manag Sci 69(4):493–498. https://doi.org/10.1002/ps.3398

Backer R, Rokem JS, Ilangumaran G, Lamont J, Praslickova D, Ricci E, Subramanian S, Smith DL (2018) Plant growth-promoting Rhizobacteria: context, mechanisms of action, and roadmap to commercialization of biostimulants for sustainable agriculture. Front Plant Sci 9:1473. https://doi.org/10.3389/fpls.2018.01473

Begum N, Qin C, Ahanger MA, Raza S, Khan MI, Ashraf M, Ahmed N, Zhang L (2019) Role of arbuscular mycorrhizal fungi in plant growth regulation: implications in abiotic stress tolerance. Front Plant Sci 10:1068. https://doi.org/10.3389/fpls.2019.01068

Bhat MA, Mishra AK, Bhat MA, Banday MI, Bashir O, Rather IA, Rahman S, Shah AA, Jan AT (2021) Myxobacteria as a source of new bioactive compounds: a perspective study. Pharmaceutics 13(8):1265. https://doi.org/10.3390/pharmaceutics13081265

Bihal R, Al-Khayri JM, Banu AN, Kudesia N, Ahmed FK, Sarkar R, Arora A, Abd-Elsalam KA (2023) Entomopathogenic fungi: an eco-friendly synthesis of sustainable nanoparticles and their nanopesticide properties. Microorganisms 11(6):1617. https://doi.org/10.3390/microorganisms11061617

Bravo A, Gómez I, Porta H, García-Gómez BI, Rodriguez-Almazan C, Pardo L, Soberón M (2013) Evolution of *Bacillus thuringiensis* cry toxins insecticidal activity. Microb Biotechnol 6:17–26. https://doi.org/10.1111/j.1751-7915.2012.00342.x

Carvalho TLG, Rosman AC, Grativol C, de Nogueira E, Baldani JI, Hemerly AS (2022) Sugarcane genotypes with contrasting biological nitrogen fixation efficiencies differentially modulate nitrogen metabolism, auxin signaling, and microorganism perception pathways. Plants (Basel) 11(15):1971. https://doi.org/10.3390/plants11151971

Chalivendra S (2021) Microbial toxins in insect and nematode pest biocontrol. Int J Mol Sci 22(14):7657. https://doi.org/10.3390/ijms22147657

Coyne KJ, Wang Y, Johnson G (2022) Algicidal bacteria: a review of current knowledge and applications to control harmful algal blooms. Front Microbiol 13:871177. https://doi.org/10.3389/fmicb.2022.871177

Deising HB, Gase I, Kubo Y (2017) The unpredictable risk imposed by microbial secondary metabolites: how safe is biological control of plant diseases? J Plant Dis Prot 124:413–419. https://doi.org/10.1007/s41348-017-0109-5

Di Marco S, Metruccio EG, Moretti S, Nocentini M, Carella G, Pacetti A et al (2022) Activity of Trichoderma asperellum strain ICC 012 and Trichoderma gamsii strain ICC 080 toward diseases of esca complex and associated pathogens. Front Microbiol 12:813410. https://doi.org/10.3389/fmicb.2021.813410

Díaz MA, Pereyra MM, Picón-Montenegro E, Meinhardt F, Dib JR (2020) Killer yeasts for the biological control of postharvest fungal crop diseases. Microorganisms 8(11):1680. https://doi.org/10.3390/microorganisms8111680

Dow L, Stock F, Peltekis A, Szamosvari D, Prothiwa M, Lapointe A et al (2020) The multifaceted inhibitory effects of an alkylquinolone on the diatom Phaeodactylum tricornutum. Chembiochem 21:1206–1216. https://doi.org/10.1002/cbic.201900612

Ebrahimi-Zarandi M, Saberi Riseh R, Tarkka MT (2022) Actinobacteria as effective biocontrol agents against plant pathogens, an overview on their role in eliciting plant defense. Microorganisms 10(9):1739. https://doi.org/10.3390/microorganisms10091739

Ffrench-Constant R, Waterfield N (2006) An ABC guide to the bacterial toxin complexes. Adv Appl Microbiol 58:169–183

Ganie SA, Bhat JA, Devoto A (2022) The influence of endophytes on rice fitness under environmental stresses. Plant Mol Biol 109(4–5):447–467. https://doi.org/10.1007/s11103-021-01219-8

Gao Y, Oppert B, Lord JC, Liu C, Lei Z (2012) *Bacillus thuringiensis* Cry3Aa toxin increases the susceptibility of Crioceris quatuordecimpunctata to Beauveria bassiana infection. J Invertebr Pathol 109(2):260–263. https://doi.org/10.1016/j.jip.2011.12.003

Gasulla F, Del Campo EM, Casano LM, Guéra A (2021) Advances in understanding of desiccation tolerance of lichens and lichen-forming algae. Plants (Basel) 10(4):807. https://doi.org/10.3390/plants10040807

Gu J, Ye R, Xu Y, Yin Y, Li S, Chen H (2021) A historical overview of analysis systems for *Bacillus thuringiensis* (Bt) cry proteins. Microchem J 165:10613. https://doi.org/10.1016/j.microc.2021.106137

Guarnizo N, Oliveros D, Murillo-Arango W, Bermúdez-Cardona MB (2020) Oligosaccharides: defense inducers, their recognition in plants, commercial uses and perspectives. Molecules 25(24):5972. https://doi.org/10.3390/molecules25245972

Guo W, Kain W, Wang P (2019) Effects of disruption of the peritrophic membrane on larval susceptibility to Bt toxin Cry1Ac in cabbage loopers. J Insect Physiol 117:103897. https://doi.org/10.1016/j.jinsphys.2019.103897

Guo X, Xu D, Li F, Bai J, Su R (2023) Current approaches on the roles of lactic acid bacteria in crop silage. Microb Biotechnol 16(1):67–87. https://doi.org/10.1111/1751-7915.14184

Hata EM, Yusof MT, Zulperi D (2021) Induction of systemic resistance against bacterial leaf streak disease and growth promotion in rice plant by *Streptomyces shenzhenesis* TKSC3 and *Streptomyces* sp. SS8. Plant Pathol J 37(2):173–181. https://doi.org/10.5423/PPJ.OA.05.2020.0083

Heil M, Bostock RM (2002) Induced systemic resistance (ISR) against pathogens in the context of induced plant defences. Ann Bot 89(5):503–512. https://doi.org/10.1093/aob/mcf076

Hernández-Melchor DJ, Mendarte C, Cortes-Pérez S, García-Sánchez C, López-Pérez PA (2023) The use of soil microorganisms in the biological control of insects: the application of microorganism toxins and their effects on agrosystems. In: Weevils: biology, behavior and ecology. Nova Science and Technology, pp 69–94

Huang Z, Liang X, Wang Y, Mo M, Qiu Y, Liu B (2023) Ginger blotches on Agaricus bisporus due to monoacetylphloroglucinol production by the pathogen pseudomonas 'gingeri'. Pest Manag Sci 79(12):5197–5207. https://doi.org/10.1002/ps.7725

Igiehon NO, Babalola OO, Aremu BR (2019) Genomic insights into plant growth promoting rhizobia capable of enhancing soybean germination under drought stress. BMC Microbiol 19:159. https://doi.org/10.1186/s12866-019-1536-1

Ismail S, Jiang B, Nasimi Z, Inam-Ul-Haq M, Yamamoto N, Danso Ofori A, Khan N, Arshad M, Abbas K, Zheng A (2020) Investigation of *Streptomycesscabies* causing potato scab by various detection techniques, its pathogenicity and determination of host-disease resistance in potato germplasm. Pathogens 9(9):760. https://doi.org/10.3390/pathogens9090760

Jacoby R, Peukert M, Succurro A, Koprivova A, Kopriva S (2017) The role of soil microorganisms in plant mineral nutrition—current knowledge and future directions. Front Plant Sci 8:1617. https://doi.org/10.3389/fpls.2017.01617

Jalmi SK, Sinha AK (2022) Ambiguities of PGPR-induced plant signaling and stress management. Front Microbiol 13:899563. https://doi.org/10.3389/fmicb.2022.899563

Khan RAA, Najeeb S, Mao Z, Ling J, Yang Y, Li Y, Xie B (2020) Bioactive secondary metabolites from *Trichoderma* spp. against phytopathogenic bacteria and root-knot nematode. Microorganisms 8(3):401. https://doi.org/10.3390/microorganisms8030401

Khan S, Srivastava S, Karnwal A, Malik T (2023) *Streptomyces* as a promising biological control agents for plant pathogens. Front Microbiol 14:1285543. https://doi.org/10.3389/fmicb.2023.1285543

Krug L, Morauf C, Donat C, Müller H, Cernava T, Berg G (2020) Plant growth-promoting Methylobacteria selectively increase the biomass of biotechnologically relevant microalgae. Front Microbiol 11:427. https://doi.org/10.3389/fmicb.2020.00427

Kumar J, Ramlal A, Mallick D, Mishra V (2021) An overview of some biopesticides and their importance in plant protection for commercial acceptance. Plants (Basel) 10(6):1185. https://doi.org/10.3390/plants10061185

Kumar C, Tomar A, Pandey S, Prasad MNV (2023) Plant growth promoting Rhizobacteria sustaining saline and metal contaminated soils. In: Prasad MNV, Chitranjan K (eds) Agroecological approaches for sustainable soil management. Willey, ISBN 978-11-1991-196-8, pp 437–455. https://doi.org/10.1002/9781119911999.ch19

Kushwah J, Somvanshi VS (2016) Photorhabdus: a microbial factory of insect-killing toxins. In: Microbial factories: biodiversity, biopolymers, bioactive molecules, vol 2. Springer India, New Delhi, pp 235–240. https://doi.org/10.1007/978-81-322-2595-9_15

Laanbroek HJ, Bär-Gilissen MJ, Hoogveld HL (2002) Nitrite as a stimulus for ammonia-starved Nitrosomonas europaea. Appl Environ Microbiol 68(3):1454–1457. https://doi.org/10.1128/AEM.68.3.1454-1457.2002

Lahlali R, Ezrari S, Radouane N, Kenfaoui J, Esmaeel Q, El Hamss H, Belabess Z, Barka EA (2022) Biological control of plant pathogens: a global perspective. Microorganisms 10:596. https://doi.org/10.3390/microorganisms10030596

Le Pioufle O, Ganoudi M, Calonne-Salmon M, Ben Dhaou F, Declerck S (2019) Rhizophagus irregularis MUCL 41833 improves phosphorus uptake and water use efficiency in maize plants during recovery from drought stress. Front Plant Sci 10:897. https://doi.org/10.3389/fpls.2019.00897

Lee CW, Efetova M, Engelmann JC, Kramell R, Wasternack C, Ludwig-Müller J, Hedrich R, Deeken R (2009) Agrobacterium tumefaciens promotes tumor induction by modulating pathogen defense in *Arabidopsis thaliana*. Plant Cell 21(9):2948–2962. https://doi.org/10.1105/tpc.108.064576

Liu Y, Feng J, Pan H, Zhang X, Zhang Y (2022a) Genetically engineered bacterium: principles, practices, and prospects. Front Microbiol 13:997587. https://doi.org/10.3389/fmicb.2022.997587

Liu Y, Han S, Yang S, Chen Z, Yin Y, Xi J, Liu Q, Yan W, Song X, Zhao F, Guo J, Liu X, Hao D (2022b) Engineered chimeric insecticidal crystalline protein improves resistance to lepidopteran insects in rice (*Oryza sativa* L.) and maize (Zea mays L.). Sci Rep 12(1):12529. https://doi.org/10.1038/s41598-022-16426-6

Llobregat B, González-Candelas L, Ballester AR (2024) Exploring the biocontrol capability of non-Mycotoxigenic strains of *Penicillium expansum*. Toxins (Basel) 16(1):52. https://doi.org/10.3390/toxins16010052

Masuda Y, Satoh S, Miyamoto R et al (2023) Biological nitrogen fixation in the long-term nitrogen-fertilized and unfertilized paddy fields, with special reference to diazotrophic iron-reducing bacteria. Arch Microbiol 205:291. https://doi.org/10.1007/s00203-023-03631-8

Maron PA, Sarr A, Kaisermann A, Lévêque J, Mathieu O, Guigue J, Karimi B, Bernard L, Dequiedt S, Terrat S, Chabbi A, Ranjard L (2018) High microbial diversity promotes soil ecosystem functioning. Appl Environ Microbiol 84(9):e02738–e02717. https://doi.org/10.1128/AEM.02738-17

Maruyama CR, Bilesky-José N, de Lima R, Fraceto LF (2020) Encapsulation of *Trichoderma harzianum* preserves enzymatic activity and enhances the potential for biological control. Front Bioeng Biotechnol 8:225. https://doi.org/10.3389/fbioe.2020.00225

Mehta PK, Yadav AK, Kumar D (2018) Secondary metabolites and functional food components: role in health and disease. In: Secondary metabolites from microorganisms. Nova Science Publishers Inc., pp 153–177

Meidani C, Savvidis A, Lampropoulou E, Sagia A, Katsifas E, Monokrousos N, Hatzinikolaou DG, Karagouni AD, Giannoutsou E, Adamakis IS, Ntalli NG (2020) The Nematicidal potential of bioactive *Streptomyces* strains isolated from Greek rhizosphere soils tested on *Arabidopsis* plants of varying susceptibility to *Meloidogyne* spp. Plants (Basel) 9(6):699. https://doi.org/10.3390/plants9060699

Mohanty P, Singh PK, Chakraborty D, Mishra S, Pattnaik R (2021) Insight into the role of PGPR in sustainable agriculture and environment. Front Sustain Food Syst 5:667150. https://doi.org/10.3389/fsufs.2021.667150

Moriyama H, Urayama SI, Higashiura T, Le TM, Komatsu K (2018) Chrysoviruses in *Magnaporthe oryzae*. Viruses 10(12):697. https://doi.org/10.3390/v10120697

Muhialdin BJ, Saari N, Meor Hussin AS (2020) Review on the biological detoxification of mycotoxins using lactic acid bacteria to enhance the sustainability of foods supply. Molecules 25(11):2655. https://doi.org/10.3390/molecules25112655

Mullis MM, Rambo IM, Baker BJ, Reese BK (2019) Diversity, ecology, and prevalence of antimicrobials in nature. Front Microbiol 10:2518. https://doi.org/10.3389/fmicb.2019.02518

Nascimento JD, Goncalves KC, Dias NP, de Oliveira JL, Bravo A, Polanczyk RA (2022) Adoption of *Bacillus thuringiensis*-based biopesticides in agricultural systems and new approaches to improve their use in Brazil. Biol Control 165:104792. https://doi.org/10.1016/j.biocontrol.2021.104792

Nazari M, Smith DL (2020) A PGPR-produced Bacteriocin for sustainable agriculture: a review of Thuricin 17 characteristics and applications. Front Plant Sci 11:916. https://doi.org/10.3389/fpls.2020.00916

Ngalimat MS, Mohd Hata E, Zulperi D, Ismail SI, Ismail MR, Mohd Zainudin NAI, Saidi NB, Yusof MT (2021) Plant growth-promoting bacteria as an emerging tool to manage bacterial rice pathogens. Microorganisms. 9(4):682. https://doi.org/10.3390/microorganisms9040682

Norton J, Ouyang Y (2019) Controls and adaptive management of nitrification in agricultural soils. Front Microbiol 10:1931. https://doi.org/10.3389/fmicb.2019.01931

Olorunsogbon T, Adesanya Y, Atiyeh HK, Okonkwo CC, Ujor VC, Ezeji TC (2022) Effects of *Clostridium beijerinckii* and medium modifications on acetone-butanol-ethanol production from switchgrass. Front Bioeng Biotechnol 10:942701. https://doi.org/10.3389/fbioe.2022.942701

Olson S (2015) An analysis of the biopesticide market now and where it is going. Outlooks Pest Manag 26(5):203–206. https://doi.org/10.1564/v26_oct_04

Peng G, Xia Y (2015) Integration of an insecticidal scorpion toxin (BjαIT) gene into *Metarhizium acridum* enhances fungal virulence towards *Locusta migratoria* manilensis. Pest Manag Sci 71(1):58–64. https://doi.org/10.1002/ps.3762

Peng Q, Yu Q, Song F (2019) Expression of cry genes in *Bacillus thuringiensis* biotechnology. Appl Microbiol Biotechnol 103(4):1617–1626. https://doi.org/10.1007/s00253-018-9552-x

Purkayastha GD, Mangar P, Saha A, Saha D (2018) Evaluation of the biocontrol efficacy of a Serratia marcescens strain indigenous to tea rhizosphere for the management of root rot disease in tea. PLoS One 13(2):e0191761. https://doi.org/10.1371/journal.pone.0191761

Qi J, Song Y, Liang J, Bai Y, Hu C, Liu H, Qu J (2021) Growth inhibition of Microcystis aeruginosa by sand-filter prevalent manganese-oxidizing bacterium. Sep Purif Technol 256:117808. https://doi.org/10.1016/j.seppur.2020.117808

Regaiolo A, Dominelli N, Andresen K, Heermann R (2020) The biocontrol agent and insect pathogen *Photorhabdus luminescens* interacts with plant roots. Appl Environ Microbiol 86(17):e00891–e00820. https://doi.org/10.1128/AEM.00891-20

Ruanpanun P, Chamswarng C (2016) Potential of actinomycetes isolated from earthworm castings in controlling root-knot nematode *Meloidogyne incognita*. J Gen Plant Pathol 82:43–50. https://doi.org/10.1007/s10327-015-0637-4

Saeed Q, Xiukang W, Haider FU, Kučerik J, Mumtaz MZ, Holatko J, Naseem M, Kintl A, Ejaz M, Naveed M, Brtnicky M, Mustafa A (2021) Rhizosphere bacteria in plant growth promotion, biocontrol, and bioremediation of contaminated sites: a comprehensive review of effects and mechanisms. Int J Mol Sci 22(19):10529. https://doi.org/10.3390/ijms221910529

Santi C, Bogusz D, Franche C (2013) Biological nitrogen fixation in non-legume plants. Ann Bot 111(5):743–767. https://doi.org/10.1093/aob/mct048

Saxena AK, Kumar M, Chakdar H, Anuroopa N, Bagyaraj DJ (2020) Bacillus species in soil as a natural resource for plant health and nutrition. J Appl Microbiol 128(6):1583–1594. https://doi.org/10.1111/jam.14506

Silva-Filha MHNL, Romão TP, Rezende TMT, Carvalho KDS, Gouveia de Menezes HS, Alexandre do Nascimento N, Soberón M, Bravo A (2021) Bacterial toxins active against mosquitoes: mode of action and resistance. Toxins 13:523. https://doi.org/10.3390/toxins13080523

Simons A, Alhanout K, Duval RE (2020) Bacteriocins, antimicrobial peptides from bacterial origin: overview of their biology and their impact against multidrug-resistant bacteria. Microorganisms 8(5):639. https://doi.org/10.3390/microorganisms8050639

Sree KS, Padmaja V (2008) Destruxin from *Metarhizium anisopliae* induces oxidative stress effecting larval mortality of the polyphagous pest *Spodoptera litura*. J Appl Entomol 132(2008):68–78. https://doi.org/10.1111/j.1439-0418.2007.01239.x

Suman J, Rakshit A, Ogireddy SD, Singh S, Gupta C, Chandrakala J (2022) Microbiome as a key player in sustainable agriculture and human health. Front Soil Sci 2:821589. https://doi.org/10.3389/fsoil.2022.821589

Ulzen J, Abaidoo RC, Mensah NE, Masso C, AbdelGadir AH (2016) *Bradyrhizobium* inoculants enhance grain yields of soybean and cowpea in northern Ghana. Front Plant Sci 7:1770. https://doi.org/10.3389/fpls.2016.01770

Valtierra-de-Luis D, Villanueva M, Berry C, Caballero P (2020) Potential for *Bacillus thuringiensis* and other bacterial toxins as biological control agents to combat dipteran pests of medical and agronomic importance. Toxins (Basel) 12(12):773. https://doi.org/10.3390/toxins12120773

Verma J, Kumar C, Sharma M, Shukla AC, Saxena S (2024) Exploitation of microbial consortia for formulating biofungicides, biopesticides and biofertilizers for plant growth promotion. In: Shukla AC (ed) Entrepreneurship with microorganisms. Elsevier. ISBN 978-0-443-19049-0. https://doi.org/10.1016/B978-0-443-19049-0.00019-0

Vishwakarma K, Kumar N, Shandilya C, Mohapatra S, Bhayana S, Varma A (2020) Revisiting plant-microbe interactions and microbial consortia application for enhancing sustainable agriculture: a review. Front Microbiol 11:560406. https://doi.org/10.3389/fmicb.2020.560406

Vodovar N, Vallenet D, Cruveiller S, Rouy Z, Barbe V, Acosta C, Cattolico L, Jubin C, Lajus A, Segurens B, Vacherie B, Wincker P, Weissenbach J, Lemaitre B, Médigue C, Boccard F (2006) Complete genome sequence of the entomopathogenic and metabolically versatile soil bacterium *Pseudomonas entomophila*. Nat Biotechnol 24(6):673–679. https://doi.org/10.1038/nbt1212

Vu NT, Oh CS (2020) Bacteriophage usage for bacterial disease management and diagnosis in plants. Plant Pathol J 36(3):204–217. https://doi.org/10.5423/PPJ.RW.04.2020.0074

Weselowski B, Nathoo N, Eastman AW et al (2016) Isolation, identification and characterization of *Paenibacillus polymyxa* CR1 with potentials for biopesticide, biofertilization, biomass degradation and biofuel production. BMC Microbiol 16:244. https://doi.org/10.1186/s12866-016-086

Yao X, Guo H, Zhang K, Zhao M, Ruan J, Chen J (2023) Trichoderma and its role in biological control of plant fungal and nematode disease. Front Microbiol 14:1160551. https://doi.org/10.3389/fmicb.2023.1160551

Zapaśnik A, Sokołowska B, Bryła M (2022) Role of lactic acid bacteria in food preservation and safety. Food Secur 11(9):1283. https://doi.org/10.3390/foods11091283

Zeffa DM, Perini LJ, Silva MB, de Sousa NV, Scapim CA, Oliveira ALM, Amaral Júnior ATD, Azeredo Gonçalves LS (2019) Azospirillum brasilense promotes increases in growth and nitrogen use efficiency of maize genotypes. PLoS One 14(4):e0215332. https://doi.org/10.1371/journal.pone.0215332

Zhang X, Li S, Li X, Song M, Ma S, Tian Y, Gao L (2023) Peat-based hairy root transformation using *Rhizobium rhizogenes* as a rapid and efficient tool for easily exploring potential genes related to root-knot nematode parasitism and host response. Plant Methods 19(1):22. https://doi.org/10.1186/s13007-023-01003-3

Zhu L, Huang J, Lu X, Zhou C (2022) Development of plant systemic resistance by beneficial rhizobacteria: recognition, initiation, elicitation and regulation. Front Plant Sci 13:952397. https://doi.org/10.3389/fpls.2022.952397. Erratum in: Front Plant Sci 2023 Jan 19; 13: 1118073

Chapter 19
Bioactive Metabolites Produced by Microorganisms and Plants with Potential Application in Agriculture and Medicine

Mohd Ali, Deepak Kumar, Sandeep Kour, Parkirti, Roohi Sharma, Vikram, Rajesh Kumari Manhas, Harish Changotra, Puja Ohri, Nitika Kapoor, and Renu Bhardwaj

Abstract Different secondary metabolites (SMs) with novel biological properties, such as antifungal, antibacterial, antiplasmodium, cytotoxic, aggregation, antiplatelet, immunostimulant, enzyme-inhibiting, algicide, antitumor, anticancer, antiviral, and phytotoxic activities, are produced by microorganisms, algae, lichens, and plants. Secondary metabolites play an important role in interacting with the environment whereas, primary metabolites are present throughout the nature and are actively involved in the growth and development of plants. Terpenes are the largest class of secondary metabolites found in plants and are crucial for defence mechanisms against diseases and herbivores. Alkaloids, phenolics, and diterpenes are examples of plant secondary metabolites that have significant anticarcinogenic potential. Carotenoids serve as a pigment to absorb light during photosynthesis and offer photoprotection. Due to their ability as an allelopathic agent and phagostimulant, carotenoids are also effective in interacting with other organisms. Similarly, flavonoids are important for UV filtration, nitrogen fixation, flower coloration, and defences against a variety of illnesses. Plants are also protected by these secondary metabolites from various biotic and abiotic stressors. However, a large number of bioactive metabolites are also present in microorganisms, though not all of them. Alkaloids, terpenoids, polyketides, isoprenoids and non-isoprenoids, quinones, and

M. Ali · D. Kumar · S. Kour · Parkirti · R. Sharma · Vikram · P. Ohri
Department of Zoology, Guru Nanak Dev University, Amritsar, Punjab, India

R. K. Manhas
Department of Microbiology, Guru Nanak Dev University, Amritsar, Punjab, India

H. Changotra
Department of Molecular Biology and Biochemistry, Guru Nanak Dev University, Amritsar, Punjab, India

N. Kapoor · R. Bhardwaj (✉)
Department of Botanical and Environmental Sciences, Guru Nanak Dev University, Amritsar, Punjab, India

M. N. V. Prasad et al. (eds.), *Biotoxins*,
https://doi.org/10.1007/978-3-031-75309-1_19

other secondary metabolites are abundant in marine microorganisms. These metabolites work to reduce inflammation, kill germs, and fight tumors. Due to their antifungal and antibacterial activity against numerous diseases, secondary metabolites from fungus have significant medicinal importance. This chapter therefore offers an insight into bioactive metabolites produced by microorganisms and plants with their potential application in agriculture and/or medicine.

Keywords Bioactive metabolites · Metabolites against abiotic stress · Metabolites against biotic stress · Metabolites against plant pathogens · Nitrogen compounds · Phenolics · Secondary metabolites · Terpenes

19.1 Introduction

Substances that are involved in the metabolic processes of an organism are known as metabolites. They are the byproducts of metabolism and often include small molecules (Bano et al. 2023; Tiwari and Rana 2015; Liu et al. 2023). Bioactive metabolites can be identified in a variety of marine species, as well as in microorganisms (Sedeek et al. 2022), fungi (Wang et al. 2023), plants (Fayaz et al. 2023), and animals (Dobrijević et al. 2023) and are mainly classified into primary and secondary metabolites (Twaij and Hasan 2022). Plant growth regulators (PGR) may serve as primary, secondary, or intermediary metabolites (Assaf et al. 2022). Terpenoids (isoprenoids) such as gibberellins, plastoquinone, brassinosteroids, ubiquinone, and carotenoids, which are primary metabolites in plants are also called secondary metabolites because of their ecological roles in defence against other organisms (Gutensohn et al. 2022). Primary metabolites are produced by combining the end result of catabolic reactions (Wang et al. 2022). These primary metabolites in turn produce essential metabolites that are important for structural and functional aspects of organisms. The precursor for secondary metabolites is provided by the primary metabolic pathway hence, factors impacting primary metabolites also affect secondary metabolites (Khan et al. 2018). Bioactive metabolite production is highly dependent on weather conditions (Singh et al. 2023a, b; Khan et al. 2017). Seasonal variation, intra- and inter-competitions, an organism's genotype, genetic regulation and gene expression, environmental factors, ecology, and biochemistry all affect their production (Afshar et al. 2022; Karuppusamy 2009; Gobbo-Neto and Lopes 2007). Plants have developed a wide range of defence mechanisms that help them handle different stress situations and encourage the buildup of SMs. Such mechanisms include receptor and sensor activation, which enables plants to recognize harmful signals and take defensive action to defend themselves (Jan et al. 2021). SMs also affect how plants react to different biotic and abiotic stresses by affecting their color, flavor, and aroma. Besides their important role in plant stress physiology and stress adaptation, SMs have significant uses in medicine, nutrition, and cosmetics. SMs have been used for medicinal purposes since ancient times and are the basis for many pharmaceutical products today (Humbal and Pathak 2023). About

80% of the world's population receives their primary health care from plant-based chemical constituents (Khan and Ahmad 2019). So, the only source of diverse bioactive chemicals for the synthesis of novel medications to treat a number of diseases is plant SMs, the metabolites are rarely synthesized outside the plant tissues or during developmental stages (Sampaio et al. 2016). Plants are exposed to numerous stressors, elicitors, or signaling molecules, which cause them to accumulate SMs. Abiotic and biotic elements, such as physical or chemical agents, as well as microbial elements, may both function as such elicitors. In the end, these elicitors can encourage plants to increase the production of SMs (Halder et al. 2019; Thakur et al. 2019). Therefore, this chapter takes into account the bioactive metabolites produced by microorganisms and plants with potential applications in agriculture and medicine.

19.2 Classification of Bioactive Metabolites

Plants and microorganisms have a wide variety of metabolites that are involved in their metabolic pathways under normal and stressful conditions. These metabolites are further divided into primary and secondary metabolites. Primary metabolites are responsible for the main metabolic pathways that are essential for the survival of plants while, secondary metabolites are involved in developing the ability of the plants to interact with their environment (Jha and Mohamed 2022; Santra et al. 2022). Secondary metabolites are metabolic intermediates or products that are not required for organisms' growth or survival but are required for organisms' interaction with their environment or are created in reaction to stress (Böttger et al. 2018). Secondary metabolites are engaged in defence over other species, symbiotic improved resistance to abiotic stresses, metal transport, attractive for other organisms to boost fertilization, or seed dispersal and are employed as stressful sexual hormones (Tiwari and Rana 2015). Terpenoids, phenolics, nitrogen-containing (alkaloids), and sulpur-containing chemicals are the most common secondary metabolites (Kumar et al. 2015). Phenolics are heterogeneous groups of natural substances having aromatic rings with one or more hydroxyl groups (Graça et al. 2005). Phenolics are involved in growth, pigmentation, reproduction, resistance to pathogens, plant–animal and plantmicrobe interactions, anti-oxidation, metal chelation, as signaling agent, and UV light screening (Lattanzio et al. 2009). Phenolics are majorly classified as flavonoids and non-flavonoids. Flavonoids are the largest class of phenolics constituting about 4500 different representatives (Tiwari and Rana 2015). They have antioxidant, antibacterial, antiallergic, antiviral, anti-inflammatory, visual attractiveness, light screening, photoreception, feeding repellents, and vasodilation properties (Cushnie and Lamb 2005). Non-flavonoids are subclasses of phenolic compounds. Their major dietary source is wine. They are colorless, but they enhance and sooth the color and flavor of red wine and show biological activities. Triterpenoids (C30) are composed of approximately 20,000 isoprene units. It has numerous medicinal and biological effects, including antiviral,

anti-inflammatory, anticancer, and antifungal properties, as well as a role in plant growth (Cmoch et al. 2008). The condensation of two isoprene units (C10) produces monoterpenoids. Monoterpenoids can be used in industries as medications, pesticides, flavoring agents, and perfumes (Mahmoud and Croteau 2002). Sesquiterpenoids (C15) are lipophilic colorless chemicals generated by farnesyl pyrophosphate (FPP) condensation of three isoprene units in the endoplasmic reticulum (Yu and Utsumi 2009). Triterpenoids (C30) are composed of approximately 20,000 isoprene units. It has numerous medicinal and biological effects, including antiviral, anti-inflammatory, anticancer, and antifungal properties, as well as a role in plant growth (Cmoch et al. 2008). Tetraterpenoids (C40) are carotenoids generated by the elimination of eight isoprene molecules (da Silveira Vasconcelos et al. 2020). Alkaloids are a structurally diverse collection of bioactive metabolites that include nitrogen (Roberts et al. 2010). Their names finish in "-ine or -in" suffixes and are created by bacteria, mammals, fungi, and plants. They are secondary metabolites with antibacterial, antioxidant, antitumor, anti-inflammatory, anti-herbivore, antimalarial, and anticancer properties (Kabera et al. 2014). Monoterpenoid alkaloids (monoterpenoidindole alkaloids) are a significant group of alkaloids, with over 3000 identified alkaloids (Pan et al. 2016). They are antifungal, antibacterial, antimalarial, anticancer, and anti-gastric ulcer (O'Connor and Maresh 2006). The diterpenoid alkaloids are a diverse group of bioactive metabolites (Owens et al. 2019). They are categorized as C18- (Lappaconitine), C19- (Lycaconitine), and C20- (aluminum chloride) diterpenoid alkaloids, all of which have the same general structure (Ren et al. 2017) (Fig. 19.1).

19.3 Plant Bioactive Secondary Metabolites

There are over 4,000,000 plant species on earth, and each one makes hundreds of thousands different metabolites (Humbal and Pathak 2023). A tiny molecule created as an interim result of metabolic secondary metabolite is known as a metabolite. A rising number of these compounds were discovered by means of improvements in scientific methods, notably chromatography in the middle of the twentieth century, which finally gave rise to the profession of physiochemists (Hussein and El-Anssary 2019). Diverse secondary metabolites are thought to be abundant in plants, or in plant sources. Secondary metabolites are vital for the survival of producing organic secondary metabolites but are not necessary for the continuation of life (Selim et al. 2021). Plant secondary metabolites are produced in unique routes and might vary in production sites depending on the component and the plant type (Li et al. 2020). Plant metabolites are made up of both simple elements such as proteins, nucleic acids, and polysaccharides as well as complex molecules such as sugars, amino acids, tricarboxylic acids, and Kreb's cycle intermediates (Kandar 2021). Plant metabolites are engaged in the essential functions of life including division of cells, develop other plant metabolites, oxygen consumption, storage facilities, and procreation which are based on their activities in core metabolic pathways (Kumar et al.

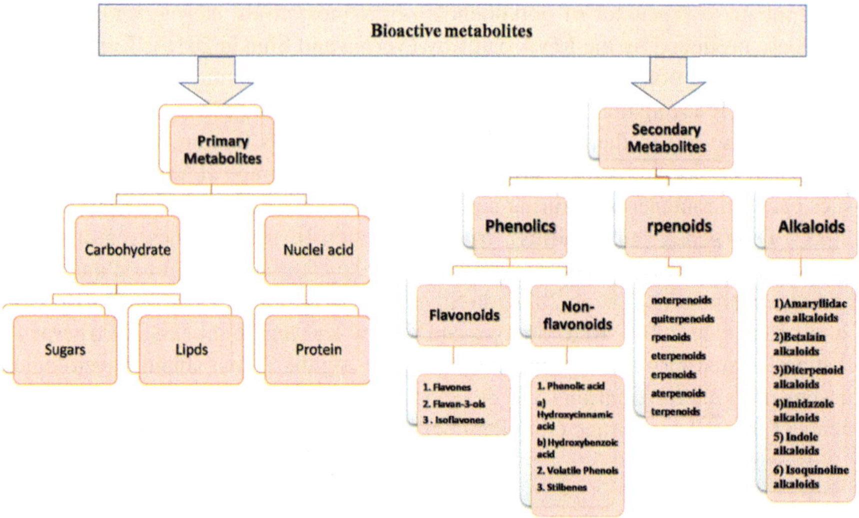

Fig. 19.1 Classification of bioactive metabolites

2022). The fundamental components of crucial procedures including glycolysis, the Krebs or citric acid cycle, photosynthesis, and related pathways are found in plant metabolites (Kumar and Dubey 2019). When the plant is exposed to a demanding and changing environment, these secondary metabolites act in defense mechanism system (Jan et al. 2021). Also, the primary metabolites create the secondary metabolite precursors (Erb and Kliebenstein 2020). The medicinal qualities of plants can be understood based on their chemical makeup. Phenolics, terpenoids, and nitrogen-containing chemicals, which are all involved in environmental adaptation and stress tolerance are the three groups into which plant secondary metabolites are primarily divided (Boncan et al. 2020).

19.3.1 Terpenes

Terpenes make up the biggest category of secondary metabolites found in plants and about 10,000 of them have been purified till date (Thirumurugan et al. 2018). The major scaffolds of several structurally distinct terpenoids are created by stereospecific carbocation cyclization, rearrangement, and elimination processes which control the chemistry of a variety of terpenoids. Terpene biosynthesis occurs in plants via two distinct pathways, one in the plastids and the other in the cytoplasm of the plant cell. Mevalonate (MVA) and methylerythritol 4-phosphate (MEP) pathways mediate terpene production (Wölwer-Rieck et al. 2014). Mono and diterpenes, isoprene, carotenoids, phytol, tocopherols, plastoquinone, phylloquinone, and other substances are produced via the MEP route. On the other hand, isopentenyl

diphosphate (a precursor of polyphenols), brassinosteroids, sterols, and sesquiterpenes are produced by the MVA pathway (Verma and Shukla 2015). Terpenes have a variety of functions in plants including plant growth regulation, pollinator attraction, and defense against various diseases (Divekar et al. 2022; Das et al. 2013). Terpenes also aid in thermotolerance through processes that are not exclusive to plants that produce terpenes (Sasaki et al. 2007). The storage of terpene in plant tissues occurs spontaneously during contact with certain insects (Tlak Gajger and Dar 2021). Diterpenes (gibberellins), triterpenes (sterol), tetraterpenes (carotenoids), and sesquiterpenes (ABA) are significant terpenes that are crucial to plant growth and development (Liu et al. 2022; da Silveira Vasconcelos et al. 2020; Bröker et al. 2018; Kermode 2005). Red, orange, and yellow carotene pigments are triterpenes that act as auxiliary hues in the course of photosynthesis and shield physiological tissues from photooxidation (Xue et al. 2019). Sterols and terpenes are key components of the cell membrane (Kolekar et al. 2019).

19.3.2 Phenolics

Around 10,000 different compounds make up the heterogeneous family of molecules known as phenols, which are solvable in both aqueous and organic solvents (Chiorcea-Paquim et al. 2020). Plants may make a variety of secondary compounds containing phenol groups, which are hydroxyl functional groups on aromatic rings. Flavonoids, tannins, styryl pyrones, aryl pyrones, lignin, lignans, stilbenes, and coumarins are some of the compounds that make up phenol compounds (Crozier et al. 2006). The formation of phenolic compounds involves the shikimic and malonic acid pathways (Dhaniaputri et al. 2022). Phenylalanine serves as a frequent intermediate in the shikimic acid pathway, which converts precursors from the pentose phosphate route and glycolysis into a variety of aromatic amino acids (Bartwal et al. 2013). It is well established that phenolics play vital roles in plants including those related to growth, reproduction, and resistance to biotic and abiotic stresses. Due to their redox properties, phenolic compounds can serve as antioxidants and detoxify singlet oxygen (Dumanović et al. 2021). Photosynthesis, protein synthesis, enzyme activity, and food absorption are some of the other crucial functions of phenolics in plants (Kołton et al. 2022). Plants employ phenolic compounds in order to protect themselves from diseases and parasites like worms such as nematodes (Kumar et al. 2023; Ali and Ohri 2023). These are among the most typical plant-derived substances discovered in veggies, fruits, cacao and tea cultivation, and other plants. They have a number of physiological properties that are beneficial for one's health, including those that are anti-inflammatory, antioxidative, and anticarcinogenic (Ahammed et al. 2023; Martău et al. 2023). They also guard against oxidative damage and associated diseases (Hajam et al. 2023).

19.3.3 Nitrogen Compounds

The class of compounds containing nitrogen includes around 12,000 substances (Gavhale et al. 2023). Botanical secondary metabolites with nitrogen in their molecular makeup include alkaloids, cyanogenic glucosides, and nonprotein amino acids (Al-Khayri et al. 2023). They are highly sought-after due to their curative properties. A few amino acids that include lysine and tyrosine are necessary for the formation of some alkaloids (Rajak et al. 2023). Alkaloids include substances such as nicotine, purine alkaloids, tropane alkaloids, terpenoids, indole alkaloids, and benzylisoquinoline alkaloids. Terpenoid indole alkaloids are made possible by the amino acid tryptophan. Several terpenoid indole alkaloids defend plants from biotic and abiotic stresses (Mohammed et al. 2021). Plants with medicinal effects, like *Catharanthus roseus* and *Rauwolfia serpentina*, are abundant sources of terpenoid indole alkaloids (Kumar et al. 2018; Pandey et al. 2014). This category includes cough suppressants such as codeine (Jahan et al. 2015), painkillers in the form of morphine (Sousa et al. 2022), antimicrobials such as berberine and sanguinarine (Khameneh et al. 2019), and medications that relax muscles such as papaverine (Ashrafi et al. 2023). Also, the building blocks for the synthesis of purine alkaloids are purine nucleotides (Ashihara et al. 2013). For instance, *Coffea arabica* plants are known to have pure purine alkaloid metabolism (Zheng and Ashihara 2004). In addition to alkaloids, plants also contain glycosides known as glucosinolates which are nitrogenous protecting compounds and cyanogenic glycosides (Tiku 2018). It is sometimes called mustard oil glycosides which degrade to release chemicals that are volatile (Chhajed et al. 2020). Plants such as radish, broccoli, and cabbage have a distinctive flavor and scent because of the presence of glucosinolates (Afsar et al. 2021; Bell et al. 2018) (Fig. 19.2).

19.4 Secondary Metabolites Produced by Microorganism

Microorganisms are beneficial in pharmaceuticals and agricultural fields. Several strains of beneficial microbes are useful in disease management, pest management, crop production, and quality and quantity enhancer of crops. These microbes synthesize secondary metabolites (SMs) that have several beneficial effects. SMs of microbes are complex biological compounds having low molecular mass and are produced after growth phase. SMs production in microbes starts during stress conditions. These metabolites do not play any vital role for the growth and reproduction of microorganisms themselves and only particular microbes synthesize a particular SM. Their production depends on growth condition factors and thus can be enhanced by optimizing growth condition factors such as light, temperature, pH, and medium composition (Kumar et al. 2022). Several microbial SMs have been reported in recent years. These SMs interfere with the potential of pathogens to cause diseases

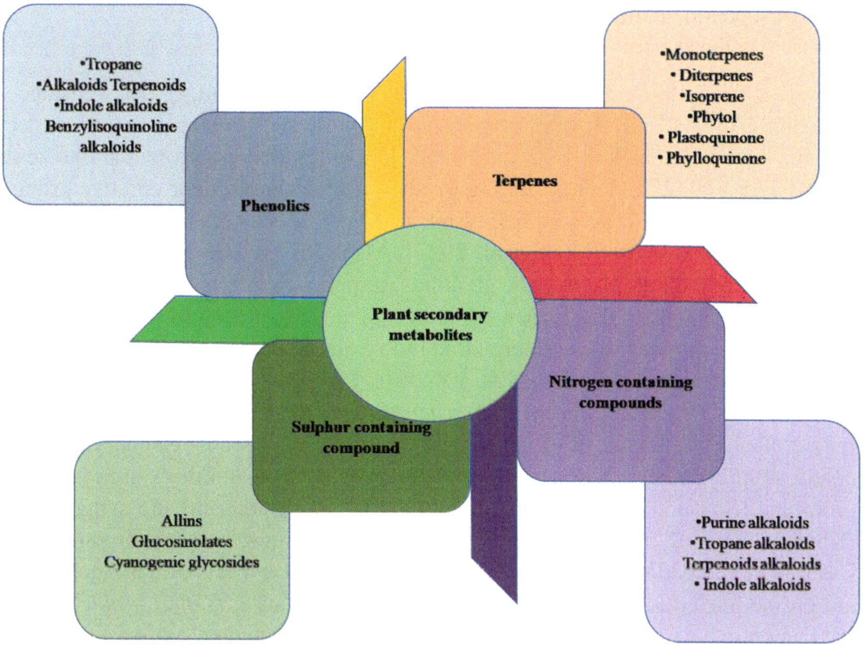

Fig. 19.2 Plant bioactive secondary metabolites

by inhibiting their growth or by enhancing the defence system. These compounds are produced by plants, animals, algae, fungi, and bacteria also.

19.4.1 Bacteria

All bacteria produce a diverse range of small bioactive molecules having agricultural potential. There are many agriculturally important bacteria such as *Pseudomonas* spp. that can produce a number of secondary metabolites, like 2,4-diacetylphloroglucinol (DAPG). DAPG-producing *Pseudomonas* spp. can protect plants from numerous pathogens and thus are efficient by used as biocontrol agents. Similarly, a polyketide pyoluteorin, which is a secondary metabolite produced by *Pseudomonas protegens* Pf-5, acts as a strong antibacterial agent against *Pantoeaananatis*. Where *P. ananatis* is a Gram-negative bacterium responsible for severe losses in crop plants such as honeydew melon, onion, pineapple, and maize. In maize, it causes brown rot disease in these plants, causes premature leaf drying, reduced kernel length, decreased weight, and shortened growth cycle. Also, it causes water-soaked leaf lesions that eventually become straw colored and necrotic (Guo et al. 2022). Other secondary metabolites produced by *Pseudomonas* spp. include phloroglucinols, pyrrolnitrin, mupirocin, hydrogen cyanide, oxyvinylglycines,

rhizoxins, peptide antibiotics, pyoluteorin, 2,5-dialkylresorcinols (DAR), and phenazines. All these compounds are important in agricultural and pharmacological fields. In another study, secondary metabolites of *Trichoderma* spp. were tested for nematicidal and antibacterial activity against *Meloidogyne incognita* and bacterial strains *Xanthomonas compestris* and *Ralstonia solanacearum,* respectively. While for fungal pathogens, *Trichoderma* spp. produces various SMs such as trichothecenes, lactone cremenolide, anthraquinones, stigmasterol, koninginins A–E, azaphilones, pyridines, peptaibols, and epipolythiodioxopiperazines (ETPs). Similarly, antifungal secondary metabolites are active biofungicides against fungal phytopathogens. The secondary metabolites produced by *Bacillus* genus have potential to act against several phytopathogens. Some main SMs of this genus are bacircine, halobacillin, pumilacidin, bacillomycin, mycosubtilin, zwittermicin A, macrolactin A, cyclo (L-Pro-L-Leu), tryptamine, and bacilysin. All these compounds act as antifungal, antibacterial, and growth promoters also. Some SMs from actinomycetes including staurosporine, ivermectin, actinocarbasin, lignin peroxidase, Baeyer-Villiger monooxygenases, etc. act as biocatalysts, antiparasitic agents, and antifungal and anticancer agents. Similarly, PGPR-based secondary metabolites such as exopolysaccharides (EPS), volatile organic compounds (VOCs), and osmoprotectants/compatible solutes were found significant in ameliorating some abiotic stresses in crops.

19.4.2 Fungi

Several beneficial fungi biosynthesize the fungal secondary metabolites. These metabolites positively affect plant metabolism, control phytopathogens, increase plant resistance against various diseases, enhance plant defence system, and also enhance plant growth. Fungal metabolites such as cerinolactone and 5-hydroxy-1-(3-methyl-3-buten-1-ynyl)-7-oxabicyclo [4.1.0]-hept-3-en-2-one are beneficial in plant growth. There are also many antifungal secondary metabolites that are produced by fungal endophytes such as cytochalasin alkaloids, terpenoids, alkaloids, polyketides, and other miscellaneous compounds that are beneficial in agricultural as well as in pharmaceutical sector. While griseofulvin and trichothecene are important fungal metabolites from agricultural point of view (Xue et al. 2019). A study was also conducted where the secondary metabolites of fungi *Colletotrichum gloeosporioides* and *Chaetomium globosum* were found beneficial for plant growth in agricultural cultivation. In a study on marine endophytic fungi, maraniol which is a secondary metabolite was found effective against cancer cells. Some metabolic compounds of *Alternanria* spp. given in a study include propanenitrile, proparglylamine, 2-butyne, cyclobutanecarbonitrile, benzeneethanamine, coumarin, etc. having properties such as antimicrobial, transcription, repression activity, and antiviral activity. Some bioactive SMs reported from *F. oxysporium* involve compounds such as fumonisins, alkaloids, cyclic depsipeptides, pyran and furan derivatives, xanthone derivatives, Glucosylceramides (GCs), jasmonates, and anthranilates.

Aspergillusniger, the most common and important species of fungus, includes a number of SMs such as pyranone derivatives, pyranonigrin derivatives, six amides, cyclopeptides, and sterols and thus has potential to act as therapeutic agents, vitamins, pesticides, and medicines and has various other applications in agriculture and food industry.

19.4.3 Lichens

Lichens rely upon a symbiotic result of an autotrophic photosynthetic component and a heterotrophic fungus with dominant role. The photosynthetic component may belong to Cyanobacteria or Chlorophyta division. The primary function of secondary metabolite in lichens is to protect thallus from various stresses. The main pathway for secondary metabolism is acetate-polymalonate pathway. While most other metabolites are of fungal origin and are derived mainly from polyketide pathway and a small amount also comes from shikimate and mevalonate biosynthetic pathways. Secondary metabolites such as carotenoids, steroids, and terpenes are synthesized via mevalonic acid pathway. While other secondary metabolites include usnolic acid, physodic acid, and physodalic acid. While usnic acid and atranorin are two secondary metabolites found effective against saprotrophic fungi (Furmanek et al. 2022). Gyrophoric acid and other metabolites such as salazinic acid, vulpinic acid, ceratinalone, ramalin, tumidulin, parmosidone, flavicansone, ohycsiosporin, and atranorin modulate several pathways related to various biomedical disorders and disease. Lichens have a number of SMs, for example, 1,2,8-trihydroxy-5-methoxy-3-methylxanthone, ethyl hematommate, evernic acid, gyrophoric acid, methyl hematommate, norstictic acid, orcinol, physodic acid, perlatolic acid, picrolichenic acid, parietin, scabrosin, sekikaic acid, and zeorin.

19.4.4 Algae

Algae possess essential secondary metabolic compounds important from pharmacological and agricultural view. The bioactive secondary metabolites present in algae contain various active biological phytochemicals such as sterols, proteins, heterocyclic carbons, terpenes, fatty acids, proteins, and polysaccharides. Thus, these metabolites have antifungal, antibacterial, antitumor, anti-inflammatory, and antioxidant capacity. Seaweeds are reported to have various secondary metabolites such as terpenes, steroids, lipids, acetogenins, phlorotannins, aromatic-like compounds, and other polymeric compounds active in response to various viruses, bacteria, insects, and other microbes. The microalgae, *Chlorella vulgaris*, was also used to alleviate drought stress in broccoli plants as algae elevated the secondary metabolites such as total flavonoids and phenols in response to drought stress. Some other SMs of marine algae include phlorotannins, fucols, phlorethols, fucophlorethols, eckols,

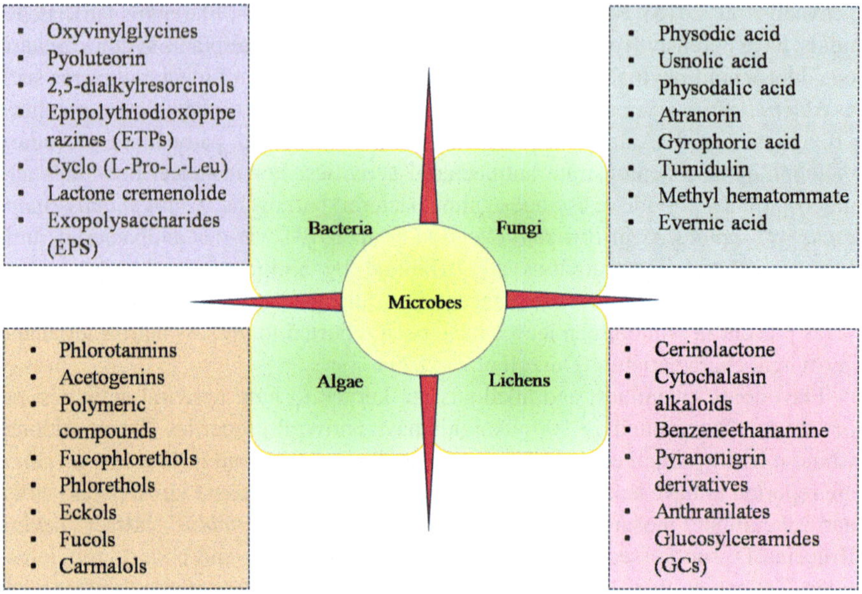

Fig. 19.3 Secondary metabolites produced by microorganism

and carmalols, which may act as antiallergens, anti-inflammatory agents, antioxidants, anticancer, and antimicrobials (Fig. 19.3).

19.5 Role of Bioactive Secondary Metabolites against Biotic Stress

Plants are exposed to a variety of pathogenic microorganisms such as bacteria, fungi, nematodes, herbivores, and viruses in the environment. Being sessile, they cannot physically avoid these biotic stressors; however, they have evolved various tactics to defend themselves against these notorious disease-causing agents. One such tactic is the production of secondary metabolites such as terpenoids, saponins, alkaloids, flavonoids, and steroids having antimicrobial properties (Ashraf et al. 2018; Anjali et al. 2023). In response to biotic stress, plants show elevated biosynthesis of secondary metabolites (Shah and Smith 2020). For instance, plants infected by the fungal pathogen, *Lupinus angustifolius,* showed significantly enhanced levels of phenolic compounds (Verma and Shukla 2015). Secondary metabolites are activated via the action of MAMPs (microbe-associated molecular patterns), which are perceived by PRRs (pattern recognition receptors) (Ahuja et al. 2012; Ranf 2018). Antimicrobial characteristics of various secondary metabolites inhibit proliferation of pathogens in the apoplast of plants. Antifungal activity of saponins (α-tomatine in tomato plants) caused oxidative burst in *Fusarium oxysporum*

(Zaynab et al. 2018). Antimicrobial actions of phytoalexins, like nonvolatile terpenoids, have defensive roles in plants (Block et al. 2019). Rice plants synthesize various diterpenoid phytoalexins that provide resistance against fungal pathogens such as *Magnaporthe oryze* and *Pyricularia oryzae*, which are responsible for rice blast (Toyomasu 2008; Azizi et al. 2019). Apart from fungal pathogens, secondary metabolites also demonstrate antibacterial activities. For instance, phenolics and glucosinolates provide resistance against bacterial pathogens, *Xanthomonas campestris* pv. *campestris* in *Brassica* (Velasco et al. 2013). Various antibacterial, antifungal, and antiviral activities are exhibited by anthraquinone-like secondary metabolites produced by certain plant extracts (Shang et al. 2019). Hydroxycinnamic acids present in wintergreen leaves have been reported to possess antibacterial and antifungal characteristics (Darzuli et al. 2021).

Flavonoids, phenolics, and alkaloids are known to have antiviral activities. As per the studies, around 18,500 alkaloids have antiviral properties in conventional Chinese herbs (Anjali et al. 2023). Bruceine-D alkaloid found in *Brassica javanica* are reported to have ameliorative effects against tobacco mosaic virus, potato virus, and cucumber mosaic virus. *Chenopodium amaaranticolor* extract having Bruceine-D, suppressed > 90% of cucumber mosaic virus and potato virus infections (Zaynab et al. 2018). Flavonoids are also produced in response to plant parasitic nematode infections. Glyceollin, a flavonoid, released against *Meloidogyne penetrans* reduced crop damage in soyabean (Chin et al. 2018). Flavonols, myricetin, kaempferol, and quercetin deterred and reduced *M. incognita* infective juvenile populations, even at micromolar concentrations (Wuyts et al. 2006). Seed and leaf extracts of *Solanum elaeagnifolium* were found to contain quercetin and kaempferol that displayed nematicidal potential (Balah and Abdel Razek 2020). Besides their antibacterial, antifungal, antiviral, and nematicidal potential, secondary metabolites also have anti-herbivore potential. Several studies have reported their insecticidal activities. Secondary metabolites like flavonoids (vitexin) in the leaf extracts of *Basella alba* restricted the growth of *Spodoptera litura* larvae (Aboshi et al. 2018). In another study, enhanced levels of quercetin were found in *B. oleracea* (cabbage) infected with *Pieris brassicae* and *Phyllotreta nemorum* (Kovalikova et al. 2019). Examples of different secondary metabolites having ameliorative roles against pathogenic microbes in plants are listed in Table 19.1.

Table 19.1 Role of secondary metabolites against plant pathogens

Secondary metabolite	Pathogen	Plant	Effect	References
Catechin, chrysin, isorhamnetin *italium*, naringin, kaempferol and quercetin	*Fusarium oxysporum* and *Alternaria alternata*	*Flourensia microphylla*	Repressed mycelial growth	Carrillo-Lomelí et al. (2022)
Flavonoids	*Penicillium italicum*	*Commelina erecta* and *Richardia brasiliensis*	Reduced total lipid and ergosterol content in the cell. Disrupted the membrane integrity of fungal pathogen	Guo et al. (2022)
Laburnetin	*Stemphylium vesicarium*	*Retama raetam*	Caused more than 50% reduction in fungal growth	Soriano et al. (2022)
Chrysoeriol	*Staphylococcus aureus*	*Artemisia rupestris*	Increased membrane permeability, inhibited NorA pump, reduced plant damage	Lan et al. (2021)
Quercetin	*Oxycarenus hyalinipennis*	*Euphorbia nivulia*	Enhanced 3rd instar larval mortality	Younus et al. (2021)
Terpenoids, flavonoids, and alkaloids	*Aspergillus* spp.	*Carthamus tinctorius*	Reduced fungal growth	Hussain et al. (2021)
Hydroxyl −3-methoxyflavone and chlorogenic acid	*Meloidogyne incognita*	*Solanum elaeagnifolium*	Enhanced infective juvenile mortality, decreased the number of egg masses and galls, increased growth of egg plant	Balah and Abdel Razek (2020)
Flavonols and Flavones	*M. incognita*	*Tagetes patula*	Elevated infective juvenile mortality	Bano et al. (2020)
Quercetin	*Colletotrichum gloeosporioide*	*Mangifera indica*	Subdued germination of conidia and growth of pathogenic fungi	Sudheeran et al. (2020)
Flavonoids	*Macrosiphum rosaeformis*	*Carica papaya*	Enhanced aphid mortality under laboratory and field conditions	Gupta et al. (2020)

(continued)

Table 19.1 (continued)

Secondary metabolite	Pathogen	Plant	Effect	References
Flavonoids, terpenoids, saponins, steroids, coumarins, and cardiac glycosides	*Alternaria solani*	*Azadirachta indica*	Decreased pathogen growth and increased crop yield	Chohan et al. (2019)
Rutin	*Rhizoctonia solani* and *F. culmorum*	*Musa paradisiaca*	Decreased the growth of pathogenic fungi	Behiry et al. (2019)
Flavonoids	*Spodoptera litura*	*Moringa oleifera*	Reduced larval biomass and enhanced plant growth	Tridiptasari et al. (2019)
Apiofuranosyl Xylopyranoside	*S. aureus*	*Graptophyllum grandulosum*	Disrupted cell membrane and caused lysis of bacterial cell	Tagousop et al. (2018)
Vitexin	*S. litura*	*Basella alba*	Decreased larval growth	Aboshi et al. (2018)

19.6 Role of Bioactive Secondary Metabolites against Abiotic Stress

Abiotic stress, which reduces average yields for the majority of primary agricultural crops by more than 50%, is the main factor in crop loss worldwide. Environmental stressors such as drought, salinity, cold, and heat have a negative impact on plant growth and agricultural output. Physiological reactions and signaling pathways, such as the buildup of suitable solutes and metabolites, activation of antioxidants, and synthesis of stress proteins, are frequently triggered by varied environmental stresses. Many plants rely heavily on their secondary metabolites to protect themselves in shifting or stressful environments. Secondary metabolites have served as a line of defence against cell damage during abiotic stress. The build-up of secondary metabolites in numerous plants enables the plants to adapt to severe abiotic stresses. Furthermore, these environmental elements, such as temperature, light, salinity, and humidity, have an impact on plant development and secondary metabolite production (Kumar et al. 2023). For instance, increased production of secondary metabolites including phenols and flavonoids has been reported during heavy metal stress in tomato plants. Similarly, enhanced accumulation of secondary metabolites has been found in drought and other stresses. Exogenous application of plant and microbial secondary metabolites can mitigate different types of stresses and improve plant growth. Application of exogenous proline alleviated the salt-mediated damage to PSII functioning in *Aloe vera.* Furthermore, treated plants exhibited decreased accumulation of hydrogen peroxide and malondialdehyde content which was attributed to increased activity of antioxidant enzymes including superoxide dismutase, catalase, and ascorbate peroxidase. In another study, application of glycine betaine

preserved nutritional quality and maintained low levels of malondialdehyde and electrolyte leakage in pomegranate fruits. Additionally, treated fruits also exhibited increased radical scavenging activity and increased levels of total phenols, flavonoids, and anthocyanin content. Emanuil et al. reported that application of caffeine in heavy metal-stressed spinach plant increased the synthesis of chlorophyll and accumulation of proline and ascorbic acid and reduced accumulation of malondialdehyde and hydrogen peroxide. In an experiment, Ahmad et al. reported that treatment with salicylic acid reduced the negative effects of salt stress in *Pisum sativum* in terms of improved growth, photosynthetic pigments, enzymatic and nonenzymatic antioxidant defense system, and content of soluble sugars and proline. Similarly, treatment of chromium stressed-*Trigonella corniculate* with quercetin increased intracellular CO_2 concentration, photosynthetic rate, transpiration rate, and stomatal conductance and activity of antioxidants including superoxide dismutase, peroxidase, ascorbate peroxidase, and catalase and improved the growth of plants. Also, application of coumarin enhanced salt tolerance in tomato plants by boosting antioxidant defense system and lowering lipid peroxidation and membrane damage. Treated plants also exhibited increased plant growth, photosynthetic pigments, mineral homeostasis, water retention, and biomass content. Foliar application of gallic acid and β-carotene significantly alleviated salt stress by reducing hydrogen peroxide and malondialdehyde content, regulating ion uptake, and increasing activity of enzymatic antioxidants and chlorophyll, glutathione, and phenolic content. Azeem et al. reported salicylic acid and ascorbic acid-mediated amelioration of salinity-induced negative impacts in *Sorghum bicolor*, where treated plants exhibited increased activity of enzymatic and nonenzymatic antioxidants and photosynthetic pigments and improved plant biomass. Furthermore, fusicoccin produced by fungus *Fusicoccum amydali* was discovered to be a potential plant growth promoter and mitotic activator when applied externally to onion bulbs that had been exposed to salt stress. Additionally, fusicoccin treatment reduced the negative effects of salt stress on the chromosomal structure and root anatomical structure and shielded cells from salt's cytotoxic and genotoxic effects. Additionally, by limiting damage to cell membranes in root cells and regulating the buildup of osmolyte substances like proline and antioxidant enzymes like superoxide dismutase and catalase, this application helped the onion plant fight off reactive oxygen species and increased salt tolerance. Similarly, the detrimental effects of salt stress on bulb germination and growth, the antioxidant defence system, anatomical structure, and cytogenetics were greatly reduced by exogenous application of coronatine a bacterial metabolite by *Pseudomonas syringae* (Table 19.2).

19.7 Role of Bioactive Secondary Metabolites in Medicine

The way of life we lead now is a major contributor to many diseases in people. Although allopathic drugs efficiently treat the disease, they may have severe side effects. Swelling of the face, body rashes, itching, headache, inflammation, and

Table 19.2 Role of bioactive secondary metabolites during abiotic stress

Secondary metabolite	Abiotic Stress	Plant	Effect	References
Quercetin and kaempferol	Heavy metal (As)	*Triticum aestivum*	Improved growth, photosynthesis and activity of chloroplastic antioxidants	
Fusicoccin	Salt	*Allium cepa*	Alleviated negative effects of salt stress on chromosomes and root anatomy and improved growth and salt tolerance	
L-carnitine	Water stress	*Eryngium foetidum*	Improved photosynthetic pigments	
Salicylic acid and ascorbic acid	Salt	*Triticum aestivum*	Enhanced photosynthetic pigments, activity of antioxidants and reduced lipid peroxidation	
Triacontanol	Salt stress	*Brassica juncea*	Increased levels of proteins and metabolites and activity of antioxidants leading to improved growth	
5-aminolevulinic acid	Drought	*Musa* spp.	Reduced oxidative stress by enhancing activity of antioxidants	
Phenol	Heavy metal (Cd)	*Avicennia marina*	Enhanced Cd tolerance and increased plant biomass	
Caffeic and ferulic acid	Salt	*Cucumis sativus*	Reduced lipid peroxidation and increased plant growth	
Rutin	Heavy metal (Cd)	*Amaranthus hypochondriacus*	Reduced Cd toxicity and improved plant growth	
Liquiritoside	Heavy metal (Pb)	*Brassica rapa*	Reduced oxidative stress and increased plant height and fresh and dry weight	
α-Lipoic acid	Drought	*Triticum aestivum*	Reduced malondialdehyde and hydrogen peroxide content and increased plant growth	
Trehalose	Drought	*Helianthus annuus*	Reduced accumulation of hydrogen peroxide and increased plant fresh weight, total soluble sugar, phenols and achene yield	

(continued)

Table 19.2 (continued)

Secondary metabolite	Abiotic Stress	Plant	Effect	References
Phenolic acid	Salt	*Brassica* sp.	Increased phenolics and proline content and photosynthetic performance	
Cinnamic acid	Salt	*Capsicum annum*	Reduced lipid peroxidation and increased flavonoid content and antioxidants activity	
Vanillic acid	Salt	*Solanum lycopersicum*	Reduced accumulation of reactive oxygen species and increased plant growth	
L-Proline	Heat	*Solanum lycopersicum*	Lowered malondialdehyde and increased fruit yield	

drug resistance are typical manifestations of the adverse effects of allopathic medications. Herbal or plant or microbial-derived medications, which have been used to treat ailments since antiquity, are a safer option (Bapat et al. 2023; Han et al. 2023). Chemical components of metabolites are wholly responsible for the diverse and extensive pharmacological actions of such metabolites. Various plants and microbes have been reported to produce metabolites that have anticancer, antimicrobial, antiviral, and anticholestromic activity (Kaur and Ahmed 2021; Singh et al. 2023a, b). For instance, effectiveness of ethanolic extracts of three medicinal herbs, *Cardiospermum halicacabum, Gomphrena celosioides,* and *Scoparia dulcis*, against the cholangiocarcinoma (CCA) cell lines KKU-100 and KKU-213 was examined in a study conducted by Promraksa et al. (2019). The research team concluded that *S. dulcis* extract significantly enhanced the BAX/Bcl-2 ratio and the permeability of cell membranes among all the extracts examined and prevented cancer cells from growing. Similarly, in another in-vivo study on effects and mechanism of the bioactive secondary metabolite eugenol, which is typically found in *Eugenia caryophyllus* (clove) essential oil and other MAPs like *Ocimum minimum* and *Melissa offcinalis*, against high cholesterol and lipids was conducted by Harb et al. (2019) in hypercholesteromic rats. Eugenol effectively decreased total cholesterol, low-density lipoproteins (LDL), and anthrogenic index in rats fed a high-cholesterol diet, according to the study's findings. Steatosis, hepatic inflammation, and hepatomegaly were also reduced. Additionally, a significant decrease in the liver damage marker enzymes alanine amino transferase (ALT) and alkaline phosphatase (ALP) as well as an improvement in the antioxidant enzymes superoxide dismutase (SOD) and catalase were seen. In another study, crude extract of *Scenedesmus obliquus* exhibited anticancerous activity. GCMS analysis revealed that 9-octadecadienoic acid, quercetin, nonadecane, and butylated hydroxytoluene were responsible for bioactivity of the plant extract (Marrez et al. 2019). Furthermore, in traditional remedies, a wide variety of phenolic substances and flavonoids directly interact with neurotransmitter systems in various animal models and produce sedative effects by physically binding to GABAA receptors. Through antagonistic binding to GABAA

receptors and cholinergic upregulation, this combination also improves cognition. It also has antidepressant effects by blocking monoamine oxidase, which raises noradrenaline levels in certain parts of the brain (Bano et al. 2023). Furthermore, different kinds of antibiotics known to be produced by actinomycetes and fungi have been a rich source of several medicinal compounds ever since penicillin, which was discovered and isolated from the fungus *Penicillium notatum*, was discovered (Abdel-Razek et al. 2020). Therefore, numerous plant and microbial species are examined for these metabolites and also isolated from microbes and plants.

19.8 Conclusion

The bioactive secondary metabolites of plants and microorganisms are essential for numerous processes in the organism. Secondary metabolites are essential for plants' defence mechanisms against various biotic and abiotic stressors. In addition, they are employed in the medical industry and aid in the interaction of plants with other living things. Particularly in the marine environment, microbes constitute a significant source of bioactive secondary metabolites. The metabolic processes of microbes are a significant source of secondary metabolites that are vital to human diet and are very profitable to pharmaceutical business. Plant growth promoters (PGR) and plant growth-promoting rhizobacteria (PGPR) are key players in the synthesis of secondary metabolites because they have unique mechanisms for boosting the production of bioactive compounds.

References

Abdel-Razek AS, El-Naggar ME, Allam A, Morsy OM, Othman SI (2020) Microbial natural products in drug discovery. Processes 8(4):470

Aboshi T, Ishiguri S, Shiono Y, Murayama T (2018) Flavonoid glycosides in Malabar spinach Basella alba inhibit the growth of *Spodoptera litura* larvae. Biosci Biotechnol Biochem 82(1):9–14

Afsar T, Razak S, Aldisi D, Shabbir M, Almajwal A, Al Kheraif AA, Arshad M (2021) *Acacia hydaspica* R. Parker ethyl-acetate extract abrogates cisplatin-induced nephrotoxicity by targeting ROS and inflammatory cytokines. Sci Rep 11(1):17248

Afshar M, Najafian S, Radi M (2022) Seasonal variation on the major bioactive compounds: total phenolic and flavonoids contents, and antioxidant activity of rosemary from shiraz. Nat Prod Res 36(16):4287–4292

Ahammed MS, Kader MA, Islam KM, Lopa SS, Alam AHMK, Sadik G (2023) Anti-inflammatory activity of the phenolic compounds from *Vanda roxburghii* R. Br. Bangladesh J Pharmacol 26(1):1–6

Ahuja I, Kissen R, Bones AM (2012) Phytoalexins in defense against pathogens. Trends Plant Sci 17(2):73–90

Ali M, Ohri P (2023) Deciphering the synergistic effect of Jasmonic acid and Spermine in mitigating root-knot nematode stress in tomato plants through enhancing growth and activity of antioxidant enzymes. S Afr J Bot 161:21–35

Al-Khayri JM, Rashmi R, Toppo V, Chole PB, Banadka A, Sudheer WN, Nagella P, Shehata WF, Al-Mssallem MQ, Alessa FM, Rezk AAS (2023) Plant secondary metabolites: the weapons for biotic stress management. Metabolites 13(6):716

Anjali, Kumar S, Korra T, Thakur R, Arutselvan R, Kashyap AS, Nehela Y, Chaplygin V, Minkina T, Keswani C (2023) Role of plant secondary metabolites in defence and transcriptional regulation in response to biotic stress. Plant Stress 100154

Ashihara H, Yokota T, Crozier A (2013) Biosynthesis and catabolism of purine alkaloids. In: Advances in botanical research, vol 68. Academic Press, pp 111–138

Ashraf MA, Iqbal M, Rasheed R, Hussain I, Riaz M, Arif MS (2018) Environmental stress and secondary metabolites in plants: an overview. In: Ahmad P, Ahanger MA, Singh VP, Tripathi DK, Alam P, Alyemeni MN (eds) Plant metabolites and regulation under environmental stress. Academic Press, London, pp 153–167

Ashrafi S, Alam S, Sultana A, Raj A, Emon NU, Richi FT, Sharmin T, Moon M, Park MN, Kim B (2023) Papaverine: a miraculous alkaloid from opium and its multimedicinal application. Molecules 28(7):3149

Assaf M, Korkmaz A, Karaman Ş, Kulak M (2022) Effect of plant growth regulators and salt stress on secondary metabolite composition in *Lamiaceae* species. S Afr J Bot 144:480–493

Azizi P, Osman M, Hanafi MM, Sahebi M, Yusop MR, Taheri S (2019) Adaptation of the metabolomics profile of rice after *Pyricularia oryzae* infection. Plant Physiol Biochem 144:466–479

Balah MA, Abdel Razek GM (2020) Pesticidal activity of *Solanum elaeagnifolium* Cav. Leaves against nematodes and perennial weeds. Acta Ecol Sin 40(5):373–379

Bano S, Iqbal EY, Lubna Zik-ur-Rehman S, Fayyaz S, Faizi S (2020) Nematicidal activity of flavonoids with structure activity relationship (SAR) studies against root knot nematode *Meloidogyne incognita*. Eur J Plant Pathol 157:299–309

Bano A, Qadri TA, Khan N (2023) Bioactive metabolites of plants and microbes and their role in agricultural sustainability and mitigation of plant stress. S Afr J Bot 159:98–109

Bapat VA, Kishor K, Penna S (2023) Heterologous production of high value metabolites in plants and microbes. Front Plant Sci 14:1223033

Bartwal A, Mall R, Lohani P, Guru SK, Arora S (2013) Role of secondary metabolites and brassinosteroids in plant defense against environmental stresses. J Plant Growth Regul 32:216–232

Behiry SI, Okla MK, Alamri SA, El-Hefny M, Salem MZ, Alaraidh IA, Ali HM, Al-Ghtani SM, Monroy JC, Salem AZ (2019) Antifungal and antibacterial activities of Musa paradisiaca L. peel extract: HPLC analysis of phenolic and flavonoid contents. Processes 7(4):215

Bell L, Oloyede OO, Lignou S, Wagstaff C, Methven L (2018) Taste and flavor perceptions of glucosinolates, isothiocyanates, and related compounds. Mol Nutr Food Res 62(18):1700990

Block AK, Vaughan MM, Schmelz EA, Christensen SA (2019) Biosynthesis and function of terpenoid defense compounds in maize (Zea mays). Planta 249:21–30

Boncan DAT, Tsang SS, Li C, Lee IH, Lam HM, Chan TF, Hui JH (2020) Terpenes and terpenoids in plants: interactions with environment and insects. Int J Mol Sci 21(19):7382

Böttger A, Vothknecht U, Bolle C, Wolf A, Böttger A, Vothknecht U, Bolle C, Wolf A (2018) Plant secondary metabolites and their general function in plants. In: Böttger A, Vothknecht U, Bolle C, Wolf A (eds) Lessons on Caffeine, Cannabis and Co: Plant-derived drugs and their interaction with human receptors. Springer, Cham, pp 3–17

Bröker JN, Müller B, van Deenen N, Prüfer D, Schulze Gronover C (2018) Upregulating the mevalonate pathway and repressing sterol synthesis in *Saccharomyces cerevisiae* enhances the production of triterpenes. Appl Microbiol Biotechnol 102:6923–6934

Carrillo-Lomelí DA, de Rodríguez DJ, Moo-Huchin VM, Ramón-Canul L, Rodríguez-García R, González-Morales S, Villarreal-Quintanilla JA, Peña-Ramos FM (2022) How does Flourensia microphylla extract affect polyphenolic composition, antioxidant capacity, and antifungal activity? Ind Crop Prod 186:115248

Chhajed S, Mostafa I, He Y, Abou-Hashem M, El-Domiaty M, Chen S (2020) Glucosinolate biosynthesis and the glucosinolate–myrosinase system in plant defense. Agronomy 10(11):1786

Chin S, Behm CA, Mathesius U (2018) Functions of flavonoids in plant–nematode interactions. Plan Theory 7(4):85

Chiorcea-Paquim AM, Enache TA, De Souza Gil E, Oliveira-Brett AM (2020) Natural phenolic antioxidants electrochemistry: towards a new food science methodology. Compr Rev Food Sci Food Saf 19(4):1680–1726

Chohan S, Perveen R, Anees M, Azeem M, Abid M (2019) Estimation of secondary metabolites of indigenous medicinal plant extracts and their in vitro and in vivo efficacy against tomato early blight disease in Pakistan. J Plant Dis Protect 126:553–563

Cmoch P, Pakulski Z, Swaczynová J, Strnad M (2008) Synthesis of lupane-type saponins bearing mannosyl and 3, 6-branched trimannosyl residues and their evaluation as anticancer agents. Carbohydr Res 343(6):995–1003

Crozier A, Jaganath IB, Clifford MN (2006) Phenols, polyphenols and tannins: an overview. In: Crozier A, Clifford MN, Ashihara H (eds) Plant secondary metabolites: occurrence, structure and role in the human diet, vol 1. Wiley, New York, pp 1–25

Cushnie TT, Lamb AJ (2005) Antimicrobial activity of flavonoids. Int J Antimicrob Agents 26(5):343–356

da Silveira Vasconcelos M, de Oliveira LMN, Nunes-Pinheiro DCS, da Silva Mendes FR, de Sousa FD, de Siqueira Oliveira L, de Aquino AC, de Souza TDFG, Silva AS, Nabavi SM, de Melo DF (2020) Analysis of tetraterpenes and tetraterpenoids (carotenoids). In: Recent advances in natural products analysis. Elsevier, pp 427–456

Darzuli N, Budniak L, Slobodianiuk L (2021) Investigation of the antibacterial and antifungal activity of the *Pyrola rotundifolia* L. leaves dry extract. Pharmacol Online 1:395–403

Das A, Lee SH, Hyun TK, Kim SW, Kim JY (2013) Plant volatiles as method of communication. Plant Biotechnol Rep 7:9–26

Dhaniaputri R, Suwono H, Amin M, Lukiati B (2022) Introduction to plant metabolism, secondary metabolites biosynthetic pathway, and in-silico molecular docking for determination of plant medicinal compounds: an overview. In: 7th international conference on biological science (ICBS 2021). Atlantis Press, pp 373–382

Divekar PA, Narayana S, Divekar BA, Kumar R, Gadratagi BG, Ray A, Singh AK, Rani V, Singh V, Singh AK, Behera TK (2022) Plant secondary metabolites as defense tools against herbivores for sustainable crop protection. Int J Mol Sci 23(5):2690

Dobrijević D, Pastor K, Nastić N, Özogul F, Krulj J, Kokić B, Bartkiene E, Rocha JM, Kojić J (2023) Betaine as a functional ingredient: metabolism, health-promoting attributes, food sources, applications and analysis methods. Molecules 28(12):4824

Dumanović J, Nepovimova E, Natić M, Kuča K, Jaćević V (2021) The significance of reactive oxygen species and antioxidant defense system in plants: a concise overview. Front Plant Sci 11:552969

Erb M, Kliebenstein DJ (2020) Plant secondary metabolites as defenses, regulators, and primary metabolites: the blurred functional trichotomy. Plant Physiol 184(1):39–52

Fayaz A, Unnisa G, Faizan M, Ahmed SM (2023) Medicinal uses of plant secondary metabolites: a brief review. Indian J Appl Pure Biol 38(1):170–175

Gavhale SS, Waghmare SA, Kamble HV (2023) A review on: alkaloids important therapeutic compound. World J Pharm Res 12(5):741–752

Gobbo-Neto L, Lopes NP (2007) Plantas medicinais: fatores de influência no conteúdo de metabólitos secundários. Química nova 30:374–381

Graça MAS, Bärlocher F, Gessner MO (2005) Methods to study litter decomposition: a practical guide. Dordrecht, Springer

Guo L, Li Y, Mao X, Tao R, Tao B, Zhou Z (2022) Antifungal activity of polymethoxylated flavonoids (PMFs)-loaded citral nanoemulsion against *Penicillium italicum* by causing cell membrane damage. J Fungi 8(4):388

Gupta G, Sharma S, Kumar NR (2020) Carica papaya aqueous leaf extract as potential botanical insecticide against rose aphids (*Macrosiphum rosaeformis* D.). J Entomol Zool Stud 8:960–964

Gutensohn M, Hartzell E, Dudareva N (2022) Another level of complexity: the role of metabolic channeling and metabolons in plant terpenoid metabolism. Front Plant Sci 13:954083

Hajam YA, Lone R, Kumar R (2023) Role of plant phenolics against reactive oxygen species (ROS) induced oxidative stress and biochemical alterations. In: Plant phenolics in abiotic stress management. Springer Nature, Singapore, pp 125–147

Halder M, Sarkar S, Jha S (2019) Elicitation: A biotechnological tool for enhanced production of secondary metabolites in hairy root cultures. Eng Life Sci 19(12):880–895

Han J, Zhao X, Zhao X, Wang Q, Li P, Gu Q (2023) Microbial-derived γ-aminobutyric acid: synthesis, purification, physiological function, and applications. J Agric Food Chem 71:14931

Harb AA, Bustanji YK, Almasri IM, Abdalla SS (2019) Eugenol reduces LDL cholesterol and hepatic steatosis in hypercholesterolemic rats by modulating TRPV1 receptor. Sci Rep 9(1):14003

Humbal A, Pathak B (2023) Influence of exogenous elicitors on the production of secondary metabolite in plants: a review ("VSI: secondary metabolites"). Plant Stress:100166

Hussain AY, Hussein HJ, Al-Rubaye AF (2021) Antifungal activity of the secondary metabolites extracted. Clin Schizophr Relat Psychoses 15(6)

Hussein RA, El-Anssary AA (2019) Plants secondary metabolites: the key drivers of the pharmacological actions of medicinal plants. In: Builders PF (ed) Herbal Medicine. IntechOpen, London

Jahan Y, Mahmood T, Bagga P, Kumar A, Singh K, Mujahid M (2015) Future prospects of cough treatment, herbal medicines v/s modern drugs. Int J Pharm Sci Res 6(9):1000–1009

Jan R, Asaf S, Numan M, Lubna, Kim KM (2021) Plant secondary metabolite biosynthesis and transcriptional regulation in response to biotic and abiotic stress conditions. Agronomy 11(5):968

Jha Y, Mohamed HI (2022) Plant secondary metabolites as a tool to investigate biotic stress tolerance in plants: a review. Gesunde Pflanzen 74(4):771–790

Kabera JN, Semana E, Mussa AR, He X (2014) Plant secondary metabolites: biosynthesis, classification, function and pharmacological properties. J Pharm Pharmacol 2(7):377–392

Kandar CC (2021) Secondary metabolites from plant sources. In: Pal D, Nayak AK (eds) Bioactive natural products for pharmaceutical applications. Springer, Cham, pp 329–377

Karuppusamy S (2009) A review on trends in production of secondary metabolites from higher plants by in vitro tissue, organ and cell cultures. J Med Plants Res 3(13):1222–1239

Kaur N, Ahmed T (2021) Bioactive secondary metabolites of medicinal and aromatic plants and their disease-fighting properties. In: Aftab T, Hakeem KR (eds) Medicinal and aromatic plants: healthcare and industrial applications. Springer, Cham, pp 113–142

Kermode AR (2005) Role of abscisic acid in seed dormancy. J Plant Growth Regul 24:319–344

Khameneh B, Iranshahy M, Soheili V, Fazly Bazzaz BS (2019) Review on plant antimicrobials: a mechanistic viewpoint. Antimicrob Resist Infect Control 8(1):1–28

Khan MSA, Ahmad I (2019) Herbal medicine: current trends and future prospects. In: Khan MSA, Ahmad I, Chattopadhyay D (eds) New look to phytomedicine. Academic Press, London, pp 3–13

Khan N, Ali S, Shahid MA, Kharabian-Masouleh A (2017) Advances in detection of stress tolerance in plants through metabolomics approaches. Plant Omics 10(3):153–163

Khan N, Zandi P, Ali S, Mehmood A, Adnan Shahid M, Yang J (2018) Impact of salicylic acid and PGPR on the drought tolerance and phytoremediation potential of Helianthus annus. Front Microbiol 9:2507

Kolekar SM, Jain BU, Kondawarkar MS (2019) A review on steroids and terpenoids (stereochemistry, structural elucidation, isolation of steroids and terpenoids). Res J Pharmaceut Dosage Forms Technol 11(2):126–130

Kołton A, Długosz-Grochowska O, Wojciechowska R, Czaja M (2022) Biosynthesis regulation of folates and phenols in plants. Sci Hortic 291:110561

Kovalikova Z, Kubes J, Skalicky M, Kuchtickova N, Maskova L, Tuma J et al (2019) Changes in content of polyphenols and ascorbic acid in leaves of white cabbage after pest infestation. Molecules 24(14):2622

Kumar P, Dubey KK (2019) Citric acid cycle regulation: back bone for secondary metabolite production. In: New and future developments in microbial biotechnology and bioengineering. Elsevier, pp 165–181

Kumar K, Kumar SR, Dwivedi V, Rai A, Shukla AK, Shanker K, Nagegowda DA (2015) Precursor feeding studies and molecular characterization of geraniol synthase establish the limiting role of geraniol in monoterpeneindole alkaloid biosynthesis in Catharanthusroseus leaves. Plant Sci 239:56–66

Kumar S, Singh A, Kumar B, Singh B, Bahadur L, Lal M (2018) Simultaneous quantitative determination of bioactive terpene indole alkaloids in ethanolic extracts of *Catharanthus roseus* (L.) G. Don by ultra high performance liquid chromatography–tandem mass spectrometry. J Pharm Biomed Anal 151:32–41

Kumar S, Saini R, Suthar P, Kumar V, Sharma R (2022) Plant secondary metabolites: their food and therapeutic importance. In: Plant secondary metabolites: physico-chemical properties and therapeutic applications. Springer Nature, Singapore, pp 371–413

Kumar D, Manhas RK, Ohri P (2023) Deciphering the growth promoting and stress curtailing role of nitric oxide in *Meloidogyne incognita* infested *Solanum lycopersicum* seedlings. Sci Hortic 319:112147

Lan JE, Li XJ, Zhu XF, Sun ZL, He JM, Zloh M, Gibbons S, Mu Q (2021) Flavonoids from Artemisia rupestris and their synergistic antibacterial effects on drug-resistant *Staphylococcus aureus*. Nat Prod Res 35(11):1881–1886

Lattanzio V, Kroon PA, Quideau S, Treutter D (2009) Plant phenolics—secondary metabolites with diverse functions. Recent Adv Polyphenol Res 1:1–35

Li Y, Kong D, Fu Y, Sussman MR, Wu H (2020) The effect of developmental and environmental factors on secondary metabolites in medicinal plants. Plant Physiol Biochem 148:80–89

Liu Y, Chen X, Zhang C (2022) Sustainable biosynthesis of valuable Diterpenes in microbes. Eng Microbiol 3(1):100058

Liu X, Zhang P, Zhao Q, Huang AC (2023) Making small molecules in plants: a chassis for synthetic biology-based production of plant natural products. J Integr Plant Biol 65(2):417–443

Mahmoud SS, Croteau RB (2002) Strategies for transgenic manipulation of monoterpene biosynthesis in plants. Trends Plant Sci 7(8):366–373

Marrez DA, Naguib MM, Sultan YY, Higazy AM (2019) Antimicrobial and anticancer activities of *Scenedesmus obliquus* metabolites. Heliyon 5(3):e01404

Martău GA, Bernadette-Emőke T, Odocheanu R, Soporan DA, Bochiş M, Simon E, Vodnar DC (2023) Vaccinium species (Ericaceae): phytochemistry and biological properties of medicinal plants. Molecules 28(4):1533

Mohammed AE, Abdul-Hameed ZH, Alotaibi MO, Bawakid NO, Sobahi TR, Abdel-Lateff A, Alarif WM (2021) Chemical diversity and bioactivities of monoterpene indole alkaloids (MIAs) from six Apocynaceae genera. Molecules 26(2):488

O'Connor SE, Maresh JJ (2006) Chemistry and biology of monoterpeneindole alkaloid biosynthesis. Nat Prod Rep 23(4):532–547

Owens KR, McCowen SV, Blackford KA, Ueno S, Hirooka Y, Weber M, Sarpong R (2019) Total synthesis of the diterpenoid alkaloid arcutinidine using a strategy inspired by chemical network analysis. J Am Chem Soc 141(35):13713–13717

Pan Q, Mustafa NR, Tang K, Choi YH, Verpoorte R (2016) Monoterpenoidindole alkaloids biosynthesis and its regulation in Catharanthusroseus: a literature review from genes to metabolites. Phytochem Rev 15:221–250

Pandey P, Kaur R, Singh S, Chattopadhyay SK, Srivastava SK, Banerjee S (2014) Long-term stability in biomass and production of terpene indole alkaloids by hairy root culture of *Rauvolfia serpentina* and cost approximation to endorse commercial realism. Biotechnol Lett 36:1523–1528

Promraksa B, Phetcharaburanin J, Namwat N, Techasen A, Boonsiri P, Loilome W (2019) Evaluation of anticancer potential of Thai medicinal herb extracts against cholangiocarcinoma cell lines. PLoS One 14(5):e0216721

Rajak N, Tiwari A, Kumar P, Garg N (2023) Overview on nutraceuticals and biotechnology. In: Nutraceuticals. Academic Press, pp 175–191

Ranf S (2018) Pattern recognition receptors—versatile genetic tools for engineering broad-spectrum disease resistance in crops. Agronomy 8(8):134

Ren MY, Yu QT, Shi CY, Luo JB (2017) Anticancer activities of C18-, C19-, C20-, and bis-diterpenoid alkaloids derived from genus *Aconitum*. Molecules 22(2):267

Roberts MF, Strack D, Wink M (2010) Biosynthesis of alkaloids and betalains. In: Wink M (ed) Annual plant reviews, volume 40: biochemistry of plant secondary metabolism. Wiley, pp 20–91

Sampaio BL, Edrada-Ebel R, Da Costa FB (2016) Effect of the environment on the secondary metabolic profile of Tithonia diversifolia: a model for environmental metabolomics of plants. Sci Rep 6(1):29265

Santra HK, Maity S, Banerjee D (2022) Production of bioactive compounds with broad spectrum bactericidal action, bio-film inhibition and antilarval potential by the secondary metabolites of the endophytic fungus *Cochliobolus* sp. APS1 isolated from the Indian medicinal herb *Andrographis paniculata*. Molecules 27(5):1459

Sasaki K, Saito T, Lämsä M, Oksman-Caldentey KM, Suzuki M, Ohyama K et al (2007) Plants utilize isoprene emission as a thermotolerance mechanism. Plant Cell Physiol 48(9):1254–1262

Sedeek AM, Ismail MM, Elsayed TR, Ramadan MA (2022) Recent methods for discovering novel bioactive metabolites, specifically antimicrobial agents, from marine-associated microorganisms. Lett Appl Microbiol 75(3):511–525

Selim MSM, Abdelhamid SA, Mohamed SS (2021) Secondary metabolites and biodiversity of actinomycetes. J Genet Eng & Biotechnol 19(1):72

Shah A, Smith DL (2020) Flavonoids in agriculture: chemistry and roles in, biotic and abiotic stress responses, and microbial associations. Agronomy 10(8):1209

Shang XF, Zhao ZM, Li JC, Yang GZ, Liu YQ, Dai LX et al (2019) Insecticidal and antifungal activities of *Rheum palmatum* L. anthraquinones and structurally related compounds. Ind Crop Prod 137:508–520

Singh AA, Ghosh A, Agrawal M, Agrawal SB (2023a) Secondary metabolites responses of plants exposed to ozone: an update. Environ Sci Pollut Res 30:1–32

Singh S, Singh G, Attri S, Kaur P, Rashid F, Bedi N, Haque S, Janahi EM, Arora S (2023b) Development and optimization of nanoparticles loaded with erucin, a dietary isothiocyanate isolated from *Eruca sativa*: antioxidant and antiproliferative activities in ehrlich-ascites carcinoma cell line. Front Pharmacol 13:1080977

Soriano G, Petrillo C, Masi M, Bouafiane M, Khelil A, Tuzi A, Isticato R, Fernández-Aparicio M, Cimmino A (2022) Specialized metabolites from the allelopathic plant Retama raetam as potential biopesticides. Toxins 14(5):311

Sousa JP, Ramos MJ, Fernandes PA (2022) Modern strategies for the diversification of the supply of natural compounds: the case of alkaloid painkillers. Chembiochem 23(10):e202100623

Sudheeran PK, Ovadia R, Galsarker O, Maoz I, Sela N, Maurer D, Feygenberg O, Oren Shamir M, Alkan N (2020) Glycosylated flavonoids: fruit's concealed antifungal arsenal. New Phytol 225(4):1788–1798

Tagousop CN, Tamokou JDD, Ekom SE, Ngnokam D, Voutquenne-Nazabadioko L (2018) Antimicrobial activities of flavonoid glycosides from *Graptophyllum grandulosum* and their mechanism of antibacterial action. BMC Complement Altern Med 18(1):1–10

Thakur M, Bhattacharya S, Khosla PK, Puri S (2019) Improving production of plant secondary metabolites through biotic and abiotic elicitation. J Appl Res. Med Aromat Plants 12:1–12

Thirumurugan D, Cholarajan A, Raja SS, Vijayakumar R (2018) An introductory chapter: secondary metabolites. In: Raja SSS, Vijayakumar R (eds) Secondary metabolites - sources and applications. InTech, pp 3–21

Tiku AR (2018) Antimicrobial compounds and their role in plant defense. In: Singh A, Singh IK (eds) Molecular aspects of plant-pathogen interaction. Springer, Singapore, pp 283–307

Tiwari R, Rana CS (2015) Plant secondary metabolites: a review. Int J Eng Res Gen Sci 3(5):661–670

Tlak Gajger I, Dar SA (2021) Plant allelochemicals as sources of insecticides. Insects 12(3):189

Toyomasu T (2008) Recent advances regarding diterpene cyclase genes in higher plants and fungi. Biosci Biotechnol Biochem 72(5):1168–1175

Tridiptasari A, Leksono AS, Siswanto D (2019) Antifeedant effect of Moringa oleifera (L.) leaf and seed extract on growth and feeding activity of *Spodoptera litura* (Fab.) (*Lepidoptera: Noctuidae*). J Exp Life Sci 9(1):25–31

Twaij BM, Hasan MN (2022) Bioactive secondary metabolites from plant sources: types, synthesis, and their therapeutic uses. Int J Plant Biol 13(1):4–14

Velasco P, Lema M, Francisco M, Soengas P, Cartea ME (2013) In vivo and in vitro effects of secondary metabolites against Xanthomonas campestris pv. campestris. Molecules 18(9):11131–11143

Verma N, Shukla S (2015) Impact of various factors responsible for fluctuation in plant secondary metabolites. J Appl Res Med Aromat Plants 2(4):105–113

Wang S, Li Y, He L, Yang J, Fernie AR, Luo J (2022) Natural variance at the interface of plant primary and specialized metabolism. Curr Opin Plant Biol 67:102201

Wang Z, Wang L, Pan Y, Zheng X, Liang X, Sheng L, Zhang D, Sun Q, Wang Q (2023) Research advances on endophytic fungi and their bioactive metabolites. Bioprocess Biosyst Eng 46(2):165–170

Wölwer-Rieck U, May B, Lankes C, Wüst M (2014) Methylerythritol and mevalonate pathway contributions to biosynthesis of mono-, sesqui-, and diterpenes in glandular trichomes and leaves of *Stevia rebaudiana* Bertoni. J Agric Food Chem 62(11):2428–2435

Wuyts N, Swennen R, De Waele D (2006) Effects of plant phenylpropanoid pathway products and selected terpenoids and alkaloids on the behaviour of the plant-parasitic nematodes *Radopholus similis*, *Pratylenchus penetrans* and *Meloidogyne incognita*. Nematology 8(1):89–101

Xue L, Chen YY, Yan Z, Lu W, Wan D, Zhu H (2019) Staphyloxanthin: a potential target for antivirulence therapy. Infect Drug Resist 12:2151–2160

Younus M, Hasan MM, Ali S, Saddq B, Sarwar G, Ullah MI et al (2021) Extracts of *Euphorbia nivulia* Buch.-Ham. showed both phytotoxic and insecticidal capacities against *Lemna minor* L. and *Oxycarenus hyalinipennis* Costa. PLoS One 16(4):e0250118

Yu F, Utsumi R (2009) Diversity, regulation, and genetic manipulation of plant mono-and sesquiterpenoid biosynthesis. Cell Mol Life Sci 66:3043–3052

Zaynab M, Fatima M, Abbas S, Sharif Y, Umair M, Zafar MH, Bahadar K (2018) Role of secondary metabolites in plant defense against pathogens. Microb Pathog 124:198–202

Zheng XQ, Ashihara H (2004) Distribution, biosynthesis and function of purine and pyridine alkaloids in *Coffea arabica* seedlings. Plant Sci 166(3):807–813

Chapter 20
Therapeutic Applications of Plant Toxins Used in Traditional Medicine

J. M. Dahanayake and P. K. Perera

Abstract Traditional herbal medicines have played a very important role in primary health care systems globally. Also, these medicines of these traditional health systems are used to treat various acute and chronic disease conditions without or least toxic effect. Plants are rich in bioactive phyto-constituents known as secondary metabolites. These secondary metabolites have been used to cure illnesses due to their various pharmacological properties. Plant toxins can be known as a large group of structurally diverse small molecules that result from the plant's secondary metabolism. These toxins may be synthesized by plants themselves or by nonpathogenic endophytic microorganisms living within the plants. Among these plant toxins, some groups show high therapeutic values which can be used to cure various illnesses. Most of these plant toxins are derivatives of important secondary metabolites such as alkaloids, terpenes, steroids, and phenolic compounds. In Ayurveda and Sri Lankan traditional system of medicine used toxic plants such as *Aconitum ferox, Strychnus nux-vomica, Nerium indicum, Datura metal, Croton tiglium, Papaver somniferum,* and *Gloriosa superba*. The usage parts of these plants are subjected to deep purification procedures before preparing the medicines. These plants were categorized under poisons and semi-poisons in Ayurveda system of medicine. Hence, the prime focus of this review is to identify toxic plants used in traditional systems of medicines and analyze their therapeutic efficacy on human body for various ailments. Also, this review aims to categorize various purification procedures according to the plant toxins and search the evidence-based modern pharmacological properties of these special phyto-constituents.

Keywords Plant toxins · Phyto-constituents · Purification procedures · Therapeutic efficacy

J. M. Dahanayake · P. K. Perera (✉)
Faculty of Indigenous Medicine, University of Colombo, Colombo, Sri Lanka
e-mail: kamalperera@fim.cmb.ac.lk

M. N. V. Prasad et al. (eds.), *Biotoxins*,
https://doi.org/10.1007/978-3-031-75309-1_20

20.1 Introduction

Medicinal plants have been used for centuries, and numerous cultures still rely on indigenous medicinal plants for their primary health care needs. The World Health Organization (WHO) estimated that 80% of the population of developing countries relies on traditional medicines, mostly plant drugs, for their primary health care practices (Reid et al. 2018). Modern pharmacopeia still contains at least 25% of drugs derived from plants and many other synthetic medicines built on prototype compounds isolated from plant materials. Among these plant materials, many plants are used in traditional medicinal systems as effective medicines and also contain some poisonous plant materials. However, these poisonous plant materials are also rich with many therapeutic values.

Correct identification and knowledge about the therapeutic actions and doses on poisonous plants are very important for the physician in the treatment of various illnesses. According to the Acharya Charaka in Ayurveda medicine, he mentioned that a strong poison can be converted to nectar (*Amruta*) if processed and administrated properly, and on the other hand, even the most useful medicine may become a poison if handled incorrectly (Acharya 2011).

The traditional systems of medicine, such as Ayurveda, Siddha, Unani, Sri Lankan traditional system, Chinese traditional medicine, etc., have been in existence for several centuries. Hence it is important to have an awareness regarding the poisonous plants when used in the treatment. Appropriate doses of these processed poisonous plant materials act as potent therapeutic agents. Poisons or toxins are molecules that are harmful to living organisms. Poisons include both naturally produced compounds and chemicals manufactured by humans. Natural poisons are produced by species of bacteria, fungi, protists, plants, and animals.

According to the scientific findings, the numbers of phyto-constituents have been isolated from the various poisonous medicinal plants that possess activities like diuretic, purgative, laxative, anti-allergic, and various other important medicinal properties. Hence poisonous medicinal plants are used for various ailments as antidiabetic, anticancer, antibacterial, antifungal, and cytogenetic agents (Tamilselvan et al. 2014).

The symptoms of herb poisoning can differ depending on the plant material ingested and the individual's sensitivity. In addition, these symptoms can occur through absorption by the skin and inhalation in the respiratory system. Some common signs and symptoms include nausea, vomiting, abdominal pain, diarrhea, and loss of appetite as gastrointestinal distresses, headaches, dizziness, confusion, hallucinations, seizures, and loss of consciousness as neurological effects, irregular heart rhythms, increased heart rate, and changes in blood pressure as cardiovascular effects and difficulty in breathing, chest tightness, and wheezing as respiratory tract issues (Willingsible 2023). Many traditional systems use these toxic plant materials with proper purification procedures before use as medicines.

The therapeutic utilizations of various poisonous plants were mentioned in literature related to traditional medicine especially in Ayurveda medicinal system. In

Ayurveda, the origin, properties, purification procedures, therapeutic uses, and toxic symptoms related to particular plant with its antidotes were well described. These details of the poisonous and semi-poisonous plants (*Visha* and *Upavisha Dravya*) can be found in *Rasa Shastra* which is known as the special branch of Ayurveda. In therapeutic point of view, these poisonous and semi-poisonous plants are highly important due to their quick effectiveness even in smaller doses after proper purification of these materials. However, at the same time, it may be harmful to human beings if used without proper purification and in higher doses. Hence, correct identification of these plant materials, purification procedures, its important uses, medicinal applications, benefits, toxic symptoms, and its management is very important in the clinical practice of traditional medicine.

20.2 Classification of Poisonous Plants in Ayurveda

The categorization of toxic materials is based on certain basic criteria like origin, base, properties, and potency of particular materials. Some of the Ayurvedic classics and texts have classified all poisons into two categories *Mahavisha* or *Visha* (poisons) and *Upavisha* (Semi-poisons) based on their toxicity and potency. Poisons (*Visha)* have been classified into two categories, based on their origin, as animal poisons (*Jangama visha*) and plant and mineral poisons (*Sthavara visha*) (Bramhanand and Dephande 2002, 2007). Semi-poisons (*Upavisha*) are group of drugs, which are less toxic in nature and not very poisonous but produce certain toxic symptoms on consumption or administration. The symptoms produced in the body due to semi-poisons are less toxic, usually not life-threatening and their toxicity can be controlled by therapeutic measures. Among the poisonous herbs-tuberous and/or root poisons are more toxic (Dahikar et al. 2023).

20.3 Concept of Purification (*Shodhana*) of Toxic Plants

Historically, *Shodhana* concept was in existence since the time of Samhita period, and this concept has further developed after the development of *Rasa Shastra* in the field of Ayurveda medicine. As in *Rasa Sahastra* generally, the metals/minerals/mercury and some plants of poisonous nature are used which contains some toxic effects (Savrikar and Ravishankar 2011). These poisonous and semi-poisonous plants are used as single drugs or formulations after applying the suitable purification procedures mentioned in authentic texts. The concept of *Sodhana* in Ayurveda is not only a process of purification or detoxification of poisonous materials, but also it is a process of reduction of adverse effects and enhances the therapeutic efficacy of the purified drugs by using different medias specific to a particular substance (Priyanka et al. 2023). Research findings revealed that these purification procedures help to minimize the level of toxicity to a sustainable limit of the body

and to reduce the toxic constituents to some extent or by potentiating their chemical transformation to nontoxic or relatively less toxic substances by enhancing their biological efficacy (Maurya et al. 2015). The classical purification procedures mentioned in the texts related to Rasa Shastra are summarized in Table 20.1 (Fig. 20.1).

Table 20.1 Purification procedures of poisonous medicinal plants mentioned in Ayurveda texts (Subodh 2018; Shastri 2012; Mitra et al. 2011; Ayurveda Pharmacopoea of Sri Lanka 1979; Sharma 2017; Mishra 2010; Sharma et al. 2019)

Medicinal plant	Usage part	Media used for purification	Process of purification
Aconitum ferox wall.	Root	Cow's urine	Keep the cut pieces of *A. ferox* roots in freshly collected cow's urine for 3 consecutive days. Each day fresh cow's urine should be added after decant the previous day cow's urine. On the 3rd day rinse with lukewarm water and dry under sunlight.
		Cow's milk or Goat's milk	Make a bundle from *A. ferox* root and boil (*Swedana*) it in cow's milk or goat's milk in *Dolayantra* for three to six hours. Then rinse with water and dry under sunlight.
Strychnos nuxvomica L.	Seeds	*Kanji*	Soak the well-grown seeds of *Strychnos nuxvomica* in *Kanji* for three days. Then remove the seeds from *Kanji* and remove the external cover and dried under sunlight.
		Ghee	Fry the well-grown seeds of *S. nuxvomica* in an open iron vessel along with the required quantity of ghee. The process of frying is continued until the color of the seeds becomes reddish yellow.
		Cow's milk	Make a bundle from seeds of *S. nuxvomica* and boil (Swedana) it in cow's milk Dolayantra for three hours. Then remove the outer cover of the seeds and rinse with water and dry under sunlight.
		Aloe vera and Ginger juice	Immerse the seeds in the exudates scraped from the fresh leaves and stems of *Aloe vera* for 15 days, followed by ginger juice (*Ārdraka svarasa*) for 7 days for purification.

(continued)

Table 20.1 (continued)

Medicinal plant	Usage part	Media used for purification	Process of purification
Semicarpus anacardium Linn.	Nuts	Powders of bricks	Remove the stalks of nuts by cutting with a sharp knife and tied these nuts in a thick cloth or sac with required quantity of the pieces of bricks. Keep the bundle for a period of one week and after one week rub the bundle thoroughly till the oil comes out. When the cloth of the bundle becomes wet, take out nuts and wash them with hot water for 3–4 times. Allow them to dry under sunlight.
		Coconut water	After removing top part of nuts cuts in two parts are placed in *Pottalli, Swedana* is done by filling it with coconut water for 1–2 hours in *Dolayantra*.
		Burning charcoal	Heat the nuts with burning Charcole in an open iron pan
Croton tiglium L.	Seeds	Cow's milk	Remove the testa of the seeds and make a bundle. Then boil (*Swedana*) the bundle in cow's milk in a *Dolayantra* for three hours. Repeat the procedure for 3 times. Rinse with water and dry under sunlight.
		Purified Tankana and Cow's milk	Remove the testa of the seeds and make a bundle along with 1/8th part of purified *Tankana*. Then boil (*Swedana*) the bundle in cow's milk in a *Dola Yantra* for three hours. Repeat the procedure for 2 times. Rinse with water and dry under sunlight.
		Cow's milk	Remove the testa of the seeds and make a bundle. Then boil (*Swedana*) the bundle in cow's milk in a *Dolayantra* for three hours. Then make a paste from the seeds and apply the paste over the top of a new mud pot and leave for absorption of oil through the pores of the new mud pot, the dry paste on the pot is scraped and collected.
Datura metel L.	Seeds	Cow's milk	Make a bundle from *D. metel* seeds and boil (*Swedana*) it in cow's milk in *Dolayantra* for three hours. Then rinse with water and dry under sunlight.
		Cow's urine	Soak the seeds in freshly collected *Gomūtra* and keep for 12 hours. Then wash the seeds and boil in a Dolayantra for 3 hours. Then rinse with water and dry under sunlight (Mitra et al. 2011; Ayurveda Pharmacopoea of Sri Lanka 1979)
Nerium indicum L.	Roots	Cow's milk	Roots of *Karveera* are purified by *Swedana* process. In *Dolayantra* using *Godugdh* (Cow's milk) for 3 hours.

(continued)

Table 20.1 (continued)

Medicinal plant	Usage part	Media used for purification	Process of purification
Cannabis sativa L.	Leaves	Water and Ghee	Soak the dry leaves of *C. sativa* in enough quantity of water for a while. Then squeeze the leaves and remove the water. Fry the leaves in a pan over mild fire by adding the required quantity of ghee. When properly fried, the leaves are taken down, and dried under the hot sun.
		Decoction prepared with *Acacia nilotica* plant	Boil the dry leaves of *C. sativa* with a decoction of *Babbula* (*Acacia nilotica*) for 24 minutes. Decant the decoction and dry the leaves.
Papaver somniferum L.	Latex	*Ardraka swarasa*	Give 7 times *Bhavana* of *Adraka Swaras* (fresh ginger juice) and then dry it
Abrus pricatorius Linn.	Seeds		Make a bundle from *Gunja* seeds and boil (*Swedana*) it in cow's milk or *kanji* (type of viniger) in *Dolayantra* for three hours. Then rinse with water and dry under sunlight.
Gloriosa superba L.	Roots	Cow's urine	Fresh roots and seeds are deep in Gomutra for 24 hour and wash with lukewarm water
Euphorbia neriifolia Linn.	Latex	Juice of Tamarind (*Tamarindus indica*)	Four parts of milk of *Snuhi* is collected and mixed with the one part of juice of *Amlika* (Tamarind) put it into a vessel and let them dry in direct sunlight, after proper drying it should be used.
		Neem leaves juice (*Azadirachta indicia*)	Mix 1 *pala* (60 ml) of *E. nerifolia* latex with 1 *kudawa* (120 ml) of Neem leave juice, kept sometime, and dry under sunlight (Sharma 2017).
		Decotions of *Mahapasmula, Vruhati (Solanum indicum)* and *Kantakari (Solanum xanthocarpum)*	Mix *E. nerifolia* latex with equal volume of 3 decoctions, respectively, and dried on heated charcoal (Sharma 2017).
		Dashamoola decoction	Mix *E. nerifolia* latex with equal volume of *Dashamoola* decoction and dried on heated charcoal. (Mishra 2010)

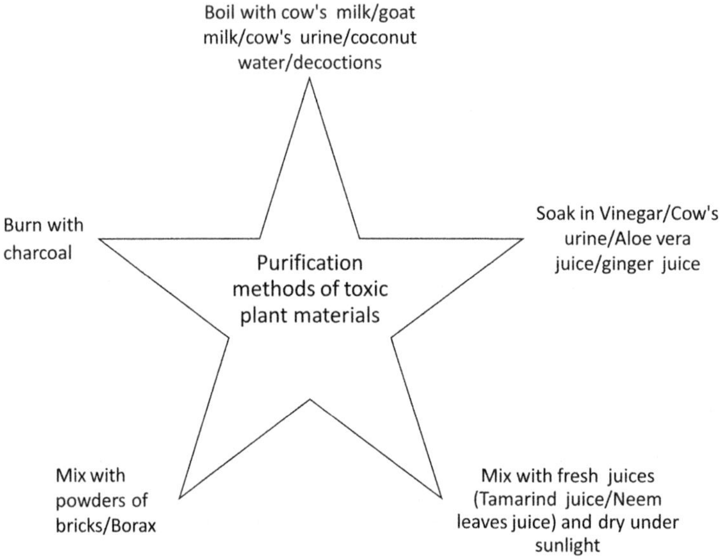

Fig. 20.1 Purification methods of toxic plant materials

20.4 *Aconitum ferox* Wall

20.4.1 Family: Rananculaceae

Aconitum ferox Wall. is commonly used herbal medicine in Ayurveda system of medicine especially for fever and rheumatoid arthritis conditions. It is named Vatsanabha in Sanskrit, resembling the umbilical code of the newborn calf. Root of this plant is the main use part, and it is categorized under poisonous drugs due to its toxic nature. This plant, *Aconitum ferox, is a well-known rejuvenator and an anti-pyretic medicine* after its proper purification or detoxification (Chandewar and Harode 2022).

Aconitum ferox is a perennial, tuberous rooted, and herbaceous shrub that grows upto 3 feet in height. Leaves are cleft or lobed, simple, alternate, and oval in shape. The tuberous roots of this plant are dark brown in color externally and internally it's yellow in color. *A. ferox* bares a beautiful blue color inflorescence.

The toxic effect of the *A. ferox* plant is due to the alkaloids present in the root and research revealed that the higher doses of this root may have cardiotoxic and neuro-toxic effects (Rachna et al. 2023; Panda and Debnath 2010). One gram of *A. ferox* root is lethal for humans and 2 mg of Aconite chemical compound can cause death of the person (Malviya and Choudhari 2018). Toxic effects can occur due to incomplete purification procedures and overdosing of herbal or herbo-mineral Ayurveda formulations that include *A. ferox* root can cause bradycardia, hypotension, or bidirectional tachycardia (Rastogi et al. 2007; Sastri 1978). Hence, the therapeutic dose of *A. ferox root is* mentioned as 8 mg to 16 mg/day in Ayurveda (Deore et al. 2013).

Research done on purification of *A. ferox* root revealed that after proper purification the total alkaloid content of the root was decreased and the contents of less toxic substances (aconine, hypoaconine, and benzylhypoaconine) were increased (Sabeena et al. 2021). Research studies suggest this may be due to the change of the toxic alkaloids to their respective amino alcohols after purification processes (Jayaweera 1980).

20.5 Phytochemicals and Pharmacological Actions

The medicinal plant species of Aconitum are a rich source of alkaloids and flavonoids, many of which exhibit a broad spectrum of activity. Also isolated and identified are various polysaccharides and free fatty acids (Sabeena et al. 2021). The symptoms may appear almost immediately, usually not later than one hour, and with large doses death is almost instantaneous. Death usually occurs within two to six hours in fatal poisoning (20 to 40 mL of tincture may prove fatal). The initial signs are gastrointestinal including nausea, vomiting and diarrhea, burning sensation, tingling, numbness in mouth, and burning in the abdomen. Cardiovascular features include hypotension, sinus bradycardia, and ventricular arrhythmias. Other features may include sweating, dizziness, difficulty in breathing, headache, and confusion. The main causes of death are ventricular arrhythmias and asystole, and paralysis of the heart or the respiratory center (Rachna et al. 2023).

20.6 *Datura metel* Linn

20.6.1 Family: Solanaceae

Datura metel is a perennial plant that contains simple, alternate, dark green leaves and extending branches and is about 1.5 m in height. Flowers are large, single, and trumpet-shaped with a pleasant fragrance and various colors such as white to yellow and dark purple color. The fruit is egg-shaped and capsulated with short spines (Al-Snafi 2017).

D. *metel* contains many phytochemicals, including alkaloids, flavonoids, tannins, phenols, cardiac glycosides, amino acids, and carbohydrates (Sharma et al. 2021). Various parts of the herb are being utilized as hallucinogenic, narcotic, anti-tussive, antispasmodic, bronchodilator, and anti-asthmatic and are used in epilepsy, diarrhea, hysteria, skin diseases, rheumatic pains, painful menstruation, hemorrhoids, wounds, burns, and skin ulcers. Due to the presence of alkaloids in the *D. metal* seeds of this plant, it shows very toxic effects and may be fatal (Islam et al. 2023; Sayyed and Shah 2014). Toxic alkaloids of *D. metel* have anticholinergic properties also. Therefore, can experience symptoms such as muscle cramps, dry mouth,

giddiness, and excessive thirst when used in high dose or an unpurified state (Li 2021; Patel et al. 2010). Ajay Kumar and team observed that significant changes have been observed in different physic-chemical parameters, HPTLC, HPLC, GC-MS, and LC-MS chromatographic profiling of before and after *purification* process in the *Datura metel* seeds. Also, in HPLC profile revealed that the peak area of all peaks of the processed samples was reduced compared with the unprocessed samples. The quantification results of the HPLC analysis demonstrated the complete depletion of toxic alkaloid scopolamine in the processed sample compared with the unprocessed *Datura metel* seeds. In this study, the complete removal of scopolamine reflects the importance of the *purification* process in the seeds of *Datura metel* L., before preparing the herbal medicines (Ajay Kumar et al. 2022).

In the traditional system of medicines, *D. metel* is used to treat epilepsy along with skin diseases, hysteria, fever with catarrh, heart diseases, and diarrhea. In China and Vietnam, this plant is used in the treatment of asthma conditions. In Bangladesh, *D. metal* plant extract is used as general anesthesia and in the treatment of chronic bronchitis (Srivastava and Srivastava 2020). *D. metel* is also used traditionally in India as a cure for madness or insanity, cerebral complaints and catarrhal infections, elephantiasis, ear discharge, skin diseases, mad dog bites, and indigestion (Wang et al. 2008). The flower of *D. metel* is called biamantuoluo in Chinese traditional medicine, where it is mentioned as useful for skin inflammation and psoriasis (Monira and Munan 2012). Seeds of *D. metel* are used as an ingredient in tea in Brazil, which provides a sedative effect, and dried flowers of the plant are used for smoking (Ishola et al. 2021). Nigerians also consider *D. metel* as a hallucinogenic agent (Rahmatullah et al. 2010). In Bangladesh, different parts of the *D. metel* plant are utilized in the treatment of scabies, eczema, and allergies (Rahmatullah et al. 2010), abnormal breathing, pupil problems, pain and swelling, breast pain, asthma, convulsions, and rheumatism (Mirgane 2021).

20.7 *Euphorbia nerifolia* Linn

20.7.1 Family: Euphorbiaceae

Euphobia nerifolia is a large succulent shrub or a small tree or which grows upto 6–20 ft. high, with jointed, cylindrical, or obscurely 5 angled branches bearing short stipule thorns, more or less confluent in vertical or slightly spiral lines. Flowers are yellowish green or greenish yellow in color. Leaves are fleshy, deciduous, 6–12 in. long. Fruits are tricolcus, consisting appearing three radiating follicles, and minute-like seeds can be seen (Chinmayi et al. 2017).

Phytochemical analyses of *E. nerifolia* plant extracts revealed the presence of various classes of secondary metabolites, including euphol, monohydroxy triterpene, nerifoliol, taraxerol, β-amyrin, glut-5-(10)-en-1-one, neriifolione, and cycloartenol, etc. (Naga et al. 2016). Furthermore, it contains phytoconstituents, such as

flavonoids, monoterpenoids, diterpenoids, triterpenoids, and alkaloids, which prove to have significant antimicrobial properties. Moreover, it consists of antianxiety, anticonvulsant and antipsychotic effects, immunomodulatory, anti-inflammatory, analgesic properties, and hepatoprotective activity.

Fatal doses of *E. nerifolia* depend on the usage parts of the plants as follows. Root Powder: 0.5-1 g, Stem Juice: 5-10 ml, Milky Latex: 125-50 mg (Mali and Panchal 2017). Toxicity of latex when applied to skin cause vesication, and inflammation involving eyesight when droop into eyes. When latex is taken inside, it causes irritation in the gastrointestinal tract, vomiting, diarrhea, burning sensation in the abdomen, convulsions, and coma (Chinmayi et al. 2017).

Latex of *E. nerifolia* is acrid, pungent, and good for tumors, abdominal troubles, and leukoderma. It is also used as a purgative, rubefacient, carminative, expectorant, whooping cough, gonorrhea, dropsy, leprosy, asthma, dypepesia, jaundice, enlargement of spleen, colic, and stone in bladder. It is used to remove cutaneous eruptions and warts. It is liable to cause dermatitis. It is used as a drastic purgative in the enlargement of the liver and spleen, syphilis, dropsy, general anasarcaleprosy, etc. (Wadhwani et al. 2021).

20.8 *Calotropis gigantea* (L) Dryand

20.8.1 Family Apocyanaceae

Genus *Calotropis* has two common species viz. *Calotropis procera* (*Rakta Arka in Sanskrit*) and *Calotropis gigantea* (*Sweata Arkain Sanskrit*), which are described as possessing vital pharmacological properties in Ayurveda toxicology and therapeutics. *Calotropis procera* is more toxic than *Calotropis gigantea* and is assumed to be even more poisonous than cobra venom. *C. gigantea* is commonly used for traditional medicinal systems than the *C. procera* (Lakshani et al. 2022).

Calotropis gigantea is an erect and highly branched and perennial shrub or small tree with pale gray color barks and young woolly hairy shoots, with milky latex throughout the plant. Leaves are simple, opposite, sub-sessile, slightly thick and fleshy. *C. gigantea* bears flowers with white, cream, pale pink, or purple in color. Fruit is a couple of fleshy avoid follicles and seeds are ovoid in shape 5–6 mm long, small, and flat (Imosemi 2016).

The latex of *C. gigantea* contains a powerful bacteriolytic enzyme, a very toxic glycoside calactin (the concentration of which is increased following insect or grasshopper attack as a defense mechanism), calotropin D I, calotropin D II, calotropin-F I, calotropin F II, and a nontoxic protealytic enzyme calotropin (2%-3%). This plant shows anticancer, antifungal, and insecticidal activity, hepatoprotective activity, anti-inflammatory, antipyretic, analgesic, and antimicrobial effects, oestrogenic functionality and larvicidal activity (Imosemi 2016).

Due to the toxicity of the latex, it may cause quick, painless vision loss, and photophobia in human eyes. Local administration of the latex of *C. gigantea* induces intense inflammatory response. Accidental exposure to latex produces contact dermatitis, keratitis, and toxic iridocyclitis (Sinha and Biswas 2016).

However, *C. gigantea* is used in traditional medicinal systems to treat a variety of conditions including leucoderma, ulcers, rheumatoid arthritis, and illnesses of the spleen, liver, and stomach. Furthermore, it is used as a purgative and antihelmintic agent and traditionally, the powdered root is used to treat bronchitis, asthma, leprosy, eczema, and elephantiasis while the latex is used to treat vertigo, baldness, hair loss, toothache, intermittent fevers, rheumatoid/joint swellings, and paralysis. The leaves are used to treat joint pains and to reduce swelling (Lakshani et al. 2022).

20.9 *Nerium indicum* L. (*Syn. N. oleander* L.)

20.9.1 Family: Apocynaceae

Nerium indicum is a small evergreen tree that grows to a height of 2 to 5 meters and is found in a variety of geographical and ecological locations. Nerium is popular for accidental poisoning especially among children and sometimes intentional suiciding too in adults. The leaves are 5 to 20 cm long, acuminate or acute, shortly petiolate, narrow, with a dark-green blade. Flowers are produced in terminal clusters about 5 cm in diameter with five petals and different colors vary from lilac, salmon, carmine, deep to pale pink, purple, copper, apricot, orange, white, and yellow. The fruit consists of a narrow follicle of 7.5 to 17.5 cm long, opens to disperse fluffy seeds, and can be propagated by seed (Bandara et al. 2010).

Nerium plant contains a wide range of phytochemicals such as phenolic, glycosides, alkaloids, tannin, and flavonoids. All parts of the plant are toxic and contain a variety of cardiac glycosides including neriifolin, thevetin A, thevetin B, and oleandrin (Dey et al. 2012). Based on the HPLC analysis, cinnamic acid was the major component and other components were epicatechine, catechin, and chlorogenic acid. Major of the fractions was pectic polysaccharide which is composed of arabinose, galacturonic acid, galactose, and rhamnose (Bandara et al. 2010). *N. indicum* consists of wide range of pharmacological activities such as antinociceptive activity, antimicrobial activity, antifungal activity¸ potent antiviral activity, larvicidal activity against Aedes aegypti, antitumor activity, cardiotonic activity, neuroprotective actions, antioxidant capacity, hepatotoxic activity, antidiabetic activity, analgesic activity, antiulcer activity, and anti-inflammatory activity (Schmelzer and Gurib-Fakim 2013). Due to the toxicity of unpurified Nerium, ingestion of any part of the plant results in nausea, vomiting, abdominal pain, diarrhea, dysrhythmias, and hyperkalemia (Dey et al. 2012).

Ethno-medicinal uses of *Nerium oleander* include the treatment of diseases such as cardiac illnesses, asthma, corn, cancer, and epilepsy. A dye obtained from the

flower is used in the treatment of skin diseases and also possesses wound healing and anti-inflammatory properties. Hot water extract of the leaves and seeds is used for upper respiratory tract and gastrointestinal tract infections in Kenya. In the Philippines, the plant is used as ethnomedicine to treat fever, headache, and dermatological problems (Schmelzer and Gurib-Fakim 2013).

20.10 *Gloriosa superba* L

20.10.1 Family: Liliaceae

Gloriosa superba is a herbaceous perennial plant that grows from a fleshy rhizome. It is scandent, climbing using modified leaf-tip, called tendrils. Flowers are axillary, solitary, bisexual, and regular. Fruit is a loculicidal, oblong capsule containing up to 20–40 seeds (Chaudhary et al. 2019).

G. superba tubers contain colchicines, benzoic and salicylic acid, sterols, and resinous substances such as colchicines, 3-demethyl colchicine, 1,2-didemethyl colchicine, 2,3-didemethyl colchicine, N-formyl, N deacetyl colchicines, colchicocide, gloriosine, tannins, and superbine (Ashokkumar 2015). The different parts of *G. superba* exhibited diverse pharmacological activities as anti-microbial and anti-cancer, antibacterial, antifungal, mutagenic, antioxidant, antihaemolytic, and anti-thrombotic/anticoagulant activity (Umavathi et al. 2020).

Sufficient ingestion of alkaloidcolchicines of this plant could cause fatalities to humans and animals. This is majorly found in tubers of *Gloriosa superba.* It slows the division of cell and sedates the bone marrow. Two hours after ingestion of the tubers, the person develops symptoms such as nausea, coldness, soreness of the throat, and diarrhea and leads to severe dehydration. Along with excessive hair loss has been reported among the patients. Symptoms of alopecia and dermatitis may develop after two to three weeks of poisoning. The toxic conditions of the *G. superba* shows the symptoms such as gastroenteritis, hematological abnormalities, acute renal failure, and cardiac arrest (Bhide and Acharya 2012).

The plant *G. superba* is commonly used to cure skin disease, snake bites, as an abortifacient, an antidote for a scorpion sting, healing of wounds, and as head lice killer (Bhide and Acharya 2012). Root of this plant is used as an ingredient in many Ayurvedic classical formulations and indicated for various clinical conditions such as inflammation/odema, wound, lymphadeinitis, skin diseases, hair loss, loss of apettite, piles and joint pains/arthritis, and many others (Datta et al. 2021).

20.11 *Cannabis sativa* L

20.11.1 Family: Cannabinaceae

Cannabis sativa L. (cannabis) is well known for its psychoactive and medicinal properties. Cannabis is considered to be the most widely used illegal drug in the world, and it was estimated that 4% of the world's adult population (160 million) were using cannabis. The two main preparations derived from cannabis are Marijuana: referring today as the dried leaves and flowers of the Cannabis plant and Hashish (Arabic name for Indian hemp) is the viscous resin of the plant (Kuddus et al. 2013).

The erected stems of the plant are growing from 3 to 10 feet or more high, very slightly branched, and having greyish-green hairs. Leaves are palmate in shape, with five to seven leaflets, numerous, on long thin petioles with acute stipules at the base, and the margins of the leaves are sharply serrate. The flowers are unisexual and the fruits are small, smooth, light brownish-grey in color, and completely filled by the seed (Kopustinskiene et al. 2022).

The number of phytochemicals isolated from *C. sativa* plant and currently 566 compounds have been identified and isolated. These phyto-constituents over 18 classes of different secondary metabolites found in the cannabis plant. Also these chemicals are mainly in the flowers and leaves of the plant (Al Ubeed et al. 2022). Among these 566 compounds, 125 are cannabinoids, 198 are non-cannabinoids and 120 are terpenes, and the rest of the substances identified in the plant include 2 other alkaloids, flavanoids, phenols, and sterols (ElSohly et al. 2017). Due to the terpenes produced by the *C. sativa* plant, it gets an aromatic odor in female plants and these terpenes have insect-repellent properties and slow down the growth of neighboring vegetation (Radwan et al. 2017).

The cannabinoids in *C. sativa* are unique substances formulated from terpenes and phenols. Approximately 100 cannabinoids are formed in epidermal trichomas of the plant. These cannabinoids contain different types of cannabinoids including tetrahydrocannabinol (THC), cannabinol, cannabidiol (CBD), cannabigerol, and cannabichromene. Among these, **two** cannabinoids are currently licensed for medicinal uses. Synthetically prepared tetrahydrocannabinol **(Marinol)** has licensed in the USA for the treatment of nausea which occurs after the cancer chemotherapy and for the appetite enhancement in patients with AIDS. **Nabilone** is the second cannabinoids which is completely synthetically derived and licensed in the UK for the treatment of nausea caused by chemotherapy (Shakil et al. 2021).

The main use of Hemp as a medicine is for lessening pain and comforting influence in nervous disorders. It is useful in gout, neuralgia, rheumatism, insanity, insomnia, etc. The action of this medicine is almost entirely on the higher nerve centers. It can produce an exhilarating intoxication with hallucinations. Variety of pharmacological actions is shown by Cannabis such as analgesic effect, anticancer activity, antidepressant-like actions, antidiuretic activity, antiemetic,

anti-inflammatory activity, antitumor activity, and appetite enhancement and bronchoconstrictor activity (Kopustinskiene et al. 2022).

Cannabis is used as a medicine in various traditional medicinal systems of the world due to its high therapeutic value. According to the ethnomedicinal uses, Cannabis was used as a medicine before the Christian era in Asia, particularly in Sri Lanka and India. In Sri Lanka, purified leaves were used to prepare many pills and pastes which indicated for diseases especially in gastro-intestinal tract, respiratory system, and nervous system (Sharma 2017). It has been found that folk medicine practitioners of Bangladesh used *Cannabis sativa* to treat various ailments such as sleep-associated problems, neuropsychiatric and CNS problems, and infections and respiratory problems followed by rheumatism, gastrointestinal, gynecological, cancer, sexual, and other ailments (Fordjour et al. 2023).

The *C. sativa* is also used in the food industry, mostly worldwide, which uses seeds and their derivatives commonly. However other plant parts such as sprouts, leaves, and flowers are consumed in raw state in dishes and drinks. These plant parts contain higher amounts of bioactive phytochemicals like polyphenols and cannabinoids than the cannabis seeds. Cannabis intake makes up 7.29% of all uses, with 58.72% corresponding to traditional meals and 41.28% to traditional beverages (Ngernsaengsaruay et al. 2023).

20.12 *Papaver somniferum* L.

20.12.1 Family: Papaveraceae

Papaver somniferum is also known as opium poppy which used as medicine since long time. ***P. somniferum*** is an annual herb of 0.27–1.5 m tall with milky white latex with characteristic odor. This white color latex turns into dark brown or black when exposed to the air. This exudate can be obtained from the pod when it is cut. Leaves are simple, cauline, spiral, sessile; lamina unlobed or pinnately lobed, lanceolate, lanceolate-ovate, ovate, or broadly ovate, and upper surface is green in color and lower surface is pale color. Flowers *are* solitary, terminal, 1–3(−6) per plant, showy, actinomorphic, fully opened and erect flowersgreen outside, turning pale green or pale green–yellow (nearly opened flowers), white inside. Fruits are capsular, poricidal (pores below stigmatic disk), green, glabrous, glaucous, turning stramineous, brown, or black–brown when dried (Butnariu et al. 2022).

Papaver somniferum produces some of the most widely used medicinal alkaloids such as morphine, codeine, thebaine, neopine and porphyroxine. Apart from these alkaloids, opium poppy produces approximately 80 alkaloids belonging to various tetrahydrobenzylisoquinoline-derived classes. Morphinan alkaloids accumulate in the latex of *Papaver somniferum* pods. *P. somniferum* was the first source of natural drugs that obtain morphine as analgesic drugs in the natural product field. Also *P. somniferum* plant is a rich source of phenolic compounds, such as anthocyanins,

flavonols, and the characteristic indole derivatives and essential oil volatiles, which are responsible for various pharmacological actions (Masihuddin et al. 2018). Pharmacological properties of this plant are antiallodynic activity, analgesics activity, antitussive activity, anti-diarrheal activity, anti-smoking activity (Addiction), antidepressant activity, anti-anxiety activity, bronchodilator activity, antibacterial activity, narcotic activity, etc. (Lal 2022). Furthermore, it shows hypnotic and sedative, astringent, expectorant, diaphoretic, and antispasmodic actions.

Opium is derived from the dried latex of the *P. somniferum* pods and can be taken intravenously injected, smoked, and in the form of pills. The opium and its components, especially the three morphine alkaloids, viz. morphine, codeine, and thebaine, are the natural (authentic) narcotics—generally referred to as "opiates." They possess narcotic property, which produces insensibility or stupor because of their depressant effects on the central nervous system, results reducing pain and inducing sleep (Verma et al. 2019). Opioids, by acting on the opioid receptors, have the advantageous therapeutic effect of blocking pain sensation. However, it also may cause number of uncomfortable effects, including nausea, constipation, sedation, and most significantly, respiratory depression as a harmful effect (Ray et al. 2006).

In Sri Lanka, dried excudate of P. somniferum plant was used to prepare traditional pills and pastes for different ailments (Sharma 2017). Opium tincture was used for several diseases by traditional physicians of India such as conjunctivitis, for relief of pain and renal colics, common cold, cough, insomnia, and as an anti-diarrheal agent (Guo et al. 2018).

20.13 *Strychnosnux vomica* L.

20.13.1 Family: Loganiaceae

Strychnosnux-vomica is a tree that is 5–25 m in height. Leaves are papery and have an opposite decussate arrangement, while the leaf blade is suborbicular, broadly elliptic, or ovate. Flowers are dull-green and white in color. Fruit is a brownish-yellow color berry that contains a gelatinous pulp in which one to five seeds are embedded. The seed has a flattened disk shape and is slightly concave and completely covered with hairs radiating from the center to the sides; thus, giving the seed a very characteristic sheen. The seed with dark gray horny endosperm, where the small embryo is housed, has no odor but tastes very bitter. These seeds can be kept for long periods (Perumal et al. 2016). The main usage part of this plant is seeds which need to be purified before preparing medicines from the seeds.

The dried seeds of Nux vomica contain 2.6%- 3% total alkaloids and out of these alkaloids strychnine and brucine are considered the major bioactive and principal toxic compounds and the rest are vomicine and igasurine (Eldahshan and Abdel-Daim 2015). Some other minor alkaloids are αcolubrine, β-colubrine, 3-methoxyicajine, protostrychnine, novacine, n-oxystrychnine, pseudostrychnine,

isostrychnine, chlorogenic acid, and glycoside (Behera et al. 2017). Strychnine and brucine act as stimulant effects on the nervous system. Also, these alkaloids increase the secretion of gastric juices and enhance sensory awareness. Furthermore, it is used as a respiratory and circulatory stimulant, especially in debilitating conditions (Kushwaha et al. 2014).

The seeds of purified Nux-vomica seeds are mainly used as aphrodisiac, appetizer, digestive, purgative, and stimulant. They are also used in anemia, asthma, bronchitis, intermittent malarial fever, and in the weakness of extremities. Purified seeds are recommended during senility as rejuvenator, which has antioxidant properties, since it is considered to be potent drug in old age problems. Unpurified Nux vomica seeds are powerful poison in large doses, producing tetanic convulsions and eventually death (Semalty et al. 2010).

It is observed that seeds of *Strychnosnux-vomica* are used as an ingredient of compound formulations of Ayurvedic medicine, indicated for nervous debility, paralysis and weakness of limbs, sexual weakness, dyspepsia, dysentery, and chronic rheumatism. Additionally, it is employed in the treatment of anemia, asthma, bronchitis, colic, intermittent fever, hysteria, etc. in a specific therapeutic dose (Semalty et al. 2010).

20.14 *Semecarpus anacardium* Linn

20.14.1 Family: *Anacardiaceae*

Semicarpus anacardium is a medium-to-large size tree, 15–25 m in height with grey bark exfoliating in small irregular flakes. Leaves of the plant are simple and alternate, obviate—oblong in shape. Flowers are greenish white in color and the fruit is a nut that is ovoid in shape and smooth lustrous black in color (Paras Jain and Sharma 2013).

Semecarpus anacardium is a well-known plant for its medicinal value in Ayurveda and Siddha systems of medicine. The most important phyto-constituents of the *S. anacardium* are alkaloids, phenolic compounds, flavonoids, sterols, glycosides, minerals, vitamins, and amino acids. An alkaloid, Bhilawanol, has been isolated from oil and seeds (Mishra and Bhavprakash 1984). Also, anacardoside, semecarpetin, nallaflavanone, jeediflavanone, semecarpuflavanone, galluflavanon, anacarduflavone, mono-olefin I, diolefin II, bhilawanol-A, and bhilawanol-Betc were isolated from the nut extracts. A variety of nut extract preparations from this source are effective against many diseases, viz., arthritis, tumors, skin diseases, and infections (**Paras Jain and Sharma** 2013). Anti-inflammatory, antiarthritic, antioxidant activity, hypolipidemic, hypoglycaemic, anti-atherogenic, anti-inflammatory, antifertility, and neuroprotective activities of *Semecarpus anacardium* nut with different solvents are also reported on experimental animal and cell lines (Mishra and Bhavprakash 1984).

Various reports on trials by Ayurvedic practitioners indicate that *Semecarpus anacardium* Linn can be given orally with milk, Ghee, peanut oil, etc. and with such type of administration toxic effects are not observed but on the contrary anabolic effects are obtained (Premalatha 2000). The toxic reactions of high dosages are diarrhea, vomiting, swelling all over the body, ulceration, and vesication of the skin. On external application, least appearance of redness or rash of the skin or an itchy or an uneasy sensation in any part of the body, hence should be discontinued immediately (Patel et al. 2020).

In Ayurveda, detoxified*Semicarpus anacardium*are is used against several diseases such as skin disease, tumors, malignant growth, fungal disease, excessive menstruation, vaginal discharge, fever, hemoptysis, constipation, and intestinal parasites. It is also used as a brain tonic, blood purifier, and hematinic tonic. The combination, *Semecarpus anacardium, Terminalia chebula,* and *Sesamum indicum L.* seeds powders with jaggery, has excellent results in chronic rheumatic disorders. In dysmenorrheal (painful menstruation) and oligomenorrhea (scanty menstruation), the medicated milk or its oil is salubrious (Semalty et al. 2008). Bhallataka is used for hair care in the traditional system of medicines. It is used for dyeing, and promoting hair growth in folk medicine (WHO and Tewari n.d.) (Fig. 20.2).

Fig. 20.2 Morphology of plant part that contain toxins

20.15 Conclusion

Medicinal plants have been used for centuries and in many cultures still rely on herbal medicines for their primary health care practices. Among these herbal medicines, preparations that contain toxic substances can be effectively used in traditional therapeutics. The aim of this review is to provide information pertaining to toxic medicinal plants used for various treatments in traditional systems. It is evident that after correct and proper purification methods mentioned in traditional systems were helpful in detoxification of the toxic plant materials. These detoxified plant materials have a potential role and when used with correct therapeutic doses, can be safely used in various disease conditions such as skin diseases, neurological disorders, diseases in gastrointestinal tract, and in infectious diseases. This review reveals that various phyto-constituents identified from the toxic medicinal plants, possess pharmacological activities such as purgative, diuretic, anticancer, antiallergic, anti-inflammatory, analgesic, and antipyretic. Also, this review categorized the various purification procedures according to the plant toxins mentioned in Ayurveda and other traditional systems. This information would be useful in future development of natural medicines for different ailments.

References

Acharya JT (2011) Agnivesa: Charaka Samhita. Chawkhambha Vidyabhawan, Varanasi, p 23

Ajay Kumar M, Poorna V, Ravindra S, Kusuma G, Narayanam S, Kartar Singh D, Mohit M, Sadhna C, Amit Kumar D (2022) Detoxification of Datura metel L. seeds using Shodhana (purifying process) and estimation of scopolamine content. J Drug Res Ayurvedic Sci 7(4):229–242. https://doi.org/10.4103/jdras.jdras_45_21

Al Ubeed HMS, Bhuyan DJ, Alsherbiny MA, Basu A, Vuong QV (2022) A comprehensive review on the techniques for extraction of bioactive compounds from medicinal cannabis. Molecules 27:604. https://doi.org/10.3390/molecules27030604

Al-Snafi AE (2017) Medical importance of Datura fastuosa (syn: Datura metel) and Datur stramonium: a review. IOSR J Pharm 7(2):43–58

Ashokkumar K (2015) *Gloriosa superba* (L.): A brief review of its phytochemical properties and pharmacology. Int J Pharmacogn Phytochem Res 7(6):1190–1193

Ayurveda Pharmacopoea of Sri Lanka (1979) Department of Ayurveda, Sri Lanka

Bandara V, Weinstein SA, White J, Eddleston M (2010) A review of the natural history, toxinology, diagnosis and clinical management of Nerium oleander (common oleander) and Thevetia peruviana (yellow oleander) poisoning. Toxicon 56(3):273–281

Behera MC, Mohanty TL, Paramanik BK (2017) Silvics, phytochemistry and ethnopharmacy of endangered poison nut tree (Strychnos nux-vomica L.): a review. J Pharmacogn Phytochem 6(5):1207–1216

Bhide B, Acharya R (2012) Uses of Langali (Gloriosasuperba Linn.): an ethnomedicinal perspective. Ayurpharm Int J Ayur Alli Sci 1(3):65–72

Bramhanand T, Dephande P (2002) Charak Samhita. Chukhambha Orientalia, Varanasi, p 746

Bramhanand T, Dephande P (2007) Asthanga Hridaya. Chukhabha Sanskrita Pratisthan, Varanasi, p 1144

Butnariu M, Quispe C, Herrera-Bravo J, Pentea M, Sarac I, Küşümler AS, Özçelik B, Painuli S, Semwal P, Imran M, Gondal TA, Emamzadeh-Yazdi S, Lapava N, Yousaf Z, Kumar M, Eid AH, Al-Dhaheri Y, Suleria HAR, Del Mar Contreras M, Sharifi-Rad J et al (2022) *Papaver* plants: current insights on phytochemical and nutritional composition along with biotechnological applications. Oxidative Med Cell Longev 2022:1. https://doi.org/10.1155/2022/2041769

Chandewar RN, Harode RS (2022) A review of Vatsanabh (Aconitum Ferox) in Ayurveda. J Ayurveda Integr Med Sci 7(4):45–48. https://jaims.in/jaims/article/view/1837

Chaudhary S, El-Shorbagi A, Shridhar B, Gupta M, Verma H (2019) A review on phytochemical and pharmacological profile of gloriosa superba Linn. Int Res J Pharm 10:1–5. https://doi.org/10.7897/2230-8407.1004113

Chinmayi U, Satish AND, S. (2017) A review on Euphorbia neriifolia plant. Int J Pharma Chem Res 3:149–154

Dahikar GK, Rathi B, Phadnis R, Kamble SB (2023) Concept of purification w.r.t. Visha Dravya (poisonous drugs). Int J Ayu Pharm Chem 19(2):106–117

Datta S, Ramamurthy PC, Anand U, Singh S, Singh A, Dhanjal DS, Dhaka V, Kumar S, Kapoor D, Nandy S, Kumar M (2021) Wonder or evil?: multifaceted health hazards and health benefits of Cannabis sativa and its phytochemicals. Saudi J Biol Sci 28(12):7290–7313

Deore SL, Moon KV, Khadabadi SS, Deokate UA, Baviskar BA (2013) Evaluation of toxicity of 'Vatsanabha' (Aconitum ferox, Ranunculaceae) before and after Shodhana. J Young Pharm 5(1):3–6

Dey P, Roy S, Chaudhuri TK (2012) A quantitative assessment of bioactive phytochemicals of Neriumindicum: an ethnopharmacological herb. Int J Res Pharm Sci 3(4):579–587

Eldahshan OA, Abdel-Daim MM (2015) Phytochemical study, cytotoxic, analgesic, antipyretic and anti-inflammatory activities of Strychnos nux-vomica. Cytotechnology 67(5):831–844. https://doi.org/10.1007/s10616-014-9723-2

ElSohly MA, Radwan MM, Gul W, Chandra S, Galal A (2017) Phytochemistry of Cannabis sativa L. Prog Chem Org Nat Prod 103:1–36

Fordjour E, Manful CF, Sey A, Javed R, Pham TH, Thomas R, Cheema M (2023) Cannabis: a multifaceted plant with endless potentials. Front Pharmacol 14:1200269. https://doi.org/10.3389/fphar.2023.1200269

Guo R, Wang T, Zhou G, Xu M, Yu X, Zhang X, Sui F, Li C, Tang L, Wang Z (2018) Botany, phytochemistry, pharmacology and toxicity of Strychnosnux-vomica L.: a review. Am J Chin Med 46(01):1–23

Imosemi IO (2016) Evaluation of the toxicity, medicinal use and pharmacological actions of Calotropisprocera. Eur J Pharm Med Res 3(9):28–36

Ishola AO, Imam A, Ajao MS (2021) Effects of datumetine on hippocampal NMDAR activity. Toxicol Rep 8:1131–1142

Islam T, Ara I, Islam T, Sah PK, de Almeida RS, Matias EFF, Ramalho CLG, Coutinho HDM, Islam MT (2023) Ethnobotanical uses and phytochemical, biological, and toxicological profiles of *Datura metel* L.: a review. Curr Res Toxicol 4:100106. https://doi.org/10.1016/j.crtox.2023.100106

Jayaweera DMA (1980) Medicinal plants (indigenous and exotic) used in Ceylon. National Science Foundation, Colombo

Kopustinskiene DM, Masteikova R, Lazauskas R, Bernatoniene J (2022) Cannabis sativa L. bioactive compounds and their protective role in oxidative stress and inflammation. Antioxidants 11:660. https://doi.org/10.3390/antiox11040660

Kuddus M, Ginawi IA, Al-Hazimi A (2013) Cannabis sativa: an ancient wild edible plant of India. Emir J Food Agric 25:736–745

Kushwaha RK, Berval R, Sharma A (2014) The therapeutic and toxicological effect of kupilu (Strychnosnux-vomica L.): a review. Ayushdhara 1(2):1–4

Lakshani AARP, De Silva NN, Dahanayake N (2022) A review on *Calotropisgigante*a (WARA): A miracle plant. J Agro-Technol Rural Sci 2(1):5–8. https://doi.org/10.4038/atrsj.v2i1.36

Lal RK (2022) The opium poppy (*Papaver somniferum* L.): historical perspectives recapitulate and induced mutation towards latex less, low alkaloids in capsule husk mutant: a review. J Med Plant Stud 10(3 Part A):19–29
Li W (2021) Toxic effects of Datura. Int J Drug Res Technol 10(11):1–2
Mali PY, Panchal SS (2017) Euphorbia neriifolia L. review on botany, ethnomedicinal uses, phytochemistry and biological activities. Asian Pac J Trop Med 10(5):430–438
Malviya N, Choudhari S (2018) Review of Vatsanabha and its toxicity. World J Pharm Med Res 4(8):115–117
Masihuddin M, Jafri MA, Siddiqui A, Chaudhary S (2018) Traditional uses, phytochemistry and pharmacological activities of papaversomniferum with special reference of unani medicine an updated review. J Drug Deliv Ther 8(5-s):110–114
Maurya SK, Seth A, Laloo D, Singh NK, Gautam DNS, Singh AK (2015) Śodhana: an Ayurvedic process for detoxification and modification of therapeutic activities of poisonous medicinal plants. Anc Sci Life 34(4):188–197
Mirgane RA (2021) Snuhi (*Euphorbia nerifolia* Linn.) Upavisha–a review
Mishra BS (2010) Yogaratnakara. Choukambha Prakashana, Varanasi, pp 167–169
Mishra B, Bhavprakash RV (1984) The kasha Sanskrit Series, 6th edn. Chaukhambha Sanskrit Sansthan, Varanasi, p 141
Mitra S, Shukla VJ, Acharya R (2011) Role of castor oil in processing (Shodhana) of kupeelu (Strychnosnux vomica Linn.) seeds: an approach of traditional Ayurveda. Int J Ayurvedic Med 2(2):62–71
Monira KM, Munan SM (2012) Review on Datura metel: a potential medicinal plant. Glob J Res Med Plant Indigenous Med 1(4):123
Naga A, Jyothi A, Rajalakshmi R (2016) A review on Upavisha-snuhi (Euphorbia neriifolia Linn). Int Ayurvedic Med J 4:1231–1234
Ngernsaengsaruay C, Leksungnoen N, Chanton P, Andriyas T, Thaweekun P, Rueansri S, Tuntianupong R, Hauyluek W (2023) Morphology, taxonomy, anatomy, and palynology of the opium poppy (Papaver somniferum L.) cultivation in Northern Thailand. Plants 12(11):2105
Panda AK, Debnath SK (2010) Overdose effect of aconite containing Ayurvedic medicine ('Mahashankha Vati'). Int J Ayurveda Res 1(3):183
Paras Jain PJ, Sharma HP (2013) A potential ethnomedicinal plant: Semecarpusanacardium Linn.: a review. Int J Res Pharm Chem 3(3):2231–2781
Patel Y, Savitha DB, Acharya R, Ashok BK, Shukla VJ (2010) Role of Shodhana on analytical parameters of Datura innoxia Mill and Datura metel Linn seeds. Int J Res Ayurveda Pharm (IJRAP) 1(2):249–254
Patel D, Inchulkar SR, Kaushik Y, Chauhan NS (2020) A toxicological review of Bhallataka. J Ayurveda Integr Med Sci 5(2):209–214
Perumal B, Velmani S, Santhosh C, Maruthupandian A (2016) Phytochemical and medicinal properties on Strychnos-nux-vomica (L). J Adv Appl Sci Res 1(3):96–99. https://doi.org/10.46947/joaasr13201624
Premalatha B (2000) Semecarpus anacardium Linn. nuts—a boon in alternative medicine. Indian J Exp Biol 38(12):1177–1182
Priyanka, Tiwari RC, Yadav A, Sharma A, Indermani (2023) Medicinal uses of poisonous plant – a brief review. Int J Ayurveda Pharma Res 10(12):48–51. https://doi.org/10.47070/ijapr.v.10i2.2614
Rachna I, Tomar BS, Chawla SK, Gusain T (2023) A critical review of Sthavara Visha 'Vatsanabha': a cardiotoxic poison. Int Res J Ayurveda Yoga 6:92–98. https://doi.org/10.47223/IRJAY.2023.6614
Radwan MM, Wanas AS, Chandra S, Elsohly MA (2017) Natural cannabinoids of cannabis and methods of analysis. Cannabis sativa l.-botany and biotechnology. Springer, New York
Rahmatullah M, Azam MNK, Mollik MAH, Hasan MM, Hassan AI, Jahan R, Jamal F, Nasrin D, Ahmed R, Rahman MM, Khatun MA (2010) Medicinal plants used by the Kavirajes of DaulatdiaGhat, Kushtia district, Bangladesh. Am-Eur J Sustain Agr 4:219–229

Rastogi S, Rastogi R, Singh RH (2007) Adverse effects of Ayurvedic drugs: an overview of causes and possibilities in reference to a case of Vatsanabha (aconite) overdosing. Int J Risk Saf Med 19(3):117–125
Ray, R., Kattimani, S., & Sharma, H. K. (2006). Opium abuse and its management: global scenario. World Health Organization Department of Mental Health and Substance Abuse Management of Substance Abuse. National Drug Dependence Treatment Centre All India Institute of Medical Sciences New Delhi, 1–13
Reid A, Oosthuizen CB, Fibrich BD, Twilley D, Lambrechts IA, de Canha MN, Rademan S, Lall N (2018) Chapter 1 - Traditional medicine: the ancient roots of modern practice. In: Medicinal plants for holistic health and well-being. Academic Press, London
Sabeena A, Chouhan R, Sultan P, Parvaiz Q, Gandhi S (2021) A comprehensive review of phytochemistry, pharmacology and toxicology of the genus Aconitum L. Adv Tradit Med 23:299. https://doi.org/10.1007/s13596-021-00565-8
Sastri A (1978) Vagbhattacharya: Rasaratna Samuchchaya, vol 590, 6th edn. Chaukhamba Sanskrit Series Office, Varanasi
Savrikar SS, Ravishankar B (2011) Introduction to 'Rasashaastra' the iatrochemistry of Ayurveda. Afr J Tradit Complement Altern Med 8(5 Suppl):66–62. https://doi.org/10.4314/ajtcam.v8i5S.1
Sayyed A, Shah M (2014) Phytochemistry, pharmacological and traditional uses of Datura stramonium L. J Pharmacogn Phytochem 2(5):123–125
Schmelzer GH, Gurib-Fakim A (2013) Plant resources of tropical Africa 11 (2): medicinal plants 2. Plant Resour Trop Africa 11(2):384
Semalty M, Semalty A, Joshi GP, Rawat MS (2008) Herbal hair growth promotion strategies for alopecia. Indian Drugs 45(9):689–700
Semalty M, Semalty A, Badola A, Joshi GP, Rawat MSM (2010) Semecarpusanacardium Linn.: A review. Pharmacogn Rev 4(7):88
Shakil SS, Gowan M, Hughes K, Azam MNK, Ahmed MN (2021) A narrative review of the ethnomedicinal usage of Cannabis sativa Linnaeus as traditional phytomedicine by folk medicine practitioners of Bangladesh. J Cannabis Res 3(1):1
Sharma KR (2017) Charaka Samhitha, vol 1. Chaukhambha Sanskrit Series Office, Varanasi, p 101
Sharma VB, Soni MK, Onkar JM, Sharma OP (2019) Benefits and uses of vatsanabh (Acontium ferox wall ex seringe): a review. World J Pharm Med Res 5(9):85–86
Sharma M, Dhaliwal I, Rana K, Delta AK, Kaushik P (2021) Phytochemistry, pharmacology, and toxicology of *Datura* species: a review. Antioxidants (Basel, Switzerland) 10(8):1291. https://doi.org/10.3390/antiox10081291
Shastri K (2012) Rasa Tarangani, 11th edn. Motilal Banarasidas, New Delhi, pp 651–652
Sinha SN, Biswas K (2016) A concise review on Nerium oleander L.—an important medicinal plant. Trop Plant Res 3:408–412
Srivastava R, Srivastava P (2020) The medicinal significance of Datura stramonium: a review. Biomed J Sci Tech Res 29(2):22223–22226
Subodh SA (2018) The effect of shodhan process on toxicity of upavisha kupeelu beeja (strychnos nuxvomica): a critical study. Ayushdhara 4(6):1430–1435
Tamilselvan N, Thirumalai T, Shyamala P, David E (2014) A review on some poisonous plants and their medicinal values. J Acute Dis 3(2):85–89
Umavathi S, Gopinath K, Manjula MS, Chinnasamy B, Ayyakannu A (2020) Gloriosasuperba L: a critical review of recent advances. Abasyn J Life Sci 3(2):48–65
Verma RK, Kumar R, Sankhla MS (2019) Forensic identifications of Papaver Somniferum (Opium) for fatal poisoning. Int Medico-legal Rep J 1(1):9
Wadhwani BD, Mali D, Vyas P, Nair R, Khandelwal P (2021) A review on phytochemical constituents and pharmacological potential of *Calotropis procera*. RSC Adv 11(57):35854–35878. https://doi.org/10.1039/d1ra06703f
Wang QH, Xiao HB, Yang BY, Yao FY, Kuang HX (2008) Studies on pharmacological actions of the effective parts for psoriasis in FlosDaturae (I). Chinese J Exp Trad Med Formulae 14:48–51

WHO, Tewari DN (n.d.) Report of the task force on conservation & Sustainable use of Medicinal Plants (Online). Available from: http://planningcommission.nic.in/aboutus/taskforce/tsk_medi.pdf

Willingsible K (2023) Understanding herb toxicity: a comprehensive overview. Med Aromat Plant 12:449

Chapter 21
Saxitoxin: An Omics Understanding with Future Ocular Applications

Sukanya Madhanmohan, Ranjith Kumar Ravi Kumar, Jayasree Balasubramanyan, Raghu Sripada, Narayanan Janakiraman, Sundara Ramprabhu, and Sailaja V. Elchuri

Abstract Saxitoxin, a notably potent neurotoxin produced by Harmful Algal Blooms, often associated with "red tides," presents a significant environmental challenge. This chapter delves into the complex nature of Saxitoxin compounds, focusing on their biosynthesis and genetic underpinnings from an omics perspective. It also examines the latest developments in diagnostic techniques for detecting this biotoxin and reviews the regulatory and sanitary standards within the marine and food industries. Additionally, this chapter explores Saxitoxin's potential therapeutic applications, particularly in ophthalmology, positioning it alongside botulinum toxin as one of the most extensively researched biotoxins. This comprehensive overview aims to encapsulate the latest scientific advancements and diverse interests surrounding Saxitoxin compounds.

Keywords Biotoxins · Harmful Algal blooms · Saxitoxin · Eye effects · Therapy

21.1 Introduction

Several algal species that produce toxins are collectively called harmful algal blooms (HAB). The biotoxins from marine organisms are classified as hydrophilic and lipophilic compounds (Fig. 21.1). Amnesic shellfish poisoning (ASP), paralytic shellfish poisoning (PSP), and pufferfish poisoning (PFP), whose toxins are not produced by HABs but by marine bacteria, are grouped in the hydrophilic marine biotoxins, while diarrhetic shellfish poisoning (DSP), yessotoxins (YTXs), azaspirazid poisoning (AZP), ciguatera poisoning (CP), neurotoxic shellfish poisoning

S. Madhanmohan · R. K. Ravi Kumar · J. Balasubramanyan · N. Janakiraman
S. V. Elchuri (✉)
Department of Nanobiotechnology, Vision Research Foundation, Sankara Nethralaya, Chennai, India

R. Sripada · S. Ramprabhu
Department of Physics, Indian Institute of Technology Madras, Chennai, India

M. N. V. Prasad et al. (eds.), *Biotoxins*,
https://doi.org/10.1007/978-3-031-75309-1_21

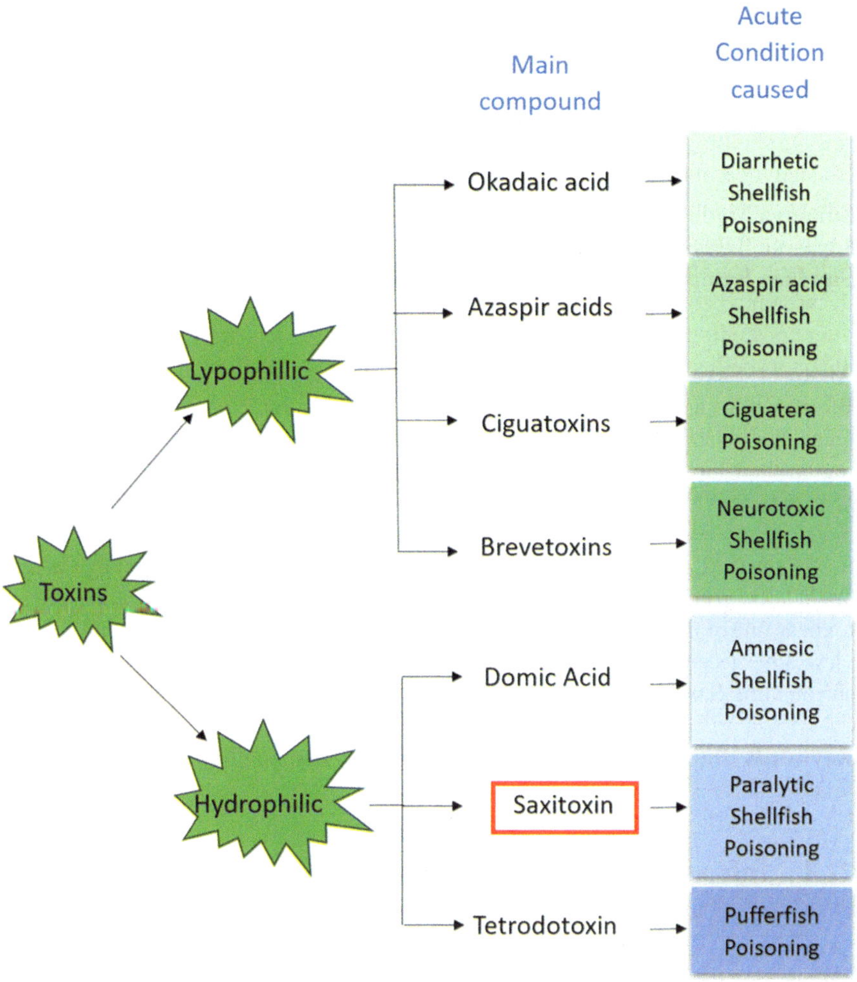

Fig. 21.1 Toxins from harmful algal blooms and their relevant acute conditions

(NSP), and cyclic imines (CIs) are included in the lipophilic marine biotoxins. This chapter focuses on Saxitoxin (STX), which is a family of >57 analogs that include guanyautoxins and STXs. They are alkaloids and can be classified as neurotoxins under the mode of action category. These are some of the most potent neurotoxins known. They are synthesized by some toxic species of cyanobacteria genera such as *Dolichospermum*, *Aphanizomenon*, *Cylindrospermopsis*, *Lynabya*, *Planktothrix, Raphidiopsis, Fischerella, Geitlerthena* and *Scytonema*. STX is also produced by planktons, *Gonyaulax*, marine dinoflagellates (*Alexandrium* sp., *Gymnodinium* sp) (Akbar et al. 2020). STX as a standalone compound from its analog has a molecular formula of $C_{10}H_{17}N_7O_4$ and, a molecular weight of 299.29 g/mol. It is a carbamate ester, a member of guanidines, a pyrrolopurine, and a ketone hydrate. Its conjugate

base is STX $^{(2+)}$. The Haz-map shows its LC50 as 0.3/m^3/24 hr. in the mouse. Previously, mouse bioassay was used to determine whether the shellfish is safe for consumption. This methodology is now replaced with analytical techniques. The EU regulations for toxicity confirmation for STX as 800 μg STX.2Hcl equivalent (eq)/kg shellfish flesh. Even though this limit is set for STX, all the analogs detected by analytical methods have to be determined. Recently, tetrodotoxin levels seem valid for the same level still it opens up the opportunity for regulation to be put forth in the food industry as and when new toxin forms or impacts on human health are reported. The analogs of STX have been metabolized to various other forms to form decarbomoyl analogs in the postmortem studies from 2 fishermen who died 3–4 hrs after consuming 7–9 ribbed mussels and had 8575 μg of STX in MBA equivalent to 100 g of shellfish meat (Finch et al. 2023). Every year about 2000 deaths are reported due to consumption of seafood contaminated with this toxin. The toxins produced in algae and Dinoflagellates end up in mammals, birds, and humans (Fig. 21.2). STX was observed in zooplankton, clams, worms, pelagic fishes, birds, and marine mammals (Lefebvre et al. 2022). STX poisoning was reported as a causal reason for the death of whales and walruses (Lefebvre et al. 2016), (Doucette et al. 2006). Research is ongoing for optimizing the conditions needed to estimate toxins in whales (Bowers et al. 2022; Frey et al. 2023). The warming of Oceans is observed which is affecting several Ocean communities and primary producers (Frey et al. 2023). Additionally, warming Ocean temperatures are causing the increased production of STX in dinoflagellates (Anderson et al. 2021). Similarly, Ocean warming has caused changes in Arctic community ecosystems by altering the planktonic community (Anderson et al. 2021). Furthermore, climate change can alter toxin-producing organisms and harmful bacteria available in the same niche leading to several diseases in humans (Chatterjee and More 2023). Therefore, the study of the toxin in changing environments and their potential role in biotechnological applications has received major interest recently. Additionally, about 895 million dollars

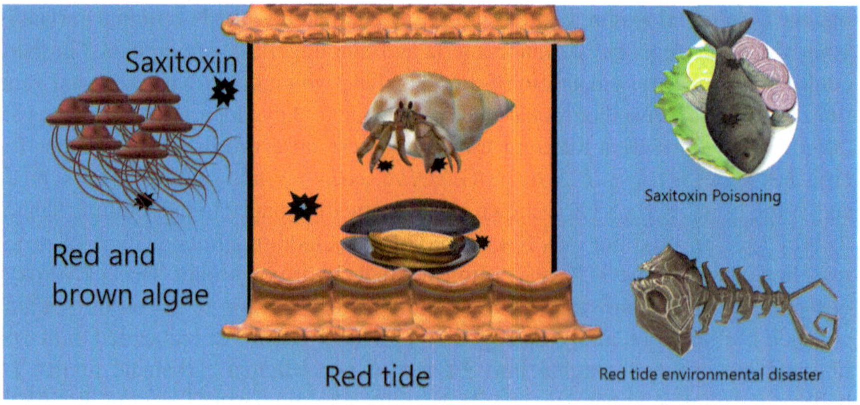

Fig. 21.2 The toxins secreted from marine algae can affect shellfishes feeding them that are in turn toxic to marine organisms and human. Drawn using Microsoft 3D Paint

are used by people worldwide to remove these toxins from their coastal waters eliciting their role in the economic aspects of the countries. More work will continue to happen in toxicology for cyanotoxins. Research on STX is emerging for food safety and industrial applications. On the other hand, for consumer safety new regulations and standards, analytical techniques, toxicity studies, new analogs, new toxins, and new food-borne carriers are ever emerging which demands new studies for public health (Amzil et al. 2023). These are a few aspects that will be discussed in this book chapter.

The world bodies have established regulatory guidelines on the toxins. The toxins-causing organisms were higher in 2000–2018 compared to 2018–2024 (Fig. 21.3). This suggests that collective regulations and the sharing of information between countries are showing promising results. EU-regulated toxins, such as STX and its analogs, and nonregulated toxins such as tetrodotoxin, and cyclic imines, have been recorded from *Perna viridis* and *Crassostrea madrasensis* from southwest Indian freshwater. They have used LC-PLD, LC-UV, and LC-MS/MS in detecting these toxins. In Canadian arctic waters along with phycotoxins (AST, PSTs), STX was also found. The author records it as a species-specific accumulation for ASTs. This opens up yet again the need to expand the regulatory guidance of biotoxins along with established STX guidelines (Pućko et al. 2023). In the toxins studied on the Italian coast (Mediterranean Sea coast) over a period of time, STX is detected but the regulatory limit set for this toxin cannot be compared to other toxins since other toxins exceed had high regulatory limit (Accoroni et al. 2024).

21.2 Toxicity of Saxitoxins

STX is synthesized by homologous *sxt* genes in Algae and dinoflagellates. From the genomic study of the cyanobacteria, it is observed that *sxtA* and *sxtG* produce toxins, the C-terminal variant conferred toxicity compared to the N-terminal variant is nontoxic. Phylogenetically cyanobacteria and dinoflagellates are unrelated or there could be an ancestral eukaryote of dinoflagellate which could have acquired gene transfer from various other prokaryotes independent of cyanobacteria (Hackett et al. 2013). There were eight different transcripts were observed in Dinoflagellates for eight key genes (*sxtA, sxtB, sxtD, sxtG, sxtH/T, sxtI, sxtS*, and *sxtU*) for STX synthesis (Wang et al. 2020). STX is shown electrostatically bonded to the sodium channel thereby blocking sodium ions from entering the cells through its guanidium structure. In neutral pH, the STX gets protonated in its hydroxyl groups. Hence hydroxyl group and guanidinium structure are important for the toxin to act on the sodium channel. Furthermore, all the analogs of STXs have a common unit called imidazoline and are grouped roughly into 3 types of non-sulfated carbamate toxins, 11 hydroxyls sulfated-carbamate, and 21 N-sulfocarbamyl toxin (Turner et al. 2017). STX is also shown to act on the potassium channel by modifying the channel gating thereby increasing the membrane depolarization and reducing overall potassium

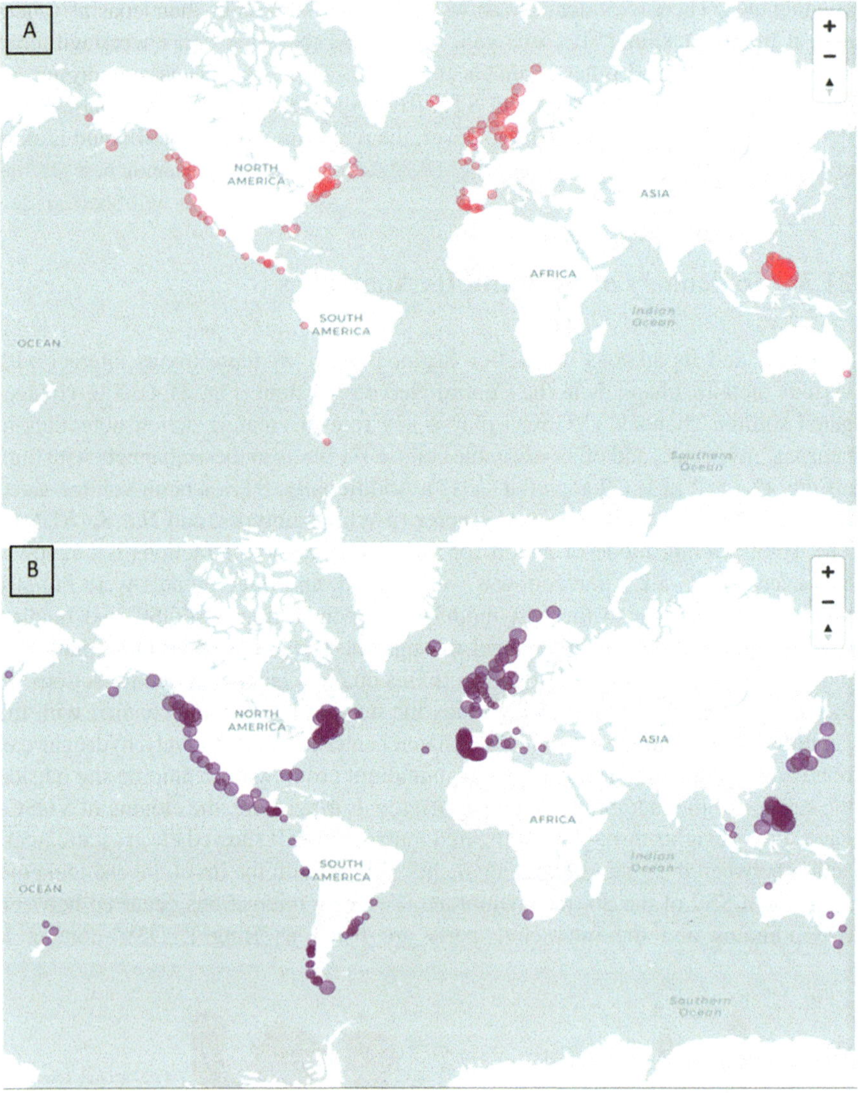

Fig. 21.3 The presence of marine toxins in the coasts of countries. (**a**). The year 2018–2024 and (**b**). 2000–2018. This map is generated from the interactive tool available for the public created by the Harmful Algal Information System (HAIS) for International Oceanographic Data and Information Exchange by the Intergovernmental Oceanographic Commission (IOC) of UNESCO). This map excludes incidences that happened to humans in large waterbodies other than coastal regions (US National Office for Harmful algal blooms 2024). The map was built using the Harmful Algal Information System (HAIS) available at https://data.hais.ioc-unesco.org/ (accessed on Jan 12th, 2024)

conductance. There is evidence to show STX acting on voltage-gated calcium channels at the extracellular site. It is well established that STXs act on voltage-gated sodium channels and other channels thereby its effect on the nervous system is established. In mice, it is seen that STX from *Acanthocardia tuberculation* exhibited oxidative stress in the liver. However, the renal function is not affected (Takati et al. 2023).

21.3 Toxicology of STX and Its Analogues

The STX and its analogs interaction are neurotoxic as these toxins interact with various metallic channels in the Central Nervous System (Fig. 21.4). The voltage-gated sodium channels (VGSCs) play a key role in creating action potentials in neurons, myocytes, and other excitable cells. STX binds to these channels with high affinity ($K_d \approx 2$ nM) (Zhang et al. 2013). Additionally, STX acts on voltage-gated K^+, Ca^{2+}, channels, acetylcholine receptor (nAchR) subtypes, and Na^+ K^+ ATPase. The exact binding mode of action for these nonspecific interactions is unknown. However, VGSC-STX interaction is well studied, and three models were initially proposed. According to the first model, STX penetration of VGSC and binding occurs at the bottom of the channel through the formation of an ionic pair. This model is rejected for its analog gonyautoxins due to a lack of steric interactions. In the second proposed model, STX binds the outside edge of the channel with the guanidium group facing the top of the channel entrance. Alternatively, hydrogen can bond with -OH, and ion pairing of the guanidium group with an anionic site outside the surface of the membrane could happen too. Furthermore, the cloning of VGSCs gave better insights through site-directed mutagenesis. It showed electrostatic interaction between the guanidium components of STX with the lip of the channel pore in segment SS2 of the S6 transmembrane. Two ring interactions occurred between two domains and the interactive parts are listed as Ring I: D384-domain I,

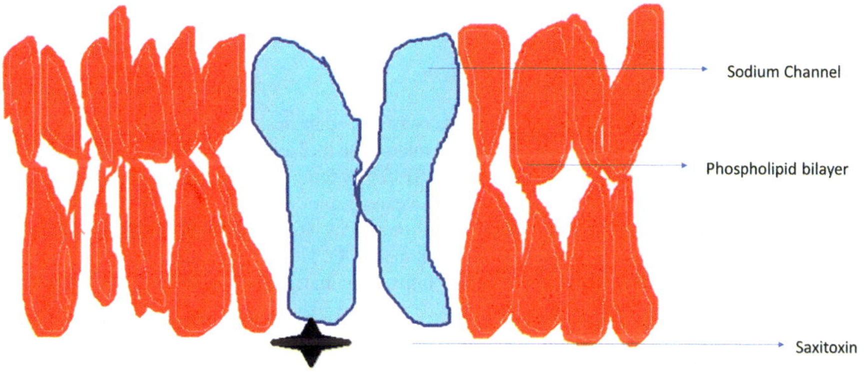

Fig. 21.4 The effect of STX binding on voltage-gated sodium channel

E392-domain II, K1423-domain III, and A1714-domain IV. Ring II: E387-domain I, E945-domain II, M1425-domain III, D1717- domain IV (Zhang et al. 2013). For Nav1.4 (VG SC)—STX interaction: Model 1 residues were P loop I-D400, E403; P loop II-E755, E758; P loop III-K1237, P loop IV. However, none of the predicted interactions were for high binding. Model 2 residues were; P loop I-D400, P loop II-C753, G754, P loop III-K1237, F1236, P loop IV. These interactions had less than <20% interactions. The author concluded that STX and its analogs interact electrostatically with Nav1.4. However, this does not correlate to the toxin's actual potency (Durán-Riveroll et al. 2016). It is also interesting to study the combined toxicology and synergistic effects of two toxins such as STXs and tetrodotoxin, STX and Gonyautoxin, Tetrodotoxin and Gonyautoxin, etc. For example, in tetrodotoxin (TTX) resistant CA1 neuronal cells which is a subpopulation of CA1 neuronal cell culture; Nav1.9 sensitivity is shown to a nanomolar concentration of STX against reversible BDNF evoked sodium currents (Gilchrist and Bosmans 2012).

21.4 STX as a Potassium and Calcium Channel Blocker

Even though studies are done on STX's binding activity on voltage-gated sodium channels, it is important to explore its activity on other channels such as Ca^{2+} channels and K^+ channels, ATPases, glutamate-gated channels, and receptors in CNS and PNS due to its high neurotoxic effect. A partial activity of STX and not TTX was observed on native L-type calcium channels in mouse ventricular myocytes and tsA-201 cell lines (Su et al. 2004). This binding effect is shown to be rapid in higher concentrations of about 10 μM. A K_d of 300 nM with a 49% inhibition was observed. A decrease in hERG K^+ currents by STX was observed. The STX here played an agonist-like role in increasing the K^+ currents for a short amount of time which they call a toxin as a gating modifier of K^+ channels (Wang et al. 2003).

21.5 Exploring the Intricacies of STX Biosynthesis: A Genomic, Environmental, and Translational Perspective

Saxitoxin, a potent neurotoxin, emerges from a complex biosynthetic web woven by aquatic microorganisms like cyanobacteria and dinoflagellates, particularly species of the *Alexandrium* genus. Research delving into the depths of this process has unveiled a myriad of genetic, environmental, and regulatory factors at play. In the world of cyanobacteria, the synthesis of STX is a story of meticulous orchestration involving a cluster of 26 genes, spanning from 25.7 kb to 35 kb in sequence length. This Biosynthetic Gene Cluster (BGC) illustrates the intricate genetic programming of organisms like cyanobacteria. However, this clustering is a stark contrast to the

genomic architecture in dinoflagellates, where such organized gene clusters are absent. Focusing on the *Alexandrium* species, such as *A. pacificum* (Abassi et al. 2023), *A. catenella*, and *A. fundyense*, researchers have dived into the translational regulation of STX biosynthesis. Here, the *sxtA* and *sxtG* genes are particularly noteworthy, regulated at the translational level, rather than during transcriptional RNA processing. This nuance in genetic control underscores the complex regulatory mechanisms dictating toxin synthesis. Nutrient conditions play a pivotal role in STX production. Microarray analyses revealed how varying nutrient ratios (C: N, C:P, N:P) and growth conditions (like mixotrophic environments) lead to the expression of previously unannotated genes in species like *Alexandrium*. Especially intriguing is the *sxtA4* gene in dinoflagellates, unique to these organisms and central to STX synthesis. Transcriptome studies have confirmed the independent evolutionary paths of STX biosynthesis in cyanobacteria and dinoflagellates. In *A. catenella*, the interplay of genes like *sxtO* and *sxtZ*, along with the regulation of the *sxtA* gene, paints a complex picture of genetic control mechanisms. Meanwhile, in *A. fundyense* exposure to soluble inorganic nitrogen triggers upregulation of the *sxtA* gene, demonstrating an environmental influence on genetic expression. Adding another layer to this ecological interplay, *A. minutum* was found to increase STX synthesis upon exposure to copepod amides from zooplankton like *Centropages typicus*. Interestingly, these zooplankton are resistant to STX, posing questions about ecological interactions and toxin resistance. In contrast, *S. trochoidea*, which does not produce STX, shows a different transcriptomic response to nitrogen depletion, suggesting species-specific adaptations to environmental factors. This narrative of STX biosynthesis in cyanobacteria and dinoflagellates, especially in *Alexandrium* species, is a testament to the intricate interplay of genetics and environment. The ongoing research not only deepens our understanding of toxic algal blooms but also sheds light on broader ecological dynamics and potential public health implications.

Light and temperature are important factors that can affect toxin production. STX production in *Cylindrospermopsis raciborskii* strains is increased when higher light and temperatures are given during the organism's growth (Mesquita et al. 2019). Interestingly, the effect of STX-producing Cyanobacteria on *Daphnia* sp. was studied recently. The organism was fed with nutritious green algae in the presence and absence of cyanobacterium *Raphidiopsis raciborskii*. The growth of the organism was affected when the STX-producing algae were present. The organism could not grow well in the presence of toxic algae even though green algal food was available indicating the effect of toxins on the organism (Dos Reis et al. 2023). The cyanobacterium was able to adapt to the grazer *Daphnia* by increasing *stx* gene expression and STX concentration. However, the growth and photosynthesis processes of the cyanobacteria were affected indicating that toxin production against the predators is achieved at the cost of the growth of the organism (Vilar et al. 2021). In another study, the STX-producing and nonproducing strains of *Cylindrospermopsis raciborskii* were studied on its grazer *Eudiaptomus gracilis* (Calanoida, Copepoda). Different culture temperatures (17, 25, and 32 °C) were used to study the interaction. Higher temperatures elicited higher STX production in the toxin-producing strain. The copepods avoided grazing on STX+ strain indicating that higher

temperatures in water bodies might favor the growth of toxic algae severely limiting the food availability in the food chain and leading to Ecosystem health (Rangel et al. 2016). Similarly, the filament length of the STX+ cyanobacterial strain was tested on Copepod grazing. This factor did not affect the grazing ability of the copepods. They avoided STX-producing algae irrespective of their filament length (Rangel et al. 2020). The increase in phosphorus and Nitrogen caused by the eutrophication of lakes has resulted in the elevation of toxic algal growth posing a threat to animals consuming these waters (Hai et al. 2009). Heavy metals and trace elements were found to influence cyanotoxin production in cyanobacteria. Extensive studies were done on *Anabena Microcystis Aphanizomenon, Anacystis*, and *Gloeotrichia* sp. in various lakes of the world. Irkutsk Reservoir near Lake Baikal in Russia exhibited alarming STX levels by the cyanobacterium *Dolichospermum lemmermannii*. An increased expression of *stxA* was identified accounting for the increased toxin production (Grachev et al. 2018). Researchers observed that metal concentrations greatly influenced cyanobacterial growth and toxin production. The toxin gene expression increases when metals are present in the aquatic bodies (Facey et al. 2019). Microcystis is known to produce two cyanotoxins and STX. These cyanotoxin genes were observed in polluted lakes. A study conducted in Songkhla Lake Basin (SLB) located in Southern Thailand, where industrial pollutants are leached into the lake waters, observed expression of microcystin and STX genes and an increase in toxin content (Naknaen et al. 2021). However, STX production was observed in *Anabaena, Aphanizomenon, Cylindrospermopsis, Lyngbya, Phormidium, Planktothrix, Raphidiopsis, Scytonema*, and *Woronichinia* (Bullerjahn et al. 2016; Schneider et al. 2020; Moustaka-Gouni et al. 2016; Bullerjahn et al. 2016), in freshwater lakes. Several algal species were observed in polluted lakes and high toxin production was observed (Podduturi et al. 2021). However, in non-polluted lakes in Sullivan Bay, Kabetogama Lake, Minnesota, USA, water was sampled for 1 year. The anatoxin-a and STX were discovered in the cyanobacteria before the visible algal blooms were observed in the region (Christensen et al. 2019). Salinity stress was tested on STX production in *Alexandrium pacificum*. An increase in salinity till an optimum level induced STX production without transcriptional regulation of *stxA4* and *sxtG* genes (Bui et al. 2021).

21.6 Diagnostic Methods for Detecting STX

The most traditional gold standard assay is the mouse bioassay which slowly is getting replaced with ELISA-based immune assays for STX after the antibody era. For research and commercial purposes, more analytical techniques for extraction and purification are emerging. A microscale sample method in extracting STX and its analogs along with liquid extraction LLE and solid phase extraction methods has been published recently (Ishak et al. 2023). Several methodologies have been tried in the past for regulatory purposes. STX detection is performed either by oxidation of compounds before separation (pre-column), which is generally used in mouse

bioassay or the analytes are separated and oxidized post-column purification (PCOX) (Fig. 21.5). This method became the AOAC (Association of Agricultural Chemists) official method for detecting toxins in collaboration with the Canadian Food and Inspection Agency and the National Research Council of Canada in 2011. Researchers analyzed STX and all its analogs together from various marine biosystems such as in mussels, clams, oysters, and scallops (Van De Riet et al. 2011).

In parallel to the AOAC mouse bioassay, in AOAC Frances M Van Dolah et al. receptor binding assay's efficiency was determined for PSTs in 2011. This method also became the AOAC method of detection. The main steps in the method are given below.

Shellfish homogenates extracted as per MBA → [^{3}H] STX compete with unlabelled STX for Na^+ channel site in rat brain membrane at 4◦C → [^{3}H] STX quantified by liquid scintillation counting (CPM)

Data analyzed with hill equation

Data analyzed with hill equation

$$Y = \min + \frac{\max - \min}{1 + 10^{(x - \log EC50\text{hill scope})}}$$

The authors observe that the RBA method using pre-column oxidation HPLC yielded better results. It is interesting to note that these methods see Na^+ channels as the only site for competitive binding of labeled and unlabeled STX. The work is in progress for other targets of STX such as K^+ channels and Ca^{2+} channels.

For drinking water in general EPA listed ELISA linked in LC-MS to be used for PST detection. For continuously monitoring the STX, a microfluidic immunoassay system was developed by Adama Marie Sesay and Laura Micheli's team in 2021

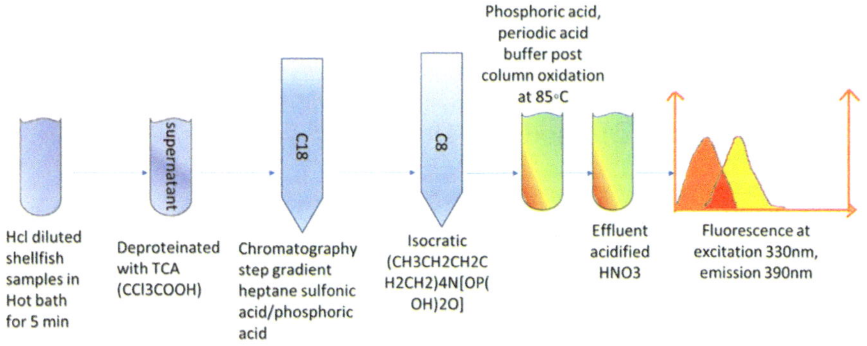

Fig. 21.5 The method for the post-column oxidation method (PCOX method), which is an AOAC Official method 2011.02 for the determination of Paralytic shellfish toxins from Mussels, Clams, Oysters, and Scallops

(Celio et al. 2021). This system can detect 3.5×10^{-7} to 2×10^{-5} ng/ml with a detection limit of $1x\ 10^{-7}$ ng/ml. This system has been optimized and tested with seawater samples with an in-situ monitoring of marine water contaminations. This method uses competitive binding of free STX and Horse radish peroxidase treated STX against polyclonal antibody for STX and uses 3,3′,5,5′-tetramethylbenzidine (TMB) as a substrate which when bound to free STX with pAb STX shows a calorimetric reading at 655 nm (Fig. 21.6). The continuous analysis data can be plotted to give a peak-like shape with height (H), width (W), and area (A). A three-parameter Gaussian distribution is used to determine the desired value. When no linear fitting is obtained from the Gaussian bell curve, the maximum absorbance (desired peak maximum) $\propto \dfrac{conc\ of\ enzyme\ antigen\ complex}{conc\ of\ injected\ analyte}$; this measure has been represented as relative standard deviation (RSD in %). For STX determination a generalized logistic function can be used which has a slow increase in activity until a rapid asymptotic maximum point is reached. For absorbance values in % = $\dfrac{A}{A_{sat}} = 100x\dfrac{A - A_o}{A_{sat} - A_0}$; (A being the absorbance in the presence of analyte, A_0 is the absorbance when there is no analyte presence, A_{sat} is the saturated absorbance). The method stringently accommodates the probability of false positives kept significantly low. Other experimental and operational parameters such as flowrate optimization, HRP concentration optimization, FI-IA optimization, binding curve and competition time, concentration determination of STX-HRP, calibration curve, and seawater matrix effect all using bioreactor, calibration curve for the full system (bioreactor and chip) is also performed here (Celio et al. 2021).

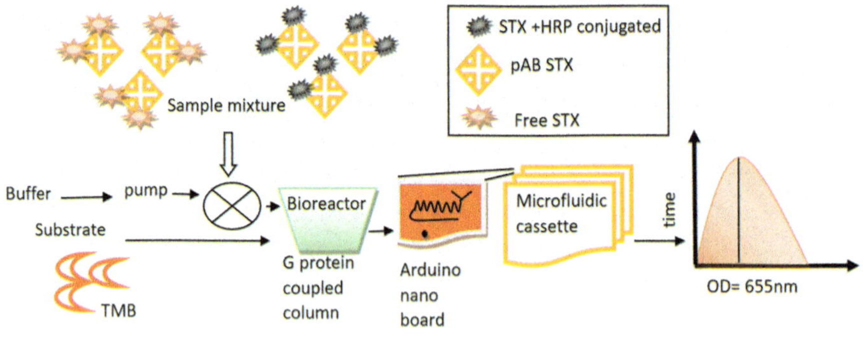

Fig. 21.6 Schematic method flow for microfluidic assay system for STX. The free STX with polyclonal anti-STX + TMB is the actual signal which has an absorbance at 655 nm. The nano board has microfluidic mixing methods such as diffusion, herringbone mixer, serpentine reaction channel, and optical sensing channel which acts mechanically through fluid dynamics for the reaction to be completed and detected using the optical density method

21.7 Aptamers and STX

Aptamers are small protein/DNA/RNA molecules that specifically bind to target molecules and are popular due to their property of ease of synthesis and eliciting less immune response in humans when they are used as drugs. These are employed as drugs and selective detection for making sensors. One example for treating drug-resistant cancer cells is as follows: The cancer cell surface overexpressed protein, Epithelial cell adhesion molecule EpCAM specific marker was selected for targeted delivery. This aptamer was conjugated to siRNA, which can silence oncoprotein, signal transducer, and activator of transcription 3 (STAT3). Specifically, drug-resistant spheroids in breast cancer cells were used as targets, this aptamer could not kill these cancer cells. Then when nanotechnology using porous carbon (synthesized from polypyrrole using pyrolysis) and polymer Polydimethylsiloxane (PDMS) was used the drug-resistant cancer spheroids were killed (Fig. 21.7). Similarly, aptamers are used against various molecules including toxins for therapy as well as detection of toxins.

G quadruplex structures are guanine-rich oligonucleotides, and through the Hoogstein base pairing principle, form stable-G tetrad structures that can be used as delivery vehicles due to their structure and nontoxic cellular uptake. G-quadruplex structures identification in the genes and in vitro assays UV visible and Raman spectroscopies are illustrated using putative aptamers for MicroRNA. These are small noncoding RNA sequences regulating RNA expression by complementary RNA binding. These are important in cell differentiation, cell cycle progression, apoptosis, and signaling pathways. The miRNA regulates important cellular

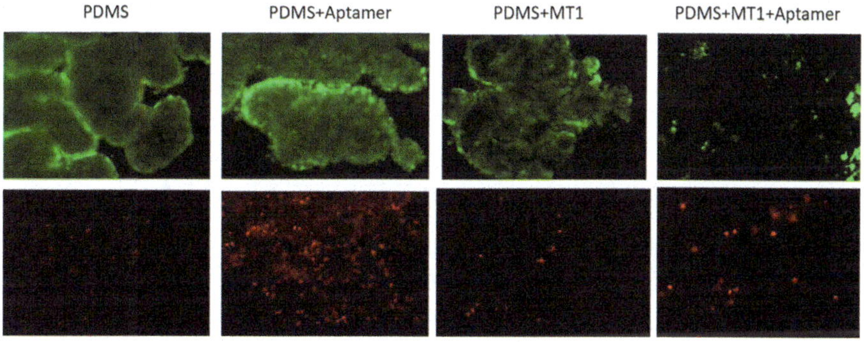

Fig. 21.7 Aptamer siRNA chimeras using porous carbon scaffolds can effectively kill drug-resistant spheroid MCF7 breast cancer cells. MT1-Porous Carbon synthesized from Polypyrrole. Aptamer-EpCAM-STAT3 siRNA chimera. *Polydimethylsiloxane* (PDMS) polymer. The breast cancer cells were grown on the scaffolds. The cells are visualized using Live and dead stains. Green cells are live cells, while red cells are cells undergoing apoptosis (programmed cell death). The cells are visualized using a Fluorescence microscope. Calcein-AM and Ethidium homodimer solutions, stained viable, and dead cells, respectively. Calcein-AM in cells can be visualized at Excitation/Emission 494/517 nm and Ethidium homodimer inside the cells was visualized at Excitation/Emission 528/617 nm, respectively

functions and dysregulation leads to disease conditions such as cancer. The Hsa-mir-373, hsa-mir-330, The hsa-mir-152, hsa-mir-196b, and the hsa-mir-675 are a few miRNAs that regulate cancer progression (Suzuki et al. 2013). A method of choosing regions from promoters of the genes that form G quadruplex structures as aptamers has been proposed as a rapid method for time-consuming SELEX-based technologies. The predicted DNA sequences that form G quadruplex structures from the above-mentioned miRNA are studied by UV visible spectroscopy and Raman spectroscopy. The putative aptamers had a 15–17 nm shift in the Soret band of the drug indicating the formation of G quadruplex structures in vitro as the $TmPyP_4$ drug preferentially binds to G quadruplex DNA structures (Fig. 21.8a and b). The G-quadruplex structures usually attain several confirmations (Fig. 21.8c). Raman spectroscopy gives characteristic bands that can be employed to identify compounds. Raman peaks at 1480 cm^{-1} and 1570 cm^{-1} (Fig. 21.9). These peaks were attributed to C_8-N_7-H_2 deformation of the G tetrad involved in Hoogstein base pairing and N_3C_4-C_4C_5 stretching of the deoxyguanosine which is involved in the guanine residue stability of the G tetrad, respectively (Pagba et al. 2010). There is recent interest in using G-quadruplex forming aptamer structures in disease therapy and diagnostics.

Systematic evolution of ligands via exponential enrichment technology (SELEX) was used to create a structurally stable DNA/G-quadruplex aptamer which binds to Gonyautoxin. The binding studies of the aptamer to the toxins to select the best aptamer is time-consuming. Therefore, Raman spectroscopy can be used as the G-quadruplex formation can be sensitively identified by Raman vibrational peaks

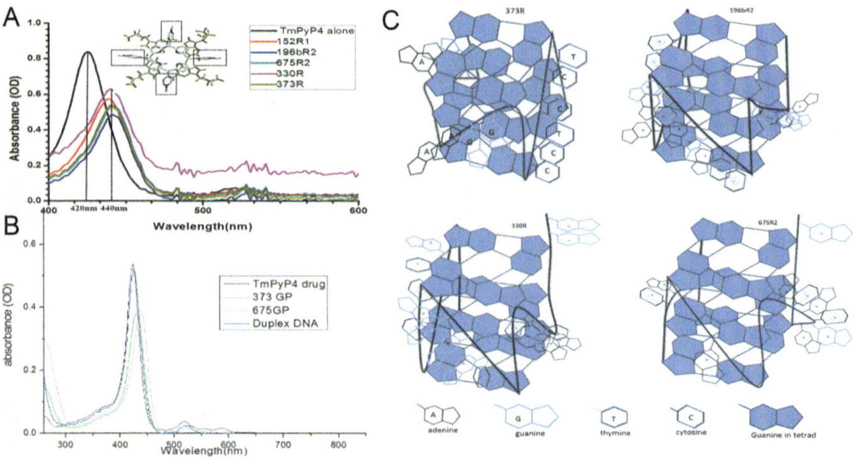

Fig. 21.8 Characterization of putative G-quadruplex structures using UV visible spectroscopy and Raman spectroscopy. The 50 μM DNA solutions were mixed with 50uM of The $TmPyP_4$ drug to a final volume of 2 μL and recorded in a UV-visible spectrophotometer at Vision Research Foundation. (**a**) The drug reacts with G quad structures in miRNA promotors compared to double-stranded DNA (**b**) The shift in drug peak in miRNA putative G-quad structures (**c**) Confirmations of G-quad structures

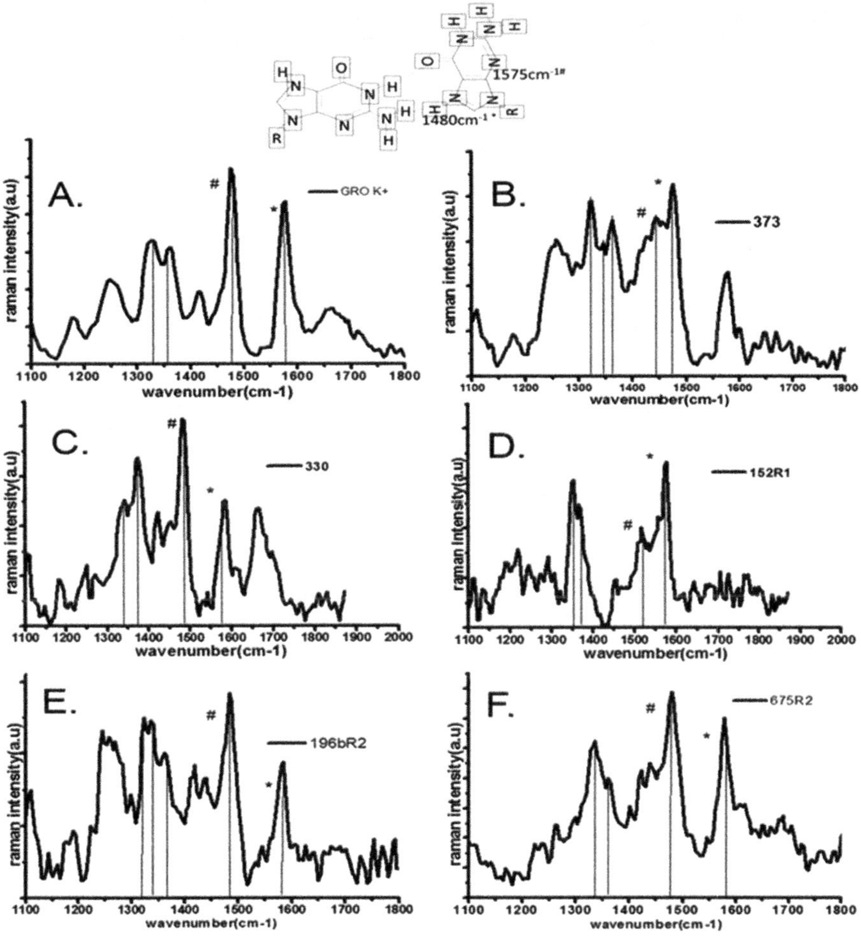

Fig. 21.9 Raman spectra from the structures indicate the formation of G-quad structures. The vibrational mode of guanines involved in the formation of G quadruplex structures was studied using Raman spectroscopy. The 50 μM DNA sequences in 100 KCl were dried on a silica chip. The background subtraction for the silica peak was done before recording the Raman spectra. The spectrum was recorded with Witec alpha 300confocal Raman system [Department of Physics, IIT Madras] with Nd: YAG laser having 532 nm [2.33 eV] excitation

that are like fingerprinting of chemical and biological moieties. The Raman spectroscopy sensitivity is increased by using metal nanoparticles such as silver and gold. Silver nanoparticles were used to enhance the Raman signal by the phenomenon of surface-enhanced Raman spectroscopy (SERS) when the aptamer binds to the toxin. An efficient sensitive toxin detection method has been developed using Raman spectroscopy (Liu et al. 2022). In the future, aptamers can be cleaning agents that are completely nontoxic to humans as a preventive measure before consuming shellfish especially shellfish and mollusks from recreational harvesting/fishing and

home-based cooking. Similarly, SELEX technology was used from the G quadruplex library for screening binding moieties to STX. The 45e aptamer was used as label free optical Biolayer interferometry (BLI) aptasensor for the detection of STX (Zhou et al. 2022).

Sensitive detection of STX was performed using aptamer binding studies (Handy et al. 2013; Zheng et al. 2015; Alfaro et al. 2015). The aptamer-STX binding at the single molecule level was probed using the Optical tweezer technique. In this method, laser light and optical trapping are used to study biomolecule conformations at a single molecule level and binding regions are identified for probable sensitive STX detection and making an aptasensor (Casanova-Morales et al. 2019). This novel strategy can mitigate blindfold risk and provides minimum safety from STX, Gonyautoxin, and its analogs which is a heat-stable toxin. This strategy paves a small-scale industrial solution and has a futuristic commercial prospect in handling food-borne poisoning. The strategy of STX detection using SELEX technology and signal enhanced by SERS could be extended to other toxin detection. Another toxin called Ochratoxin A also has a G quadruplex binding aptamer to it (Xinhui et al. 2018). For the treatment of contaminated water, green nanoparticles were used against STX and its derivatives. Two different magnetic nanostructures were created. Fe_3O_4NPs magnetite nanoparticle (in magnetic phase) and carbon prepared through a step solvothermal process with a diameter of 400 nm. The carbon-coated Fe_3O_4NPs were coated on mono-porous silica by soft template method with a diameter of 700 nm. It was observed that carbon-coated nanostructure eliminated adsorbent 70% of marine toxins which does not have sulfate groups in their structure and up to 84.7% of freshwater toxins with cyclic peptide structures (González-Jartín et al. 2020).

21.8 Current Therapy Methods for STX

Currently, there is no antidote for PSPs especially STX and its analogs even though extensive research is going on to find therapy molecules. There is no single approved drug or compound to treat PSPs. The treatment is only symptomatic-based. The investigation trends in British Columbia, Canada show increase in shell fish poisoning to humans from 1940 to 2020. The research studies indicated rise in poisoning due to altered climatic conditions. A storm "Beast from the East" brought in new benthic organisms which were consumed by dogs leading to PSPs with exceeding levels of 14,000 µg STX/eq kg. In one wave, dicarbonyl STX and the other STX and gonyautoxin were identified across Northeast England (Dean et al. 2020). In fatal/severe cases of STX poisoning, humans use life support systems such as oxygen support until the toxin gets removed mostly through urinary clearance. Some survivors show fatigue. A PSP case series for seven patients in Washington, experienced peripheral paresthesia, nausea, vomiting, diarrhea, ataxia, weakness, and shortness of breath. A 62-year-old female experienced dysarthria and was kept in ICU having low K^+ levels. The STX levels from their shellfish had an average of 6250 meg/100gm

of shellfish. All seven patients here were treated for their symptoms, observed for their recovery, and discharged (William et al. 2014). From the current equivalency standards of measurement, STX and its analogs are always grouped as PSPs, and they are always relatively measured. The toxicity equivalency is always relatively measured. The toxicity equivalency factors (TEPs) are used which is the "Toxicity ratio of a compound from a chemical group that shares the same mode of action of a reference compound in the same group" which means there is no absolute measuring standard to the analog compounds without its reference; example STX. Due to various environmental factors, new strains, and new levels, new symptoms post threat either directly or indirectly to humans which makes the therapeutic world even more challenging. Nevertheless, new technologies and methodologies tend to bridge the gap and bring new treatment options.

21.9 International Regulatory Bodies and their Sanitation Program for Preventing PSPs

The United States Food and Drug Administration (FDA) plays a critical role in ensuring the safety and sanitation of shellfish through the National Shellfish Sanitation Program (NSSP) and the Interstate Shellfish Sanitation Committee (ISSC). These programs are instrumental in monitoring and preventing shellfish-related illnesses, particularly Paralytic Shellfish Poisoning (PSP). The collaboration involves various stakeholders, including shellfish-producing and nonproducing states, the FDA, the Environmental Protection Agency (EPA), the National Oceanic and Atmospheric Administration (NOAA), and the shellfish industry. Internationally, the FDA allows participation from other national governments in the NSSP under national Memorandums of Understanding (MoUs). This international cooperation indicates the program's global significance in shellfish sanitation and safety. The NSSP guide, known as the model ordinance, is crucial in guiding the sanitation control for raw and post-harvest processed shellfish. It sets standards and provides guidelines for the safe handling and processing of these products. However, certain categories of shellfish, such as cooked shellfish and packaged raw shellfish intended to be cooked, fall outside the scope of NSSP. These categories are regulated under the seafood Hazard Analysis Critical Control Point (HACCP) regulations (21 CFR 123), ensuring comprehensive coverage of all types of shellfish products. The scope of the NSSP is broad, covering regulation of shellfish growing areas, harvesting, processing procedures, labeling, storage, handling, packing, and logistics. It encompasses the entire supply chain, from production to distribution, to ensure that shellfish are safe for consumption (Anon 2020). Certification is a critical aspect of the NSSP. The program offers various types of certifications and permits, covering a wide range of operations including shucker-packer, re-packer, shell stock shipper, re-shipper, and aquaculture activities. These certifications ensure that those handling shellfish are trained and comply with safety standards. Risk management is a

cornerstone of the NSSP, encompassing measures for quality control and assurance. The FDA enforces and oversees these measures, relying on inspections and reports from NSSP inspectors. These measures are essential for maintaining the safety and quality of shellfish products. The program also addresses shellfish-related illnesses and the presence of harmful substances in shellfish meats. Guidance for controlling toxins and pathogens like *Vibrio parahaemolyticus*, and toxic substances in shellfish meats, is provided. The NSSP establishes control plans as preventive measures against these health hazards. Laboratory testing and quality assurance are integral to the NSSP. The program mandates the use of validated laboratory methods, such as those from AOAC, BAM, and EPA. These methods ensure the reliability and accuracy of testing, which is crucial for monitoring and maintaining shellfish safety. In terms of biotoxin control, the NSSP has established limits for various toxins, such as PSP-80 μg STX equivalents/100grams, NSP-20MU/10gram/0.8 mg, DSP-0.16 mg okadaic acid (OA) equivalents/kg (0.16 ppm), AZP-0.16 mg azaspiracid-1(AZA-1) equivalent/kg (0.16 ppm), and ASP-2 mg domoic acid (DA)/100grams (20 ppm). These limits are based on extensive research and are critical for protecting public health. Furthermore, the NSSP's functions under the model ordinance are comprehensive, covering aspects like shell stock relaying, wet storage in approved areas, control of shellfish harvesting, and transportation. These functions are essential for maintaining the integrity and safety of shellfish from production to consumption. FDA's National Shellfish Sanitation Program, in collaboration with the ISSC and other stakeholders, provides a robust framework for ensuring the safety and sanitation of shellfish. Through comprehensive regulations, certifications, risk management, and quality control measures, the program plays a pivotal role in protecting public health and maintaining the integrity of the shellfish industry (Anon 2023d).

Canada has a very similar sanitation program Canadian Shellfish Sanitation Program (CSSP) administered by the Canadian Inspection Food Agency (CFIA), Fisheries and Oceans Canada (DFO), and Environmental and Climate Change Canada (ECCC). This program releases its manual which is linked to preventative control plan requirements as per the Safe Food for Canadians Regulations (SFCR) (Anon 2023a). In India, the Food Safety Standards and Authorities of India (FSSAI) guide shellfish processing together as a food safety management system (FSMS)—a guide to implement GMP/GHP (fssai 2023). EU as per (EC) No 854/2004, the EU monitors and controls microbiological monitoring (EURL) and classification of bivalve mollusk harvesting areas. They classify harvesting areas into different classes with class A being the most stringent. Nick Taylor and his colleagues in the Centre for Environment Fisheries and Aquaculture Science (CEFAS) did a comparison document for EU and NSSP protocols and found it to be equivalent to most of the class A measures. To enhance trade relationships better, EU accepts NSSP standards (Anon 2023c). Japan has sanitation agreements for shellfish with the US and therefore follows both NSSP and CSSP (Anon 1964). South Korea has a similar program to CSSP called as Korean Shellfish Sanitation Program (KSSP) and has agreements with Canada for virtually audited by CFIA through CSSP since 1974 (Anon 2023b).

Apart from regulatory sanitation programs from various governing bodies, different strategies are adopted in treating shellfish aquaculture and reducing the impact of its benefactor the blue-green algae which provides nutrients to these aquaculture-carrying cyanobacteria and dinoflagellates causing PSPs. Prevention of bloom formation through early detection and hydrological manipulation, water body treatments through conventional novel methods including nanotechnology and genetic manipulation, application of cyanocides, chemicals, and efficient water treatments are needed. Commercially available Phoslock® (lanthanum modified bentonite) is used for nutrient depletion, cyanocides for chemical treatments hydrogen peroxide, synthetic surfactants such as ammonium compounds, alkyl trimethyl ammonium compounds, clays made of hexadecyl trimethyl ammonium bromide, granulated composites of bentonite with micelles of octadecyl trimethyl ammonium bromides (ODTMA), etc. For toxin adsorption, activated carbon removes organic contaminants (Oh et al. 2023; Tseytlin et al. 2024; Ng et al. 2023).

21.10 Toxins for Pain Management

A well-known toxin that is in the market as a therapeutic agent is botulinum toxin commercially called Botox. This toxin was named Onabotulinum toxin A and Dr. Alan Scott in the 1970s used it first for muscle surgery to weaken extraocular eye muscles in patients with strabismus. Its importance knocked on the commercial doors and the FDA started to monitor the licensing, and quality control procedures which were sold to Allergen to become what is known today as Botox (Scott et al. 2023). Botox binds to nerve terminals, internalized, and cleaves synaptosomal associated protein (SNAP-25), which is a part of the SNARE complex. This SNARE complex plays an important role in synapses during vesicle docking and fusion. Successful clinical studies with Onobotulinum toxin A have been done to reduce muscle contractions by inhibiting acetylcholine in the motor terminals. This toxin is now having a promising future for its benefits in major depression, and cosmetic applications such as skin quality apart from pain management (Brin and Burstein 2023). Similar to Botox, it is interesting to see the potential of STX and its analogs. In 2020, the FDA published an open collaboration opportunity for licensing purified STX for pain management. Pure and synthesized STX and its derivatives, analogues can be used in disease conditions related to their mechanism of action. To understand the disease condition related to STX's target molecular protein (receptor), the very first group of diseases that is highly related are channelopathies and their impact. Sodium channel targeting is involved in pain management and toxins can be used as therapy. The nervous system's response to therapy for pain is categorized as tetrodotoxin-sensitive currents and tetrodotoxin-resistant currents. STX and its analogs are found in general as sodium channel inhibitors. The use of sodium channels for channelopathies is summarised in Table 21.1.

Furthermore, voltage-gated Nav1.7 is shown to be effectively inhibited by ST-2530 an analog of STX for pain studies in a preclinical study (Beckley et al.

Table 21.1 Sodium channel's role in various neuropathic conditions

Sodium channel subtypes	Channelopathy conditions with references	Subtype gene involved in the human genome
Nav1.3	• Sciatic nerve CCI (Leffler et al. 2002) • Diabetic neuropathy (Yang, 2016) • Trigeminal neuralgia (Gambeta et al. 2020) • Postherpetic neuralgia (Attal and Bouhassira 2006) • Peripheral nerve injury (Hains et al. 2004)	SCN3A
Nav1.6	• Diabetic neuropathy (Schreiber et al. 2015) • Infraorbital nerve injury (Henry et al. 2007)	SCN8A
Nav1.7	• Diabetic neuropathy (Yang et al. 2016) • Inflammatory pain (Hameed 2019) • Trigeminal neuralgia (Kwon et al. 2021)	SCN9A
Nav1.8	• Diabetic neuropathy (Schreiber et al. 2015) • Painful neuropathy (Hameed 2019) • Visceral pain (Lv et al. 2023) • Chronic nerve compression (Yang et al. 2021)	SCN10A
Nav1.9	• Inflammatory pain (Amaya et al. 2006) • Trigeminal neuralgia (Luiz et al. 2015)	SCN11A

2021). Different toxins are discussed here along with STX such as neoSTX, gonyautoxin, and Tetrodotoxin (guanidium group of toxins) binding to site 1 sodium channels; 4,9 anhydrous tetrodotoxin specifically binds to Nav1.6. Nav1.6 is shown to play an important role in diabetic neuropathy, trigeminal neuralgia, etc. (Cardoso and Lewis 2018). Similarly, calcium channels, potassium channels, and other channels such as chloride channels, etc. also play important roles in channelopathies in the nervous system, and it would be an insight to study them from the toxicology of these natural toxins.

21.11 Prospects of Using Toxins for Ocular Diseases

Even though there is not much research work done on the interaction of STX in ocular diseases or conditions similar to Botox, there are high relations and prospects for ocular application. STX's direct impact as Nav channel blockers allows us to see if ocular myotonia-related symptoms during channelopathy can be a solution. Being a natural Nav1.4 blocker, STX and its analogs such as NeoSTX (Clinical trial: NCT00273065 for local anesthetic properties, Chile), can be used for ocular

myotonia currently treated with an anticonvulsant, sodium, and potassium channel blocker carbamazepine which is a first-line therapy is used (Lee and Lee 2022). Similarly, these natural blockers (STX, NeoSTX, TTX) can also be used. In another study genetic inheritance of ocular neuromyotonia involves SCN4A mutation which encodes the Nav1.4 channel (Sampaio et al. 2020). TNF-α enhances Nav1.6 channels in retinal ganglion cells leading to apoptosis in experimental glaucoma models. It induces hyperexcitability in these cells. Since the guanidium group of toxins has the potential to bind Nav1.6, though opposing the hyperexcitability of the cells, it is important to use these toxins as negative markers and confirm the effect with purified toxins for glaucoma to see if chemical alternatives can be done for the synergistic effect (Cheng et al. 2021). Another ocular disease called dry eye disease involves increased expression of Nav1.8 (Ashok et al. 2023). As a general natural Nav blocker, STX can be used in conditions where the currents continue to get hyperexcited as tetrodotoxin-resistant currents. In a promising study, phase II clinical trial was done in the UK, NCT01451593, for neuroprotection with phenytoin (a sodium channel blocker) in optic neuritis in 2015. This study gives natural sodium channel blockers hope to be reengineered for drug screening in optic neuritis conditions. This study shows the activity of phenytoin against the Nav1.6 channel (Raftopoulos et al. 2016). Apart from disease condition and therapeutic application, fundamental research related to the eye shows few potassium channel blockers like tetraethylammonium (TEA), 3,4-diaminopyridine (DAP) reduces the amplitude of dark current in the inner segments of rod photoreceptor (Drachman and Strichartz 1991). This fundamental study can also give insight into the interactor of potassium channel gating modification by STX which has a unique property of delaying the action potential in neurons when applied. This delay could also have any therapeutic potential for potassium channelopathy which requires a delay in excitation of neurons (Wang et al. 2003). STX binds to L-type Calcium channels and blocks their function. The L-type calcium channels play an important role along with T-type calcium channels in experimental glaucoma models. Apoptosis of Retinal ganglion cells is one of the pathological characteristics of glaucoma, which leads to irreversible blindness. Again $Ca_v1.2$ of L-type calcium channels were downregulated in glaucoma condition and $Ca_v3.3$ is over-expressed in the same which is one of the T-type calcium channels (Wang et al. 2023). Though STX blocks L-type calcium, it is important to see in this condition how STX impacts the L-type Calcium currents (Su et al. 2004). In uveal melanoma (leading cancer that kills adults), L type Calcium blockage is beneficial as shown in melanoma for M17 and JG cell lines there is an upregulation of $Ca_v1.2$ or the L-type calcium channel, and if STX binds and is shown to block L type calcium channels which makes it worth investigating the therapeutic property on uveal melanoma (Shaughnessy et al. 2019). L-type Calcium channels especially $Ca_v1.3$ are shown to have a "light rise" phenomenon using nifedipine which inhibits the gradual Ca^{2+} influx to generate the light rise. This phenomenon was found in the pathologic condition called degenerative retinitis pigmentosa to delay the gradual visual impairment but this study has failed to give more causal relation of the photoreceptor protection effect and blockage of retinal L-type Calcium channels. However, this experiment gives a lead for the future to more

thoroughly investigate other toxin analogs for hereditary rod and cone cell death (Zamponi et al. 2015). $Na_v1.3$, Na_v 1.7, and Na_v 1.8 have roles in diabetic neuropathy, it is important to look into diabetic retinopathy conditions to look into diabetic retinopathy conditions in the retina. Sodium currents and L-type calcium channels importance in release of calcium play an important role in photoreceptors for glutamate release in the retina (Kern and Berkowitz 2015). The toxin and its derivatives can be studied in the future as a therapeutic target for diabetic retinopathy (Schmitz and Witkovsky 1997). The direct use for STX and Tetrodotoxin is being explored for corneal topical analgesia along with other Site 1 Sodium Channel Blockers (S1SCB). Tetrodotoxin and STX with and without dexmedetomidine, and proparacaine were studied in rat cornea, and tactile sensitivity, and blink response with Cochet-Bonnet esthesiometer were studied. In vitro cytotoxicity and in vivo corneal healing with corneal debridement due to drug administration was measured. Local anesthetics cause delays in epithelial healing, corneal ulceration, and occasional neurotrophic keratopathy. By combining TTX and or STX with dexmedetomidine this effect was not seen which could be a potential method to prevent pain during longer surgical procedures. The team has also successfully shown this synergistic action of these compounds against widely used amino ester ocular anesthetic, 0.5% (weight/volume) proparacaine (McAlvin et al. 2015). Although like Botox every naturally occurring toxin has therapeutic potential; it is important to overcome challenges, especially for highly water soluble STX, and its analogues is its systemic toxicity. This is addressed by creating a supramolecular mimetic system of sodium channels by self-assembling peptides as nanofibers which is shown to safely deliver STX and Tetrodotoxin in the sciatic nerves of the rats. This delivery vehicle is shown to overcome this systemic toxicity with benign local tissue reaction. These site-1 sodium channel blockers (S1SCBs) are engineered as a potent local anesthetic agent (Ji et al. 2021).

21.12 Future Directions in STX Research

Oyster farming co-cultivating with mangrove trees is a natural solution. Mangrove sediments harbored novel bacteria, *Mangrovimonas yunxiaonensis* LY01 that could regulate *Alexandrium sp* growth and toxin production. The genome of this algicidal bacterium has a genome size of about 2,673,510 base pairs. A total of 846 genes and 1781 genes were assigned to COG classes and KEGG pathways (Li et al. 2014). Several bioremediation technologies could be researched further to sequester the toxins. However, Mangrove health should be monitored and periodically PSPs should be monitored below the acceptable limits. Therefore, it is important to consider the basic sanitation bio-geochemical cycles of Mangroves a priority to monitor PSP in farming (Alves et al. 2022). Efforts should be made by countries to prioritize environmental and ecosystem health. The nutrient reduction accompanied by Eutrophication lowering methods by regulating the water quality at the groundwater level significantly has a greater impact on HAB reduction, especially in

coastal ecosystems (Green et al. 2021). Water quality management should be prioritized by Governments to reduce toxin poisoning by marine organisms. As an unusual strategy to resist STX, frogs use saxiphilin a 91 kDa transferrin homolog protein from American bullfrog. A novel STX binding site in the saxiphilin STX complex was identified through crystallographic methods where both target (voltage-gated sodium channel) and saxiphilin recognize STX through both the glutamates and use aromatic ring to engage with STX through cation π interaction. Therefore, saxiphilin core which is built from the transferrin fold acts as dual function protein with high affinity interactions with STX and cysteine proteases which along with the cation π interaction of the carboxylates side chain makes the protein unique in binding with STX. Hence frogs' resistance to STX might be attributed to this saxiphilin protein which acts like a molecular sponge (Chen et al. 2022). Several natural mechanisms present in Nature should be researched and used as natural sponges for biotoxin reduction. Additionally, this gives us an opportunity for further research to be done with other derivatives of STX using sax Phillip protein since there is no study performed against neo-STX, gonyautoxins, dicarbamoyl derivatives, N-sulforcarbomyl, Brevotoxins, and tetrodotoxins. Another mussel *Mytilus chilensis* Hemocytes is known for its resistance to PSPs such as STX. The antioxidant enzymes and subunits like Cytocrome Oxidase subunits I and II (COI, COIII), cytochrome B, ATPase 6 were expressed highly along with increased expression of K^+ and Ca^{2+} channel genes. K^+ channel genes increased four-fold increase against the toxin (Astuya et al. 2015). Immune system modulation was observed in the mussels that can detoxify the STX (De la Ballina et al. 2022). Future studies on immunomodulation against the toxins could develop immunotherapy against the toxins similar to cancer immunotherapy available presently. The toxins from *Alexandrium sp* are tested as antiviral and anticancer agents (Montuori et al. 2024). They can be brought from bench to bedside. Another avenue of research could be using changes in Environmental parameters, nutrient recycling, Government regulations, and genomics of organisms by Artificial intelligence to find solutions to biotoxin regulation (Marrone et al. 2023).

Acknowledgments Authors thank the Department of Biotechnology under program support for research on retinoblastoma Grant No. BT/01/CEIB/11/V/16, Department of Biotechnology, Government of India (Grant no. NO.BT/PR3580/PID/6/625/2011) and SERBEMR/2015/000607. We thank Dr. Krishnakumar Subramanyan for providing the necessary facilities for experimental work.

References

Abassi S, Kim HS, Bui QTN, Ki JS (2023) Effects of nitrate on the saxitoxins biosynthesis revealed by sxt genes in the toxic dinoflagellate *Alexandrium pacificum* (group IV). Harmful Algae 127:102473. https://doi.org/10.1016/j.hal.2023.102473

Accoroni S, Cangini M, Angeletti R, Losasso C, Bacchiocchi S, Costa A, Di Taranto A, Escalera L, Fedrizzi G, Garzia A, Longo F, Macaluso A, Melchiorre N, Milandri A, Mil S (2024) Marine

phycotoxin levels in shellfish—14 years of data gathered along the Italian coast. Harmful Algae 131:102560. https://doi.org/10.1016/j.hal.2023.102560

Akbar MA, Mohd Yusof NY, Tahir NI, Ahmad A, Usup G, Sahrani FK, Bunawan H (2020) Biosynthesis of Saxitoxin in marine dinoflagellates: an omics perspective. Mar Drugs 18(2):103. https://doi.org/10.3390/md18020103

Alfaro K, Bustos P, O'Sullivan C, Conejeros P (2015) Facile and cost-effective detection of Saxitoxin exploiting aptamer structural switching. Food Technol Biotechnol 53(3):337–341. https://doi.org/10.17113/ftb.53.03.15.3911

Alves FADS, de Sousa EB, Martins MP, da Silva Rocha CC, Faustino SMM, Mendes RA, de Oliveira LM, Schneider MPC (2022) Evaluation of paralytic shellfish toxins in marine oyster farming and microalgae in the Atlantic Amazon evidences safety but highlights potential risks of shellfish poisoning. Toxins (Basel) 14(10):654. https://doi.org/10.3390/toxins14100654

Amaya F, Wang H, Costigan M, Allchorne AJ, Hatcher JP, Egerton J, Stean T, Morisset V, Grose D, Gunthorpe MJ, Chessell IP, Tate S, Green PJ, Woolf CJ (2006) The voltage-gated Sodium Channel Nav1.9 is an effector of peripheral inflammatory pain hypersensitivity. J Neurosci 25, 50:12852–12860. https://doi.org/10.1523/JNEUROSCI.4015-06.2006

Amzil Z, Derrien A, Terre Terrillon A, Savar V, Bertin T, Peyrat M, Duval A, Lhaute K, Arnich N, Hort V, Nicolas M (2023) Five years monitoring the emergence of unregulated toxins in shellfish in France (EMERGTOX 2018-2022). Mar Drugs 21(8):435. https://doi.org/10.3390/md21080435

Anderson DM, Fachon E, Pickart RS, Lin P, Fischer AD, Richlen ML, Uva V, Brosnahan ML, McRaven L, Bahr F, Lefebvre K, Grebmeier JM, Danielson SL, Lyu Y, Fukai Y (2021) Evidence for massive and recurrent toxic blooms of *Alexandrium catenella* in the Alaskan Arctic. Proc Natl Acad Sci USA 118(41):e2107387118. https://doi.org/10.1073/pnas.2107387118

Anon (1964) Japan shelfish sanitation aggreement [Online]. Available at https://www.fda.gov/international-programs/cooperative-arrangements/fda-japan-shellfish-sanitation-agreement. Accessed 09 Jan 2024

Anon (2020) Purified Saxitoxin for food safety applications [Online]. Available at: https://www.fda.gov/science-research/licensing-and-collaboration-opportunities/purified-saxitoxin-food-safety-applications. Accessed 09 Jan 2024

Anon (2023a) Canadian shellfish sanitation program [Online]. Available at: https://inspection.canada.ca/food-guidance-by-commodity/fish/canadian-shellfish-sanitation-program/eng/1527251566006/1527251566942?chap=0. Accessed 09 Jan 2024

Anon (2023b) Canadian-virtual-audit-of-the-republic-of-Korea [Online]. Available at: https://inspection.canada.ca/importing-food-plants-or-animals/food-imports/foreign-systems/audits/canadian-virtual-audit-of-the-republic-of-korea/eng/1680788569196/1680788569821. Accessed 09 Jan 2024

Anon (2023c) Comparison of the European Union and US NSSP in relation to EU regulations 2015_2285 [Online]. Available at: https://www.cefas.co.uk/media/pvsmmwjr/comparisons-of-the-european-union-and-us-nssp-in-relation-to-eu-reg-2015_2285-dj-passed.pdf. Accessed 09 Jan 2024

Anon (2023d) US FDA National shellfish Sanitation Program (NSSP) [Online]. Available at: https://www.fda.gov/food/federalstate-food-programs/national-shellfish-sanitation-program-nssp. Accessed 09 Jan 2024

Ashok N, Khamar P, D'Souza S, Gijs M, Ghosh A, Sethu S, Shetty R (2023) Ion channels in dry eye disease. Indian J Ophthalmol 71(4):1215–1226. https://doi.org/10.4103/IJO.IJO_3020_22

Astuya A, Carrera C, Ulloa V, Aballay A, Núñez-Acuña G, Hégaret H, Gallardo-Escárate C (2015) Saxitoxin modulates immunological parameters and gene transcription in *Mytilus chilensis* Hemocytes. Int J Mol Sci 16(7):15235–15250. https://doi.org/10.3390/ijms160715235

Attal N, Bouhassira D (2006) Translating basic research on sodium channels in human neuropathic pain. J Pain 7:S31–S37. https://doi.org/10.1016/j.jpain.2005.09.005

Beckley JT, Pajouhesh H, Luu G, Klas S, Delwig A, Monteleone D, Zhou X, Giuvelis D, Meng ID, Yeomans DC, Hunter JC, Mulcahy JV (2021) Antinociceptive properties of an isoform-selective

inhibitor of Nav1.7 derived from saxitoxin in mouse models of pain. Pain 162:1250–1261. https://doi.org/10.1097/j.pain.0000000000002112

Bowers EK, Stimmelmayr R, Hendrix A, Lefebvre KA (2022) Stability of Saxitoxin in 50% methanol fecal extracts and raw feces from bowhead whales (*Balaena mysticetus*). Mar Drugs 20:547. https://doi.org/10.3390/md20090547

Brin MF, Burstein R (2023) Botox (onabotulinumtoxinA) mechanism of action. Medicine (Baltimore) 102(S1):e32372. https://doi.org/10.1097/MD.0000000000032372

Bui QTN, Kim H, Park H, Ki JS (2021) Salinity affects Saxitoxins (STXs) toxicity in the dinoflagellate *Alexandrium pacificum*, with low transcription of SXT-biosynthesis genes sxtA4 and sxtG. Toxins (Basel) 13(10):733. https://doi.org/10.3390/toxins13100733

Bullerjahn GS, McKay RM, Davis TW, Baker DB, Boyer GL, D'Anglada LV, Doucette GJ, Ho JC, Irwin EG, Kling CL, Kudela RM, Kurmayer R, Michalak AM, Ortiz JD, Otten TG, Paerl HW, Qin B, Sohngen BL, Stumpf RP, Visser PM, Wilhelm SW (2016) Global solutions to regional problems: collecting global expertise to address the problem of harmful cyanobacterial blooms., a Lake Erie case study. Harmful Algae 54:223–238. https://doi.org/10.1016/j.hal.2016.01.003

Cardoso FC, Lewis RJ (2018) Sodium channels and pain: from toxins to therapies. Br J Pharmacol 175(12):2138–2157. https://doi.org/10.1111/bph.13962

Casanova-Morales N, Figueroa NL, Alfaro K, Montenegro F, Barrera NP, Maze JR, Wilson CAM, Conejeros P (2019) Structural characterization of the saxitoxin-targeting APTSTX1 aptamer using optical tweezers and molecular dynamics simulations. PLoS One 14:11. https://doi.org/10.1371/journal.pone.0222468

Celio L, Ottaviani M, Cancelliere R, Di Tinno A, Panjan P, Sesay AM, Micheli L (2021) Microfluidic flow injection immunoassay system for algal toxins determination: a case of study. Front Chem 9:626630. https://doi.org/10.3389/fchem.2021.626630

Chatterjee S, More M (2023) Cyanobacterial harmful algal bloom toxin microcystin and increased vibrio occurrence as climate-change-induced biological co-stressors: exposure and disease outcomes via their interaction with gut-liver-brain Axis. Toxins (Basel) 15(4):289. https://doi.org/10.3390/toxins15040289

Chen Z, Zakrzewska S, Hajare HS, Alvarez-Buylla A, Abderemane-Ali F, Bofgan M, Ramirez D, O'Connell LA, Du Bois J, Minor DL Jr (2022) Definition of a saxitoxin (STX) binding code enables discovery and characterization of the anuran saxiphilin family. Proc Natl Acad Sci USA 119(44):e2210114119. https://doi.org/10.1073/pnas.2210114119

Cheng S, Wang H-N, Xu L-J, Li F, Miao Y, Lei B, Sun X, Wang Z (2021) Soluble tumor necrosis factor-alpha-induced hyperexcitability contributes to retinal ganglion cell apoptosis by enhancing Nav1.6 in experimental glaucoma. J Neuroinflammation 18:182. https://doi.org/10.1186/s12974-021-02236-6

Christensen VG, Maki RP, Stelzer EA, Norland JE, Khan E (2019) Phytoplankton community and algal toxicity at a recurring bloom in Sullivan Bay, Kabetogama Lake, Minnesota, USA. Sci Rep 9(1):16129. https://doi.org/10.1038/s41598-019-52639-y

De la Ballina NR, Maresca F, Cao A, Villalba A (2022) Bivalve Haemocyte subpopulations: a review. Front Immunol 13:826255. https://doi.org/10.3389/fimmu.2022.826255

Dean KJ, Hatfield RG, Lee V, Alexander RP, Lewis AM, Maskrey BH, Teixeira Alves M, Hatton B, Coates LN, Capuzzo E, Ellis JR, Turner AD (2020) Multiple new paralytic shellfish toxin vectors in offshore North Sea Benthos, a deep secret exposed. Mar Drugs 18(8):400. https://doi.org/10.3390/md18080400

Dos Reis GC, de Carvalho GHA, Vilar MCP, Azevedo SMFOE, Ferrão-Filho ADS (2023) Saxitoxin-producing Raphidiopsis raciborskii (cyanobacteria) constrains daphnia fitness and feeding rate despite high nutritious food availability. Toxics 11(8):693. https://doi.org/10.3390/toxics11080693

Doucette GJ, Allan Cembella JL, Martin JM (2006) Paralytic shellfish poisoning (PSP) toxins in North Atlantic right whales *Eubalaena glacialis* and their zooplankton prey in the Bay of Fundy, Canada. Mar Ecol Prog 306:303–313. https://doi.org/10.3354/meps306303

Drachman D, Strichartz G (1991) Potassium channel blockers potentiate impulse inhibition by local anesthetics. Anesthesiology 1051–1061:1051. https://doi.org/10.1097/00000542-199112000-00018

Durán-Riveroll LM, Cembella AD, Band-Schmidt CJ, Bustillos-Guzmán JJ, Correa-Basurto J (2016) Docking simulation of the binding interactions of Saxitoxin analogs produced by the marine dinoflagellate *Gymnodinium catenatum* to the voltage-gated sodium channel Nav1.4. Toxins 8(5):129. https://doi.org/10.3390/toxins8050129

Facey JA, Apte SC, Mitrovic SM (2019) A review of the effect of trace metals on freshwater cyanobacterial growth and toxin production. Toxins (Basel) 11(11):643. https://doi.org/10.3390/toxins11110643

Finch SC, Webb NG, Boundy MJ, Harwood DT, Munday JS, Sprosen JM, Somchit C, Broadhurst RB (2023) A sub-acute dosing study of Saxitoxin and Tetrodotoxin mixtures in mice suggests that the current paralytic shellfish toxin regulatory limit is fit for purpose. Toxins (Basel) 15(7):437. https://doi.org/10.3390/toxins15070437

Frey KE, Comiso JC, Stock LV, Young LNC, Cooper LW, Grebmeier JM (2023) A comprehensive satellite-based assessment across the Pacific Arctic distributed biological observatory shows widespread late-season sea surface warming and sea ice declines with significant influences on primary productivity. PLoS One 18(7):e0287960. https://doi.org/10.1371/journal.pone.0287960

fssai (2023) Guidance document [Online]. Available at: https://fssai.gov.in/upload/uploadfiles/files/Guidance_Document_Fish_01_06_2018.pdf. Accessed 09 Jan 2024

Gambeta E, Chichorro JG, Zamponi GW (2020) Trigeminal neuralgia: an overview from pathophysiology to pharmacological treatments. Mol Pain 16:1744806920901890. https://doi.org/10.1177/1744806920901890

Gilchrist J, Bosmans F (2012) Animal toxins can alter the function of Nav1.8 and Nav1.9. Toxins (Basel) 4(8):620–632. https://doi.org/10.3390/toxins4080620

González-Jartín JM, de Castro L, Alves AA, Piñeiro Y, Vilar SY, Rodríguez I, Gomez MG, Osorio ZV, Sainz MJ, Vieytes MR, Rivas J, Botana LM (2020) Magnetic nanostructures for marine and freshwater toxins removal. Chemosphere 256:127019. https://doi.org/10.1016/j.chemosphere.2020.127019

Grachev M, Zubkov I, Tikhonova I, Ivacheva M, Kuzmin A, Sukhanova E, Sorokovikova E, Fedorova G, Galkin A, Suslova M, Netsvetayeva O, Eletskaya E, Pogadaeva T, Smirnov V, Ivanov A, Shagun V, Minaev V, Belykh O (2018) Extensive contamination of water with Saxitoxin near the dam of the Irkutsk Hydropower Station reservoir (East Siberia, Russia). Toxins (Basel) 10(10):402. https://doi.org/10.3390/toxins10100402

Green L, Magel C, Brown C (2021) Management pathways for the successful reduction of nonpoint source nutrients in coastal ecosystems. Reg Stud Mar Sci 45:1–15. https://doi.org/10.1016/j.rsma.2021.101851

Hackett JD, Wisecaver JH, Brosnahan ML, Kulis DM, Anderson DM, Bhattacharya D, Plumley FG, Erdner DL (2013) Evolution of saxitoxin synthesis in cyanobacteria and dinoflagellates. Mol Biol Evol 30(1):70–78. https://doi.org/10.1093/molbev/mss142

Hai X, Paerl HW, Qin B, Zhu G, Gaoa G (2009) Nitrogen and phosphorus inputs control phytoplankton growth in eutrophic Lake Taihu, China. Limnol Oceanogr 55(1):420–432. https://doi.org/10.4319/lo.2010.55.1.0420

Hains BC, Saab CY, Klein JP, Craner MJ, Waxman SG (2004) Altered sodium channel expression in second-order spinal sensory neurons contributes to pain after peripheral nerve injury. J Neurosci 24(20):4832–4839. https://doi.org/10.1523/JNEUROSCI.0300-04.2004

Hameed S (2019) Nav1.7 and Nav1.8: role in the pathophysiology of pain. Mol Pain 15:1744806919858801. https://doi.org/10.1177/1744806919858801

Handy SM, Yakes BJ, DeGrasse JA, Campbell K, Elliott CT, Kanyuck KM, Degrasse SL (2013) First report of the use of a saxitoxin-protein conjugate to develop a DNA aptamer to a small molecule toxin. Toxicon 61:30–37. https://doi.org/10.1016/j.toxicon.2012.10.015

Henry MA, Freking AR, Johnson LR, Levinson S (2007) Sodium channel Nav1.6 accumulates at the site of infraorbital nerve injury. BMC Neurosci 8:6. https://doi.org/10.1186/1471-2202-8-56

Ishak SM, Yahaya N, Loh SH, Kamaruzaman S, Zain NNM, Waras MN, Abdullah WN, Miskam M, Raoov M, Aziz NA, Hassan Y (2023) Research progress on extraction and analytical methods for Saxitoxin and its congeners. Chromatographia 86:349–373. https://doi.org/10.1007/s10337-023-04251-6

Ji T, Li Y, Deng X, Rwei AY, Offen A, Hall S, Zhang W, Zhao C, Mehta M, Kohane DS (2021) Delivery of local anaesthetics by a self-assembled supramolecular system mimicking their interactions with a sodium channel. Nat Biomed Eng 5:1099–1109. https://doi.org/10.1038/s41551-021-00793-y

Kern TS, Berkowitz BA (2015) Photoreceptors in diabetic retinopathy. J Diabetes Invest 6(4):371–380. https://doi.org/10.1111/jdi.12312

Kwon M, Jung IY, Cha M, Lee BH (2021) Inhibition of the Nav1.7 channel in the trigeminal ganglion relieves pulpitis inflammatory pain. Front Pharmacol 12. https://doi.org/10.3389/fphar.2021.759730

Lee SK, Lee MS (2022) Ocular neuromyotonia: a review of diagnosis and treatment. Curr Opin Ophthalmol 33(6):465–470. https://doi.org/10.1097/ICU.0000000000000894

Lefebvre KA, Quakenbush L, Frame E, Huntington KB, Sheffield G, Stimmelmayr R, Bryan A, Kendrick P, Ziel H, Goldstein T, Snyder JA, Gelatt T, Gulland F, Dickerson B, Gill V (2016) Prevalence of algal toxins in Alaskan marine mammals foraging in a changing arctic and subarctic environment. Harmful Algae 55:13–24. https://doi.org/10.1016/j.hal.2016.01.007

Lefebvre KA, Fachon E, Bowers EK, Kimmel DG, Snyder JA, Stimmelmayr R, Grebmeier JM, Kibler S, Ransom Hardison D, Anderson DM, Kulis D, Murphy J, Gann JC, Cooper D, Eisner LB, Duffy-Anderson JT, Sheffield G, Pickart RS, Mounsey A, Willis ML, Stabeno P (2022) Paralytic shellfish toxins in Alaskan Arctic food webs during the anomalously warm ocean conditions of 2019 and estimated toxin doses to Pacific walruses and bowhead whales. Harmful Algae 114:102205. https://doi.org/10.1016/j.hal.2022.102205

Leffler A, Cummins TR, Dib-Hajj SD, Hormuzdiar WN, Black JA, Waxman SG (2002) GDNF and NGF reverse changes in repriming of TTX-sensitive Na(+) currents following axotomy of dorsal root ganglion neurons. J Neurophysiol 88(2):650–658. https://doi.org/10.1152/jn.2002.88.2.650

Li Y, Zhu H, Li C, Zhang H, Chen Z, Zheng W, Xu H, Zheng T (2014) Draft genome sequence of the Algicidal bacterium *Mangrovimonas yunxiaonensis* strain LY01. Genome Announc 2(6):e01234–e01244. https://doi.org/10.1128/genomeA.01234-14

Liu Y, Jiang C, Song M, Cao Y, Huang Q, Feng L (2022) Optimization of Gonyautoxin1/4-binding G-Quadruplex aptamers by label-free surface-enhanced Raman spectroscopy. Toxins 14(9):622. https://doi.org/10.3390/toxins14090622

Luiz AP, Kopach O, Santana-Varela S, Wood JN (2015) The role of Na1.9 channel in the development of neuropathic orofacial pain associated with trigeminal neuralgia. Mol Pain 11:72. https://doi.org/10.1186/s12990-015-0076-4

Lv MD, Wei YX, Chen JP, Cao MY, Wang QL, Hu S (2023) Melatonin attenuated chronic visceral pain by reducing Nav1.8 expression and nociceptive neuronal sensitization. Mol Pain 19:17448069231170072. https://doi.org/10.1177/17448069231170072

Marrone BL, Banerjee S, Talapatra A, Gonzalez-Esquer R, Pilania G (2023) Toward a predictive understanding of cyanobacterial harmful algal blooms through AI integration of physical, chemical, and biological data. ACS ES&T Water 4(3):844–858. https://doi.org/10.1021/acsestwater.3c00369

McAlvin JB, Zhan C, Dohlman JC, Kolovou PE, Salvador-Culla B, Kohane DS (2015) Corneal anesthesia with site 1 Sodium channel blockers and Dexmedetomidine. Invest Ophthalmol Vis Sci 56(6):3820–3826. https://doi.org/10.1167/iovs.15-16591

Mesquita MCB, Lürling M, Dorr F, Pinto E, Marinho MM (2019) Combined effect of light and temperature on the production of Saxitoxins in Cylindrospermopsis raciborskii strains. Toxins (Basel) 11(1):38. https://doi.org/10.3390/toxins11010038

Montuori E, De Luca D, Penna A, Stalberga D, Lauritano C (2024) Alexandrium spp.: from toxicity to potential biotechnological benefits. Mar Drugs 22(1):31. https://doi.org/10.3390/md22010031

Moustaka-Gouni M, Hiskia A, Genitsaris S, Katsiapi M, Manolidi K, Zervou S-K, Christophoridis C, Triantis TM, Kaloudis T, Orfanidis S (2016) First report of Aphanizomenon favaloroi occurrence in Europe associated with saxitoxins and a massive fish kill in Lake Vistonis, Greece. Mar Freshwater Res 68(4):793–800. https://doi.org/10.1071/MF16029

Naknaen A, Ratsameepakai W, Suttinun O, Sukpondma Y, Khan E, Pomwised R (2021) Microcystis Sp. co-producing microcystin and Saxitoxin from Songkhla Lake Basin, Thailand. Toxins (Basel) 13, 9:631. https://doi.org/10.3390/toxins13090631

Ng PH, Cheng TH, Man KY, Huang L, Cheng KP, Lim KZ, Chan CH, Kam MH, Zhang J, Marques AR, St-Hilaire S (2023) Hydrogen peroxide as a mitigation against *Microcystis* sp. bloom. Aquaculture (Amsterdam, Netherlands) 577:739932. https://doi.org/10.1016/j.aquaculture.2023.739932

Oh JW, Pushparaj SSC, Muthu M, Gopal J (2023) Review of harmful algal blooms (HABs) causing marine fish kills: toxicity and mitigation. Plants (Basel) 12(23):3936. https://doi.org/10.3390/plants12233936

Pagba CV, Lane SM, Wachsmann-Hogiu S (2010) Conformational changes in quadruplex oligonucleotide structures probed by Raman spectroscopy. Biomed Opt Express 2(2):207–217. https://doi.org/10.1364/BOE.2.000207

Podduturi R, Schlüter L, Liu T, Osti JAS, Moraes MAB, Jørgensen NOG (2021) Monitoring of saxitoxin production in lakes in Denmark by molecular, chromatographic and microscopic approaches. Harmful Algae 101:101966. https://doi.org/10.1016/j.hal.2020.101966

Pućko M, Rourke W, Hussherr R, Archambault P, Eert J, Majewski AR, Niemi A, Reist J, Michel C (2023) Phycotoxins in bivalves from the western Canadian Arctic: the first evidence of toxigenicity. Harmful Algae 127:102474. https://doi.org/10.1016/j.hal.2023.102474

Raftopoulos R, Hickman SJ, Toosy A, Sharrack B, Mallik S, Paling D, Altmann DR, Yiannakas MC, Malladi P, Sheridan R, Sarrigiannis PG, Hoggard N, Koltzenburg M, Gandini Wheeler-Kingshott CA, Schmierer K, Giovannoni G, Miller DH, Kapoor R (2016) Phenytoin for neuroprotection in patients with acute optic neuritis: randomizeded, placebo-controlled, phase 2 trial. Lancet Neurol 15(3):259–269. https://doi.org/10.1016/S1474-4422(16)00004-1

Rangel LM, Ger KA, Silva LH, Soares MC, Faassen EJ, Lürling M (2016) Toxicity overrides morphology on *Cylindrospermopsis raciborskii* grazing resistance to the Calanoid copepod *Eudiaptomus gracilis*. Microb Ecol 71(4):835–844. https://doi.org/10.1007/s00248-016-0734-8

Rangel LM, Silva LHS, Faassen EJ, Lürling M, Ger KA (2020) Copepod prey selection and grazing efficiency mediated by chemical and morphological defensive traits of cyanobacteria. Toxins (Basel) 12(7):465. https://doi.org/10.3390/toxins12070465

Sampaio F, Soares S, Pereira S, Lemos JA, Mota Á (2020) Sodium Channel Myotonia and a novel Gly701Asp mutation in the SCN4A gene: from an ophthalmological symptom to a familial disease. Neuroophthalmology 45(1):41–44. https://doi.org/10.1080/01658107.2020.1779316

Schmitz Y, Witkovsky P (1997) Dependence of photoreceptor glutamate release on a dihydropyridine-sensitive calcium channel. Neuroscience 78(4):1209–1216. https://doi.org/10.1016/s0306-4522(96)00678-1

Schneider T, Simpson C, Desai P, Tucker M, Lobner D (2020) Neurotoxicity of isomers of the environmental toxin L-BMAA. Toxicon 184:175–189. https://doi.org/10.1016/j.toxicon.2020.06.014

Schreiber AK, Nones CF, Reis RC, Chichorro JG, Cunha JM (2015) Diabetic neuropathic pain: physiopathology and treatment. World J Diabetes 6(3):432–444. https://doi.org/10.4239/wjd.v6.i3.432

Scott AB, Honeychurch D, Brin MF (2023) Early development history of Botox (onabotulinumtoxinA). Medicine (Baltimore) 102:e32371. https://doi.org/10.1097/MD.0000000000032371

Shaughnessy M, Lamuraglia G, Klebanov N, Ji Z, Rajadurai A, Kumar R, Flaherty K, Tsao H (2019) Selective uveal melanoma inhibition with calcium channel blockade. Int J Oncol 55(5):090–1096. https://doi.org/10.3892/ijo.2019.4873

Su Z, Sheets M, Ishida H, Li F, Barry WH (2004) Saxitoxin blocks L-type ICa. J Pharmacol Exp Ther 308(1):324–329. https://doi.org/10.1124/jpet.103.056564

Suzuki H, Maruyama R, Yamamoto E, Kai M (2013) Epigenetic alteration and microRNA dysregulation in cancer. Front Genet 4:258. https://doi.org/10.3389/fgene.2013.00258

Takati N, Azeddoug H, Benchekroun MN, Blaghen M, Ennaji MM (2023) Biochemical responses in mice induced by saxitoxins extracted from the cockles *Acanthocardia tuberculatum*. Cell Mol Biol (Noisy-le-grand) 69(6):95–100. https://doi.org/10.14715/cmb/2023.69.6.15

Tseytlin IN, Antrim AK, Gong P (2024) Nanoparticles for mitigation of harmful cyanobacterial blooms. Toxins 16(1):41. https://doi.org/10.3390/toxins16010041

Turner AD, Dhanji-Rapkova M, Rowland-Pilgrim S, Turner LM, Rai A, Venugopal MN, Karunasagar I, Godhe A (2017) Assessing the presence of marine toxins in bivalve molluscs from southwest India. Toxicon 140:147–156. https://doi.org/10.1016/j.toxicon.2017.11.001

US National Office for Harmful Algal Blooms (2024) IOC-UNESCO harmful algae information system [Online]. Available at: https://data.hais.ioc-unesco.org/. Accessed 12 Jan 2024

Van De Riet J, Gibbs RS, Muggah PM, Rourke WA, MacNeil JD, Quilliam MA (2011) Liquid chromatography post-column oxidation (PCOX) method for the determination of paralytic shellfish toxins in mussels, clams, oysters, and scallops: collaborative study. J AOAC Int 94(4):1154–1176

Vilar MCP, Rodrigues TFCP, Silva LO, Pacheco ABF, Ferrão-Filho AS, Azevedo SMFO (2021) Ecophysiological aspects and sxt genes expression underlying induced chemical defense in STX-producing *Raphidiopsis raciborskii* (cyanobacteria) against the zooplankter Daphnia gessneri. Toxins (Basel) 13(6):406. https://doi.org/10.3390/toxins13060406

Wang J, Salata JJ, Bennett PB (2003) Saxitoxin is a gating modifier of HERG K+ channels. J Gen Physiol 121(6):583–598. https://doi.org/10.1085/jgp.200308812

Wang H, Ruoyu G, Weol-Ae L, Andrew AE, Ki J-S (2020) Comparative transcriptomics of toxin synthesis genes between the non-toxin producing dinoflagellate *Cochlodinium polykrikoides* and toxigenic *Alexandrium pacificum*. Harmful Algae 93:101777. https://doi.org/10.1016/j.hal.2020.101777

Wang HN, Qian WJ, Zhao GL, Li F, Miao YY, Lei B, Sun XH, Wang ZF (2023) L- and T-type Ca2+ channels dichotomously contribute to retinal ganglion cell injury in experimental glaucoma. Neural Regen Res 18(7):1570–1577. https://doi.org/10.4103/1673-5374.360277

William H, Cameron W, Ryan MD, Debora S (2014) Paralytic shellfish poisoning: a case series. Western J Emerg Med 15, 4:378–381. https://doi.org/10.5811/westjem.2014.4.16279

Xinhui Y, Lin Y, Wang X, Liangjun X, Wang Z, Feng FF (2018) Exonuclease-assisted multicolor aptasensor for visual detection of ochratoxin a based on G-quadruplex-hemin DNAzyme-mediated etching of gold nanorod. Microchim Acta 185:259. https://doi.org/10.1007/s00604-018-2811-9

Yang L, Li Q, Liu X, Liu S (2016) Roles of voltage-gated Tetrodotoxin-sensitive sodium channels NaV1.3 and NaV1.7 in diabetes and painful diabetic neuropathy. Int J Mol Sci 17(9):1479. https://doi.org/10.3390/ijms17091479

Yang F, Zou Y-Q, Li M, Luo W-J, Chen G-Z, Xiao-Zhi W (2021) Intervertebral foramen injection of plerixafor attenuates neuropathic pain after chronic compression of the dorsal root ganglion: possible involvement of the down-regulation of Nav1.8 and Nav1.9. Eur J Pharmacol 908:174322. https://doi.org/10.1016/j.ejphar.2021.174322

Zamponi GW, Striessnig J, Koschak A, Dolphin AC (2015) The physiology, pathology, and pharmacology of voltage-gated calcium channels and their future therapeutic potential. Pharmacol Rev 67(4):821–870. https://doi.org/10.1124/pr.114.009654

Zhang F, Xu X, Li T, Liu Z (2013) Shellfish toxins targeting voltage-gated sodium channels. Mar Drugs 11, 12:4698–4723. https://doi.org/10.3390/md11124698
Zheng X, Hu B, Gao SX, Liu DJ, Sun MJ, Jiao BH, Wang LH (2015) A saxitoxin-binding aptamer with higher affinity and inhibitory activity optimized by rational site-directed mutagenesis and truncation. Toxicon 101:41–47. https://doi.org/10.1016/j.toxicon.2015.04.017
Zhou R, Gao Y, Yang C, Zhang X, Hu B, Zhao L, Guo H, Sun M, Wang L, Jiao B (2022) Novel SELEX based on immobilizing libraries enables screening of Saxitoxin aptamers for BLI Aptasensor applications. Toxins (Basel) 14(3):228. https://doi.org/10.3390/toxins14030228

Chapter 22
Cytotoxicity and Medicinal Applications of Plant Lectins

Nihal Gören-Sağlam and Pelin Arda

Abstract Lectins are a wide variety of proteins or glycoproteins that have the ability to bind and recognize carbohydrate molecules. They are found in various types of organisms, from viruses to humans. They also have many functions in the organisms, including immune response and pathogen recognition. In particular, plant lectins play a role in the plant defense system by recognizing specific sugar moieties on the surface of pathogenic organisms such as fungi or bacteria. In addition, their activities contribute to the regulation of plant growth and development as well as to symbiotic interactions with beneficial microorganisms in plants. There is information that plant lectins have protective effects on plants against abiotic stress factors such as drought, salinity, and temperature. Despite these diverse functions of the lectins in plants, their consumption can have toxic effects on animals, including humans. Due to their functional properties, which are antimicrobial, anticancer, immunomodulatory, anti-inflammatory, and antiviral, they have also attracted attention in medicine. All of these properties are revealed by their specific binding to cell surface carbohydrates. They are therefore used as diagnostic tools to detect carbohydrate antigens on the cell surface and as therapeutic tools that bind to cancer cell membranes carrying abnormally glycosylated residues of glycoconjugates. Their ability to bind to specific carbohydrate motifs found in certain cell types, including cancer cells, also makes them a potential candidate for targeted drug delivery. The medical applications of plant lectins are still an active field of research and further studies are needed. In addition, their activity and specificity need to be modified for their medical use or their toxic effects need to be investigated much more thoroughly.

N. Gören-Sağlam (✉)
Faculty of Science, Department of Biology, Division of General Biology, Istanbul University, Istanbul, Turkey
e-mail: gorenn@istanbul.edu.tr

P. Arda
Faculty of Science, Department of Biology, Division of Molecular Biology, Istanbul University, Istanbul, Turkey
e-mail: parda@istanbul.edu.tr

M. N. V. Prasad et al. (eds.), *Biotoxins*,
https://doi.org/10.1007/978-3-031-75309-1_22

Keywords Plant lectins · Abiotic stress · Medical application · Toxic effects · Plant defence

22.1 Introduction

When we look at lectins from a historical perspective, we see that Peter Hermann Stillmark discovered ricin in 1888. Ricin is an important lectin as it is obtained from the extracts of the castor oil plant and is the first lectin discovered (Olsnes 2004; Polito et al. 2019). Shortly after the discovery of ricin, another lectin called abrin was discovered (Olsnes 2004). Lectins were initially called agglutinins, hemagglutinins, or phytoagglutinins due to their ability to agglutinate blood cells (André et al. 2015). After Sumner and Howell (1936) demonstrated the carbohydrate specificity of concanavalin A (Con A) and Watkins and Morgan (1952) discovered blood type agglutination, the idea that their selectivity was based on carbohydrate recognition emerged. When it was revealed that most of the lectins had properties that could distinguish different human blood groups, they were named lectins, based on the Latin word "lego" meaning chosen (Boyd and Reguera, 1949; Renkonen 1948; Boyd and Shapleigh, 1954). Due to their functional properties, which are antimicrobial, anticancer, immunomodulatory, anti-inflammatory, and antiviral, they have also attracted attention in medicine (Fig. 22.1) (Chettri et al. 2021).

Plant lectins are plant proteins that have at least one non-catalytic domain and bind inversely to mono- and oligosaccharides. Herbal lectins taken orally can bind to undigested food particles remaining in the intestine, various cell membranes and glycoconjugates of the intestinal and colonic mucosa, and may damage the mucosa itself, the bacterial flora in the intestine, and other internal organs (Rüdiger 1998; Peumans and Van Damme 1995). Thus, dietary lectins are generally considered toxic and antinutritional factors. However, many lectins found in tomatoes, lentils, peas, chickpeas, broad beans, and other common foods are non-toxic (Campos-Vega et al. 2009). Research on lectins has increased in recent years. Studies

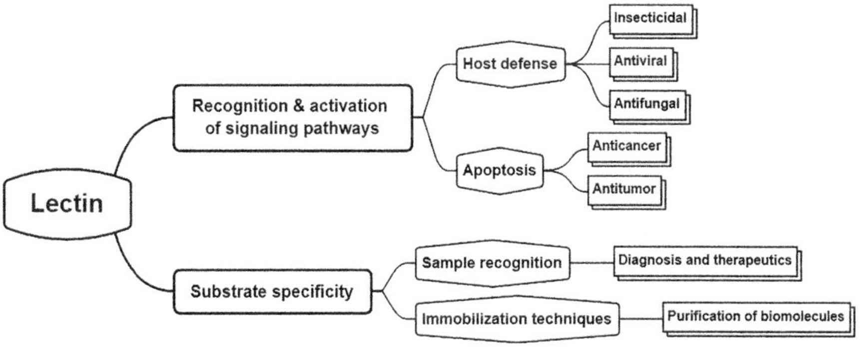

Fig. 22.1 Application areas of lectins (Chettri et al. 2021)

conducted on experimental animals have shown that they prevent certain types of cancer, activate innate defense mechanisms, and control and prevent obesity (Roy et al. 2010; Pryme et al. 2006; Hartmann and Meisel 2007). The effect of lectins in tumors is thought to be to reduce cell division. Lectins can bind to cell membranes and receptors, thereby causing cytotoxicity and apoptosis. Some studies suggest that lectins have the ability to increase the number of macrophages. Thus, the sensitivity of tumor cells to attacks increases. Immunomodulatory effects of legumes in peas are based on their activation of spleen lymphocytes in mice. In human studies, lentil lectins have shown a robust effect in reducing the onset of Merkel skin carcinomas (Sames et al. 2001) and human hepatoma [Wang et al. 2000). Vicia faba agglutinin (VFA), a lectin found in broad beans, reduced colon cancer cells that aggregated, promoted morphological differentiation, and became malignant (Jordinson et al. 1999). In mice fed a diet containing PHA (phytohemagglutinin), a lectin found in raw beans, it significantly reduced the growth of non-Hodgkin lymphoma tumors, tumors that cause intra-abdominal fluid collection, and subcutaneous hard tumors. The degree of reduction is related to the dose used (Pryme and Bardocz 2001). The number of Kerbs II tumor cells in the acidic fluid of mice fed a control diet for 8 days was three times higher than that of mice with PHA in their diet. Feeding mice with PHA for less than 8 days after injection of tumor cells also resulted in a reduction in tumor cell growth (Bardocz et al. 1997). The inclusion of raw beans in the diet of obese mice reduced lipid accumulation due to the insulin-lowering effect of lectins. Even at high doses, no body or muscle protein losses occurred in normal rats, suggesting that the use of lectins as therapeutic agents to treat obesity would be beneficial (Pusztai et al. 1998).

22.2 Cytotoxicity of Plant Lectins and Medicinal Potentials

Plant lectins are proteins in plants that bind reversibly to carbohydrates and glycoconjugates, contributing to the bioactive properties of many medicinal plants. Many lectins isolated from medicinal plants are known to have antioxidant, antibacterial, antiviral, anti-inflammatory, immunomodulatory, and antitumor effects (Wu and Bao 2013; Jandú et al. 2017; Padiyappa et al. 2022; Bektas and Kaptan 2023). Dietary lectins with biological activity have been mostly characterized by purification from dicotyledonous seeds. The therapeutic activities of dietary lectins obtained from non-seed plant parts, such as tubers and leaves, have been the subject of many studies in recent years (Van Damme 2014). The best-studied lectin whose cytotoxic effect is known is ricin, which was discovered in the seeds of *Ricinnus communis*. Ricin is a very potent toxin and this toxic effect can be attributed to the inhibition of protein synthesis due to its catalytic action on ribosomes (Hartley and Lord 2004). For this reason, ricin is widely used in traditional medicine in various forms and is also frequently used for criminal purposes (Olsnes 2004). Ricin enters mammals via endocytosis. Since ricin binds to different glycoconjugates in different cellular modules, different endocytic compartments can be used for entry (Hartley and Lord

2004). The ability of ricin to inactivate the ribosome and interfere with eukaryotic maintenance has been investigated for its therapeutic use in cancer treatment. The nonspecific affinity of the β-chain of ricin for glycoproteins and glycolipids, which can be found in almost all cell types, may limit its selective binding to cancer cells as an anticancer agent. The challenges associated with the use of ricin as a therapeutic agent can be addressed by improving the nanocarriers that mediate the transport of the drug to cancer cells. Various types of nanocarriers (liposomes, nanoparticles, carbon nanotubes, etc.) that have been used systemically so far are effective both in vitro and in vivo (Tyagi et al. 2015).

Increased sialylation and changes in protein glycosylation on the cancer cell surface are hallmarks of cancerous tissues. Many studies focus on the usability of plant lectins in differentiating between benign and malignant tumors (Lalwani et al. 1996; Gao et al. 2019; Mazalovska and Kouokam 2020). With these properties, lectins can be used in histochemical studies to reveal the stages of carcinogenesis and also can be utilized as a therapeutic target using specific lectins. It has been reported that the expression of growth enzymes such as α-2,3-sialyltransferase (ST3Gal1) is responsible for sialyl Lewis x and a (sLex/sLea) composition in carcinogenesis (Koike et al. 2004). sLex serves the metastatic process by interacting with selectins found in endothelial cells and platelets (Varki and Varki 2007). The β-1 integrin hypersialylated with α-2,6-sialic acid, which has been identified as a parameter of colon adenocarcinoma, stimulates cell migration by allowing the carcinoma cells to change their affinity preferences with the extracellular matrix (Seales et al. 2005). Although sialic acid plays an important role in many biological processes, hypersialylation makes an important contribution to determining the phenotype of metastatic cells (Litynska et al. 2001). It has been reported that the treatment of Maackia amurensis lectin (MAL-II), which is extracted from the Amur tree and has a specific binding capacity for these residues, significantly suppressed the tumorigenic and malignant properties of these cells. MAL-II exposure not only suppressed the proliferative and invasive properties of anaplastic thyroid cancer cells but also significantly reduced tumor cell-endothelial interaction, which is one of the key steps of the metastatic process (Kaptan et al. 2018). Additionally, treatment of anaplastic thyroid cancer cells with MAL-II altered the expressions of numerous genes involved in key signal transduction pathways related to development, proliferation, stem cell regulation, and carcinogenesis (Bektas and Kaptan 2022).

Plant lectins interact with various glycan ligands found in animal tissues that serve as targets for lectin-based markers and can be used as glycoprobes in cancer diagnosis. For example, ovarian cancer biomarker (CA125), prostate cancer biomarker (PSA), and breast cancer biomarker (Her2/neu) are glycoproteins that can be easily recognized by lectins with selective sugar-binding potential (Majee and Biswas 2013). Concanavalin A (ConA) and Peanut Agglutinin (PNA), as demonstrated by immunohistochemical analysis, are promising lectin markers for identifying the malignant phenotype of the cell surface of the head and neck squamous cell carcinomas (Lalwani et al. 1996). Gao and colleagues suggested that *Aleuria aurantia* lectin (AAL), a galactose-specific lectin, *Ulex europaeus I* (UEA-I), a fucose-binding lectin, and *Phaseolus vulgaris* erythroagglutinin (PHA-E) show different

degrees of binding properties to metastatic and non-metastatic pancreatic cancer cells. They suggest that these lectins can be used as markers in processes related to metastasis (Gao et al. 2019). It was also reported a decrease in cell survival in papillary thyroid cancer cells following AAL treatment (Kaptan et al. 2018). This indicates that AAL not only serves as a diagnostic marker but also has potential as a therapeutic agent.

Various preclinical and clinical studies have shown that some plant lectins may have therapeutic efficacy in the treatment of cancer and AIDS (Ji et al. 2005; An et al. 2006; Liu et al. 2010; Wang et al. 2011). The Galanthus nivalis agglutinin (GNA)-related lectin family, which is commonly obtained from monocotyledonous plants includes mannose-binding specific lectins. The first lectin isolated from snowdrop bulbs belonging to this family was *Galanthus nivalis* agglutinin (GNA), a model lectin that binds to terminal mannose groups on N-glycans (Van Damme et al. 1987). GNA-related plant lectins are the subject of many preclinical studies for therapeutic uses due to their anticancer, microbicidal, and antiviral activities. GNA-related plant lectins are the subject of many preclinical studies testing their therapeutic uses due to their anticancer, microbicidal, and antiviral activities (Balzarini et al. 2004; Ji et al. 2005; Wu and Bao 2013). Several GNA lectins have been reported to induce apoptosis and autophagy in various cancer cell lines in vitro, exhibiting cytotoxic effects (Wu and Bao 2013). Garlic lectin, a plant lectin from the GNA family isolated from *Allium sativum* L. belonging to the Liliaceae family, has been shown to have many pharmacological properties (Sharma and Singh 2022). Garlic lectin has been reported to have cytotoxic effects on human U937 and HL60 cells by inhibiting cell proliferation and inducing apoptosis (Karasaki et al. 2001). Padiyappa et al. (2022) partially purified the Glc/Man-specific garlic lectin from garlic bulbs and applied it to in vitro human melanoma (A375) cell lines and an in vivo murine Ehrlich ascites carcinoma (EAC) tumor model. In vivo experiments in this study revealed the antiproliferative activity of garlic lectin. Additionally, this lectin exhibited anti-tumor and anti-angiogenic effects by regressing tumor progression and inhibiting new blood vessel formation. Garlic lectin, a natural immune-stimulating biomolecule, is of great importance for its antioxidant effects and potential use in cancer therapy.

Polygonatum cyrtonema lectin (PCL), a member of the GNA-related lectin family, is an important lectin used in traditional Chinese medicine, particularly in treating hypertension and diabetes for many years (An et al. 2006). Experimental studies have shown that PCL exhibits anti-tumor effects by stimulating apoptosis and autophagy in human melanoma (A375) cells, human breast adenocarcinoma (MCF-7) cells, human cervical cancer (HeLa) cells, and murine fibrosarcoma (L929) cells. This lectin, which binds both mannose and sialic acid, has been reported to exhibit lower cytotoxicity for normal cells compared to other mannose-binding lectins isolated from Amaryllidaceae, Liliaceae, and Orchidaceae (Liu et al. 2010; Wang et al. 2011; Bektas and Kaptan 2023). An et al. (2006) reported that this lectin exhibited 10–100 times stronger anti-HIV activity on Human T lymphocytes (MT-4 cells) and T lymphoid leukemia (CEM cells) than other mannose-binding lectins. The antiviral activity of PCL is suggested to occur through high-affinity

binding to α-(1–3)-D-mannose oligomers on gp120, an envelope protein of HIV, thereby blocking the virus's binding to target cells (Balzarini et al. 2004; Ji et al. 2005). Investigating the toxicity and therapeutic effectiveness of the GNA-related lectin family through clinical studies is important for these lectins to be considered promising agents in the treatment of cancer and HIV infections.

22.3 The Role of Plant Lectins in Immunomodulation and Inflammatory Processes

Some plant lectins exhibit strong immunomodulatory properties by interacting with glycans on the surface of various immune cells. The discovery that plant lectins exert mitogenic effects on lymphocytes has revealed the importance of the interaction between lectins and lymphocytes in immunomodulation (Souza et al. 2013; da Silva and Correia 2014; Jandú et al. 2017). The first studies on this subject attracted attention in 1960 when Peter Carey Nowell observed that kidney bean (*Phaseolus vulgaris*) phytohemagglutinin (PHA) had the potential to stimulate mitosis in lymphocytes (Sharon and Lis 2007). Such an interaction can stimulate the production of cytokines, reactive oxygen, and reactive nitrogen species, and increase the phagocytic activity of immune cells, triggering immune defense mechanisms involved in the response to microbial infections (Souza et al. 2013; da Silva and Correia 2014).

Silva et al. (2016) reported that the application of mannose-binding lectin isolated from *Canavalia brasiliensis* and *Cratylia argentea* protected mice from a systemic infection caused by Salmonella by regulating the release of pro-inflammatory cytokines and nitric oxide. Some reports suggest that certain plant lectins can combat protozoan infections through immunomodulatory activity. Lectins such as Jacalin, *Synadenium carinatum* (ScLL), and *Artocarpus heterophyllus* (ArtinM) have been successfully applied in murine models to treat infections caused by *Toxoplasma gondii*, *Trypanosoma cruzi,* and *Leishmania* spp. (Albuquerque et al. 1999; Panunto-Castelo et al. 2001; de Souza et al. 2016). In these studies, lectins were also used as adjuvants to create a more effective immunization.

Concanavalin A (ConA), one of the most studied plant lectins, is a legume lectin with mannose/glucose-binding activity obtained from the seeds of the jack bean (*Canavalia ensiformis*). Various researchers have reported that ConA causes the proliferation of immune cells and increases the expression of toll-like receptors (Sodhi et al. 2007; da Silva and Correia 2014). Kuo et al. (2007) administered *Klebsiella pneumoniae* bacteria intragastrically to mice, causing organ damage and sepsis in an experimental model. In this study, single-dose ConA treatment (1 mg/kg or 2 mg/kg) had no effect on bacterial growth in vitro but increased the animal survival by regulating immune responses against the bacterial infection (Kuo et al. 2007). Con A pretreatment has been reported to exhibit antifungal effects by enhancing the phagocytic ability of macrophages and neutrophils in experimental animals infected with *Candida albicans*. (Loyola et al. 2002; Moresco et al. 2002). However, it is also known that Con A triggers inflammation by stimulating the release of

pro-inflammatory cytokines at high doses. Con A, which plays a role in T cell-mediated inflammatory response formation and immunomodulation, is widely used to create autoimmune or viral hepatitis experimental models in vivo (Ye et al. 2018; Hao et al. 2022).

After Con A injection into experimental animals, macrophages and CD4+ T cells are activated, stimulating the release of proinflammatory cytokines such as TNF-α, IL-1β, IL-6, and IFN-γ. Intracellular signaling pathways stimulated by these cytokines can cause apoptosis and autophagy in hepatocytes. Concanavalin A triggers acute inflammation by causing inflammatory cell infiltration and necrosis in the liver, depending on the dose and time (8–12 hours after tail vein injection at a dose of 20 mg/kg) (Ye et al. 2018). When the applied ConA concentration is too high or the application time is prolonged, irreversible liver damage occurs, and mice die in a short time. Ye et al. (2018) reported that a dose of 15 mg/kg ConA caused 20% of C57BL/6 mice to die within 18 hours, and doses above 20 mg/kg were lethal within 12 hours after administration.

Plant lectins can exhibit both pro-inflammatory and anti-inflammatory effects, depending on the dose and application time. There is strong evidence that plant lectins, such as *Canavalia boliviana* lectin, lectin isolated from butterfly pea (*Clitoria fairchildiana*) seeds, monocot lectin from *Canna limbata*, and soybean agglutinin, have anti-inflammatory effects (Benjamin et al. 1997; Leite et al. 2012; Araujo et al. 2013; Bezerra et al. 2014). Lectins can inhibit the binding of selectins to leukocytes by competitively binding to their glycosylated ligands on leukocytes, thus preventing vascular inflammation by inhibiting cell migration to the inflammation area. They may also prevent the release of pro-inflammatory cytokines and chemokines from immune cells (Jandú et al. 2017).

22.4 The Role of Plant Lectins in Plants

Lectins are protein groups found in most plants that have binding properties of glycoproteins, glycolipids, and polysaccharides (Öktem 2001). Plant lectins comprise heterogeneous and biochemically complex groups of proteins. It is believed that the main role of lectins is to recognize the pathogen, detect stress, and form the defense mechanism. Many plant lectins play a defensive role because they bind glycans such as chitin, which are rare or absent in plants (Peumans and Van Damme 1995). Studies on lectins are generally concentrated in the medical field. Studies show the use and effects of lectins in the field of health, such as the classification of red blood cells, the treatment of acute wounds, and the cure of incurable diseases such as cancer, AIDS, and Coronavirus and human immunity (Furukawa et al. 1995; Bird and Wingham 1983; De Mejía and Prisecaru 2005; Mazalovska and Kouokam 2020; Keyaerts et al. 2007; Lepenies and Lang 2019).

From a plant study perspective, in agriculture, lectins have been used to protect lectin-rich plants, such as legumes, from microbial infection by recognizing microbes and stimulating the immune system to confront them (Gautam et al. 2018).

Scientists at the "Plant Protection Institute" in Kiev have managed to prepare a commercial preparation called "Azolec" that works both as a fertilizer and a natural pesticide. This preparation consists of a combination of Azotobacter bacteria cells and some lectin molecules (Sergienko et al. 2008; Kyrychenko 2021).

Accumulation of plant lectins in storage organs also indicates the role of a passive defense mechanism. Lectin genes stimulate the plant to respond to many pathogens, including plant parasitic nematodes (Oka et al. 1997; Jammes et al. 2005; Fuller et al. 2007). Lectin genes can act by interacting with glycoproteins found in the intestines or with those other than those that bind to glycoproteins found on nematodes (Fuller et al. 2008). The first study in this field was conducted by Marban-Mendoza et al. (1987) and they showed that the application of Concanavalin A reduced the *M. incognita* population in tomato plant roots by 75%. After lectins perceive external signals and warnings by plant parasitic nematodes, the gene products are directed for cellular or root release by expression of the appropriate gene in plant roots, thus creating significant resistance.

22.4.1 The Role of Lectins against the Effects of Abiotic and Biotic Stress Factors in Plants

Abiotic and biotic stress factors are the most important factors limiting plant growth, development, and production. With the effect of increasing global warming in recent years, drought, salinity, and heat stresses negatively affect plant growth and development. Pathogens such as bacteria, viruses, and fungi also cause diseases in plants, limiting plant growth and development and crop yield. In recent years, researchers have accelerated their work on understanding stress-related responses in plants and developing strategies to produce stress-resistant varieties. Molecular networks encompassing numerous genes and functional proteins have been shown to play a crucial role in generating responses to combat various stresses. Recently, there has been renewed interest in investigating the role of lectins in modulating various biological responses in plants (Fig. 22.2). Many plant lectins have been identified and functionally characterized, but their role in stress tolerance has not been analyzed in detail or comprehensively. However, the development of new analysis methods has added a new dimension to plant lectin research (Marothia et al. 2023).

Studies on the effects of lectins on plant growth and development show that lectins have effects against abiotic and biotic stress in plants. Due to their versatile and multifunctional roles, lectins can increase plant survival efficiency under different stress conditions (Van Damme 2020) (Fig. 22.3). Recently, there has been increasing recognition of the functional role of lectins in orchestrating abiotic stress responses in plants. Plant lectins play an important role as mediators and effectors in various abiotic stress-related signaling pathways (Tsaneva and Van Damme 2020; Liu et al. 2023).

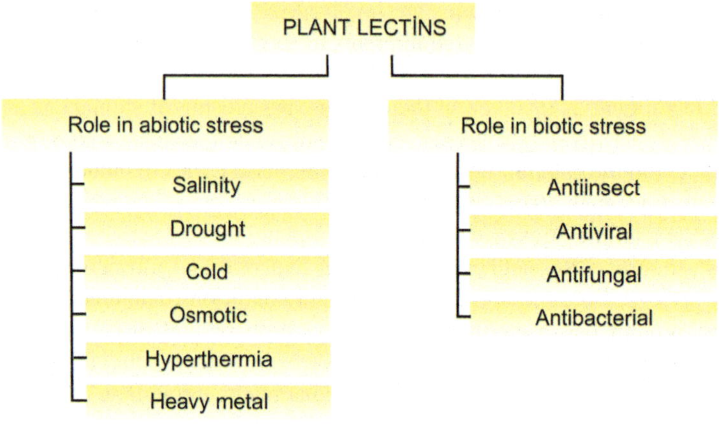

Fig. 22.2 Role of plant lectins in plant stress

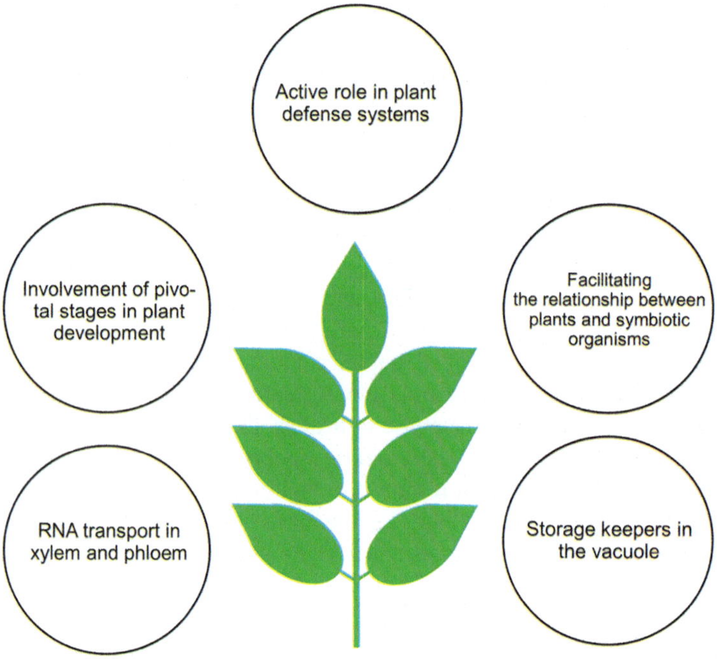

Fig. 22.3 Plant lectins play an important roles in plant defense system

Drought is an important factor limiting crop production in many of the world's agricultural areas. Drought stress triggers numerous physiological, biochemical, and molecular events in plants, which can lead to the development of tolerance mechanisms that enable adaptation to constrained environmental conditions (Kaur et al. 2022; Takahashi et al. 2020). Many stress-related genes, transcription factors,

signaling molecules, and proteins are involved in the regulation of drought stress. Lectin proteins are also expressed in plants in response to drought stress (Naithani et al. 2021; Mahmood et al. 2019; Marothia et al. 2023). In Arabidopsis thaliana, the *EULS3* lectin gene is involved in the opening and closing of stomata as a primary defense response to drought stress in an ABA-dependent manner. *EULS3* is believed to interact with the calcium-dependent protein kinase *CPK3* to induce stomatal closure (Van Hove et al. 2015). *R40* proteins are lectins primarily responsible for controlling the osmotic stress response in many plant species and are upregulated in response to stress. Researchers have shown that rice and tobacco plants overexpressing *Osr40c1* are more tolerant to drought conditions than control plants. The *Osr40c1* protein has been shown to interact with several proteins that regulate drought response (Sahid et al. 2020). Another study showed that the *OsSalT* gene confers tolerance to drought stress in Nicotiana plants through an ABA-dependent pathway via osmoregulation. The data obtained indicate that the expression of *AREB* (*ABA-responsive element-binding*), *NCED* (*9-cis epoxycarotenoid dioxygenase*), and *ERD* (early responsive to dehydration) is increased in transgenic lines. Similarly, under both stress and normal conditions, OsSalT transgenic lines have shown increased levels of the *LEA* (*late embryogenesis-abundant*) gene compared to wild-type plants (Kaur et al. 2022). The obtained data provide some information on the participation of lectins in the drought stress tolerance process in plants; however, the role of lectins in drought stress requires detailed study.

Because plants are sessile organisms, they are affected by global warming. Temperature stress is another type of stress related to climate change. The negative effects of extreme temperature increases on plants are alarming (Balfagón et al. 2020). To protect themselves from these effects, plants develop defense mechanisms against the stressful conditions they face. Generally, these mechanisms involve modulating their physiology, metabolism, and gene expression (Dos Santos et al. 2022). It has been determined that the amount of lectins increases in response to heat stress in plants (Jiang et al. 2010). Researchers have shown that the *DB57* lectin protein accumulates in response to heat shock in cell suspension cultures of the plant *Dolichos biflorus*. In wheat and rice embryos, the levels of lectin proteins have been found to increase in response to high temperatures (Spadoro-Tank and Etzler 1988; Shakirova et al. 1996). Lectins have been observed to affect antioxidant enzyme activity in wheat roots under high- and low-temperature stresses. Sp7 and Sp245 lectins from Azospirillum brasilense exhibited increased peroxidase and SOD activity but decreased catalase activity in wheat roots when exposed to hypo- and hyperthermic stress (Alenkina et al. 2018). Furthermore, in response to low temperature, the activity of cell wall-bound lectin increased in wheat, conferring freezing tolerance in a calcium-dependent manner. Cell wall-bound lectins serve as receptors that detect stress stimuli and transmit these signals to various cellular compartments in response to stress (Timofeeva et al. 2010; Gataulina et al. 2021).

Salt stress is one of the major factors limiting crop productivity in many areas around the world (Kaur et al. 2022, 2020). The response of plants to salt stress is a complex process that involves regulating various physiological, molecular, and biochemical events (Gupta and Huang 2014). Lectins have been reported to play a role

in plant adaptation to salt stress. Researchers have shown that the expression of the rice protein OsJRL, a jacalin-related lectin belonging to the mannose-binding lectin superfamily, is upregulated under saline conditions. Additionally, lectins have been found to regulate the levels of many stress-regulated proteins, including LEA proteins, which play a protective role in response to damage caused by salt stress (Bhardwaj et al. 2013; He et al. 2017). This suggests that lectins may play an indirect role in protecting cellular mechanisms in response to salt stress. Overexpression of chickpea lectin in Brassica juncea has been found to mitigate the effects of salt stress by regulating water retention capacity, chlorophyll content, and membrane integrity (Kumar et al. 2015). Although lectin proteins are known to play a critical role in responding to salt stress, their mechanism of action remains unresolved.

Biotic factors are stress factors caused by the infection of microorganisms (fungi, bacteria, and viruses) and attacks by harmful animals. Plants may be subject to attacks by other harmful organisms that limit or affect their development. Furthermore, the application of chemical treatments to combat these can also be stressful, depending on the conditions in which they are applied. Recently, a large number of lectins have been identified for their roles in plant innate defense systems against various biotic stresses. Lectins are abundant in storage organs, where plants store some of their nitrogen in the form of proteins that bind to carbohydrates. These can later be used as passive defense proteins in immune responses (Van Damme 2014). Lectin proteins are known to possess various biological activities, including anti-insect, antibacterial, antifungal, and antiviral properties. They protect plants against insect attacks by inhibiting insect development (Macedo et al. 2015). Lectins cause insects attacking the plant to lose weight, have difficulty absorbing nutrients, reduce fertility, and experience other physiological problems (Vandenborre et al. 2011). Lectins can bind to various glycan structures within insect bodies, which is believed to cause a toxic effect leading to their death. However, there are still unresolved issues regarding the insecticidal properties of lectins (Macedo et al. 2015). Recent studies suggest that lectins attach to glycoproteins in the insect gut, ultimately inhibiting food absorption and resulting in death (Mishra et al. 2019). Additionally, lectins have been demonstrated to counteract fungi that cause significant crop damage in plants (Holle and Van Damme 2019; Esch and Schaffrath 2017). Using a proteomic approach, a mannose-binding lectin (JRL) was detected in the extracellular secretions of *Helianthus annuus*. This JRL exhibits antifungal activity against pathogenic fungi from the genera *Candida* and *Pichia*. The results showed that the JRL affected membrane permeability and induced cell aggregation in the tested fungi (Regente et al. 2014). Additionally, susceptibility of wheat plants to the facultative fungal disease Fusarium graminearum and the biotrophic fungus Blumeria graminis was reduced by TaJRLL1, a JRL expressed in the stems and spikes of wheat. Furthermore, *A. thaliana* modified with the gene encoding this protein exhibited improved resistance to *F. graminearum* and *Botrytis cinerea*, leading to a reduction in conidial growth of both fungi (Xiang et al. 2011). In addition, lectins exhibit antiviral activity when bound to high-mannose or complex carbohydrates. Examples of such lectins include jacalin, Con A, GNA, HHA, and *Urtica dioica* agglutinin (UDA) (El-Maradny et al. 2021). In fact, targeting viral glycans

with lectins is considered a promising antiviral strategy due to the number of glycosylated sites on the virus capsid (Van Damme 2020). Several recent reports have provided detailed information on various lectins and their roles in biotic stress management (Van Damme 2020; Holle and Van Damme 2019; El-Maradny et al. 2021).

22.5 Conclusion

The anticancer properties of lectins have been the subject of extensive research and have been tested in preclinical and clinical studies. In recent years, the discovery of immunomodulatory and anti-inflammatory effects of plant lectins has encouraged further research into their use in the treatment and prevention of infectious and inflammatory diseases. Depending on the type, dose, and route of administration of the lectin, serious side effects such as kidney diseases, intestinal damage, and hepatitis have been reported in vivo animal models. Considering these drawbacks, there are serious concerns about the therapeutic applicability of lectins in various human diseases. Many mechanistic and interdisciplinary studies need to be carried out on this subject.

Plants contain lectin proteins that serve various functions, including protection against biotic and abiotic stresses at morphological, biochemical, physiological, and molecular levels. Furthermore, plants overexpressing lectin genes demonstrate increased resistance to extreme environmental conditions. Current studies highlight the significance of employing lectins to combat stress in plants, focusing primarily on classical lectins. Nonetheless, there is limited data on inducible lectins, suggesting that future research should concentrate on exploring the roles of these proteins. It is important to elucidate the mechanisms of action and regulatory roles of lectins under various stress stimuli. Detailed studies will shed light on the functional properties of lectin proteins, which will be crucial in developing stress-resistant plants.

References

Albuquerque DA, Martins GA, Campos-Neto A, Silva JS (1999) The adjuvant effect of jacalin on the mouse humoral immune responseto trinitrophenyl and *Trypanosoma cruzi*. Immunol Lett 68:375–381

Alenkina SA, Romanov NI, Nikitina VE (2018) Regulation by *Azospirillum lectins* of the activity of antioxidant enzymes in wheat seedling roots under short-term stresses. Rev Bras Bot 41:579–587

André S, Kaltner H, Manning JC, Murphy PV, Gabius HJ (2015) Lectins: getting familiar with translators of the sugar code. Molecules 20(2):1788–1823

An J, Liu JZ, Wu CF, Li J, Dai L, Van Damme E et al (2006) Anti-HIV I/II activity and molecular cloning of a novel mannose/sialic acid-binding lectin from rhizome of Polygonatum cyrtonema Hua. Acta Biochim Biophys Sin 38(2):70–78

Araujo TS, Teixeira CS, Falcao MA, Junior VR, Santiago MQ, Benevides RG, Delatorre P, Martins JL, Alexandre-Moreira MS, Cavada BS, Campesatto EA, Rocha BA (2013) Anti-inflammatory

and antinociceptive activity of chitin-binding lectin from *Canna limbata* seeds. Appl Biochem Biotechnol 171:1944–1955

Balfagón D, Zandalinas SI, Mittler R, Gómez-Cadenas A (2020) High temperatures modify plant responses to abiotic stress conditions. Physiol Plant 170(3):335–344

Balzarini J, Hatse S, Vermeire K, Princen K, Aquaro S, Perno CF, Clercq ED et al (2004) Mannose-specific plant lectins from the Amaryllidaceae family qualify as efficient microbicides for prevention of human immunodeficiency virus infection. Antimicrob Agents Chemother 48:3858–3870

Bardocz S, Grant G, Duguid TJ, Brown D, Pusztai A, Pryme IF (1997) Intracellular levels of polyamines in Krebs II lymphosarcoma cells in mice fed phytohaemagglutinin-containing diets are coupled with altered tumour growth. Cancer Lett 121:25–29

Bektas S, Kaptan E (2022) RNA-Seq transcriptome analysis reveals Maackia amurensis leukoagglutinin has antitumor activity in human anaplastic thyroid cancer cells. Mol Biol Rep 49(10):9257–9266

Bektas S, Kaptan E (2023) Therapeutic potential of lectins in the treatment of breast cancer: a review. Int J Biol Macromol 126073:126073

Benjamin CF, Figueiredo RC, Henriques MG, Barja-Fidalgo C (1997) Inflammatory and anti-inflammatory effects of soybean agglutinin. Braz J Med Biol Res 30:873–881

Bezerra GA, Viertlmayr R, Moura TR, Delatorre P, Rocha BAM, do Nascimento KS et al (2014) Structural studies of an anti-inflammatory lectin from Canavalia boliviana seeds in complex with dimannosides. PLoS One 9(5):e97015

Bhardwaj R, Sharma I, Kanwar M, Sharma R, Handa N, Kaur H et al (2013) LEA proteins in salt stress tolerance. In: Ahmad P, Azooz MM, Prasad MNV (eds) Salt stress in plants: signalling, omics and adaptations. Springer, New York, pp 79–112

Bird GWG, Wingham J (1983) "New" lectins for the identification of erythrocyte cryptantigens and the classification of erythrocyte polyagglutinability: medicago disciformis and Medicago turbinata. J Clin Pathol 36(2):195–196

Boyd WC, Reguera RM (1949) Hemagglutinating substances for human cells in various plants. J Immunol 62(3):333–339

Boyd WC, Shapleigh E (1954) Specific precipitating activity of plant agglutinins (lectins). Science 119(3091):419–419

Campos-Vega R, Reynoso-Camacho R, Pedraza-Aboytes G, Acosto-Gallegos JA, Guzman-Maldonado SH, Paredes-Lopez O, Oomah BD, Loarca-Pina G (2009) Chemical composition and in-vitro polysaccharide fermentation of different beans (*Phaseolus vulgaris* L.). J Food Sci 74:59–65

Chettri D, Boro M, Sarkar L, Verma AK (2021) Lectins: Biological significance to biotechnological application. Carbohydr Res 506:108367

De Mejía EG, Prisecaru VI (2005) Lectins as bioactive plant proteins: a potential in cancer treatment. Crit Rev Food Sci Nutr 45(6):425–445

da Silva LC, Correia MT (2014) Plant lectins and toll-like receptors: implications for therapy of microbial infections. Front Microbiol 5:20

de Souza LPF, Ramos EL, Santana SS, Silva MV, Santiago FM, Mineo TW et al (2016) Lectins from *Synadenium carinatum* (ScLL) and *Artocarpus heterophyllus* (ArtinM) are able to induce beneficial immunomodulatory effects in a murine model for treatment of *Toxoplasma gondii* infection. Front Cell Infect Microbiol 6:164

Dos Santos TB, Ribas AF, de Souza SGH, Budzinski IGF, Domingues DS (2022) Physiological responses to drought, salinity, and heat stress in plants: a review. Stresses 2(1):113–135

El-Maradny YA, El-Fakharany EM, Abu-Serie MM, Hashish MH, Selim HS (2021) Lectins purified from medicinal and edible mushrooms: insights into their antiviral activity against pathogenic viruses. Int J Biol Macromol 179:239–258

Esch L, Schaffrath U (2017) An update on jacalin-like lectins and their role in plant defense. Int J Mol Sci 18(7):1592

Fuller VL, Lilley CJ, Atkinson HJ, Urwin PE (2007) Differential gene expression in Arabidopsis following infection by plant-parasitic nematodes *Meloidogyne incognita* and *Heterodera schachtii*. Mol Plant Pathol 8:595–609
Fuller VL, Lilley CJ, Urwin PE (2008) Nematoderesistance. New Phytol 180:27–44
Furukawa T, Duguid WP, Kobari M, Matsuno S, Tsao MS (1995) Hepatocyte growth factor and Met receptor expression in human pancreatic carcinogenesis. Am J Pathol 147(4):889-895
Gao HF, Wang QY, Zhang K et al (2019) Overexpressed N-fucosylation on the cell surface driven by FUT3, 5, and 6 promotes cell motilities in metastatic pancreatic cancer cell lines. Biochem Biophys Res Commun 511(2):482–489
Gataulina GG, Medvedeva NV, Pilipenko SE, Shitikova AV (2021) Formation of seedlings of white lupin, *Lupinus albus* L. and blue lupin, *Lupinus angustifolius* L. in laboratory and field experiments. Caspian J Environ Sci 19(4):771–776
Gautam A, Srivastava N, Sharma B, Bhagyawant S (2018) Current scenario of legume lectins and their practical applications. J Crop Sci Biotechnol 21(3):217–227
Gupta B, Huang B (2014) Mechanism of salinity tolerance in plants: physiological, biochemical, and molecular characterization. Int J Genomics 2014:701596
Hao J, Sun W, Xu H (2022) Pathogenesis of Concanavalin A induced autoimmune hepatitis in mice. Int Immunopharmacol 102:108411
Hartley MR, Lord JM (2004) Cytotoxic ribosome-inactivating lectins from plants. Biochim Biophys Acta Proteins Proteomics 1701(1–2):1–14
Hartmann R, Meisel H (2007) Food-derived peptides with biological activity: from research to food applications. Curr Opin Biotechnol 18:163–169
He X, Li L, Xu H, Xi J, Cao X, Xu H et al (2017) A rice jacalin-related mannose-binding lectin gene, Os JRL, enhances Escherichia coli viability under high salinity stress and improves salinity tolerance of rice. Plant Biol 19(2):257–267
Holle SV, Van Damme EJM (2019) Messages from the past: new insights in plant lectin evolution. Front Plant Sci 10:36
Jammes F, Lecomte P, de Almeida Engler J, Bitton F, Martin-Magninette ML, Renou JP, Favery B, Abad P (2005) Genome-wideexpressionprofiling of thehostresponsetoroot-knotnematodeinfection in Arabidopsis. Plant J 44:447–458
Jandú JJ, Moraes Neto RN, Zagmignan A, de Sousa EM, Brelaz-de-Castro MC, dos Santos Correia MT, da Silva LC (2017) Targeting the immune system with plant lectins to combat microbial infections. Front Pharmacol 8:285566
Ji X, Gewurz H, Spear GT (2005) Mannose binding lectin (MBL) and HIV. Mol Immunol 42(2):145–152
Jiang SY, Ma Z, Ramachandran S (2010) Evolutionary history and stress regulation of the lectin superfamily in higher plants. BMC Evol Biol 10:1–24
Jordinson M, El-Hariry I, Calnan D, Calam J, Pignatelli M (1999) Vicia fabaagglutinin, the lectin present in broad beans, stimulates differentiation of undifferentiated colon cancer cells. Gut 44:709–714
Kaptan E, Sancar-Bas S, Sancakli A, Bektas S, Bolkent S (2018) The effect of plant lectins on the survival and malignant behaviors of thyroid cancer cells. J Cell Biochem 119(7):6274–6287
Karasaki Y, Tsukamoto S, Mizusaki K, Sugiura T, Gotoh S (2001) A garlic lectin exerted an antitumor activity and induced apoptosis in human tumor cells. Food Res Int 34(1):7–13
Kaur N, Sagar A, Sharma P, Ashish, Pati PK (2020) Structural insights into rice SalTol QTL located SALT protein. Sci Rep 10(1):16589
Kaur N, Prashanth KH, Bhatti MS, Pati PK (2022) OsSalT gene cloned from rice provides evidence of its role in salinity and drought stress tolerance. Plant Sci 320:111306
Keyaerts E, Vijgen L, Pannecouque C, Van Damme E, Peumans W, Egberink H et al (2007) Plant lectins are potent inhibitors of coronaviruses by interfering with two targets in the viral replication cycle. Antivir Res 75(3):179–187

Koike T, Kimura N, Miyazaki K, Yabuta T, Kumamoto K, Takenoshita S et al (2004) Hypoxia induces adhesion molecules on cancer cells: a missing link between Warburg effect and induction of selectin-ligand carbohydrates. Proc Natl Acad Sci USA 101:8132–8137

Kumar D, Shekhar S, Bisht S, Kumar V, Varma A, Kumar M (2015) Ectopic overexpression of lectin in transgenic Brassica juncea plants exhibit resistance to fungal phytopathogen and showed alleviation to salt and drought stress. J Bioeng Biomed Sci 5(1):147

Kuo CF, Wang YH, Lei HY, Wang CH, Tsao N (2007) Concanavalin A protects mice from a lethal inoculation of intragastric Klebsiella pneumoniae and reduces the induced liver damage. Antimicrob Agents Chemother 51:3122–3130

Kyrychenko OV (2021) The effectiveness of phytolectins and lectin compositions application for spraying plants during vegetation. Biotechnol Acta 14(1):57–68

Lalwani AK, Carey TE, Goldstein IJ, Peters BP (1996) Lectin binding characteristics of squamous cell carcinomas of the head and neck. Acta Otolaryngol 116(1):125–131

Leite JF, Assreuy AM, Mota MR, Bringel PH, Lacerda RR, Gomes Vde M, Cajazeiras JB, Nascimento KS, Pessoa Hde L, Gadelha CA, Delatorre P, Cavada BS, Santi-Gadelha T (2012) Antinociceptive and anti-inflammatory effects of a lectin-like substance from Clitoria fairchildiana R. Howard seeds. Molecules 17:3277–3290

Lepenies B, Lang R (2019) Lectins and their ligands in shaping immune responses. Front Immunol 10:2379

Litynska A, Przybyuulo M, Pochec E, Hoja-uu Lukowicz D, Ciouulczyk D, Laidler P, Gil D (2001) Comparison of the lectin-binding pattern in different human melanoma cell lines. Melanoma Res 11:205–212

Liu B, Wu JM, Li J, Liu JJ, Li WW, Li CY, Xu HL, Bao JK (2010) Polygonatum cyrtonema lectin induces murine fibrosarcoma L929 cell apoptosis and autophagy via blocking Ras-Raf and PI3K-Akt signaling pathways. Biochimie 92(12):1934–1938

Liu S, Jiang Y, Guo X, Xu L, Lei P, Luo Q et al (2023) The lectin gene TRpL1 of tetraploid Robinia pseudoacacia L. response to salt stress. J For Res 34(2):497–505

Loyola W, Gaziri DA, Gaziri LC, Felipe I (2002) Concanavalin A enhances phagocytosis and killing of Candida albicans by mice peritoneal neutrophils and macrophages. FEMS Immunol Med Microbiol 33:201–208

Macedo MLR, Oliveira CF, Oliveira CT (2015) Insecticidal activity of plant lectins and potential application in crop protection. Molecules 20(2):2014–2033

Mahmood T, Khalid S, Abdullah M, Ahmed Z, Shah MKN, Ghafoor A, Du X (2019) Insights into drought stress signaling in plants and the molecular genetic basis of cotton drought tolerance. Cells 9(1):105

Majee SB, Biswas GR (2013) Exploring plant lectins in diagnosis, prophylaxis and therapy. J Med Plant Res 7(47):3444–3451

Marban-Mendoza N, Jeyaprakash A, Jansson HB, Damon RA, Zuckerman BM (1987) Control of root-knotnematodes on tomatobylectins. J Nematol 19:331–335

Marothia D, Kaur N, Jhamat C, Sharma I, Pati PK (2023) Plant lectins: classical molecules with emerging roles in stress tolerance. Int J Biol Macromol 244:125272

Mazalovska M, Kouokam JC (2020) Plant-derived lectins as potential cancer therapeutics and diagnostic tools. Biomed Res Int 2020:1631394

Mishra A, Behura A, Mawatwal S, Kumar A, Naik L, Mohanty SS et al (2019) Structure-function and application of plant lectins in disease biology and immunity. Food Chem Toxicol 134:110827

Moresco TR, Gaziri LC, Yasumoto Y, Felipe I (2002) Phagocytic and candidacidal activities of macrophages from suckling and adult mice pretreatedwith concanavalin-A. Med Mycol 40:393–397

Naithani S, Komath SS, Nonomura A, Govindjee G (2021) Plant lectins and their many roles: carbohydrate-binding and beyond. J Plant Physiol 266:153531

Oka Y, Chet I, Spiegel Y (1997) An immunoreactive protein to wheat-germ agglutinin antibody is induced in oat roots following invasion of the cereal cyst nematode Heterodera avenae, and by jasmonate. Mol Plant Microbe Interact 10:961–969
Öktem HA (2001) Böceklere Dayanıklı Transgenik Bitkilerin Geliştirilmesi. Bitki Biyoteknolojisi Edt.; Sebahattin Özcan, Ekrem Gürel, Mehmet Babaoğlu, Türkiye
Olsnes S (2004) The history of ricin, abrin and related toxins. Toxicon 44:361–370
Padiyappa SD, Avalappa H, Venkatesh YP, Parisara N, Prabhakar BT, SN, P. (2022) Characterization of antioxidant, anti-cancer, and immunomodulatory functions of partially purified garlic (*Allium sativum* L.) lectin. Biomedicine 42(4):703–712
Panunto-Castelo A, Souza MA, Roque-Barreira MC, Silva JS (2001) KMC, A lectin from Artocarpus integrifolia, induces IL-12 p 40 production by macrophages and switches from type 2 to type 1 cell-mediated immunity against Leishmania major antigens, resulting in BALB/c mice resistance to infection. Glycobiology 11:1035–1042
Peumans WJ, Van Damme EJ (1995) Lectins as plant defense proteins. Plant Physiol 109:347–352
Polito L, Bortolotti M, Battelli MG, Calafato G, Bolognesi A (2019) Ricin: an ancient story for a timeless plant toxin. Toxins 11(6):324
Pryme IF, Bardocz S (2001) Anti-cancer therapy: diversion of polyamines in the gut. Eur J Gastroenterol Hepatol 13:1041–1046
Pryme IF, Bardocz S, Pusztai A, Ewen SW (2006) Suppression of growth of tumor cell lines in-vitro and tumors in vivo by mistletoe lectins. Histol Histopathol 21(3):285–299
Pusztai A, Grant G, Buchan WC, Bardocz S, De Carvalho AF, Ewen SW (1998) Lipid accumulation in obese Zucker rats is reduced by inclusion of raw kidney bean (Phaseolus vulgaris) in the diet. Br J Nutr 79:213–221
Regente M, Taveira GB, Pinedo M, Elizalde MM, Ticchi AJ, Diz MS et al (2014) A sunflower lectin with antifungal properties and putative medical mycology applications. Curr Microbiol 69:88–95
Renkonen KO (1948, January) Studies on hemagglutinins present in seeds of some representatives of the family of Leguminoseae. In Annales medicinae experimentalis et biologiae fenniae (Vol. 26, No. 1, pp. 66-72). Kalevankatu 11 A, 00100 Helsinki, Finland: Finnish Medical Soc Duodecim
Roy AF, Boye JI, Simpson BK (2010) A review bioactive proteins and peptides in pulse crops: pea, chickpea, lentil. Food Res Int 43:432–442
Rüdiger H (1998) Plant lectins-more than just tools for glycoscientists: occurrence, structure, and possible functions of plant lectins. Acta Anal 161:130–152
Sahid S, Roy C, Paul S, Datta R (2020) Rice lectin protein r40c1 imparts drought tolerance by modulating S-adenosylmethionine synthase 2, stress-associated protein 8 and chromatin-associated proteins. J Exp Bot 71(22):7331–7346
Sames K, Shumacher U, Halato Z, Van Damme ES, Peumans WJ, Asmus B (2001) Lectins as bioactive plant proteins: a potential in cancer treatment. Crit Rev Food Sci Nutr 45:425–445
Seales EC, Jurado GA, Brunson BA, Wakefield JK, Frost AR, Bellis SL (2005) Hypersialylation of β1 integrins, observed in colon adenocarcinoma, may contribute to cancer progression by up-regulating cell motility. Cancer Res 65(11):4645–4652
Sergienko VG, Kyrychenko OV, Perkovska GY (2008) Method of plant lectins using to protect vegetable crops from disease. Zayavnyk i Patentovlasnyk Instytut Zakhystu Roslyn UAAN; Appl., 28.10.2008; Publ. 10.06.2009, Biul. N11. (In Ukrainian)
Shakirova FM, Bezrukova MV, Shayakhmetov IF (1996) Effect of temperature shock on the dynamics of abscisic acid and wheat germ agglutinin accumulation in wheat cell culture. Plant Growth Regul 19:85–87
Sharma S, Singh DD (2022) Investigations on the biological activity of Allium sativum agglutinin (ASA) isolated from garlic. Protein Pept Lett 29(6):555–566
Sharon N, Lis H (2007) Lectins, 2nd edn. Springer, Dordrecht
Silva AFB, Matos MP, Ralph MT, Silva DL, de Alencar NM, Ramos MV, Lima-Filho JV (2016) Comparison of immunomodulatory properties of mannose-binding lectinsfrom